Dinosaur Tracks

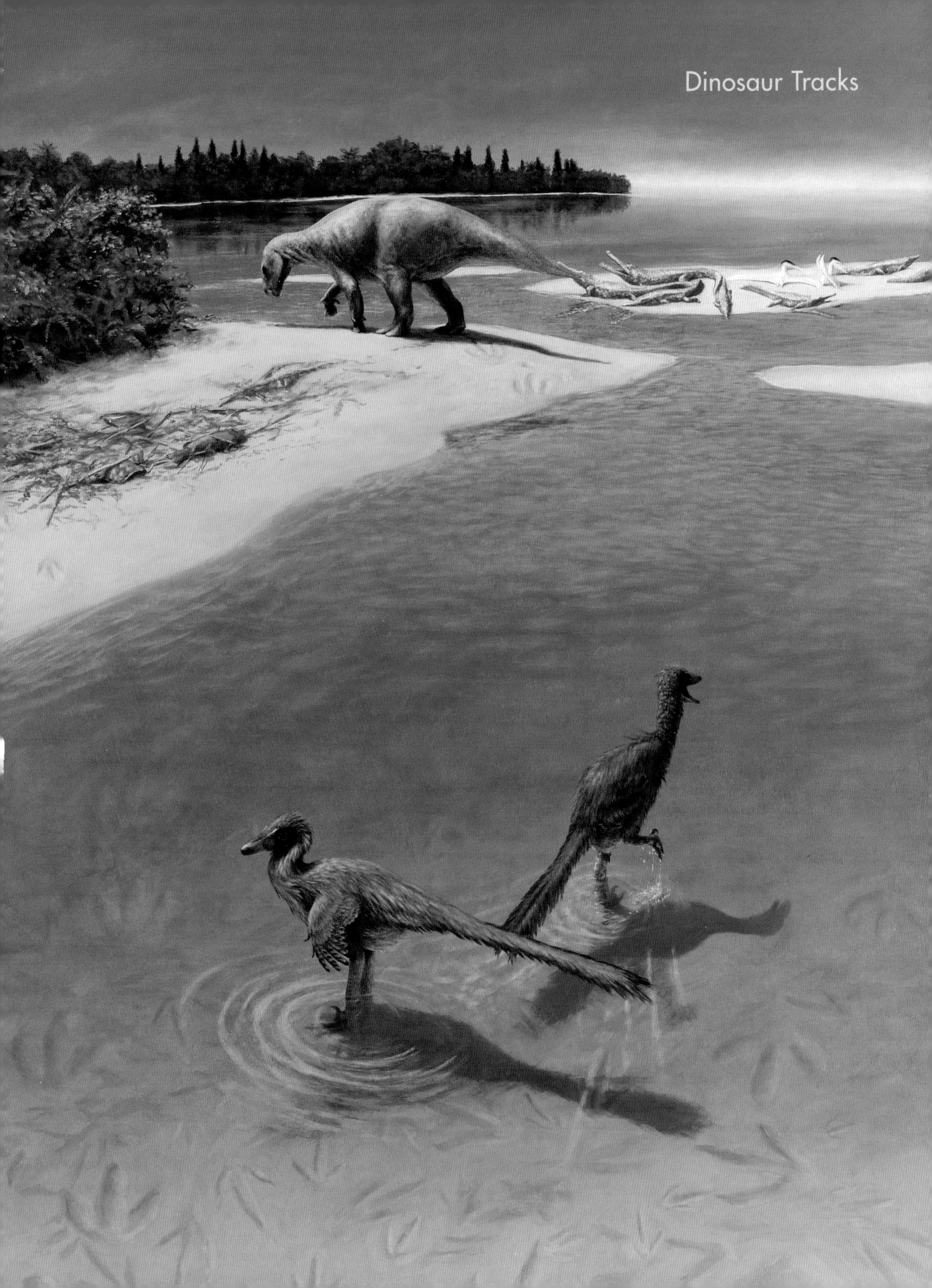

LIFE OF THE PAST   James O. Farlow, editor

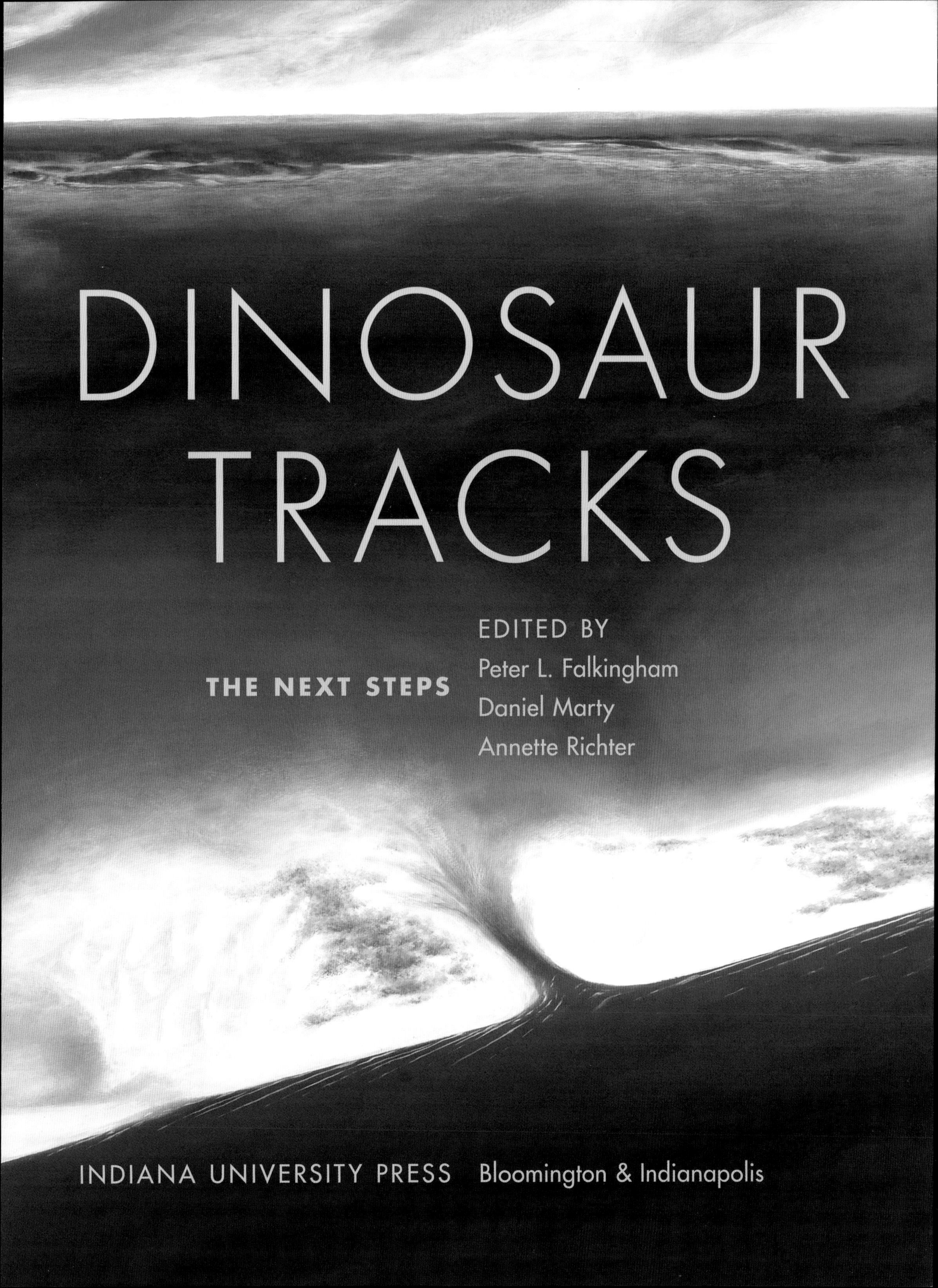

DINOSAUR TRACKS

THE NEXT STEPS

EDITED BY

Peter L. Falkingham

Daniel Marty

Annette Richter

INDIANA UNIVERSITY PRESS   Bloomington & Indianapolis

This book is a publication of

Indiana University Press
Office of Scholarly Publishing
Herman B Wells Library 350
1320 East 10th Street
Bloomington, Indiana 47405 USA

iupress.indiana.edu

The paper used in this publication meets the minimum requirements of the American National Standard for Information Sciences–Permanence of Paper for Printed Library Materials, ANSI Z39.48-1992.

Manufactured in China

Library of Congress Cataloging-in-Publication Data

Names: Falkingham, Peter L., editor. | Marty, Daniel, 1973– editor. | Richter, Annette, editor.
Title: Dinosaur tracks : the next steps / edited by Peter L. Falkingham, Daniel Marty, and Annette Richter.
Description: Bloomington : Indiana University Press, [2016] | Series: Life of the past | Includes bibliographical references and index.
Identifiers: LCCN 2016011885 (print) | LCCN 2016015807 (ebook) | ISBN 9780253021021 (cloth) | ISBN 9780253021144 (ebook)
Subjects: LCSH: Dinosaur tracks–Congresses. | Footprints, Fossil–Congresses.
Classification: LCC QE861.6.T72 D45 2016 (print) | LCC QE861.6.T72 (ebook) | DDC 567.9–dc23
LC record available at https://lccn.loc.gov/2016011885

1 2 3 4 5 21 20 19 18 17 16

Published with the generous support of:

 **Niedersächsisches Ministerium
für Wissenschaft und Kultur**

 **Landesmuseum
Hannover**
Das WeltenMuseum

Klosterkammer
Hannover

Contents

Acknowledgments

A

THIS BOOK WAS DEVELOPED FROM A DINOSAUR TRACK symposium that was organized and held in April 2011 in Obernkirchen, Germany, on behalf of the Niedersächsisches Landesmuseum Hannover (Lower Saxony State Museum Hannover). The enthusiasm generated during the short span of the symposium resulted in the idea for a new up-to-date dinosaur track book. Many of the symposium participants–leading researchers in the field of dinosaur ichnology–authored chapters in this book. We heartily acknowledge all of the authors for their excellent papers and patience throughout the process of bringing this wide-ranging book to publication, as well as the numerous reviewers that have contributed to the high quality of the peer-reviewed chapters. Thanks also to the Niedersächsisches Ministerium für Wissenschaft und Kultur (Lower Saxony Ministry for Science and Culture), which has underwritten a substantial portion of the costs associated with the publication of this book, notably the color figures throughout the book. The Klosterkammer Hannover also deserves our gratidtude for financing extra color paintings, including the cover picture. Finally, our thanks go to Jim Farlow and Bob Sloan (both of Indiana University Press) for their outstanding support for the project from its earliest inception.

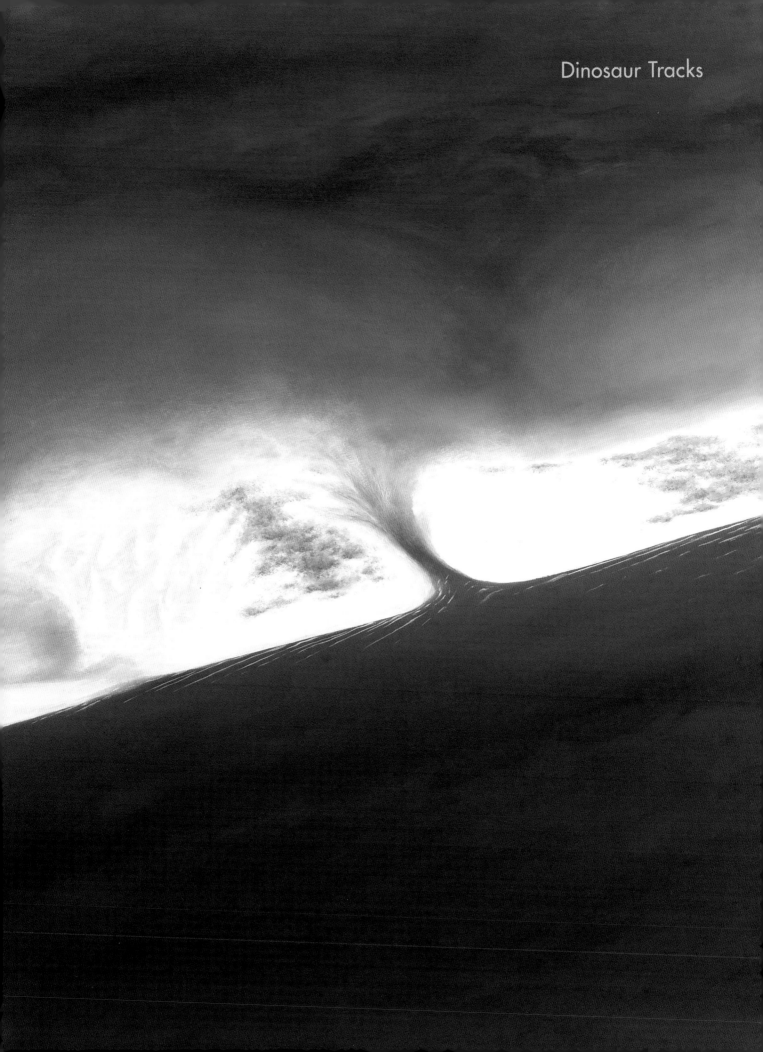

Dinosaur Tracks

Introduction

Peter L. Falkingham, Daniel Marty, and Annette Richter

THE DINOSAURIA ARE ONE OF THE MOST MORPHOLOGI-cally diverse groups of terrestrial vertebrates (Alexander, 1989), spanning several orders of magnitude in size from the smallest hummingbird to the largest sauropods. Ancestrally bipedal, groups within the Dinosauria evolved into a range of habitually and facultatively bipedal and quadrupedal animals. Their skeletons have been found on every continent (Weishampel, Dodson, and Osmólska, 2004), and their fossilized footprints are known from all except Antarctica.

The public perception of dinosaurs comes almost exclusively via their skeletons, and much of our knowledge about how these enigmatic animals looked and lived comes from osteological information. But the bones can only reveal so much, being as they are the product of a dead animal. Footprints and traces, on the other hand, are made by an animal during its life and can therefore shed light on paleobiological aspects that are not preserved in osteological remains – aspects such as behavior, locomotion, or paleoecology.

Vertebrate tracks are biogenic sedimentary structures and not body fossils or biological objects in the common sense. They result from the complex interaction of three factors: the sediment (its consistency and resistance to deformation), the foot dynamics (i.e., the kinematics and kinetics, or motions and forces, of the distal-most limb), and the anatomy of the foot (Padian and Olsen, 1984; Minter, Braddy, and Davis, 2007; Falkingham, 2014). Once formed, both pre- and postlithification they are subject to all of the taphonomic processes that affect other sedimentary structures (Scott et al., 2007; Marty, Strasser, and Meyer, 2009). A track is an intricate structure, existing in three dimensions (3-D) both at and below the foot-sediment interface (i.e., there is both a 3-D surface and a 3-D volume component to the track). As a field, vertebrate ichnology has grown to accommodate this complex nature by becoming increasingly interdisciplinary, interfacing with other fields such as sedimentology, soil mechanics, and biomechanics, as well as more traditional taxonomic and paleontological fields.

EARLY DINOSAUR ICHNOLOGY

Dinosaur ichnology has existed as a field for over 150 years, with fossil tracks being documented earlier than any osteological dinosaur material. The first recorded fossil vertebrate tracks were discovered in the 1820s in the Permian of Scotland, and (incorrectly) interpreted as turtle tracks by means of experimental ichnology by William Buckland (Pemberton, 2010). Shortly after, the famous Triassic archosaur tracks, named *Chirotherium* (Greek: hand animal) due to their obvious resemblance with the human hand (Kaup, 1835), stimulated great interest and controversy regarding trackmaker identification and reconstruction. The first dinosaur tracks were described by Edward Hitchcock in 1836, six years before Owen formally named the Dinosauria. A decade later, a large tridactyl track from the Early Cretaceous of southern England was discovered (Tagart, 1846) and attributed to *Iguanodon* in 1862 (Jones, 1862).

It remained Hitchcock, however, who produced the largest and most significant contribution during this time with a large body of work in a series of publications (Hitchcock, 1848, 1858, 1865). Hitchcock described a plethora of forms from the Early Jurassic of the Connecticut Valley, erecting nearly 100 ichnogenera, and more than 200 ichnospecies, many of which are still in use today, though he attributed them to ancient birds and lizards rather than to dinosaurs.

Following Hitchcock's death in 1864, research in his field declined, and whereas our knowledge of dinosaurs began to increase dramatically, ichnology was generally neglected (Lockley and Gillette, 1989). In 1962, Lapparent reported only 27 or 28 tracksites worldwide. However, these sites received comparatively little attention, because fossil bones formed the focus of most dinosaur research. The importance of dinosaur tracks only began to be recognized again in the early 1980s with the beginning of the "dinosaur track renaissance" (Lockley, 1986; Lockley and Gillette, 1987; Lockley, 1991a, 1991b). Since then hundreds of dinosaur tracksites have been

0.1. (Top) Nocturnal view of the Early Cretaceous moderately to heavily dinoturbated Chicken Yard level at the Obernkirchen tracksite. (Bottom) Group photo of the congress attendants during the conference at the Renaissance Castle of Hülsede.

and continue to be discovered all over the world. D'Orazi Porchetti et al. (chap. 20) report 211 tracksites only from Jurassic and Cretaceous shallow marine carbonate depositional environments. Today, there are so many dinosaur tracksites that evaluating the dinosaur track record has become an important part within the field of research that may be called "geoconservation," especially because many tracksites must be protected in situ.

THE DINOSAUR TRACK RENAISSANCE

The mid-1980s was considered to be the beginning of a renaissance in dinosaur ichnology, heralded by the First International Symposium on Dinosaur Tracks and Traces, convened in Albuquerque, New Mexico, in May 1986 (Lockley and Gillette, 1987). The symposium was described as notable for focusing on tracks as a means of "understanding the paleobiology and habits of dinosaurs" rather than ichnotaxonomic studies (Lockley and Gillette, 1987:247). Dinosaur tracks were forming the basis of studies interested not merely in describing a new morphology but also in adding to our understanding of paleobiology, paleoecology (Lockley, 1986, 1987), biostratigraphy, and locomotion (Alexander, 1977, 1989; Padian and Olsen, 1989). The field also saw the introduction of new methods for discriminating tracks (Moratalla, Sanz, and Jimenez, 1988) and early attempts at documenting them in 3-D (Ishigaki and Fujisaki, 1989).

There was also a growing awareness that tracks are 3-D, extending beneath the tracking surface. Descriptions of tracks in cross-section (Loope, 1986) were supported by the experimental work carried out by Allen (Allen, 1989, 1997), who used colored plasticine to observe undertrack formation. The renaissance continued into the 1990s (Lockley, 1991a), by which time several important books had been written on dinosaur tracks that to this day form the core of a vertebrate ichnologist's library (Leonardi, 1987; Lockley and Gillette, 1989; Lockley and Hunt, 1995; Thulborn, 1990; Lockley, 1991c).

21ST-CENTURY DINOSAUR ICHNOLOGY

Documenting and Communicating Tracks

In the early part of the century, laser scanning saw some uptake in the study of dinosaur tracks (Bates, Breithaupt, et al., 2008; Bates et al., 2009; Falkingham et al., 2009; Adams et al., 2010; Bates et al., 2010; Platt, Hasiotis, and Hirmas, et al., 2010). Although the method enabled recording of 3-D data from fossil tracks, the cost of the hardware and the logistics associated with it (transporting of delicate, bulky machinery, power sources, user expertise required for data capture and processing) prevented laser scanning from becoming mainstream in ichnology.

At around the same time that laser scanners were entering the field, photogrammetry saw some limited use (Breithaupt et al., 2001, 2006; Breithaupt and Matthews, 2001; Breithaupt, Matthews, and Noble, 2004; Matthews et al., 2005; Matthews, Noble, and Breithaupt, 2006; Bates, Breithaupt, et al., 2008). Widespread adoption was hindered by expensive software that required substantial user interaction and powerful computers. This changed with the development of free, open-source photogrammetry software in which matching algorithms and model generation required almost no user input (Falkingham, 2012). The software could run on reasonably powerful but otherwise common computer hardware and required little in the way of expertise. Technological progress continues and photogrammetry software can now process in minutes what took hours only a few years ago. What is more, this can be done on extremely modest hardware such as the laptop an ichnologist might use to write a manuscript or edit photographs on.

Complementing the rise in 3-D data acquisition techniques has been a parallel advancement in communication methods. Ichnologists can present data as raw 3-D files in supplemental information, as 3-D PDFs, or as videos. In doing so, far more information about a track or tracksite can be communicated, and to a wider audience, than ever before. It almost seems like magic that one ichnologist in the field can take a few photos with the phone in his pocket, and e-mail the data to a colleague in the lab at the other side of the world who can view the track digitally in 3-D.

In 1990, Thulborn stated that "the basic equipment comprises: notebook, graph paper, pens or pencils, compass, clinometer, camera with tripod and plenty of film, a stiff brush, hammer and cold chisels, tape measure, ruler and chalk" (67). One might add a computer to that list, for many of the documentation and analysis methods you will see in this book, and one might replace "plenty of film" with "a high-capacity SD card," but otherwise the list remains unchanged.

A Mechanistic Understanding of Track Formation

Accompanying the rise in fossil digitization methods was a desire to understand the track-forming process experimentally. Manning (2004) and later Jackson, Whyte, and Romano (2009, 2010) used artificial indenters to produce "footprints" in strongly controlled sediments, layered with plaster of paris to enable recovery of subsurface deformations. Milàn and Bromley (2006, 2008) carried out similar experiments with emu, using both living animals and severed feet to produce tracks in colored cement, whereas Marty, Strasser, and Meyer (2009) made neoichnological experiments with human footprints in microbial mats of modern tropical supratidal flats.

The impact of these studies was most evident in how they highlighted the importance of undertracks, or transmitted tracks. Though the phenomenon had been known since the work of Hitchcock, and illustrated clearly in Allen's (1989, 1997) indenter experiments, these new studies illustrated the transmission of deformation in a way that resonated with many track workers, reminding them that the footprints exposed on any given bedding plane were not necessarily a true track and an accurate representation of the foot morphology.

Later studies explored the process of track formation using computer-simulation techniques such as finite element analysis (Margetts et al., 2005; Falkingham et al., 2009; Falkingham, 2010; Falkingham, Margetts, and Manning, 2010; Falkingham et al., 2011a, 2011b; Schanz et al., 2013; Falkingham, Hage, and Bäker, 2014), the discrete element method (Falkingham and Gatesy, 2014), and 3-D modeling approaches (Henderson, 2003, 2006a, 2006b; Sellers et al., 2009).

Quantitative Studies of Tracks

Coupled with the increase in objective data acquisition, workers have begun utilizing that data to attempt to categorize and describe track morphology in meaningful, quantitative ways. One of the key aspects of science is repeatability, and there has been a movement to apply numerical techniques to describing and comparing tracks. Such a movement traces its roots into the dinosaur track renaissance with Moratalla, Sanz, and Jimenez's (1988) application of multivariate analysis to discriminating theropods and ornithopods. Those methods are still in practice today (Romilio and Salisbury, 2011), though there is some contention as to their utility (Thulborn, 2013). Other quantitative track studies have used the 3-D digital data to derive objective comparisons between tracks, calculating parameters such as length and width repeatably from the data (Bates, Manning, et al., 2008; Castanera et al., 2013; Razzolini et al., 2014).

KEY APPLICATIONS AND BENEFITS OF DINOSAUR TRACKS

The fossil tracks of vertebrates are a diverse, abundant source of data that supplement the fossil record. Both osteological and ichnological records are incomplete, but their completeness differs (bones and tracks are rarely found in the same rocks [Crimes and Droser, 1992]), making the two lines of evidence complimentary (Carrano and Wilson, 2001; Falkingham, 2014). Tracks not only present evidence from different paleoenvironments, but they can also reveal aspects of paleobiology that are absent from the body fossil material,

notably tracks and trackways are direct records of locomotion and, in turn, behavior. Footprints can therefore often provide the only test of biomechanical hypotheses derived from the skeletons and present a dynamic, vivid impression of a dinosaur as a living creature (Farlow et al., 2012).

The paleoecological aspect of dinosaur tracks may be equally as important as the paleobiological one. The immutable association of track and sediment means that interpretations of the environment based on sedimentology directly tell us of the environment in which the trackmaker lived. Contrarily, the case of body fossils only provides information on where the animal's body came to finally rest, which may be great distances from where the animal actually lived or may be reworked from older deposits with little evidence to the fact (Behrensmeyer, 1982).

Integral to the utility of tracks is their abundance. A single animal only leaves one skeleton, but it can produce many thousands of tracks throughout its life time (Lockley, 1998). Large terrestrial vertebrates only sparsely populate any given area, living at low population densities and having relatively large ranges. It takes special circumstances to generate a bone accumulation, and the processes responsible for concentrating the skeletal remains into fossil localities also bias representation of species, individuals, and body parts relative to the original populations (Behrensmeyer, 1991). The mud beside a watering hole, on the other hand, has the potential to act as a record of the majority of animals in a given area (Cohen et al., 1991; Cohen et al., 1993), potentially more completely recording the wider ecosystem.

THE DIFFICULTIES OF DINOSAUR ICHNOLOGY—ERRORS AND MISINTERPRETATIONS

Despite the wealth of information to be found within a track, trackway, or tracksite—or perhaps because of it—confident interpretation of footprints can prove to be particularly difficult. Understanding the strengths and weaknesses (uncertainties) of data, and the potential biases, is an integral part of all paleobiological research, and this is no truer than for ichnology.

Trackmaker Identification

One of the first questions often asked of a new track or tracksite is "What made them?" This desire to identify the trackmaker is not unwarranted—as Carrano and Wilson (2001:567) stated, "the level to which the trackmaker can be identified affects nearly all types of ichnological analysis." Hitchcock (1858) argued that if the anatomists of the time could identify an animal from a single bone (Cuvier's principle), why then should it not be possible to reconstruct the whole animal based on its track? After all, a track records anatomy from much more than just a single bone.

Unfortunately, the task is not so straightforward. Final track morphology only partially reflects foot anatomy. The remaining contributions come from the substrate properties and the kinematics of the foot, yet often these factors are not addressed or are poorly understood. Compounding this difficulty, many clades retain conservative foot morphology across species and genera—many theropods, for instance, tend to have very similar tridactyl feet. As such, tracks have a relatively low taxonomic resolution and are usually not attributable below the family level. The exception to this are taxa with particularly distinctive morphology, which may leave equally distinctive impressions, such as didactyl tracks left by deinonychosaurian dinosaurs (see Lockley et al., chap. 11, for a review).

Ichnotaxonomy

The assignation of dinosaur tracks into new or existing ichnotaxa—that is, giving a scientific name to the track—forms a large majority of the track literature. This is especially true historically, where almost all publications were concerned with descriptions of new specimens. To some workers outside the field, it can seem odd to give a binomial name to a track, the terms "ichnogenus" and "ichnospecies" might appear to imply some hierarchical relatedness. But, just as for body fossils, these names are assigned purely based on morphology, and the genus/species distinction is subjectively made according to the morphological similarity or disparity. Ichnotaxonomy offers a means of communicating the complex forms of tracks without needing to devote entire paragraphs or pages to descriptive text, and such was the reasoning that Hitchcock (1858:4) gave for assigning names: "Without some such designations, it is nearly impossible, since they have become so numerous, to describe the different sorts of tracks."

However, whereas ichnotaxonomy is a kind of shorthand for complex 3-D morphologies, the temptation to link a track to the supposed trackmaker is ever present and often manifests itself in the ichnotaxonomic names. *Camptosaurichnus*, *Iguanodonichnus*, *Hadrosaurichnus*, or *Tyrannosauripus* are just a few examples of track names that imply a specific trackmaker. The practice can be traced back to Hitchcock, who by 1858 had ceased naming the tracks themselves and was instead attempting to name the trackmakers directly: "for several years I merely gave names to these tracks with reference to their supposed affinities; such as *Ornithichnites*, or stony bird-tracks. . . . But more recently, I have named the animals that made the tracks" (Hitchcock, 1858:4). Confusion can arise when tracks are subsequently reinterpreted as having been made by a different animal entirely but must retain the original ichnotaxa according to the rules of the ICZN (of the four ichnogenera listed above, two have since been reinterpreted as being produced by a theropod and *Iguanodonichnus* as being made by a sauropod (Lockley, Nadon, and Currie, 2004).

The semantics of naming aside, ichnotaxonomy suffers from the same problems that organismal taxonomy suffers from. First, the means by which morphological differences are defined can affect how large a difference is required for the erection of a new ichnogenus or ichnospecies. Even simple metrics such as track length and width can be hard to define accurately (Falkingham, chap. 4). Second, as more and more intermediate ichnospecies or ichnogenera are found, the distinction between taxa can become highly blurred. However, where the body fossil record is relatively sparse, recording as it does only a tiny fraction of the evolutionary tree, the ichnofossil record is far more continuous. Tracks can vary within a trackway due to changes in substrate consistency, the way the animal moves, or due to differential weathering/erosion either before or after lithification. Whereas the body fossil record therefore only presents a relatively discrete subset of the continuum of life, tracks can present an almost infinite range of morphologies. In 2012, Farlow et al. asked, "What are ichnotaxa for? Or, what is the goal, and what is to be gained, by giving names to vertebrate trace fossils?" (739). If we consider ichnotaxa as simply a means to distinguish different morphologies, there should be no problem assigning multiple ichnotaxa to a single trackway in which morphology varies for one reason or another. But given the knowledge that the tracks were produced by a single organism, and that some studies attempt to gauge diversity based on ichnotaxa, is this the correct approach? Sadly, there are no simple answers to this question.

Track Formation

Most of the confusion related to the understanding of tracks is certainly related to the fact that a single foot impact may leave tracks not only on the superficial level (true tracks) but also on (several) underlying levels and in different forms (undertracks, underprints). The track-forming process may also lead to the formation of a variety of associated (extramorphological) track features such as displacement rims and downfolding. After track formation, taphonomic processes may lead to blurring of tracks and thus to the formation of modified true tracks and/or (internal) overtracks (Marty, Strasser, and Meyer, 2009).

The importance of the distinction between true tracks and undertracks has been recognized as early as 1858 by Hitchcock, and today it is generally agreed that the prerequisite for meaningful ichnotaxonomic and paleoecological studies is the correct identification and exclusive use of true tracks among all these different kind of tracks. Only fine anatomical details such as toe marks, claw marks, or skin impressions generally identify true tracks with any level of certainty. Unfortunately, such details are often not recorded because the trackmaker's feet are not suitable to leave such traces, because the substrate properties do not favor preservation of such fine details, or because such details have been lost during the taphonomic process (Padian and Olsen, 1984; Cohen et al., 1991; Nadon, 2001; Henderson, 2006b; Milàn and Bromley, 2006; Scott et al., 2007). Generally, exposed tracks degrade rapidly after formation and have a low preservation potential, even though some processes such as early cementation (e.g., of the sediment or within a microbial mat), rapid covering by sediment, and overgrowth by microbial mats may potentially preserve tracks (Phillips et al., 2007; Marty, Strasser, and Meyer, 2009; Carmona et al., 2011; Carvalho, Borghi, and Leonardi, 2013). The amount of time between track formation and burial affects their preservation potential (Laporte and Behrensmeyer, 1980), as well as the degree of time-averaging of the ichnoassemblage (Cohen et al., 1993), making confident interpretation difficult.

DINOSAUR TRACKS: THE NEXT STEPS

The dinosaur track renaissance (Lockley, 1986, 1987; Lockley and Gillette, 1987; Lockley, 1991a) made considerable inroads toward unifying the study of tracks and the study of bones into a more complete paleobiological framework. The field has remained strong since the turn of the century, and many advancements have been made during that time.

In 2011, a dinosaur track symposium was organized and held in Obernkirchen, Germany, on behalf of the Niedersächsisches Landesmuseum Hannover (Lower Saxony State Museum Hannover) and the main organizing foundation Schaumburger Landschaft (Richter and Reich, 2012), to bring together active researchers in dinosaur ichnology. More than 90 participants from around the world who were working on dinosaur ichnology attended the symposium. During this symposium, many important aspects of dinosaur ichnology were addressed and discussed. This was complemented by field trips to the amazing Lower Cretaceous Obernkirchen and Münchehagen tracksites, including a spectacular nocturnal view of the unique Chicken Yard level (Fig. 0.1). This symposium was sponsored by the Schaumburger Landschaft, Stiftung Niedersachsen, the Sparkassenstiftung Schaumburg, and the Klosterkammer Hannover, enabling invitations to 20 specialists from more than 10 countries. The Niedersächsisches Ministerium für Wissenschaft und Kultur (Lower Saxony Ministry for Science and Culture)

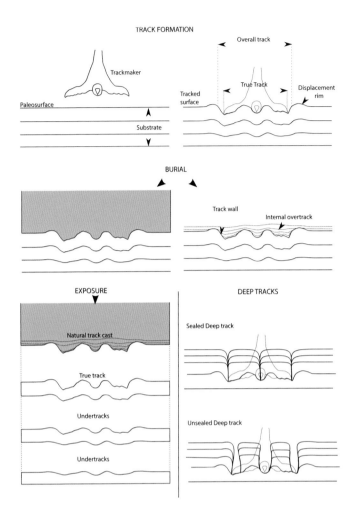

0.2. Illustration of some of the most important terms used to describe fossil tracks.

has also sponsored a substantial part of the costs linked to the publication of this book, notably for the color figures throughout the book. We heartily acknowledge all of these contributions. It was also during this meeting when the idea for a new dinosaur track book was born.

This book comprises 20 contributions that discuss or apply the recent advancements in dinosaur ichnology. The contributions are from active researchers and research groups, and they cover a wide range of topics within the study of dinosaur tracks.

The chapters that follow are arranged loosely into four broad themes. The first of these, approaches and techniques, contains chapters that review the state of the art in the field or introduce new methods for studying fossil tracks. Milàn and Falkingham start us off with a review of experimental ichnology, covering in more detail previous studies that have used indenters, living animals, and computer simulation to study the formation of dinosaur footprints. Matthews, Noble, and Breithaupt then provide a historical and modern account of photogrammetry, the digitization technique that has become a mainstay of vertebrate ichnologists. To

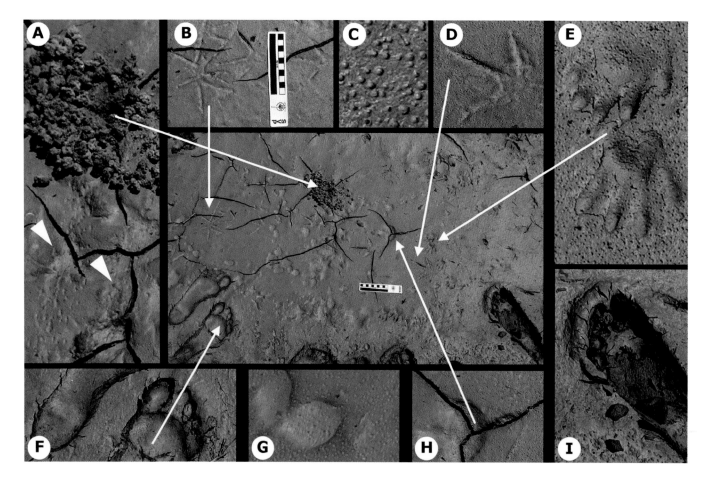

0.3. The concept of vertebrate ichnoassemblage demonstrated by an example from a supratidal, microbial mat-covered flat south of San Pedro Town (Ambergris Caye, Belize). The pictures were taken after a heavy rainfall, when the flat was susceptible for track recording. Time-averaging is minor and the assemblage is likely to record at least a part of the biocoenosis of the surrounding area. (A) Burrow with pellets of a decapod crustacean, note trails heading toward and away from the burrow (arrows). (B) Faint shorebird tracks left in an area with firm sediment. (C) Gas-bubble escape structures related to the decay of organic matter within the sediment. (D) Reasonably defined bird track. (E) Well-defined tracks of *Iguana* with digit impressions, organized in a trackway. (F) Well-defined footprints with toe impressions left by the photographer (D.M.). (G) Raindrop impact impressions. (H) Desiccation cracks. (I) Human track with displacement rims.

illustrate this, Wings, Lallensack, and Mallison (chap. 3) apply photogrammetry to the Early Cretaceous dinosaur trackways in Münchehagen, Germany, as a basis for geometric morphometric shape analysis and evaluation of material lost during excavation. Falkingham follows this in chapter 4 with a discussion about objectively defining track outlines for such purposes and the difficulties therein. Chapter 5, by Gatesy and Ellis, applies new techniques in the form of biplanar X-rays to study 3-D sediment motion beneath the surface as a track is formed. Identifying the quality of track preservation is important when defining ichnotaxa or making paleobiological inferences, and Belvedere and Farlow introduce a new scale of preservation for this purpose in chapter 6. The section concludes with Alcalá et al.'s evaluation of the dinosaur track record, both for scientific and conservation/management ("geoheritage") purposes.

The second theme of the book is interpreting paleobiology and evolution from tracks and begins with Castanera et al. exploring the variations in Iberian sauropod tracks through

time. Hall, Fragomeni, and Fowler take a more focused approach and attempt to use the fossil trackway record to test the hypothesis that sauropods used the pedal unguals for gripping the substrate. Not all trackways are formed during walking or running, and Milner and Lockley review the field's current understanding of dinosaur swim track assemblages. In chapter 11, Lockley et al. examine the fossil track record of two-toed dinosaurs through time. Following this, Hornung et al. apply a morphometric approach to understanding whether track diversity among German ornithopod tracks is down to diversity, ontogeny, or both. Stevens, Ernst, and Marty then use computer modeling to discuss the uncertainty and ambiguity in the interpretation of sauropod trackways. Finally, Cobos et al. use exceptionally preserved tracks to attempt to reconstruct the foot motions of different dinosaur species in chapter 14.

No ichnological text would be complete without contributions concerning ichnotaxonomy and trackmaker identification. The first of these is by Buckley, McCrea, and

Lockley, who use multivariate statistical analyses to resolve the complex Cretaceous avian ichnotaxonomy. This is joined by chapter 16 by Hübner, who discusses the elusive ornithischian tracks from the amazing Chicken Yard level at the Obernkirchen tracksite (Germany).

The final theme includes depositional environments in a broader sense. The heavily dinoturbated Chicken Yard level at the Obernkirchen tracksite is the focus of chapter 17 by Richter and Böhme, who provide a description of this unique ichnoassemblage and a review on dinoturbation. Loope and Milàn (chap. 18) review dinosaur tracks in eolian strata, desert-like environments often considered as devoid of much life and that thus might seem to be a poor place to look for animal tracks. In chapter 19, Schanz et al. analyze desiccation crack patterns with experimental test series and finite element analysis and discuss their potential use for the quantitative interpretation of fossil tracks. Finally, in chapter 20, d'Orazi Porchetti et al. provide a thorough review of the dinosaur track record from Jurassic and Cretaceous shallow marine carbonate depositional environments in the form of a detailed database that may be used in the future for new large-scale evolutionary studies based on track data.

The high diversity of these up-to-date essays emphasizes that dinosaur ichnological research is alive and kicking, that new important discoveries are continuously made, and new methods are being developed, applied, and refined. This book also highlights the importance of interdisciplinary scientific research in earth sciences and biosciences. It demonstrates that ichnology has an important contribution to make toward a better understanding of dinosaur paleobiology. Tracks and trackways are among the best sources of evidence to understand and reconstruct the daily life of dinosaurs. They are windows on past lives, dynamic structures produced by living, breathing, moving animals now long extinct, and they are every bit as exciting and captivating as the skeletons of their makers.

We have tried, where possible, to present the following chapters with a common terminology. To that effect, Figures 0.2 and 0.3 demonstrate many of the important terms used throughout this book.

ACKNOWLEDGMENTS

Editing a scientific book is not an easy task, and the editors thank the authors for providing interesting chapters that could easily and quickly have been submitted to peer-reviewed journals and for their patience during the noticeably longer editing process of a book. This book would not have been possible without the great help of all of the conscientious reviewers, which we particularly acknowledge. Jim Farlow and Robert Sloan have been an invaluable source of help throughout the entire process of producing this book since the very beginning, and we extend our thanks to them. And we acknowledge the funders that made this possible: the Kllosterkammer Hannover, the Niedersächsisches Ministerium für Wissenschaft und Kultur, and Schaumburger Landschaft.

REFERENCES

Adams, T., C. Strganac, M. J. Polcyn, and L. L. Jacobs. 2010. High resolution three-dimensional laser scanning of the type specimen of *Eubrontes(?) glenrosensis* Shuler, 1935, from the Comanchean (Lower Cretaceous) of Texas: implications for digital archiving and preservation. Palaeontologia Electronica 13: 1T.

Alexander, R. M. 1977. Mechanics and scaling of terrestrial locomotion; pp. 93–110 in T. J. Pedley (ed.), Scale Effects in Animal Locomotion. Academic Press, London.

Alexander, R. M. 1989. Dynamics of Dinosaurs and Other Extinct Giants. Columbia University Press, New York, New York, 167 pp.

Allen, J. R. L. 1989. Fossil vertebrate tracks and indenter mechanics. Journal of the Geological Society 146: 600–602.

Allen, J. R. L. 1997. Subfossil mammalian tracks (Flandrian) in the Severn Estuary, S. W. Britain: mechanics of formation, preservation and distribution. Philosophical Transactions of the Royal Society B: Biological Sciences 352: 481–518.

Bates, K. T., P. L. Manning, B. Vila, and D. Hodgetts. 2008. Three dimensional modelling and analysis of dinosaur trackways. Palaeontology 51: 999–1010.

Bates, K. T., B. H. Breithaupt, P. L. Falkingham, N. A. Matthews, D. Hodgetts, and P. L. Manning. 2008. Integrated LiDAR and photogrammetric documentation of the Red Gulch dinosaur tracksite (Wyoming, USA); pp. 101–103 in S. E. Foss, J. L. Cavin, T. Brown, J. I. Kirkland, and V. L. Santucci (eds.), Proceedings of the Eighth Conference on Fossil Resources, St. George, Utah, May 19–21. BLM Regional Paleontologist, Salt Lake City, Utah.

Bates, K. T., P. L. Falkingham, F. Rarity, D. Hodgetts, A. Purslow, and P. L. Manning. 2010. Application of high-resolution laser scanning and photogrammetric techniques to data acquisition, analysis and interpretation in palaeontology. International Archives of Photogrammetry, Remote Sensing and Spatial Information Sciences 38: 68–73.

Bates, K. T., P. L. Falkingham, D. Hodgetts, J. O. Farlow, B. H. Breithaupt, M. O'Brien, N. Matthews, W. I. Sellers, and P. L. Manning. 2009. Digital imaging and public engagement in palaeontology. Geology Today 25: 134–139.

Behrensmeyer, A. K. 1982. Time resolution in fluvial vertebrate assemblages. Paleobiology 8: 211–227.

Behrensmeyer, A. K. 1991. Terrestrial vertebrate accumulations; pp. 291–335 in P. Allison and D. E. G. Briggs (eds.), Taphonomy: Releasing the Data Locked in the Fossil Record. Plenum, New York.

Breithaupt, B. H., and N. A. Matthews. 2001. Preserving paleontological resources using photogrammetry and geographic information systems; pp. 62–70 in D. Harmon (ed.), Crossing Boundaries in Park Management: Proceedings of the 11th Conference on Research and Resource Management in Parks and Public Lands. The George Wright Society, Hancock, Michigan.

Breithaupt, B. H., N. Matthews, and T. Noble. 2004. An integrated approach to three-dimensional data collection at dinosaur tracksites in the Rocky Mountain West. Ichnos 11: 11–26.

Breithaupt, B. H., E. H. Southwell, T. Adams, and N. A. Matthews. 2001. Innovative documentation methodologies in the study of the most extensive dinosaur tracksite in Wyoming; pp. 113–122 in V. L. Santucci and L. McClelland (eds.), Proceedings of the 6th Fossil Research Conference. National Park Service D-2228.

National Park Service, Geological Resources Division, Lakewood, Colorado.

Breithaupt, B. H., E. H. Southwell, T. Adams, and N. A. Matthews. 2006. The Red Gulch dinosaur tracksite: public participation in the conservation and management of a world-class paleontological site. Bulletin New Mexico Museum of Natural History and Science 34: 10.

Carmona, I. S., C. N. Bournod, J. J. Ponce, and D. G. Cuadrado. 2011. The role of microbial mats in the preservation of bird footprints: a case study from the mesotidal Bahia Blanca Estuary. SEPM Special Publication 101: 37–45.

Carrano, M. T., and J. A. Wilson. 2001. Taxon distributions and the tetrapod track record. Paleobiology 27: 564–582.

Carvalho, I. D. S., L. Borghi, and G. Leonardi. 2013. Preservation of dinosaur tracks induced by microbial mats in the Sousa Basin (Lower Cretaceous), Brazil. Cretaceous Research 44: 112–121.

Castanera, D., C. Pascual, N. L. Razzolini, B. Vila, J. L. Barco, and J. I. Canudo. 2013. Discriminating between medium-sized tridactyl trackmakers: tracking ornithopod tracks in the base of the Cretaceous (Berriasian, Spain). PLoS One 8: e81830.

Cohen, A., J. Halfpenny, M. G. Lockley, and E. Michel. 1993. Modern vertebrate tracks from Lake Manyara, Tanzania and their paleobiological implications. Paleobiology 19: 433–458.

Cohen, A., M. Lockley, J. Halfpenny, and A. E. Michel. 1991. Modern vertebrate track taphonomy at Lake Manyara, Tanzania. Palaios 6: 371–389.

Crimes, T. P., and M. L. Droser. 1992. Trace fossils and bioturbation: the other fossil record. Annual Review of Ecology and Systematics 23: 339–360.

Falkingham, P. L. 2010. Computer simulation of dinosaur tracks. Ph.D. dissertation, University of Manchester, Manchester, U.K., pp. 305.

Falkingham, P. L. 2012. Acquisition of high resolution three-dimensional models using free, open-source, photogrammetric software. Palaeontologia Electronica 15: 1T: 15p.

Falkingham, P. L. 2014. Interpreting ecology and behaviour from the vertebrate fossil track record. Journal of Zoology 292: 222–228.

Falkingham, P. L., and S. M. Gatesy. 2014. The birth of a dinosaur footprint: subsurface 3D motion reconstruction and discrete element simulation reveal track ontogeny. Proceedings of the National Academy of Sciences USA 111: 18279–18284.

Falkingham, P. L., J. Hage, and M. Bäker. 2014. Mitigating the Goldilocks effect: the effects of different substrate models on track formation potential. Royal Society Open Science 1: 140225.

Falkingham, P. L., L. Margetts, and P. L. Manning. 2010. Fossil vertebrate tracks as paleopenetrometers: confounding effects of foot morphology. Palaios 25: 356–360.

Falkingham, P. L., K. T. Bates, L. Margetts, and P. L. Manning. 2011a. The "Goldilocks" effect: preservation bias in vertebrate track assemblages. Journal of the Royal Society: Interface 8: 1142–1154.

Falkingham, P. L., K. T. Bates, L. Margetts, and P. L. Manning. 2011b. Simulating sauropod

manus-only trackway formation using finite-element analysis. Biology Letters 7: 142–145.

Falkingham, P. L., L. Margetts, I. Smith, and P. L. Manning. 2009. Reinterpretation of palmate and semi-palmate (webbed) fossil tracks; insights from finite element modelling. Palaeogeography, Palaeoclimatology, Palaeoecology 271: 69–76.

Farlow, J. O., R. E. Chapman, B. H. Breithaupt, and N. Matthews. 2012. The scientific study of dinosaur footprints; pp. 713–759 in M. K. Brett-Surman, T. R. Holtz, and J. O. Farlow, (eds.), The Complete Dinosaur. 2nd edition. Indiana University Press, Bloomington and Indianapolis, Indiana.

Henderson, D. M. 2003. Footprints, trackways, and hip heights of bipedal dinosaurs: testing hip height predictions with computer models. Ichnos 10: 99–114.

Henderson, D. M. 2006a. Burly gaits: centers of mass, stability, and the trackways of sauropod dinosaurs. Journal of Vertebrate Paleontology 26: 907–921.

Henderson, D. M. 2006b. Simulated weathering of dinosaur tracks and the implications for their characterization. Canadian Journal of Earth Sciences 43: 691–704.

Hitchcock, E. 1836. Ornithichnology–description of the foot marks of birds (Ornithichnites) on New Red Sandstone in Massachusetts. American Journal of Science 29: 307–340.

Hitchcock, E. 1848. An attempt to discriminate and describe the animals that made the fossil footmarks of the United States, and especially of New England. Memoirs of the American Academy of Arts and Science 3: 129–256.

Hitchcock, E. 1858. Ichnology of New England. A Report on the Sandstone of the Connecticut Valley, Especially Its Fossil Footmarks. W. White, Boston, Massachusetts [reprinted 1974 by Arno Press, New York, New York], 232 pp.

Hitchcock, E. 1865. Supplement to the Ichnology of New England. Wright and Potter State Printers, Boston, Massachusetts, 96 pp.

Ishigaki, S., and T. Fujisaki. 1989. Three dimensional representation of eubrontes by the method of moiré topography; pp. 421–425 in D. D. Gillette and M. Lockley (eds.), Dinosaur Tracks and Traces. Cambridge University Press, Cambridge, UK.

Jackson, S. J., M. A. Whyte, and M. Romano. 2009. Laboratory-controlled simulations of dinosaur footprints in sand: a key to understanding vertebrate track formation and preservation. Palaios 24: 222–238.

Jackson, S. J., M. A. Whyte, and M. Romano. 2010. Range of experimental dinosaur (Hypsilophodon foxii) footprints due to variation in sand consistency: how wet was the track? Ichnos 17: 197–214.

Jones, T. R. 1862. Tracks trails and surface markings. Geologist 5: 128–139.

Kaup, J. J. 1835. Über Thierfährten bei Hildburghausen. Neues Jahrbuch für Mineralogie, Geologie und Paläontologie 1835: 227–228.

Laporte, L. F., and A. K. Behrensmeyer. 1980. Tracks and substrate reworking by terrestrial vertebrates in Quaternary sediments of Kenya. Journal of Sedimentary Petrology 50: 1337–1346.

Lapparent, A. F. 1962. Footprints of dinosaur in the Lower Cretaceous of Vestspitsbergen-Svalbard. Arbok Norsk Polarinstitutt 1960: 14–21.

Leonardi, G. 1987. Glossary and Manual of Tetrapod Footprint Palaeoichnology. República Federativa do Brasil, Ministério das Minas e Energia, Departamento Nacional da Produção Mineral, Brasilia, Brazil, 137 pp.

Lockley, M. G. 1986. The paleobiological and paleoenvironmental importance of dinosaur footprints. Palaios 1: 37–47.

Lockley, M. G. 1987. Dinosaur trackways and their importance in paleoenvironmental reconstruction; pp. 81–95 in S. Czerkas and E. C. Olson (eds.), Dinosaurs Past and Present. Los Angeles County Museum, Los Angeles, California.

Lockley, M. G. 1991a. The dinosaur footprint renaissance. Modern Geology 16: 139–160.

Lockley, M. G. 1991b. Dinosaur tracking revolution: new applied dimensions for geology. AAPG Bulletin–American Association of Petroleum Geologists 75: 1807.

Lockley, M. G. 1991c. Tracking Dinosaurs. Cambridge University Press, Cambridge, UK.

Lockley, M. G. 1998. Philosophical perspectives on theropod track morphology: blending qualities and quantities in the science of ichnology. Gaia 15: 279–300.

Lockley, M. G., and D. D. Gillette. 1987. Dinosaur tracks symposium signals a renaissance in vertebrate ichnology. Paleobiology 13: 246–252.

Lockley, M. G., and D. D. Gillette. 1989. Dinsoaur tracks and traces: an overview; pp. 3–10 in M. G. Lockley and D. D. Gillette (eds.), Dinosaur Tracks and Traces. Cambridge University Press, Cambridge, UK.

Lockley, M. G., and A. P. Hunt. 1995. Dinosaur Tracks and Other Fossil Footprints of the Western United States. Columbia University Press, New York, New York, 338 pp.

Lockley, M. G., G. Nadon, and P. J. Currie. 2004. A diverse dinosaur-bird footprint assemblage from the Lance Formation, Upper Cretaceous, eastern Wyoming: implications for ichnotaxonomy. Ichnos 11: 229–249.

Loope, D. B. 1986. Recognizing and utilizing vertebrate tracks in cross section: Cenozoic hoofprints from Nebraska. Palaios 1: 141–151.

Manning, P. L. 2004. A new approach to the analysis and interpretation of tracks: examples from the Dinosauria. Geological Society, London, Special Publications 228: 93–123.

Margetts, L., I. M. Smith, J. Leng, and P. L. Manning. 2005. Simulating dinosaur track formation, pp. 1–4 in E. Oñate and D. R. J. Owen (eds.), Extended Abstracts, VIII International Conference on Computational Plasticity (COMPLAS). CIMNE, Barcelona, Spain.

Marty, D., A. Strasser, and C. A. Meyer. 2009. Formation and taphonomy of human footprints in microbial mats of present-day tidal-flat environments: implications for the study of fossil footprints. Ichnos 16: 127–142.

Matthews, N. A., T. A. Noble, and B. H. Breithaupt. 2006. The application of photogrammetry, remote sensing and geographic information systems (GIS) to fossil resource management. Bulletin New Mexico Museum of Natural History and Science 34: 119–131.

Matthews, N. A., B. H. Breithaupt, T. Noble, A. Titus, and J. Smith. 2005. A geospatial look at the morphological variation of tracks at the Twentymile Wash dinosaur tracksite, Grand Staircase-Escalante National Monument, Utah. Journal of Vertebrate Paleontology 25: 90A

Milàn, J., and R. G. Bromley. 2006. True tracks, undertracks and eroded tracks, experimental work with tetrapod tracks in laboratory and field. Palaeogeography, Palaeoclimatology, Palaeoecology 231: 253–264.

Milàn, J., and R. G. Bromley. 2008. The impact of sediment consistency on track and undertrack morphology: experiments with emu tracks in layered cement. Ichnos 15: 19–27.

Minter, N. J., S. J. Braddy, and R. B. Davis. 2007. Between a rock and a hard place: arthropod trackways and ichnotaxonomy. Lethaia 40: 365–375.

Moratalla, J. J., J. L. Sanz, and S. Jimenez. 1988. Multivariate analysis on Lower Cretaceous dinosaur footprints: discrimination between ornithopods and theropods. Geobios 21: 395–408.

Nadon, G. C. 2001. The impact of sedimentology on vertebrate track studies; pp. 395–407 in D. H. Tanke and K. Carpenter (eds.), Mesozoic Vertebrate Life. Indiana University Press, Bloomington and Indianapolis, Indiana.

Padian, K., and P. E. Olsen. 1984. Footprints of the Komodo monitor and the trackways of fossil reptiles. Copeia 3: 662–671.

Padian, K., and P. E. Olsen. 1989. Ratite footprints and the stance and gait of Mesozoic theropods; pp. 232–241 in D. D. Gillette and M. Lockley (eds.), Dinosaur Tracks and Traces. Cambridge University Press, Cambridge, UK.

Pemberton, S. G. 2010. History of ichnology: the Reverend William Buckland (1784–1856) and the fugitive poets. Ichnos 17: 246–263.

Phillips, J. P. L., G. A. Ludvigson, R. Matthew Joeckel, L. A. Gonzalez, R. L. Brenner, and B. J. Witzke. 2007. Sequence stratigraphic controls on synsedimentary cementation and preservation of dinosaur tracks: example from the lower Cretaceous, (Upper Albian) Dakota Formation, southeastern Nebraska, U.S.A. Palaeogeography, Palaeoclimatology, Palaeoecology 246: 367–389.

Platt, B. F., S. T. Hasiotis, and D. R. Hirmas. 2010. Use of low-cost Multistripe Laser Triangulation (MLT) scanning technology for three-dimensional, quantitative paleoichnological and neoichnological studies. Journal of Sedimentary Research 80: 590–610.

Razzolini, N. L., B. Vila, D. Castanera, P. L. Falkingham, J. L. Barco, J. I. Canudo, P. L. Manning, and À. Galobart. 2014. Intra-trackway morphological variations due to substrate consistency: the El Frontal dinosaur tracksite (Lower Cretaceous, Spain). PLoS One 9: e93708.

Richter, A., and M. Reich (eds.). 2012. Dinosaur Tracks 2011: An International Symposium, Obernkirchen, April 14–17, 2011 Abstract Volume and Field Guide to Excursions. Universitätsverlag Göttingen, Göttingen, Germany.

Romilio, A., and S. W. Salisbury. 2011. A reassessment of large theropod dinosaur tracks from the mid-Cretaceous (late Albian-Cenomanian) Winton Formation of Lark Quarry, central-western Queensland, Australia: a case for mistaken identity. Cretaceous Research 32: 135–142.

Schanz, T., Y. Lins, H. Viefhaus, T. Barciaga, S. Läbe, H. Preuschoft, U. Witzel, and P. M. Sander. 2013. Quantitative interpretation of tracks for determination of body mass. PLoS One 8: e77606.

Scott, J. J., R. W. Renaut, R. B. Owen, and W. A. S. Sarjeant. 2007. Biogenic activity, trace formation, and trace taphonomy in the marginal sediments of saline, alkaline lake Bogoria, Kenya Rift Valley; pp. 311–332 in R. G. Bromley, L. A. Buatois, G. Mángano, J. F. Genise, and R. N. Melchor (eds.), Sediment-Organism Interactions: A Multifaceted Ichnology. SEPM Special Publication. Society for Sedimentary Geology, Tulsa, Oklahoma.

Sellers, W. I., P. L. Manning, T. Lyson, K. Stevens, and L. Margetts. 2009. Virtual palaeontology: gait reconstruction of extinct vertebrates using high performance computing. Palaeontologia Electronica 12: 11A.

Tagart, E. 1846. On markings in the Hastings sands near Hastings, supposed to be the footprints of birds. Quarterly Journal of the Geological Society of London 2: 267.

Thulborn, R. A. 1990. Dinosaur Tracks. Chapman and Hall, London, UK., 410 pp.

Thulborn, R. A. 2013. Lark Quarry revisited: a critique of methods used to identify a large dinosaurian track-maker in the Winton Formation (Albian-Cenomanian), western Queensland, Australia. Alcheringa: An Australasian Journal of Palaeontology 37: 312–330.

Weishampel, D. B., P. Dodson, and H. Osmólska. 2004. The Dinosauria. 2nd edition. University of California Press, Berkeley, California, 880 pp.

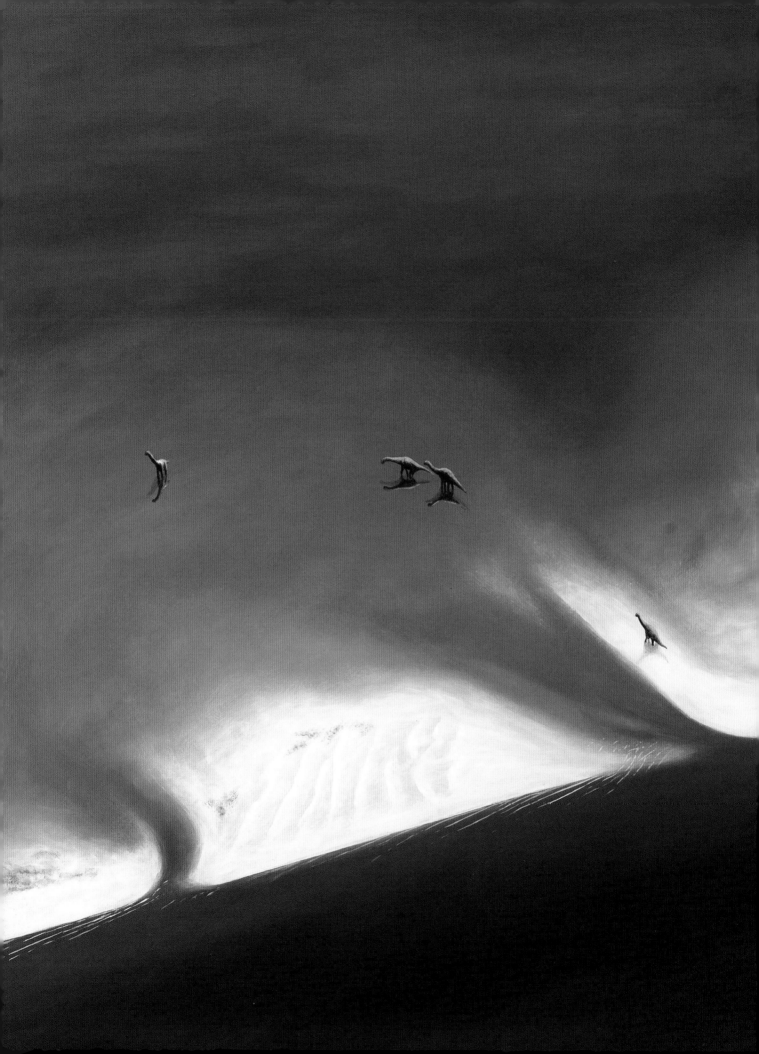

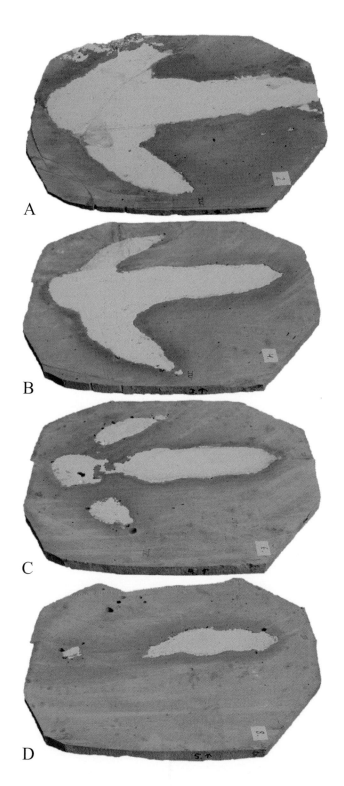

A

B

C

D

1.1. The morphological changes in a tridactyl track exposed to different degrees of erosion. The example is a plaster cast of an emu track emplaced in soft mud and afterward sectioned horizontally to simulate erosion of a track with the sedimentary infill still in place. (A) Section cut just below the tracking surface. (B) Section cut 14 mm below the tracking surface. (C) Section cut 25 mm below the tracking surface. (D) Section cut 38 mm below the tracking surface. Notice how the overall dimensions of the track become smaller with depth and that the individual parts of the track become separated with depth, until only the most deeply impressed parts are present, in this case, the distal part of the impression of the middle digit and the pad covering the metatarsal joint. *Figure based on experimental data from Milàn and Bromley (2006).*

Experimental and Comparative Ichnology

Jesper Milàn and Peter L. Falkingham

ONE OF THE MAIN PROBLEMS FACED IN PALEOICHNOL-ogy is the delicate relationship between the organism and the sediments it leaves its tracks and traces in. Since the first scientific report of comparisons between fossil and modern tracks, researchers have turned to making experiments and comparing tracks and trackways of modern animals in order to interpret fossil tracks and traces. The easiest experimental approach is simply to make living analogues to the fossil animals walk through soft sediment and directly study the tracks they produce. Modern, more sophisticated experimental procedures include laboratory-controlled settings with sediments of different properties and model feet and indenters impressed into the sediment to various degrees. When cement or plaster is used as a tracking medium in laboratory settings, it is possible to cut vertical sections through the tracks after hardening and to study the formation and morphology of undertracks along the subjacent horizons below the foot. Complementing physical experimentation is computer simulation, in which both substrate- and indenter-specific variables can be precisely, independently, and systematically controlled. Resultant virtual tracks can be visualized completely in three dimensions, together with a time component. Experimental ichnology is an important tool for people working with tracks because the experimental settings are able to provide important data about the variations in track morphologies that can occur as a result of erosion, gait, undertrack formation, ontogeny, and individual behavior of the track maker.

INTRODUCTION

A fossil vertebrate track is much more than just the mere impression of the trackmaker's foot in the substrate. In reality, a track is a complex three-dimensional structure extending into the substrate, the morphology of which is dependent on the local sedimentary conditions, the anatomy of the foot, plus any foot movements exercised by the trackmaker during the time of contact between the animal and the substrate (e.g., Padian and Olsen, 1984a; Allen, 1989, 1997; Gatesy et al., 1999; Manning, 2004; Milàn, Clemmensen, and Bonde, 2004; Milàn, 2006; Milàn and Bromley, 2006; Milàn et al.,

2006; Minter, Braddy, and Davis, 2007; Falkingham, 2014). Tracks and trackways are biogenic sedimentary structures, and as such, the taphonomic processes that influence their preservation are different from those that influence body-fossil preservation. Tracks are therefore likely to be preserved in sedimentary environments where no body fossils are preserved and, furthermore, cannot be transported from the sedimentary environment in which they are made. Tracks are thus very important sources of additional information about past biodiversity and animal behavior. Tracks made in particularly compliant substrates may record soft tissue morphology and distribution in the pedal parts and may even record the motion path of the foot (Gatesy et al., 1999; Gatesy, 2001; Avanzini, Piñuela, and García-Ramos, 2012; Cobos et al., 2016); information that in many instances is unobtainable from the study of skeletons alone.

Unfortunately, it is rarely simple to read all of this information directly from a track. Tracks emplaced in deep, soft substrates may not record the shape of the foot, or do so only poorly, but instead their morphology can be strongly determined by kinematics of the lower parts of the limb, creating an elongation of the track at the tracking surface (Gatesy et al., 1999). In some sediments, the different parts of the foot can also penetrate the sediment to different depths due to differentiated weight loads on the different parts of the foot (Falkingham, Bates, et al., 2011b); this can be particularly evident as part of the step cycle, because toe-down and kick-off phases present a smaller area and thus higher pressure to the substrate than during the weight-bearing phase (Thulborn and Wade, 1984; Thulborn, 1990). A track whose morphology is strongly influenced by foot anatomy and motion, as well as by the sediment collapsing or transmitting force, will vary considerably according to the depth at which it is exposed either by excavation or by weathering and erosion (Fig. 1.1).

One important factor to take into consideration when studying fossil tracks and trackways is erosion. When a track becomes exposed to subaerial erosion, the shape will gradually disintegrate and fine anatomical details will be lost (Henderson, 2006a). Tracks exposed to severe erosion can be hard to distinguish from undertracks (Milàn and Bromley, 2006). In fact, the act of erosion may destroy the surface

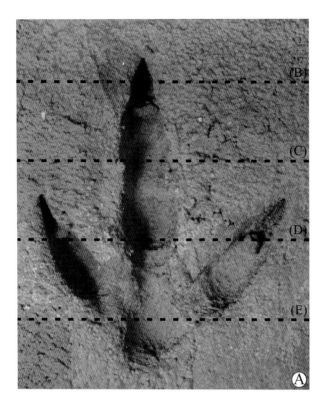

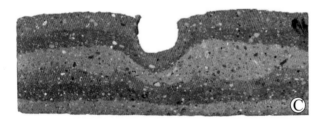

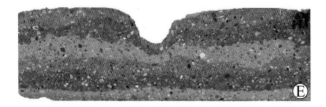

track and instead reveal some blurred fusion of subsurface undertracks at the surface. Furthermore, erosion can alter the total size and morphology of a track. This is especially the case with tracks originally emplaced in deep soft substrates where the trackmaker's foot has sunk to a considerable depth below the tracking surface, as the track may have a longer period of degeneration before being entirely destroyed.

Another factor able to strongly alter the appearance of a track is the phenomenon of undertracks. When a track is emplaced, the weight of the trackmaker's foot can be transmitted down and outward into the surrounding sediment. In cases where the rock is layered and breaks within the volume deformed by the transmitted force, a stacked succession of undertracks can be exposed (Fig. 1. 2). The phenomenon of undertracks was first noted by Hitchcock (1858), who depicts the same track exposed at successive, subjacent sediment surfaces. If not recognized for what they are, undertracks can be a source of confusion and misinterpretation, because they can make the track seem larger, less detailed and more rounded than the true track. Experimental work with track and undertrack formation has helped to illuminate morphologic variation of undertracks (Allen, 1997; Manning, 2004; Milàn and Bromley, 2006) (Fig. 1. 3).

When all of these factors are taken into consideration – that the track encountered can be the combined result of (1) the foot morphology of the animal, (2) the foot movements exercised by the animal, (3) the consistency of the substrate at the time of track formation, and that (4) the visible track may not represent a real "surface" track (due to where and how it is exposed) – then it becomes increasingly difficult to interpret and determine the original morphology and origin of the true track and subsequently the likely trackmaker.

In cases with modern animals, individuals can be identified from their tracks by a sufficiently trained eye (Speakman, 1954; Sharma, Jhala, and Sawarkar, 2005). This can be taken to extremes with modern trackers such as neoichnologist Tom Brown (1999), who teaches the tracking skills of Native Americans in his Tracker School, and who claims to be able not only to identify individual animals but also their

1.2. (A) Emu track emplaced in a package of layered colored cement. The used foot was freshly severed and still retained full flexibility of all joints. The broken lines indicate the location of sections in B–E. (B) Section cut through the impression of the claw of digit III. The sharp edges of the claw have cut down through the upper layers and formed a shallow undertrack in the lower layers. (C) Section through the middle of the impression of digit III. The shape of the digit is well preserved in the deformed surface layer, and undertracks are formed along the subjacent layers, becoming successively shallower and wider downward. (D) Section through the impression of digits II, III, and IV. The tridactyl pattern is recognizable in the undertracks. (E) The rounded pad covering the metatarsal/phalangeal joint has left a rounded impression recognizable in the undertracks. The length of the emu track is 19.5 cm. *Figure after Milàn and Bromley (2008).*

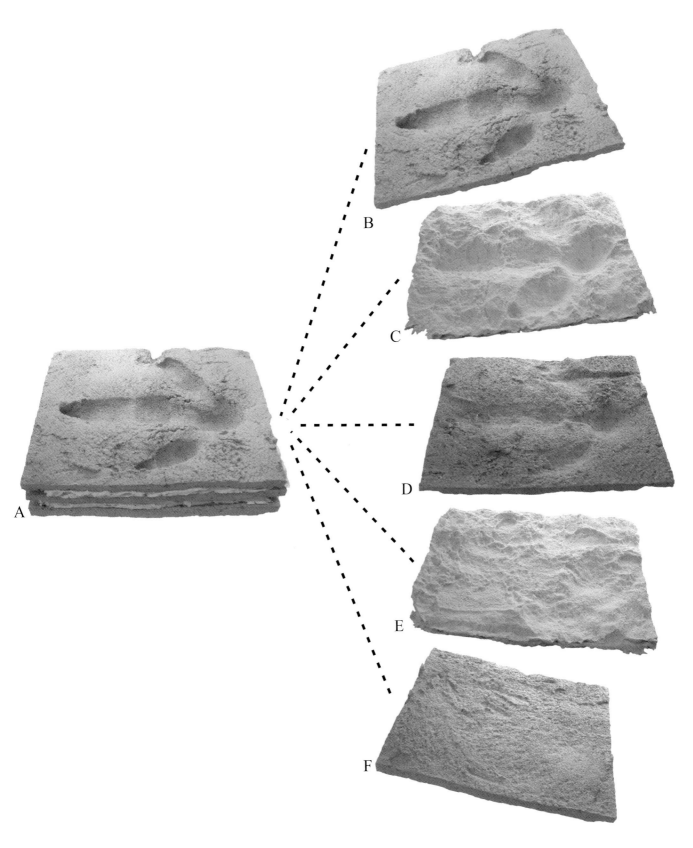

1.3. True track and undertracks. (A) Experimental setting with an emu track emplaced in an artificial layered heterogeneous substrate, allowing the package to be split along several subjacent horizons. (B) The true track at the surface is the direct impression of the trackmaker's foot and has preserved fine anatomical details like number and arrangement of digital pads, claw imprints and skin texture. (C) The track is still easily recognizable as an undertrack along the horizon 1 cm below the tracking surface but appears more rounded and less detailed, and it has a shallower relief (D–F). The undertracks become successively shallower and less detailed downward along each subjacent horizon until the track is unrecognizable. *Modified from Milàn and Bromley (2008).*

behaviors, sex, and intentions from subtle variations in the tracks.

When dealing with fossil footprints, however, such exquisite details are only rarely preserved – and even when they are, the trackmaker may remain unknown to science. In most cases it is, at best, only possible to assign tracks to higher taxonomic groups.

In order to obtain a better understanding of the factors affecting the morphology of a track, experimental and comparative work with track formation has been shown to be an important tool. This chapter will give an historical account of experimental ichnology from the scientific literature, explore some results and approaches of recent experimental work with living animals, and end in the modern and future age of computer-aided virtual ichnology.

HISTORY OF EXPERIMENTAL ICHNOLOGY

Throughout the history of ichnology, researchers have used extant animals with an inferred comparable anatomy and lifestyle to help understand and interpret fossil tracks and traces made by long extinct taxa. The first documented example of experimental ichnology where a scientist used the tracks of extant animals to compare with fossil tracks was that of the Reverend William Buckland, who in 1828 made crocodiles and tortoises walk through soft pie-crust, wet sand, and soft clay in order to identify the origin of fossil tracks and trackways from Permian sandstones in Scotland (Sarjeant, 1974). Buckland was also the first to describe some of the potential problems of experimental ichnology as his tortoise got stuck in the drying clay and had to be freed manually (full story transcribed in Tresise and Sarjeant, 1997). Later, Hitchcock (1836, 1858) compared the Triassic tracks and trackways of the Connecticut Valley with tracks of ratite birds and concluded that the fossil tracks had originated from large ground-dwelling birds (an interpretation that in light of modern understanding of bird evolution was not far wrong!), which led him to coin the name *Ornithichnites* to the tracks. A similar approach was used by Sollas (1879), who used casts of footprints from emus, rheas, and cassowaries to compare with the slender-toed tridactyl theropod footprints in the Triassic conglomerates of south Wales. Based on the close similarities, he suggested that the footprints could originate from ancestors of ratite birds (dinosaurs were only known from very sparse material at that time). He further noticed that the track morphology of the emu changed with the mode of progression, so that in tracks where the emu was accelerating, the metatarsal pad was less impressed into the sediment than when the emu was walking, and thus the morphology of the tracks from the same trackmaker changed with mode of progression.

In order to interpret the rich amphibian ichnofauna of the Permian Coconino Sandstone in northern Arizona, a substantial amount of comparative work with salamander and reptilian trackways has been conducted through time. McKee (1944, 1947) performed a series of experiments in which he made different kinds of reptiles, mostly lizards, walk up the slopes on simulated dune forests, similar to those found in the Coconino Sandstone. By varying the angle of the slope and the water content of the sand, from dry to saturated, McKee (1944, 1947) made convincing analogies to the different track morphologies found in the Coconino Sandstone. Peabody (1959) made detailed research on the trackways of living salamanders for comparison with Tertiary salamander tracks from California. Brand and Tang (1991) used subaqueous salamander trackways to argue for an underwater origin for the Coconino Sandstone, otherwise considered aeolian, but their arguments were heavily disputed (Lockley, 1992; Loope, 1992). Similar experiments with salamanders in substrates ranging from muddy to fine sand, level or sloping and with moisture contents from dry to submerged, clearly showed that the condition of the substrate is an important factor for the trackway morphology (Brand, 1979, 1996). McKeever and Haubold (1996) reclassified several Permian vertebrate trackways by demonstrating that several of the different ichnogenera erected through time were, in reality, sedimentological variations of no more than four valid ichnogenera.

MODERN FIELD AND LABORATORY EXPERIMENTS

The use of comparative and experimental ichnology, established in the early 1800s, remains to this day the ichnologist's most useful tool in the interpretation of dinosaur tracks. Modern experimental ichnology can largely be divided into those studies using extant taxa, building upon the earlier experimental work or more constrained and controlled laboratory-based indenter experiments.

Using Extant Taxa as Analogues

Following the work of Sollas (1879), ratite birds have been used especially for comparison with small bipedal dinosaurs. Padian and Olsen (1989) used the tracks and trackway pattern of a rhea (*Rhea americana*) to infer stance and gait of Mesozoic theropods, and Farlow (1989) examined the footprints and trackways of an ostrich (*Struthio camelus*) and compared them with theropod tracks and trackways. Diminutive theropod trackways from Zimbabwe were compared with a trackway from an ostrich chick to demonstrate the juvenile nature of the theropod trackmakers (Lingham-Soliar and Broderick,

1.4. Field experiments with emus. The foot of the extant emu is similar to that of nonavian theropods and is ideal for comparative track work. (A, B) Emus walking back and forth and making tracks and trackways in prepared lanes of sediment of different consistencies. (C) A track emplaced in damp sand preserves impressions of anatomical details such as number and arrangement of digital pads, claws, and even the faint texture of the skin. (D) In wet sand, the foot sinks down, creating steep track walls from the bottom of the track to the tracking surface. Radiating fractures are formed in the sediment around the track. (E) Track emplaced in deep firm mud. After withdrawal of the foot, the track walls slowly converge and destroy the shape of the track. An amount of sediment is sticking to the sole of the foot and is transported to the next step. (F) In deep semiliquid mud, the track flows together immediately after the foot is lifted, leaving only an amorphous depression in the tracking surface. Notice the trace from the claw of digit III scraping the surface as the foot is lifted forward.

A

B

1.5. Plaster casts of emu tracks emplaced in moist sand. When walking on firm substrates, the emu occasionally carries little or no weight on digit II. (A) Normal tridactyl track with all digits evenly impressed into the substrate. (B) Didactyl variant of emu track obtained from the same trackway as in A. Only a faint imprint of the claw of digit II is visible in the track.

2000). Gatesy et al. (1999) compared peculiar, partly collapsed theropod tracks emplaced in deep mud from Jameson Land, East Greenland, with the tracks of a turkey walking in similar deep substrate and found close similarities in the track morphologies. Gatesy et al. (1999) therefore concluded that the foot movement of theropods during walking exhibited close similarities to the foot movements of modern birds. Despite the gross overall similarities, the footprints of large ratites differ significantly from each other when examined in detail. This phenomenon was investigated by Farlow and Chapman (1997) and Farlow, McClain, and Shearer (1997), who used field observations and casts of tracks from emu, ostrich, cassowary, rhea, and the extinct moa to demonstrate how tracks from even closely related forms exhibit differences so significant as to warrant assignation to different ichnotaxa had they been found as fossil footprints.

Among ratite birds, the emu (*Dromaius novaehollandiae*) has in particular been the subject of much recent experimental and comparative ichnological work due to the close resemblance of its feet to those of nonavian theropods. The emu foot is tridactyl, consisting of digits II, III, and IV. Digit I, the hallux, which in modern birds is posteriorly directed and used for grasping branches and which occurs uncommonly as posterolateral traces in dinosaur footprints (Irby,

1995; Gatesy et al., 1999), is absent in the emu and the other ratites, except for the kiwi (Davies, 2003). The emu foot is 18–20 cm long, and the digits have all the anatomical details found in theropod tracks, including the small tubercles covering the ventral side of the foot, the configuration of the digital pads around the phalangeal joints, and the prominent claws.

Recent field experiments with emus conducted at a private emu farm in Denmark have provided important information about the span of variation within trackway parameters and range of track morphologies likely to be produced by the same animal. A study singularly carried out on tracks from emus walking in sediment of different consistencies (Milàn, 2006), demonstrated that even the tracks from the same animal could appear dramatically different if emplaced in substrates of different consistencies. The substrates were prepared using local soil mixed with different quantities of water to produce consistencies from firm to liquid mud and sand ranging from dry to saturated. Subsequently, the emus were encouraged to walk through the prepared areas to leave their tracks in the different sediments (Fig. 1.4).

One peculiar result obtained from the emu trackways was that the birds, when walking on firm substrates such as damp sand, impressed digit II to a lesser extent in substrate than digits III and IV, presumably due to a lower pressure under that digit compared with the rest of the foot. Often only a very faint impression hinted at the existence of the digit, and in some cases, digit II left no trace in the sediment at all, leaving perfectly didactyl traces (Milàn, 2006) (Fig. 1.5).

The divarication angle between the outer digits in a track is a parameter often used to characterize fossil tracks. However, by measuring 30 random tracks from trackways from the same emu, the divarication angle ranged from 61° to 102° with a mean value of 77° (Milàn, 2003). Even within the same track, the measured angle of divarication between the outer digits can differ significantly depending on where in the track the measurements are conducted (see Falkingham, 2016). A track obtained while the emu was accelerating to run showed the foot being forced down into the sediment to a depth of approximately 7 cm. During this process, the angle of divarication between the outer toes increased from 62° at the tracking surface to 77° at the bottom of the track. Not only did the overall divarication angle between the two outer digits, digits II and IV, increase while the foot was impressed into the substrate, but also the angle between the individual digits did not change evenly during the process. The angle between digits II and III decreased by 11° from the tracking surface to the bottom of the track, whereas the angle between digits III and IV increased by 27° (Table 1.1) (Milàn, 2006). Such a track would, if fossilized, and encountered eroded to different depths, show varying angles of divarication from 62° to 77°, according to the extent of erosion.

Although dinosaurs are the focus of this chapter, and indeed this book, the use of modern analogues to study track formation extends well beyond the Dinosauria. Turtle-like trackways from the Late Jurassic of Asturias, Spain, were compared with trackways from living turtles walking on sands and muds with different contents of moisture and some covered with 5–10 cm of water, making the turtle semisubmerged. The resulting trackway morphologies were strikingly similar to the morphologies of the suspected Jurassic turtle trackways (Avanzini et al., 2005).

The fossil trackway, *Pteraichnus*, described by Stokes (1957) as the trackway from a pterosaur, were reinterpreted by Padian and Olsen (1984a), who used a recent caiman walking on soft clay to demonstrate that *Pteraichnus* could as well be of crocodilian origin, though later research (Lockley et al., 1995; Mazin, Billon-Bruyat, and Padian, 2009) suggested that at least some *Pteraichnus* trackways are of pterosaur origin. Furthermore, Padian and Olsen (1984b) conducted experimental work with Komodo monitors, the tracks and trackways of which are similar to those of Triassic pseudosuchian thecodonts and to a lesser extent Early Jurassic crocodiles.

Diedrich (2002) demonstrated that in Triassic rhynchosaurid tracks there were several different preservational

Table 1.1. The changes in divarication angle and interdigital angle from the tracking surface to the bottom of the same track

Interdigital angle	Digits II–III	Digits III–IV	Digits II–IV
Tracking surface	35°	27°	62°
Bottom of track	24°	53°	77°
Difference	−11°	27°	15°

Note: The overall divarication angle increases from 62° to 77° from the tracking surface to the bottom of the track. That the divarication angle can differ according to depth in the track is important to bear in mind when interpreting fossil tracks exposed to erosion. The track is obtained when an emu is accelerating to run on moist sand.

variants caused by differences in water content of the sediments. Tracks made in dry subaerial sediments consisted of little more than faint claw imprints. With increasing water content of the sediment, shallow tracks were found having skin texture preserved. In more water-rich and thus softer sediments, the tracks became deeper and more blurred in shape until finally subaquatic tracks produced by a swimming trackmaker were found as elongated parallel scratch traces.

The tracks left by early hominids can be compared with those left by modern humans (Bates et al., 2013), and in these cases, it is rarely similarities but rather differences that the experimenter is looking for. The closeness between fossil and subfossil hominid tracks and modern human tracks also allows researchers to explore how tracks can rapidly deteriorate, and how they can be best recorded; Bennett et al. (2013) did just this using comparable laser scanning and photogrammetric techniques to record and compare hominid and human footprints.

Experimental work has also played a significant part in the identification of invertebrate trackways. Davis, Minter, and Braddy (2007) carried out an exhaustive study using five different types of extant terrestrial arthropods that produced a series of very detailed comparative schemes for arthropod trackways.

Experiments with Indenters and Artificial Substrates

Whereas field experiments with living animals are extremely useful because they include all the kinematic and behavioral aspects of a live animal, it can be difficult to fully control, or even record, the sediment properties and foot motion. In this case, laboratory-controlled experimental settings with model feet impressed into preprepared sediment packages can offer valuable insight into the connection between substrate consistency and track morphology.

It was the important work by Allen (1989, 1997) that first showed experimentally the formation of undertracks and subsurface deformations through indentation in layered plasticine. By forcing slotted and flat indenters into the plasticine, and then sectioning the resultant track volumes both vertically and horizontally, Allen was able to show with clarity the motion of the substrate at depth beneath the indenter. However, the foot of an animal generally does not perform a static up and down movement during its contact with the sediment (though see Milàn, Christiansen, and Mateus, 2005), rather the force vector applied through the foot varies throughout the step cycle as the animal's center of mass moves anteriorly (Alexander, 2003; Biewener, 2003). This movement was incorporated in later experiments with a tridactyl model foot that was brought into contact with the sediment at an angle, then rolled forward and lifted upward at angles similar to what was expected to have been performed by theropod dinosaurs (Manning, 2004). These experiments were carried out in a package of sand, clay, and plaster of paris. After oven-drying, the sand could be brushed away, leaving individual layers of plaster, each recording an undertrack. This enabled the study of subsurface deformations occurring beneath the tracks, as well as the formation of extramorphological features occurring in the sediment around and below the tracks. More recently, Jackson, Whyte, and Romano (2009, 2010) continued Manning's work, systematically altering sediment conditions in order to explore the variability in undertracks in light of substrate, particularly with regard to water content and the level of saturation.

The foot of a living animal, however, is not a static unit but a dynamic unit, with several joints each performing their characteristic part of the stride. Furthermore, the morphology and distribution of the soft fleshy parts of the foot affects the morphology of the tracks. Well-preserved theropod tracks from the Triassic of East Greenland show a pronounced lateral flattening of the digital pads during the foot contact phase of the step cycle (Gatesy, 2001). To incorporate as much as possible of the actual foot movements to the experimental settings, Milàn (2003) and Milàn and Bromley (2006, 2008) carried out experiments using a fresh emu foot impressed into packages of layered cement. During the contact between the foot and the sediment, the movements of the foot of a living emu were mimicked closely. The experimental tracks were then sectioned horizontally and vertically. Aside from showing that the tracks increased in horizontal dimensions with depth (while simultaneously decreasing in vertical dimensions), the authors also demonstrated that in particularly wet substrate, the greatest preservation of detail was found in undertracks rather than surface tracks.

In field studies, emus are observed to suddenly stop mid-stride and to continue after a while without this action being evident from the trackways (Milàn, pers. obs., 2006). Furthermore, emus occasionally produce erratic trackways with pace angulations in excess of 180°, overcrossing steps, which would be hard to interpret if only observed in a fossil trackway (Thulborn and Wade, 1984; Breithaupt, Southwell, and Matthews, 2006; Romillio and Salisbury, 2011). Also, studies of emu tracks from birds of different ontogenetic ages have produced important growth curves for their tracks, curves that can be directly compared with previously established growth rates for nonavian dinosaurs (Breithaupt, Southwell, and Matthews, 2007).

One of the greatest limitations of studying track formation physically is that the foot-substrate interaction is hidden from view (by both the foot and the substrate). The result is that a track can only be studied after it has been made. Although this is comparable to the way in which we can study fossil tracks, a greater insight may be gained if track formation can be observed, as it happens, at the foot-substrate interface. To this end, recent work by Ellis and Gatesy (2013) and Gatesy and Ellis (2016) used biplanar X-rays and X-ray reconstruction of moving morphology (XROMM) techniques (Gatesy et al., 2010; Brainerd et al., 2010) to visualize subsurface foot motion (from the bones) and sediment motion (from metal beads within the sediment). By manually manipulating a severed guineafowl leg, these authors were able to create realistic tracks in a very soft clay and track the interacting motion of the foot and substrate. The result is a data set providing a direct correlation between foot motions and track formation.

Computer Simulation

The setup, running, and analysis of physical experiments, whether laboratory indenter–based or live animal field–based, can be a long, laborious process. Arranging access to and time with animals requires cooperation and willingness on behalf of their keepers, whereas laboratory setups must be meticulously prepared and often involve long periods of oven-drying and preparation in order to observe subsurface deformation (Manning, 2004). Added to these difficulties, aside from the study of Gatesy and Ellis (2016), physical experimentation has thus far lacked an important aspect of track formation—that of observing real-time subsurface deformation. In all of these regards, computer simulation of tracks offers an alternative and complementary experimental method for studying and interpreting fossil tracks.

Initial uses of computer simulation in ichnology explored track placement within a trackway as a result of gait (Henderson, 2006b; Sellers et al., 2009) or weathering effects on track morphology (Henderson, 2006a). More recently, computer simulation has been employed to study track formation, particularly exploring subsurface deformation and the effects of varying load according to foot dynamics and body mass distribution. Much of this research up to the time of writing

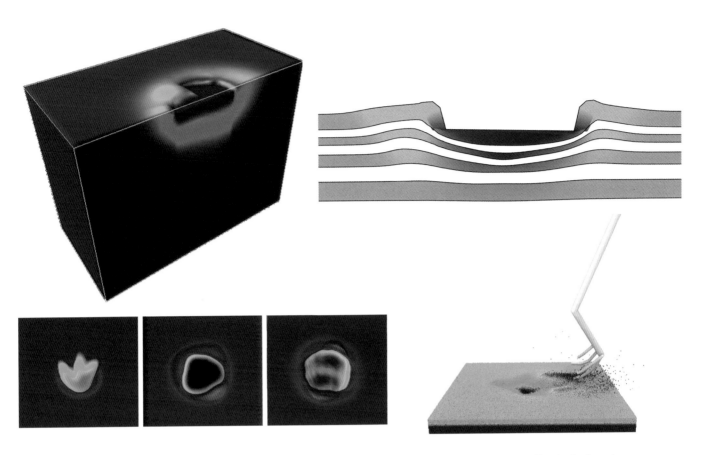

1.6. Various ways in which computer-simulated tracks can be visualized. (Upper left) A theropod track simulation sectioned longitudinally and coloured according to displacement. (Upper right) A simulated sequence of undertracks. (Bottom left) The flexibility of simulation enables a range of tracks to be made while controlling all parameters. (Bottom right) A discrete element simulation of guineafowl foot motions creating a track.

has used finite element analysis (FEA) (Margetts et al., 2006; Falkingham, Margetts, et al., 2009; Falkingham, Margetts, and Manning, 2010; Falkingham, Bates, et al., 2011a, 2011b; Schanz et al., 2013), an engineering methodology used for investigating stress and strain in materials under load (see Rayfield, 2007, for review of FEA applied to other paleontological areas). Computer simulation is particularly suited to isolating the effects of individual variables of the substrate, the shape of the foot, the force applied through the foot, and the motion of the foot. Each variable in a computer simulation can be kept constant, while the variable of interest can be systematically altered, even when in physical experiments such variables may be intrinsically linked. By taking advantage of high-performance computing, many simulations can be carried out simultaneously, and trends resulting from the variation of parameters can be observed.

Such systematic variation of parameters was used by Falkingham, Bates, et al. (2011a, 2011b) to show how track depth varied for multiple trackmakers over a range of substrates. This work led to two major findings; the first was that any given sediment exhibited what those authors termed a "Goldilocks Effect" such that the substrate consistency had to be "just right" – too soft and it an animal of a given size

would be unable to walk on it, too firm and that same animal would leave no tracks (Falkingham, Bates, et al., 2011b). The Goldilocks' range was remarkably narrow, though only homogeneous muds were considered in that work (Falkingham, Bates, et al., 2011b). Upcoming work by Falkingham and Baeker (in review) expand on this finding, and though more complex, homogeneous substrates widen the effective track bearing range of consistencies, the effect remains.

Related to that required specificity of substrate, Falkingham, Bates, et al. (2011a) showed that the position of the animals' center of mass (and thus loading over the forefeet and hind feet), in conjunction with the relative sizes of the fore and hind feet, could lead to cases where the manus exerted a substantially higher underfoot pressure to the pedes (or vice versa). In such cases, a substrate may be of such a consistency that it only records the impressions of the forefeet. That work offered an alternative to the "punting" hypothesis long associated with sauropod manus-only trackways.

More recently, Schanz et al. (2013) used FEA to back-calculate the mass of an animal from its tracks. Specifically, Schanz et al. used an elephant to produce tracks in moist sand. By characterizing the substrate mechanically and then simulating the track formation, they were able to link track depth with the weight of the animal. The logical next step, as those authors noted, will be to apply that methodology to fossil tracks, specifically those of sauropods, in order to make estimations of mass independently from those derived by measuring skeletal remains.

Finite element analysis, as used in the studies discussed has proved to be an incredibly useful tool for understanding the relationship between sediment and track morphology. However, there are limitations inherent in the method for simulating particularly complex track formation scenarios. Because the finite element mesh is a continuum, it is difficult to model extreme deformations, which occur when a foot sinks deeply into a substrate. Falkingham and Gatesy (2014) faced such a scenario in their study of guineafowl traversing dry sand-like poppy seeds. That study used XROMM (Brainerd et al., 2010; Gatesy et al., 2010) to capture the limb kinematics as the bird moved over and sank into the substrate.

In order to model the interaction of the foot and sediment, and particularly the collapse of the dry substrate, they used the discrete element method, in which particles representing individual grains were simulated. By transferring the motions captured by XROMM into the discrete element method simulation, a virtual track closely matching the real thing was produced, which could split along virtual bedding planes. Falkingham and Gatesy (2014) were then able to use the simulated guineafowl track to interpret enigmatic features of 200 million year–old dinosaur tracks.

Visualization is a major advantage to computer modeling of track formation. Any simulated track can be sectioned multiple times, in any direction while retaining the original–something that is impossible with physical modeling. In addition, tracks can be color mapped according to displacements or stresses occurring in the deformed substrate, or they can be peeled away to reveal virtual undertracks at any level (Fig. 1.6). However, confidence in the validity of simulations and applicability to real-world scenarios can only come from physical experiments with indenters and/or live animals, making virtual experiments a complementary, rather than replacement approach.

DISCUSSION

Experimental ichnology has proved itself to be a useful tool for obtaining a better understanding of the many factors involved in all the different processes that influence the morphology of the track, from the moment it is originally emplaced to its discovery as a fossil. There are pros and cons associated with both animal-/field-based experiments and those carried out in the laboratory. Whereas experiments using model feet and indenters (Allen, 1989, 1997; Manning, 2004) or computer simulations (Falkingham, 2011b; Schanz

et al., 2013) are easier to conduct and, importantly, much easier to document because there is total control over all parameters, the experiments can lack the realism afforded by live animals. By using live animals, all the primary factors that determine track morphology–anatomy, locomotor kinematics, and substrate consistency–can be included simultaneously. However, extrapolating which of these variables is responsible for the morphological variation in the resultant tracks can be extremely difficult, and for this reason, laboratory or computational experiments remain a vital companion for experimental work with extant animals.

Because the pattern of a trackway is the result of all the actions exercised by the animal during the walk, observations of extant animals with comparable lifestyles are important when fossil trackways are interpreted. The extant emu, *Dromaius novaehollandiae*, and other large cursorial birds are the best living analogues to medium- to large-sized Mesozoic theropods. The emu and the rhea (*Rhea americana*), in having pedal skeleton and footprint morphology resembling that of nonavian theropods, are especially obvious candidates for comparative ichnological work. The ostrich (*Struthio camelus*), although larger than both the emu and the rhea, has a specialized highly reduced didactyl foot that has lost the resemblance with the tracks of nonavian theropods. For studying tracks made by smaller theropods, modern analogues may include turkeys (*Meleagris* sp.) and other similar predominantly ground-based birds. Birds, being the direct descendants of theropod dinosaurs, offer a logical choice as an extant analogue, but what of other extinct animals? There is a much more restricted choice for the large sauropods, ceratopsians, and hadrosaurs–elephants may be the closest living animals in terms of size but likely differ so considerably in locomotor biomechanics and behavior as to be of limited use for detailed studies beyond weight-based interpretations (as in Schanz et al., 2013). Other extant taxa beyond the dinosaurs offer a similar challenge; pterosaurs, pelycosaurs, and early tetrapods are other examples for which a useful modern analogue would be difficult to justify.

The field of experimental ichnology has rapidly expanded within the last decade or so, as an increasing number of researchers have discovered the applications of experimental work with track formation. As more people become involved in the topic, new and exciting experimental methods are constantly being invented. Computer-based methods are providing a new approach to experimental ichnology, an approach based on quantitative values, and one that will integrate strongly in the future with the rise in digitization techniques employed by ichnologists including laser scanning (Bates, Manning, et al., 2008; Bates, Rarity, et al., 2008; Farlow et al., 2010; Bennett et al., 2013) and photogrammetry (Breithaupt, Matthews, and Noble, 2004; Bates et al., 2009; Remondino et al., 2010; Falkingham, 2012; Bennett et al., 2013; Falkingham, Bates, and Farlow, 2014; Razollini et al., 2014).

Although some wider trends and phenomena have been elucidated by experimental ichnology, work so far has predominantly been highly specialized, looking at specific case studies of either taxa or substrate. What remains are broader challenges for which experimental ichnology must be tied to descriptive and taxonomic work. More systematic laboratory and simulation work may open the door to a comprehensive understanding of how aspects of morphological variation between tracks are linked to the trackmaker's anatomy, its behavior, and the substrate it walks through. Currently, disentangling these factors from the final morphology is incredible difficult. Such work may eventually provide a definitive means of identifying subsurface undertracks or whether a track has been altered significantly from its original form by weathering and erosion. More experimental work with live animals will enable a closer linkage between track morphologies and limb motions and will pave the way for a standardized approach for inferring biomechanics from tracks—particularly studies involving different taxa or different ontogenetic stages of a single taxon. In the future, we envisage that experimental ichnology will be entirely integral to the descriptive and taxonomic part of the science. Interpretations of tracks and trackways will be accompanied by experimental data used to support hypotheses or disprove null hypotheses about how the tracks were made and the identity of what made them. Experimental ichnology therefore, despite its 150 year age, remains a modern, exciting field of research within paleontology.

ACKNOWLEDGMENTS

Part of this research was supported by a Ph.D. grant to Milàn from the Faculty of Science, University of Copenhagen. Falkingham was supported by a Marie Curie International Outgoing Fellowship within the Seventh European Framework Programme. Richard G. Bromley kindly read, commented, and improved the language of an early version of the manuscript. Karin Holst, Mønge, Denmark, kindly provided access to her domesticated emus. We wish to thank Daniel Marty, Don Henderson, and Nic Minter for their highly useful comments and suggestions in improving this essay.

REFERENCES

Alexander, R. M. 2003. Principles of Animal Locomotion. Princeton University Press, Princeton, New Jersey, 371 pp.

Allen, J. R. L. 1989. Fossil vertebrate tracks and indenter mechanics. Journal of the Geological Society, London 146: 600–602.

Allen, J. R. L. 1997. Subfossil mammalian tracks (Flandrian) in the Severn Estuary, S.W. Britain: mechanics of formation, preservation and distribution. Philosophical Transactions of the Royal Society of London, B 352: 481–518.

Avanzini M., L. Piñuela, and J. C. García-Ramos. 2012. Late Jurassic footprints reveal walking kinematics of theropod dinosaurs. Lethaia 45: 238–252.

Avanzini, M., J. C. Garcia-Ramos, J. Lires, M. Menegon, L. Piñuela, and L. A. Fernandez. 2005. Turtle tracks from the Late Jurassic of Asturias, Spain. Acta Palaeontologica Polonica 50: 743–755.

Bates, K. T., P. L. Manning, B. Vila, and D. Hodgetts. 2008. Three dimensional modelling and analysis of dinosaur trackways. Palaeontology 51: 999–1010.

Bates, K. T., B. H. Breithaupt, P. L. Falkingham, N. A. Matthews Neffra, D. Hodgetts, and P. L. Manning. 2009. Integrated LiDAR and photogrammetric documentation of the Red Gulch Dinosaur Tracksite (Wyoming, USA); pp. 101–103 in S. E. Foss, J. L. Cavin, T. Brown, J. L. Kirkland, and V. L. Santucci (eds.), Proceedings of the Eighth Conference on Fossil Resources, St. George, Utah, May 19–21. BLM Regional Paleontologist, Salt Lake City, Utah.

Bates, K. T., F. Rarity, P. L. Manning, D. Hodgetts, B. Vila, O. Oms, À. Galobart, and R. Gawthorpe. 2008. High-resolution LiDAR and photogrammetric survey of the Fumanya dinosaur tracksites (Catalonia): implications for the conservation and interpretation of geological heritage sites. Journal of the Geological Society, London 165: 115–127.

Bates K. T., D. Collins, R. Savage, J. McClymont, E. Webster, T. C. Pataky, K. D'Aout, W. I. Sellers, M. R. Bennett, and R. H. Crompton. 2013. The evolution of compliance in the human lateral mid-foot. Proceedings of the Royal Society B: Biological Sciences 280: 20131818.

Bates, K. T., P. L. Falkingham, D. Hodgetts, J. O. Farlow, B. H. Breithaupt, M. O'Brien, N. Matthews, W. I. Sellers, and P. L. Manning. 2009. Digital imaging and public engagement in palaeontology. Geology Today 25: 134–139.

Bennett, M. R., P. L. Falkingham, S. Morse, K. Bates, and R. H. Crompton. 2013. Preserving the impossible: conservation of soft-sediment hominin footprint sites and strategies for three-dimensional digital data capture. PLoS One 8: e60755.

Biewener, A. A. 2003. Animal Locomotion. Oxford University Press, Oxford, UK.

Brainerd, E. L., D. B. Baier, S. M. Gatesy, T. L. Hedrick, K. A. Metzger, S. L. Gilbert, and J. J. Crisco. 2010. X-ray reconstruction of moving morphology (XROMM): precision, accuracy and applications in comparative biomechanics research. Journal of Experimental Zoology 313(5): 262–279.

Brand, L. 1979. Field and laboratory studies on the Coconino Sandstone (Permian) vertebrate footprints and their palaeoecological implications. Palaeogeography, Palaeoclimatology, Palaeoecology 28: 25–38.

Brand, L. R. 1996. Variations in salamander trackways resulting from substrate differences. Journal of Paleontology 70: 1004–1010.

Brand, L. R., and T. Tang 1991. Fossil vertebrate footprints in the Coconino Sandstone (Permian) of northern Arizona: evidence for underwater origin. Geology 19: 1201–1204.

Breithaupt, B. H., N. Matthews, and T. Noble. 2004. An integrated approach to three-dimensional data collection at dinosaur tracksites in the Rocky Mountain West. Ichnos 11: 11–26.

Breithaupt, B. H., E. H. Southwell, and N. A. Matthews. 2006. Walking with emus: insight into dinosaur tracking in the 21st century. Geological Society of America Abstracts 38: 537.

Breithaupt, B. H., E. H. Southwell, and N. A. Matthews. 2007. Growing up in the middle Jurassic: ichnological evidence for family groups of theropods in Wyoming; comparison of footprints and growth rates of emus and dinosaurs. In J. J. Liston (ed.), 55th Symposium of Vertebrate Palaeontology and Comparative Anatomy and the 16th Symposium of Palaeontological Preparation and Conservation held at the University of Glasgow, 28th August–1st September 2007: Abstracts of Presentations. 1 p.

Brown, T., Jr. 1999. The Science and Art of Tracking. Berkley Books, New York, New York, 240 pp.

Cobos, A., F. Gascó, R. Royo-Torres, M. G. Lockley, and L. Alcalá. 2016. Dinosaur tracks as "four-dimensional phenomena" reveal how different species moved; chap. 14 in P. L. Falkingham, D. Marty, and A. Richter (eds.), Dinosaur Tracks: The Next Steps. Indiana University Press, Bloomington, Indiana.

Davies, S. J. J. F. 2003. Kiwis; pp. 89–90 in M. Hutchins (ed.), Grzimek's Animal Life Encyclopedia. 8 Birds I Tinamous and Ratites to Hoatzins. 2 edition. Gale Group, Farmington Hills, Michigan.

Davis, R. B., N. J. Minter, and S. J. Braddy. 2007. The neoichnology of terrestrial arthropods. Palaeogeography, Palaeoclimatology, Palaeoecology 255: 284–307.

Diedrich, C. 2002. Vertebrate track bed stratigraphy at new megatrack sites in the Upper Wellenkalk member and orbicularis member (Muschelkalk, Middle Triassic) in carbonate tidal flat environments of the western Germanic Basin. Palaeogeography, Palaeoclimatology, Palaeoecology 183: 185–208.

Ellis, R. G., and S. M. Gatesy. 2013. A biplanar X-ray method for three-dimensional analysis of track formation. Palaeontologia Electronica 16(1): 16.

Falkingham, P. L. 2012 Acquisition of high resolution three-dimensional models using free, open-source, photogrammetric software. Palaeontologia Electronica 15: 1T: 15p.

Falkingham, P. L. 2014 Interpreting ecology and behaviour from the vertebrate fossil track record. Journal of Zoology 292: 222–228.

Falkingham, P.L., and S. M. Gatesy. 2014. The birth of a dinosaur footprint: subsurface 3D motion reconstruction and discrete element simulation reveal track ontogeny. PNAS 111(51): 18279–18284.

Falkingham, P. L., L. Margetts, and P. L. Manning. 2010. Fossil vertebrate tracks as palaeopenetrometers: confounding effects of foot morphology. PALAIOS 25: 356–360.

Falkingham, P. L., L. Margetts, I. M. Smith, and P. L. Manning. 2009. Reinterpretation of palmate and semi-palmate (webbed) fossil tracks: insights from finite element modelling. Palaeogeography, Palaeoclimatology, Palaeoecology 271: 69–76.

Falkingham, P. L., K. T. Bates, L. Margetts, and P. L. Manning. 2011a. Simulating sauropod manus-only trackway formation using finite-element analysis. Biology Letters 7: 142–145.

Falkingham, P. L., K. T. Bates, L. Margetts, and P. L. Manning. 2011b. The 'Goldilocks' effect: preservation bias in vertebrate track assemblages. Journal of the Royal Society Interface 8: 1142–1154.

Falkingham, P. L., K. T. Bates, and J. O. Farlow. 2014. Historical photogrammetry: Bird's Paluxy River dinosaur chase sequence digitally reconstructed as it was prior to excavation 70 years ago. PLoS One 9: e93247.

Farlow, J. O. 1989. Ostrich footprints and trackways: implications for dinosaur ichnology; pp. 243–248 in D. D. Gillette and M. G. Lockley (eds.), Dinosaur Tracks and Traces. Cambridge University Press, Cambridge, UK.

Farlow, J. O., and R. E. Chapman. 1997. The scientific study of dinosaur footprints; pp. 519–552 in J. O. Farlow and M. K. Brett-Surman (eds.), The Complete Dinosaur. Indiana University Press, Bloomington, Indiana.

Farlow, J. O., J. McClain, and K. Shearer 1997. Intraspecific and interspecific variability in foot and footprint shapes in ground birds: implications for the ichnology of bipedal dinosaurs. Journal of Vertebrate Paleontology 17: 45A.

Farlow, J. O., M. O'Brien, G. J. Kuban, B. F. Dattilo, L. Pinuela, K. T. Bates, P. L. Falkingham, A. Rose, A. Freels, C. J. Kumagai, C. Libben, J. Smith, and J. Whitcraft. 2010. Dinosaur tracksites of the Paluxy River (Glen Rose Formation, Lower Cretaceous), Dinosaur Valley State Park, Somervell County, Texas, USA; pp. 15–16 in 5th Jornadas Internacionales sobre Paleontology a de Dinosaurios y su Entorno. Salas de los Infantes, Burgos, Spain.

Gatesy, S. M. 2001. Skin impressions of Triassic theropods as records of foot movement. Bulletin of the Museum of Comparative Zoology 156: 137–149.

Gatesy, S. M., and R. Ellis. Beyond surfaces: a particle-based perspective on track formation; chap. 5 in P. L. Falkingham, D. Marty, and A. Richter (eds.), Dinosaur Tracks: The Next Step. Indiana University Press, Bloomington, Indiana.

Gatesy, S. M., D. B. Baier, F. A. Jenkins, and K. P. Dial. 2010. Scientific rotoscoping: a morphology-based method of 3-D motion analysis and visualization. Journal of Experimental Zoology 313(5): 244–261.

Gatesy, S. M., K. M. Middleton., F. A. Jenkins Jr., and N. H. Shubin. 1999. Three-dimensional preservation of foot movements in Triassic theropod dinosaurs. Nature 399: 141–144.

Henderson, D. M. 2006a. Simulated weathering of dinosaur tracks and the implications for their characterization. Canadian Journal of Earth Sciences 43: 691–704.

Henderson, D. M. 2006b. Burly gaits: centers of mass, stability, and the trackways of sauropod dinosaurs. Journal of Vertebrate Paleontology 26(4): 907–921.

Hitchcock, E. 1836. Ornithichnology: description of the toot marks of birds, (Ornithichnites) on New Red Sandstone in Massachusetts. American Journal of Science 29: 307–340.

Hitchcock, E. 1858. Ichnology of New England: A Report on the Sandstone of the Connecticut Valley, Especially Its Fossil Footmarks. William White, Boston, Massachusetts, 296 pp.

Irby, G. V. 1995. Posterolateral markings on dinosaur tracks, Cameron Dinosaur tracksite, Lower Jurassic Moenave Formation, Northeastern Arizona. Journal of Paleontology 69(4): 779–784.

Jackson, S. J., M. A. Whyte, and M. Romano. 2009. Laboratory-controlled simulations of dinosaur footprints in sand: a key to understanding vertebrate track formation and preservation. Palaios 24(3–4): 222–238.

Jackson, S. J., M. A. Whyte, and M. Romano. 2010. Range of experimental dinosaur (Hypsilophodon foxii) footprints due to variation in sand consistency: how wet was the track? Ichnos 17: 197–214.

Lingham-Soliar, T., and T. Broderick. 2000. An enigmatic Early Mesozoic dinosaur trackway from Zimbabwe. Ichnos 7: 135–148.

Lockley, M. G. 1992. Comment and reply on "Fossil vertebrate footprints in the Coconino Sandstone (Permian) of Northern Arizona: evidence for underwater origin." Geology 20: 666–667.

Lockley, M. G., T. J. Logue, J. J. Moratella, A. P. Hunt, R. J. Schultz, and J. W. Robinson. 1995.

The fossil trackway Pteraichnus is pterosaurian, not crocodillian: implications for the global distribution of pterosaur tracks. Ichnos 4: 7–20.

Loope, D. B. 1992. Comment on "Fossil vertebrate footprints in the Coconino Sandstone (Permian) of northern Arizona: evidence for underwater origin." Geology 20: 667–668.

Manning, P. 2004. A new approach to the analysis and interpretation of tracks: examples from the Dinosauria; pp. 93–123 in D. McIlroy (ed.), The Application of Ichnology to Palaeoenviromental and Stratigraphic Analysis. Special Publications 228. Geological Society, London, UK.

Margetts, L., I. M. Smith, J. Leng, and P. L. Manning. 2006. Parallel three-dimensional finite element analysis of dinosaur trackway formation; pp. 743–749 in H. F. Schweiger (ed.), Numerical Methods in Geotechnical Engineering. Taylor & Francis, London, UK.

Mazin, J.-M., J.-P. Billon-Bruyat, and K. Padian. 2009. First record of a pterosaur landing trackway. Proceedings of the Royal Society B: Biological Sciences 276: 3881–3886.

McKee, E. D. 1944. Tracks that go uphill. Plateau 16: 61–72.

McKee, E. D. 1947. Experiments on the development of tracks in fine cross-bedded sand. Journal of Sedimentary Petrology 17: 23–28.

McKeever, P. J., and H. Haubold. 1996. Reclassification of vertebrate trackways from the Permian of Scotland and related forms from Arizona and Germany. Journal of Paleontology 70: 1011–1022.

Milàn, J. 2003. Experimental Ichnology, experiments with track and undertrack formation using emu tracks in sediments of different consistencies, with comparisons to fossil dinosaur tracks. Ms.C. thesis, Geological Institute, University of Copenhagen, Copenhagen, Denmark, 124 pp.

Milàn, J. 2006. Variations in the morphology of emu (Dromaius novaehollandiae) tracks, reflecting differences in walking pattern and substrate consistency: ichnotaxonomical implications. Palaeontology 49: 405–420.

Milàn, J., and R. G. Bromley 2006. True tracks, undertracks and eroded tracks, experimental work with tetrapod tracks in laboratory and field. Palaeogeography, Palaeoclimatology, Palaeoecology 231: 253–264.

Milàn, J., and R. G. Bromley. 2008. The impact of sediment consistency on track- and undertrack morphology: experiments with emu tracks in layered cement. Ichnos 15: 18–24.

Milàn, J., P. Christiansen, and O. Mateus. 2005. A three-dimensionally preserved sauropod manus impression from the Upper Jurassic of Portugal: implications for sauropod manus shape and locomotor mechanics. Kaupia 14: 47–52.

Milàn, J., L. B. Clemmensen, and N. Bonde. 2004. Vertical sections through dinosaur tracks (Late Triassic lake deposits, East Greenland): undertracks and other subsurface deformation structures revealed. Lethaia 37: 285–296.

Milàn, J., M. Avanzini, L. B. Clemmensen, J. C. Garciá-Ramos, and L. Piñuela. 2006. Theropod foot movement recorded from Late Triassic, Early Jurassic, and Late Jurassic fossil footprints; pp. 352–364 in J. D. Harris, S. G. Lucas, J. A. Spielmann, M. G. Lockley, A. R. C. Milner, and J. I. Kirkland (eds.), The Triassic-Jurassic Terrestrial Transition. Bulletin 37. New Mexico

Museum of Natural History and Science, Albuquerque, New Mexico.

Minter, N. J., S. J. Braddy, and R. B. Davis. 2007. Between a rock and a hard place: arthropod trackways and ichnotaxonomy. Lethaia 40: 365–375.

Padian, K., and P. E. Olsen 1984a. The fossil trackway *Pteraichnus*: not pterosaurian, but crocodillian. Journal of Paleontology 58: 178–184.

Padian, K., and Olsen, P. E. 1984b. Footprints of Komodo monitor lizard and the trackways of fossil reptiles. Copeia 1984: 662–671.

Padian, K., and P. E. Olsen. 1989. Ratite footprints and the stance and gait of Mesozoic theropods; pp. 231–242 in D. D. Gillette and M. G. Lockley (eds.), Dinosaur Tracks and Traces. Cambridge University Press, Cambridge, UK.

Peabody, F. E. 1959. Trackways of living and fossil salamanders. Publications in Zoology 63: 1–72.

Rayfield, E. J. 2007. Finite element analysis and understanding the biomechanics and evolution of living and fossil organisms. Annual Review of Earth and Planetary Sciences 35: 541–576.

Razzolini, N. I., B. Vila, D. Castanera, P. L. Falkingham, J. L. Barco, J. I. Canudo, P. L. Manning, and À. Galobart. 2014.

Intra-trackway morphological variations due to substrate consistency: the El Frontal dinosaur tracksite (Lower Cretaceous, Spain). PLoS One 9: e93708.

Remondino, F., A. Rizzi, S. Girardi, F. M. Petti, and M. Avanzini. 2010. 3D ichnology: recovering digital 3D models of dinosaur footprints. Photogrammetric Record 25: 266–282.

Romilio, A., and S. W. Salisbury. 2011. A reassessment of large theropod dinosaur tracks from the mid-Cretaceous (late Albian-Cenomanian) Winton Formation of Lark Quarry, central-western Queensland, Australia: a case for mistaken identity. Cretaceous Research 32: 135–142.

Sarjeant, W. A. S. 1974. A history and bibliography of the study of fossil vertebrate footprints in the British Isles. Palaeogeography, Palaeoclimatology, Palaeoecology 16: 265–378.

Schanz, T., Y. Lins, H. Viefhaus, T. Barciaga, S. Läbe, H. Preuschoft, U. Witzel, and P. M. Sander. 2013. Quantitative interpretation of tracks for determination of body mass. PLoS One 8: e77606.

Sellers, W. I., P. L. Manning, T. Lyson, K. Stevens, and L. Margetts. 2009. Virtual palaeontology: gait reconstruction of extinct vertebrates using

high performance computing. Palaeontologia Electronica 12: 11A.

Sharma, S., Y. Jhala, and V. B. Sawarkar. 2005. Identification of individual tigers (*Panthera tigris*) from their pugmarks. Journal of the Zoological Society of London 267: 9–18.

Sollas, W. J. 1879. On some three-toed footprints from the Triassic conglomerate of southern Wales. Quarterly Journal of the Geological Society of London 35: 511–517.

Speakman, F. J. 1954. Tracks, Trails and Signs. G. Bell and Sons, London, UK., 154 pp.

Stokes, W. L. 1957. Pterodactyl tracks from the Morrison Formation. Journal of Paleontology 31: 952–954.

Thulborn, R. A. 1990. Dinosaur Tracks. Chapman and Hall, London, UK., 410 pp.

Thulborn, R. A., and M. Wade. 1984. Dinosaur trackways in the Winton Formation (mid-Cretaceous) of Queensland. Memoirs of the Queensland Museum 21(2): 413–517.

Tresise, G., and W. A. S. Sarjeant. 1997. The Tracks of Triassic Vertebrates. The Stationary Office, London, UK., 216 pp.

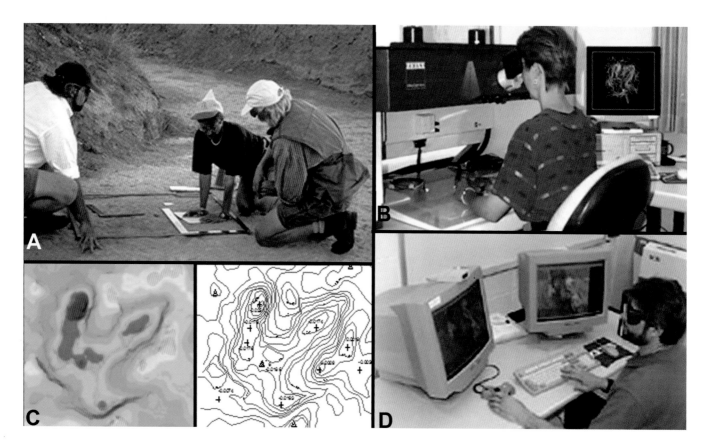

2.1. (A) Placing photogrammetric control on the track surface at the Red Gulch Dinosaur Tracksite (RGDT), Wyoming, summer 1998. (B) Compiling topographic contour maps using an analytical stereoplotter, fall 1998. (C): Topographic contour map of single, Middle Jurassic theropod track from RGDT (right) and depth map (left) interpolated from the contour elevations. (D) Digital Softcopy Photogrammetric workstation circa 2001. See Breithaupt and Matthews (2001) and Breithaupt et al. (2001).

Close-Range Photogrammetry for 3-D Ichnology: The Basics of Photogrammetric Ichnology

2

Neffra Matthews, Tommy Noble, and Brent Breithaupt

INTRODUCTION

VERTEBRATE TRACE FOSSILS REFLECT THE COMPLEX interrelationship between an animal's activities and the substrate (Manning, 2004; Falkingham, 2014), which is well represented in the ichnofaunal record of Mesozoic dinosaurs (Thulborn, 1990; Lockley, 1991; Lockley and Meyer, 2000; Wright and Breithaupt, 2002). As such, these unique three-dimensional (3-D) fossils warrant detailed recordation that captures their multidimensional features to fully understand formation and preservation of the ichnofossils, as well as dinosaur community dynamics (Lockley, 1986; Falkingham, 2014). Currently, the most cost-efficient and high-resolution mechanism to collect 3-D digital data of trace fossils is through the proper use of photogrammetry. Digital ichnological and spatial data capture a large portion of the incredible wealth of information provided at tracksites and are the basis for photogrammetric ichnology. As such, close-range photogrammetry (CRP) can assist in the proper documentation, preservation, and assessment of ichnological resources of any size at any location no matter the orientation of the track surface. A properly executed ichnological photogrammetric project has the quality, reliability, and authenticity necessary for scientific use. Three-dimensional image data sets created from stereoscopic digital photography provide permanent digital records of fossil tracks, including the creation of digital type specimens, or "Digitypes" (Adams et al., 2010). CRP is a noninvasive, objective recording and analysis method, which provides a visual, quantifiable baseline to evaluate track-bearing surfaces. The CRP data sets support accurate visualization of the fossils and can be used to create a digital archive from tracksites worldwide, allowing researchers to conduct detailed scientific studies on these paleontological resources. Imagery that is correctly taken now can be used in software developments and remain relevant into the future. Fortunately, the tools to conduct photogrammetric documentation (e.g., digital camera, scale bar) are already part of any good ichnologist's tool kit. Although conducting photogrammetric documentation need not be difficult, there are concepts and complexities that exist. A better understanding of these concepts may be reached by briefly reviewing the history of photogrammetry.

HISTORY OF PHOTOGRAMMETRY

The history of photogrammetry is a journey of conceptual developments punctuated by advances in technology that began with early man's observations of the natural world and continues through the technological advancements of today. Highlights of a very few of these developments and advancements (and in some cases technology limitations) will be discussed in this chapter. For a more complete chronicling of the history of photogrammetry, see Gruner (1977), Thompson and Gruner (1980), Ghosh (2005), and Albertz (2007).

The concepts of projected geometry and perspective viewing that form the basis of modern photogrammetry have been utilized since Babylonian and Egyptian times. These concepts were put to use for mapmaking in the late 1400s and early 1500s by Leonardo da Vinci and Albrecht Duerer (Gruner, 1977; Thompson and Gruner, 1980). Both of these individuals utilized the concepts of changes in scale and perspective to illustrate and measure a subject without directly touching the subject. Included in their "tool kit" was the pinhole camera or camera obscura used to project a scene onto a piece of vellum, where it could then be drawn and measured.

For much of its early history, advancements in the discipline that would become photogrammetry were made without photography (capturing images on a light-sensitive plate), a component that today is considered essential. In 1839, after several years of experimentation, a partnership between Louis-Jacques-Mandé Daguerre and Joseph Nicéphore Niépce produced an effective method of capturing an image on a light-reactive plate—the daguerreotype. This was a 'wet process' as it was necessary to expose the plate to light while it was coated with a solution applied just before exposure. Despite difficulties of early photography, Frenchman Gaspard-Félix Tournachon (nicknamed Nadar) took the first aerial photography in 1855 using a hot air balloon to take himself and his camera equipment aloft. At about the same time, Aimé Laussedat utilized kite and balloon photography for topographic map compilation and called his technique iconometry (icon, image; metry, to measure) (Gruner, 1977; Thompson and Gruner, 1980).

The term photogrammetry (making measurements from photographs) was first published in 1867 in the *Wochenblatt des Architekten-Vereins zu Berlin* (Berlin Architectural Society-Weekly Journal) to describe the technique introduced by the German civil engineer Albrecht Meydenbauer. Meydenbauer realized that direct measurements of a structure could be replaced by indirect measurements made from photographic images. He referred to this method as plane table photogrammetry and recognized the following requirements for success: (1) a camera with a wide-angle lens, (2) a defined principal distance (or calibrated focal length), (3) an image coordinate system, and (4) two or more images required to plot a point. Much of Meydenbauer's career was a struggle to prove the accuracy and utility of his 'Photogrammetrie' method for architectural documentation. A mark of his eventual success was his appointment by the Prussian Ministry of Culture as the head of the Königlich Preußische Meßbildanstalt (Royal Prussian Photogrammetric Institute) in 1885. The institute was responsible for the documentation of cultural monuments and was the first photogrammetric institution in the world (Albertz, 2002). Another visionary was Sebastian Finsterwalder, who wrote a dissertation in 1899 on the basics of spatial resection and orientation of stereopairs of images, titled "the Fundamental Geometry of Photogrammetry." This formed the basis of what would become aerotriangulation, although it was not fully integrated into the discipline until the use of the computer in the 1960s (Gruner, 1977; Thompson and Gruner, 1980).

Prior to the early 1900s, the use of photogrammetry required laborious manual computation of measurements and plotting of bearings and distances in a point-to-point progression. Between 1900 and 1920, advancements in technology and instrumentation converted the conceptual advancements of the past century into mechanical realization. The Orel-Zeiss Stereoautograph allowed for the first time continuous computation, abandoning the laborious point-to-point method. This invention enabled the use of photogrammetry for map-making to flourish and its military applications to be realized (Gruner, 1977; Thompson and Gruner, 1980). At almost the same time (1926), the German inventor Professor Reinhard Hugershoff conceived of the first universal stereoplotter. This instrument utilized the concept of observation of imagery through the perspective of the camera lens at the time the photo was taken. This allowed for the use of sequences of stereopairs of imagery, which required orientation parameters for each pair. To advance beyond the individual stereopair setup, Otto von Gruber laid down the foundation of the theory of analog restitution of a sequence of photos in 1924, making the first attempts of spatial aerotriangulation. These foundational elements laid the groundwork for the innovations that would continue through the 20th century. The stereoplotter (a mechanical or analog workstation) held a stereopair of film images in the exact orientation as the moment of capture. When the images were brought into alignment, a point was projected so that it could be manually plotted on a map compilation. The block adjustment process allowed hundreds of thousands of photos to be linked to each other by the Van Gruber points. In addition to providing spatial orientation of photos one to the other in a large block of aerial imagery coverage, it was also possible to extend a ground control survey network of coordinates over hundreds of kilometers. This represented significant savings in the time and costs of topographic mapping, which had previously required surveyors to triangulate in features and control points on the ground alidade and plane table (Gruner, 1977).

To this point, photogrammetry was applied fairly equally to both terrestrial (or close range) and aerial strategies for image collection. However, through much of the early to mid-1900s, the innovations in powered flight and large format (250 mm × 250 mm negative) mapping cameras advanced aerial collection for wartime reconnaissance and topographic mapping (Matthews, Noble, Brady, et al., 2014). By the end of the 1960s, advances in electronics and computational devices had ushered in the new era of analytical photogrammetry. The analytical stereoplotter greatly reduced the time necessary to set up the large format film images of a stereopair and supported compilation of data directly into a digital drawing environment. In addition, the computational process for aerial block adjustment and aerotriangulation were automated, greatly reducing the time between acquiring imagery and data collection. Micro (PC) computers, semiconductors, and microelectronics led to state-of-the-art analytical photogrammetric workstations in the 1980s (Ghosh, 2005). Whereas the analytical stereoplotter greatly streamlined the process of aerial photogrammetric compilation of data, rigorous requirements from the past century were still present. Accurate measurements could be recorded from aerial photographic images, only when the following conditions were met: (1) nadir stereoscopic image pairs (two or more overlapping photographs) cover the object to be analyzed; (2) accurate x, y, z coordinates are known for at least three defined object points in the overlapping photographs; and (3) a calibrated mapping or metric camera is used to take the photographs. The standard outputs of the time were hardcopy (paper) topographic contour map plots and dimensioned drawings (Thompson and Gruner, 1980).

HISTORY OF PHOTOGRAMMETRIC ICHNOLOGY

In 1998, research on Middle Jurassic theropod footprints at the Red Gulch Dinosaur Tracksite (RGDT) in the Bighorn Basin of Wyoming was the initiation of the use of CRP for the documentation of fossil sites by the United States Department of Interior, Bureau of Land Management (BLM). BLM scientists were among the first to use CRP for the documentation, management, and interpretation of fossil resources (Breithaupt et al., 2001; Breithaupt and Matthews, 2001; Adams and Breithaupt, 2003). Other early examples included photogrammetric documentation of isolated Late Jurassic dinosaur footprints from the Picketwire Canyonlands Tracksite in southern Colorado in 1997 (Breithaupt, Matthews, and Noble, 2004; Matthews, Noble, and Breithaupt, 2006), Late Jurassic dinosaur tracks at the Moutier dinosaur disco tracksite in the Jura Mountains of Switzerland (Lockley and Meyer, 2000), and individual, Pliocene hominid footprints from the Pliocene Laetoli site in Tanzania in the 1970s (Jones, 1987) and 1990s (Musiba et al., 2012). Subsequently, more thorough documentation was done at some of these sites, and photogrammetric analyses were performed by the authors (Breithaupt, Matthews, and Noble, 2004; Matthews et al., 2006; Musiba et al., 2012). In addition to photogrammetric documentation at the RGDT, other data collected included traditional ichnological measurements, maps, hand-sketches, and Mylar tracings. All of these data were integrated together utilizing geographic information systems (GIS) analysis (Breithaupt et al., 2001, 2006; Breithaupt, Matthews, and Noble, 2004; Matthews, Noble, and Breithaupt, 2006).

The decision to use the best science to capture the paleontological values of RGDT led it to become one of the most thoroughly documented fossil tracksites (Matthews, Noble, Brady, et al., 2014). The unique documentation technologies (especially low- and high-level photogrammetry) used at this site have subsequently spawned companion studies at various localities around the world. Prior to the late 1990s, BLM photogrammetric projects mainly consisted of traditional aerial topographic mapping for internal resource management uses. During the early days of 3-D photodocumentation at RGDT (1998–2000), the process was very labor-intensive and could require as much as a week to get a final data set for a single fossil footprint, as there were no automated image correlation processes at that time and traditional "aerial-type" photogrammetric workflows and requirements were being followed. Photogrammetric practice of the time dictated that ground control data must be three times more accurate than the resolution of the imagery. Thus, for imagery with a final resolution of 0.1 mm, ground control coordinates better than 0.033 mm (33 μm) in all three dimensions would be

required. Unfortunately, both traditional total station and GPS surveying equipment of the day were only accurate to the subcentimeter. Thus, it was necessary to place a control mark at what would be an origin point and measure out with calipers to locate the other three or more points in Cartesian space. The overlapping stereophotos were taken using a 35-mm Rollei Metric Surveying film camera with a factory calibration report to satisfy the remaining two requirements. Due to this very laborious capture process, only a fraction of the thousands of tracks preserved at RGDT and the surrounding area (i.e., Sundance Vertebrate Ichnofaunal Province; see Breithaupt et al., 2006; Adams, Breithaupt, and Matthews, 2014) were photographed. Upon return to the photogrammetry lab, it was then necessary to send the positive slide film out for processing, inspect it upon return, and choose the best photos for stereoscopic compilations. The two slides of the stereopair were placed in the analytical photogrammetric workstation, the fiducial markings on the images were read, and the camera lens calibration values were appropriately applied. Lines of equal elevation were captured by a time-consuming, manual tracing process to capture the surface data in a digital 3-D format (Fig. 2.1). As a result, many days were needed to establish control, photograph the subject, and compile the surface data for one individual footprint (Matthews, Breithaupt, and Southwell, 2000; Breithaupt et al., 2001; Matthews and Breithaupt, 2001, 2012; Matthews, Noble, Brady, et al., 2014).

21ST-CENTURY INNOVATIONS

The first decade of the 21st century introduced truly affordable consumer digital cameras and laptop computers. Photogrammetry embraced this innovation in technology by incorporating scans of aerial film or digital imagery into a softcopy workstation environment. The transition from analytical to digital photogrammetry brought many advantages (such as automated image matching and surface generation); however, a dependence on specialized workstations still remained for topographic compilation, as did the traditional requirements of consistent nadir stereo, ground control, and a calibrated mapping camera (Konecny, 1985). Advancements in close-range photogrammetry also occurred, as three-dimensional measuring and mapping (3DMM) software, such as PhotoModeler, came on the scene. At the time, PhotoModeler supported point-to-point measurements, development of a robust camera calibration, semiautomated target recognition, and the ability to establish a user-defined coordinate system, but it lacked an automated surface generation function.

The BLM National Operations Center in Denver, Colorado, had expertise in traditional photogrammetry and a

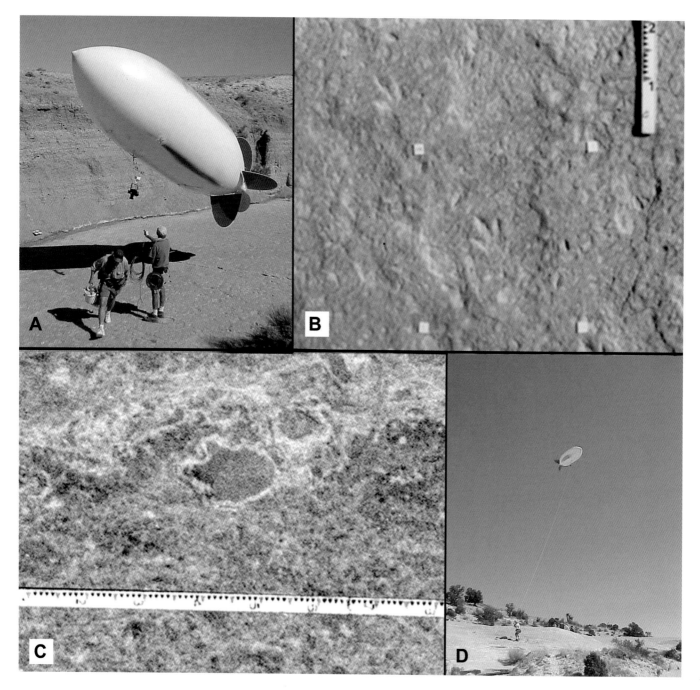

2.2. (A) The Aerial Camera Blimp System (ACBS) used at the Red Gulch Dinosaur Tracksite, Wyoming; note camera mount and tether to operator, summer 2000. (B) Imagery of the Middle Jurassic, Sundance Formation track surface taken from the ACBS, stadia rod interval is 1 foot, square porcelain tiles are 2.5 inches.

(C): ACBS imagery of the Middle Jurassic, Entrada Sandstone track surface from Twentymile Wash Dinosaur Tracksite, Grand Staircase Escalante National Monument, Utah (one of the first tracksites to be entirely documented using photogrammetry). Stadia rod interval is 1 foot. (D) ACBS capturing stereoscopic imagery

at the Twentymile Wash Dinosaur Tracksite, Grand Staircase Escalante National Monument, Utah. See Breithaupt et al. (2001); Breithaupt and Matthews (2001); Matthews et al. (2002); Breithaupt, Matthews, and Noble (2004); and Matthews, Noble, and Breithaupt (2006).

high-end, professional, softcopy workstation environment, along with 3DMM software. In addition, there was a wealth of projects at a variety of scales ranging from traditional large format aerial- to ground-based, very large-scale, close-range projects available for extensive testing of the camera calibration function of the 3DMM software (Fig. 2.2). As a result of the testing, an understanding of the critical importance of

the camera calibration and the need for a robust field calibration procedure was developed. The ability to internally and accurately apply scale to a project without reliance on ground control (which often did not meet accuracies of very close-range projects) was also realized (Breithaupt et al., 2001; Matthews and Breithaupt, 2001; Matthews, Noble, and Breithaupt, 2004).

During the documentation of the RGDT, BLM expanded the traditional aerial block photogrammetry methods, with terrestrial photogrammetry or CRP, forming a hybrid method that combined photogrammetry and surveying. The hybrid method streamlined digital CRP capture techniques for field use (Breithaupt and Matthews, 2001; Breithaupt, Matthews, and Noble, 2004; Matthews, Noble, and Breithaupt, 2005, 2006; Matthews and Breithaupt, 2011b, 2012). This hybrid process focused on taking a combined series of photos that satisfied the requirements for camera calibration and coordinate system definition of the 3DMM software and the stereopairs necessary for automated surface generation of the softcopy photogrammetry workstation. The hybrid method was flexible enough for field capture and significantly reduced the time needed for CRP. For example, a 10-m-long trackway could be completed from image acquisition to final 3-D data production in approximately 1 week. The output products included a 3-D surface data set and an orthorectified image mosaic. The hybrid process facilitated the documentation of a wide variety of natural and cultural resources located in the United States and included the RGDT (Breithaupt et al., 2001, 2006; Breithaupt, Matthews, and Noble, 2004; Matthews and Breithaupt, 2006; Matthews, Noble, and Breithaupt, 2006), Twentymile Wash Dinosaur Tracksite (Matthews et al., 2002; Matthews and Breithaupt, 2006), St. George Dinosaur Discovery Site at Johnson Farm (Matthews, Noble, and Breithaupt, 2005; Milner et al., 2009; Matthews and Breithaupt, 2011b) Red Rock Canyon Dinosaur Tracksite (Rowland et al. 2014), and Moccasin Mountain Tracksite (Matthews et al., 2008; Matthews and Breithaupt, 2011b).

During the second decade of the 21st century, amazing advancements in digital camera equipment, computer architecture, and hardware have taken place. These advancements coupled with breakthroughs in computer vision, structure from motion (SfM) and image-matching algorithms, and software; dramatically changed digital photogrammetry (Fraser and Stamatopoulos, 2014). Today's graphic processors and cloud computing make it possible to take hundreds of digital photos and produce dense point clouds of 3-D data in a matter of minutes. The new generation of photogrammetric software implements said breakthroughs and algorithms and automatically connects photographs based on perspective geometry. From these connected photographs, surfaces are reconstructed, meshes are derived, and point data is triangulated. Not only can this 3-D data surface contain hundreds of thousands of very precise x, y, z coordinate locations (accurate to the subpixel level), each data point can also carry a red, green, blue color model value depicting the natural color of the subject.

These robust 3-D data sets can be generated without the dependence on the specialized equipment of the softcopy photogrammetric workstation, making CRP an incredibly efficient means of 3-D data capture. The requirement for only minimal equipment and capability to do initial field processing on a laptop gives the photographer almost immediate feedback on the success of image capture, making CRP an excellent tool for both field and laboratory situations. Often the use of photogrammetry can be more efficient, less labor-intensive, and more cost-effective than other types of field 3-D data collection (Breithaupt, Noble, and Matthews, 2012). Currently, photogrammetric software ranges in cost from freeware (Falkingham, 2012) and online services to expensive professional mapping suites. However, not all software is created equal. Image size limitations, extent to which the camera-lens system is calibrated, dependence on outside coordinate system control, and scaling are some of the major differences that can be seen among software. Of these differences, the degree to which the camera is calibrated has the greatest impact on final project accuracies. Photogrammetric software (such as PhotoScan, ADAM Technology 3DM Analyst), support the orientation of hundreds (even thousands) of images, making it possible to integrate both ground-based and aerial imagery (Mudge et al., 2010; Breithaupt and Matthews, 2012; Breithaupt, Noble, and Matthews, 2012; Matthews and Breithaupt, 2012; Matthews, Noble, and Breithaupt, 2012, 2014a, 2014b; Breithaupt, Matthews, and Noble, 2014; Matthew, Noble, Brady, et al., 2014; Matthews, Pond, and Breithaupt, 2014).

The authors have utilized a variety of platforms to capture imagery for photogrammetric processing at dinosaur tracksites in western North America, as well as other countries. These include tripods of various heights (1 to 10 m) and monopods (extended overhead up to 3 m) with remote triggers. Another ground-based option for getting very high-resolution stereoscopic images is the use of telephoto lenses and tripod heads designed to capture panoramas (e.g., Gigapan robotic head). The resulting panoramas, captured with proper geometry in relation to each other, can be stitched into very large (several hundred megapixels, even gigapixels) images. Specialized software is needed for processing these stereoscopic Gigapan pairs to remove lens distortions and create a virtual stereo image (Mudge et al., 2010, 2012; Matthews, Noble, Brady, et al., 2014). In addition to ground-based camera platforms, a variety of aircraft have been used to capture nadir imagery at RGDT and other BLM-managed paleontological sites These platforms include manned aircraft, such as helicopters, ultralights, and single-engine fixed wing aircraft, and unmanned platforms such as blimps and unmanned aircraft systems (Breithaupt, Matthew, and Noble, 2004; Matthews, Noble, and Breithaupt, 2006; Matthews and Breithaupt, 2011a, 2011b, 2012; Chapman et al., 2012; Matthews, Noble, Brady, et al., 2014b) (Fig. 2.3). SfM-based

2.3. (A) Neffra Matthews in preparation for low-level imagery collection using DSLR camera mounted below a Bell Ranger helicopter of the Moccasin Mountain Tracksite near Kanab, Utah, summer 2008. (B) Ultralight image collection over the Red Gulch Dinosaur Tracksite (RGDT), Wyoming, summer 1999. (C) Monopod mounted camera with remoter trigger, Laetoli Tracksite, Tanzania, spring 2011. (D) Tommy Noble setting up the Gigapan robotic mount to collect high-resolution imagery at RGDT, summer 2012. (E) A 10-m tripod used to collect imagery at RGDT, summer 1999. (F) Tripod and control grid setup for imagery collection at RGDT from 1999 to 2001. The control grid was constructed to align with the 1-m mapping units; in addition, three-dimensional control was provided by calibrated bars extended from the grid. Extensions along the base of the grid assisted in maintaining consistent stereoscopic overlap. See Matthews et al. (2002, 2008); Matthews and Breithaupt (2011a, 2011b); Musiba et al. (2012); and Matthews, Noble, Brady, et al. (2014).

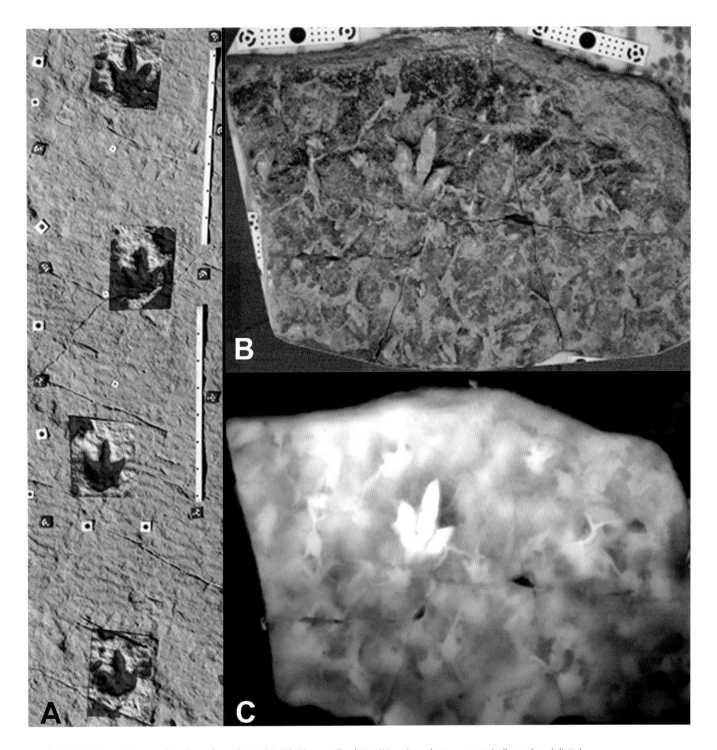

2.4. (A) Middle Jurassic theropod trackway from the Red Gulch Dinosaur Tracksite, Wyoming, photogrammetrically produced digital orthoimage with color depth map of individual fossil footprints; scale bar is 36 inches. (B) Middle Jurassic Kilmaluag Formation track block (with multiple-sized theropod footprints) from Isle of Skye, Scotland, summer 2006; calibrated scale 30 cm. (C) Photogrammetrically generated digital surface with grayscale depth map. See Clark, Ross, and Booth (2005) and Breithaupt et al. (2006).

photogrammetric software also provides the ability to utilize film-based imagery, thus historic aerial or ground-based photos can be scanned and processed. If the photos were captured with enough overlap, they may be utilized on their own or incorporated with recent project photos (Breithaupt et al., 2004; Matthews, Noble, and Breithaupt, 2006; Matthews and Breithaupt, 2012; Falkingham, Bates, and Farlow, 2014).

Historically, CRP was treated much in the same fashion as traditional aerial photogrammetry, in that photos were taken over a subject (such as a tracksite) from a nadir position to the surface and in a line-of-flight type configuration. This strategy is still an efficient method for capturing information about a relatively flat surface (such as a single fossil footprint or an entire trackway). In addition, the use of

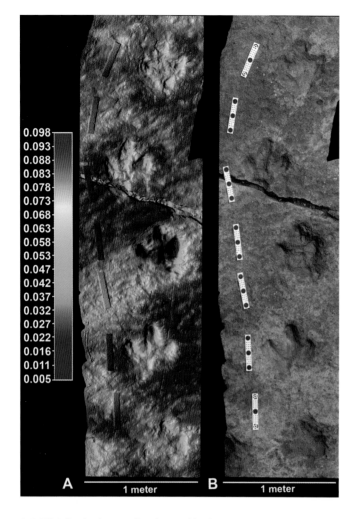

2.5. (A) Color depth map of trackway with relative depth legend in meters. (B) Photogrammetric orthophoto image of ornithopod trackway from Early Cretaceous Obernkirchen Sandstone, Obernkirchener Sandsteinbrueche Quarry, Germany, spring 2011. See Hornung et al. (2012) and Richter et al. (2012).

photogrammetry may be applied to dimensionally complex subjects, such as overhanging or tilted bedding planes, quarries, outcrops, skeletal elements, high-relief or deep dinosaur tracks (unsealed penetrative tracks), and museum mounts. Photogrammetric documentation may take place in any situation where quality, consistent photographs can be taken, including the field, laboratory, or museum. When capturing dimensionally complex subjects, it is often necessary to combine a number of strategies for camera location in relation to the subject (Mudge et al., 2010; Breithaupt, Matthews, and Noble, 2014; Mallison and Wings, 2014; Matthews, Noble, and Breithaupt, 2014a).

Photogrammetric point cloud data can be exported into a variety of file formats, including products traditionally associated with aerial photogrammetry (such as orthoimage maps, topographic contour maps, and color-coded elevation maps), as well as a variety of digital outputs. There are a

number of analytical tools that support direct comparison of 3-D point cloud data of morphologic features, such as those between individual tracks, trackways, or tracksites (Belvedere et al., 2013; Castanera, Pascual, et al., 2013a; Matthews, Pond, and Breithaupt, 2014; Razzolini et al., 2014; Castanera et al., 2015; Wings, Lallensack, and Mallison, 2016) (Figs. 2.4 and 2.5). As a scientific community, we can now build a library of photogrammetric image data sets (Pond, Belvedere, and Dyke, 2012; Pond et al., 2014) (Fig. 2.6). These 3-D digital surrogates can be utilized in a virtual environment or "printed" as hardcopy replicas for research, management, preservation, and interpretation. Fortunately, the basic equipment (i.e., scale bar and camera) necessary to successfully create photogrammetric point cloud data digitally is easily available to scientists in the field, lab, or museum, giving the ability to capture our natural world in 3-D at any time in any place.

Associated with the various technological advancements over the last decade, a marked increase in the use of 3-D data capture for the purpose of documentation, evaluation, and preservation of paleontological resources can be seen. Subjects now vary from an isolated tooth to an entire bonebed and from a single fossil footprint to an entire tracksite. Along with photogrammetry, the other most widely used method for capturing 3-D data of paleontological subjects is LiDAR. In comparison, photogrammetric point cloud data contain both the exterior physical dimensionality of a subject and a high-quality, natural color, image texture derived from a geometrically linked set of photographs adjoined to form a digitally reconstructed surface mesh. LiDAR point clouds are composed of independently reflected 3-D data points accompanied by a value signifying the strength of return. Any natural color information is added as a secondary process when imagery is combined with the point data (Breithaupt, Matthews, and Noble, 2014).

The authors began experimenting with the integration of CRP and LiDAR in 2001 with projects in Colorado on Dinosaur Ridge and Skyline Drive (Breithaupt, Matthews, and Noble, 2004; Matthews, Noble, and Breithaupt, 2006); subsequent studies included work at RGDT (Bates et al., 2009). Close-range photogrammetry and LiDAR can be effective methods to document very large, complex, and difficult to access track surfaces (Bates et al., 2008). Recent comparisons by the authors utilizing SfM-based photogrammetry and LiDAR focused on the documentation of architectural structures that provided intriguing results. Great effort was taken to ensure that best practices in photogrammetry and LiDAR capture were followed and that point accuracies and density benchmarks were stated in advance. Metrics such as internal versus external point precision statistics, completeness of coverage, fidelity of surface rendering, and image

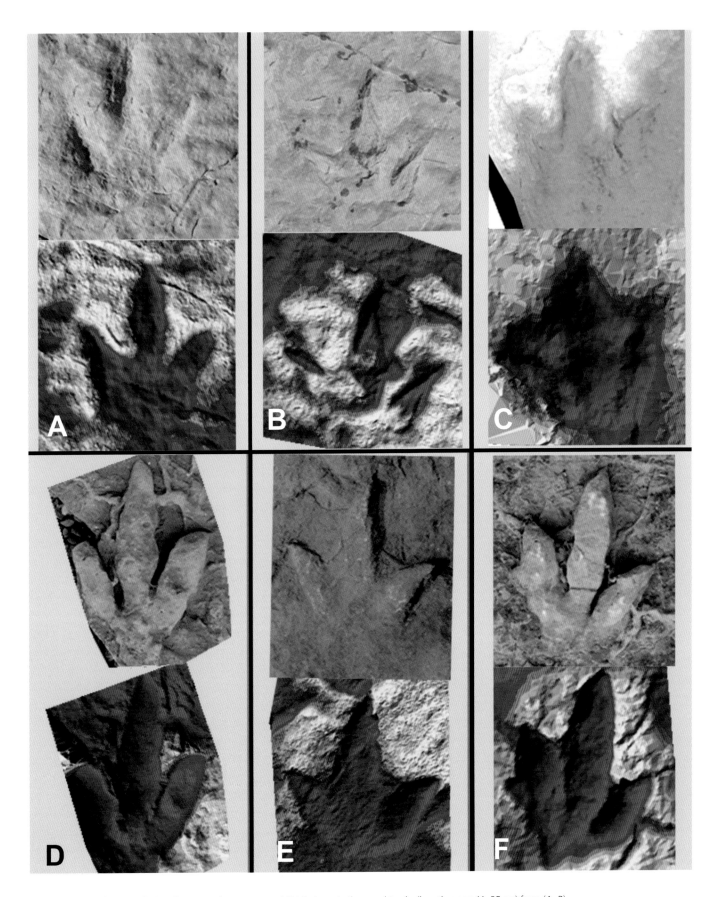

2.6. The matrix of images depicts for comparison purposes Middle Jurassic theropod tracks (length range 11–25 cm) from (A, B) Sundance Ichnofaunal Province, Wyoming (Sundance Formation); (C) *Carmelopodus* type locality, Utah (Carmel Formation); and (D, E, F) Trotternish Peninsula, Isle of Skye, Scotland (Valtos and Kilmaluag formations). Photogrammetrically derived orthoimage and color depth maps. See Adams and Breithaupt (2003), Clark and Barco Rodriguez (1998), Clark et al. (2005), and Breithaupt et al. (2006).

correlation; time factors for gathering and processing the digital data; cost of labor; and hardware/software needs were considered. Other concepts such as project planning, deliverables and derived products, data management planning, and site assessment were considered along with harder to quantify factors such as quality checks and correcting errors in the data. Open source software for manipulating, scaling, and comparing point cloud data were used. The comparison results demonstrated that today photogrammetric point clouds can be generated at a level that meets or exceeds the instrument specifications for the LiDAR unit used in the comparison (Breithaupt, Noble, and Matthews, 2012). See Petti et al. (2008) and Belvedere et al. (2012) for other comparison studies.

PHOTOGRAMMETRIC BASICS

The integration of multiview matching and SfM algorithms is a significant milestone in the development of photogrammetry. These robust, economic, workflow (or expert)-driven software programs have opened photogrammetry to an extremely wide user base well beyond the traditional photogrammetry, GIS, and mapping fields. SfM-based photogrammetric software allows even novice users to take digital photos and process them into a 3-D digital model. Based on the end use of the data set, an image acquisition and processing workflow may be used to capture a visually pleasing (but low accuracy and resolution) 3-D model or a high-fidelity, highly accurate 3-D recreation of a subject. The uses of these data sets can vary from online models for visualization to scientific research to solid model printouts (Matthews, Noble, and Breithaupt, 2014a).

Error

As discussed earlier, the discipline of photogrammetry has a very long history that includes the identification, quantification, and removal of sources of error from the process. Early photogrammetrists worked very diligently to prove the technique as a reliable and accurate means of data capture leading to a very rigorous approach that could be conducted with confidence. Utilization by the military and civilian agencies to provide reconnaissance and topographic maps is an example of this confidence. The process of error detection and minimization is a routine part of the traditional photogrammetric process conducted by a professional aerial photogrammetric operator. As noted herein, the tools used to minimize error include an aerial surveying camera and wide-angle lens with factory calibration report, properly overlapping aerial imagery, high-quality ground control survey, aerotriangulation and bundle adjustment algorithms, and attention to root mean square and sigma statistical results (Matthews, Noble, and Breithaupt, 2014b).

Error may be introduced at any point during the photogrammetric process (from image capture to product generation) and propagate quickly by tens or even hundreds of times. Thus, depending on the size of the subject and the resolution of capture imagery, error from a few millimeters to tens of meters could be present. In all likelihood, the occurrence of error will not be discernible in the resulting point cloud, but it will only be observable when it is sought during the processing phase or when compared with other models of the same surface. The extent to which SfM-based photogrammetric software manages, eliminates, and reports error varies widely. Although some software programs provide in-depth and rigorous error detection, minimization, and reporting (e.g., PhotoScan, ADAM Technology, PhotoModeler), these measures are not always incorporated as part of the expert workflow. It is thus incumbent on the user to understand the errors inherent in the process and take the necessary steps to remove error. Error in the photogrammetric process falls into two categories—direct error and indirect error (Matthews, Noble, and Breithaupt, 2014a, 2014b).

Direct error refers to those phases of the process over which the operator has influence based on the decisions made during project execution. These phases include the following: (1) taking a set of quality photos at a consistent (or fixed) focus and focal length and aperture with a wide-angle (20 to 30 mm) lens, (2) taking this image set with geometrically appropriate stereoscopic overlap (i.e., 66% overlap), (3) adding to this set images taken with the camera turned at 90° and 270° (for camera calibration), and (4) including at least two objects of known dimension captured in the stereo overlap of at least two photos. Indirect error refers to those factors in which error can only be limited by the operator and not removed, basically the lens distortion and alignment of the lens to the sensor during camera construction. To minimize camera error, it is paramount that a set of images be provided to the SfM-based photogrammetric software to take full advantage of the software's ability to correctly determine and apply camera and lens distortions. It is also important to ensure that a strategy of removal of poorly matched points and reprocessing is repeated within the software, so that a significant amount of error can be removed and the resulting root mean square error (RMSE) falls to an acceptable level. The RMSE expresses the average magnitude of the error between an actual and a predicted value and gives a high value for large errors. With properly taken photographs using a high-quality DSLR camera with wide-angle lens, a RMSE of 0.13 to 0.17 of a pixel (low error) should be achievable.

Minimizing and quantifying the error in a CRP project can quantify the resulting data in terms of RMSE, regardless of the use of externally collected geographic coordinate control. Defining a confidence level allows the data to be utilized in a wide variety of scientific, resource documentation, and monitoring applications. An initial investment in time to properly master the requirements in a logical progression of first achieving good photographs, knowing the requirements of scaling the subject, achieving good geometry for a relatively flat project, and understanding the software-processing workflow prior to moving to more dimensionally complex subjects will certainly pay off in the long run. Once these processes are mastered, virtually any subject can be photogrammetrically documented (Matthews, Noble, and Breithaupt, 2014b).

Taking Good Photos

The role the camera and lens play in a quality photogrammetry project cannot be overstated, because when a project is executed properly they become the survey instrument. Whereas a high-resolution DSLR camera with a wide-angle (20 to 30 mm) prime lens is the best for scientific photogrammetry, a variety of other types of cameras may be used if care is taken. Should it be necessary to use a point and shoot or other non-DSLR camera, it is important to determine whether the camera is equipped with a manual focus and aperture mode so that these may be fixed. It is also important to know the size of the sensor, its impact on the field of view, and thus the stereo overlap. In photogrammetric terms, good quality refers to sharp pictures that have uniform exposure and high contrast and that fill the frame with the subject. The camera should be set to aperture priority (preferably F8) and the ISO, shutter speed, white balance, and other settings should be adjusted to achieve properly exposed images. To obtain the highest order results, it is necessary to ensure that focal distance, physical distance, and zoom do not change for a given sequence of photos. This can be achieved by taking a single photo at the desired distance using the autofocus function, then turning the camera to manual focus and taping the focus ring (to restrict accidental movement) in place. A set of photos taken in which the focal distance is set are referred to as a calibration group. When using a camera that lacks the ability to fix these settings, it becomes more important to maintain a consistent distance from the subject so that the camera elements do not move drastically from picture to picture in a calibration group. When the camera elements are not fixed, very low RMSE may not be achievable, thus higher reported error and less precision may occur. BLM Technical Note 428 (Matthews, 2008) provides documentation on

the use of CRP (Matthews and Breithaupt, 2009, 2011a, 2012; Mudge et al., 2010).

Image Geometry and Stereoscopic Overlap

Capturing photographs for stereoscopic photogrammetric processing may be accomplished in as few as six photos for a small subject and can provide extremely dense, high-resolution, geometrically and orthometrically correct, 3-D, digital data sets (Matthews, 2008; Mudge et al., 2010). Because of the flexibility of this technique, it is possible to obtain highly accurate 3-D data from subjects that are at almost any orientation (horizontal, vertical, above, or below) to the camera position. However, it is important to keep the plane of the sensor and lens as parallel as possible to the subject and to maintain a consistent height (or distance) from the subject. A geometrically sound network of camera locations results in complete coverage of the subject and an automated workflow. Matching algorithms in SfM-based photogrammetric software work by evaluating sequences of images, finding groups of spatially and spectrally similar pixels, and matching them to like groups from other images. Abrupt changes in scale, changes in orientation, distortions caused by highly oblique photos, or drastic changes in lighting can all cause the matching algorithms to fail to correlate points from those images. The best results in image matching occur when the software is provided with a consistent framework of stereoscopic images (Mudge et al., 2010;Breithaupt, Noble, and Matthews, 2012; Matthews and Breithaupt, 2012; Matthews, Noble, and Breithaupt, 2012, 2014a, 2014b; Breithaupt, Matthews, and Noble, 2014; Matthews, Noble, Brady, et al., 2014; Matthews, Pond, and Breithaupt, 2014).

The first consideration when designing the stereophoto image framework is the needed precision and therefore the photo scale that is required to adequately represent the subject. The final accuracy of the resulting, dense surface model is governed by the image resolution, or ground sample distance. The ground sample distance is a result of the resolution of the camera sensor (higher is better), the focal length of the lens, and the distance from the subject (closer = higher resolution) (Matthews, 2008). Next is achieving good camera location geometry. Once the needed resolution is determined, the distance or "height" of the camera from the subject is determined. The distance the camera must move to create stereoscopic overlap is the "base," the geometry that governs a good photogrammetric framework is the "base-to-height" ratio (BtH ratio). A BtH ratio of 1:1 distributes error equally between the base (x- and y-axis) and the height (z-axis); however, photos taken with this ratio would most likely not have enough locatable points in common to successfully

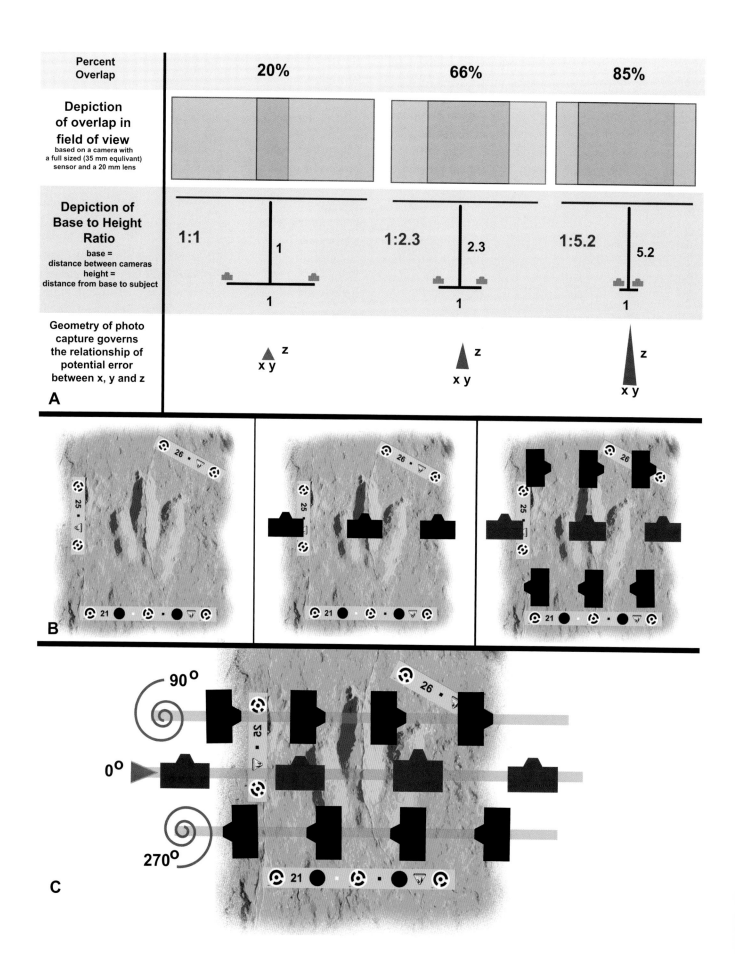

align in SfM-based photogrammetric software. Therefore, photos taken with a BtH ratio of 1:3 provide enough overlap between photos for successful orientation in the software, while keeping the opportunity for z error to a minimum. When BtH ratios range from 1:7 to 1:10, or greater, the opportunity for error in z increases dramatically, so much so that it is not outweighed by the high amount of points that may be generated in the software. When using a wide-angle lens of 20 to 30 mm on a DSLR camera with a full-size sensor, images with a BtH ratio of 1:2 to 1:5 will produce an ideal stereo overlap of 66%. However, when using a lens of 50 mm equivalent or longer, achieving a stereo overlap of 66% forces a higher BtH ratio and could introduce significant error in the z-axis. See Figure 2.7A. The result may be an unnecessarily rough 3-D surface model, unless care is taken during the processing phase. The smaller sensor size of many point and shoot cameras will have a similar effect, as will a longer lens, which forces a higher BtH and thus increases the potential error in z or roughness of the surface. When using less than a full-size sensor, ensuring that there is good overlap and that the BtH is as low as possible is important.

Camera Calibration

All camera lens systems have distortions due to the curvature of the lens and the alignment of the lens with respect to the sensor. When images are used in the photogrammetric process with little or no camera or lens distortion information, significant error may be introduced to any resulting measurements or surface data. SfM-based software programs provide camera calibration functions as part of the project workflow. A robust field calibration may be accomplished most effectively when there are a large number of autocorrelated points in common among a group of photos taken with proper camera settings and geometric framework. At least four additional photos are required; two taken with the camera physically rotated 90° to the previous line of stereoscopic photos and two additional photos with the camera rotated 270°. The additional camera calibration photos may be taken at any location along the line of stereo photographs,

but the best results occur in areas where the greatest number of autocorrelated points can be generated (Matthews, 2008; Matthews and Breithaupt, 2009, 2011a, 2012; Mudge et al., 2010; Breithaupt, Noble, and Matthews, 2012; Matthews, Noble, and Breithaupt, 2012, 2014a, 2014b; Breithaupt, Matthews, and Noble, 2014; Matthews, Noble, Brady, et al., 2014; Matthews, Pond, and Breithaupt, 2014). See Figure 2.7B.

Adding Measurability

Many SfM-based photogrammetry software programs (Agi-Soft PhotoScan, ADAM Technology, PhotoModeler, etc.) provide the ability to introduce real-world values (or scale) to a project (Matthews, 2008; Matthews and Breithaupt, 2009; 2011a, 2012; Mudge et al., 2010; Breithaupt, Noble, and Matthews, 2012; Matthews, Noble, and Breithaupt, 2012, 2014a, 2014b; Breithaupt, Matthews, and Noble, 2014; Matthews, Noble, Brady, et al., 2014; Matthews, Pond, and Breithaupt, 2014) This is accomplished by simply adding an object of known dimension (meter stick or other object) that is visible in at least two stereo models (three photos). It is preferable to have two or more such objects to ensure visibility and for accuracy assessment. Calibrated target sticks may be used in addition to (or in place of) the object of known dimension. Many software are equipped with the ability to detect and decode particular configurations of pixels such as circular barcodes, survey crosses, or the centers of circles. Utilizing these features can greatly streamline the process of adding scale, especially when these targets are incorporated in the calibrated target sticks at known intervals. These objects may then be assigned their proper lengths during processing, and most photogrammetrically based software packages conduct a mathematical procedure known as a bundle adjustment. Once an object length is established, the bundle adjustment passes those measurements to all photos and reduces error in the project. High accuracy may be extended for a long distance along a series of photos when the framework geometry and camera calibration are observed. These steps allow the object of know dimension to be placed so as to not detract visually from the subject.

2.7. (A) Comparison of changes in image overlap with changes in distance between camera positions (base) and distance to subject (height). Field of view based on a DSLR camera with a full-size sensor and a 20-mm lens. Base to height ratios and thus quality of derived surface will change with a different-size sensor and longer lens. For example, when using a 50-mm lens, an overlap of 66% will give a 1:5 ratio and could result in a bumpy three-dimensional surface model. (B) Image on the left illustrates a single theropod track with ideal placement of 3 scale bars. Middle image shows schematic of camera placement for landscape orientation to capture proper stereo overlap. Right image illustrates the addition of two more lines of photographs, taken in portrait mode rotated at 90° and 270°. (C) Conceptual layout illustrating proper geometry for camera locations needed to capture the depicted area of interest. All photos should be taken with consistent focus, focal length, and aperture. The lines of imagery that overlap each other by 66% provide complete stereoscopic coverage of the area of interest, as well as providing redundancy and an opportunity to physically rotate the camera as indicated, so that the optical center and lens distortion may be computed. Objects of known length are positioned around the area of interest and will be used to add real-world units to the photogrammetric project during processing. See Matthews, Noble, and Breithaupt (2014a, 2014b).

Archiving

With all these advances and opportunities available for 3-D data capturing and processing (both now and into the future), it is important to consider the basic components and what data can and should be archived. While there are numerous formats for the output and utilization of 3-D data, there are none that are considered as universally agreed upon archival formats. Many 3-D file formats contain x, y, z coordinate locations, the surface mesh, and even the original image texture. It is important to understand how the final 3-D data set will be used and in what software it will be handled to determine the best output file format. For general purposes, .obj is the most widespread format and maintains a map to the texture in relation to its position on the mesh. The .ply format is less widely used than the .obj is, although it contains much the same data and has the additional advantage of a supported standard binary version, which can minimize file sizes. Fortunately, photogrammetry provides an option with regard to archiving of data not available to other types of 3-D data capture. Many archival standards accept images in the form of .tif or .dng (digital negatives). Thus, it is suggested that the original photographic set of images used to process 3-D data can themselves be the archival unit, along with a basic ASCII text file. This text file should contain relevant metadata such as the name of the project and location, the reason the project was conducted, what camera was used, who was involved, measurements and units of the objects of known length, geographic coordinates, and other information pertinent to the project (e.g., expected RMSE and other statistical information related to processing) (Breithaupt, Noble, and Matthews, 2012; Matthews, Noble, and Breithaupt, 2014b).

PHOTOGRAMMETRIC ICHNOLOGY OVERVIEW AND ICHNOLOGY BEST PRACTICES

A properly done ichnological photogrammetric project can produce permanent 3-D digital image data sets that have the quality, reliability, and authenticity necessary for scientific use, including the creation of digital type specimens (Adams et al., 2010). CRP is a noninvasive, objective recording and analysis method, which provides a visual, quantifiable baseline to evaluate track-bearing surfaces. The following are 20 key points to consider regarding photogrammetric ichnology. Based on the experience of the authors, when images are taken in accordance to the criteria set out herein, they may be successfully processed in both traditional and multiview matching photogrammetric software. Imagery that is correctly taken now can be used in software and stay relevant for use as future advances take place. As long as proper

imagery sets and information are saved, the 3-D output can be reproduced virtually on the fly, making photogrammetric documentation a technology that will stretch well beyond our current understanding.

1. Take 3-D documentation and photogrammetry of trace fossils seriously. Ichnology is a rigorous discipline and trace fossils should be carefully dealt with, managed, and documented (Breithaupt and Matthews, 2014a). As such, develop a project documentation plan prior to collecting data. A well-conducted photogrammetric project can provide as much good data as any other site documentation technique can. Track measurements (e.g., length, width, and depth), stride length, straddle, morphology, substrate deformation, and many other features can be derived from a well-constructed 3-D data set. Photogrammetry should be done at the beginning of the documentation phase of a project and not as an afterthought, because a good primary photogrammetric data set does not weather and degrade with time, primary measurements can still be made from it long after the original surface is degraded. A tracksite can be recorded to monitor changes, including field excavations and sites in dynamic environments that might suffer erosion or other factors.

2. The 3-D surface derived from photogrammetry can be very precise and have submillimeter resolution and high morphometric fidelity. Therefore, anything that is on or over the surface will be modeled in 3-D and become part of the data set. Thus, it is best to conduct the photogrammetric documentation when the surface is cleanest and freshest. Gently clean all foreign material (including water) from the track surface and from within track depressions. Test all tools used to prepare the surface (even stiff brushes), to ensure they do not damage the track-bearing surface and add unwanted morphological features during cleaning. If it is permissible, remove vegetation that is on or overhanging the track surface so it will not cast shadows or obscure the track surface. Utilizing a team of volunteers for assistance with cleaning and preparing a track surface in a short amount of time is invaluable and engages the public in understanding the scientific and educational values of paleontological resources.

3. In most cases, an unexposed track-bearing surface buried in its original stratigraphic sequence is at its most stable and best-preserved state. Reburial does not protect a site as well as the original stratigraphy does (Musiba et al., 2012). Once excavated and

2.8. (A) In situ view of the first dinosaur track from Denali National Park, Alaska (Early Cretaceous Cantwell Formation). Photogrammetric documentation was conducted on a nearly vertical facing outcrop, prior to collection and molding of the specimen, summer 2005. (B) Photogrammetric documentation after collection and transportation. The BLM hybrid method was utilized in 2005. At this time, numerous scale bars were needed to provide field camera calibration (scale bars = 30 cm). (C) A digital elevation model (DEM) was generated using an autocorrelation function of the softcopy photogrammetric workstation. Topographic contours were interpolated from the DEM. (D) Orthorectified photo image of theropod track. See Fiorillo et al. (2007, 2014).

exposed, the surface will start to degrade (García-Ortiz, Fuertes-Gutiérrez, and Fernández-Martínez, 2014). The type of strata and the physicality of the site will play a large part in determining whether the degradation will take place over days, months, years, or hundreds of years. It is the authors' opinion that a tracksite should not be exposed until it can be documented thoroughly, which includes digital documentation. If photogrammetry or some other form of digital documentation cannot be conducted as soon as the surface is exposed and cleaned, then excavation and exposure of the site should not be undertaken until the equipment and expertise are available to conduct proper documentation. If the site is large and will take multiple days to clear, then predetermined areas should be cleaned and then photographed. Choose an area that can easily be prepared in one day, clean it, and while the light is good, photograph that area. Overlap the photographic project strips taken each day (ideally at the same time of day), so there is continuity for the photogrammetric software and so that all component strips align successfully.

4. If a track-bearing surface is being excavated for the first time and infillings are found in association with the tracks, they should be kept in place and photographed together.

5. Perform photogrammetric documentation before collecting in situ specimens (if authorized to do so). An excellent example of following this procedure was the photogrammetric documentation done prior to molding and collecting of the first dinosaur track

found in Denali National Park, Alaska (Fiorillo et al., 2007, 2014). See Figure 2.8.

6. All photogrammetric ichnology projects should be well planned prior to the start of documentation. Proper equipment (e.g., camera, monopod, tripod, remote triggering devices, scales) should be acquired and in good working order before documentation. Understanding the camera settings and how to get good quality photographs, as well as proper photogrammetric techniques prior to exposing and/or cleaning a surface is advised. A freshly excavated and cleaned track surface is not the place to learn for the first time how to do photogrammetry. In addition, once the surface is exposed, the degradation clock is ticking.

7. Photogrammetric ichnology may be conducted on a single track, a trackway, or an entire tracksite; however, the approaches may vary slightly. A good methodology to photograph a single track (or a small portion of a trackway) is to lay a scale bar along the base (e.g., width) and height (e.g., length) of a track. Make sure to place the scales far enough away from the track center so as not to obscure any associated anatomical or morphometric information. Position the camera at an appropriate distance to the track, autofocus, and keeping that focus, set the camera to manual. Keeping consistent distances, take a series of photos along the long axis of the track. Turn 90° and take another series along the base, then turn and take another series at 90° to that. Tipping the camera inward or slightly oblique to the surface for each line is permissible. Make sure that the photographer's feet are not visible in the images, otherwise they will need to be masked out later. Take note of sun orientation and avoid casting shadows on the surface. If a consistent height is maintained, several tracks may be done in a row by this method without refocusing. If camera settings and focal distance remain the same, hundreds of photos can be taken without refocusing, allowing for large areas (even entire tracksites) to be documented as part of one photogrammetric project. It is good practice to note the photo exposure number associated with a particular footprint within a trackway. In addition, a break (e.g., context or "tourist") picture may be

taken between each sequence of track projects or if it is necessary to refocus. See Matthews (2008) and Matthews, Noble, and Breithaupt (2014b).

8. When photogrammetrically documenting a long portion of an entire trackway, the best results will be realized when photos are taken from a nadir perspective to the surface. Overlapping photos by 66% both within and between strips will produce excellent results and, with the exception of very deep tracks, will result in very little occlusion of the surface. For larger areas, it is recommended to use a DSLR camera with a wide-angle lens mounted on a monopod. Many new digital cameras are equipped to communicate directly with smart devices, such as phones and tablets. For many older model DSLR cameras, several tools (such as a CamRanger) are available. Not only does their use allow for remote adjustment of camera settings and triggering, but they also provide real-time viewing of the captured image on a smart device, making stereoscopic image acquisition from the monopod very efficient. Other remote triggering mechanisms (either radio or infrared controlled) are also available. In addition, setting the camera on interval mode and walking and stopping in sync with the picture taking is another strategy for extended reach imagery capture. For large areas, it is advisable to measure the area to be photographed and grid off the photo strips or lines. Ceramic tiles, casino chips, pin flags, small cones, spent CDs, scale bars, or other small, easy to see items may be used to mark the beginning and end of the photographic strips. If utilizing objects of known length with software-recognizable targets, take care not to reuse any uniquely coded targets, as these can cause difficulties later during processing (e.g., either with incorrect decoding or erroneous image alignment) (Fig. 2.9). See Matthews (2008) and Matthews, Noble, and Breithaupt (2014b).

9. A well-planned set of photos taken with proper geometry/overlap at consistent focus and camera setting will provide an excellent framework, yielding a robust data set and site coverage. A recent example of this type of work was done in the main active quarry of the Obernkirchen Sandstone quarries in Obernkirchen, Germany (Fig. 2.10). Additional

2.9. (A) Brent Breithaupt conducting systematic stereoscopic imagery acquisition of the main track-bearing surface at the Mill Canyon Dinosaur Tracksite (MCDT), Utah (Early Cretaceous Cedar Mountain Formation), summer 2014. (B) Schematic of image framework (blue rectangles denote camera positions). Red polygons highlight overlapping blocks of imagery captured on sequential days. (C) Orthorectified image mosaic composed of over 1000 images. (D) Inset showing detail of a small portion of the surface as a color depth map. (E) A single theropod track from MCDT: (left) topographic contour map; scale bar = 25 cm (Lockley, Gierlinski, Dubicka, et al. 2014); (center) orthorectified image; (right) color depth map, with relative depths recorded in meters. See Lockley, Gierlinski, and Dubicka, et al. (2014); Lockley, Gierlinski, Houck, et al. (2014); and Matthews, Noble, and Breithaupt (2014a, 2014b).

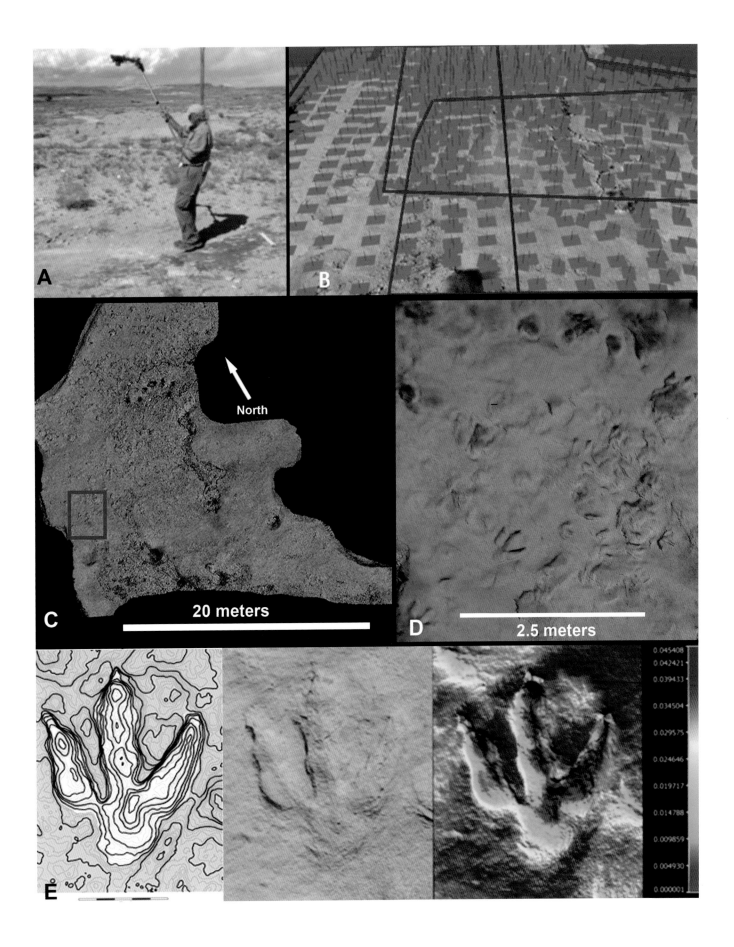

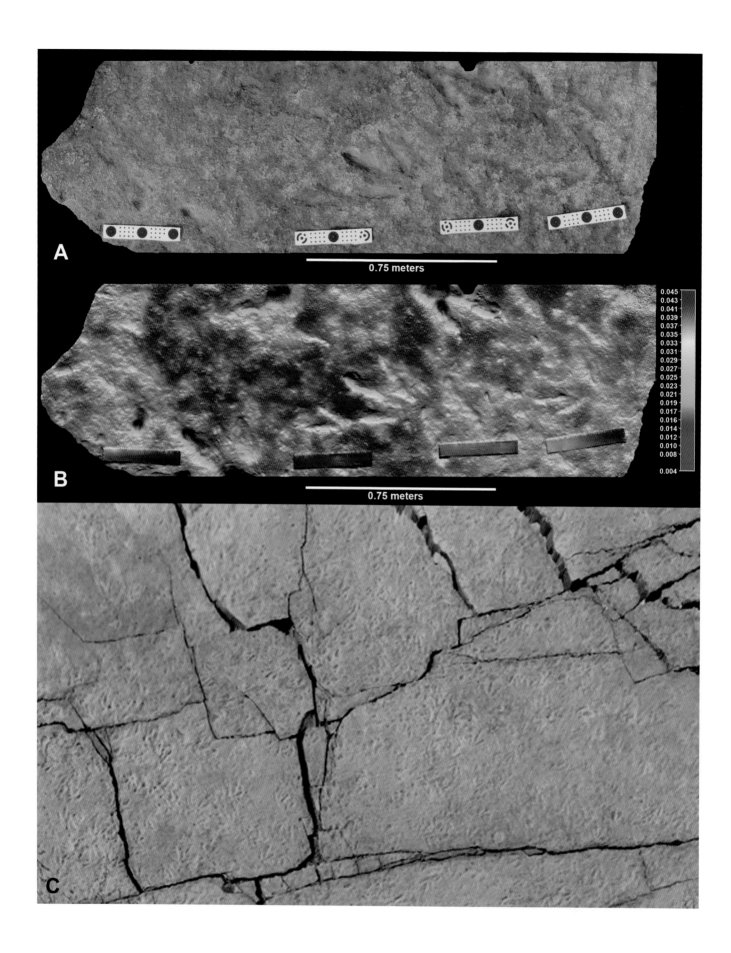

photos may be added to this framework set to ensure that important details are not missed. These additional photo sets can be particularly important if there are areas where tracks are deep or raised, have overhangs or undercuts, or have textural details (e.g., skin impressions or claw marks). In these cases, additional photos (taken in sets of stereopairs) may be shot at a high oblique angle, at different focal distances, with a flash, or zoomed into a specific detail. However, it is important to ensure that there is continuity with the existing overall framework set of photos. When changing focus or focal distance (e.g., zooming), either by moving closer to get more detail or farther out to get more site context, it is important to do so in distances of ½ or 2 times. For example, if the framework set of photos is taken at 2-m distance from the surface, the higher resolution photos of an individual track should be taken at 1 m. Conversely, an additional set of photos could be taken at 4 m to put the track layer in context with the outcrop. By using the ½ or 2 times rule, most SfM-based photogrammetric software (e.g., AgiSoft PhotoScan) will automatically align all of these photos together when processed in the same chunk, but grouped as distinct camera calibration sets. Similarly, a set of photos taken at an oblique angle that have continuity with the framework set, may serve to enhance coverage and ensure that areas are not lost due to depth of features or overhanging edges (Matthews and Breithaupt, 2011a; Matthews, Noble, and Breithaupt, 2014b).

10. Shadows, or drastic changes in surface illumination, such as that produced by using a flash or morning versus evening sun angle may impede the matching function of the processing software, causing failure or the erroneous interpretation of shadows as morphologic characters. Photographing with flat light (e.g., at high noon with low shadows or a time with complete overcast conditions) and at a consistent time for multiple day shoots are best practices. Checking photos during the documentation process for exposure and consistency is important. For relatively small areas (e.g., 1–2 m), tarps or other equipment can be used to create consistent shade on the surface if needed.

11. Incorporating other documentation methods (e.g., LiDAR and traditional ichnological data collection, such as maps and tracings) in a GIS along with your photogrammetric data increases the value of the data set (see Breithaupt and Matthews, 2001; Breithaupt, Matthews, and Noble, 2004; Matthews, Noble, and Breithaupt, 2006; Marty et al., 2010). Ichnocartography is a traditional method for two-dimensional measuring and mapping of tracks (Lockley, Gierlinski, Houck, et al., 2014). It is a very useful tool, especially when combined with photogrammetry. Just as standard photographs should accompany traditional ichnocartographic data, natural color orthophotos or digital photographs should accompany color depth maps, color contour diagrams, and digital elevation models that are derived from the photogrammetric data. Using multiple data collecting methods for the documentation of a tracksite has been shown to have extensive merits.

12. Marking high-quality GPS, total station, or other datum/grid points with visible targets (such as casino chips or spent CDs painted white) prior to conducting photogrammetric documentation is useful (Fig. 2.2). Make sure to include these marked points in the stereo overlap of at least two photos. This technique will provide high-quality reference points to incorporate the photogrammetric data with the traditional maps, tracings, and field notes.

13. Although it is possible to have apparent good success by randomly photographing a subject with hundreds of photos taken at oblique angles, in fans or panoramas, in descending spirals, or any number of random configurations; this type of approach can also lead to less than desirable results due to gaps in coverage and inability for the software to orient all groups of photos into the project as a whole and may require excessive manual intervention. Systematic overlapping coverage that satisfies the basic photogrammetric requirements can be processed in a variety of software programs, both today and into the future. Randomly shot photos, panoramic fans of photos, and lots of low oblique images that do not have image continuity with each other will likely not improve a project. They most certainly will increase the processing time and reduce automated workflow options.

2.10. (A) Orthorectified image mosaic of Chicken Yard track horizon at the Early Cretaceous Obernkirchen Sandstone, Obernkirchener Sandsteinbrueche Quarry, Germany, spring 2011. Note the optimal lighting conditions for photogrammetry resulted in minimal shadows within the track features. (B) Color depth map highlights track locations and morphology of tridactyl and didactyl theropod tracks in a heavily dinoturbated area. Relative depth legend in meters on right. (C) Orthorectified image mosaic created from high oblique photos taken from the high wall. See Hornung et al. (2012), Richter et al. (2012), and Richter and Böhme (2016).

14. Carry out photogrammetric documentation before applying any type of molding compound (if authorized to do so). Molding compounds almost always affect the track-bearing surface, enhancing mechanical and/or chemical weathering. It is also advisable to photogrammetrically document molds, as all molding compounds will degrade over time (Leite et al., 2007). In the past, prior to digital data collection techniques, the best 3-D documentation of a track may have been a mold. In addition, some unique preservational scenarios today require that molds be created because the actual tracks are voids preserved in the subsurface (McCrea et al., 2014).

15. As each tracksite presents a different set of unique conditions and resources, ingenuity and various documentation technologies may have to be tailored to individual sites (Breithaupt et al., 2001). However, any tracksite can be photogrammetrically documented.

16. Take camera RAW in addition to JPG or TIFF photos. Digital camera sensors have a "best" ISO setting that should be chosen. All lenses have an f-stop that will produce the sharpest images, usually around f8. As mentioned, take good, clear, in-focus, well-lit photographs with an ideal overlap of 66%, including an object of known dimension in at least two of the stereophotos, with consistent focus and focal length. Include a redundant set of images taken with the camera turned at 90° and 270° for camera calibration. Even if the research or documentation team does not currently have access to photogrammetric software (or the expertise for processing), it is of vital importance that a framework set of photos be taken. Properly taken photos will serve to protect and enhance the value of the research and the site into the future.

17. In those cases where a track has great depth, is a natural cast, or is reflected on both the top and bottom of the rock layer, it may be necessary to photogrammetrically document this specimen in-the-round. Advances to software and cameras allow this technique to be used on paleontological specimens of all shapes and sizes in the field, lab, and collections. As with relatively flat objects, the final data set is still dependent on the camera-lens system used, the distance from the subject, proper image geometry (i.e., 66% stereoscopic overlap), and a redundant set of images taken with the camera turned at 90° and 270°. Again, a set of good quality images; with proper exposure, good contrast, and sharp focus are a must. An advantage to capturing

dimensionally complex subjects in-the-round is that the redundancy mentioned is satisfied when completely encircling a subject with photographs taken at positions from 10° to 15° around the subject. In-the-round photogrammetry can be accomplished for specimens of virtually any size, from small specimens mounted on a turntable to large subjects lying on a table, or those that must be captured by walking around them. When capturing in-the-round subjects, a variety of considerations must be made, including scale, proper background, lighting (of subject and background), and appropriate turntable and specimen mounting, as well as the processing software utilized. The new generation of computational power supports the simultaneous processing of hundreds of photographs, resulting in an integrated point cloud allowing for efficient documentation of subjects in-the-round. Now researchers, curators, collections managers, and preparators can document material in the field and collections for research, management, and preparation purposes (Breithaupt, Matthews, and Noble et al., 2014; Mallison and Wings, 2014; Matthews, Noble, and Breithaupt, 2014a).

18. Although the use of a monopod or ladder can be effective for capturing stereoscopic imagery over a large or hard to access track-bearing surface, the use of unmanned aircraft can also be an effective tool for photogrammetric imagery collection (Breithaupt, Matthews, and Noble, 2004; Matthews, Noble, Brady, et al., 2014). Unmanned aircraft systems (UAS) are remotely piloted vehicles and fall into two broad categories—fixed wing and rotary. Depending on the size and configuration of the vehicles, camera payloads can range from a GoPro-type action camera to a DSLR. Often erroneously referred to as "drones" (a vehicle that is navigated autonomously without human control or beyond line of sight), UAS are becoming a more widely used tool for capturing imagery over a variety of paleontological resources, including tracksites and quarries. As with any photogrammetric project, good, blur-free images with proper stereoscopic (66%) overlap and proper geometry are paramount to obtaining a high-quality data set. Depending on the type of UAS and onboard camera used, meeting the preceding requirements may be difficult due to several factors. Image blur occurs due to motion during image capture and can be caused by low-shutter speeds (less than 1/2000 seconds) or the failure to isolate the camera from aircraft vibration. Consistent overlapping

stereoscopic coverage may be difficult to obtain without an expensive onboard flight management system and is a function of aircraft speed, image capture speed, and image download time. Camera battery life, digital storage capacity, and environmental considerations (such as elevation above sea level, temperature and humidity levels, wind speed and sheer) must all be taken into consideration when conducting a UAS photogrammetry project. Just as with a ground-based project, including several objects of known length, monumented ground control, and datum points visible in the imagery are important practices. In many cases, low-cost UAS have low-resolution onboard cameras and result in very poor imagery. A good point-and-shoot camera taped on a broom handle, set to interval mode will give better results. In addition, video or images from a GoPro camera will require special processing software and lens algorithms, and care should be taken that these images are processed correctly in photogrammetric software. It is also important to be aware of any governmental regulations that apply to the use of UAS. Although currently there is open use in many European countries (although it is likely that some countries will institute laws to cover UAS usage in the future), there are regulations that govern the use of any UAS in the United States National Airspace. See Matthews, Noble, Brady, et al. (2014).

19. When processing the photogrammetric image set, it is important to choose software that has the algorithms to accurately model the type of camera and lens that are being used for documentation. For example, if using a UAS with a GoPro-type camera, make sure that the photogrammetric software can model a fisheye-type lens, or if using a focal stacking setup, ensure that the resulting lens model and distortions are solved for correctly. If the software does not have the capability to correctly model the distortions of the lens/camera system, the resulting 3-D model will be inaccurate. In addition, ensure the software used can produce a robust camera/lens distortion model and that steps are taken to apply it correctly to a particular data set. When integrating a framework set of images with fixed focal distance, these photos should be grouped together and assigned a unique camera in the processing software. Any other associated groups of images taken with fixed settings should in turn be assigned to their own unique camera. These steps will allow the software to independently model the lens distortion for each set and thus remove most possible errors

from the resulting surface (Matthews, Noble, and Breithaupt, 2014b).

20. Be aware of the land status and associated rules and regulations for working in an area (Breithaupt and Matthews, 2014a). Some areas of federal public lands in the United States require specific authorizations before any paleontological surveys and documentation are engaged, even if no collection or excavation is done. Make sure that you are aware of the rules in the country you are working in. Additional authorizations may be required for collecting and/or molding of trace fossils (both in the field as well as in museum collections). Keep locality information for tracksite proprietary, because on US federal public lands this information is often kept confidential to protect the sites from theft and vandalism.

MANAGEMENT/CONSERVATION OF ICHNOFOSSILS AND PHOTOGRAMMETRY

Some of the most valuable clues to Earth's history may be represented by trace fossils. As such, these paleontological resources should be managed utilizing scientific principles and expertise to safeguard their priceless values. In 2009, the US Congress passed legislation (Omnibus Public Land Management Act) that recognized the value of and provided the authority to protect paleontological resources, using best practices on US federal public lands, as natural and irreplaceable parts of America's heritage. Many other countries around the world also have laws and mandates that protect fossil resources, including the traces of past animals and plants. In some cases, management strategies for trace fossils vary from those for body fossils (Breithaupt and Matthews, 2014a). Unlike fossilized bones and teeth (which once discovered are often removed from the ground), best practices in ichnology often encourage that the tracks and traces of vertebrate animals be left in situ and not collected. In the United States, loose specimens or in situ tracks require authorization prior to collection from public lands. Any destructive analysis (e.g., coring, thin sections) of tracks, whether done in the field or lab, also may require additional authorization. In addition, exposing a track-bearing surface prior to documentation needs to be considered early in the research planning process. Although a person may wish to expose an entire track surface for research, a responsible decision needs to be made related to the longevity and the multiple uses of the ichnological resources present in the area. Thus, it may be determined that it is not in the best interest of the resource to expose a track-bearing surface at a particular time, as paleontological resources that remain naturally

2.11. (A) BLM staff members showing visitors at the Red Gulch Dinosaur Tracksite (RGDT), Wyoming, the steps and strides of dinosaurs, summer 2012. (B) Tommy Noble and Neffra Matthews showcasing photogrammetric documentation techniques with Randy Hayes filming for BLM informational podcast. (C) RGDT theropod track (i.e., *Carmelopodus*) orthophoto (left) and color depth map (right). See Breithaupt and Matthews (2014a, 2014b).

buried are naturally protected. Once exposed, trace fossils are susceptible to natural erosion (even if reburied), unintentional or intentional damage by humans, and impacts by various animals and plants. However, once exposed, track surfaces should be digitally documented, especially prior to any collection or molding (Breithaupt and Matthews, 2014a).

No matter how carefully it is done, molding of trace fossils can damage a track-bearing surface. This activity also often requires additional authorization. State-of-the-art molding techniques utilize some type of liquid rubber (e.g., latex, silicone) or soft putty applied to the surface. Because these materials are pliable when cured, they can often be easily removed from the trace-bearing surface, providing that the surface is properly prepared (e.g., cleaned, stabilized, cracks/overhangs filled, and separator applied) prior to molding. However, even in the best cases, this activity may inevitably affect the surface chemically, mechanically, or biologically. In most situations, only experienced researchers should perform this activity on scientifically significant trace fossils (such as dinosaur footprints). In addition, a tradition (from neoichnological studies) exists where materials that cure to a hardened state (e.g., plaster, resin) are applied directly onto the track-bearing surface (Farlow et al., 2012). To be successful and not damage the fossil track, usually a great deal of

preparation needs to be done to the surface, as well as making sure the lithology and track preservation is stable enough to withstand the impact. Unfortunately, in many cases, this activity has resulted in the track-bearing surface being permanently scarred or lost, or remnants of cured material have been left in place. In some incidents, entire footprints have been accidently removed from trackways, as a result of this procedure. Although relatively cheap and easy to do, "hard molding" (i.e., casting) of trace fossils is an archaic method for collecting 3-D data for scientific purposes and should rarely if ever be done today, and only after careful consideration in extreme circumstances where original specimens cannot be preserved (e.g., the tracksite will be destroyed because of mining or road construction). In addition, unauthorized molding (especially that which results in damage to a trace fossil) may be considered vandalism, which on US federal public lands may result in criminal or civil penalties (Breithaupt and Matthews, 2014a).

Trace fossils are unique in that they provide valuable scientific information about the activities and behaviors of animals beyond the knowledge gained from body fossils. In addition, very significant fossil footprint discoveries have changed the scientific understanding of evolution and biostratigraphy. For these reasons, one of the most important

2.12. (A) Composite image showing camera positions (in lavender) of helicopter aerial image capture at Moccasin Mountain Tracksite (MMT) near Kanab, Utah (Early Jurassic Navajo Sandstone), summer 2007. (B) Interpretive map of the MMT created from photogrammetric products. (C) Contextual image of Early Jurassic Navajo Sandstone tracksite preserved on fossil dune face, Paria Canyon-Vermilion Cliffs Wilderness, Utah-Arizona border. (D) Sparse point cloud of outcrop generated from ground base images (track horizon highlighted by black line). See Matthews et al. (2008); Breithaupt and Matthews (2010); and Matthews, Noble, Brady, et al. (2014).

aspects of vertebrate ichnology is the context of the tracks in their preservational environment, as well as their relationship to other tracks and traces in the area. Removing a single footprint from its context often reduces greatly its scientific, educational, and interpretative value, as well as those of other associated tracks (such as in a trackway) and diminishes the information value. The collection of entire trackways or tracksites is often unpractical due to size, weight, material failure, and space for proper collection and storage. Thus, for many tracksites, tracks are left in place to be documented and studied by paleontologists, as well as visited and enjoyed by the general public (Marty et al., 2004; Breithaupt and Matthews, 2014b).

Fortunately, as described herein, with the arrival of the digital age, there are a variety of techniques other than molding and specimen collection for capturing and preserving the 3-D data associated with trace fossils (Chapman et al., 2002,

2012). Although laser scanning and LiDAR have been experimented with by various researchers; currently, the easiest, most efficient, low-cost, high-resolution mechanism to collect digital data of trace fossils in the field is photogrammetry. Associated with ichnology best practices and US legislation mandating that appropriate plans be developed for inventorying, monitoring, and scientifically and educationally using paleontological resources on federal public lands, the BLM pioneered many advancements in photogrammetry for state-of-the-art, noninvasive, digital data capture for 3-D data of trace fossils of all shapes and sizes (Breithaupt and Matthews, 2014a). As the photogrammetric data yield high-resolution topographic maps and orthophoto images, these photos and maps (along with microtopographic profiles) can be used for measurement and analysis of trace fossils at a submillimeter level. In addition to its value to researchers, this resolution of data allows land managers to make scientific-based management decisions regarding use, as well as monitor impacts to these paleontological resources over time.

In addition, as per current US legislation, programs to increase public awareness about the significance of paleontological resources have been developed by the BLM, National Park Service, and United States Forest Service. In many cases, vertebrate tracksites (especially those created by dinosaurs) are excellent forums for the public to experience these resources. Walking alongside the footprints of prehistoric beasts that once roamed the very same area millions of years ago is an exhilarating experience (Fig. 2.11). Examples of developed public tracksites can be found in Wyoming, Colorado, Utah, and New Mexico. These tracksites are some of the premier public paleontology sites currently managed by the federal government and are excellent examples of providing access and information about America's natural heritage on US public lands. In addition, various dinosaur tracksites have been well documented and preserved as public attractions in other countries around the world. As such, use of state-of-the-art documentation technology allows ichnologists to better interpret the formation, preservation, and location of vertebrate footprints in the context of their unique paleoenvironments. Once properly studied, tracks left in situ can become wonderful outdoor museums (Breithaupt and Matthews, 2014b). Showcasing these paleontological resources to the public requires that important management decisions be made (Fig. 2.12). Unfortunately, increased visitation may result in human erosion (sometimes including vandalism), which reinforces the need to permanently capture and preserve the 3-D data of tracks and tracksites.

DISCUSSION/SUMMARY

Associated with the actual observations of track-bearing surfaces, the key to more fully understanding formation and preservation of footprints (as well as the scientific information that they provide) is detailed 3-D documentation, along with the creation of digital archives from sites worldwide. Because cameras are currently standard pieces of field equipment, CRP is one of the easiest and most cost-effective digital data collection techniques. Fortunately, thorough documentation of the world's dinosaur tracks and tracksites no longer requires an enormous effort or expense, because combining traditional techniques with photogrammetry has proven to be an easy and highly accurate method of collecting 3-D digital data. Photogrammetry can be used on any size track (ranging from a couple of centimeters to those over 100 cm) in any orientation. It can be used for heavily bioturbated areas, capturing high-resolution ichnological data of thousands of footprints and their preservational context in a relatively

short amount of time (Richter and Böhme, 2016). In addition, tracksites of any size and location can be successfully documented. Large track localities including megatracksites (regionally extensive track-bearing horizons) can be successfully documented using various photogrammetric platforms to achieve submillimeter precision. Photogrammetry is especially useful for those areas that would be the most challenging to document via traditional methods: in particular, remote areas or areas of varying exposure (e.g., tidal or high-elevation areas). As dinosaur trace fossils reflect the complex interrelationship between an animal's activities and the substrate, CRP can assist in the proper documentation, preservation, and assessment of these ichnological resources, utilizing stereo images that have the quality, reliability, and authenticity necessary for scientific use. Three-dimensional image data sets created from stereoscopic digital photography provide permanent digital records of fossil tracks, including the creation of digital-type specimens. CRP is a noninvasive, objective recording and analysis method, which provides a visual, quantifiable baseline to evaluate track-bearing surfaces, which has proven very useful in locations around the world (Bates et al., 2008; Petti et al., 2008; Marty et al., 2010; Remondino et al., 2010; Castanera, Pascual, et al., 2013; Castanera, Vila, et al., 2013; Pond et al., 2014; McCrea et al., 2015).

As 3-D terrain surfaces or point clouds created from photogrammetric documentation may contain thousands of very accurate x, y, z coordinates, researchers can measure various ichnological dimensions at a submillimeter level. In addition to traditional ichnological measurements, unbiased, higher-level, mathematical analyses may be conducted on the 3-D data. Software algorithms can automatically quantify areas of surface curvature, roughness, slope, and other morphometric characteristics (Matthews, Pond, and Breithaupt, 2014; Wings, Lallensack, and Mallison, 2016). Photogrammetric ichnology allows for objective morphological correlations of various ichnofaunas to be made and data normalized (e.g., converting convex hyporelief forms to concave epirelief) for comparison purposes. In various parts of the world, this level of geospatial documentation has been conducted on various dinosaur tracksites; some of which are quite extensive or complex. To better understand the meaning of the ichnomorphologic characters, track formation, as well as the taphonomic, ontogenetic, and behavioral implications of fossil footprint data, detailed 3-D data in GIS is used along with computer-modeled simulations and neoichnological studies (Breithaupt and Matthews, 2012). Information derived from this research is being used to help unravel numerous ichnological complexities and provide a unique glimpse of the paleoecology, paleobiology, and paleoethology of dinosaur communities.

ACKNOWLEDGMENTS

Thanks to Annette Richter and the editors of this volume for their hard work and patience compiling the chapters of this book. In addition, gratitude is extended to Dr. Richter and the organizers of the Dinosaur Track Symposium Obernkirchen 2011, as well as all of the participants at that meeting, which spawned this volume. In addition, appreciation is extended to the Niedersächsisches Landesmuseum Hannover for its support for travel and attendance at that meeting. Thank you to the BLM for support of the management and documentation on the various ichnology projects throughout North America. Thanks to Alan Bell for his continued assistance,

encouragement, and support. And finally, thank you to all of the ichnologists old and new, traditional and technological, professional and volunteer who have assisted in the proper documentation of dinosaur tracks around the world for over 200 years. This is an exciting time for the authors, with over 25 years of experience in photogrammetry, ranging from analog (film) workstations to current multiview matching software, bridging a pivotal time in the development of the discipline. There is no doubt that the current ease of use, flexibility, and adaptability of the new breed of photogrammetric software will continue making photogrammetry a standard practice in the science of ichnology.

REFERENCES

Adams, T. L., and B. H. Breithaupt. 2003. Documentation of Middle Jurassic dinosaur tracks at the Yellow Brick Road Dinosaur Tracksite, Bighorn Basin, Wyoming. Wyoming State Geological Survey. Wyoming Geo-Notes 76: 28–32.

Adams, T. L., C. Stragnac, M. J. Polcyn, and L. L. Jacobs. 2010. High resolution three-dimensional laser-scanning of the type specimen of *Eubrontes* (?) *glenrosensis* Shuler, 1935, from the Comanchean (Lower Cretaceous) of Texas: implications for digital archiving and preservation. Palaeontologia Electronica 13(3): 1T: 1–11.

Adams, T. L., B. H. Breithaupt, and N. A. Matthews. 2014. Paleoecology and paleoenvironment of Middle Jurassic dinosaurs (Yellow Brick Road Dinosaur Tracksite), Bighorn Basin, northern Wyoming. New Mexico Museum of Natural History and Science Bulletin 62: 225–233.

Albertz, J. 2002. Albrecht Meydenbauer: pioneer of photogrammetric documentation of the cultural heritage. International Archives of Photogrammetry Remote Sensing and Spatial Information Sciences 34(5/C7): 19–25.

Albertz, J. 2007. A look back: 140 years of 'photogrammetry,' some remarks on the history of photogrammetry. Photogrammetric Engineering and Remote Sensing 53(5): 505–506.

Bates, K. T., B. H. Breithaupt, P. L. Falkingham, N. A. Matthews, D. Hodgetts, and P. L. Manning. 2009. Integrated LiDAR and photogrammetric documentation of the Red Gulch Dinosaur Tracksite (Wyoming, USA); pp. 101–103 in S. E. Foss, J. L. Cavin, T. Brown, J. I. Kirkland, and V. L. Santucci (eds.), Proceedings of the Eighth Conference on Fossil Resources, St. George, Utah, May 19–21. BLM Regional Paleontologist, Salt Lake City, Utah.

Bates, K. T., F. Rarity, P. L. Manning, D. Hodgetts, B. Vila, O. Oms, À. Galobart, and R. L. Gawthorpe. 2008. High-resolution LiDAR and photogrammetric survey of the Fumanya dinosaur tracksites (Catalonia): implications for the conservation and interpretation of geological heritage sites. Journal of the Geological Society, London 165: 115–127.

Belvedere, M., A. Baucon, C. Neto de Carvalho, C. Venturini, F. Felletti, and G. Muttoni. 2012. To

(laser) scan or not to scan? Hamletic question at the Pramollo ichnolagerstätte (Carboniferous, Italy-Austria). DigitalFossil 2012, Museum für Naturkunde, Berlin, Germany, 1 p.

Belvedere, M., N-E Jalil, A. Breda, G. Gattolin, H. Bourget, F. Khaldoune, and G. J. Dyke. 2013. Vertebrate footprints from the Kem Kem beds (Morocco): a novel ichnological approach to faunal reconstruction. Palaeogeography, Palaeoclimatology, Palaeoecology 383–384: 52–58.

Breithaupt, B. H., and N. A. Matthews. 2001. Preserving paleontological resources using photogrammetry and geographic information systems; pp. 62–70 in D. Harmon (ed.), Crossing Boundaries in Park Management: Proceedings of the 11th Conference on Research and Resource Management on Parks and Public Lands. The George Wright Biennial Conference, Denver. The George Wright Society, Hancock, Michigan.

Breithaupt, B. H., and N. A. Matthews. 2010. An Early Jurassic desert ichnofauna: paleontological resources in the Vermillion Cliffs National Monument and Paria Canyon-Vermilion Cliffs Wilderness. Society of Vertebrate Paleontology, Seventieth Anniversary Meeting. Program and Abstracts 30(2, Supplement): 64A.

Breithaupt, B. H. and Matthews, N. A. 2012. Neoichnology and photogrammetric ichnology to interpret theropod community dynamics; p. 17 in A. Richter and M. Reich (eds.), Dinosaur Tracks 2012. An International Symposium, Obernkirchen, April 14–17, 2011, Abstract Volume and Field Guide to Excursions. Universitätsverlag, Universitätsdrucke, Göttingen.

Breithaupt, B. H., and N. A. Matthews. 2014a. Fossil tracks and future science: managing ichnological resources on BLM's public land; pp. 21–28 in V. L. Santucci, G. A. Liggett, B. A. Beasley, H. G. McDonald, J. S. Tweet, J. Chilstrom, S. E. Foss, and S. Y. Shelton (eds.), Protecting the Past, Managing the Present, Planning for the Future: Proceedings of the Tenth Conference on Fossil Resources. Dakoterra 6.

Breithaupt, B. H., and N. A. Matthews. 2014b. Podcasts and the public: paleontological

education at the Red Gulch Dinosaur Tracksite, Wyoming; pp. 29–31 in V. L. Santucci, G. A. Liggett, B. A. Beasley, H. G. McDonald, J. S. Tweet, J. Chilstrom, S. E. Foss, and S. Y. Shelton (eds.), Protecting the Past, Managing the Present, Planning for the Future: Proceedings of the Tenth Conference on Fossil Resources. Dakoterra 6.

Breithaupt, B. H., N. A. Matthews, and T. A. Noble. 2004. An integrated approach to three-dimensional data collection at dinosaur tracksites in the Rocky Mountain West. Ichnos 11: 11–26.

Breithaupt, B. H., T. A. Noble, and N. A. Matthews. 2012. Photogrammetry vs. LiDAR scanning: different digital data collecting tools for the 3D data preservation of natural subjects. DigitalFossil 2012. Museum für Naturkunde, Berlin, Germany.

Breithaupt, B. H., N. A. Matthews, and T. A. Noble. 2014. Collecting quality, 360°, three-dimensional, digital data with photogrammetry: fast and furious vs. slow and steady. DigitalSpecimen 2014. Museum für Naturkunde, Berlin, Germany, 1 p.

Breithaupt, B. H., E. H. Southwell, T. L. Adams, and N. A. Matthews. 2001. Innovative documentation methodologies in the study of the most extensive dinosaur tracksite in Wyoming; pp. 113–122 in Santucci and McCelland (eds.), Proceedings of the 6th Fossil Resources Conference. United States Department of Interior, National Park Services, Geological Resources Division, Lakewood, Colorado.

Breithaupt, B. H., E. H. Southwell, T. L. Adams, and N. A. Matthews. 2006. Myths, fables, and theropod community dynamics of the 'Sundance Vertebrate Ichnofaunal Province'; pp. 1–4 in P. M. Barrett and S. E. Evans (eds.), Ninth International Symposium on Mesozoic Terrestrial Ecosystems and Biota. Natural History Museum, London, UK.

Castanera, D., J. Colmenar, V. Sauque, and J. L. Canudo. 2015. Geometric morphometric analysis applied to theropod tracks from the Lower Cretaceous (Berriasian) of Spain. Paleontology 58(1): 183–200.

Castanera, D., C. Pascual, N. L. Razzolini, B. Vila, J. L. Barco, and J. I. Canudo. 2013.

Discriminating between medium-sized tridactyl trackmakers: tracking ornithopod tracks in the base of the Cretacous (Berriasian, Spain). PLoS One 8(11): e81830.

Castanera, D., B. Vila, N. L. Razzolini, P. L. Falkingham, J. L. Canudo, P. L. Manning, and À. Galobart. 2013. Manus track preservation bias as a key factor for assessing trackmaker identity and quadrupedalism in basal ornithopods. PLoS One 8(1): e54177.

Chapman, R. A., A. Anderson, B. H. Breithaupt, and N. A. Matthews. 2012. Technology and the study of dinosaurs; pp. 247–272 in M. K. Brett-Surman, T. R. Holtz Jr., and J. O. Farlow (eds.), The Complete Dinosaur. 2nd edition. Indiana University Press, Bloomington, Indiana.

Chapman, R. E., N. A. Matthews, M. H. Schweitzer, and C. C. Horner. 2002. Applying 21st century technology to very old animals; pp. 137–144 in J. G. Scotchmoor, D. A. Springer, B. H. Breithaupt, and A. R. Fiorillo (eds.), Dinosaurs: The Science behind the Stories. American Geological Institute, Alexandria, Virginia.

Clark, N. D. L., and J. L. Barco Rodriguez. 1998. The first dinosaur trackway from the Valtos Sandstone Formation (Bathonian, Jurassic) of the Isle of Skye, Scotland, UK. Geogaceta 24: 79–82.

Clark, N. D. L., D. A. Ross, and P. Booth. 2005. Dinosaur tracks from the Kilmaluag Formation (Bathonian, Middle Jurassic) of Score Bay, Isle of Skye, Scotland, UK. Ichnos 12: 93–104.

Falkingham, P. L. 2012. Acquisition of high resolution three-dimensional models using free, open-source, photogrammetric software. Palaeontologia Electronica 15(1): 1T: 15 p.

Falkingham, P. L. 2014. Interpreting ecology and behaviour from the vertebrate fossil track record. Journal of Zoology 292: 222–228.

Falkingham, P. L., K. T. Bates, and J. O. Farlow. 2014. Historical photogrammetry: Bird's Paluxy River dinosaur chase sequence digitally reconstructed as it was prior to excavation 70 Years Ago. PLoS One 8(4): e93247.

Farlow, J. O., R. E. Chapman, B. H. Breithaupt, and N. A. Matthews. 2012. The scientific study of dinosaur footprints; pp. 713–759 in M. K. Brett-Surman, T. R. Holtz Jr., and J. O. Farlow (eds.), The Complete Dinosaur. 2nd edition. Indiana University Press, Bloomington, Indiana.

Fiorillo, A. R., P. J. McCarthy, B. H. Breithaupt, and P. Brease. 2007. Dinosauria and fossil Aves footprints from the Lower Cantwell Formation (latest Cretaceous), Denali Park and Preserve, Alaska. Alaska Park Science 6: 41–43.

Fiorillo, A. R., M. Contessi, Y. Kobayashi, and P. J. McCarthy. 2014. Theropod tracks from the lower Cantwell Formation (Upper Cretaceous) of Denali National Park, Alaska, USA with comments on theropod diversity in an ancient, high-latitude terrestrial ecosystem. New Mexico Museum of Natural History and Science Bulletin 62: 429–439.

Fraser, C. S., and C. Stamatopoulos. 2014. Automated target-free camera calibration; pp. 23–28 in Proceedings of the ASPRS 2014 Annual Conference, Louisville, Kentucky, March 23–28.

García-Ortiz, E., I. Fuertes-Gutiérrez, and E. Fernández-Martínez. 2014. Concepts and terminology for the risk of degradation of geological

heritage sites: fragility and natural vulnerability, a case study. Proceedings of the Geologist's Association 125: 463–479.

Ghosh, S. K. 2005. Fundamentals of Computational Photogrammetry. Concept Publishing Company, New Delhi, India, 254 pp.

Gruner, H. 1977. Photogrammetry: 1776–1976. Photogrammetric Engineering and Remote Sensing 43(5): 569–574.

Hornung, J. J., A. Bohme, T. Lubbe, M. van der Reich, and A. Richter. 2012. Vertebrate tracksites in the Obernkirchen Sandstone (late Berriasian, Early Cretaceous) of northwest Germany: their stratigraphical, palaeogeographical, palaeoecological and historical context. Paläontologische Zeitschrift 86: 231–267.

Jones, P. 1987. Recording the hominid footprints in Laetoli, a Pliocene site in northern Tanzania; pp. 551–558 in M. D. Leakey and J. M. Harris (eds.), Laetoli: a Pliocene Site in Northern Tanzania. Oxford University Press, Oxford, UK.

Konecny, G. 1985. The International Society for Photogrammetry and Remote Sensing–75 years old, or 75 years young, keynote address. Photogrammetric Engineering and Remote Sensing 51(7): 919–933.

Leite, M. B, H. E. LaGarry, B. H. Breithaupt, and N. A. Matthews. 2007. Direct and photogrammetric replication of latex-molded trackways from Toadstool Geologic Park, Northwestern Nebraska, U.S.A. 2007 GSA Denver Annual Meeting. Geological Society of America Abstracts with Programs 39(6): 557.

Lockley, M. G. 1986. The paleobiological and paleoenvironmental importance of dinosaur footprints. Palaios 1: 37–47.

Lockley, M. G. 1991. Tracking Dinosaurs: A New Look at an Ancient World. Cambridge University Press. Cambridge, UK., 238 pp.

Lockley, M., and C. Meyer. 2000. Dinosaur Tracks and Other Fossil Footprints of Europe. Columbia University Press, New York, New York, 360 pp.

Lockley, M. G., G. D. Gierlinski, Z. Dubicka, B. H. Breithaupt, and N. A. Matthews. 2014a. A preliminary report on new dinosaur tracksite in the Cedar Mountain Formation (Cretaceous) of eastern Utah. New Mexico Museum of Natural History and Science Bulletin 62: 279–285.

Lockley, M. G., G. D. Gierlinski, K. Houck, J. D. Lim, S. K. Kim, D. Y. Kim, T. H. Kim, S.-H. Kang, R. Hunt Foster, R. Li, C. Chesser, R. Gay, Z. Dubicka, K. Cart, and C. H. Wright. 2014b. New excavations at the Mill Canyon dinosaur tracksite (Cedar Mountain Formation, Lower Cretaceous) of eastern Utah. New Mexico Museum of Natural History and Science Bulletin 62: 287–300.

Mallison, H., and O. Wings. 2014. Photogrammetry in paleontology: a practical guide. Journal of Paleontological Techniques 12: 1–31.

Manning, P. L. 2004. A new approach to analysis and interpretation of tracks: examples from the Dinosauria. Geological Society of London Special Publications 228: 93–123.

Marty, D., L. Cavin, W. A. Hug, P. Jordan, M. G. Lockley, and C. A. Meyer. 2004. The protection, conservation and sustainable use of the Courtedoux dinosaur tracksite, Canton Jura, Switzerland. Revue de Paléobiologie Volume Spécial 9: 39–49.

Marty, D. M., C. A. Belvedere, C. M. Meyer, P. Mietto, C. Paratte, C. Lovis, and B. Thuring. 2010. Comparative analysis of Late Jurassic sauropod trackways from the Jura Mountains (NW Switzerland) and the central High Atlas Mountains (Morocco); implications for sauropod Ichnotaxonomy. Historical Biology 22: 109–133.

Matthews, N. A. 2008. Resource documentation, preservation, and interpretation: aerial and close-range photogrammetric technology in the Bureau of Land Management. Technical Note 428. U.S. Department of the Interior, Bureau of Land Management. National Operations Center. Denver, Colorado.

Matthews, N. A., and B. H. Breithaupt. 2001. Close-range photogrammetric experiments at Dinosaur Ridge. Mountain Geologist 38(3): 147–153.

Matthews, N. A., and B. H. Breithaupt. 2006. Terrestrial tracks trace theropods through time and terrain; pp. 80–83 in P. M. Barrett and S. E. Evans (eds.), Ninth International Symposium on Mesozoic Terrestrial Ecosystems and Biota. Manchester, UK, Abstracts and Proceedings. Natural History Museum, London, UK.

Matthews, N. A., and B. H. Breithaupt. 2009. Close-range photogrammetric technology for paleontological resource documentation, preservation, and interpretation; pp. 94–96 in S. E. Foss, J. L. Cavin, T. Brown, J. I. Kirkland, and V. L. Santucci (eds.), Proceedings of the Eighth Conference on Fossil Resources, St. George, Utah, May 19–21.

Matthews, N. A., and B. H. Breithaupt. 2011a. 3D image capture and close-range photogrammetry: an overview of field capture methods; pp. 32–39 in J. W. Bonde and A. R. C. Milner (eds.), Field Trip Guide Book 71st Annual Meeting of the Society of Vertebrate Paleontology. Paleontological Papers Number 1. Nevada State Museum, Carson City, Nevada.

Matthews, N. A., and B. H. Breithaupt. 2011b. Moccasin Mountain tracksite; pp. 55–62 in J. W. Bonde and A. R. C. Milner (eds.), Field Trip Guide Book 71st Annual Meeting of the Society of Vertebrate Paleontology. Paleontological Papers Number 1. Nevada State Museum, Carson City, Nevada.

Matthews, N. A., and B. H. Breithaupt. 2012. Taking measure of geospatial technology: innovations in close-range photogrammetry for 3D ichnology; p. 41 in A. Richter and M. Reich (eds.), Dinosaur Tracks 2012. An International Symposium, Obernkirchen, April 14–17, 2011, Abstract Volume and Field Guide to Excursions. Universitätsverlag, Universitätsdrucke, Göttingen.

Matthews, N. A., B. H. Breithaupt, and E. H. Southwell. 2000. Photogrammetric documentation of a Middle Jurassic dinosaur tracksite in the Sundance Formation of the Bighorn Basin of Wyoming. Journal of Vertebrate Paleontology Abstracts with Programs 20(3): 57A.

Matthews, N. A., T. A. Noble, and B. H. Breithaupt. 2004. From dinosaur tracks to dam faces: a new method for collecting three-dimensional data. Geological Society of America Abstracts 36(5): 384.

Matthews, N. A., T. A. Noble, and B. H. Breithaupt. 2005. Microtopographic documentation of a sitting dinosaur from the early

Jurassic of Utah; p. 16 in Tracking Dinosaur Origins: The Triassic/Jurassic Terrestrial Transition, Abstract Volume. Dixie State College, St. George, Utah.

Matthews, N. A., T. A. Noble, and B. H. Breithaupt. 2006. The application of photogrammetry, remote sensing, and geographic information systems (GIS) to fossil resource management; pp. 119–131 in S. G. Lucas, J. A. Spielmann, M. H. Hester, J. P. Kenworthy, and V. L. Santucci (eds.), America's Antiquities: 100 Years of Managing Fossils on Federal Lands: Proceedings of the 7th Federal Fossil Resources Conference. Bulletin 34. New Mexico Museum of Natural History and Science, Albuquerque, New Mexico.

Matthews, N. A, T. A. Noble, and B. H. Breithaupt 2012. Photogrammetry: a perspective on the past and projection for the future. DigitalFossil 2012. Museum für Naturkunde, Berlin, Germany, 1 p.

Matthews, N. A., T. A. Noble, and B. H. Breithaupt. 2014a. Focusing around with photogrammetry: capturing 3D data on dimensionally complex subjects; pp. 59–61 in V. L. Santucci, G. A. Liggett, B. A. Beasley, H. G. McDonald, J. S. Tweet, J. Chilstrom, S. E. Foss, and S. Y. Shelton (eds.), Protecting the Past, Managing the Present, Planning for the Future: Proceedings of the Tenth Conference on Fossil Resources. Dakoterra 6.

Matthews, N. A, T. A. Noble, and B. H. Breithaupt. 2014b. The photogrammetric data set: a look at capturing imagery, assessing accuracy, archiving strategies, and sharing projects. DigitalSpecimen 2014. Museum für Naturkunde, Berlin, Germany, 1 p.

Matthews, N., S. Pond, and B. Breithaupt. 2014. A multi-dimensional look at morphological variation in ichnofauna: tracking changes within and between ichnotaxa. Journal of Vertebrate Paleontology Program and Abstracts 2014: 182.

Matthews, N. A., T. A. Noble, L. R. Brady, and B. H. Breithaupt. 2014. Up, up, and away: the use of unmanned aircraft systems for paleontological documentation; pp. 231–239 in V. L. Santucci, G. A. Liggett, B. A. Beasley, H. G. McDonald, J. S. Tweet, J. Chilstrom, S. E. Foss, and S. Y. Shelton (eds.), Protecting the Past, Managing the Present, Planning for the Future: Proceedings of the Tenth Conference on Fossil Resources. Dakoterra 6.

Matthews, N. A., M. Lockley, B. H. Breithaupt, A. Titus, and T. Noble. 2008. The Moccasin Mountain Tracksite, Utah: where science, technology and recreation meet paleontological resource management. Journal of Vertebrate Paleontology 28(3): 114A.

Matthews, N. A., T. A. Noble, A. L. Titus, J. R. Foster, J. A. Smith, B. H. Breithaupt, and D. Kett, 2002. Tracking dinosaurs using low-altitude aerial photography at the Twentymile Wash Dinosaur Tracksite, Grand Staircase-Escalante National Monument, Utah. 54th Annual Meeting, Rocky Mountain Section,

Geological Society of America Abstracts with Programs 34(3): A-6.

McCrea, R. T., L. G. Buckley, J. O. Farlow, M. G. Lockley, P. J. Currie, N. A. Matthews, and S. G. Pemberton. 2014. A 'terror of tyrannosaurs': the first trackways of tyrannosaurids and evidence of gregariousness and pathology in Tyrannosauridae. PLoS One 9 (7): e103613.

McCrea, R. T., L. G. Buckley, A. G. Plint, M. G. Lockley, N. A. Matthews, T. A. Noble, L. Xing, and J. R. Krawetz. 2015. Vertebrate ichnites from the Boulder Creek Formation (Lower Cretaceous: middle to ?upper Albian) of northeastern British Columbia, with a description of a new avian ichnotaxon, *Paxavipes babcockensis* ichnogen. et isp. nov. Cretaceous Research 55: 1–18.

Milner, A. R. C., J. D. Harris, M. G. Lockley, J. I. Kirkland, and N. A. Matthews. 2009. Bird-like anatomy, posture, and behavior revealed by an Early Jurassic Theropod dinosaur resting trace. PLoS One 4(3): e4591.

Mudge, M., C. Schroer, G. Earl, K. Martinez, H. Pagi, C. Toler-Franklin, S. Rusinkiewicz, G. Palma, M. Wachowiak, M. Ashley, N. Matthews, T. Noble, and M. Dellepiane. 2010. Principles and practices of robust, photography-based digital imaging techniques for museums; in A. Artusi, M. Joly-Parvex, G. Lucet, A. Ribes, and D. Pitzalis (eds.), Principles and Practices of Robust, Photography-based Digital Imaging Techniques for Museums. The 11th International Symposium on Virtual Reality, Archaeology and Cultural Heritage VAST.

Mudge, M., C. Schroer, T. Noble, N. Matthews, S. Rusinkiewicz, and C. Toler-Franklin. 2012. Robust and scientifically reliable rock art documentation from digital photographs; pp. 644–659 in J. McDonald and P. Veth (eds.), A Companion to Rock Art. John Wiley & Sons, Chichester, UK.

Musiba, C., A. Mabulla, M. Mutakyahwa, F. Masao, V. Runyoro, J. L. Kim, M. Dominguez-Rodrigo, N. Mbwana, F. Ndunguru, S. Odunga, J. Washa, J. Paresso, and V. Stepanek. 2012. Tanzania and the outstanding universal value of its paleoanthropology: approaches at Laetoli and lessons learned; pp. 115–126 in N. Sanz (ed.), Human origin sites and the World Heritage Convention in Africa. UNESCO Publication Series 33, United Nations Educational, Scientific and Cultural Organization, New York, New York.

Petti, F. M., M. Avanzini, M. Belvedere, M. Degasperi, P. Ferretti, S. Girardi, F. Remondino, and R. Tomasoni. 2008. Digital 3D modelling of dinosaur footprints by photogrammetry and laser scanning techniques: integrated approach at the Coste dell'Anglone tracksite (Lower Jurassic, Southern Alps, northern Italy). Studi Trentini di Scienze Naturali Acta Geologica 83: 303–315.

Pond, S., M. Belvedere, and G. Dyke. 2012. The ichnologist's guide to 3D models: from the field to the cloud. DigitalFossil 2012. Museum für Naturkunde, Berlin, Germany, 1 p.

Pond, S., M. G. Lockley, J. A. F. Lockwood, B. H. Breithaupt, and N. A. Matthews. 2014. Tracking dinosaurs on the Isle of Wight: a review of tracks, sites, and current research. Biological Journal of the Linnean Society 113: 737–757.

Razzolini, N. L., B. Vila, D. Castanera, P. L. Falkingham, J. L. Barco, J. Canudo, P. L. Manning, and À. Galobart. 2014. Intra-trackway morphological variations due to substrate consistency: the El Frontal dinosaur tracksite (Lower Cretaceous, Spain). PLoS One 9 (4): e93708.

Remondino, F., A. Rizzi, S. Girardi, F. Petti, and M. Avanzini. 2010. 3D ichnology: recovering digital 3D models of dinosaur footprints. Photogrammetric Record 25 (131): 266–282.

Richter, A., and A. Böhme. 2016. Too many tracks: preliminary description and interpretation of the diverse and heavily dinoturbated Early Cretaceous 'Chicken Yard' ichnoassemblage (Obernkirchen tracksite, northern Germany); chap. 17 in P. L. Falkingham, D. Marty, and A. Richter (eds.), Dinosaur Tracks: The Next Steps. Indiana University Press, Bloomington, Indiana.

Richter, A., J. J. Hornung, A. Böhme, and U. Stratmann. 2012. Excursion guide A1: Obernkirchen sandstone quarries: a natural workstone lagerstaette and a dinosaur tracksite; pp. 73–100 in A. Richter and M. Reich (eds.), Dinosaur Tracks 2012. An International Symposium, Obernkirchen, April 14–17, 2011, Abstract Volume and Field Guide to Excursions. Universitätsverlag, Universitätsdrucke, Göttingen.

Rowland, S. M., B. H. Breithaupt, H. M. Stoller, N. A. Matthews, and M. Saines. 2014. First report of dinosaur, synapsid, and arthropod tracks in the Aztec Sandstone (Lower-Middle Jurassic) of Red Rock Canyon National Conservation Area, southern Nevada. New Mexico Museum of Natural History and Science Bulletin 62: 249–259.

Thompson, M. M., and H. Gruner. 1980. Foundations of photogrammetry; pp. 10–36 in C. C. Slama, C. Theurer, and S. W. Henriksen (eds.), Manual of Photogrammetry. 4th edition. American Society of Photogrammetry, Falls Church, Virginia.

Thulborn, R. A. 1990. Dinosaur Tracks. Chapman and Hall, London, UK.

Wings, O., J. N. Lallensack, and H. Mallison. 2016. The Early Cretaceous Dinosaur Trackways in Münchehagen (Lower Saxony, Germany): 3D photogrammetry as basis for geometric morphometric analysis of shape variation and evaluation of material loss during excavation; chap. 3 in P. L. Falkingham, D. Marty, and A. Richter (eds.), Dinosaur Tracks: The Next Steps. Indiana University Press, Bloomington, Indiana.

Wright, J. L., and B. H. Breithaupt. 2002. Walking in their footsteps and what they left us: dinosaur tracks and traces; pp. 117–126 in J. G. Scotchmoor, D. A. Springer, B. H. Breithaupt, and A. R. Fiorillo (eds.), Dinosaurs: The Science behind the Stories. American Geological Institute. Alexandria, Virginia.

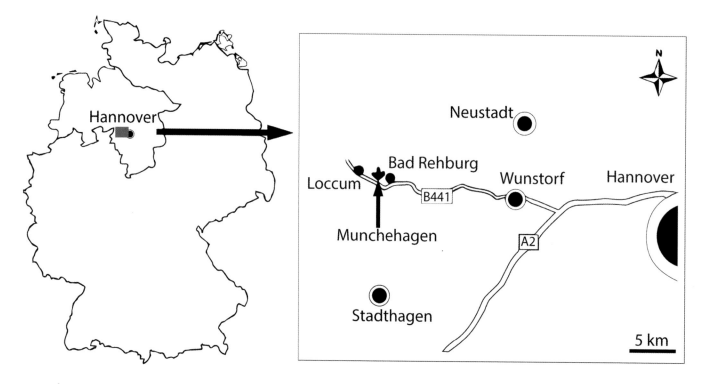

3.1. The Münchehagen locality in Lower Saxony, Germany.

The Early Cretaceous Dinosaur Trackways in Münchehagen (Lower Saxony, Germany): 3-D Photogrammetry as Basis for Geometric Morphometric Analysis of Shape Variation and Evaluation of Material Loss during Excavation

3

Oliver Wings, Jens N. Lallensack, and Heinrich Mallison

LOWER CRETACEOUS SANDSTONES IN LOWER SAXONY, northern Germany, are well known for their abundant fossil dinosaur tracks. One of the most productive sites is Münchehagen, which is well known for the only German Cretaceous sauropod trackways and hundreds of tracks of ornithopods and theropods, often forming long individual trackways with dozens of consecutive footprints. The largest theropod trackway T3 from the layer that has produced the best preserved true tracks (Lower Level) shows variations in the footprint morphology that allow use of this data as an example for studying the variability of tridactyl dinosaur track measurements.

Photogrammetric high-resolution three-dimensional (3-D) digital models of the tracks were the basis for geometric morphometrics and shape-difference analyses of true tracks and natural track casts. Geometric morphometrics was used to analyze footprint variability. The greatest variability in footprints of the T3 theropod trackway can be found in the positions of the hypex points and in the heel. The position of the medial and lateral hypices behave independently from each other, and we suggest that the relative position of the hypices is a questionable character in ichnotaxonomy. Shape-difference analyses indicate only minimal differences in shape between true tracks and their natural track casts. Furthermore, it is possible to estimate a minimum thickness for any sediment layer that separated the rocks in the field by shifting best-fit aligned models of true tracks and natural track casts until all overlap between the models is removed.

INTRODUCTION

Lower Cretaceous sandstones in Lower Saxony, northern Germany are well known for their abundant fossil dinosaur tracks. Beside Obernkirchen (Richter and Böhme, 2016), one of the most important localities for fossil tracks in the region is Münchehagen (Fig. 3.1). The site has provided the only

evidence for Cretaceous sauropod dinosaurs in Germany and features exquisitely preserved tridactyl tracks, including unusually long trackways. These trackways, with high numbers of footprints that we can be 100% sure stem from one individual, allow us to study and answer several questions of general importance in vertebrate ichnology: (1) How can we quantify shape variation of footprints within a trackway? (2) How much alteration occurs during careful excavation of tracks (i.e., how much original material is lost)?

Dinosaur footprints can exhibit a great deal of variation, even within a single trackway (e.g., Thulborn, 1990; Razzolini et al., 2014). A good understanding of variation patterns in individual trackways is of great importance for any description of fossil footprints, especially when attempting to distinguish separate trackmaker species based on a small sample size. Traditionally, footprint shape is quantified by taking sets of length and angular measurements from the footprints (Demathieu, 1990). According to Lockley (1998), measurements commonly employed to distinguish trackmaker species of mesaxonic, tridactyl dinosaur footprints can include, among others

- footprint length, width, and span
- projection of digit III past digits II and IV
- length and width of the digit impressions
- distance between heel and medial and lateral hypex
- interdigital angles
- relative position of the hypices

Detailed studies on the variability of these measurements within tridactyl dinosaur trackways, however, remain rare (but see Weems, 1992). Traditional attempts have used the coefficient of variation to quantify the variability of certain measurements within a trackway (Leonardi, 1987; Weems, 1992). Such attempts, however, provide insufficient information on where in the footprint the recorded shape differences

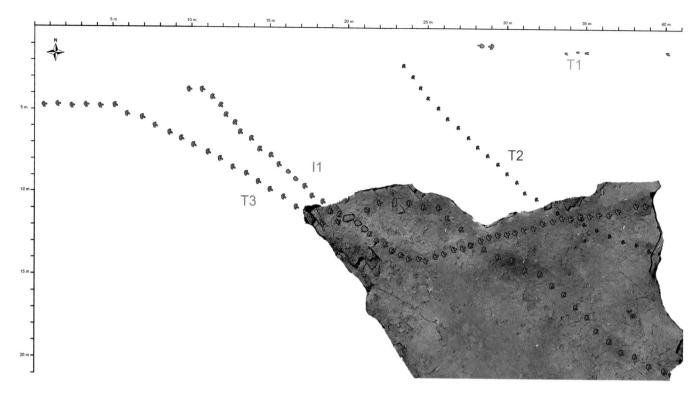

3.2. Map of the excavated trackways during 2009–2011, partially based on a photogrammetric orthophoto. Note the long trackways and the curved walking direction of the trackmakers. Green: ornithopod trackways, Red: theropod trackways.

actually do occur. For example, variability in footprint length can be explained by variations in the length of the digits, variations in the extension of the heel, or both. A second example is the variability in digit length, which can be caused by both variations in the lengths of the digits and variations in the position of the respective hypices. Belvedere (2008) was, to our knowledge, the first to apply geometric morphometrics to analyze shape variability within a trackway, concluding that the greatest variability can be found in the position of the hypices. Based on photogrammetric 3-D data, we attempt a similar approach: calculating the mean shape of the footprints by using six landmarks and multiple semilandmarks and carrying out a principal components analysis (PCA) based on six landmarks. PCA reveals whether shape variations in different parts of the footprint depend on, or behave independently, from each other.

Locality

Berriasian sandstones are exposed in several hilly areas in middle and southern Lower Saxony in northern Germany. All these exposures represent overall similar depositional environments in a large paleogeographic system at the southern rim of the Lower Saxony Basin (Hornung et al.,

2012). Münchehagen (Fig. 3.1), a part of the town of Rehburg-Loccum, has several quarries that are located on the western flank of the Rehburg Mountains. Stratigraphically, the strata in the Münchehagen quarries belong to the Obernkirchen Sandstone, which is a part of the Bückeberg Formation (Hornung et al., 2012).

All of these quarries yielded dinosaur tracks. The most well-known sauropod trackways were found in an abandoned quarry. They are protected as a Natural Monument by the Federal State of Lower Saxony and are now included in the Dinosaur Park Münchehagen. Since 2004, new tridactyl dinosaur trackways are found almost continuously in the active Wesling-Quarry adjacent to the Dinosaur Park (Figs. 3.2 and 3.3).

Sedimentology

The Münchehagen quarries yield predominantly brownish to yellow-gray, fine to medium quartz sandstones that are strongly siliceously cemented. The beds dip slightly 3°–6° west-southwest. Slack water-deposited siltstones and mudstones are intercalated with sandstones, indicating temporarily higher-flowing regimes (e.g., storm events) deposited in a predominantly brackish environment with freshwater as well

3.3. Photograph of a section of theropod trackway T3 during excavation in 2011. Note the small, fractured siltstone layer around the foremost track (T3/35)–the true track layer. The tracks were covered by a 6–12-cm-thick sandstone layer that can be seen at the far left.

as marine influences (Mutterlose, 1997). Individual beds can be traced on a centimeter-to-decimeter scale but are often laterally variable and were therefore assigned to 19 lithological units (LU) (Wings et al., 2012).

Ripple marks are present on most bedding planes. Classified as small-scale ripples (Allen, 1968), they are straight crested and sinuous in phase (Collinson, Thompson, and Mountney, 2006), indicating wave ripples, which are generated by an unidirectional undulating water flow (foreset laminae) in shallow water areas with a temporary influence of currents (Wings et al., 2012). The ripple marks are symmetrical with mostly rounded crests, which are parallel, partly bifurcated and mostly oriented in a northsouth direction. There are no indications for subaerial exposure such as mud cracks. Minimal water depth ranged from 2.7 to 5.4 cm, whereas the maximum water depth cannot be determined (Schwennicke, 1998). The identical orientation of crest lines on different bedding planes reveals that flow and wave direction were consistent during the deposition of several beds. The sandstone grain sizes of 0.063 mm to 0.2 mm (fine sand to medium sand) characterize current velocities of 10–30 cm/s (Allen, 1968; Reineck and Singh, 1980). The bedding planes can be very rough. Some of them are partially covered with coaly layers or inhomogenously embedded coaly particles, resulting from plant detritus. Drainage structures are present on few bedding planes. They have a maximum length of several meters and show a paleoflow direction to the west. Occasionally occurring small (1–3 cm) oval impressions are interpreted as shell imprints. Common colorful dark browngray sutures with a distorted and wavy-jagged outline were caused by grain margin dissolution during diagenesis.

Especially interesting regarding dinosaur tracks are LU0, LU7, and LU16 (Wings et al., 2012). The quarry base (LU0) is a fine-grained sandstone bed with ripple marks and bioturbation by invertebrates and dinosaurs. This bed yields the well-known sauropod trackways (Fischer, 1998; Schwennicke, 1998).

In the upper part of LU7, the sandstone is overlain by laterally very variable, alternating small-scale (millimeter-sized) layers of mudstone, siltstone, and fine-grained sandstone with a combined thickness of up to 3 cm. These fluctuating thin-bedded fine-grained layers constitute the tracking surface (i.e., the surface on which the animals walked). They contain the best preserved tracks in the profile (Fig. 3.3). The true tracks show prominent vertical reliefs, well-defined outlines with displacement rims, and anatomical details such as pad impressions. There is no sedimentological evidence for erosion of a complete layer; natural track casts are found on the underside of LU8. Undertracks of the footprints in LU7 are visible in LU6 and LU5.

THE NATURAL MONUMENT "DINOSAURIERFÄHRTEN MÜNCHEHAGEN," OLD WESLING QUARRY

Originally described as the new ichnotaxon *Rotundichnus muenchehagensis* (Hendricks, 1981), 256 unambiguous sauropod tracks have been described by Fischer (1998). These trackways represent the first and until now the only confirmation for sauropod dinosaurs in the German Wealden (Fischer, 1998). The tracks consist of oval pes impressions overstepping rather roundish manus imprints. The footprints lack claw impressions and morphological details and are not diagnostic of any particular type of sauropod track, rendering the ichnotaxon *Rotundichnus* a nomen dubium (Wright, 2005). However, the footprints show several similarities to the ichnotaxon *Brontopodus* (Hornung et al., 2012).

Hendricks (1981) considered the diplodocid *Apatosaurus* a possible trackmaker, whereas (Lockley, Wright, and Thies, 2004) suggested brachiosaurids or titanosaurids (i.e., members of the clade Macronaria) as likely trackmakers. The trackways have been classified as wide gauge (Lockley, Wright, and Thies, 2004) or as medium gauge (Romano, Whyte, and Jackson, 2007), supporting the attribution of the tracks to macronarians (Wilson and Carrano, 1999). Strikingly, the heteropody of the trackways is among the lowest in any known sauropod trackways, with the manus imprints measuring approximately half the size of the pes imprints (Lockley, Wright, and Thies, 2004). At the other extreme, the size difference between manus and pes tracks can be as high as ¼ or ⅕ in other sauropod tracks (Santos et al., 1994). Henderson (2006) suggested that proportionally larger areas of manus tracks imply an anterior shift of the center of mass of the trackmaker. An anterior position of the center of mass is characteristic for macronarians (Henderson, 2006), again supporting a macronarian affinity of the trackmakers.

Most well-preserved tracks belong to at least seven trackways. Leading to approximately the same direction, these trackways have been interpreted as evidence for a herd (Fischer, 1998; Fischer and Thies, 2000). However, it cannot be ruled out that the tracksite represents a time-averaged assemblage generated by independently traveling individuals (e.g., Myers and Fiorillo, 2009).

Poorly preserved tridactyl dinosaur tracks can be found in the eastern part of the protection hall (Fischer, 1998; Lockley, Wright, and Thies, 2004). Though there is no clear argument for a theropod trackmaker, the tracks do not resemble all other known ornithopod tracks in the German Wealden (Fischer, 1998). Recent studies revealed the presence of more tridactyl tracks, most of them not well preserved but clearly attributable to ornithopods (Wings et al., 2012).

Upper Level, Active Wesling Quarry

Almost all tracks on the Upper Level of the active Wesling Quarry are undertracks that can be referred to pes imprints of ornithopod dinosaurs. The associated true tracks are preserved in a very fragile mudstone layer which cannot be preserved. Only one theropod trackway was identified. The Upper Level yielded the longest individual ornithopod trackway known from the German Berriasian, consisting of 57 well-preserved and another two weakly preserved imprints (Richter et al., 2007; Wings et al., 2012). Another trackway includes very well-preserved manus imprints with a pronounced crescent, kidney-shaped outline, with their long axis being steeply angled inward to the walking direction axis. A more detailed description of the trackways in this level can be found elsewhere (Wings et al., 2012).

Lower Level, Active Wesling Quarry

Abundant and excellently preserved tridactyl dinosaur trackways, which all show bipedal gaits, were found on the bedding surface of LU7 in summer 2004 and excavation and documentation of these tracks was commenced (Wings et al., 2012). From 2004 to 2011, a total area of approximately 1300 m² was uncovered, documented, and excavated (Figs. 3.2 and 3.3). Due to irregular progress in quarrying and discontinuities in the excavation schedule during the following years, the exact number of trackways found in this level is difficult to determine. Rapid quarrying progress on short notice resulted in breaks in documentation and hence difficulties in correlation and assignment of trackways found during different years of the excavations. This is further complicated by the fact that each trackway extends in a different direction, often crossing each other, and none of them represents a straight line of walking.

At least seven iguanodontid ornithopod trackways, corresponding to small-sized, possibly subadult animals, and eight theropod trackways have been documented in LU7 (Wings, Broschinski, and Knötschke, 2005; Wings et al., 2012; Troelsen and Wings, 2014). One ornithopod trackway consists of 53 consecutive footprints and was still continuing at the end of the excavation in 2011.

In total, more than 100 theropod footprints were found in this layer (Troelsen and Wings, 2014). The theropod trackways represent two size classes; all but one can be attributed to medium-sized trackmakers. The excavated section of the longest trackway T3 consists of 48 consecutive footprints. With a footprint length greater than 35 cm and a mean pace length of about 114 cm, these tracks also belong to the largest theropod known from that layer (Troelsen and Wings, 2014).

Variations in the morphology of the individual T3 footprints make this trackway an ideal candidate for our study.

Material and Methods

Some of the best-preserved dinosaur footprints from the Münchehagen locality were studied. All tracks belong to a single trackway of a large-sized theropod dinosaur (trackway number: T3) from the Lower Track Level in LU7. Digital photographs of the tracks were taken by O.W. following methods outlined by Mallison and Wings (2014) with an 8-megapixel Canon EOS 30D and a Canon EF 17–40 mm/1:4,0 L USM during the 2009–2011 excavations. Textured 3-D models of the tracks were produced in the photogrammetric software Agisoft PhotoScan Professional 1.0.4 build 1847 (64 bit) with generally high settings. An example, the dense point cloud of the natural track cast of T3/44, is shown in Figure 3.5.

FOOTPRINT SHAPE VARIATION

This analysis is based on these photogrammetric data. Only a part of the T3 trackway (starting with footprint 23), which at this time was still preserved in situ, is analyzed herein. A single photogrammetric model of the whole trackway section was built (Fig. 3.4). The horizontal plane was defined by placing three marker points on the surface of the model; the accuracy of the assignment of the plane was achieved by maximizing the distance between the marker points. This way, the horizontal plane is the same in each of the footprint analyses. Only the best preserved 13 footprints (footprints 23, 26, 29, 34, 35, 36, 37, 39, 40, 43, 44, 46, 47) (Fig. 3.4) were selected for this analysis. Elevation maps including contour lines were produced using the Open Source software Paraview (http://www.paraview.org/). Based on these maps, interpretative outline drawings were made using the Open Source software Inkscape (http://www.inkscape.org/).

Depending on the individual footprint, the degree of subjectivity of an outline drawing often becomes problematic, especially in footprints with rounded margins that grade imperceptibly into the surrounding rocks (Thulborn, 1990). Dimensions of an outline can vary significantly depending on the chosen height level (Falkingham, 2016). Additional interpretational bias can also be introduced by extramorphological features such as cracks or, as in the Münchehagen site, ripple marks crossing the footprint. As pointed out by Falkingham (2016), the subjectivity of published outlines does not allow the derivation of objective results. To minimize subjectivity, we traced outlines following three criteria: the steepest slope, a consistent elevation, and the maximization of digit length. Exceptions were made to exclude

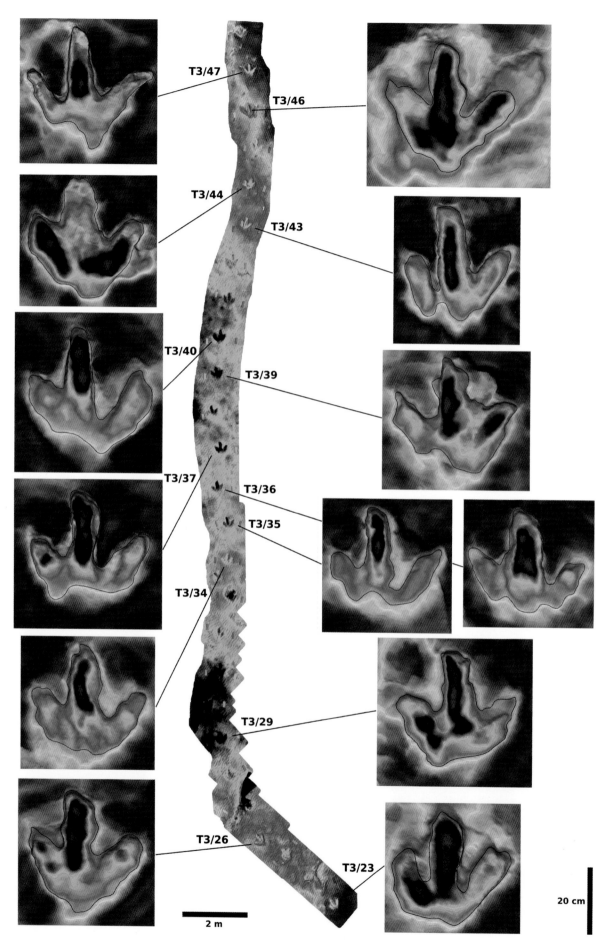

T3/47

T3/46

T3/44

T3/43

T3/40

T3/39

T3/37

T3/36

T3/35

T3/34

T3/29

T3/26

T3/23

2 m

20 cm

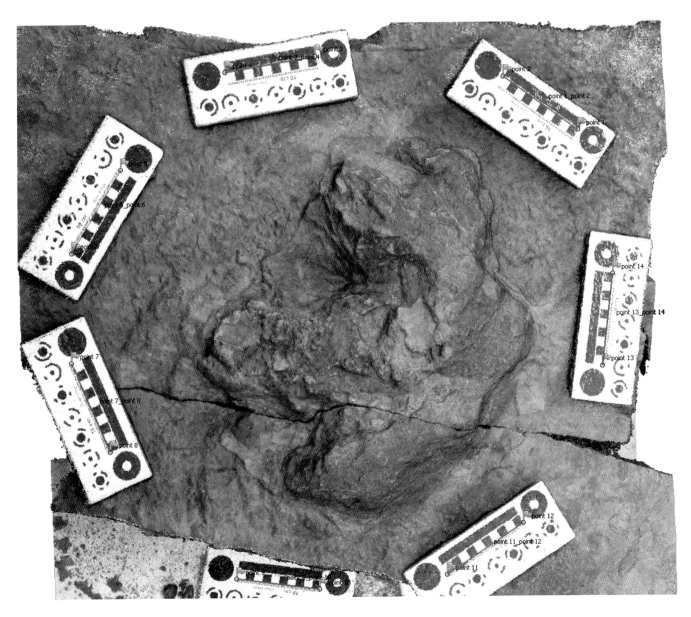

3.5. Dense point cloud of the natural track cast of T3/44 with physical and in-program scales as well as alignment points. Toes point to the left.

extramorphological features and to include partly filled digit impressions.

Six landmark points were defined on the outlines, located on the tips of each of the three digits, on the two hypices, and on the heel region. A landmark on the tip of a digit impression is defined as the distal end of the digital axes, which is a line dividing the digit impression in two approximately even-sized halves. A landmark on a hypex is defined as the midpoint between the enclosing digit impressions. The landmark on the heel is defined as the intersection of an extension of the digital axis of digit III with the outline of

the heel, which, given the straight morphology of digit III in the analyzed footprints, roughly corresponds to the midline of the footprint when measured from the tips of digit impressions II and IV.

Landmarks were digitized using the software tpsDig 2.17 (Rholf, 2013). To derive a mean shape and to visualize the overall shape variation of the analyzed set of footprints, six curves consisting of equally spaced semilandmarks were placed between the landmarks in each footprint outline using tpsDig. To allow for an equal distribution of the semilandmarks across all six curves, the number of semilandmarks

3.4. Depth-color maps of the analyzed sequence of the T3 trackway (footprints T3/23 to T3/48) and depth-color contour maps of individual footprints selected for geometric morphometric analysis. The interpreted outlines are drawn within each footprint. As the depth-color contour maps were created separately for each footprint, the depth indicated by the colors and the contours is different in separate images. All footprint maps are scaled according to the scale bar in the lower right.

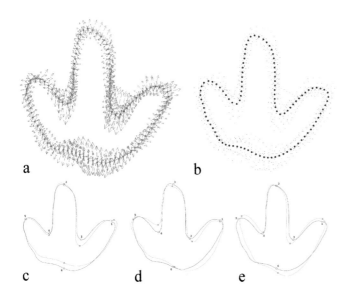

3.6. (A) Diagram showing the overall shape variability present in the 13 analyzed footprints. The vector arrows show the deviations of the landmarks and semilandmarks of each of the individual footprints from the procrustes mean shape. Major shape variability can be found in the hypex points and in the heel. (B) Diagram showing the procrustes fitted landmarks and semilandmarks for each of the 13 analyzed footprints (small black dots) and the procrustes mean shape (large blue dots). As the procrustes analysis removes position, rotation, and scaling from the footprints, the diagram is without scale. (C) Diagram showing shape changes described by the first principal component (PC1), which accounts for 46% of the total variance. The most important shape change occurs in a more posterior position of the lateral hypex (recorded by landmark 2) as well as in a more anterior position of the margin of the heel (recorded by landmark 6). (D) Diagram showing shape changes described by the second principal component (PC2), which accounts for 23% of the total variance. The most important shape change can be seen in the medial hypex position. As the main variation of the two hypices is described by separate principal components (PC1 for the lateral hypex and PC2 for the medial hypex), it can be inferred that the hypex positions change independently from each other. (E) Diagram showing shape changes described by the third principal component (PC3), which accounts for 17% of the total variance. This principal component records a smaller digit divarication.

for each curve was calculated by measuring the relative lengths of the curves for each footprint. The curve points then were converted into landmarks using tpsUtil 1.85 (Rholf, 2013), and subjected to a generalized procrustes analysis using tpsRelw 1.54 (Rholf, 2014). The generalized procrustes analysis is a statistical analysis that provides a best fit between the footprint shapes by translating, rotating, scaling, and eliminating any information but the mere shape of the footprints (Rohlf and Slice, 1990). The procrustes mean can be regarded as the average shape of the analyzed set of footprints. The deviations of each footprint from the procrustes mean have been plotted using vector arrows to visualize the amount of variation around the outline (see Fig. 3.6A).

A PCA was carried out using MorphoJ 1.06c (Klingenberg, 2011). For the PCA, only the six landmarks were used, not the semilandmarks. The procrustes mean shape produced with

tpsRelw was imported in MorphoJ in order to make use of the warped outline drawing function to allow for a better visualization of shape deformations. Due to the relatively small sample size, only the first three principal components were taken into account, which account for 86% of the total variance.

Shape Difference True Track versus Natural Track Cast

Photogrammetric documentation of three very well-preserved consecutive true tracks of the T3 trackway (T3/44–T3/46) and their natural casts were used to calculate and compare the shapes of both track fossils in order to classify and quantify the loss of material while exposing the fragile trackway layers.

Sets of photographs (between 52 and 89 images per set) were taken of all three tracks and natural casts. From these photo sets, subsets of images were used to create photogrammetric models (T3/44 true track: 12 images, T3/44 natural track cast: 13 images, T3/45 true track: 8 images, T3/45 natural track cast: 10 images, T3/46 true track: 18 images, T3/46 natural track cast: 33 images).

The image sets were imported into PhotoScan Professional (www.agisoft.com), one project file per track: both the set of a track and of its natural cast were loaded into separate "chunks" of one file. On each set of images, a 3-D reconstruction was performed with the main steps alignment, dense point cloud generation, and mesh generation. Manual intervention was limited to an optimization of the alignment by a more stringent setting of parameters and cropping of the dense point cloud. Any errors and inaccuracies in the models are thus preserved in the final model, and there was no human interpretation directly influencing the final shapes of the 3-D models.

The models were scaled via scale bars that had been placed around the tracks for photography. Markers were manually placed on the images on the scale bars, and in-program scale bars were created from them. The program-calculated errors for the 10-cm scales were all lower than 0.1 mm.

True track and natural track cast models were then pre-aligned in Photoscan Pro via three manually placed markers. Then, the models were exported as polymesh *.ply files with color information and imported into McNeel Associates Rhinoceros 5.0 (www.rhino3d.com). There, a first visual inspection of the alignment was made, which in all cases was found to be sufficiently accurate that the steps described herein could be undertaken directly, without further manual alignment of the models. Now, the models were rotated pair-wise (i.e., both track and natural track cast models of one track) so that the sediment surface coincided with the system x-y plane as well as could be judged visually and exported as polymesh *.stl files, which incidentally removed color information. Figure 3.5 shows the dense point cloud of the natural track cast of T3/44 with physical and in-program scales as well as alignment points.

The files were now imported pair-wise into Geomagic 8.0 (www.geomagic.com). There, both the natural track cast and the true track models of the same footprint were cropped together in view along the z-axis (i.e., in top view) so that the physical scale bars in both models and the corresponding areas of the respective other model were removed, as their shapes would otherwise influence the alignment. Also, parts of the track 3-D model that had no corresponding area in the natural track cast models were cropped. This process resulted in an exact correlation of x-y extent of the models (i.e., in surface extent).

Now, the automated best-fit alignment option of Geomagic 8.0 was used to improve the alignment of the two 3-D models to each other. The resulting shift of relative position was minimal in all cases, amounting to less than 5-mm shifts. The rotations induced by the "best-fit" alignment process cannot be determined exactly, but visual inspection of the discrepancies of the previously exactly superimposed borders of the models indicated only minimal rotations.

To visualize the shape differences, the "3-D compare" tool in Geomagic was used to create maps that color-code the 3-D distance between the two models. The best-fit alignment option cannot be limited to positive deviation alone; therefore, the two models penetrate each other. Accordingly, deviations are shown both as positive and negative values. In reality, two pieces of rock cannot penetrate each other, and they did not in the field. Therefore, to recreate the physical conditions in the field, the 3-D model of the natural track cast was now shifted along the z-axis in 0.5 mm steps until the two 3-D models no longer penetrated each other. Then, a new 3-D shape difference map was created, again in top view (i.e., along the z-axis). Additionally, when the shape difference maps indicated that other motions than a pure z-axis shift could result in better alignment, such motions were tested manually.

As a further measure of shape difference, the volumes of one track and its natural cast were compared (T3/44). To this end, the pair of 3-D models was intersected with a horizontal plane after best-fit alignment, and the resulting top rim was closed with a flat cap. Volumes were calculated in Rhinoceros 5.0.

Footprint Shape Variation

The procrustes mean shape (Fig. 3.6B) exhibits the major features of the footprints, including a straight digit impression III with a medially displaced apex, a lateral hypex located somewhat deeper than the medial hypex, and an asymmetrical heel that shows its greatest extension at the base of digit IV. The greatest variation is seen to occur in both hypices as well as in the heel region, whereas the shapes of the digit impressions are relatively constant (Fig. 3.6A). An examination of the footprints shows that the variation in the heel can be attributed to both changes in extension and morphology.

First principal component (PC1) accounts for 46% of the total variance. In PC1, the greatest variation from the mean shape occurs in the position of the lateral hypex (landmark 2), which is situated deeper than in the mean shape, followed by the position of the distal margin of the heel (landmark 6), which is situated more anteriorly (Fig. 3.6C). Smaller variations occur in the position of the tip of the digit impressions III and IV (landmarks 1 and 3) and in the medial hypex (landmark 4), while the tip of digit II (landmark 5) represents the most stable feature.

Second principal component (PC2), accounting for 23% of the total variance, shows its greatest variation in the position of the medial hypex, which is situated more deeply compared to the mean shape (Fig. 3.6D). Simultaneously, the digit divarication is smaller, whereas the extension of the heel region is greater. Third principal component (PC3), accounting only for 17% of the total variance, again shows great variation in the hypices (Fig. 3.6E), which are displaced in opposite directions, with the lateral hypex situated deeper and the medial hypex situated higher. The heel region is extended asymmetrically in its lateral half.

*Shape Difference True Track
versus Natural Track Cast*

For track T3/44 the best-fit alignment results in an average positive deviation of 1.023 mm, an average negative deviation of 1.122 mm, and a standard deviation of 1.487 mm (Fig. 3.7A). At a vertical offset of 17 mm, the shape difference map (Fig. 3.7B) shows clearly that the best-fit alignment resulted in a horizontally suboptimal position, as there remained one small area with strong negative overlap (i.e., the natural track cast model intersects the true track model) on the medial side of toe II, whereas the remaining areas of the models were already well separated. Further vertical displacement was

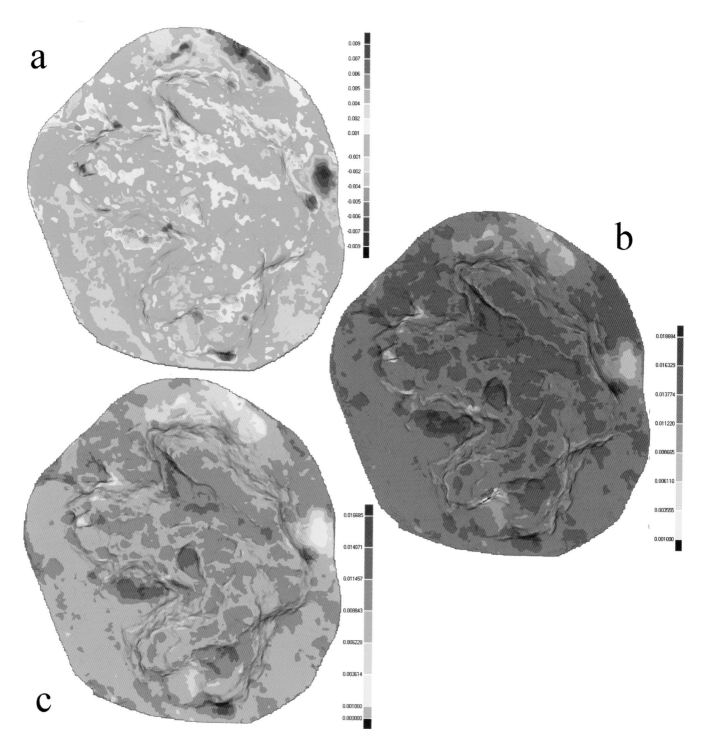

3.7. Shape difference maps of the best-fit alignment results of Track T3/44. (A) Map with an average negative deviation of 1.122 mm and a standard deviation of 1.487 mm. (B) As in A, but with an additional vertical offset of 17 mm between the three-dimensional (3-D) models of true track and natural track cast. (C) As in A, but with an additional vertical offset of only 11.5 mm between the 3-D models of true track and natural track cast and an additional shift along the y-axis (medially) by 2 mm. Red colors indicate large positive offset, and blue colors indicate large negative offset. Offset scale in meters.

halted, and a horizontal shift tested, because the overlapping area contains topography that is subparallel to the z-axis, so that small shifts in the xy-plane could lead to nonoverlap of the models without further vertical shift. A shift along the y-axis (medially) by 2 mm not only resulted in a removal of all overlap but allowed reduction of the vertical displacement to 11.5 mm (Fig. 3.7C). Compared to the purely vertical (z-axis) offset of 17 mm, the average deviation shrank from 12.45 mm to 8 mm, and the standard deviation from 1.84 mm to 1.67 mm. Further reduction of the vertical distance caused overlap in areas of the surrounding sediment, indicating that the minimum distance of the two rocks was reached at 11.5 mm z-offset. The difference in volume between the two models pertaining to T3/44 is low, at slightly over 3% (~1.657 liters for the cast versus 1.708 liters for the track).

In track T3/45, the average positive and negative deviations of the best-fit alignment at 1.144 mm and 0.694 mm, respectively, were similar to those of T3/44 (Fig. 3.8A). The vertical offset required from best-fit autoalignment to remove all overlap was 6 mm. At 7.29 mm and 1.39 mm, track T3/45 shows the smallest standard and average deviations, respectively, between the true track and natural track cast models in nonoverlapping positions (Fig. 3.8B).

Track T3/46 (Fig. 3.9A) has the largest deviation values of all three tracks, with the average positive and negative deviations being 2.412 mm and 1.435 mm, respectively, and the standard deviation 2.910 mm. A vertical displacement of 20 mm was required to remove all overlap (Fig. 3.9B). The best-fit alignment resulted in a large part of the sediment surfaces being fairly level to each other but an area on the left of Figure 3.9A to be significantly out of alignment. This discrepancy indicated that a better alignment might be achieved by tilting the natural track cast model versus the true track model. The vertical distance could be reduced to 17 mm by rotating the model 0.5° around both the x and y axes (i.e., diagonally down and right in Figure 3.9C). The rotations reduced the average deviation of the nonoverlapping positions from 17.75 mm to 14.98 mm, and the standard deviation from 3.13 mm to 2.826 mm.

DISCUSSION

Footprint Shape Variation

Both the deviations from the procrustes mean shape (Fig. 3.6A) and PC1 (Fig. 3.6C) indicate that the greatest variability occurs in the lateral hypex. This feature depends on the extent of the sediment bar separating digit III and digit IV, which (in its extremes) can approach the distal margin of the heel region of the footprint, or can be lacking completely. To reduce subjectivity, hypex points were defined at a similar height to the remaining outline where the steepness of the slope is maximized. Obviously, this steepest slope is in many cases not related to the foot anatomy of the trackmaker (i.e., the anatomical hypex). Nevertheless, the observed high variability of the lateral hypex (and, to a lesser degree, the medial hypex) is in accordance with Belvedere (2008), who identified the hypices as the most variable features in well-preserved theropod footprints of a trackway from the Late Jurassic of Morocco.

The PCA shows whether shifts in the positions of certain landmarks are correlated with shifts in the position of other landmarks. PC1 shows both a strong posterior displacement of

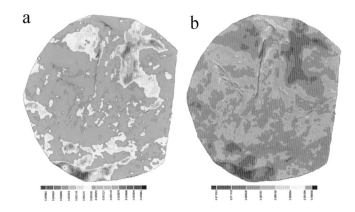

3.8 Shape difference map of the best-fit alignment results of Track T3/45. (A) Map showing positive and negative deviations of 1.144 mm and 0.694 mm, respectively. (B) Map with an additional vertical offset of 6 mm, showing standard and average deviations of 7.29 mm and 1.39 mm between true track and natural track cast models. Red colors indicate large positive offset, and blue colors indicate large negative offset. Offset scale in meters.

the lateral hypex and an anterior displacement of the margin of the heel (Fig. 3.6C), indicating that these shape changes co-occur in several of the analyzed footprints. Interestingly, the greatest variability of the medial hypex does not show up in PC1 but in PC2, indicating that major shifts in the position of the medial hypices in general are unrelated to shifts in position of the lateral hypex. This indicates that the relative position of the hypices does not always represent the anatomy of the trackmaker's foot and, therefore, can be a misleading feature in ichnological descriptions. Hypex positions are strongly affected by variations of the substrate, foot kinematics, or a combination of both (cf. Falkingham, 2014).

Shape Difference True Track versus Natural Track Cast

The automatic best-fit alignment proves to be a useful approach for determining how well two 3-D models match each other in shape, but as the examples of T3/44 and T3/46 show, human input can sometimes still improve on the algorithm's results.

The best-fit shape deviation maps for the three tracks (Figs. 3.7A, 3.8A, 3.9A) and the deviation values highlight the overall high similarity of shape between each true track and its natural cast. The natural cast of T3/45 is most similar to the track, but in all cases, the shape differences are negligible, as they amount on average to a few percent of the track length. It is unsurprising that T3/46 has a higher standard and average deviation than T3/44 and T3/45, because the natural cast of T3/46 has large breaks (i.e., fissures) running through it, which directly result in larger shape differences than an unbroken surface would, and facilitate the loss of small bits of rock that further reduce the congruence of the two surfaces.

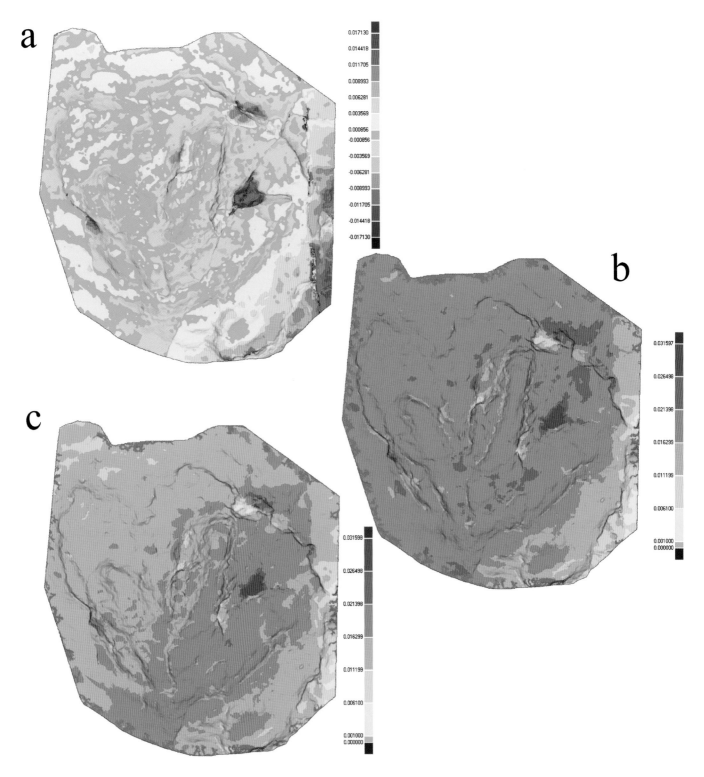

3.9. Shape difference map of the best-fit alignment results of track T3/46. (A) Map showing positive and negative deviations of 2.412 mm and 1.435 mm, respectively, and a standard deviation of 2.910 mm. (B) As in A but with an additional vertical offset of 20 mm, showing an average deviation of 17.75 mm and a standard deviation of 3.13 mm. (C) As in A but with an additional vertical offset of only 17 mm between the three-dimensional models of true track and natural track cast and additional rotations around the x and y axes (i.e., diagonally down and right) of 0.5° each. The average deviation is 14.98 mm and the standard deviation 2.826 mm. Red colors indicate large positive offset, and blue colors indicate large negative offset. Offset scale in meters.

Nevertheless, most of the track shape is preserved highly accurately in the natural cast (Figure. 3.9). Problematic areas with larger divergences are located mostly along the cracks in the rock, where the potential unreliability of the natural cast can be easily recognized. Such areas likely also account for the small difference in volume (3%) we found for T3/44. We therefore conclude that at least in the Münchehagen quarry, natural casts without visible damage can be used instead of the true tracks for ichnological studies.

The deviation values for the model positions without overlap can be taken as minimum thicknesses for parts of the small-scale layers between the preserved tracks and the natural casts that were lost during excavation. The three values used produce an average offset of 11.5 mm. Field observations by O.W. indicate that this value is realistic. The thickness of the combined section of the laterally very variable mudstone, siltstone, and sandstone layers measured usually between 2 and 30 mm. The high variability in grain sizes within the true track horizon was possibly enhanced by diagenetic and tectonic processes (Wings et al., 2012).

SUMMARY

The Münchehagen tracksites are highly significant because of the following: (1) the large number and partially excellent preservation of theropod and ornithopod tracks, (2) the only preserved Cretaceous sauropod trackways in Germany, (3) the otherwise general scarcity of long trackways in the Early Cretaceous of Germany, and (4) tridactyl trackways with unusually curved walking directions of the track-producers.

Geometric morphometrics is shown to be of great utility for the analysis of footprint variability and may become a new standard in vertebrate ichnology. The greatest variability in footprints of the T3 theropod trackway can be found in the positions of the hypex points and in the heel. The position of the medial and lateral hypices are shown to behave independently from each other, rendering the relative position of the hypices a questionable character to be used in ichnotaxonomy.

Shape-difference analyses on the basis of high-resolution 3-D digital models indicate that in the Münchehagen quarry natural track casts can be substituted for true tracks in ichnological studies, as there are only minimal differences in shape between them and the real tracks. We assume that this is a general rule for well-preserved tracks, but this has to be confirmed by future studies. Also, by shifting best-fit aligned models of true tracks and natural track casts until all overlap between the models is removed, it is possible to estimate a minimum thickness for any missing soft sediment layers that separated the rocks in the field. Comparing similar best-fit aligned photogrammetric 3-D models also is considered a useful tool for the evaluation of the erosion rate of other tracksites around the world.

ACKNOWLEDGMENTS

We wish to express our gratitude to Ferdinand Wesling Sr., Bernd Wolter, and Franz-Josef Dickmann, for permission for and support during excavations. We very gratefully acknowledge all staff members, volunteers, and students, without whom this project would have been impossible. Especially, Nils Knötschke and Annette Richter were essential during fieldwork. We would like to thank Anneke van Heteren for helpful comments and support on our geometric morphometrics approach. Financial support for the fieldwork was provided by the Dinosaurier-Park Münchehagen, the Verein zur Förderung der niedersächsischen Paläontologie e.V., and the MWK of Lower Saxony, Hannover. O.W. is currently funded by the Volkswagen Foundation and H.M. was funded by the DFG through grant SCHW 1452/3-1.

REFERENCES

Allen, J. R. L. 1968. Current Ripples: Their Relation to Patterns of Water and Sediment Motion. North-Holland Publishers, Amsterdam, Holland, 433 pp.

Belvedere, M. 2008. Ichnological researches on the Upper Jurassic dinosaur tracks in the Iouaridène area (Demnat, central High-Atlas, Morocco). Ph.D. dissertation, Università degli Studi di Padova, Padua, Italy. 121 pp.

Collinson, J. D., D. B. Thompson, and N. P. Mountney. 2006. Sedimentary Structures. Terra Publishing, Hertfordshire, U.K., 292 pp.

Demathieu, G. R. 1990. Problems in discrimination of tridactyl dinosaur footprints, exemplified by the Hettangian trackways, the Causses, France. Ichnos 1: 97–110.

Falkingham, P. L. 2014. Interpreting ecology and behaviour from the vertebrate fossil track record. Journal of Zoology 292: 222–228.

Falkingham, P. L. 2016. Applying objective methods to subjective track outlines; chap. 4 in P. L. Falkingham, D. Marty, and A. Richter (eds.), Dinosaur Tracks: The Next Steps. Indiana University Press, Bloomington, Indiana.

Fischer, R. 1998. Die Saurierfährten im Naturdenkmal Münchehagen. Mitteilungen aus dem geologischen Institut der Universität Hannover 37: 3–59. [German].

Fischer, R., and D. Thies. 2000. Das Dinosaurier-Freilichtmuseum Münchehagen und das Naturdenkmal 'Saurierfährten Münchehagen.' 2nd edition. Weserdruckerei, Stolzenau, Germany, 125 pp.

Henderson, D. M. 2006. Burly gaits: centers of mass, stability, and the trackways of sauropod dinosaurs. Journal of Vertebrate Paleontology 26: 907–921.

Hendricks, A. 1981. Die Saurierfährte von Münchehagen bei Rehburg-Loccum (NW-Deutschland). Abhandlungen aus dem Landesmuseum für Naturkunde zu Münster in Westfalen 43: 3–22.

Hornung, J., A. Böhme, T. van der Lubbe, M. Reich, and A. Richter. 2012. Vertebrate tracksites in the Obernkirchen Sandstone (late Berriasian, Early Cretaceous) of northwest Germany: their stratigraphical, palaeogeographical, palaeoecological, and historical context. Paläontologische Zeitschrift 86: 231–267.

Klingenberg, C. P. 2011. MorphoJ: an integrated software package for geometric morphometrics. Molecular Ecology Resources 11: 353–357.

Leonardi, G. 1987. Glossary and Manual of Tetrapod Footprint Palaeoichnology. Departemento Nacional da Produção Mineral Brasil, Brasília, Brazil, 137 pp.

Lockley, M. 1998. Philosophical perspectives on theropod track morphology: blending qualities and quantities in the science of ichnology. Gaia 15: 279–300.

Lockley, M. G., J. C. Wright, and D. Thies. 2004. Some observations on the dinosaur tracks at Münchehagen (Lower Cretaceous), Germany. Ichnos 11: 261–274.

Mallison, H., and O. Wings. 2014. Photogrammetry in paleontology: a practical guide. Journal of Paleontological Techniques 12: 1–30.

Mutterlose, J. 1997. 2. 15 Münchehagen quarry; pp. 123–134 in J. Mutterlose, M. G. E. Wippich, and M. Geisen (eds.), Cretaceous Depositional Environments of NW Germany. Bochumer Geologische und Geotechnische Arbeiten, Bochum, Germany.

Myers, T. S., and A. R. Fiorillo. 2009. Evidence for gregarious behavior and age segregation in sauropod dinosaurs. Palaeogeography, Palaeoclimatology, Palaeoecology 274: 96–104.

Razzolini, N. L., B. Vila, D. Castanera, P. L. Falkingham, J. L. Barco, J. I. Canudo, P. L. Manning, and À. Galobart. 2014. Intra-trackway morphological variations due to substrate consistency: the El Frontal Dinosaur Tracksite (Lower Cretaceous, Spain). PLoS One 9: e93708.

Reineck, H.-E., and I. B. Singh. 1980. Depositional Sedimentary Environments. Springer, Berlin, Germany, 551 pp.

Richter, A., and Böhme, A. Too many tracks: preliminary description and interpretation of the diverse and heavily dinoturbated Early Cretaceous 'Chicken Yard' ichnoassemblage (Obernkirchen tracksite, northern Germany); chap. 17 in P. L. Falkingham, D. Marty, and A. Richter (eds.), Dinosaur Tracks: The Next Steps. Indiana University Press, Bloomington, Indiana.

Richter, A., O. Wings, U. Richter, and N. Knötschke. 2007. 'Happy Feet': a new Lower Cretaceous dinosaur tracksite from Münchehagen, Germany. Journal of Vertebrate Paleontology 27: 134A.

Rohlf, F. J. 2013. Tps series. Department of Ecology and Evolution, State University of New York at Stony Brook, New York. Available at http://life.bio.sunysb.edu/morph/. Accessed October 22, 2015.

Rohlf, F. J. 2014. Tps series. Department of Ecology and Evolution, State University of New York at Stony Brook, New York. Available at http://life.bio.sunysb.edu/morph/. Accessed October 22, 2015.

Rohlf, F. J., and D. Slice. 1990. Extensions of the procrustes method for the optimal superimposition of landmarks. Systematic Biology 39: 40–59.

Romano, M., M. A. Whyte, and S. J. Jackson. 2007. Trackway ratio: a new look at trackway gauge in the analysis of quadrupedal dinosaur trackways and its implications for ichnotaxonomy. Ichnos 14: 257–270.

Santos, V. F., M. G. Lockley, C. A. Meyer, J. Carvalho, A. G. De Carvalho, and J. J. Moratalla. 1994. A new sauropod tracksite from the Middle Jurassic of Portugal. Gaia 10: 5–13.

Schwennicke, T. 1998. Sedimenttexturen und Lebensspuren der Sohlbank des Naturdenkmals 'Saurierfährten Münchehagen.' Mitteilungen aus dem Institut für Geologie und Paläontologie der Universität Hannover 37: 61–102.

Thulborn, T. 1990. Dinosaur Tracks. Chapman and Hall, London, U.K., 410 pp.

Troelsen, P., and O. Wings. 2014. New data from the Lower Cretaceous theropod trackways in Muenchehagen (Lower Saxony, Germany). Journal of Vertebrate Paleontology, Program and Abstracts 2014: 241.

Weems, R. E. 1992. A re-evaluation of the taxonomy of Newark Supergroup saurischian dinosaur tracks, using extensive statistical data from a recently exposed tracksite near Culpeper, Virginia; pp. 113–127 in P. C. Sweet (ed.), Proceedings 26th Forum on the Geology of Industrial Minerals. Virginia Division of Mineral Resources, Charlottesville, Virginia.

Wilson, J. A., and M. T. Carrano. 1999. Titanosaurs and the origin of 'wide-gauge' trackways: a biomechanical and systematic perspective on sauropod locomotion. Paleobiology 25: 252–267.

Wings, O., A. Broschinski, and N. Knötschke. 2005. New theropod and ornithopod dinosaur trackways from the Berriasian of Münchehagen (Lower Saxony, Germany). Kaupia: Darmstädter Beiträge zur Naturgeschichte Heft 14: 105.

Wings, O., D. Falk, N. Knötschke, and A. Richter. 2012. Excursion Guide B1: The Early Cretaceous Dinosaur Trackways in Münchehagen (Lower Saxony, Germany)–the Natural Monument 'Saurierfährten Münchehagen' and the adjacent Wesling Quarry; pp. 113–142 in A. Richter, T. R. Hübner, and M. Reich (eds.), Dinosaur Tracks 2011. An International Symposium, Obernkirchen, April 14–17, 2011. Abstract Volume and Field Guide to Excursions. Universitäts-Verlag, Göttingen, Germany.

Wright, J. L. 2005. Steps in understanding sauropod biology; pp. 252–285 in K. Curry Rogers, and J. A. Wilson (eds.), The Sauropods: Evolution and Paleobiology. University of California Press, Berkeley, California.

Direct track

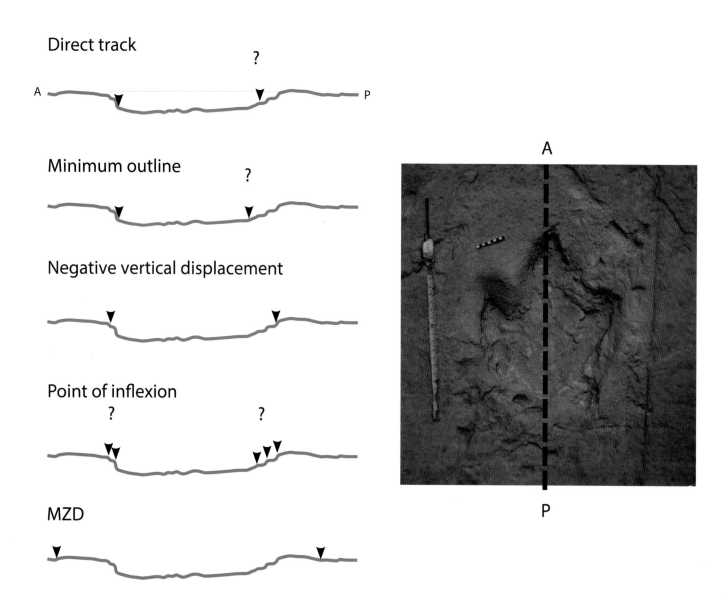

Minimum outline

Negative vertical displacement

Point of inflexion

MZD

4.1. Various methods of defining track edges and recording track length. Question marks indicate where it is particularly difficult to define track extents according to the method.

Applying Objective Methods to Subjective Track Outlines

Peter L. Falkingham

4

FORMALLY COMMUNICATING THE MORPHOLOGY OF A track generally occurs via a two-dimensional (2-D) medium (i.e., paper). For this reason, track outlines are often used to convey the geometry and morphology of a track. However, these track outlines are routinely subjective, based on the interpreter's opinion of where the track ends and the surrounding undeformed substrate begins. Although such outlines are not a problem themselves, any subsequent application of numerical objective methods such as multivariate analyses or equations using track parameters can be strongly influenced by the subjective nature of the outline. This effect is compounded in deeper tracks with sloping sides. However, although there are numerous ways in which to define track extents objectively (horizontal plane intercept, maximum inflexion, direct track impression, etc.), none are applicable to all tracks and there is no universally "correct" objective definition of a track outline.

In order to illustrate this point, outlines were produced from a digital model of a large tridactyl dinosaur track. In order to produce objective outlines, a 'base-line' surface was produced by fitting a plane to the undisturbed surrounding sediment surface. This virtual plane acted as a reference, approximating the tracking surface prior to the animal walking on it. From this plane, outlines were produced from height isolines taken at regular intervals within the track. Track parameters varied considerably depending on which outline was used, with track length shown to vary by as much as 27%. This variation in numerical parameters directly affected the results and potential interpretations of multivariate analyses, trackmaker identity, and speed.

Rather than carrying out such objective analyses on single outlines, it is suggested that a "best guess and bracketing" approach is adopted, where minimum and maximum outlines are used to constrain, rather than specify interpretations and conclusions. This study also highlights the importance of ichnologists taking advantage of modern digitizing techniques in order to communicate and share full three-dimensional (3-D) data, rather than just interpreted outlines.

INTRODUCTION

Ichnology is an important aspect of paleontology, providing an independent source of information that may be used to support or reject hypotheses based on osteological remains or to provide information that would be otherwise unavailable from the body fossil record such as soft tissue anatomy (Thulborn 1990; Currie, Nadon, and Lockely, 1991; Lockley, 1991; Falkingham et al., 2009), speed (Alexander, 1976; Farlow, 1981; Thulborn and Wade, 1984; Manning, 2008), behavior (Thulborn and Wade, 1979, 1984; Farlow et al., 2012), or paleoenvironment (Lockley and Conrad, 1989; Avanzini et al., 1997; Falkingham et al., 2011b; Falkingham, Bates, and Mannion, 2012).

Over the past few decades, paleontology has seen an increasing prevalence of objective numerical methods to support or reject hypotheses. These methods are varied and numerous, ranging from morphometric analyses of bones to computer simulations and even extending to areas such as cladistics, which have become more reliant on efficient algorithms. Despite originating over 150 years ago (Hitchcock, 1836, 1858), dinosaur ichnology remained almost exclusively a subjective science, based on interpretations of track outlines and comparisons between those interpretations, for far longer.

One of the first numerical approaches adopted by dinosaur ichnology was calculating speed from trackways using Alexander's dimensionless speed formula (Alexander, 1976). Despite containing several approximations (such as hip height from foot length), and known error bounds, the equation for speed was objective and could provide a specific answer for any given input. Since that time, multiple objective numerical techniques have entered the field, including morphometrics, landmark analysis (Belvedere, 2008), multivariate analyses (Moratalla, Sanz, and Jimenez, 1988), etc. Each of these methods and techniques provide dinosaur ichnologists with consistent tools with which to study tracks. This consistency between scientists and the methods they use is vital in order for ichnology to adhere to a fundamental tenet of science: repeatability of results. Unfortunately, in many cases, these numerical objective methods have their objectivity severely compromised by the input data they use – the parameters measured from the tracks themselves.

This compromise does not come from the ichnologists themselves as such; it is neither a consequence of inaccuracy nor human error but in fact semantics. Defining where a track begins and ends is not trivial due to the continuous nature of track and surrounding substrate, particularly in fine-grained, cohesive substrates that exhibit large areas of deformation, and several authors have made attempts over the years to specifically define where the extents of a track lie (Fig. 4.1). Measuring the length of a track is not as straightforward a task as, for instance, measuring the length of a bone. Yet defining where a track begins and ends is the very crux of producing a track outline and deriving the input variables such as length, width, interdigital angle, or landmark points, which the objective numerical techniques require. The difficulty in acquiring these input data with any real level of objectivity means that objective methods are usually applied to subjective data, making the resultant outcome subjective, rather than objective.

The smallest values of length and width of a track (assuming no collapse) will come from measuring the direct track, defined by Gatesy (2003) as the grains of sediment that contacted the foot. In tracks with no major complex deformation occurring, measuring the direct track will result in metrics closest to the size and shape of the trackmaker's foot, which are arguably most useful when attempting to make paleobiological interpretations such as speed, gait, and trackmaker identity. However, without the presence of skin impressions or scale drag marks, which preserve only rarely, it can be almost impossible to define with any confidence the extents of the direct track impression. Also, only recording the dimensions of the direct track impression serves to ignore a wealth of information contained in the deformation of the surrounding substrate – information that can be useful in identifying limb kinetics and kinematics (Gatesy et al., 1999; Avanzini, Piñuela, and García-Ramos, 2012) or substrate rheology and associated paleoenvironmental indicators from the time of track formation (Manning, 2004; Falkingham et al., 2011b). Attempting to measure a track based on the direct contact between sediment grains and the foot becomes problematic when considering undertracks or tracks that have been subject to weathering and/or erosional processes, because in these cases, the "direct track" sensu stricto no longer exists because the grains that contacted the foot have since been removed.

In order to maintain a consistent way of measuring track parameters throughout surface and subsurface series, Manning (1999) used the inflexion point of the track sides

to define the track edges, though determined this "by eye" in specimens. Bates (2006) repeated this method but used random remeasuring to quantify repeatability. The inflexion point is the location at which the slope of the edge of the track impression changes, usually from curving upward to curving downward (Fig. 4.1). If we consider an indenter with sharp edges penetrating a compliant substrate (as in Allen, 1989, 1997b), we can see that at the surface, the point of inflexion becomes analogous to the extents of the direct track. However, if the indenter or foot lacks sharp corners (as would, in reality, usually be the case), or if looking at a subsurface or weathered/eroded track, this point of inflexion would move upward to some point between the base of the track and the top of the displacement rim. In these cases, it can be difficult to ascertain with any degree of precision where this point occurs, particularly in tracks formed in an uneven surface, where the track sides may have multiple poorly defined points of inflexion. Therefore, although the point of inflexion remains objectively defined, potentially providing a degree of consistency between measurements made by different workers, the point is difficult to measure and varies in position between tracks depending on substrate conditions. This makes measurements based on this point useful for comparisons of separate layers in a track volume but less useful for comparisons between different tracks.

Martin et al. (2012), following Brown (1999) and Halfpenny and Bruchac (2002), employed a minimum-outline approach to define the outlines and subsequent metrics of tracks representing a polar dinosaur track assemblage. Martin et al. described the outline as being defined "where the lowermost concave interior ('floor') of each track met the ascending surface ('wall')" (2012:174). This definition of track extents can remove the effects of extensive sediment deformation through cohesion (though suffers where tracks walls have slumped), thereby potentially defining an outline that most closely resembles the trackmaker's foot. In tracks where the walls and floor meet low in the track, the minimum-outline approach and the inflexion-point approach may result in very similar outlines. However, as with the point of inflexion method, determining where the floor and wall meet can be difficult to determine in an objective manner, particularly if the track has a bowl-like form either as a consequence of track formation conditions (substrate, anatomy, kinematics) or weathering/erosion.

Rather than measuring the "inside" of the track, some authors have explored measuring the total deformation caused by the foot as it indented the substrate (Fig. 4.1). Manning (1999, 2004) referred to this as the maximum zone of deformation (MZD) and noted that a correlation between MZD and the length of the track indentation could potentially relate to other factors such as depth (Manning, 1999). Although such a measure is an important factor in describing a track, and subsequently understanding and interpreting the sediment mechanics involved, MZD can often bear little correlation with the "true" track length and can extend several times the length of the trackmaker's foot (Graversen, Milàn, and Loope, 2007).

In an attempt to retain consistency in measurements for comparison of computer-simulated tracks, both as track and undertrack series and as separate experimental runs, Falkingham (2010) defined the track edges as the points where sediment had been displaced in a negative vertical direction, that is, the track length was defined as the distance on the original surface that had been displaced downward (Fig. 4.1). In the same work, Falkingham made the case that the method for defining track extents could also be applied to real tracks, if those tracks were first digitized (e.g., through laser scanning or photogrammetry [Bates, Breithaupt, et al., 2009; Falkingham, 2012]) and then a plane representing the original sediment surface were fitted. However, this method of measuring tracks is variable depending on substrate consistency, because as will be shown, the slope of the track wall will directly influence the "length" at the top of the track, in much the same way as the point of inflexion would.

The varying ways in which to measure the seemingly simple metrics such as track length and width are not problematic in and of themselves. Each method provides an objective, specifically defined way to measure a track that provides consistency both between tracks and between track workers. Indeed, no one method is the correct method, just as no one method is inherently wrong; rather, each way incorporates to different degrees the contributions of foot morphology, substrate consistency, and limb motion to final track morphology (Padian and Olsen, 1984; Minter, Braddy, and Davis, 2007)—all of which are factors that any ichnologist must consider, though priorities will vary between workers according to personal interests.

The result of this difficulty in defining track extents is that published track outlines are inherently subjective, regardless of how they were determined. That is not to say in any way that published track outlines are incorrect but rather that one worker may (or indeed, is likely to) produce a different outline than another would, either by using a different definition of track extents or by interpreting the same method in a different way. Given that many comparative studies acquire data from the outlines and measurements published in the literature, any numerical objective methods employed will ultimately produce only subjective output based on the subjective input.

This chapter aims to demonstrate that the results from objective numerical methods can be drastically different with even small changes in the way track extents are defined and measured and attempts to provide a means to constrain results from objective analyses, rather than accepting single value results. In order to demonstrate the potential subjectivity present in any given track outline, this study will use as an example a track from Lark Quarry, Australia. The track site at Lark Quarry consists of many hundreds of small tridactyl tracks and trackways, and a single large tridactyl trackway. The large tridactyl tracks were originally attributed to a theropod trackmaker and were noted to possess V-shaped or tapering digits (Thulborn and Wade, 1979, 1984). Recently, this interpretation was questioned by Romilio and Salisbury (2011), who applied the multivariate analysis described by Moratalla, Sanz, and Jimenez (1988) to the original outlines produced by Thulborn and Wade (1979). The conclusions of Romilio and Salisbury were that the large tridactyl tracks were of ornithopod origin, contrasting with the original interpretation. The purpose of this chapter is not to support or reject hypotheses about the trackmaker's identity but rather to use this track to illustrate the problems inherent in defining and using 2-D outlines when describing and interpreting tracks that are fundamentally 3-D

MATERIALS AND METHODS

The track used in this study is the third track in the large, tridactyl trackway at Lark Quarry (Thulborn and Wade, 1984; Thulborn, 1990). Because the interpretation of the trackway has been called into question based on a numerical objective analysis of tracks in this trackway, the site provides an ideal setting with which to illustrate how objective methods may be flawed or ambiguous if applied to subjective data such as track outlines.

To enable accurate, repeatable measurements, a digital model of the track was produced using photogrammetry (Falkingham, 2012). The raw point cloud was meshed using the Poisson surface reconstruction algorithm in Meshlab (Fig. 4.2).

In order to provide a reference from which to measure the track, a plane was fitted to approximate the original pretracking surface. This was accomplished using the freely available program Cloud Compare, by selecting the undeformed surface around the track, and applying the plane orientation function. This provides a translation matrix with which to move the point cloud or mesh such that the undeformed surface around the track is orientated in the xy plane, and placed at a height of $z = 0$. With the digital model aligned in this way, the vertical coordinate $z < 0$ for points within the track, whereas displacement rims and uplifted structures are located where $z > 0$ (Fig. 4.2).

In order to generate a series of outlines from the digital model of the track, isolines of height (z coordinate) were

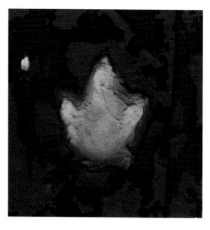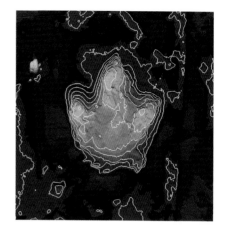

Elevation (m)
-0.08   -0.04   0   0.04   0.08

4.2. Photogrammetric model phototextured (left) and colored according to height (middle) with isolines of height (right).

generated using the free software Paraview. Four isolines were produced at 2, 3, 4, and 5 cm beneath the original tracking surface; each contour representing an outline of the track derived from the intersection of the track walls with a plane parallel to the sediment surface at different depths (Fig. 4.3). These isolines were then treated as independent track outlines and were subjected to a series of measurements and analyses commonly carried out on track outlines as reported in the literature.

RESULTS

Variations in Size

Track length and width along with interdigital angle were measured from each isoline, and these values are presented in Table 4.1. Each of these basic metrics decrease when measured from lower isolines compared with the uppermost outline. However, the rate at which these change with depth varies, with length increasing by 27% (12 cm) between 5 and 2 cm below the tracking surface, and width increasing by 9% (4 cm) over the same zone, giving foot length–to–foot width ratios of 1.07 in the lower parts of the track and 1.25 nearer the tracking surface.

Calculation of Trackmaker Hip Height and Speed

The variation in foot length directly affects the calculation of hip height, and subsequently speed, two of the most common applications of numerical methods to trackways. The most common calculation of hip height is four times foot length (Alexander, 1976). In this instance, using an outline derived from near to the tracking surface gives a hip height of 2.3 m, whereas an outline located deeper in the track provides a hip height of 1.8 m. Variations on this equation that use a factor other than 4 (e.g., Thulborn, 1990) will show a similar variation in hip height calculation.

Integral to calculating speed using the dimensionless speed equation (Alexander, 1976) are hip height (from foot length) and stride length. The effect noted for track 3 can also be seen in track 5 (and the other tracks in this trackway), though a full analysis of other tracks is not presented here. However, while track length may change, this has no bearing on stride length because the tracks are altering consistently relative to each other – that is, the center, and any landmarks used, are remaining consistently spaced (contra to Manning, 2008:fig. 12.11). However, the change in calculated hip height does affect predicted speed. In this case, the mean stride length reported by Thulborn and Wade (1984) was 3.3 m. Using this stride length to calculate speed produces results ranging from 2.19 m/s to 2.88 m/s (7.88 to 10.37 km/h) depending on which outline is used to determine foot length and subsequently hip height.

Multivariate Analyses

In their analysis of the large tridactyl tracks from Lark Quarry, Romilio and Salisbury (2011) applied the multivariate analysis described by Moratalla, Sanz, and Jimenez (1988). The same analysis was carried out on the multiple outlines produced here, and the results can be observed in Figure 4.4. In almost all metrics, as depth increases, the ratios used in the analysis increase, indicating a transition to a more theropod-like outline. The exception is in the track length-to-width ratio, which decreases with depth. Note that outlines produced higher in the track, closer to the original

Table 4.1. Length, width, and interdigital angle of outlines derived by the intersection of a horizontal plane with the track at various depths

Outline Depth (cm)	Length (cm)	Width (cm)	IDA II–III	IDA III–IV
2	57.45	46.16	35.32	50.28
3	50.95	43.90	30.54	52.34
4	49.05	43.63	32.89	46.64
5	45.08	42.28	32.94	37.64
		64	56	

Source: Romilio and Salisbury (2011)

Notes: There is an overall reduction in size with depth within the track, with much of this reduction occurring in track length.

IDA – interdigital angle.

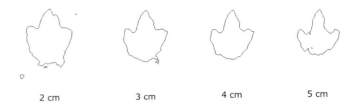

4.3. Outlines derived by a horizontal plane intersecting the track walls at 2, 3, 4, and 5 cm below the undeformed tracking surface.

tracking surface, produce values closest to those of Romilio and Salisbury (2011) as inferred from Thulborn and Wade's (1984) original outline.

DISCUSSION

It is immediately apparent from the examples shown that variations in track outline can result in widely differing results in both quantitative measures (speed, size, etc.) and qualitative interpretations (i.e., trackmaker identity). Indeed, in the example track used here, outlines separated in height by only 1 cm yield variation in track length of up to 13% (Table 4.1). This is a significant change in size that is highly dependent upon where the outline of the track is interpreted to be. This considerable variation in track length, dependent on where and how the outline is determined, should not be taken lightly. Many interpretations and inferences made from tracks and trackways are directly related to the length of the track (or rather, from this the inferred length of the foot), and errors here can propagate into calculations of speed and determination of trackmaker identity, which are further used for paleoecological and behavioral studies.

Other metrics in addition to track length are also plastic as the outline is moved deeper into the track, as illustrated by the multivariate analysis (Fig. 4.4). As the outline is derived in lower portions of the track, it becomes more gracile, the digits becoming relatively longer and thinner. There is also a reduction in interdigital angle, particularly between digits III and IV. In this particular case, it is track length that shows the greatest change depending on where the track outline is placed. That is because of the highly sloped track walls at the anterior of the track, most likely resulting from an anterior or posterior motion of the foot and ground reaction force during locomotion. It is worth noting therefore that the three factors determining track morphology (foot anatomy, kinematics, and substrate) may affect different parameters of the track to differing extents, even within a single footfall (Falkingham, 2014). There is also a shortening of length because the track floor is not parallel to the tracking surface – it is deeper under the distal digits than under the metatarsal pad, due to increased pressure at toe-off. This means that the outlines generated parallel to the tracking surface begin to lose the most posterior part of the track.

Implications for Quantitative Studies

The effects seen in the track from Lark Quarry are not specific to that specimen or even site. Any track where the track walls slope (and that is likely to be any track that is not penetrative in nature) will present a surface that can be measured in multiple ways. Castanera et al. (2012) in their report on sauropod trackways from Spain noted a 20% difference in track length compared to a prior study on the same trackways. Castanera et al. (2012) considered this variation an artefact of where the two studies had determined the "edge" of the track to be.

It is plain to see that the application of quantitative methods to any single outline from a given track may be severely misguided and result in widely variable values. An objective analysis will not produce objective results if the input data are subjective, as is the case with a track outline. A track is a product of foot anatomy, distal limb motion, and substrate properties, and these factors will be expressed to different extents within the track. Any given outline will be a function of these factors, with lower outlines near the floor of the track most closely reflecting the foot morphology, and the upper parts of the track being most directly affected by foot motion and sediment properties. It is imperative therefore to consider the whole 3-D surface of the track, not just a single outline. Note that here I do not refer to the 3-D volume, which extends beneath the visible surface (Allen, 1989, 1997a; Manning, 2004b; Milàn and Bromley, 2006; Falkingham et al., 2009; Falkingham et al., 2011a) but rather to what is exposed – a surface with depth or what we shall term here to be the 2.5-D track.

Unfortunately, as noted in the introduction, the communication of data concerning the morphology of tracks takes

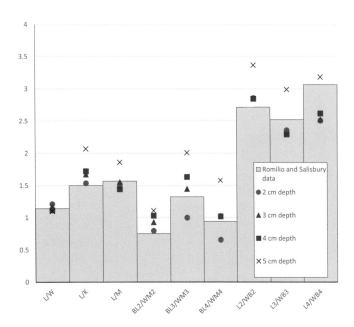

4.4. Graph showing the values of various metrics used a morphometric analysis, and how these values alter depending on how the outline is determined. Abbreviations: BL2–4, basal digit lengths; K,M; interdigital (hypex) point; L, track length; L2–4, digit lengths (digits II–IV); W, track width; WB2–4, basal digit widths.

place in a predominantly 2-D medium. Therefore, comparative work that uses the primary literature for input data to objective methodologies is in fact incorporating the subjective assessments of multiple workers. Attempts to quantify variation in tracks based on published outlines may be comparing two very different aspects of two different tracks—perhaps one outline heavily biased by sedimentary or kinematic factors with one that more closely matches the shape of the trackmaker's foot.

There is little that can be done for much of the data residing in publications from the past century and a half of dinosaur ichnology, other than cautioning that single outlines are unsuitable data for objective analyses, and this must be acknowledged in any such study. In some cases, multiple outlines are drawn for a single track, indicating the internal and external extents of the track walls (e.g., Marty, 2008), or the particular way in which an outline was determined is stated (e.g., Martin et al., 2012), enabling other researchers to account for this, although such cases tend to be sparse in the literature.

However, the documentation of tracks in the future can be modernized to account for the subjectivity present in selecting an outline. Adopting a "best guess and bracketing" approach to track outlines is the simplest method by which to alleviate the issues associated with using a single outline. Rather than presenting only the investigator's interpretation of the outline, a maximum and minimum outline can be presented. These outlines would be analogous to isolines produced where the track intersects the tracking surface and around the floor of the track, respectively. Within these maxima and minima, a third outline representing the investigator's interpretation of where the track ends and begins can be presented. Numerical studies can then be carried out on both the "maximum" and "minimum" outlines. Consistency in results from both outlines will provide confidence in interpretations, whereas wildly differing values imply that other lines of evidence are required in order to support or reject the hypothesis (e.g., trackmaker identity based on variable multivariate analyses may need additional support from features such as claw marks or pad impressions). This method mirrors recent research into dinosaur body mass estimates, in which maximum and minimum masses are used either to frame an investigator's best guess (Bates, Falkingham, et al., 2009; Bates, Manning, et al., 2009) or are presented without intermediate values (Hutchinson et al., 2011).

These maximum and minimum outlines could be determined manually in the field, but more objective repeatable outlines can be produced using the methods described here in conjunction with 3-D digital models. Techniques for the acquisition of digital models have now become so low cost (even free), and require far less expertise to use than in the past (Falkingham, 2012), that they are available to all workers. The generation of a digital model, and ideally the sharing and distribution of that model, will enable a level of repeatability and an access to primary data that is unprecedented in dinosaur ichnology. With options including self-hosting or depositing digital models as electronic supplementary material with publications, ichnologists need not be constrained by the limits of communication via paper. The implications for conducting large meta-analyses using digital models from multiple sites and specimens, rather than from outlines produced in a multitude of ways, speak for themselves.

CONCLUSIONS

This work has demonstrated that reliance upon a single outline of a track for use in objective numerical techniques is fundamentally flawed, and such analyses may be misleading at best. This is not to say that any one outline is incorrect but rather that individual investigators will define an outline based on their own interests and biases toward substrate, foot anatomy, or kinematics, and this will introduce an element of subjectivity into any related objective analyses. It is therefore of the utmost importance that outlines used encompass the full morphology of the track as a 2.5-D surface. In order to accomplish this, maximum and minimum outlines should be produced where the track meets the tracking surface and where the track walls meet the track floor, respectively.

Ideally this should be carried out using a repeatable method on 3-D digital data, which should be made available for other workers when possible. For analyses requiring a 2-D outline, confidence can be ensured when results are consistent with both outlines. When results differ depending on how an outline is defined, additional independent data must be used for supporting or rejecting hypotheses.

ACKNOWLEDGMENTS

This work was supported by a Marie Curie International Outgoing Fellowship within the Seventh European Framework Programme. I wish to offer sincere thanks to Scott Hocknull (Queensland Museum) for assistance at Lark Quarry and in providing access to data. I also wish to thank Karl Bates (Liverpool University) and Stephen Gatesy (Brown University) for comments on an early draft.

REFERENCES

Alexander, R. M. 1976. Estimates of speeds of dinosaurs. Nature 261(5556): 129–130.

Allen, J. R. L. 1989. Fossil vertebrate tracks and indenter mechanics. Journal of the Geological Society 146: 600–602.

Allen, J. R. L. 1997a. Subfossil mammalian tracks (Flandrian) in the Severn Estuary, S. W. Britain: mechanics of formation, preservation and distribution. Philosophical Transactions of the Royal Society B: Biological Sciences 352: 481–518.

Allen, J. R. L. 1997b. Subfossil mammalian tracks (Flandrian) in the Severn Estuary, SW Britain: mechanics of formation, preservation and distribution. Philosophical Transactions of the Royal Society of London Series B-Biological Sciences 352(1352): 481–518.

Avanzini, M., L. Piñuela, and J. C. García-Ramos. 2012. Late Jurassic footprints reveal walking kinematics of theropod dinosaurs. Lethaia 45(2): 238–252.

Avanzini, M., S. Frisia, K. V. D. Driessche, and E. Keppens. 1997. A dinosaur tracksite in an Early Liassic tidal flat in northern Italy: paleoenvironmental reconstruction from sedimentology and geochemistry. Palaios 12(6): 538–551.

Bates, K. T. 2006. The application of Light Detection and Range (LIDAR) imaging to vertebrate ichnology and geoconservation. M.Phil. thesis, University of Manchester, Manchester, UK, 437 pp.

Bates, K. T., P. L. Manning, D. Hodgetts, and W. I. Sellers. 2009. Estimating body mass properties of dinosaurs using laser imaging and 3D computer modelling. PLoS One 4(2): e4532.

Bates, K. T., B. H. Breithaupt, P. L. Falkingham, N. A. Matthews, D. Hodgetts, and P. L. Manning. 2009. Integrated LiDAR and photogrammetric documentation of the Red Gulch dinosaur tracksite (Wyoming, USA); pp. 101–103 in E. Foss, J. L. Cavin, T. Brown, J. I. Kirkland, and V. L. Santucci (eds.), Proceedings of the Eighth Conference on Fossil Resources, St. George, Utah, May 19–21. BLM Regional Paleontologist, Salt Lake City, Utah.

Bates, K. T., P. L. Falkingham, B. H. Breithaupt, D. Hodgetts, W. I. Sellers, and P. L. Manning. 2009. How big was "Big Al": quantifying the effect of soft tissue and osteological unknowns on mass predictions for Allosaurus (Dinosauria: Theropoda). Palaeontologia Electronica 12: 14A: 33p.

Belvedere, M. 2008. Ichnological researches on the Upper Jurassic dinosaur tracks in the Iouaridène area (Demnat, central High-Atlas,

Morocco). Ph.D. dissertation, Università degli Studi di Padova, Padua, Italy, 128 pp.

Brown, T., Jr. 1999. The Science and Art of Tracking. Berkley Books, New York, New York, 240 pp.

Castanera, D., C. Pascual, J. I. Canudo, N. Hernandez, and J. L. Barco. 2012. Ethological variations in gauge in sauropod trackways from the Berriasian of Spain. Lethaia 45(4): 476–489.

Currie, P. J., G. C. Nadon, and M. G. Lockley. 1991. Dinosaur footprints with skin impressions from the Cretaceous of Alberta and Colorado. Canadian Journal of Earth Sciences 28(1): 102–115.

Falkingham, P. L. 2010. Computer simulation of dinosaur tracks. Ph.D. dissertation, University of Manchester, Manchester, UK, 304 pp.

Falkingham, P. L. 2012. Acquisition of high resolution three-dimensional models using free, open-source, photogrammetric software. Palaeontologia Electronica 15: 1T: 15p.

Falkingham, P. L. 2014. Interpreting ecology and behaviour from the vertebrate fossil track record. Journal of Zoology 292: 222–228.

Falkingham, P. L., K. T. Bates, and P. D. Mannion, P. D. 2012. Temporal and palaeoenvironmental distribution of manus- and pes-dominated sauropod trackways. Journal of the Geological Society 169(4): 365–370.

Falkingham, P. L., K. T. Bates, L. Margetts, and P. L. Manning. 2011a. The 'Goldilocks' effect: preservation bias in vertebrate track assemblages. Journal of the Royal Society: Interface 8(61): 1142–1154.

Falkingham, P. L., K. T. Bates, L. Margetts, and P. L. Manning. 2011b. The 'Goldilocks' effect: preservation bias in vertebrate track assemblages. Journal of the Royal Society Interface 8(61): 1142–1154.

Falkingham, P. L., L. Margetts, I. Smith, and P. L. Manning. 2009. Reinterpretation of palmate and semi-palmate (webbed) fossil tracks; insights from finite element modelling. Palaeogeography, Palaeoclimatology, Palaeoecology 271(1–2): 69–76.

Farlow, J. O. 1981. Estimates of dinosaur speeds from a new trackway site in Texas. Nature 294: 747–748.

Farlow, J. O., M. O'Brien, G. J. Kuban, B. F. Dattilo, K. T. Bates, P. L. Falkingham, L. Piñuela, A. Rose, A. Freels, C. Kumagai, C. Libben, J. Smith, and J. Whitcraft. 2012. Dinosaur tracksites of the Paluxy River (Glen Rose Formation, Lower Cretaceous), Dinosaur Valley State Park, Somervell County, Texas; pp. 41–69

in P. Huerta, F. T. Fernández-Baldo, and J. I. Canudo Sanagustín (eds.), Proceedings of the V International Symposium about Dinosaur Palaeontology and Their Environment.Colectivo Arqueológico y Paleontológico Salense, Burgos, Spain.

Gatesy, S. M. 2003. Direct and indirect track features: what sediment did a dinosaur touch? Ichnos 10: 91–98.

Gatesy, S. M., Middleton, K. M., Jenkins, F. A. and Shubin, N. H. 1999. Three-dimensional preservation of foot movements in Triassic theropod dinosaurs. Nature 399(6732): 141–144.

Graversen, O., J. Milàn, and D. B. Loope. 2007. Dinosaur tectonics: a structural analysis of theropod undertracks with a reconstruction of theropod walking dynamics. Journal of Geology 115: 641–654.

Halfpenny, J., and J. Bruchac. 2002. Scats and Tracks of the Southeast. Falcon Press, Guilford, Connecticut, 192 pp.

Hitchcock, E. 1836. Ornithichnology–description of the foot marks of birds (Ornithichnites) on New Red Sandstone in Massachusetts. American Journal of Science 29: 307–340.

Hitchcock, E. 1858. Ichnology of New England. A Report on the Sandstone of the Connecticut Valley, Especially its Fossil Footmarks. W. White, Boston, Massachusetts [reprinted 1974 by Arno Press, New York, New York], 232 pp.

Hutchinson, J. R., K. T. Bates, J. Molnar, V. Allen, and P. J. Makovicky. 2011. A computational analysis of limb and body dimensions in Tyrannosaurus rex with implications for locomotion, ontogeny, and growth. PLoS One 6(10): e26037.

Lockley, M. G. 1991. Tracking Dinosaurs. Cambridge University Press, Cambridge, UK, 238 pp.

Lockley, M. G., and K. Conrad. 1989. The palaeoenvironmental context, preservation, and palaeoecological significance of dinosaur tracksites in the western USA; pp. 121–134 in D. D. Gillette and M. G. Lockley (eds.), Dinosaur Tracks and Traces. Cambridge University Press, Cambridge, UK.

Manning, P. L. 1999. Dinosaur track formation, preservation and interpretation: fossil and laboratory simulated dinosaur track studies. Ph.D. dissertation, University of Sheffield, Sheffield, UK.

Manning, P. L. 2004. A new approach to the analysis and interpretation of tracks: examples from the dinosauria; pp. 93–123 in D. McIlroy (ed.), The Application of Ichnology to

Palaeoenvironmental and Stratigraphic Analysis. Special Publication 228. Geological Society, London, UK.

Manning, P. L. 2008. *T. rex* speed trap; pp. 205–231 in K. Carpenter and P. L. Larson (eds.), *Tyrannosaurus rex:* The Tyrant King. Indiana University Press, Bloomington, Indiana.

Martin, A. J., T. H. Rich, M. Hall, P. Vickers-Rich, and G. Vazquez-Prokopec. 2012. A polar dinosaur-track assemblage from the Eumeralla Formation (Albian), Victoria, Australia. Alcheringa: An Australasian Journal of Palaeontology 36: 171–188.

Marty, D. 2008. Sedimentology, taphonomy, and ichnology of Late Jurassic dinosaur tracks from the Jura carbonate platform (Chevenez-Combe Ronde tracksite, NW Switzerland): insights into the tidal flat palaeoenvironment and dinosaur

diversity, locomotion, and palaeoecology. GeoFocus 21: 278.

Milàn, J., and R. G. Bromley. 2006. True tracks, undertracks and eroded tracks, experimental work with tetrapod tracks in laboratory and field. Palaeogeography Palaeoclimatology Palaeoecology 231: 253–264.

Minter, N. J., S. J. Braddy, and R. B. Davis. 2007. Between a rock and a hard place: arthropod trackways and ichnotaxonomy. Lethaia 40: 365–375.

Moratalla, J. J., J. L. Sanz, and S. Jimenez. 1988. Multivariate analysis on Lower Cretaceous dinosaur footprints: discrimination between ornithopods and theropods. Geobios 21(4): 395–408.

Padian, K., and P. E. Olsen. 1984. The fossil trackway pteraichnus: not pterosaurian,

but crocodilian. Journal of Paleontology 58: 178–184.

Romilio, A., and S. W. Salisbury. 2011. A reassessment of large theropod dinosaur tracks from the mid-Cretaceous (late Albian-Cenomanian) Winton Formation of Lark Quarry, central-western Queensland, Australia: a case for mistaken identity. Cretaceous Research 32(2): 135–142.

Thulborn, R. A., and M. Wade. 1979. Dinosaur stampede in the Cretaceous of Queensland. Lethaia 12(3): 275–279.

Thulborn, R. A., and M. Wade. 1984. Dinosaur trackways in the Winton Formation (Mid-Cretaceous) of Queensland. Memoirs of the Queensland Museum 21: 413–517.

Thulborn, T. 1990. Dinosaur Tracks. Chapman & Hall, London, UK, 410 pp.

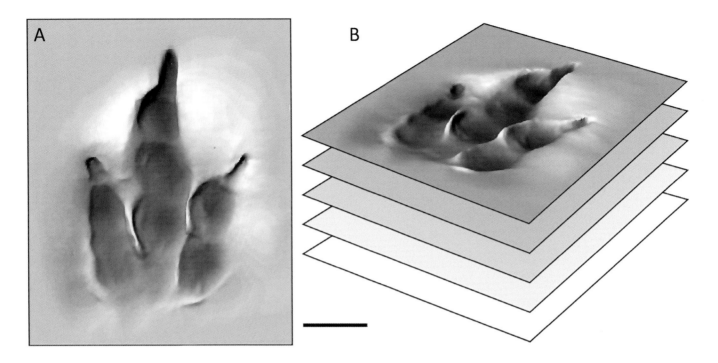

5.1. Tracks as surfaces. Computer renderings of a shallow true track (based on a photogrammetric reconstruction of a natural track cast SS.00 in Gatesy, 2001) (A) from above and (B) in perspective views. Grayscale is mapped from white (highest) to black (lowest) depth. Scale bar for A = 5 cm.

Stephen M. Gatesy and Richard G. Ellis

FOSSIL FOOTPRINTS RECORD UNRIVALED EVIDENCE OF behavior in long extinct species. For students of dinosaur locomotion, tracks offer clues about gait, speed, limb posture, foot motion (kinematics), foot loading (kinetics), and social behavior (e.g., Ostrom, 1972; Alexander, 1976; Thulborn and Wade, 1984, 1989; Padian and Olsen, 1989; Gatesy et al., 1999; Milàn, 2006; Graversen, Milàn, and Loope, 2007; Pérez-Lorente and Herrero Gascón, 2007; Ishigaki and Lockley, 2010; Avanzini, Piñuela, and Garcia-Ramos, 2011; Falkingham, 2014). Yet tracks must be studied differently from body fossils. Although complementary to skeletal remains, footprints are purely sedimentary structures that preserve traces of anatomy only indirectly. Tracks are thus neither organism nor environment but emergent features documenting their dynamic, coupled interaction (e.g., Baird, 1980; Padian and Olsen, 1984a; Allen, 1997; Falkingham and Gatesy, 2014).

Key to unlocking this exceptional evidence is a mechanistic understanding of track formation. Even so-called elite tracks are not simply molds of static pedal anatomy. Effectively extracting functional data preserved in tracks requires a geometric appreciation of their genesis. How does track morphology emerge from foot/substrate dynamics? How does track shape vary with depth below the exposed surface? How does variation in substrate consistency generate track diversity? Such fundamental questions can be approached from multiple perspectives.

Our current perception of dinosaur tracks is solidly rooted in a layer-based paradigm that emphasizes surfaces (e.g., Thulborn, 1990; Lockley, 1991; Gatesy, 2003). But what if layers and their bounding surfaces are not always the best lens through which to view footprint creation? An overreliance on this prevailing conceptual model could induce a misunderstanding of a track's origin as well as misinterpretation of its biological meaning.

In this chapter, we take a closer look at tracks as surface phenomena at layer boundaries. We then discuss an alternative perspective—tracks as volumes of displaced particles. Results from a new experimental approach reveal the benefits of quantifying the three-dimensional (3-D) trajectory of individual sedimentary particles during track formation, particularly for undertracks. We present examples from fossil tracks that illustrate how a particle based point of view can be a valuable supplement to the prevailing model for understanding the origin of track morphology.

TRACKS AS SURFACES AND LAYERS

Footprints are easily perceived as surfaces (Fig. 5.1A). We typically observe modern-day terrestrial footprints as depressions and elevations of the exposed tracked surface (Forños et al., 2002) over which the animal moved. During the ground contact phase of locomotion, loads are transmitted from foot to substrate. Such loads can permanently alter the visible surface, leaving behind a track that might one day become part of the fossil record. Paleoichnologists must consider the depth dimension as well (e.g., Manning, 2004; Boutakiout et al., 2006; Milàn and Loope, 2007; Thulborn, 2012). Researchers are now well-aware that in layered sediment a single footfall can deform not only the uppermost surface but a series of deeper bedding planes as well (Fig. 5.1B). Splitting along any one of these distorted interlayer boundaries reveals an undertrack. The true track, equivalent to the modern-day example, is only one of many possible tracks that may be produced, preserved, uncovered, and analyzed.

Integrating superficial (true) and deep (undertrack) surfaces into a single theoretical model has both intuitive and practical appeal. The volumetric nature of footprints is far better embodied by a nested set of stacked surfaces than by any single surface alone. Because the model predicts that that a footfall can generate multiple track surfaces in layered sediments, we are much less likely to be fooled by a diversity of undertrack morphologies. Studies of undertrack variation with depth should help reduce confusion stemming from errors in ichnotaxon identification, faunal analysis, and behavioral interpretation (Allen, 1997; Manning, 2004; Milàn and Bromley, 2006, 2008; Falkingham et al., 2011). Yet despite such obvious benefits, we find a surface- or layer-based paradigm lacking in several key respects.

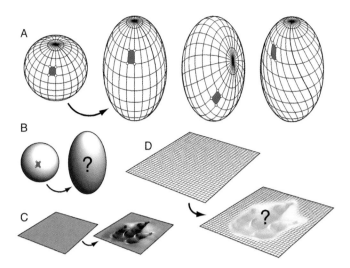

5.2. Surface point homology. (A) Locations on the surface of a globe (white dots) are designated by longitude and latitude. When modified into an ovoid, homologous points on the new surface can be identified by this two-dimensional parameterization, which reveals the nature of the deformation. (B) Without a coordinate system, point homology among surfaces is lost. (C) Track surfaces suffer the same ambiguity as the sphere and ovoid. (D) A uniform grid can be hypothesized for the hypothetical planar starting surface, but the fate of these points within the deformed track is unpreserved. The specific deformation history giving rise to track morphology remains unknown.

SURFACE DEFORMATION AND POINT HOMOLOGY

Surfaces are two-dimensional (2-D). Although contoured – different points can be at different heights – surfaces lack thickness. Thus, the location of any point on the surface can be designated by just two coordinates. For example, longitude and latitude proscribe each location on the surface of a globe (Fig. 5.2A). When a surface changes shape, such a 2-D parameterization allows homologous positions (white dots) and regions of surface (gray patches) to be identified, thereby revealing the nature of the deformation.

If we distort the sphere into an ovoid, longitude and latitude contours depict each globe point's new location on the modified surface (Fig. 5.2A). For example, a simple scaling deformation is recorded in the left ovoid by contour lines similar to the original, but stretched vertically. The nonterminal poles and inclined lines of latitude in the ovoid second from the right are evidence of surface rotation as well as scaling. A twisting deformation can be inferred from spiraling lines of longitude in the right ovoid. Unfortunately, an infinite number of such deformations can give rise to an identically shaped surface. Just describing the shapes of the original (sphere) and final (ovoid) forms is insufficient to discern homologous points (Fig. 5.2B). A coordinate system (longitude/latitude or some alternative) is essential to verify positional correspondence and thus discern the point movements responsible for overall shape change.

Like the spherical globe, a track is also 2-D. Neither the tracked surface (air-sediment interface atop the uppermost layer) nor any undertrack surface (sediment-sediment interface between layers) has thickness. Our interpretation of both true tracks and undertracks is founded on the postulate that footfall-associated changes to the surface can be identified as such. We recognize deformation by presuming a more planar starting condition (Fig. 5.2C). This hypothesis of original planarity is what makes splitting rocks along bedding planes so special.

Sadly, our ignorance about point homology between untrod and trod surfaces (Fig. 5.2C) is no less than that between sphere and ovoid (Fig. 5.2B). Simply characterizing the shape of the interface (as with a 3-D polygonal model) is not enough. A hypothetical initial plane can be parameterized with Cartesian coordinates (Fig. 5.2D), but how are these points redistributed during track formation? Peripheral regions are typically presumed to remain relatively undisturbed, but within the track areas of greatest interest, any simple homology is disrupted. Because locations on the track surface lose one-to-one correspondence to locations on the original surface, the fates of points on the plane remain conjectural.

We believe this is a major, unrecognized problem when trying to understand the formation of track morphology. A track surface only records the altered depths of points that once shared a common depth. Because evidence of horizontal movement is not preserved, surfaces give clues about only one (vertical) of the three dimensions. Such an interface-based approach is akin to trying to understand water currents by only studying the surface relief of the ocean.

SURFACE AREA AND LAYERS

To make our concerns about surfaces and layers less abstract, we provide an example. We created a 3-D model of a natural cast of a theropod track preserving skin impressions from Greenland (SS.oo in Gatesy, 2001) using photogrammetry (Falkingham, 2012). A virtual track surface (Fig. 5.1) was reconstructed by expanding the surrounding plane beyond the collected block and repairing broken claw impressions. If we assume that the tracked surface began as a plane (Fig. 5.2C), any distortion by foot contact must engender an increase in surface area. For a given scale of sampling (e.g., Scott et al., 2005), the change in area can be quantified.

This relatively shallow track has 5.8% more surface area than a flat plate of equal dimensions. Increased interface can also be quantified in 2-D by taking a transverse section (Fig. 5.3A). Elevations and depressions make the deformed curve 11.8% longer than a straight line (Fig. 5.3B). Focusing on just the portion of the relief attributable to digit III reveals a 21.8% increase in length during track formation (Fig. 5.3C). We can certainly measure such changes, but how do we account for them? Merely treating a surface as a distensible rubber sheet lacks physical reality. Expansion must entail a spreading of originally neighboring surface points and the incorporation of new material to fill the gaps (Gatesy, 2003).

The problem is not only the source of this additional material but also the fate of those surface points originally exposed. At least some points must spread to increase interface, but how? Even if we assume deformation only in the plane of transverse section, which is extremely unlikely, what happened to points on the initial plane contacted by digit III? Did all move vertically up and down (Fig. 5.3D) so that expansion was concentrated along the track walls? Or was there a more complex combination of spreading and compression that opened up a new interface in the middle of the pad impression (Fig. 5.3E)? Surfaces alone are insufficient to answer these questions.

One could argue that layers have thickness, but the issue of uncertain positional homology applies equally in 3-D. We can reasonably assume that a layer started out with uniform thickness and maintained a relatively constant volume. But what was the homologous location of any point within the deformed layer prior to footfall (Fig. 5.3F)? All we know is that the points began somewhere between the two bounding surfaces. Therefore, we know even less about point displacement within layers than we do within surfaces. Horizontal motion remains unknown, whereas the possible range of vertical displacement is even larger.

Our point is not that track surfaces and layers are not valuable; they clearly are. As the primary means of exposing footprints in the fossil record, undertrack and tracked surfaces are how we "see" tracks. Indeed, most of our data either are surfaces or are derived from surfaces. Interface shapes are simplified into outline drawings, recorded with photographs, or reconstructed in 3-D from point clouds. However, we should also consider whether such data are best or even adequate for deciphering how tracks form. In the next section, we offer an alternative look at the dynamic interaction between foot and substrate.

A PARTICLE-BASED PERSPECTIVE

At a fundamental level, track formation entails the redistribution of sedimentary particles. Prior to foot contact, each particle has a specific starting location within the initial substrate configuration. A footfall event then mobilizes a subset of these particles, which become displaced. Although some particles may return to their original positions, others travel

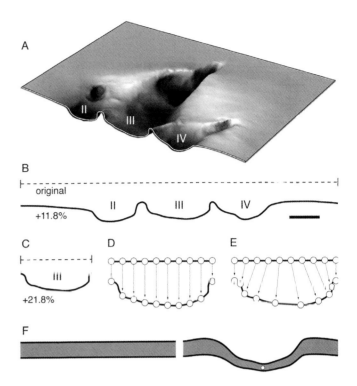

5.3. Ambiguity in track surface and layer homology. (A) Transverse section of the SS.00 model reveals track topography. (B) Deformation by digits II–IV creates a curve that is 11.8% longer than a planar section, whereas the curve associated with this portion of digit III (C) is 21.8% longer. The fate of 10 points on the original surface (white dots) is unknown even if they stay in section, potentially moving vertically (D) or spreading and compressing (E) as well. (F) Layers have thickness, but this added dimension only adds to the uncertainty about point homology. Scale bar for A, B = 2 cm.

along unique 3-D trajectories before stopping at new locations. As a sedimentary structure, a footprint can be viewed as the cumulative effect of all these trajectories. The new particle configuration forms a "track volume" encompassing the region of permanent substrate deformation.

We use the generic term "particles" in order to keep our unit of analysis flexible. In coarse substrates, particles may equate with individual sedimentary grains. In finer material, particles may consist of multiple smaller grains that move together as a relatively cohesive unit. Our main concern is that plastic substrates, no matter how fine-grained, be treated as an aggregate of physical elements. From this perspective, changes to the substrate are seen as expressions of particle motion rather than as unspecified distortions. Thus, "stretching" of a surface or "bending" of a layer are viewed as larger scale outcomes of finer scale particle rearrangement.

Some contrasts between paradigms are obvious. Whereas a surface-based approach analyzes changes in the interfaces between packets of sediment (layers), a particle-based perspective emphasizes the sediment itself. Most significantly, a reliance on surfaces cannot resolve the redistribution of material within a layer. A particle's motion (change in position

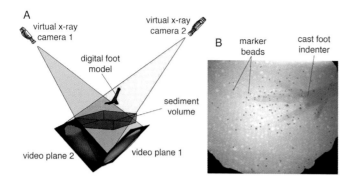

5.4. Biplanar X-ray analysis of track formation. (A) Virtual scene in the three-dimensional animation program for reconstructing movement of an indenter and marker beads at and below the substrate's surface. Calibration allows viewing the scene from the perspective of each X-ray beam. Beam sources are ~1 m from detectors. (B) Frame from one video camera showing cast foot model indenter and metal beads for tracing sediment flow. Digit III of model ~9 cm long. *Modified from Ellis and Gatesy (2013).*

through time) is studied in every dimension, revealing details of its relationship to the moving foot and to neighboring particles. Thus, particles document how sediment flows throughout the formation of a track and are key to unraveling the dynamic origin, modification, and interaction of preserved features that we recognize as tracks (e.g., Allen, 1997; Falkingham and Gatesy, 2014).

Yet if a particle-based paradigm is so desirable, why are surfaces and layers still the standard representation in most descriptions and analyses? Unfortunately, treating tracks as redistributed particles is extremely challenging, especially in fossils. Even if the entire track volume is collected, a specimen records only the final sedimentary configuration. To properly interpret such evidence from a particle-based perspective, we need to learn how patterns of displacement arise from foot-particle, particle-particle, particle-fluid (air, water), and gravitational interactions. Analyzing tracks as they are being made, particularly under controlled conditions, holds the greatest promise.

EXPERIMENTAL ICHNOLOGY

Experimental ichnologists have taken two basic approaches to studying footprint formation. One entails inducing living animals to create tracks. Examples include salamanders (Brand, 1979, 1996), horses (Lingen and Andrews, 1969), Komodo dragons (Padian and Olsen, 1984b), and ostriches (Farlow, 1989) to name a few. A subset of such in vivo studies also varied substrate consistency (McKee, 1947; Brand, 1979, 1996; Gatesy et al., 1999; Milàn, 2006). The advantage of live experiments is realism. Track shape and depth result from real feet being moved and loaded to support body weight and propel the animal forward. A big disadvantage is a lack

of control and repeatability. Animals are not easy to train and are often unpredictable.

The second approach uses indenters driven into substrates in more prescribed ways. Indenters range from severed feet to realistic models to flat plates of varying shape to cylindrical punches (Allen, 1997; Manning, 2004; Milàn and Bromley, 2006, 2008; Jackson, Whyte, and Romano, 2009, 2010). Simple indenter models let us ask "what if?" to try to separate out the effects of morphology, motion, substrate consistency, or some other parameter. The price for repeatability and control is always falling short on lifelike anatomy, movement, loading, and interaction among multiple limbs.

A major obstacle to all experimentalists is substrate opacity. Motion (or lack of motion) of the vast majority of particles remains hidden from view. Of the small minority visible on the exposed tracking surface, those particles of greatest interest are quickly concealed by the foot. Moreover, in soft substrates the limb can penetrate so deeply that toe and even metatarsal kinematics are also occluded. How can we hope to study particle redistribution if the elements of analysis are unobservable during track formation?

A BIPLANAR X-RAY METHOD OF
PARTICLE-BASED TRACK ANALYSIS

We have developed a new experimental approach based on two X-ray systems (Ellis and Gatesy, 2013) that offers a very different picture of how tracks arise (Fig. 5.4). Metal beads seeded throughout a substrate volume act as radiopaque sedimentary particles, which we record with two video cameras. In the region of beam overlap, each bead's 3-D coordinates are measured at submillimeter resolution many times a second. Using an animation-based workflow originally designed for capturing skeletal motion (Brainerd et al., 2010; Gatesy et al., 2010), we reconstruct particle trajectories and indenter motion throughout the development of a footprint. For the first time, previously invisible subterranean phenomena are directly visualized and quantified.

Results from one of our case studies (Ellis and Gatesy, 2013) are shown in Figure 5.5. In this series of experiments, a cast model of a turkey foot was mounted on a linear actuator and repeatedly plunged down and forward into a sloppy artificial sediment of sand, silica flour, ball clay, and water containing 50–190 metal beads (2 mm lead shot). We animated the 3-D locations of each bead for 45 video frames (1.43 s) spanning the interval of sediment movement. We also calculated the translations and rotations of the foot model from the 3-D coordinates of four embedded markers. Although linear penetration of a rigid foot is obviously oversimplified compared to that of a live animal, this proof of concept offers

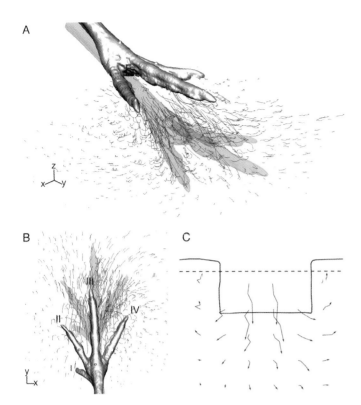

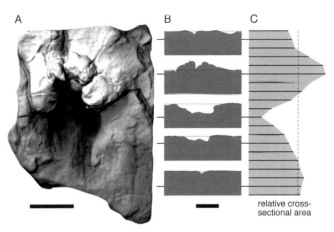

5.6. Forward sediment displacement. (A) A medium-depth theropod track from Greenland (MGUH VP 3389 [Gatesy et al., 1999]) viewed from above. (B) Five sample sections show variation in cross-sectional area. (C) Measured cross-sectional areas of 20 serial sections reveal a net transfer of sediment from back to front but no detailed accounting of displacement. Scale bars = 5 cm.

5.5. Ensemble particle flow field reconstruction. (A) Perspective and (B) dorsal views of 810 three-dimensional marker trajectories from biplanar X-ray. The foot cast (opaque render) penetrated down and forward before stopping (transparent render) 4.5 cm below the initial surface. Axes are 1 cm long (modified from Ellis and Gatesy, 2013). (C) Transverse two-dimensional displacement field assembled from a series of 8.3 mm square plasticine prism preparations (modified from Allen, 1997).

a first quantitative look at how particles in a volume of real sediment respond in real time to an anatomically correct indenter.

To get an overall picture, we illustrate 810 bead trajectories compiled from 14 of these trials as an ensemble displacement field (Figs. 5.5A, 5.5B). We achieved an aggregate density of one bead per ~4.4 cm³ within the biplanar sediment volume, but the density was higher (one bead per ~1.9 cm³) directly beneath the foot cast. From the length of each bead's path, the most affected sediment is readily apparent. Peripheral and deep beads had limited movement, whereas those directly in the model's path exhibited the longest and most complex trajectories.

Our renderings are similar to the displacement field (Fig. 5C) generated by Allen (1997:fig. 8). To document displacement by a vertically driven punch, Allen arranged colored plasticine prisms in a checkerboard pattern. However, because plasticine is opaque, he could only visualize distortion in the originally square grid by sectioning. By comparing three sets of prisms punched to different depths to an unpunched volume a coarse pseudosequence was reconstructed in 2-D. By contrast, our X-ray approach allowed

us to nondestructively trace marker particles in 3-D at very high spatial and temporal resolution.

FORWARD PARTICLE DISPLACEMENT

Our initial biplanar X-ray tests revealed that forward, rather than downward, particle motion dominated in our semiliquid substrate (Fig. 5.5A, 5.5B). This might not be surprising given the relatively low angle of penetration (39° below horizontal) and lack of foot withdrawal. However, such anterior displacement of sediment appears quite common in tridactyl tracks, particularly those that penetrate more deeply (Hitchcock, 1858; Gatesy et al., 1999; Razzolini et al., 2014). A medium-depth track from Greenland (MGUH VP 3389 [Gatesy et al., 1999]) cut into 20 transverse serial sections reveals net patterns of bulk transfer (Fig. 5.6). Rather than try to discern layer boundaries, which are extremely confusing in such sections, we compare the relative cross-sectional area of adjacent slices by measuring to a uniform depth (Fig. 5.6B). Assuming an initially level surface having sections of equal area, sections from the middle of the track show a relative paucity of sediment relative to the less disturbed end locations. A relative excess of material more anteriorly appears to make up for this deficit (Fig. 5.6C). Clearly, material driven forward by the foot accumulated to form the raised mounds so characteristic of these types of tracks (Fig. 5.6A).

Despite our ability to quantify variation in cross-sectional area, we remain ignorant about the exact nature of this final pattern. How does a foot moving through a volume of sediment give rise to such track morphology? For example, are elevated regions formed by the toes plowing forward during

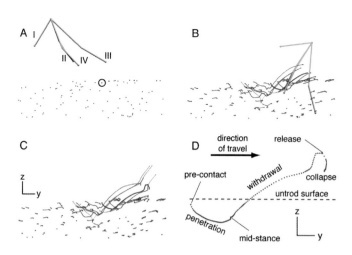

5.7. (A–C) Biplanar X-ray analysis of a full motion sequence made by manipulating an actual turkey foot. The path of one bead (circled) is shown in red. Axes = 2 cm. (D) This bead's trajectory documents major events in its motion history during track ontogeny. Axes = 1 cm. *Modified from Ellis and Gatesy (2013).*

entry early in the stance phase? Or is sediment raised only later as the foot is withdrawn? Did some features first rise much higher and then collapse down? Are all three mechanisms responsible for the ultimate track morphology? What we really would like to know is where each particle started, where it ended, and how this transit related to foot movement. How far forward did particles in the zone of net loss travel? Where did all the extra particles in the elevated zone come from, and how did they get there? We may never be able to achieve such high-resolution answers from specific fossils, but additional data from experiments can help.

By manipulating a severed turkey foot, we also recorded biplanar X-rays of more complete stepping movements through the same substrate. Markers implanted in the metatarsus and phalanges allowed us to reconstruct both bead and foot motion (Fig. 5.7), albeit for fewer markers due to ambiguity in identifying individual beads. Individual bead trajectories reveal the dramatic motion history of some markers. Beads once again moved down and forward during penetration as with the linear actuator. During midstance, motion was quite limited. Sediment was then lifted and dragged between the converging digits II–IV as they withdrew from the substrate (Figs. 5.7B, 5.7C). Some beads were lifted away in sediment stuck to the foot, whereas others collapsed back down.

The trajectory of the circled bead (Fig. 5.7A) is shown in red. Its circuitous 11.2 cm transit from ~0.3 cm below the surface to ~1.5 cm above the initial surface (a net displacement of 7.8 cm) records major phases of ground contact (Fig. 5.7D). Such detailed changes in the direction and velocity of a single particle do not answer the question of how

track morphology emerges. Yet when combined with data from other particles and from the foot, we can begin to appreciate the 3-D and temporal complexities that give rise to track variation and look for such patterns in fossils.

UNDERTRACKS IN A PARTICLE CONTEXT

The importance of undertracks in the fossil record has justifiably motivated analysis of this phenomenon in the laboratory. In such experiments, layered sediment is carefully prepared so that deformations at each interface can be visualized by sectioning or disassembly following indentation. Most protocols use alternately colored layers of uniform material (plasticine, cement, or sand with cement) to highlight interfaces (Allen, 1997; Milàn and Bromley, 2006, 2008; Jackson, Whyte, and Romano, 2009, 2010). Some workers interleaved layers of plaster (Manning, 2004) or sand (Milàn and Bromley, 2008) with their test sediment, but substrate heterogeneity was not the goal of these or other studies. Indeed, experimental and computational researchers are struggling to understand homogeneous substrates before increasing complexity with differential rheology.

The fact that both surface-based and particle-based experiments are done in homogeneous substrates invites a hypothetical question. Consider two large containers of sediment having identical mechanical properties. The first is divided into thin layers of different colors and the second is a single, massive bed of uniform color. Now imagine an animal walks over both volumes, creating a track in the exposed surface of each. Because both substrates have the same physical response to the foot, these true tracks should be almost indistinguishable. In the bedded container, we would predict the presence of undertracks. Splitting along bedding planes between colors should sample from the set of stacked surfaces at one or more sediment-sediment interfaces.

What of the massive substrate? Can a track volume without laminations or beds have undertracks? Our point is not to ask if a tree falling in the forest makes a sound if no one is there to hear it. Rather, it is to emphasize that particle displacement, which we deem most fundamental, should be essentially identical in the two containers. In the absence of significant rheological heterogeneity, bedding and laminae are irrelevant to particle movement. The particle configuration is rearranged even if deep interfaces are not there to reveal it.

The air-sediment interface is the only essential surface in this context. Undertrack surfaces are merely planes of weakness within the lithified substrate that allow us to sample (albeit crudely) the volumetric particle redistribution. Despite

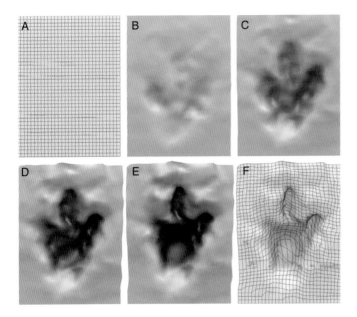

5.8. Virtual undertrack forming from interpolated particles starting at a depth of 1.25 cm below the original surface. The initial 31×41 vertex grid (A) is deformed in each dimension each frame in three dimensions to produce an estimated final undertrack (F). Starting plane = 15×20 cm.

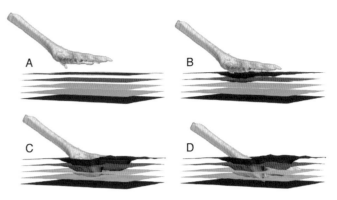

5.9. Virtual undertracks reconstructed from bead trajectories, at four increasing depths of foot penetration (A–D). For each 1/30-s time interval, we interpolated the displacement field within the 6.25-cm-deep volume to estimate the fate of particles starting at different depths. Bedding planes beginning at 1.25–5.25 cm below the original surface are shown deforming relative to the penetrating foot model and one another. Starting planes = 15×20 cm.

the dominance of surface-based perspective, this point of view should not be new or controversial. For example, Bates et al. noted that variation in surface relief "provides clues to the 3-D distribution of failure within the sediment volume during track formation" (2008:1003). Although intended for the tracking surface, this statement applies equally to undertrack surfaces as well.

Finally, consider a third container of unlayered sediment, which we seed with a 3-D grid of evenly spaced beads and image with biplanar X-rays. Aside from not needing to destroy the sample to see inside, what would be the advantage of using such markers over using colored layers? We suggest that the unique benefit would be direct quantification of 3-D motion. Not only could beads starting at the same level provide depth evidence akin to bedding planes, such markers would simultaneously reveal bead convergence/divergence and the timing of such intrabed motion. Moreover, the initial grid would serve as a consistent coordinate system (akin to the grids in Fig. 5.2) by which to assess the underlying particle reorganization responsible for large-scale deformations of the entire track volume.

VIRTUAL UNDERTRACKS

The arbitrary (and mechanistically superfluous) nature of undertracks can be demonstrated with data from our X-ray case study. By documenting the trajectories of enough real particles, the changing sediment flow field can be used to predict the fate of virtual particles starting at any location. For each 33.3 ms timestep, we calculate the x, y, and z displacements of all 810 beads. From these data, we interpolate a displacement field for each dimension at each timestep in Matlab (Simulink). The trajectory of any particle within the interpolated volume can be estimated by sequentially inputting its coordinates into the trio of displacement field functions and predicting an ending position, which serves as the starting point for the subsequent timestep.

We can thus use our experimental data to construct "virtual undertracks" within the unlayered test sediment (Fig. 5.8). To create such a virtual interface, we seed the initial displacement fields at a constant depth with a uniform grid of 1271 virtual particles, which are meshed into a horizontal bedding plane surface (Fig. 5.8A). As individual particles move through time, a virtual undertrack develops (Figs. 5.8A–5.8F). Note that such a dynamic sequence of deformable surfaces is not just a fixed grid with only the vertical dimension changing (a height field as in GIS data sets). Rather, particles defining the originally planar grid move (Fig. 5.8A) in 3-D as real sediment does (up-down, fore-aft, and transversely) based on the biplanar X-ray data (Fig. 5.8F), thus maintaining point homology.

Our method of reconstructing virtual undertracks is by no means perfect. Quality is affected by the density of the original data, displacement field artifacts arise in close proximity to the indenter, and reconstruction suffers near the tracked surface because there are no air particle trajectories to interpolate. However, our main point is that undertracks can be simulated by seeding the original volume at whatever, and as many, initial depths we choose (Fig. 5.9). Undertracks merely provide a 2-D glimpse at a 3-D, volumetric phenomenon. The presence of absence of planes of weakness in the lithified track volume is not essential to the inherent redistribution of sedimentary grains.

CONCLUSIONS

In this chapter we offer an alternative view of track formation—the genesis of track morphology. Rather than seeing tracks as alterations of the tracked and undertrack surfaces, we challenge others to consider the underlying particle redistribution responsible for changes in shape. Surface deformations come about through particle displacement. Likewise, layers that appear to have been stretched or squeezed or bent are merely large-scale reflections of the relative motion of their constituent parts. Our biplanar X-ray method of motion analysis gives a first glimpse at these previously hidden trajectories.

We believe that there is a need for both surface-/layer-based and particle-based perspectives in dinosaur ichnology. While we continue to interact with physical layers and their bounding surfaces, we should try to envision the distorted volume of sediment that is being sampled by such

subdivisions. Understanding how an initial configuration of particles is transformed into its final state should be a fundamental goal of ichnology. Only by studying track "ontogeny" can we discern how foot-sediment and sediment-sediment interactions generate track morphology (Falkingham and Gatesy, 2014). These spatial and temporal relationships will, in turn, foster the functional interpretation of such unique evidence of dinosaur locomotion in the fossil record. Indeed, track morphology and track formation are two sides of the same coin.

ACKNOWLEDGEMENTS

We thank B. Brainerd, J. Cheney, M. Dawson, P. L. Falkingham, F. A. Jenkins Jr., K. Middleton, N. Shubin, and J. Whiteside for assistance; K. Wellspring at the Beneski Museum of Natural History, Amherst College for help; two reviewers for critical comments; A. Richter, P. L. Falkingham, and D. Marty for making this volume a reality; and US National Science Foundation grants (IOS-0925077, DBI-0552051, EAR-1452119), the W. M. Keck Foundation, and the Bushnell Family Research and Education Fund for support.

REFERENCES

Alexander, R. McN. 1976. Estimates of speeds of dinosaurs. Nature 261: 129–103.

Allen, J. R. L. 1997. Subfossil mammalian tracks (Flandrian) in the Severn Estuary, S.W. Britain: mechanics of formation, preservation and distribution. Philosophical Transactions of the Royal Society, London B 352: 481–518.

Avanzini, M., L. Piñuela, and J. C. Garcia-Ramos. 2011. Late Jurassic footprints reveal walking kinematics of theropod dinosaurs. Lethaia 45(2): 238–252.

Baird, D. 1980. A prosauropod dinosaur trackway from the Navajo Sandstone (Lower Jurassic); pp. 219–230 in L. L. Jacobs (ed.), Aspects of Vertebrate History. Museum of Northern Arizona Press, Flagstaff, Arizona.

Bates, K. T., P. L. Manning, B. Vila, and D. Hodgetts. 2008. Three-dimensional modeling and analysis of dinosaur trackways. Palaeontology 51(4): 999–1010.

Boutakiout, M., M. Hadri, J. Naouri, S. Caro, and F. Pérez-Lorente. 2006. The syngenetic structure suite of dinosaur footprints in finely laminated sandstones: site n°1 of Bin el Ouidane (1BO; Central Atlas, Morocco). Ichnos 13: 69–79.

Brainerd, E. L., S. B. Baier, S. M. Gatesy, T. L. Hedrick, K. A. Metzger, S. L. Gilbert, and J. J. Crisco. 2010. X-ray reconstruction of moving morphology (XROMM): precision, accuracy and applications in comparative biomechanics research. Journal of Experimental Zoology 313A: 262–279.

Brand, L. R. 1979. Field and laboratory studies on the Coconinon Sandstone (Permian) vertebrate footprints and their

paleoecological implications. Palaeogeography, Palaeoclimatology, Palaeoecology 28: 25–38.

Brand, L. R. 1996. Variations in salamander trackways resulting from substrate differences. Journal of Paleontology 70(6): 1004–1010.

Ellis, R. G., and S. M. Gatesy. 2013. A biplanar X-ray method for three-dimensional analysis of track formation, Palaeontologia Electronica 16(1): 1T.

Falkingham, P. L. 2012. Acquisition of high resolution 3D models using free, open-source, photogrammetric software. Palaeontologia Electronica 15(1): 1–15.

Falkingham, P. L. 2014. Interpreting ecology and behaviour from the vertebrate fossil track record. Journal of Zoology 292(4): 222–228.

Falkingham, P. L., K. T. Bates, L. Margetts, and P. L. Manning. 2011. The 'Goldilocks effect' preservation bias in vertebrate track assemblages. Journal of the Royal Society Interface 8: 1142–1154.

Falkingham, P. L., and S. M. Gatesy. 2014. The birth of a dinosaur footprint: subsurface 3D motion reconstruction and discrete element simulation reveal track ontogeny. Proceedings of the National Academy of Sciences 111(51): 18279–18284.

Farlow, J. O. 1989. Ostrich footprints and trackways: implications for dinosaur ichnology; pp. 243–248 in D. D. Gillette and M. G. Lockley (eds), Dinosaur Tracks and Traces. Cambridge University Press, Cambridge, UK.

Fornós, J. J., R. G. Bromley, L. B. Clemmensen, and A. Rodriguez-Perea. 2002. Tracks and trackways of *Myotragus balearicus* Bate (Artiodactyla, Caprinae) in Pleistocene

aeolianites from Mallorca (Balearic Islands, western Mediterranean). Palaeogeography, Palaeoclimatology, Palaeoecology 180: 277–313.

Gatesy, S. M. 2001. Skin impressions of Triassic theropods as records of foot movement. Bulletin of the Museum of Comparative Zoology 156: 137–149.

Gatesy, S. M. 2003. Direct and indirect track features: what sediment did a dinosaur touch? Ichnos 10: 91–98.

Gatesy, S. M., D. B. Baier, F. A. Jenkins Jr., and K. P. Dial. 2010. Scientific rotoscoping: a morphology-based method of 3-D motion analysis and visualization. Journal of Experimental Zoology 313A: 244–261.

Gatesy, S. M., K. M. Middleton, F. A. Jenkins Jr., and N. H. Shubin. 1999. Three-dimensional preservation of foot movements in Triassic theropod dinosaurs. Nature 399: 141–144.

Graversen, O., J. Milàn, and D. B. Loope. 2007. Dinosaur tectonics—a structural analysis of theropod undertracks with a reconstruction of theropod walking dynamics. Journal of Geology 115(6): 641–654.

Hitchcock, E. 1858. Ichnology of New England: A Report on the Sandstone of the Connecticut Valley, Especially Its Fossil Footmarks, Made to the Government of the Commonwealth of Massachusetts. W. White, Boston, Massachusetts, 220 pp.

Ishigaki, S., and M. G. Lockley. 2010. Didactyl, tridactyl and tetradactyl theropod trackways from the Lower Jurassic of Morocco: evidence of limping, labouring and other irregular gaits. Historical Biology 22(1–3): 100–108.

Jackson, S. J., M. A. Whyte, and M. Romano. 2009. Laboratory-controlled simulations of dinosaur footprints in sand: a key to understanding vertebrate track formation and preservation. Palaios 24: 222–238.

Jackson, S. J., M. A. Whyte, and M. Romano. 2010. Range of experimental dinosaur (*Hypsilophodon foxii*) footprints due to variation in sand consistency: how wet was the track? Ichnos 17: 197–214.

Lingen, G. J. Van der, and P. B. Andrews. 1969. Hoof-print structures in beach sand. Journal of Sedimentary Research 39(1): 350–357.

Lockley, M. G. 1991. Tracking Dinosaurs: A New Look at Our Ancient World. Cambridge University Press, New York, New York, 238 pp.

Manning, P. L. 2004. A new approach to the analysis and interpretation of tracks: examples from the Dinosauria; pp. 93–123 in D. McIlroy (ed.), The Application of Ichnology to Paleoenvironmental and Stratigraphic Analysis. Special Publication 228. Geological Society of London, London, UK.

McKee, E. D. 1947. Experiments on the development of tracks in fine cross-bedded sand. Journal of Sedimentary Petrology 17(1): 23–28.

Milàn, J. 2006. Variations in the morphology of emu (*Dromaius novaehollandiae*) tracks reflecting differences in walking pattern and substrate consistency: ichnotaxonomic implications. Palaeontology 49(2): 405–420.

Milàn, J., and R. G. Bromley. 2006. True tracks, undertracks and eroded tracks, experimental work with tetrapod tracks in laboratory and field. Palaeogeography, Palaeoclimatology, Palaeoecology 231: 253–264.

Milàn, J., and R. G. Bromley. 2008. The impact of sediment consistency on track and undertrack morphology: experiments with emu tracks in layered cement. Ichnos 15(1): 19–27.

Milàn, J., and D. B. Loope. 2007. Preservation and erosion of theropod tracks in eolian deposits; examples from the Middle Jurassic Entrada Sandstone, Utah, USA. Journal of Geology 115(3): 375–386.

Ostrom, J. H. 1972. Were some dinosaurs gregarious? Palaeogeography, Palaeoclimatology, Palaeoecology 11(4): 287–301.

Padian, K., and P. E. Olsen. 1984a. The fossil trackway *Pteraichnus*: not pterosaurian, but crocodilian. Journal of Paleontology 58(1): 178–184.

Padian, K., and P. E. Olsen. 1984b. Footprints of the Komodo dragon and the trackways of fossil reptiles. Copeia 1984: 662–671.

Padian, K., and P. E. Olsen. 1989. Ratite footprints and the stance and gait of Mesozoic theropods; pp. 232–241 in D. D. Gillette and M. G. Lockley (eds.), Dinosaur Tracks and Traces. Cambridge University Press, Cambridge, UK.

Pérez-Lorente, F., and J. Herrero Gascón. 2007. El movimiento de un dinosaurio deducido de una rastrillada terópoda con estructuras de inmersión de los pies en barro y de arrastre de cola (Formación Villar Del Arzobispo. Galve, Teruel, España). Revista Española de Paleontología 22: 157–174.

Razzolini, N. L., B. Vila, D. Castanera, P. L. Falkingham, J. L. Barco, J. I. Canudo, P. L. Manning, and À Galobart. 2014. Intra-trackway morphological variations due to substrate consistency: the El Frontal dinosaur tracksite (Lower Cretaceous, Spain). PLoS One 9(4): e93708.

Scott, R. S., P. S. Ungar, T. S. Bergstrom, C. A. Brown, F. E. Grine, M. F. Teaford, and A. Walker. 2005. Dental microwear texture analysis shows within-species diet variability in fossil hominins. Nature 436(4): 693–695.

Thulborn, R. A. 1990. Dinosaur Tracks. Chapman and Hall, New York, New York, 410 pp.

Thulborn, R. A., and M. Wade. 1984. Dinosaur trackways in the Winton Formation (mid-Cretaceous) of Queensland. Memoirs of the Queensland Museum 21: 413–517.

Thulborn, R. A., and M. Wade. 1989. A footprint as a history of movement; pp. 51–56 in D. D. Gillette and M. G. Lockley. (eds.), Dinosaur Tracks and Traces. Cambridge University Press, Cambridge, UK.

Thulborn, T. 2012. Impact of sauropod dinosaurs on lagoonal substrates in the Broome Sandstone (Lower Cretaceous), Western Australia. PLoS One 7(5): e36208.

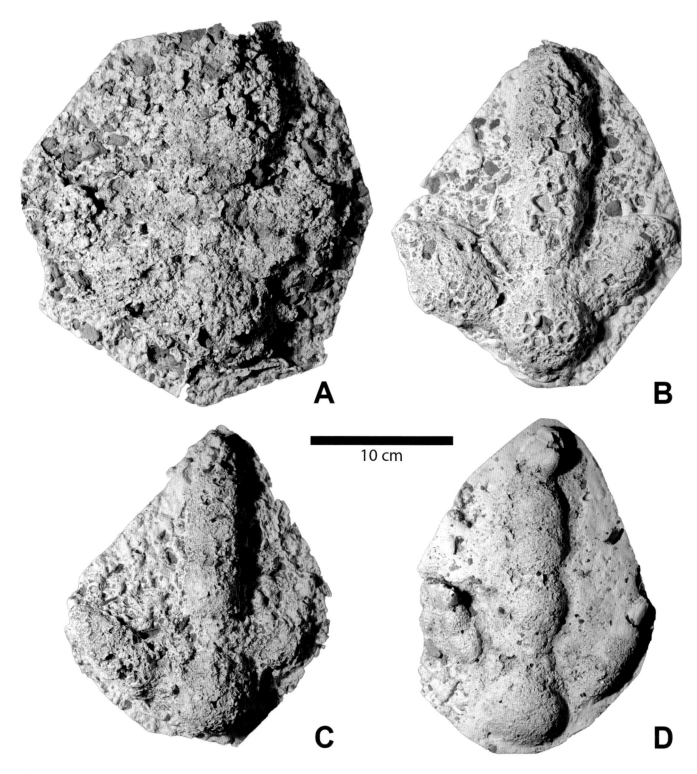

A

B

10 cm

C

D

A Numerical Scale for Quantifying the Quality of Preservation of Vertebrate Tracks

Matteo Belvedere and James O. Farlow

FROM ITS BEGINNING, VERTEBRATE ICHNOLOGY HAS described fossilized footprints in a qualitative, descriptive way. At the same time, considerable effort has gone into illustrating footprint morphology. In recent years, new technologies (e.g., laser-scanning and close-range photogrammetry) and methods (e.g., geometric morphometrics) have allowed more objective, quantitative approaches to vertebrate ichnology. However, quantitative shape analyses need to be based on data of high quality, and comparisons are best made between tracks comparable in quality of preservation. Thus, determining which footprints constitute the most reliable sample for quantitative analyses is fundamental for the progress of ichnology.

We introduce here a numerical scale to quantify the quality of preservation of vertebrate footprints, based on morphological details present in the prints. The numerical grades conceptually arise from theoretical assumptions about the fossil record, and also from experimental observations of footprints of the extant emu (*Dromaius novaehollandiae*). The numerical scale is ordinal in character, reflecting progressively better preservation of morphological details. Values range from zero (the worst prints, occurring only as aligned depressions) to three (the best preserved prints), plus some coded letters to specify the occurrence of certain morphological or preservational features (e.g., skin impressions).

Introducing this numerical scale will facilitate easier and faster comparisons among footprints, reducing the need to refer to drawings or photographs of the traces to assess their quality. This will make possible more accurate evaluation of the track record, with comparisons made only from the best preserved footprints. This scale could potentially be used when creating new ichnotaxa, which should be based on well preserved tracks (such as those with numerical values of at least two, for simple-shaped tracks). Morphometric analyses might likewise be restricted to prints of the same quality, because the results of such analyses are highly dependent on the preservation of the specimens analyzed.

INTRODUCTION

Vertebrate ichnology has usually been based on qualitative descriptions of specimens, from its beginning (e.g., Duncan, 1821; Hitchcock, 1836) to the most recent papers (e.g., Milàn, 2011; Lockley et al., 2012). This approach resulted in a very detailed lexicon (e.g., Leonardi, 1987; Thulborn, 1990) for morphological description of tracks and trackways, including some standardization of illustrations of morphological features through outline drawings, although some ambiguities and differences in techniques still remain.

In recent years, many research groups have introduced new, quantitative methods in order to update ichnology as a more mathematical, less descriptive discipline (Falkingham, 2012, and references therein); this has progressed through increasingly more common use of digitized models of specimens made with laser scanners and/or photogrammetric methods. In spite of their demonstrated weaknesses and strengths, these new methods result in comparable levels of resolution and accuracy (Remondino et al., 2010; Adams et al., 2010; Petti et al., 2011; Belvedere et al., 2012). These methods allow collection of digital data that are virtually identical replicas of the true specimen (submillimeter resolution is nowadays readily achievable). They also permit ichnologists to share objective three-dimensional (3-D) data and not just photos (whose objectivity is limited by lighting, the angle of the shot and, above all, interpretation) or interpretative outline drawings, the accuracy of which depends on many subjective factors (e.g., the experience of the person making the drawing, and light conditions).

Geometric morphometrics (i.e., landmark analysis [Rasskin-Guttman et al., 1997; Azevedo-Rodriguez and Santos, 2004; Belvedere, 2008]) is also taking root in ichnology, but, although its methods can be applied in new studies, its use with previously published data is difficult, because most of the figures and drawings in literature are generally too small and of too low resolution to permit recognition of clear

6.1. Photos of the emu tracks used to define Farlow's (unpubl.) preservation scale. (A) Grade 1; (B) grade 2; (C) grade 3; (D) grade 4.

Table 6.1. Schematic descriptions and possible uses of the preservation grades

	Description	Use	Figure
3	All digit impressions completely sharp and clear; digit walls well defined, all ungual marks clearly preserved, distinct digital pads present. In quadrupedal animals, both manus and pes perfectly preserved.	Excellent prints upon which to base new ichnotaxa, even at the level of ichnospecies. These are the only footprints that allow reliable landmark identifications for shape analyses.	2A,B
2	Toe marks fairly clear and sharp, at least over substantial portions; ungual marks and some digital pads recognizable.	More exact information on the trackmaker. Some higher, at best at generic level, ichnotaxonomy possible but not certain. Can be used for general shape comparisons. It can be used to determine precise heteropody for quadrupeds.	2C,D
1	Toe marks faint, blurred, or distorted but recognizable. Some ungual marks recognizable; for quadrupeds, manus prints distinguishable from pes prints. Only general outline preserved.	The print provides good information about the kind of trackmaker but only poor information about the shape of the autopodium. Determination of movement direction and possibly body carriage posture (plantigrade vs. digitigrade) feasible. Assignment to previously defined ichnogenus likely possible but should not be used for new ichnotaxon.	2E,F
0	No visible morphological details.	Provides only a general indication of the passage of the animal; if organized in trackway, possibly allows determining some parameters, e.g., print alignment, probable posture, probable direction of travel	3A

landmarks. Moreover, the description of the track may not accurately represent what is figured, making it thus very dubious that prints of similar preservation are being compared. In fact, the main problem in the application of landmark analysis is identifying the same reference points among different tracks. Ideally this method should be applied to very well preserved tracks ("elite tracks," sensu Gatesy, 2003) that preserve internal morphology (e.g., digital pad impressions) and not just a general footprint outline; comparing tracks of different quality can gives unreliable results. The study of the autopodium morphology, recorded in a footprint, gives numerous insights for the identification of the trackmaker, and thus for the evaluation of the (ichno)biodiversity. Through ichnotaxonomy, autopodium morphology may provide clues about the global distribution of certain taxa that can be used for paleobiological and paleogeographical considerations.

In this light, by "preservation," we mean the record of morphological features that can be related to the anatomy of the trackmaker's autopodium. Whether the diagnosable features of a footprint are due to the rheological features of the substrate or to the weathering of the lithified surface is not the aim of this scale; the purposes of the application of quantitative data for morphological and taxonomical purposes, in fact, are independent from the variable occurring in the track formation.

For the same reason, footprints with a great potential for biomechanical studies may not reach a high value in the proposed scale, as the features preserved are not sufficient for a quantitative approach to morphological studies. For

example, the tracks analyzed by Gatesy et al. (1999) in his biomechanical studies are a perfect example of what we mean: they preserve a lot of valuable and remarkable information that allowed creation and validation of a model of the foot movement through the sediment, but, from a taxonomical point of view, they cannot be of any use; indeed, the determination of a possible trackmaker is made through the assignment to an ichnogenus (e.g., *Grallator*) of other tracks present in the site (Jenkins et al., 1995). At the same site, other tracks also occur (Gatesy, 2001) that present clear outlines together with skin impressions; these would achieve a higher score in the scale because of their potential use for morphological identification, and they are also extremely important as records of foot movement. Unfortunately none of the quantitative methods of comparison can work if applied to footprints with different preservation grades; such comparisons would only increase confusion, compromising the validity of the results. Thus far, the quality of preservation of footprints has only been described verbally, without any codification or rule, and consequently depends on the opinions and descriptive skills of the authors.

For these reasons, we here introduce an ordinal numerical scale to characterize the preservation quality of footprints, with values ranging from 0 to 3, plus the possibility of adding letter codes to highlight specific features that may occur independently from the overall track preservation. The aim of this numerical scale is to provide a consistent means of assessing the preservation quality of footprints, both for future descriptions and for retroactive characterization of

6.2. Examples of the preservation scale for tridactyl and sauropod tracks. (A) Biped grade 1, Late Triassic, Dolomia Principale Formation, Italy (Leonardi and Mietto, 2000). (B) Quadruped grade 1, Early Cretaceous, Twannbach Formation, Switzerland (Les Grattes-S1, LP4-MP4, modified from Marty et al., 2013). (C) Biped grade 2, Early Cretaceous, Glen Rose Formation, Texas (Dattilo et al., 2014). (D) Quadruped grade 2, Early Cretaceous, Villar del Arzobispo Formation, Spain (specimen LCR14.6p, Castanera et al., 2011:fig. 5B). (E) Biped grade 3, Late Jurassic, Iouaridène Formation, Morocco (specimen Deio CXVIII/16, Belvedere, Mietto, and Ishigaki, 2010). (F) Quadruped grade 3, Early Cretaceous, Glen Rose Formation, Texas (Dattilo et al., 2014).

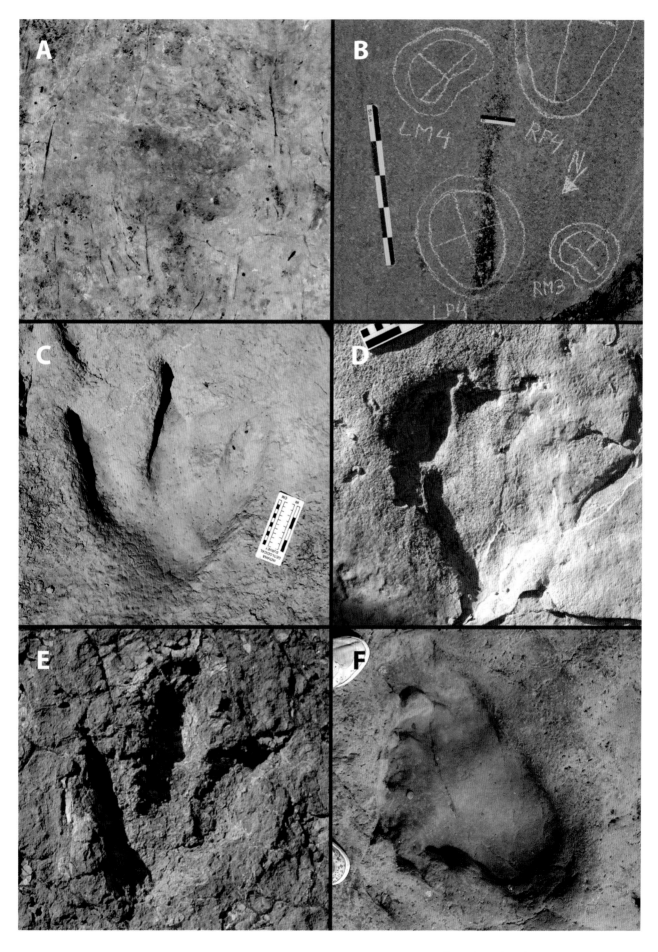

6.3. Examples of the preservation scale. (A) Grade 0, Mt. Pelmetto, Late Triassic, Dolomia Principale Formation, Italy (Leonardi and Mietto, 2000). (B) Grade 3s: "elite track" with all anatomical details and scales impression preserved. Early Jurassic, Franklin County, Massachusetts (specimen 56/1, Beneski Museum of Natural History at Amherst College, courtesy of the Trustees of Amherst College). (C) Grade 2.5pst: most of the phalangeal pad impression well preserved, partial (one digit is missing) with scale striations along the print walls. Early Cretaceous, Gara Sbaa, Morocco (Belvedere et al., 2013).

previously published material, thus facilitating application of new, quantitative methods of footprint shape analysis to previously described dinosaur tracks. The use of such a scale is not intended to replace qualitative descriptions of morphology but rather to integrate it with a numerical value, resulting in improved understanding of the morphological features present in the specimens. This numerical value will make the identification of comparably preserved prints easier, enhancing the reproducibility of comparisons of specimens.

THE SCALE

The proposed scale is modified from one initially created by Farlow (unpubl.) during a study of footprint formation by emus (*Dromaius novaehollandiae*) (Fig. 6.1), the details of which will be published elsewhere. One preservation grade has been added at the bottom and one deleted from the top and the scale is now extended to prints beyond the digitigrade and tridactyl prints of ratites. Some coded letters are added to specify the occurrence of certain details or, when clearly recognizable, some preservational features. Footprint quality continues to be expressed as an ordinal variable, with larger numbers reflecting better preservation, without specifying how much better footprints of higher numerical rank are than those of lower numerical rank.

As defined, the scale is completely independent of substrate characteristics, or gait and behavior of the trackmaker, and is based only on the morphological characteristics visible in the footprint. This permits comparisons of the largest number of footprints, from the broadest range of sedimentary paleoenvironments and modern localities, subject to different weathering conditions. This scale is intended to provide a tool to quantify the preservation quality of a track without impacting ichnotaxonomic analyses of the footprint.

In Farlow's (unpubl.) emu footprint study, values of the scale ranged 1–4, with intermediate values (e.g., 1.5, 2.5) permitted when the quality of preservation of a print did not neatly fall into the defined footprint values. For fossil tracks, further aspects have to be considered, such as diagenetic processes, weathering, and mechanical erosion, which could completely obliterate the morphological details of a footprint. However, even where footprints lack any morphological detail, they cannot be ignored and need to be characterized. For this reason, a footprint quality value of 0 has been added to the scale. Grade 4 of Farlow's (unpubl.) original scale was eliminated to allow the use of a single scale for digitigrade, plantigrade, and semiplantigrade footprints. Although very different in morphology, it is possible to define common key features to discriminate the preservation grade (Fig. 6.2). A series of letter codes added to the scale's value is introduced to specify the occurrence of certain details or, when clearly recognizable, some preservational features: s for skin/scale impressions; st for skin/scale striations; p for partial preservation (e.g., broken/missing toe impressions); u if preserved as an undertrack or transmitted track; o if preserved as an overtrack (sensu Lockley, 1991); e for epirelief; h for hyporelief.

As with Farlow's (unpubl.) original emu footprint quality scale, intermediate footprint quality categories can be assigned to tracks that have characteristics that fall between the defined categories. The detailed scale and the identification features are presented in Table 6.1, with footprints illustrating each numerical value illustrated in Figure 6.2 and examples of the letter use in Figure 6.3. Three-dimensional models of most of the key steps can be downloaded from http://dx.doi.org/10.6084/m9.figshare.1332454 and http://dx.doi.org/10.6084/m9.figshare.1322110.

DISCUSSION

Introducing a scale for numerical characterization of footprint quality provides an easy-to-use tool for sharing data, which would improve application of shape analysis to ichnology by ensuring that prints of comparable quality are employed in comparisons. Even though our scale is originally defined on the basis of true tracks (sensu Lockley, 1991), the fact that it is based on morphological characteristics present in a footprint means that the scale can be applied to any kind of footprint, such as true tracks, undertracks, overtracks, although these would have a lower score in the scale.

Poorly and only moderately well preserved (values of 0–1) footprints, nonetheless, still retain great relevance. At the very least, even the poorest (category 0) prints indicate the presence of a trackmaker. Prints of higher quality yield information of progressively greater value about the nature of trackmakers and the ichnotaxa to which they should be assigned.

We suggest that use of the scale is appropriate any time tracks from a new site are described, or previously published prints are redescribed, and should be part of the characterization of any new ichnotaxa. Usually only the best-preserved tridactyl tracks (grade 3 in our scale) should be used as holotypes to avoid generating confusion and misinterpretations.

However, certain tracks, such as those of large theropods, seldom show well-defined digital pad impressions, even though the quality of preservation of other features is quite good. In such cases new taxa can justifiably be defined from prints with a slightly lower quality grade (grade ≥2). For the same reasons, only high-score prints should be used for making reliable identifications of trackmakers. A similar reasoning can be applied also to sauropod tracks, although, in this case, the fewer morphological details commonly preserved in footprints of sauropods as opposed to tridactyl footprints of bipedal dinosaurs (due to the more compact construction of sauropod fore and hind feet) mean that grade values ≥2 would probably be adequate for describing new sauropod ichnotaxa.

"Elite tracks" were originally defined by Lockley and Hunt (1995:22–23) as "the most recently formed (youngest) tracks on the same surface" that "will stand out clearly" among the other prints in the surface, without focusing on the overall quality of the footprint. However, even though the term is still intended to identify the relatively best track(s) of a site (e.g., Wagensommer et al., 2012; Matsukawa et al., 2014), the term "elite track" can also be used to describe very well preserved tracks more generally, those which "reflect plantar morphology with digital pad, claw, and skin impressions" (Gatesy, 2003:97, fig. 6). Although the two meanings are not very different, the numerical values of elite tracks as interpreted on the basis of the quoted passages could range from grades 2 to 3 in our scale. We therefore suggest, following the definition of Gatesy (2003), that "elite track" should be used to describe only those tracks with a value of 3 in our scale (Fig. 6.3). Finally, as a general suggestion, we recommend that the verbal descriptions of footprints jointly reflect the quality of the footprints under consideration and should be always accompanied by high-quality illustration and, when possible, 3-D models of the specimens.

ACKNOWLEDGMENTS

We wish to thank the Beneski Museum of Natural History at Amherst College, and The Trustees of Amherst College for the permission to the publication of the image in Figure 6.2F; Ignacio Díaz Martínez (University of La Rioja, Spain) whose comments to an the early version have been very useful for the perfection of the manuscript; Heinrich Mallison for his accurate review; and Peter Falkingham for providing some of the three-dimensional models and for his constructive discussion, review, and suggestions that deeply improved the manuscript and the definition of the scale here proposed.

REFERENCES

Adams, T., C. Strganac, M. J. Polcyn, and L. L. Jacobs. 2010. High resolution three-dimensional LaserScanning of the type specimen of *Eubrontes* (?) *glenrosensis* Shuler, 1935, from the Comanchean (Lower Cretaceous) of Texas: implications for digital archiving and preservation. Palaeontologia Electronica 13: 1T: 11p.

Azevedo Rodrigues, L., and V. F. dos Santos. 2004. Sauropod tracks: a geometric morphometric study; pp. 129–142 in A. M. T. Elewa (ed.), Morphometrics: Applications in Biology and Paleontology. Springer-Verlag, Heidelberg, Germany.

Belvedere, M. 2008. Ichnological researches on the Upper Jurassic dinosaur tracks in the Iouaridène area (Demnat, central High-Atlas, Morocco). Ph.D. dissertation, Università degli Studi di Padova, Padua, Italy, 128 pp.

Belvedere, M., P. Mietto, and S. Ishigaki. 2010. A Late Jurassic diverse ichnocoenosis from the siliciclastic Iouaridène Formation (Central High Atlas, Morocco). Geological Quarterly 54(3): 367–380.

Belvedere, M., A. Baucon, C. Neto de Carvalho, S. Venturini, F. Felletti, and G. Muttoni. 2012. To (laser) scan or not to scan? Hamletic question at the Pramollo ichnolagerstätte (Carboniferous, Italy-Austria); Abstract 3 in Abstract Book of DigitalFossi12012, September 24–26, Berlin, Germany.

Belvedere, M., N.-E. Jalil, A. Breda, G. Gattolin, H. Bourget, F. Khaldoune, and G. J. Dyke. 2013. Vertebrate footprints from the Kem Kem beds (Morocco): a novel ichnological approach to faunal reconstruction. Palaeogeography, Palaeoclimatology, Palaeoecology 383–384: 52–58. doi: 10.1016/j.paleo.2013.04.026.

Castanera D., J. L. Barco, I. Díaz-Martínez, J. Herrero-Gascón, F. Pérez-Lorente, and J. I. Canudo. 2011. New evidence of a herd of titanosauriform sauropods from the lower Berriasian of the Iberian range (Spain). Palaeogeography, Palaeoclimatology, Palaeoecology 310: 227–237. doi: 10.1080/10420940490428805.

Dattilo, B. F., S. C. Howald, R. Bonem, J. O. Farlow, A. J. Martin, M. O'Brien, M. G. Blair, G. Kuban, L. K. Mark, A. R. Knox, W. N. Ward, and T. Joyce. 2014. Stratigraphy of the Paluxy River tracksite in and around Dinosaur Valley State Park, Lower Cretaceous Glen Rose Formation, Somervell County, Texas; pp. 307–338 in M. G. Lockley and S. G. Lucas (eds.), Fossil Footprints of Western North America. Bulletin 62. New Mexico Museum of Natural History and Science, Albuquerque, New Mexico.

Duncan, H. 1831. An account of the tracks and footprints found impressed on sandstone in the Quarry at Corncockle Muir in Dumfries-shire. Transaction of the Royal Society of Edinburgh 11: 194–209.

Falkingham, P. L. 2012. Acquisition of high resolution three-dimensional models using free, open-source, photogrammetric software. Palaeontologia Electronica 15(1): 1T: 15p.

Falkingham, L. P., and J. O. Farlow. 2015. Latex peel of theropod tracks from the Paluxy River. Figshare. Available at http://dx.doi.org/10.6084/m9.figshare.1322110. Accessed October 22, 2015.

Gatesy, S. M. 2001. Skin impressions of Triassic theropods as records of foot movement. Bulletin of the Museum of Comparative Zoology 156: 137–149.

Gatesy, S. M. 2003. Direct and indirect track features: what sediment did a dinosaur touch? Ichnos 10: 91–98.

Gatesy, S. M., K. M. Middleton, F. A. Jenkins, and N. H. Shubin. 1999. Three-dimensional preservation of foot movements in Triassic theropod dinosaurs. Nature 399: 141–144.

Hitchcock, E. 1836. Ornithichnology: description of the foot marks of birds (Ornithichnites) on New Red sandstones in Massachusetts. American Journal of Science 29: 305–339.

Jenkins, F. A., Jr., N. H. Shubin, W. W. Amaral, S. M. Gatesy, C. R. Schaff, L. B. Clemmensen, W. R. Downs, A. R. Davidson, N. Bonde, and F. Osbaeck. 1994. Late Triassic continental vertebrates and depositional environments of the Fleming Fjord Formation, Jameson Land, East Greenland. Meddelelser om Gronland, Geoscience 32: 1–25.

Leonardi, G. 1987. Glossary and Manual of Tetrapod Footprint Palaeoichnology: Departamento Nacional da Produçao Mineral, Brasilia, Brazil, 137 pp.

Leonardi, G., and P. Mietto, eds. 2000. Dinosauri in Italia. Accademia Editoriale, Pisa, Italy, 494.

Lockley, M. G. 1991. Tracking Dinosaurs: A New Look at an Ancient World. Cambridge University Press, Cambridge, U.K., 238 pp.

Lockley, M. G., and A. P. Hunt. 1995. Dinosaur Tracks and Other Fossil Footprints of the Western United States. Columbia University Press, New York, New York, 338 pp.

Lockley, M. G., J. Li, M. Matsukawa, and R. Li. 2012. A new avian ichnotaxon from the Cretaceous of Nei Mongol, China. Cretaceous Research 34(1): 84–93.

Marty, D., C. A. Meyer, M. Belvedere, J. Ayer, and K. L. Schäfer. 2013. Rochefort-Les Grattes: an Early Tithonian dinosaur tracksite from the Canton Neuchâtel, Switzerland. Revue de Paléobiologie 32(2): 373–384.

Matsukawa, M., M. G. Lockley, K. Hayashi, K. Korai, C. Peiji, and Z. Haichun. 2014. First report of the ichnogenus Magnoavipes from China: new discovery from the Lower Cretaceous inter-mountains basin of Shangzhou, Shaanxi Province, central China. Cretaceous Research 47: 131–139.

Milàn, J. 2011. New theropod, thyreophoran, and small sauropod tracks from the Middle Jurassic Bagå Formation, Bornholm, Denmark. Bulletin of the Geological Society of Denmark 59: 51–59.

Petti F. M., M. Bernardi, R. Todesco, and M. Avanzini. 2011. Dinosaur footprints as ultimate evidence for a terrestrial environment in the late Sinemurian Trento Carbonate Platform. Palaois 26(10): 601–606.

Rasskin-Gutman, D., G. Hunt, R. E. Chapman, J. L. Sanz, and J. J. Moratalla. 1997. The shapes of tridactyl dinosaur footprints: procedures, problems and potentials; pp. 377–383 in D. L. Wolberg, E. Stump, and G. D. Rosenberg (eds.), Dinofest International Proceedings. Academy of Natural Sciences, Philadelphia, Pennsylvania.

Remondino, F., A. Rizzi, S. Girardi, F. M. Petti, and M. Avanzini. 2010. 3D Ichnology−recovering digital 3D models of dinosaur footprints. Photogrammetric Record 25(131): 266–282.

Thulborn, T. 1990. Dinosaur Tracks. Chapman and Hall, London, U.K., 410 pp.

Wagensommer, A., M. Latiano, G. Leroux, G. Cassano, and S. D'Orazi Porchetti. 2012. New dinosaur tracksites from the Middle Jurassic of Madagascar: ichnotaxonomical, behavioural and palaeoenvironmental implications. Palaeontology 55(1): 109–126.

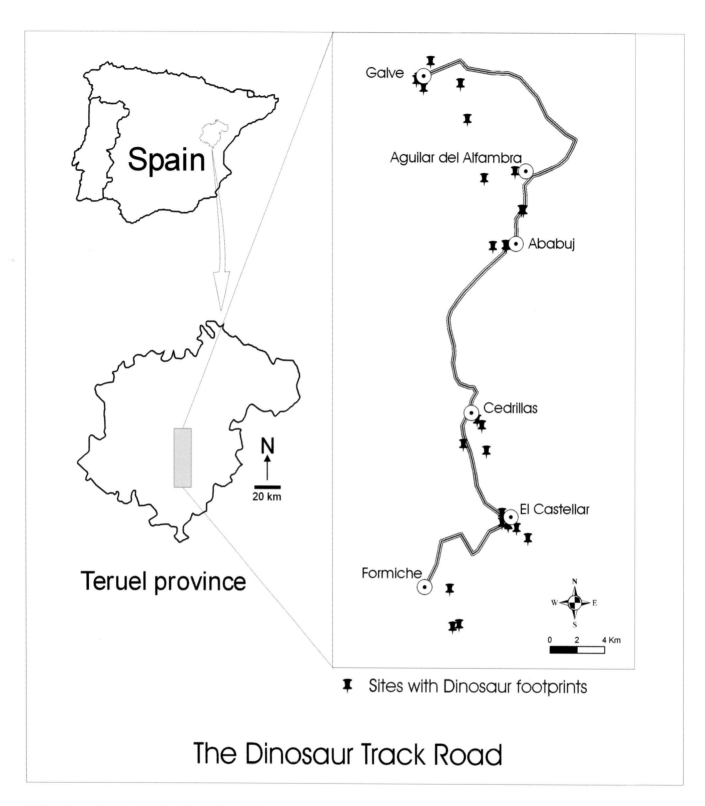

Galve

Aguilar del Alfambra

Ababuj

Cedrillas

El Castellar

Formiche

N

20 km

N
W   E
S

0   2   4 Km

🔱 Sites with Dinosaur footprints

The Dinosaur Track Road

7.1. The Dinosaur Track Road in Teruel (Spain) footprint sites.

Evaluating the Dinosaur Track Record: An Integrative Approach to Understanding the Regional and Global Distribution, Scientific Importance, Preservation, and Management of Tracksites

7

Luis Alcalá, Martin G. Lockley, Alberto Cobos, Luis Mampel, and Rafael Royo-Torres

MANY PAPERS ON FOSSIL TRACKS, FROM MANY REGIONS of the world have been published in the last two decades, and this rapid increase in documentation has itself generated the idea of a dinosaur "footprint renaissance" marked by a landslide of new discoveries and documentation. Many of these papers mention the significance of these sites in terms of selected variables such as size of site, number of tracks, new or unknown ichnotaxa, new stratigraphic or geographic occurrence, trackmaker behavioral implications, and so forth. However, the significance of fossil tracksites is often not comprehensively discussed or evaluated in such a way as to address all relevant criteria and facilitate comparison with other sites. In this chapter we describe an approach for evaluating tracksites.

In addition to criteria that deal with intrinsic scientific value, other aspects of tracksites are important for evaluation purposes. These include the aesthetic and pedagogic values for tourism and public education, historical value, vulnerability, access, and so on. Each of these can be broken down into further categories and evaluated objectively by using various consistent quantitative and qualitative measures. Thus, criteria of evaluation could be included in the following categories: scientific criteria, the most important when considering ichnites as a heritage; cultural criteria, the most important when planning regional development; and deterioration risk, crucial in terms of conservation.

One of the challenges of such management of heritage resources is that every tracksite in a given region may potentially require protection and, in many cases, interpretation. It is also important to note that when sites are nominated for national or global registers such as World Heritage List (WHL) or Global Geoparks, the evaluation process requires a comparative analysis. In theory such comparative analyses result in objective evaluation of the importance of sites that may have global applications. The main subject of this chapter is to discuss the process of evaluating tracksites in order to provide useful objective measures of their importance, and appropriate management models to help establish investment priorities. The impetus for this chapter arose from a number of practical considerations pertaining to the evaluation of ichnological sites in well-defined regional and national scale areas and comparison between sites and regions on a global scale. Case studies from Europe (Teruel, Spain) and North America (United States) are discussed.

PREVIOUS STUDIES

There have been a few previous studies, which attempt a coherent and comprehensive evaluation of tracksites in terms of all relevant criteria (Alcalá, 2002; Cobos, 2004; FCPTD, 2009). However, the simultaneous interest of some countries (Spain, Portugal, Bolivia, and South Korea) in submitting proposals to include dinosaur trackways in UNESCO's WHL fostered evaluations based on purely ichnological comparisons between sites in terms of broadly based quantifiable data such as number and diversity of tracks, trackways, size of site, and so on.

As a spin-off from these efforts, many questions have been raised about the importance of tracksites relative to other fossil sites and geological sites in general. Within the vertebrate ichnology community, interest in the criteria used in assigning value to potential WHL sites has led many in the field to begin to wonder about the value of various sites from other regions. To this end, in a dossier coordinated by the Fundación Conjunto Paleontológico de Teruel-Dinópolis (FCPTD) in 2009, a comparative method for scoring the Iberian vertebrate tracksites was described, and Lockley (2010) published an evaluation of 11 well-known tracksites in the United States. These approaches used several variables, which can be measured using objective numbers (number of tracks, publications, type ichnotaxa, and so on) and others that are less easily measured (quality of preservation, historical significance, and so on).

However, the challenge, which this chapter attempts to address, is to extend the evaluation to a well-chosen global sample of tracksites and establish variables that all in the field would consider objective, comprehensive, and of practical application. Though desirable in principle, these analyses point to a number of challenges and factors that need to be taken into consideration. The most important of these is perhaps that the data matrix compiled during the evaluation process is in a constant state of flux. This is not only due to the discovery and documentation of new sites but also to the development of existing sites as they are documented in more detail, subject to increased visitation or improved protection, or otherwise upgraded (or downgraded).

In the early days of the current dinosaur footprint renaissance, Lockley (1991) and Lockley and Gillette (1989) attempted to identify important ichnological regions in the Americas based on an estimate of the number of sites documented. The figure of about 200 tracksites for the western United States was revised to more than 550 by Lockley and Hunt (1995) and subsequently was estimated at about 1000 (Lockley, 2005a, 2005b). Leonardi (1994) has also done an admirable job in cataloging hundreds of South American sites, and the number of sites reported from China has also grown rapidly (Lockley et al., 2014, 2015).

Lockley and Gillette (1989) also listed a dozen sites, which fell into the category of sites with more than 1000 footprints and/or 100 trackways. However, there are no regional, national, or international agencies or academic groups responsible for compiling databases on trace and body fossil sites. Such compilations are usually undertaken locally by land management agencies responsible for local or regional areas (Lockley, 2005a, 2005b). Thus, the problem of "keeping track or losing track" of available sites and building useful databases usually falls to individual researchers or land management officials who choose to compile such information. Other incentives for compiling such information may arise from national or regional efforts to establish WHL sites, Global Geoparks, or other to protected areas.

A precise method of evaluating multiple dinosaur track-bearing sites was proposed by Cobos (2004) modifying the methodology proposed by Alcalá (2002) for the evaluation of Neogene mammal bone sites. They divided evaluation criteria into three basic categories, simplified here as scientific value (S_v), cultural value (C_v) and deterioration risk (D_v):

1. Scientific value. Of primary importance for research: knowledge of past life. Main interest group: paleontologists.
2. Cultural value. Of interest for touristic, educational, and economic benefits. Principal audience: society, especially students, tourists, and persons seeking leisure or public education activities.
3. Deterioration risk. State of preservation and risk of deterioration. Interested community: government guardians of local and regional heritage.

In the case of the geoheritage value of dinosaurs, it is probably not possible to distinguish between the importance of trace and body fossil sites without some subjectivity, and it is our position that the two records are complementary. Anyway, some significant practical differences exist in terms of the physical characteristics of sites and the preservation and management challenges they pose. These differences can be quantified and evaluated using standardized criteria (Alcalá, Cobos, Huh, et al., 2012). While dinosaur bones compose an inert record that helps decipher who they were, their footprints show dinosaurs as living beings and provide us with information about what they did, where they did it, and how they did it, registering evocative evidence of their activity in the soft sediments that time and geological processes have converted into enduring paleontological messages.

In recent years, there has been marked progress in understanding the global significance of dinosaur tracksites. This led to the need to standardize criteria used in evaluation (comparative analysis) and fostered a collaboration between various specialists in establishing a dialogue and creating the concept of a Global Network of Dinosaur (Tetrapod) Trackways (Meyer et al., 2008; Alcalá, 2012; Huh et al., 2012).

SITE VALUE AND EVALUATION PROCESSES

Most dinosaur body fossil discoveries are excavated and transferred to museums. This means that the discovery sites may be emptied of a previously ric fossil content. Thus, many remain as holes in the ground with little or no value for further excavation. This is not always the case, especially where large fossil deposits are developed as museum focal points (e.g., Dinosaur National Monument, United States; Zigong, China; Zhucheng, China). Likewise, even in the case of many "emptied" sites, some still retain historical importance as type localities. So, although excavation may decrease the importance of a body fossil site to some degree, the extracted fossils typically have their value enhanced by preparation, description, and interpretation, especially when new scientific material is involved. Nevertheless, from the point of view of comparative evaluation, one must take into account the disposition and significance of the extracted fossils and also treat them objectively. For example, how many are recorded (scientifically documented)? How many represent new species (holotypes, paratypes, topotypes, and so on)? Is exceptional preservation involved? And so forth.

The situation with tracks is generally quite different. Sites, many of which represent single, or multiple surfaces, as large as or larger than football fields, cannot be excavated and removed to museums, except in very rare cases (Bird, 1985; Li et al., 2012; Royo-Torres, Mampel, and Alcalá 2013). Thus, the sites must be preserved in situ. In Spain, Korea, the United States, and other areas, there are various dinosaur tracksites benefiting from the highest legal heritage categories (Cobos, 2011). The most important tracksites in the United States are state parks (Rocky Hill, Connecticut, and Dinosaur Valley State Park, Texas), whereas Dinosaur Ridge and Fossil Trace are designated as national natural landmarks (under the National Park Service umbrella). In Korea, all five sites proposed for WHL inscription are national parks (Natural Monuments 294, 411, 418, 434, and 847): four of these are tracksites and one is an egg site (Lockley et al., 2012). Likewise, numerous sites in Spain fall in the legally designated Bien de Interés Cultural (BIC, asset of cultural interest). Where large protected areas such as national parks already exist, tracksites may be found within their borders as the results of surveys.

One of the challenges of such management of heritage resources is that every tracksite in a given region may potentially require protection, and in many cases, interpretation. The main subject of this chapter is to discuss the process of evaluating tracksites in order to provide objective measures of their importance. In detail, therefore, the objective is to describe a useful evaluation model that uses sensibly prioritized evaluation criteria/data. These data in turn should be easily accessible from available publications or readily obtained from in situ sites. Thus, the evaluation can be compiled in standardized form, without the necessity of appraisals or donation of proprietary data by specialists or persons with special interest in particular local sites.

In this study, we draw attention to the inherently complex nature of any process that seeks to evaluate a natural resource. Science has provided us with valuable insight into the value of certain sites for providing novel scientific information, which in turn feeds back into broader areas of public education. However, such studies do not necessarily provide comprehensive evaluations of a site that help with long-term management or full understanding of the long-term significance of the site as part of the local, regional, national, or global natural heritage. Such values must be measured by a broader range of value and heritage criteria.

Many of these diverse criteria, which go beyond assessment of purely scientific value are of major interest to land management agencies, who routinely ask researchers to provide evaluations in ostensibly nonscientific categories such as vulnerability, public access, and tourism potential. Some funded projects may require such evaluations. For example, some agencies ask researchers to rank sites in importance and sensitivity on a scale of 1–5.

In this type of study, we could in principle compare all important vertebrate tracksites (Mampel et al., 2009). However, to make the study manageable, and to develop precedents set in previous evaluations, we confine ourselves to sites representative of the Mesozoic age of dinosaurs (Late Triassic–Late Cretaceous). Generally speaking, these belong to what Meyer et al. (2008) labeled the Global Network of Dinosaur Tracksites (or Global Network of Tetrapod Tracksites, Huh et al., 2012). In an even broader sense, such sites represent part of the global track record that has become an ichnological concept that complements, parallels, and is subsumed by the fossil record. Thus, sites are included that may reveal tracks of birds, pterosaurs, mammals, and other vertebrates in addition to dinosaurs. Our primary goal is to define and apply a coherent, manageable, and objective set of variables for evaluating tracksites that can be applied both globally and on different geographic scales. Such methodology allows for the objective comparison of sites within regions but also between regions. Moreover, it allows comparison of individual sites or contextually related groups of sites representing coherent geological units (formations, basins, regions, and so on). We can also provide quantitative measures for defining regions, so that regions of the same or differing sizes can also be compared objectively using consistent measures, based on agreed categories pertaining to scientific value/merit, cultural potential, and deterioration risk.

It is important for our evaluations to be as understandable and accessible as possible so that resource managers, local politicians, and nonspecialists can adequately comprehend the evaluations and comparisons made between sites, as well as the validity of the results and the methods used. Moreover, our proposed model could have general applications, allowing one to change the variables as necessary depending on the area of paleontological heritage under consideration, or the geologic, geographic, and evolutionary scope of the evaluation.

Having evaluated tracksites in a region and given them quantitative scores on their individual merit, we can also compare their importance in the context of the corresponding body fossil sites and the fossil record of the region in general. This broader comparative analysis of the combined trace and body fossil record represents a detailed evaluation process that is convergent with the scheme outlined herein (Lockley and Hunt, 1994; Weishampel, Dodson, and Osmólska, 2004).

Table 7.1. Variables for the evaluation of dinosaur tracksites (modified from Mampel et al., 2009)

	0	1
Scientific Value (Sv)		
S1. Dating methods	No dating available	Dubious
S2. Relationship with other vertebrate indirect remains, according to Mampel et al. (2009: table 10)	>100 km	100 km–50 km
S3. Knowledge level	Not studied	1 publication
S4. Footprint preservation (sharpness/definition)	Bad (in subimprints or not visible)	Bad (in true imprints)
S5. Number of footprints	1–20	21–200
S6. Variety of ichnogroups and/or ichnotypes	Not defined	1
S7. Length of the trackway	Nonexistent or 1 with <5 steps	≥1 trackway with 5 steps or several with fewer
S8. Number of trackways	0	1–4
S9. Variety of trackways	There are no trackways	An ichnogroup or morphotype
S10. Classification of footprints	Exist but are not classified	Can be classified
S11. Relationship with dinosaur's bones of similar age	>100 km	100 km–50 km
Cultural Value (Cv)		
C1. Relationship with dinosaur's bones	>100 km	100 km–50 km
C2. Tourist and educational potential	None	Only for specialists (microfossil site, historical site aimed for very restricted visits)
C3. Nearby spreading possibilities	Do not exist (hint/sign)	Just sign posts
C4. Accessibility	No path 1 km from last junction, road in good condition	Path (only for FWD vehicles) (1 km from last junction, road in good condition)
C5. Municipal infrastructures	None	Bar (drinks)
C6. Natural resources not >50 km	None	1 natural resource
C7. Cultural resources not >50 km	None	1 cultural resource
C8. Tourist influence scope	0	1

Natural and cultural resources are considered excellent when the site is placed inside or very near an institutionally protected zone (natural monument, cultural park, national park, etc.).

Tourist influence scope: it would be 0 when within the scope of 300 km around the site/resource the potential number of visitors is 100,000 people; 1 if 300,000; 2 if 500,000; 3 if 1,000,000; and 4 if >1,000,000.

	0	1
Deterioration Risk Value (Dv)		
D1. Accessibility	Tarmac road, site close to a town or village (<1 km)	Track in good condition (all types of vehicles), site remote from a town or village (1–5 km)
D2. Natural erosion*	None (or not known)	Weak
D3. Legal protection degree	Highest international legal protection degree	Highest national legal protection degree
D4. Physical protection	Museum in situ	Roofed and protected by fence
D5. Urban vulnerability	Land designated as protected and not for building (subject to special protection due to its scientific and cultural value)	Land designated as protected and not for building (subject to special protection for other reasons: e.g., agricultural, forestry)
D6. Civil works/mining vulnerability	None	Civil works/mining resources are scarce

Note: * Value calculated from adding up water erosion, wind, and biologic and anthropogenic agents.

In principle, the best evaluation of paleontological resources would be based on the most thorough evaluation of the broadest range of available and "objective" criteria. In practice, however, given the number of tracksites known and the variability in the quality of documentation available, it is impracticable to compile excessively detailed data or propose too many finely distinguished criteria. For example, the single criterion of number of tracks (Lockley and Gillette, 1989) is clearly insufficient, whereas the efficacy of 25 or 16 variables has yet to be tested outside the regions where they were first used (see FCPTD, 2009, for Spain and Portugal; and Lockley, 2010, for the United States).

We recognize that some variables, such as number of tracks, are more amenable to objective quantification than more subjective criteria such as quality of preservation are. However, the methodology described in Mampel et al. (2009)

2	3	4
By stratigraphic correlation	Biostratigraphic	Absolute
<50 km–26 km	<26 km and ≥1 km	At the site and surroundings (1 km)
2 publications	≥3 publications	≥1 published in SCI journals
Good (in subimprints)	Good (in true imprints)	Exceptional
201–500	501–1000	>1000
2	3	≥4
≥1 trackway with 5–10 imprints	≥1 trackway with >10 prints	>1 trackway with >5 imprints
5–10	11–20	>20
2 different ichnogroups or morphotypes	3 different ichnogroups or morphotypes	4 different ichnogroups or morphotypes
Well defined but not published	A new ichnospecies, ichnogenus, etc., is defined	Several new ichnotypes are defined
<50 km–26 km	<26 km and ≥1 km	At the same site and surroundings (1 km)
<50 km–26 km	<26 km	At the same site and surroundings
Scientific tourism (sites already excavated, without fossils and closed)	Site with materials on the spot being excavated, visitors allowed, not suitably arranged	Sites currently opened for visitors
Sign posts + triptychs and flyers	Sign posts, triptychs, and flyers + fence	Sign posts, triptychs-flyers, fence nearby + museum(s) (< 5 km)
Track in bad condition (only for FWD vehicles) (1 km from last junction, road in good condition)	Track in good condition (all types of vehicles)	Tarmac road
Bar + camping and/or hostel without meal service	Bar, camping and/or hostel + hotel, meal service	<1 (bar, camping, hostel, hotel, meal service)
2–4 natural resources	>4 natural resources	>4 natural resources and/or inside a protected zone (natural park, geopark, national monument, etc.)
2–4 cultural resources	>4 cultural resources	>4 cultural resources
2	3	4
Track in bad condition (FWD vehicles), site remote from a town or village (>3 km)	Path suitable only for FWD vehicle, site remote from a town or village (>3 km)	Without a path, remote from a town or village (>3 km)
Active and weak	Active and moderate	Very high (severe)
Highest regional legal protection degree	Highest provincial legal protection degree	Basic or none
Fence	'Caution!' signs	None
I and designated as not for building but not protected	Land suitable for urban development	Building land
Placed on a zone aimed at civil works/mining	Placed on civil works/mining area	Site found in the active civil works/mining

allows for modification of selected criteria and quantification methods and the introduction of new subcategories if more detailed evaluation criteria are desired. To represent and manage spatial data and associated databases related with paleontological sites, a geographic information system–based management tool was designed. Thanks to a list of proposed indicators, it was possible to evaluate and compare (even in a graphical way) several sites as well as create routines to help the decision-making process based on query consultation functions. In general terms, it is possible to estimate a single total value (T_v) that encompasses all the variables proposed in the three great categories: scientific value (S_v), cultural value (C_v) and deterioration risk value (D_v) through the conceptual formula (simplified from Mampel et al., 2009):

$$T_v = S_v + C_v - D_v$$

Table 7.2. Sites identified as Bien de Interés Cultural (BIC, asset of cultural interest) by the Aragón Government in Teruel

Site	Locality	Diversity	Formation	Reference
Ríos Bajos	Galve	T	Higueruelas Tithonian	Pérez-Lorente and Romero-Molina (2001)
Barranco Luca	Galve	S	Villar del Arzobispo; Tithonian-Berriasian	Blanco et al. 2000
El Cantalar	Galve	S (or C)	Villar del Arzobispo; Tithonian-Berriasian	Pérez-Lorente and Ortega (2003); Lockley (2009)
Las Cerradicas	Galve	O, S	Villar del Arzobispo; Tithonian-Berriasian	Pérez-Lorente et al. (1997); FCPTD (2009); Castanera et al. (2011, 2013)
Corrales del Pelejón	Galve	O, S, T	Villar del Arzobispo; Tithonian-Berriasian	Cuenca et al. (1993); Royo-Torres et al. (2009)
Miravete 1 and 2	Miravete de la Sierra	O, S	Villar del Arzobispo; Tithonian-Berriasian	Pérez-Lorente and Romero-Molina (2001)
Ababuj	Ababuj	O, S, St	Villar del Arzobispo; Tithonian-Berriasian	Alcalá and Martín (1995); Alcalá, Cobos, Espílez et al. (2012)
El Castellar	El Castellar	St, T	Villar del Arzobispo; Tithonian-Berriasian	Cobos et al. (2010, 2014); Alcalá, Pérez-Lorente, et al. (2014)
El Pozo	El Castellar	S, T	Villar del Arzobispo; Tithonian-Berriasian	Alcalá, Mampel, et al. (2014); Alcalá, Pérez-Lorente, et al. (2014)
Camino El Berzal	El Castellar	S, T	Villar del Arzobispo; Tithonian-Berriasian	Alcalá, Pérez-Lorente, et al. (2014)
El Hoyo	El Castellar	T	Camarillas; Lower Barremian	Alcalá, Cobos, and Royo-Torres (2003)
Puente del río Escuriza	Ariño	O, S	Alacón; Upper Barremian	Canudo et al. (2000)
Abenfigo	Castellote	T	Mosqueruela; Cenomanian	Meléndez et al. (2002)

Note: Track type diversity: C, crocodilian; O, ornithopod; S, sauropod; St, stegosaur; T, theropod.

With the system proposed, each group of variables constitutes a numeric value with which to easily compare indicators, categories, or even a total value estimated for a site. But the proposed formula is not to be considered as a universal evaluation system but as a logical method to synthesize, treat, and analyze major databases for different uses.

Nevertheless, in practice, each additional category that is used adds to the complexity of the model and the difficulty of its consistent application. Thus, in order to strike a balance, the model/method should only be as complex as is necessary to execute an effective evaluation. The tables introducing the evaluation categories (scientific categories, cultural value, and deterioration risk) are laid out using a scoring scheme, 0–4 (or 1–5), which provides measures of importance in various categories (Table 7.1). These scores can be compiled to provide more objective measures of the overall importance of a site.

The scientific value involves a number of variables that provide information and content relevant to research and research potential. For example, of primary interest would be the number of type specimens (ichnotypes) or the variety (diversity) of named track types, and of secondary interest would be the number of tracks, length of trackways, quality of preservation, resistance to erosion, and so on. Also one would include information on other associated fossils and outcrops at or near the tracksites. The number and quality of scientific publications pertaining to the site and immediate region is also a measure of scientific value. These criteria are most useful in so far as they provide a substantive, quantitative database.

The cultural value is calculated according to a series of indicators that include touristic (and educational) potential,

accessibility, infrastructure available, natural and cultural resources, and so on. A site has greater potential to be visited if it can be reached by any means of transport (at least to a nearby point) by roads or tracks in good condition (Gray, 2004) and vice versa if the site lacks these. Equally, a site with a good infrastructure (accommodation services, hiking trails, etc.) has a greater chance of succeeding from a cultural and social perspective than does a site with little or no infrastructure available. Included in such infrastructure are interpretative resources such as guidebooks, brochures, signage, and so on. Likewise, a site located close to other significant natural and cultural resources is more likely to be of greater interest to potential visitors than a site isolated from other cultural spots.

Deterioration risk is defined using different indicators. A site is considered vulnerable if it is located on unprotected land available for development that could endanger the site (pressures and threats; Gray, 2004). For instance, a tracksite situated in close proximity to (or even inside) a quarry, or in an area awaiting urban development, is deemed highly vulnerable (e.g, San Cristóbal, Galve, Spain; Royo-Torres, Mampel, and Cobos, 2013). Many important sites reported from Cretaceous Coal Mines in Canada (McCrea, Lockley, and Meyer, 2001; McCrea et al., 2013) have suffered significant deterioration or complete loss. Likewise, the spectacular Cal Orck'o site situated in a Cretaceous limestone quarry in Sucre, Bolivia, suffered a major collapse on February 2010.

For important Spanish sites, results were obtained from a matrix of 25 variables. The values of the matrix must be assessed by paleontologists who assign values from 0 to 4 (worst to best estimation) in the case of scientific and cultural

ACRON	LOCAL	AGE	Sv	Cv	Dv	Tv
LC	Galve	Tithonian- Berriasian	3,01	2,972	0,339	0,700
EO	El Castellar	Tithonian- Berriasian	3,389	2,597	1,655	0,526
CP	Galve	Tithonian- Berriasian	2,635	2,853	1,339	0,491
EP	El Castellar	Tithonian- Berriasian	2,254	2,921	2,149	0,328
ECT	Galve	Tithonian- Berriasian	1,87	2,526	1,477	0,327
RB	Galve	Tithonian	1,471	2,744	1,339	0,313
BL1	Galve	Tithonian- Berriasian	1,45	2,526	1,339	0,287
BL2	Galve	Tithonian- Berriasian	1,45	2,526	1,339	0,287
CB	El Castellar	Tithonian- Berriasian	1,773	2,597	1,793	0,275
AB-1	Ababuj	Tithonian- Berriasian	2,157	2,088	2,011	0,245
EH	El Castellar	Barremian	1,43	2,472	1,655	0,233
AG-3	Aguilar del Alfambr	Tithonian- Berriasian	2,182	1,974	2,68	0,140
CD-1	Cedrillas	Tithonian- Berriasian	1,303	2,215	1,984	0,140
CD-3-A	Cedrillas	Tithonian- Berriasian	1,272	2,215	1,984	0,136
CD-3-B	Cedrillas	Tithonian- Berriasian	1,272	2,215	1,984	0,136
CD-3-C	Cedrillas	Tithonian- Berriasian	1,272	2,215	1,984	0,136
BA	Formiche Alto	Tithonian- Berriasian	1,693	2,06	2,3	0,134
AB-5	Ababuj	Tithonian- Berriasian	1,612	1,708	1,984	0,130
CT-31	El Castellar	Tithonian- Berriasian	1,674	2,205	2,518	0,116
CD 4	Cedrillas	Tithonian- Berriasian	1,109	2,215	1,984	0,112
CT-44	El Castellar	Tithonian- Berriasian	1,107	2,205	1,984	0,111
AB-4	Ababuj	Tithonian- Berriasian	1,216	1,757	2,202	0,047
AG-1	Aguilar del Alfambra	Tithonian- Berriasian	1,481	1,974	2,64	0,046
EM	Formiche Alto	Tithonian- Berriasian	1,579	2,372	3,128	0,033
AB-3	Ababuj	Tithonian- Berriasian	1,498	1,595	2,518	0,025
CC	Formiche Alto	Tithonian- Berriasian	1,19	1,946	2,478	0,025
SC	Galve	Barremian	1,007	2,242	2,648	0,006
CT-32	El Castellar	Tithonian- Berriasian	1,189	2,205	2,834	0,001
AB-2	Ababuj	Tithonian- Berriasian	1,108	1,488	3,684	0,000
AG-4	Aguilar del Alfambr	Tithonian- Berriasian	1,108	1,488	3,684	0,000
CD-2	Cedrillas	Tithonian- Berriasian	1,109	2,566	3,204	0,000

Total Value (Tv)

7.2. Scientific values (S$_v$), cultural values (C$_v$), and deterioration risk values (D$_v$) estimated for the sites of the Dinosaur Track Road in Teruel. (Right) Total values (T$_v$) ranked according to previous evaluation. Abbreviations: Acron., site designation; Local, locality.

values and from 0 to 4 (best to worst estimation) to calculate deterioration risk (Dv). Deterioration risk values have a negative effect on the total value of the site. A low (or, ideally, 0) deterioration risk value would not alter the heritage value (intrinsic value) of the site. A high deterioration risk value, on the contrary, would deduct points from the total value.

LOCAL CASE STUDIES

Teruel Province (Spain) Dinosaur Tracksites (the Dinosaur Track Road)

The province of Teruel (Aragón, Spain) is characterized by a large number and variety of paleontological sites, especially in the case of dinosaurs. Among dinosaur tracksites, before the appraisal of 2009 (FCPTD, 2009; Mampel et al., 2009), 13 important heritage sites had been identified as assets of cultural interest (BIC) by the government of Aragón (Alcalá, 2006) (Table 7.2). This rich heritage stimulated various initiatives aimed at the local development and popularization of

paleontological sites (Cobos and Alcalá, 2007; Alcalá, 2006, 2011; Cobos, 2011). For this reason, Teruel is a location where the application of an evaluation model is fundamental in order to ensure preservation of sites and determine the priorities for action/intervention in each case.

In this study, we apply the tracksite evaluation model previously outlined by Mampel et al. (2009) in Teruel, to a regional grouping of sites we have named the Dinosaur Track Road in Teruel (Alcalá and Cobos, 2011) (Figure 7.1). Driving about 70 km along a rural road, you can visit up to six villages with almost 30 dinosaur tracksites: 8 in Galve (Maestrazgo UNESCO Global Geopark), 3 in Aguilar del Alfambra, 5 in Ababuj, 4 in Cedrillas, 6 in El Castellar, and 3 in Formiche Alto. All of them belong to the Villar del Arzobispo Formation (Tithonian-Berriasian), except for Ríos Bajos (Higueruelas Formation, Lower Tithonian) in Galve and El Hoyo (Camarillas Formation, Lower Barremian) in El Castellar.

For a specific site, the total value will range between the worst possible evaluation and the best one with the best situation: S$_v$ = 4, C$_v$ = 4, and D$_v$ = 0. The total value of a tracksite

7.3. (A, B) Tracksites of the Dinosaur Track Road that bear the highest (Las Cerradicas, Galve) and (C, D) lowest (CD-2, Cedrillas) total value according to the evaluation method proposed.

could be calculated in general terms as: $T_v = S_v + C_v - D_v$ (simplified from the methodology of Mampel et al., 2009).

Following this evaluation of sites along the Dinosaur Track Road (Figs. 7.1 and 7.2), it is evident that El Castellar, at the site of the same name, and Las Cerradicas (Galve) have the greatest scientific value.

However, these are not necessarily the best sites on the basis of all evaluation criteria. For example, from the viewpoint of cultural value, the sites of El Pozo in El Castellar and Las Cerradicas in Galve are better for their good accessibility. Moreover, the Las Cerradicas site is little affected by the risk of deterioration because it is covered and well-protected thanks to geoconservation efforts adopted since 2003 (Figs.

7.3A, 7.3B, 7.4), whereas El Pozo and El Castellar are threatened by risk of deterioration due to the lack of physical protective measures (Fig. 7.4).

A Second Case Study: The 'Top Listed'
Dinosaur Tracksites in the United States

While the aforementioned Spanish study (FCPTD, 2009) was being conducted and the nominations for the inscription of potential UNESCO WHL sites in Spain and Portugal, Korea, and Bolivia were being prepared and reviewed, an independent evaluation of 11 dinosaur tracksites in the United States was published (Lockley, 2010). This comparative analysis

(Table 7.3) was undertaken using 16 measurable categories consistent, in principle, with those used in these UNESCO proposals.

These 16 categories are similar to those used in the Spanish study, although they were scored on a scale of 1–5, rather than 0–4, and not broken into the same three groups (S_v, C_v, D_v). The method presented here is the same as in Lockley (2010), although the category numbers have been rearranged so as to fall into two groups: quantitative measures (1–8) and nonquantitative (9–16). The important criteria used in this study are summarized in Tables 7.4 and 7.5. The first listed quantitative categories 1–8 (Table 7.4) are as follows:

1. Site size, in square meters
2. Number of tracks
3. Number of trackways
4. Number of holotypes (type ichnospecies)
5. Number of general vertebrate track types
6. Number of track levels
7. Number of science publications
8. Visitation

In contrast to categories 1–8, which are measured in absolute numbers, categories 9–16 (Tables 7.4 and 7.5) as follow require more nuanced categorization:

9. Preservation
10. Historic value
11. Educational value
12. Access
13. Management ownership
14. Protection
15. Other geologic, natural, and/or historic features
16. Nearby sites

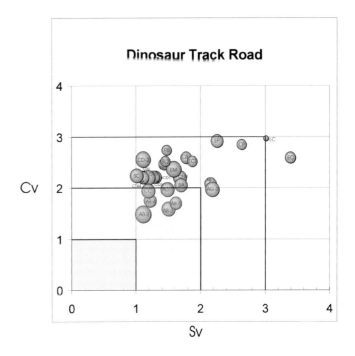

7.4. Evaluation of sites of the Dinosaur Track Road in Teruel according to their scientific values (S_v) versus their cultural values (C_v). Circle radii are proportional to the deterioration risk values, for example, low risk in Las Cerradicas (LC) and high risk in El Pozo (EP), El Castellar (EC) (acronyms of sites identified in Figure 7.2).

As described by Lockley (2010), prior to publication, the results of this comparative analysis were shown to ichnologists and paleontologists familiar with the sites evaluated (see acknowledgments), and in general, there was agreement about the scores, the rankings, and the criteria used. Thus, the results of this analysis indicate that Dinosaur Ridge, Colorado, the St. George Discovery Site, Utah, and the Fossil Trace site, Colorado, rank first, second, and third, respectively. In comparing these higher-scoring sites with the lower scoring locations, we note that the former have higher accessibility

Table 7.3. Eleven dinosaur tracksites from the United States and criteria used in their evaluation

Site Name	C_1	C_2	C_3	C_4	C_5	C_6	C_7	C_8	C_9	C_{10}	C_{11}	C_{12}	C_{13}	C_{14}	C_{15}	C_{16}	Total Score
Dino Ridge, Colorado	2	3	4	2	3	5	5	5	4	5	5	5	—	4	5	5	62
St. George, Utah	2	5	5	0	4	5	3	4	5	3	5	5	—	4	3	4	57
Fossil Trace, Colorado	3	2	3	3	5	5	3	2	2	4	3	4	—	3	4	5	51
Rocky Hill, Connecticut	3	4	2	0	2	1	2	4	5	2	5	5	—	5	2	1	43
Purgatoire, Colorado	4	4	5	1	3	3	3	2	2	4	3	2	—	3	2	2	42
Dino Valley, Texas	2	3	3	1	2	1	3	5	1	5	3	4	—	3	4	2	42
Moab Mega Quarry, Utah	4	4	5	0	1	1	2	2	2	3	2	2	—	2	4	5	36
Peacock Can, New Mexico	3	4	4	2	4	5	2	1	2	3	1	1	—	1	1	1	35
Red Gulch, Wyoming	3	4	5	0	1	1	2	2	2	2	3	3	—	3	1	1	33
Clayton Lake, New Mexico	2	3	2	0	2	1	2	4	2	2	3	3	—	3	1	1	31
Mosquero, New Mexico	2	3	4	0	2	2	2	2	2	2	2	2	—	1	1	1	24

Source: Modified after Lockley (2010)

Note: Categories C_1–C_{16} are listed in Table 7.4 and the scoring scale (1–5) is explained in Table 7.5. Note that no numerical score is possible for category C_{13} (land ownership). The maximum possible score is 75.

Table 7.4. Quantitative (1–8) and nonquantitative (9–16) categories used in comparative analysis of U.S. dinosaur tracksites

1	2	3	4	5
<250 m²	250–999 m²	1000–4999 m²	5000–20000 m²	>20,000 m²
<50	50–249	250–999	1,000–2,499	>2,500
1–25	26–50	51–75	76–100	>100
1	2	3	4	>5
1	2 or 3	4 or 5	6 or 7	>8
1	2 or 3	4 or 5	6 or 7	>8
<2	2–5	6–9	10–19	>20
>1000	1000–4999	5000–24999	25000–99999	>10,000
poor	fair	good	very good	excellent
1	2	3	4	5
1 aspect	2 aspects	3 aspects	4 aspects	>5 aspects
1	2	3	4	5
1	2	3	4	5
1	2	3	4	5
1 aspect	2 aspects	3 aspects	4 aspects	>5 aspects
1–3	4–6	7–9	10–12	>12

Source: Modified after Lockley (2010)

and visitation scores, as well as more scientific publications, whereas the latter are generally less visited, less accessible, and less-extensively documented. It would be instructive to subject the 11 US sites to a comparative analysis using the criteria developed for the analysis of the Spanish sites. Using this evaluation method, the two most outstanding sites along the Dinosaur Track Road in Teruel, Las Cerradicas and El Castellar, give values similar to those of the Purgatoire site in the United States.

In our opinion, it is important to independently evaluate sites on the basis of these different categories (S_v, C_v, D_v) in order to help the decision-making process (Mampel et al., 2009). Thus, a simple summary of all evaluation criteria may be deceptive because a scientifically important site might have its total value reduced, relative to others that lack such scientific importance but nevertheless have high value for visitation or interpretative value.

GOING GLOBAL: THE BIGGEST AND "BEST OF THE BEST"

When the proceedings of the First International Symposium on Dinosaur Tracks and Traces were published (Gillette and Lockley, 1989), the 10 largest dinosaur tracksites were listed, based on the sites with the most reported tracks (Lockley and Gillette, 1989): more than 1000 tracks or 100 trackways. At that time, the number of tracks reported at these sites was only verified by published maps in a small number of cases. Since then, many more large sites have been reliably documented. Many of these were listed in the comparative analyses included in the nominations for WHL inscription presented by Bolivia, Spain and Portugal, and Korea. In the former nomination, a series of tables were presented in which sites were listed in order of decreasing size or decreasing number of tracks and trackways. Other tables listed a small number of additional criteria. However, due to the difficulties of obtaining consistent, high-quality data in multiple (16–25) categories as was done in the case studies we have summarized, when dealing with a disparate and widely distributed global simple of sites, it has not yet proved feasible to assemble reliable data for all the world's more important sites, say the top 30–50, in multiple categories. In this section, therefore, we present some very basic statistics on large sites with large numbers of tracks and trackways based on an update of the tabulations presented in the Korea application (Table 7.6).

Table 7.5. Nonquantitative criteria used in comparative US tracksite study

Criteria	Category 1 (low score)	Category 2
9 Preservation	Poor. Most or all tracks eroded or indistinct. Trackways unclear	Fair. Some clear tracks, others indistinct, eroded. Some trackways clear
10 Historic value	Recently found, one of many similar sites, slight relation to scientific tradition	Site more or less representative. Adds minor information to database
11 Educational value	Site is named, locally known and visited	Site also has signage, trails and guide pamphlets
12 Access	Remote, difficult terrain, access not advertised by owners dangerous for unadventurous	Remote but trail access allowed by owners/managers. Moderate physical challenge
13 Management-ownership	Private owners	Local, municipal
14 Protection	Private land, owner not involved in formal management	Local municipal owners have some protection strategy, signs, fences, etc.
15 Other geological, natural and/or historic features at site	General geologic or natural scenery of contextual significance	Other scientific features (fossils, faults, landforms) already documented
16 Nearby features in 25 km radius	1–3 sites in radius	4–6 sites in radius

Source: Modified after Lockley (2010)
Note: Compare with Tables 7.3 and 7.4

The main purpose of this brief outline is to draw attention to the variable nature of some of the dinosaur tracksites being compared and the different information they yield. We also point to the need to subject this global list to the same type of standardized comparative analysis as demonstrated herein.

Another factor that has a significant impact on comparative analyses pertains to density of tracks. Sites such as Lark Quarry, Australia, and Yangshan, China, have very high densities of small tracks in small areas (Thulborn and Wade, 1984; Matsukawa, Lockley, and Jianjun, 2006). This produces impressive counts of tracks, but the density of tracks makes the identification and counting of trackways difficult. Finally, another factor pertains to the way tracks and trackways are counted. In some cases, multiple sites, reported from multiple levels in small areas, have been included in the analyses. This is the case in many of the Korean sites. In contrast, the Salt Valley Anticline site in Utah (Lockley and Hunt, 1995) is associated with a single surface; this is just one site forming part of a megatracksite complex associated with a single surface that extends for several tens of kilometers. If all the tracksite sizes, tracks, and trackway statistics for mapped areas were combined, or extrapolated, across this large region, the result would be statistics of a different order of magnitude. For example, 22 of the 30 listed sites with available data (Table 7.6) reveal a total documented area of track-bearing surfaces of only about 0.42 km^2. Although a relatively small area, these surfaces have yielded a total of about 44,000 documented tracks, which in turn represent about 2900 individual animals. That latter number is almost certainly an underestimate because high track densities and overlap prevent counting all trackways. Such numbers give a general indication of track abundance as registered on representative surfaces at some of the world's more significant dinosaur tracksites. Until recently, such useful data has been scattered in the literature, and its potential for paleobiology has yet to be more fully explored.

Having stressed the differences evident in comparisons between the sites listed in Table 7.6, we conclude that it would be possible to compile more data in multiple categories for these and other sites, with the help of specialists, and following the method of Mampel et al. (2009), used in the previous sections. Therefore, we present this table only as a work in progress that updates and corrects some of the data presented in recent WHL nominations. Once done, the results would help us more precisely to determine which sites on the global list have the most value in the three categories (scientific, cultural, and deterioration), as well as based on total scores.

A GLOBAL CASE: ASPIRATIONS AND PITFALLS INVOLVED IN COMPARISON OF THE WORLD'S MOST IMPORTANT DINOSAUR TRACKSITES

Although vertebrate ichnologists know that the best dinosaur tracksites represent a unique part of the fossil record, it is necessary to demonstrate what dinosaur ichnites explain about the history of life (of dinosaurs) that we cannot know from other dinosaur evidence. This concern has been voiced by those who evaluate WHL nominations. In short, they are asking why dinosaur tracks are important? For WHL consideration, it is evidently not enough to demonstrate, what ichnologists already know, that tracks sometimes show aspects of behavior that we cannot deduce from skeletal remains. Rather, WHL sites must be of outstanding universal value.

Category 3	Category 4	Category 5 (high score)
Good. Many tracks obvious to nonspecialists. Various clear trackways	Very good. Most tracks visually obvious, clear trackways	Excellent. Almost all tracks and trackways very clear. Some special features (e.g., skin)
Site important to science. More not less representative. Adds much to historic database	Site very important to science. More than representative Adds new info to historic database	One of first found sites, seminal to the science for novelty and significance
Signage, good guidebooks, information on Internet, tours often locally well-known	Site very widely known good signs, visitors center, website and guidebooks heavily toured	Major on site exhibits, visitors center, big tour destination, with all aspects 1–4
Vehicular access to site, by back roads encouraged by owners / managers Vehicular access to site by main roads, near major population centers. Limited food water	Direct access to visitors' facility, with handicap access, food water, public toilets, etc.	
County	State, province	National
County owners' protection strategy approaches level 4 management. Partial cover or shelter	State park (or province) level legal protection and management. Cover or shelter	National park level legal register protection and management. Cover or building
Other published scientific or natural features used as destinations in their own right	Another historic or registered natural feature or destination known in its own right	Multiple historic or registered features standing in their own right
7–9 sites in radius	10–12 sites in radius	>12 sites in radius

Table 7.6. Biggest and best dinosaur tracksites: a list of the 30 largest dinosaur tracksites for which reliable data has been compiled on site size, number of tracks, and trackways

Site	Age	Size (m²)	Tracks / trackways	Diversity	References
Cal Orck'o, Bolivia	Late Cretaceous	65,000	5055 / 465	A, O, S, T, t	Meyer, Hippler, and Lockley (2001)
Fumanya, Spain	Late Cretaceous	64,560	3500 / 30	S	FCPTD (2009); Schulp and Brokx (1999)
Lavini de Marco, Italy	Early Jurassic	58,400	— / 49	O, S, t	Dal Sasso (2001)
Altamura, Italy	Late Cretaceous	40,700	4000 / 00	A, S, T	Dal Sasso (2001)
Khodja-Pil-Ata, Turkmenistan	Late Jurassic	40,000	800 / 32	T	Lockley and Gillette (1989)
Ardley, UK	Middle Jurassic	30,646	— / 42	S, T	Day et al. (2002)
Fatima, Portugal	Middle Jurassic	28,000	1100 / 20	S	Santos et al. (1994); Santos, Moratalla, and Royo-Torres (2009)
Briar, Alaska	Early Cretaceous	25,000	— / 10	S	Pittman and Gillette (1989)
Munchehagen, Germany	Early Cretaceous	18,000	— / 10	O, S, T	Lockley, Wright, and Thies (2004)
Obernkirchen, Germany	Early Cretaceous	—	— / —	O, T	Lubbe, Richter, and Böhme (2009)
Yeosu, Korea	Late Cretaceous	15,500	3853 / 115	O, S, T, t	Lockley et al. (2012)
Goseong, Korea	Early Cretaceous	13,000	5000 / 412	O, S, T	Lockley et al. (2006)
Haenam, Korea	Late Cretaceous	8000	823 / 52	O	Hwang et al. (2008)
Hwasun, Korea	Late Cretaceous	4500	1800 / 73	O, T, t, S	Huh et al. (2006)
Purgatoire, Colorado	Late Jurassic	5600	1300 / 100	T, t, S	Lockley, Houck, and Prince (1986)
Salt Valley Anticline, Utah	Middle Jurassic	8000	2000 / —	T	Lockley and Hunt (1995)
Red Gulch, Wyoming	Middle Jurassic	4000	1000 / 125	t	Breithaupt, Matthews, and Nolan (2004)
St George, Utah	Early Jurassic	—	— / —	O, T	Milner, Lockley, and Johnson (2006)
Toro Toro, Bolivia	Late Cretaceous	2000	500 / 60	A	Leonardi (1984)
Kirmenjak, Croatia	Late Jurassic	4000	~1000 / —	S, T	Pikelj and Mezga (2007)
El Peladillo, Spain	Early Cretaceous	4231	3000 / 348	O, S, T, t	FCPTD (2009)
Lark Quarry, Australia	'Middle' Cretaceous	—	4000 / 500	O, t	Thulborn and Wade (1984)
Dinosaur Ridge, Colorado	'Middle' Cretaceous	540	~400 / 77	O, t	Lockley and Hunt, (1995)
Fossil Trace, Colorado		—	— / 60	Cer, O, T, t	Lockley and Hunt (1995)
Yanguoxia, China	Early Cretaceous	—	286 / 30	O, S, T, t	Li et al. (2006); Xing et al (2013)
Huanglonggou, China	Early Cretaceous	~2000	2000 / —	S, T, t	Li et al. (2012); Lockley et al. (2015)
Chabu, China	Early Cretaceous	—	1000 / —	T	Li et al. (2009)
Qijiang site, China	Early Cretaceous	—	300 / 50?	O, t	Xing, Wand, and Pan (2007)
Yangshan, China	Early Cretaceous	~200	1170 / —	t	Matsukawa, Lockley, and Jianjun (2006)
Peace River, Canada	Early Cretaceous	—	— / —	O, T	Currie (1989)
TOTALS (minimum)	Jurassic and Cretaceous	441,877	43,887 / 2860		31 citations

Note: Track type diversity: A, Ankylosaur; Cer, ceratopsian; O, ornithopod; S, sauropod; T, large theropod; t, small theropod. Nondinosaur tracks are not listed. Note that a total of about 441,877 m² (= 0.442 km²) of track-bearing surface was documented, at 22 sites, to produce a total count of about 44,000 tracks, comprising about 2,900 trackways.

Thus, the question arises, would our understanding of the history of life be incomplete or inadequate if the sites of paleontological significance on the WHL did not include representative tracksites? This is the difficult "universal" question addressed in this section, with reference to world's most significant sites, listed in the previous section.

It is first necessary to consider what significant or outstanding gaps on the WHL the dinosaur ichnites could fill. The "age of dinosaurs" represents a major component of the vertebrate record of life on planet Earth. In fact, it represents one-third (one of only three eras) covering the evolution of vertebrate life. Of the 12 WHL sites representing the Phanerozoic, only two partially represent the terrestrial age of dinosaurs, one focused on the very beginning (Ischigualasto/Talampaya Natural Parks in Argentina) and the other on the very end (Dinosaur Provincial Park in Canada). Thus, there is a large gap. These two dinosaur WHL sites also deal primarily with a specific type of evidence (i.e., bones). Modern developments in the science show that other areas of study (especially tracksites, eggs, and nest sites) add essential knowledge that supplements the bone record.

Despite this scientific importance and social interest, there are only three properties relating to dinosaurs in the WHL: Ischigualasto/Talampaya, Dinosaur Provincial Park, and the Dorset and East Devon Coast in United Kingdom (whc.unesco.org; FCPTD, 2009). None of these feature many tracks or tracksites. By contrast, the aforementioned Iberian, Bolivian, and Korean nominations for the WHL are among the world's most significant from an ichnological viewpoint.

The first of those three sites to be inscribed on the WHL (Dinosaur Provincial Park) is mostly known for abundant Upper Cretaceous dinosaur skeletons representing 35 different species. Surprisingly, despite the large number of skeletons there are hardly any dinosaur tracks. At Ischigualasto/Talampaya there is a complete sequence of fossiliferous continental sediments representing the entire Triassic period (45 million years). There are a few (about five) small sites with archosaur prints, but they have not been thoroughly studied since 1931, nor are they considered very important for the global track record of dinosaurs. Several isolated tracks or small outcrops have been documented along the Dorset and East Devon Coast, a location known primarily for its geological and stratigraphic importance from both a scientific and historic point of view.

So we can argue that dinosaur tracks are very poorly represented at sites on the current WHL. Does this mean that the natural heritage properties on the WHL have a serious omission? And does this mean suitable tracksites should be added? We argue "yes!" Both bones and ichnites are crucial and complementary when deciphering dinosaur evolution and behavior: see Mampel et al. (2009) for discussion of the scientific value of both types of dinosaur site evidence.

Four criteria are very significant in establishing the importance of ichnites:

1. Ichnites fossilize in rocks that form part of a specific sedimentary environment, always providing in situ information about the original ecosystem. On the contrary, dinosaur bones have usually been transported from their original habitats before, during, or after the fossilization processes acted.
2. Ichnites are normally preserved at in situ open-air sites, thus they need exceptional preservation measures as part of the natural environment (on the contrary, bones are usually well curated in city museums). In other words, tracksites are ancient landscapes that must be integrated with the modern landscape. Moreover, tracksites encourage local and regional development in rural areas (e.g., Marty et al., 2004).
3. It is also well established that dinosaur tracks are extraordinarily abundant in certain deposits, and moreover, in many cases they occur in deposits where skeletal remains are rarely if ever found. Such occurrences fill important gaps in the record of life.
4. Lastly, in a very concrete sense, tracks are a record of life in the dynamic sense that they represent living, rather than dead animals. This record of life aspect is stressed in WHL criteria pertaining to

paleontological sites. Any fossil is a record of life, but while bones represent life indirectly, after death, tracks represent life more directly and immediately. They are a record of life in two senses. Thus, eggs and nests are a record of "life before life" (or early in the life cycle). Tracks are records of 'life during life,' including the building of nests (a type of trace fossil) by parents during life, whereas bones and skeletons are a record of 'life after death.' These three categories are complementary and necessary to represent the dinosaur life cycle in a full and organic way.

So, now that we have seen the methods and criteria developed to compare different sites in Spain and the United States, we can attempt to compare the other globally important sites that stand out for the wealth of paleoichnological evidence available. As noted, the question arises as to how many criteria we might use in a realistic global comparative analysis. Again the most obvious criteria appear to be the number of tracks and trackways, and the diversity of track types. Although these numbers are in part proportional to track density, generally the size of outcrops is the most important factor, so in general larger sites (Table 7.6) provide more information (see Lockley et al., 2014, for a recent example). However, for various reasons, different sites may reveal quite different paleoenvironmental information, depending on history of study and their contextual relationship to other sites.

Here again the concept of a geosite (Wimbledon, 1996) is important because it is not only the global significance (ranking) of the site in terms of size, number of tracks, and track types that is significant or of "value." Related contextual information on the geological and paleontological environment is also very important. It is for this reason we ask to what extent many of the large sites listed herein offer more than an impressive set of statistics. One question we wish to answer is whether some of these ichnological sites fill a gap in our understanding of dinosaurs that helps us better understand the Mesozoic record of life. So when it comes to evaluating an important site, say one with 1000 or more footprints, we should go beyond simple quantitative lists and show the tracks in the fullest geological context possible.

The IUGS Geosites Programme, with the support of UNESCO, is aiming to avoid descriptions of isolated sites evaluated on ad hoc criteria and instead is proposing to compile lists of similar geological sites that can be used as a comparative standard or guideline in order to identify candidates best suited for the WHL. This again raises the question of what sites are most representative and necessary to include on the WHL in order to understand the history of life, and specifically that of the Mesozoic age of dinosaur. Would the tentative list be complete, balanced, and representative if all proposed sites were adopted?

Perhaps we cannot answer this definitively because new evidence may come to light at any time that would revolutionize our understanding of the age of dinosaurs. However, we can say that tracks have already revolutionized our understanding of the age of dinosaurs, and that a number of sites around the world have become as famous and representative of the age of dinosaurs as the better-known bone sites. Therefore, one or more of the most scientifically important and informative sites should be on the WHL as representative of the record of life during the Mesozoic era.

It is encouraging that a number of predominantly ichnological sites, especially the Spanish-Portuguese and Korean properties, have been nominated for the WHL. However, there is still a certain prejudice that tracks are less important than bones, that tracks are "only" trace fossils rather than real body fossils. However, the argument that tracks represents the living, rather than the dead animals, is also important to balance otherwise biased perceptions. A less polarizing position is that tracks and bones are complementary and that the WHL would benefit from including sites representative of both categories of evidence. At the beginning of this section, we gave four cogent reasons for considering ichnites an important category of evidence, without which the dinosaur record in the Mesozoic would be incomplete.

It remains for ichnologists to refine the site evaluation criteria developed for the Spanish-Portuguese, Korean, and American sites, into a single standardized methodology and apply it globally. The list of sites given in Table 7.6 is a useful starting point.

CONCLUSIONS

Dinosaur ichnology has made rapid advances in recent years as literally thousands of sites have been documented. During this time, many large sites have been identified, each with more than 1000 footprints (or more than 100 trackways). In the last decade, representative tracksites have been proposed for the WHL. As these nominations have been submitted and reviewed, various methods of comparative analysis have been developed and the world's largest and most significant sites have been identified. As new sites have been discovered or uncovered, this list has grown. The 30 top-listed sites represent a very significant ichnological database that is only the tip of an iceberg comprising thousands of smaller and perhaps less spectacular sites. However, some of these lesser sites may score highly in certain categories and be important for specific reasons.

Although workers evaluating these sites have generally agreed on the main criteria used, the two case studies presented here pertain to different countries (continents), and we are still working toward a completely standardized methodology that can be applied globally. We nevertheless recognized that evaluation criteria fall into certain definite categories (scientific, cultural, and deterioration). Moreover, we argue that the Spanish study comes closest to a standardized methodology, based on a comprehensive and balanced list of variables (25 as compared with 16 used in the US analysis). This methodology is efficient and relatively easy to apply. It does not necessarily require the input of specialized ichnologists, assuming published data are available, and/or the evaluators have direct access to the sites evaluated. Once results have been obtained, more objective decisions can be made regarding the best use or development of the sites and the best strategies to develop the sites proposing this method as an integrative quantitative approach to evaluate the dinosaur track record in the future.

ACKNOWLEDGMENTS

This contribution is part of the Departamento de Educación, Universidad, Cultura y Deporte and Departamento de Industria e Innovación (Aragón Government and Fondo Social Europeo); Spanish R&D projects CGL2009-07792 DINOSARAGÓN and CGL2013-41295-P DINOTUR (Ministerio de Economía y Competitividad, Spanish Government); Research Group E-62 FOCONTUR project and Instituto Aragonés de Fomento. Spanish and Portuguese heritage national authorities and Spanish Communities of Aragón, Asturias, Castilla y León, Cataluña, La Rioja and Valencia heritage authorities entrusted some of us with the scientific evaluation of the Dinosaur Ichnites of the Iberian Peninsula WHL candidacy dealing with tracksites (FCPTD, 2009). A draft of the US study (Lockley, 2010) was shown to selected specialists with knowledge of specific sites. In this regard, we thank James Farlow (Indiana University), Andrew Milner (St. George Dinosaur Discovery site), Bruce Schumacher (US Forest Service), and Annette Richter (Niedersächsisches Landesmuseum Hannover) for their comments on the methodology and results obtained.

Alcalá, L. 2002. Los yacimientos de vertebrados fósiles de la fosa de Teruel; pp. 227–242 in G. Meléndez and E. Peñalver (eds.), El patrimonio paleontológico de Teruel, Instituto de Estudios Turolenses, Teruel, Spain.

Alcalá, L. 2006. Los yacimientos turolenses de icnitas de dinosaurios y el Conjunto Paleontológico Dinópolis; pp. 265–285 in F. Torcida Fernández-Baldor (ed.), Huellas que perduran. Icnitas de dinosaurios: patrimonio y recurso. Fundación del Patrimonio Histórico de Castila y León, Valladolid, Spain.

Alcalá, L. 2011. Territorio Dinópolis. Her&Mus 3(3): 96–106.

Alcalá, L. 2012. The IDPI (Icnitas de Dinosaurios de la Península Ibérica) WHL candidacy in perspective; pp. 255–258 in M. Huh, H. J. Kim, and J. Y. Park (eds.), The 11th Symposium on Mesozoic Terrestrial Systems: Biota and Ecosystem, and Their Global Correlation, Kimdaejung Convention Center, Gwangju City, South Korea, August 15–18, 2012.Abstract book. Korea Dinosaur Research Center, Chonnam National University, Gwangju, Korea.

Alcalá, L., and A. Cobos. 2011. The dinosaur track road in Teruel (Spain); p. 13 in A. Richter and M. Reich (eds.), Dinosaur Track Symposium 2011. Abstracts Volume and Field Guide to Excursions, Obernkirchen. Universitätsverlag Göttingen, Göttingen, Germany.

Alcalá, L., and C. Martín. 1995. Huellas de dinosaurio en el Jurásico Superior de Ababuj (Teruel). Geogaceta 17: 19–22.

Alcalá, L., A. Cobos, and R. Royo-Torres. 2003. Icnitas de dinosaurio en El Castellar (Teruel); pp. 28–29 in M. V. Pardo Alonso and R. Gozalo (eds.), Libro de Resúmenes XIX Jornadas de la Sociedad Española de Paleontología, Morella.

Alcalá, L., L. Mampel, R. Royo-Torres, and A. Cobos. 2014. On small quadrupedal ornithopod tracks in Jurassic-Cretaceous transition intertidal deposits (El Castellar, Teruel, Spain). Spanish Journal of Palaeontology 29(2): 183–190.

Alcalá, L., A. Cobos, M. Huh, M. Lockley, and R. Royo-Torres. 2012. Comparing bones and traces: evaluating scientific value of the dinosaur record; pp. 263–266 in M. Huh, H. J. Kim, and J. Y. Park (eds.), The 11th Symposium on Mesozoic Terrestrial Systems: Biota and Ecosystem, and Their Global Correlation, Kimdaejung Convention Center, Gwangju City, South Korea, August 15–18, 2012.Abstract book. Korea Dinosaur Research Center, Chonnam National University, Gwangju, Korea.

Alcalá, L., F. Pérez-Lorente, L. Luque, A. Cobos, R. Royo-Torres, and L. Mampel. 2014. Preservation of dinosaur footprints in shallow intertidal deposits of the Jurassic-Cretaceous transition in the Iberian Range (Teruel, Spain). Ichnos 21 (1): 19–31.

Alcalá, L., A. Cobos, E. Espílez, F. Gascó, L. Mampel, C. Martín Escorza, and R. Royo-Torres. 2012. Icnitas de dinosaurios en la Formación Villar del Arzobispo de Ababuj (Teruel, España). Geogaceta 51: 31–34.

Bird, R. T. 1985. Bones for Barnum Brown: Adventures of a dinosaur hunter. Texas Christian University Press, Fort Worth, Texas.

Blanco, M., S. Caro, F. Pérez-Lorente, L. E. Requeta, and M. Romero. 2000. Un nuevo yacimiento jurásico con icnitas saurópodas en la Cordillera Ibérica. Geogaceta 28: 23–26.

Breithaupt, B. H., N. A. Matthews, and T. A. Nolan. 2004. An integrated approach to three dimensional data collection at dinosaur tracksites in the Rocky Mountain West. Ichnos 11: 11–26.

Canudo, J. I., S. Alberto, J. L. Barco, R. Royo-Torres, and J. I. Ruiz-Omeñaca. 2000. La dinoturbación de Ariño. Cauce 4: 9–15.

Castanera, D., J. L. Barco, I. Díaz-Martínez, J. Herrero-Gascón, F. Pérez-Lorente, and J. I. Canudo. 2011. New evidence of a herd of titanosauriform sauropods from the Lower Berriasian of the Iberian range (Spain). Palaeogeography, Palaeoclimatology, Palaeoecology 310: 227–237.

Castanera, D., B. Vila, N. L. Razzolini, P. L. Falkingham, J. I. Canudo, P. I. Manning, and À. Galobart. 2013. Manus track preservation bias as a key factor for assessing trackmaker identity and quadrupedalism in basal Ornithopods. PLoS One 8(1): e54177. doi:10.1371/journal. pone.0054177.

Cobos, A. 2004. Valoración patrimonial de los yacimientos de icnitas de dinosaurio de la provincial de Teruel. Geogaceta 36: 191–19.

Cobos. A. 2011. Los dinosaurios de Teruel como recurso para el desarrollo territorial. Ph.D. dissertation, Universidad del País Vasco, Bilbao, Spain, 584 pp.

Cobos, A., and L. Alcalá. 2007. Los itinerarios de icnitas de El Castellar. Prames-Departamento de Educación, Cultura y Deporte, Gobierno de Aragón, Zaragoza, Spain, 23 pp.

Cobos, A., M. G. Lockley, F. Gascó, R. Royo-Torres, and L. Alcalá. 2014. Megatheropods as apex predators in the typically Jurassic ecosystems of the Villar del Arzobispo Formation (Iberian Range, Spain). Palaeogeography, Palaeoclimatology, Palaeoecology 399: 31–41.

Cobos, A., R. Royo-Torres, L. Luque, L. Alcalá, and L. Mampel. 2010. An Iberian stegosaurs paradise: the Villar del Arzobispo Formation (Tithonian-Berriasian) in Teruel (Spain). Palaeogeography, Palaeoclimatology, Palaeoecology 293: 223–236.

Cuenca, G., R. Ezquerra, F. Pérez, and A. R. Soria. 1993. Las huellas de dinosaurios (icnitas) de los Corrales del Pelejón. Departamento de Cultura y Educación, Gobierno de Aragón, Zaragoza, Spain, 14 pp.

Currie, P. J. 1989. Dinosaur tracksites of western Canada; pp. 293–300 in D. D. Gillette and M. G. Lockley (eds.), Dinosaur Tracks and Traces. Cambridge University Press, Cambridge, UK.

Dal Sasso, C. 2001. Update on Italian dinosaurs; p. 27 in 6th European Workshop on Vertebrate Palaeontology, Florence, September 19–22. Abstract volume. Università degli Studi di Firenze, Florence, Italy.

Day, J. J., P. Upchurch, D. Norman, A. S. Gale, and H. P. Powell. 2002. Sauropod trackways, evolution, and behavior. Science 296: 1659.

FCPTD (Fundación Conjunto Paleontológico de Teruel-Dinópolis) (coord). 2009. Dinosaur Ichnites of the Iberian Peninsula. World Heritage Candidacy. Ministerio de Cultura de España, Spain; and Ministerio do Ambiente, do Ordenamento do Território e do Desenvolvimiento Regional, Portugal.

Gillette, D. D., and M. G. Lockley. 1989. Dinosaur Tracks and Traces. Cambridge University Press, Cambridge, UK, 480 pp.

Gray, M. 2004. Geodiversity: Valuing and Conserving Abiotic Nature. Wiley, Chichester, UK, 448 pp.

Huh, M., M. Lockley, L. Alcalá, A. Cobos, and C. Meyer. 2012. The Global Network on Dinosaur Tracksites (GNDT) concept: comparing major tracksites around the world; pp. 259–262 in M. Huh, H. J. Kim, and J. Y. Park (eds.), The 11th Symposium on Mesozoic Terrestrial Systems: Biota and Ecosystem, and Their Global Correlation, Kimdaejung Convention Center, Gwangju City, South Korea, August 15–18, 2012.Abstract book. Korea Dinosaur Research Center, Chonnam National University, Gwangju, Korea.

Huh, M., S. L. Paik, M. G. Lockley, K. G. Hwang, and S. K. Kwak. 2006. Well-preserved theropod tracks from the Upper Cretaceous of Hwasun County, Southwestern Korea and their paleobiological implications. Cretaceous Research 27: 123–138.

Hwang, K. G., M. G. Lockley, M. Huh, and I. S. Paik. 2008. A reinterpretation of dinosaur footprints with internal ridges from Cretaceous Uhangri Formation, Korea. Paleogeography, Paleoclimatology, Paleoecology 258: 59–70.

Leonardi, G. 1984. Le Impronti Fossili di Dinosauri; pp. 165–186 in Sulle ormi di dinosauri. Editio Editrice, Venice, Italy.

Leonardi, G. 1994. Annotated Atlas of South American Footprints (Devonian-Recent). Companhia de Pesquisa de Recursos Minerais, Brasilia, Brazil, 247 pp.

Li, D., Y. Azuma, M. Fujita, Y. N. Lee, and Y. Arakawa. 2006. A preliminary report on two new vertebrate tracksites including dinosaurs from the Early Cretaceous Hekou Group, Gansu Province, China. Journal of the Paleontological Society of Korea 52: 29–49.

Li, J., M. G. Lockley, Y. Zhang, M. Matsukawa, S. Hu, and Z. Bai. 2012. An important Ornithischian tracksite in the Early Jurassic of the Shenmu Region, Shaanxi, China. Acta Geologica Sinica 86: 1–10.

Li, J., M. Lockley, Z. Bai, L. Zhang, Q. Wei, Y. Ding, M. Matsukawa, and K. Hayashi. 2009. New bird and small theropod tracks from the Lower Cretaceous of Otog Qi, Inner Mongolia, P. R. China. Memoirs of the Beijing Museum of Natural History 61: 51–79.

Lockley, M. G. 1991. The dinosaur footprint renaissance. Modern Geology 16: 139–160.

Lockley, M. G. 2005a. Loosing track or keeping track: the challenges of compiling a fossil footprint database for the western U.S.A.; p. 5 in 57th Annual Meeting Rocky Mountain Section Abstracts with Programs. Geological Society of America, Boulder, Colorado.

Lockley, M. G. 2005b. The vertebrate ichnology database and landscape: a story of two decades of rapid global growth of collections and scientific education; pp. 46–47 in J. Le Loeuff (ed.), International Symposium on Dinosaurs and Other Vertebrates Paleoichnology. Fumanya-St. Corneliu, Cercs.

Lockley, M. G. 2009. Some comparisons between dinosaur-dominated footprint assemblages in North America and Europe; pp. 121–130 in P. Huerta Hurtado and F. Torcida Fernández-Baldor (eds.), Actas IV Jornadas Internacionales sobre Paleontología de Dinosaurios y su entorno. Colectivo Arqueológico y Paleontológico de Salas, Salas de los Infantes, Burgos, Spain.

Lockley, M. G. 2010. Dinosaur Ridge ranks tops among America's largest dinosaur tracksites! Dinosaur Ridge Report 22(3): 16–19.

Lockley, M. G., and A. P. Hunt. 1994. A review of vertebrate ichnofaunas of the western interior United States: evidence and implications; pp. 95–108 in M. V. Caputo, J. A. Peterson, and K. J. Franczyk (eds.), Mesozoic Systems of the Rocky Mountain Region, United States. The Rocky Mountain Section, SEPM (Society for Sedimentary Geology), Denver, Colorado.

Lockley, M. G., and A. P. Hunt. 1995. Dinosaur Tracks and Other Fossil Footprints of the Western United States. Columbia University Press, New York, New York, 338 pp.

Lockley, M. G., and D. D. Gillette. 1989. Dinosaur tracks and traces: an overview; pp. 3–10 in D. D. Gillette and M. G. Lockley (eds.), Dinosaur Tracks and Traces. Cambridge University Press, Cambridge, UK.

Lockley, M. G., K. Houck, and N. K. Prince. 1986. North America's largest dinosaur tracksite: implications for Morrison Formation paleoecology. Geological Society of America Bulletin 97(10): 1163–1176.

Lockley, M. G., J. L. Wright, and D. Thies. 2004. Some observations on the dinosaur tracks at Münchehagen (Lower Cretaceous), Germany. Ichnos 11: 261–274.

Lockley, M. G., J. Holbrook, R. Kukihara, and M. Matsukawa. 2006. An ankylosaur-dominated dinosaur tracksite in the Cretaceous Dakota Group of Colorado and its paleoenvironmental and sequence stratigraphic context. Bulletin New Mexico Museum of Natural History and Science 35: 95–104.

Lockley, M. G., L. Xing, J. Y. Kim, and M. Matsukawa. 2014 Tracking Early Cretaceous dinosaurs in China: a new database for comparison with ichnofaunal data from Korea, the Americas and Europe. Biological Journal of the Linnean Society 113: 770–789.

Lockley, M. G., M. Huh, J. Y. Kim, J. D. Lim, and K. S. Kim. 2012. Recent Advances in Korean Vertebrate ichnology: the KCDC comes of age. Ichnos 19: 1–5.

Lockley, M. G., R. Li, M. Matsukawa, L. Xing, J. Li, M. Liu, and X. Xing. 2015. Tracking the yellow dragons' implications of China's largest dinosaur tracksite (Cretaceous of the Zhucheng area, Shandong Province, China). Palaeogeography, Palaeoclimatology, Palaeoecology 423: 62–79.

Lubbe, van der T., A. Richter, and A. Böhme. 2009. Velociraptor's sisters: first report of troodontid tracks from the Lower Cretaceous of northern Germany. Journal of Vertebrate Paleontology 29(3): 194A–195A.

Mampel, L., A. Cobos, L. Alcalá, L. Luque, and R. Royo-Torres. 2009. An integrated system of heritage management applied to dinosaur sites in Teruel (Aragón, Spain). Geoheritage 1(2–4): 53–73.

Marty, D., L. Cavin, W. A. Hug, P. Jordan, M. G. Lockley, and C. A. Meyer. 2004. The protection, conservation and sustainable use of the Courtedoux dinosaur tracksite, Canton Jura, Switzerland. Revue de Paléobiologie Special Volume 9: 39–49.

Matsukawa, M., M. G. Lockley, and L. Jianjun. 2006. Cretaceous terrestrial biotas of East Asia, with special reference to dinosaur-dominated ichnofaunas: towards a synthesis. Cretaceous Research 27: 3–21.

McCrea, R., M. G. Lockley, and C. A. Meyer. 2001. Global distribution of purported Ankylosaur track occurrences; pp. 413–454 in K. Carpenter (ed.), The Armored Dinosaurs. Indiana University Press, Bloomington, Indiana.

McCrea, R. T., L. G. Buckley, A. G. Plint, P. J. Currie, C. W. Helm, J. W. Haggart, and S. G. Pemberton. 2103. A review of vertebrate track-bearing formations from the Mesozoic and earliest Cenozoic of western Canada with a description of a new theropod ichnospecies and reassignment of an avian ichnogenus. Bulletin New Mexico Museum of Natural History and Science 62: 5–93.

Meléndez, A., R. Ezquerro, E. Zurita, and F. Pérez-Lorente. 2002. El yacimiento de huellas de Abenfigo (Teruel, España); p. 34 in F. Pérez-Lorente (ed.), Resúmenes Congreso Internacional de Dinosaurios y otros Reptiles Mesozoicos de España. Instituto de Estudios Riojanos, Logroño, Spain; and Universidad La Rioja, Logroño, Spain.

Meyer, C. A., D. Hippler, and M. G. Lockley. 2001. The Late Cretaceous vertebrate ichnofacies of Bolivia: facts and implications; pp. 133–139 in VII International Symposium on Mesozoic Terrestrial Ecosystems. Publicación Especial. Asociación Paleontológica Argentina, Buenos Aires, Argentina.

Meyer, C. A., B. Thüring, L. Alcalá, M. Huh, M. G. Lockley, and A. Cobos. 2008. The Global Network on Dinosaur Tracksites: towards a strategy of protecting important vertebrate tracksites; p. 81 in Ichnia, The Second International Congress on Ichnology, Cracow, Abstract book. Asociación Paleontológica Argentina, Buenos Aires, Argentina.

Milner, A. R. C., M. G. Lockley, and S. Johnson. 2006. The Story of the St. George Dinosaur Discovery Site at Johnson Farms: an important new Lower Jurassic dinosaur tracksite in the Moenave Formation of southwestern Utah. New Mexico Museum of Natural History and Science Bulletin 37: 329–345.

Pérez-Lorente, F., and F. Ortega. 2003. Dos nuevos rastros de arcosaurios en el Cretácico inferior de Galve (Teruel, España). El yacimiento del Cantalar; pp. 129–136 in F. Pérez-Lorente (ed.), Dinosaurios y otros reptiles mesozoicos de España. Instituto de Estudios Riojanos 26. Instituto de Estudios Riojanos, La Rioja, Spain.

Pérez-Lorente, F., and M. M. Romero-Molina. 2001. Nuevas icnitas de dinosaurios terópodos y saurópodos en Galve y Miravete de la Sierra (Teruel, España). Geogaceta 30: 111–114.

Pérez-Lorente, F., G. Cuenca-Bescós, M. Aurell, J. I. Canudo, A. R. Soria, and J. I. Ruiz-Omeñaca. 1997. Las Cerradicas tracksite (Berriasian, Galve, Spain): growing evidence for quadrupedal ornithopods. Ichnos 5: 109–120.

Pikelj, K., and A. Mezga. 2007. Origin and composition of Upper Tithonian marley layers from Kirmenjak Quarry (Istria, Croatia); p. 289 in A. Zeililids (ed.), 25th IAS Meeting of sedimentology. Abstract book.

Pittman, J., and D. D. Gillette. 1989. The Briar site: a new sauropod dinosaur tracksite in Lower Cretaceous beds of Arkansas; pp. 313–332 in D. D. Gillette and M. G. Lockley (eds.), Dinosaur Tracks and Traces. Cambridge University Press, Cambridge, UK.

Royo-Torres, R., L. Mampel, and L. Alcalá. 2013. Icnitas de dinosaurios del yacimiento San Cristóbal 3 de la Formación Camarillas en Galve (Teruel, España). Geogaceta 53: 5–8.

Royo-Torres, R., A. Cobos, L. Luque, A. Aberasturi, E. Espílez, I. Fierro, A. González, L. Mampel, and L. Alcalá. 2009. High European sauropod dinosaur diversity during Jurassic–Cretaceous transition in Riodeva (Teruel, Spain). Palaeontology 52: 1009–1027.

Santos, V. F., J. J. Moratalla, and R. Royo-Torres. 2009. New sauropod trackways from the Middle Jurassic of Portugal. Acta Paleontologica Polonica 54(3): 409–422.

Santos, V. F. dos, M. G. Lockley, C. A. Meyer, J. Carvalho, A. M. Galopim de Carvalho, and J. J. Moratalla. 1994. A new sauropod tracksite from the Middle Jurassic of Portugal. Gaia 10: 5–14.

Schulp, A. S., and W. A. Brokx. 1999. Maastrichtian sauropod footprints from the Fumanya site, Bergedà, Spain. Ichnos 6: 239–250.

Thulborn, R. A., and M. Wade. 1984. Dinosaur trackways in the Winton Formation (mid-Cretaceous) of Queensland. Memoirs Queensland Museum 21: 413–517.

Weishampel D. B., P. Dodson, and H. Osmólska. 2004. The Dinosauria. 2nd edition. University of California Press, Berkeley, California.

Wimbledon, W. A. P. 1996. GEOSITES – a new IUGS initiative to compile a global comparative site inventory, an aid to international and national conservation activity. Episodes 19: 87–88.

Xing, L. D., F. P. Wang, and S. G. Pan. 2007. The discovery of dinosaur footprints from the Middle Cretaceous Jiaguan Formation of Qijiang County, Chongqing City. Acta Geologica Sinica (Chinese edition) 81: 1591–1602.

Xing, L., D. Li, J. D. Harris, P. R. Bell, Y. Azuma, M. Fujita, Y. N. Lee, and P. J. Currie. 2013. A new Dromaeosauripus (Dinosauria: Theropoda) ichnospecies from the Lower Cretaceous Hekou Group, Gansu Province, China. Acta Palaeontologica Polonica 58: 723–730.

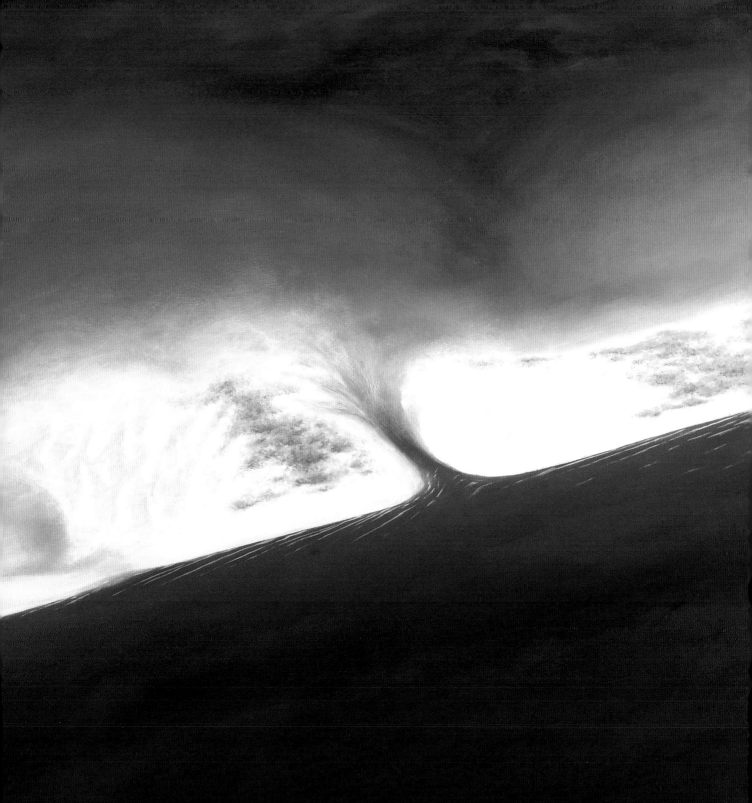

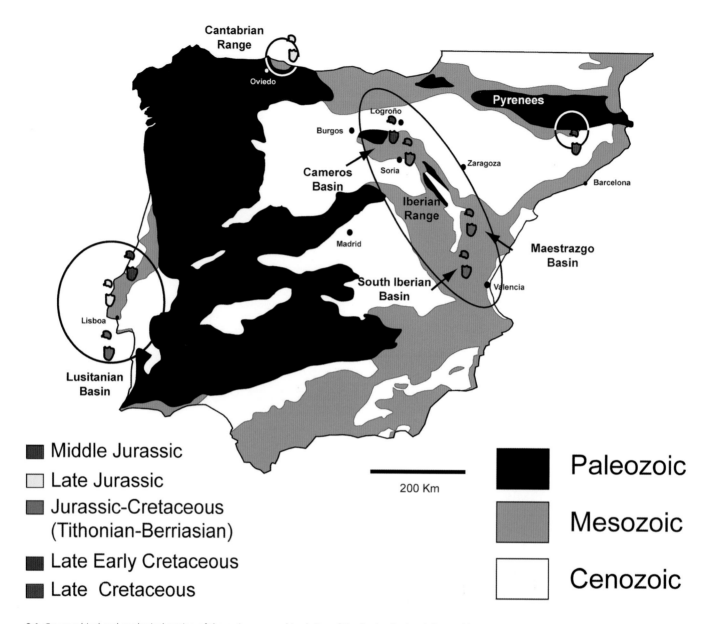

8.1. Geographical and geological setting of the main sauropod tracksites of the Iberian Peninsula located in four broad areas: Lusitanian Basin, Cantabrian Range, Iberian Range, and the Pyrenees.

Iberian Sauropod Tracks through Time: Variations in Sauropod Manus and Pes Track Morphologies

8

Diego Castanera, Vanda F. Santos, Laura Piñuela, Carlos Pascual, Bernat Vila, José I. Canudo, and José Joaquin Moratalla

THE IBERIAN SAUROPOD TRACK RECORD HAS YIELDED more than 100 sauropod tracksites ranging in age from the Middle Jurassic (Bathonian) to the Late Cretaceous (Maastrichtian). During this wide range of time, four different types of manus prints can be differentiated, changing in morphology from (1) speech-bubble–shaped with a prominent claw mark in digit I (Middle Jurassic), (2) kidney-shaped with a claw mark in digit I or (3) without a claw mark in digit I (Late Jurassic and Early Cretaceous), to (4) horseshoe-shaped (Cretaceous). Pes prints are slightly more conservative in morphology through the Mesozoic and are generally subtriangular. They can mainly be differentiated on the basis of the number and orientation of the claw marks, although the presence of a lateral notch behind digit V and the heel can be useful as well. There seems to be a lateralization of the claw marks after the Middle Jurassic, where the pes have four claw marks, two of them oriented anteriorly and two laterally. Subsequently, pes prints have three (Late Jurassic–Early Cretaceous) or four (Late Cretaceous) claw marks oriented anterolaterally and decreasing in size. The variation in the manus and pes morphology in the Iberian sauropod tracks is a reflection of the changes in the sauropod faunas over time. The different types of manus prints suggest that the forelimbs should play a major role in sauropod ichnotaxonomy.

INTRODUCTION

The Iberian sauropod track record ranges from the Middle Jurassic (Bathonian) to the Late Cretaceous (Maastrichtian). There are more than 100 tracksites where sauropod footprints have been described (see Pérez-Lorente, 2003; Hernández-Medrano et al., 2008; Royo-Torres, 2009; Castanera et al., 2014). The quantity of data and the wide range of time can provide a significant overview and allow us to improve our knowledge about the evolution of sauropod manus and pes print morphologies throughout an entire geographic area such as the Iberian Peninsula. Despite the fact that there are a great number of tracksites, not many of them preserve tracks showing clear anatomical details, some tracks being just oval to rounded marks. In the global record, there is a

similar scenario, with few tracks (either manus or pes) preserving anatomical details, especially claw marks (Wright, 2005; Hall, Fragomeni, and Fowler, 2016). The consequence is that as yet there are few sauropod ichnotaxa considered valid, due to the absence of clear anatomical features (Lockley, Farlow, and Meyer, 1994; Wright, 2005; Marty et al., 2010; Kim and Lockley, 2012). Since the earliest papers on the subject, sauropod tracks have been classified on the basis of the trackway gauge type (Farlow, 1992; Lockley, Farlow, and Meyer, 1994). Nonetheless, in the last few years, some papers have questioned the utility of this parameter in sauropod ichnotaxonomy (Wright, 2005; Moratalla, 2009; Santos, Moratalla, and Royo-Torres, 2009; Castanera et al., 2012). Moreover, the description of some quadrupedal wide-gauge trackways as being stegosaur in origin has further increased the complexity of the task (Cobos et al., 2010).

The aims of this chapter are to provide an overview of the best-preserved Iberian sauropod tracks showing anatomical features and put them in chronological and stratigraphical order to show the manus and pes evolution of this group of dinosaurs on the basis of their tracks. Moreover, we draw comparisons between the footprints and the skeletal record. Finally, we discuss the variations in the manus and pes morphologies and their implications in sauropod ichnotaxonomy.

GEOGRAPHICAL AND GEOLOGICAL SETTING

The Iberian sauropod tracksites are located in four broad geographic areas (Fig. 8.1): the Lusitanian Basin (Santarém and Setúbal districts, Central Portugal) in Portugal, and the Cantabrian Range (province of Asturias), the Iberian Range (provinces of Burgos, La Rioja, Soria, Teruel and Valencia), and the Pyrenees (province of Barcelona) in Spain.

Lusitanian Basin (Portugal)

The main Portuguese sauropod tracksites are Jurassic in age and are preserved in shallow marine limestones. Outstanding among them and classified as a natural monument is the Monumento Natural das Pegadas de Dinossáurios da Serra de Aire (MNDPDSA, also known as Galinha tracksite, lower Bathonian, Serra de Aire Formation) at Ourém–Torres Novas, as well as the Pedra da Mua tracksite (Tithonian, Espichel unit–?Lourinhã Formation) in the Sesimbra region, which contains the best-preserved sauropod footprints that are preserved as true tracks (Lockley, Meyer, and Santos, 1994; Meyer et al., 1994; Santos et al., 1994; Santos, Moratalla, and Royo-Torres, 2009). The Galinha track level was deposited in lacustrine, paralic, and very shallow, restricted marine conditions (Santos et al., 1994; Azeredo, 2007; Santos, Moratalla, and Royo-Torres, 2009), whereas the Pedra da Mua tracksite was deposited in shallow marine conditions (Lockley, Farlow, and Meyer, 1994; Santos, 2003). Furthermore, isolated sauropod casts in the Lourinhã area and the locality of Porto Dinheiro (Lourinhã Formation, Kimmeridgian-Tithonian) have been recovered as well. In these cases, the tracks have been preserved as sandstone casts in fluvial environments (Milàn, Christiansen, and Mateus, 2005; Mateus and Milàn, 2010).

Cantabrian Range (Asturias)

The tracksites of the Cantabrian Range are located in the cliffs of the Asturian Coast, where impressive outcrops of Late Jurassic (Kimmeridgian) materials are exposed. The best-preserved sauropod tracks are from the Lastres Formation, which was deposited in a fluvial-dominated deltaic system (García-Ramos, Piñuela, and Lires, 2006; Lockley et al., 2008). The tracks are preserved as sandstone casts; some of them were found isolated (MUJA-1899 and MUJA-1896 [housed in Museo del Jurásico de Asturias]) and others as one trackway that preserved pes (JVLCS-261) and manus (JVLCS-262). (JVLCS is the Jurásico, Municipio de Villaviciosa, Formación Lastres, Cuadrúpedo, Suelto, which is translated as "Jurassic, municipality of Villaviciosa, Lastres Formation, Quadrupedal, Isolated.") The isolated tracks MUJA-1899 and MUJA-1896 are from localities in the villages of Tazones (Villaviciosa) and Luces (Colunga), whereas the sauropod trackway is from Quintueles (Villaviciosa).

Iberian Range (Soria, La Rioja, Teruel)

The Iberian Range yields the great majority of the sauropod tracksites showing anatomical features. These tracksites are preserved in different stratigraphic units and range in age from the Tithonian (Late Jurassic) to the Aptian (late Early Cretaceous).

The tracksites of José María Herrero and Las Cerradicas in the village of Galve are located in the Maestrazgo Basin (Villar del Arzobispo Formation, Tithonian-Berriasian) in the province of Teruel (Castanera, Canudo, et al., 2010; Castanera et al., 2011). The El Pozo tracksite is located in the South Iberian Basin (Villar del Arzobispo Formation, Tithonian-Berriasian); it is situated in the village of El Castellar, also in the province of Teruel (Cobos, 2011; Alcalá et al., 2014). The Villar del Arzobispo Formation is composed of carbonates and terrigenous materials representing transitional environments between a shallow marine carbonate platform and more terrestrial settings. Thus, the tracks have been preserved in different facies and in different modes, such as sandstone casts (José María Herrero tracksite; Castanera, Canudo, et al., 2010), true tracks and natural casts preserved in sandstone layers (Las Cerradicas; Castanera et al., 2011), and true tracks preserved in carbonate layers (El Pozo tracksite; Cobos, 2011; Alcalá et al., 2014).

The largest sample of sauropod tracks is from the Cameros Basin, where different stratigraphic units have yielded sauropod tracksites. Thus, the tracksites of Las Cuestas I (Santa Cruz de Yanguas village, Soria province), Miraflores I (Fuentes de Magaña village, Soria province) and Salgar de Sillas (Los Campos village, Soria province) (Castanera, Barco, et al., 2010; Pascual-Arribas and Hernández-Medrano, 2011; Castanera et al., 2012) in the Huérteles Formation (Berriasian), El Majadal (Yanguas village, Soria province) (Castanera, Pascual, and Canudo, 2013) in the Urbión Group (Barremian-Aptian) and Los Cayos S (Cornago village, La Rioja province) (Moratalla, Hernán, and Jiménez, 2003; Moratalla and Hernán, 2008) in the Enciso Group (Aptian) are the tracksites that record the best-preserved sauropod tracks. The tracksites from the Huérteles Formation (Las Cuestas I, Miraflores I, and Salgar de Sillas) are preserved in siliciclastic materials (mainly sandstones) that were deposited in a tidal flat environment associated with a fluvial system (Quijada et al., 2013). Here, the tracks are preserved as concave epireliefs (true tracks or shallow undertracks?) in the surface of the outcrops (Miraflores I and Salgar de Sillas) and as true tracks and natural casts (Las Cuestas I). In the El Majadal tracksite, the track has been preserved in a sandstone level as a true track, with part of the natural cast preserved inside the footprint. This material was deposited in a braided fluvial system (Fernández Barrenechea, 1993). In the Los Cayos S tracksite, the track has been preserved as a true track in siliciclastic materials deposited in a low-gradient lacustrine environment (Moratalla, Hernán, and Jiménez, 2003; Moratalla and Hernán, 2008).

Pyrenees (Barcelona)

The Late Cretaceous Fumanya tracksites ('Grey Unit' of the Tremp Formation, early Maastrichtian), located in the village of Vallcebre in Barcelona province, record the largest sample of sauropod trackways from the Iberian Peninsula (Vila et al., 2008). Some long trackways with manus and pes prints showing good preservation and anatomical characters have been reported from the tracksites of Mina Esquirol and Fumanya Sud (Schulp and Brokx, 1999; Vila, Oms, and Galobart, 2005, Vila et al., 2008). The tracks have been preserved as concave epireliefs in the surface of thin marly beds deposited within a lagoonal (mudflat) setting (Vila et al., 2008; Marmi et al., 2014).

MATERIAL AND METHODS

We have selected the sauropod tracks on the basis of their stratigraphic position and state of preservation, considering those tracks that preserve clear morphologies and anatomical details and are exempt from taphonomic bias such as overprinting or mud collapse. The footprints under comparison occur in different states of preservation (mainly true tracks or natural casts) and in different substrates (mainly sandstones, siltstones, marsh, and limestones). Despite the different substrates and modes of preservation of the tracks, comparison has been possible because we have mainly analyzed the general track outline of each footprint, distinguishing broad categories, especially for the manus prints (mainly speech-bubble, kidney, and horseshoe shapes). Thus, we have considered the differences between them to be a consequence of anatomy and not due to the substrate or other extramorphological factors. In order to show the variations in the track morphologies in the figure, all the tracks have been drawn as if they were left ones. Nonetheless, some of them are actually right tracks, whereas others are preserved as a natural casts, so in these cases, they have been mirrored. Within the Lusitanian Basin tracksites, we have selected single tracks from trackways MNDPDSA-G1 (manus) and MNDPDSA-G5 (manus and pes) at Galinha tracksite; trackways PM5-1 (manus) and PM5-2 (pes) at Pedra da Mua level 5 tracksite (a gypsum plaster cast of the latter, MNHN-MG-P270, is housed in the Museu Nacional de História Natural e da Ciência da Universidade de Lisboa); and isolated manus (ML 965) and pes casts (ML 1151) described by Milàn, Christiansen, and Mateus (2005) and Mateus and Milàn (2010) and housed in the Museu da Lourinhã. Among the Cantabrian Range tracksites, we have selected the isolated tracks MUJA-1899 (manus cast) and MUJA-1896 (pes cast) housed in the Museo del Jurásico de Asturias in the village of Colunga and a sauropod trackway preserved as natural casts in the Quintueles tracksite (JVLCS-261, pes; JVLCS-262, manus). Among the Iberian Range tracksites, the selected tracks are a manus track (JMH1.7m) from Jose María Herrero tracksite, which is housed in the Museo de Galve in Teruel; a manus track (1CT-2) from El Pozo tracksite described by Cobos (2011) and Alcalá et al. (2014); a manus track (LCR 9.5m) and a pes track (LCR14.6p) from Las Cerradicas tracksite (the latter is provisionally housed at the Museo Paleontológico de

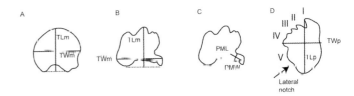

8.2. Measurements taken from the tracks. (A) Scheme showing the measurements taken in the horseshoe-shaped manus tracks redrawn from Vila et al. (2008). (B, C) Scheme showing the measurements taken in the speech-bubble–shaped (or kidney-shaped) manus tracks redrawn from Santos, Moratalla, and Royo-Torres, (2009). (D) Scheme showing the measurements taken in a pes track including digits I–V redrawn from Lockley, Meyer, and Santos, (1994). Abbreviations: m, manus; p, pes; PML, pollex mark length; PMW, pollex mark width; TL, track length; TW, track width.

la Universidad de Zaragoza); a manus (LCU-I-37-24m) and a pes (LCU-I-R37-12p) track from Las Cuestas I tracksite housed at the Museo Numantino de Soria; a manus track (MI-VIII-B-1m) from Miraflores I tracksite; a manus track (SS1-R1-4 m) from Salgar de Sillas tracksite; a manus track (LCS9-R1) from Los Cayos S tracksite; and a pes track (EMajS2.1) from El Majadal tracksite. Among the Pyrenean tracksites, we have selected the manus track morphology from trackway ME#42 at Mina Esquirol tracksite and the pes morphology from trackway FS#29 at Fumanya Sud tracksite.

The specimens have been described morphologically and photographed. The values given in Tables 8.1 and 8.2 represent data taken from perpendicular photographs of the tracks using the software ImageJ, with measurements taken by the authors on the specimens and data from the original papers. Measurements were taken of the track length (TL) and track width (TW), and the pollex mark length (PML) and pollex mark width (PMW) (Fig. 8.2), taking the maximum length and width. The TL/TW ratio has been calculated for each imprint in order to quantify the metacarpal arrangement in the case of manus prints (TLm/TWm) and the differences in the pes dimensions (TLp/TWp). The letters m and p refers to manus and pes, respectively. The tracks have also been compared using the heteropody ratio (Lockley, Farlow, and Meyer, 1994). The tracks were qualitatively considered shallow or deep in consideration of their overall size.

DESCRIPTION OF SAUROPOD
TRACK MORPHOLOGIES

Middle Jurassic Tracksites

Of the Iberian tracksites under study, the Galinha tracksite is the only one from the Middle Jurassic. Here, at least two trackways (MNDPDSA-G1 and MNDPDSA-G5) with well-preserved tracks have been described (Santos et al., 1994; Santos, Moratalla, and Royo-Torres, 2009). In MNDPDSA-G5, the manus prints (track 1 in Fig. 8.3; see also Fig. 8.4A) are wider

Table 8.1. Measurements taken from the studied sauropod manus tracks

Tracksite	Location	Footprint	Age	Shape
Galinha*	Ourém-Torres Novas (Santarém)	MNDPDSA-G1	Bathonian	speech bubble
Galinha*	Ourém-Torres Novas (Santarém)	MNDPDSA-G5	Bathonian	speech bubble
Tazones+	Villaviciosa (Asturias)	MUJA-1899	Kimmeridgian	kidney
Quintueles+	Villaviciosa (Asturias)	JVLCS-262	Kimmeridgian	kidney
Pedra da Mua* level 5	Sesimbra (Setúbal)	PM5-1	Tithonian	kidney
Jose María Herrero +	Galve (Teruel)	JMH1.7m	Tithonian/Berriasian	kidney
El Pozo*	El Castellar (Teruel)	1CT-2	Tithonian/Berriasian	kidney
Las Cerradicas+	Galve (Teruel)	LCR9.5m	Berriasian	kidney
Las Cuestas I+	Santa Cruz de Yanguas (Soria)	LCU-I-37-24m	Berriasian	kidney
Miraflores I+	Fuentes de Magaña (Soria)	MI-VIII-B-1m	Berriasian	kidney
Salgar de Sillas+	Los Campos (Soria)	SS1-R1-4 m	Berriasian	kidney
Los Cayos S+	Cornago (La Rioja)	LCS9-R1	Aptian	Horseshoe
Mina Esquirol* (Fumanya)	Vallcebre (Barcelona)	ME#42	Maastrichtian	Horseshoe

Notes: Values given in centimeters. Asterisk indicates that tracksites have been preserved in carbonate beds (marsh-limestones). Plus sign indicates that tracksites have been preserved in siliciclastic beds (sandstones-siltstones). CT, El Pozo; JMH, Jose María Herrero; JVLCS, Jurásico, Municipio de Villaviciosa, Formación Lastres, Cuadrúpedo, Suelto; LCR, Las Cerradicas; LCS, Los Cayos S; LCU-I, Las Cuestas I; m, manus; ME, Mina Esquirol; MI, Miraflores I; MNDPDSA, Monumento Natural das Pegadas de Dinossáurios da Serra de Aire; MUJA, Museo del Jurásico de Asturias; p, pes; PL, pollex mark length; PM, Pedra da Mua; PW, pollex mark width; SS, Salgar de Sillas; TL, track length; TL/TW, track length/track width ratio; TW, track width.

than long (TLm/TWm ratio = 0.68) and are characterized by an asymmetrical speech-bubble shape with a large digit I impression (17 cm long × 13 cm wide) oriented in a medial direction. Moreover, they reveal a large, posteriorly oriented, triangular claw I mark and the impressions of individual digits (II–V). By contrast, trackway MNDPDSA-G1 has slightly different manus morphology (track 2 in Fig. 8.3; see also Fig. 8.4B). The manus tracks also have a speech-bubble shape, but they are more symmetrical, and no individual digits can be discerned. They are wider than long (TLm/TWm ratio = 0.75) but slightly longer in proportion. They have a rounded lateral and medial margin and also a large digit I impression (17 cm long × 12 cm wide) projected posteromedially from the center of the track's rear margin.

The pes prints in MNDPDSA-G5 (track 3 in Fig. 8.3; Santos, Moratalla, and Royo-Torres, 2009:fig. 8K) are longer than wide (90 cm × 60 cm, TLp/TWp ratio = 1.5), with four claw marks oriented anteriorly (I–II) and laterally (III–IV). The

heel impression is rounded. In MNDPDSA-G1, the pes prints are oval in shape (see Santos, Moratalla, and Royo-Torres, 2009:fig. 7), longer than wide (95 cm × 70 cm, with a TLp/TWp ratio of 1.35), and the number of claw marks is not easily discernible, probably due to preservational/taphonomical factors. The impression of the heel is rounded as well and is separated from the anterior part of the footprint by a shallower area. MNDPDSA-G1 and MNDPDSA-G5 are also characterized by a low heteropody value (manus/pes area ratio of 1:2).

Late Jurassic Tracksites

In the Late Jurassic, the best-preserved track morphologies have been described in various Spanish (Cantabrian Range) and Portuguese localities. In the former case, different isolated manus and pes tracks recovered from the cliffs have been described (García-Ramos, Lires, and Piñuela, 2002,

Table 8.2. Measurements taken from the studied sauropod pes tracks.

Tracksite	Location	Footprint	Age	TLp (cm)	TWp (cm)	TLp/TWp
Galinha*	Ourém-Torres Novas (Santarém)	MNDPDSA-G5	Bathonian	90	60	1.5
Quintueles+	Villaviciosa (Asturias)	JVLCS-261	Kimmeridgian	67	61	1.1
Luces+	Colunga (Asturias)	MUJA-1896	Kimmeridgian	28	26	1.08
Porto Dinheiro+	Porto Dinheiro (Lisboa)	ML 1151	Kimmeridgian	94	88	1.06
Pedra da Mua* level 5	Sesimbra (Setúbal)	PM5-2/E11	Tithonian	80	58	1.37
Las Cerradicas+	Galve (Teruel)	LCR14.6p	Berriasian	27	22	1.2
Las Cuestas I+	Santa Cruz de Yanguas (Soria)	LCU-I-37-12p	Berriasian	73	56	1.3
El Majadal+	Yanguas(Soria)	EMajS2.1	Barremian/Aptian	54	50	1.08
Fumanya Sud*	Vallcebre (Barcelona)	FS#29	Maastrichtian	71	58	1.2

Note: Values given in centimeters. EMaj, El Majadal; FS, Fumanya Sud; ML, Museum Lourinhã. Please see Table 8.1 for other abbreviations and explanation of symbols.

TLm (cm)	TWm (cm)	TLm/ TWm	Digit I Claw marks (PML × PMW)	References
50	66	0.75	17 × 12	Santos et al. (1994); Santos, Moratalla, and Royo-Torres (2009)
43	63	0.68	17 × 13	Santos, Moratalla, and Royo-Torres (2009)
14	21	0.66	None	García-Ramos, Piñuela, and Lires (2006); Lockley et al., (2008)
21	36	0.58	about 10 cm	García-Ramos, Piñuela, and Lires (2006); Lockley et al., (2008)
23	33	0.69	None	Santos (2003)
40	55	0.72	10 × 8	Castanera, Canudo, et al. (2010); Castanera and Canudo (2011)
29	38	0.76	?	Alcalá et al. (2014)
13	19	0.68	None	Castanera et al. (2011)
29	43	0.67	Reduced	Pascual-Arribas et al. (2008); Pascual-Arribas and Hernández-Medrano (2011)
58	82?	0.70?	15 × 10?	Latorre-Macarrón et al. (2006); Castanera et al. (2010)
42	58?	0.72	none?	Castanera et al., (2012)
35	37	0.94	None	Moratalla, Hernán, and Jiménez (2003); Moratalla and Hernán (2008)
25	29	0.86	None	Vila, Oms, and Galobart (2005); Vila et al. (2008)

García-Ramos, Piñuela, and Lires, 2006; Lockley et al., 2007, 2008). The manus track MUJA-1899 is wider than long (TLm/ TWm ratio = 0.66) and kidney-shaped in morphology (track 4 in Fig. 8.3; see also Fig. 8.4C). The manus track JVLCS-262 is also wider than long (TLm/TWm ratio = 0.58) and kidney-shaped but has the impression of a prominent digit I claw mark (about 10 cm in length) that is oriented medially (track 5 in Fig. 8.3). Neither of the manus tracks shows any indication of any other digits.

The pes tracks from the Asturian localities (MUJA-1896, JVLCS-261, tracks 6 and 7 in Fig. 8.3; see also Figs. 8.5A, 8.5B) are subtriangular in shape (TLp/TWp ratio = 1.1 and 1.08, respectively), with impressions of digits and claws. They have marks of five digits, with claw marks oriented anterolaterally for digits I–III and the rounded marks of digits IV–V oriented laterally. The impressions of the heels are rounded, and in JVLCS-261, the heel impression is separated from the anterior part of the footprint by a shallower area. MUJA-1896 shows a

lateral notch behind digit V. Among the Portuguese localities, at Porto Dinheiro locality, pes casts showing four digit marks (track 8 in Fig. 8.3) have also been reported (Mateus and Milàn, 2010). ML 1151 shows blunt claw marks and division into digital pads. The heel impression is rounded and is separated from the anterior part of the footprint by a shallower area. Milán, Christiansen, and Mateus (2005) described a "semi-lunate" manus cast from the Lourinhã Formation with a small claw mark in digit I oriented posteromedially (see Milàn, Christiansen, and Mateus, 2005:fig. 2). There is no indication of any other digit.

Jurassic-Cretaceous Interval (Tithonian-Berriasian)

At the Pedra da Mua cliffs and in the Iberian Range, various sauropod manus and pes morphotypes from the Tithonian-Berriasian interval have been described (Lockley, Meyer, and Santos, 1994; Meyer et al., 1994; Santos, 2003; Castanera,

Claw marks	Lateral notch	Heel	References
anteriorly (I–II) and laterally (III–IV)	None	rounded	Santos, et. al (1994); Santos, Moratalla, and Royo-Torres (2009)
laterally (I–III)	present	rounded	García-Ramos, Piñuela, and Lires (2006)
laterally (I–III)	present	rounded	Lockley et al. (2008)
laterally	None	rounded	Mateus and Milàn (2010)
laterally	Present	rounded	Meyer et al. (1994)
laterally	Present	subtriangular	Castanera et al. (2011)
laterally	None	rounded	Pascual-Arribas et al. (2008)
anteriorly-anterolaterally (I–II), laterally (III)	Present	subtriangular	Castanera, Pascual, and Canudo (2013)
laterally (I–IV)	None	rounded	Schulp and Brokx (1999); Vila et al. (2008)

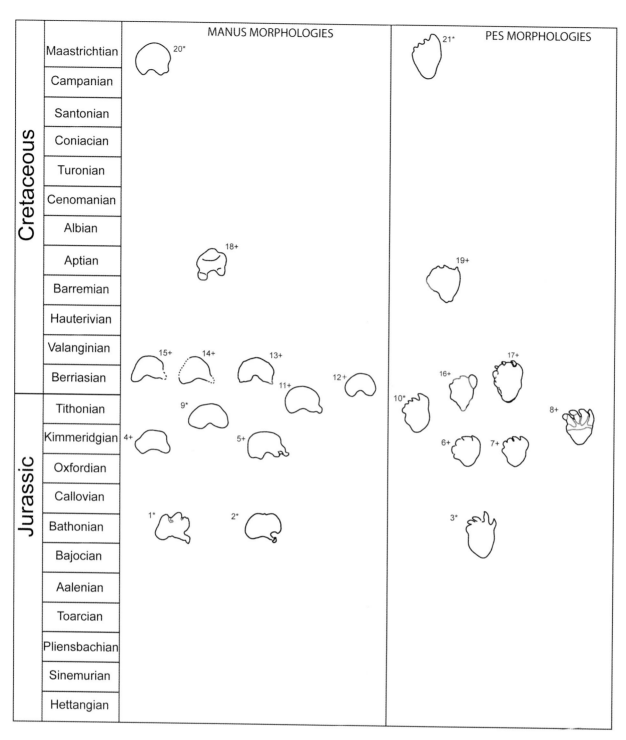

8.3. Iberian manus and pes track morphologies through time. Tracks 3, 6, 7, 11, and 13 have been mirrored. Tracks 4–8, 11, 13, 16, and 17 have been preserved as natural casts. Drawings not to scale. Asterisk indicates tracks that have been preserved in carbonate beds (marsh-limestones). Plus sign indicates tracks that have been preserved in siliciclastic beds (sandstones-siltstones). (1) MNDPDSA-G5, Monumento Natural das Pegadas de Dinossáurios da Serra de Aire, Galinha tracksite. (2) MNDPDSA-G1, Galinha tracksite. (3) MNDPDSA-G5, Galinha tracksite. (4) MUJA-1899 [housed in the Museo del Jurásico de Asturias], Tazones, Villaviciosa locality. (5) JVLCS-262, Jurásico, Municipio de Villaviciosa, Formación Lastres, Cuadrúpedo, Suelto, Quintueles, Villaviciosa locality. (6) MUJA-1896, Luces, Colunga locality. (7) JVLCS-261, Quintueles, Villaviciosa locality. (8) ML 1511, Porto Dinheiro locality. (9) PM5-1, Pedra da Mua tracksite. (10) PM5-2/E11, Pedra da Mua tracksite. (11) JMH1.7m, José María Herrero tracksite. (12) LCR9.5m, Las Cerradicas tracksite. (13) LCU-I-37-24m, Las Cuestas I tracksite. (14) MI-VIII-B-1m, Miraflores I tracksite. (15) SS1-R1-4 m, Salgar de Sillas tracksite. (16) LCR14.6p, Las Cerradicas tracksite. (17) LCU-I-37-12p, Las Cuestas I tracksite. (18) LCS9-R1, Los Cayos S tracksite. (19) EMajS2.1, El Majadal tracksite. (20) ME#42, Mina Esquirol tracksite. (21) FS#29, Fumanya Sud tracksite. Tracks 1–3 redrawn from Santos, Moratalla, and Royo-Torres (2009); 4–6 redrawn from Lockley et al. (2008); 7 redrawn from García-Ramos, Piñuela, and Lires (2006); 8 redrawn from Mateus and Milàn (2010); 10 redrawn from Lockley, Meyer, and Santos, (1994); 11 and 14 redrawn from Castanera and Canudo (2011); 12 and 16 redrawn from Castanera et al. (2011); 13 and 17 redrawn from Pascual Arribas et al. (2008); 15 redrawn from Castanera et al. (2012); 18 redrawn from Moratalla, Hernán, and Jiménez (2003); 19 redrawn from Castanera et al. (2013); and 20 and 21 redrawn from Vila et al. (2008).

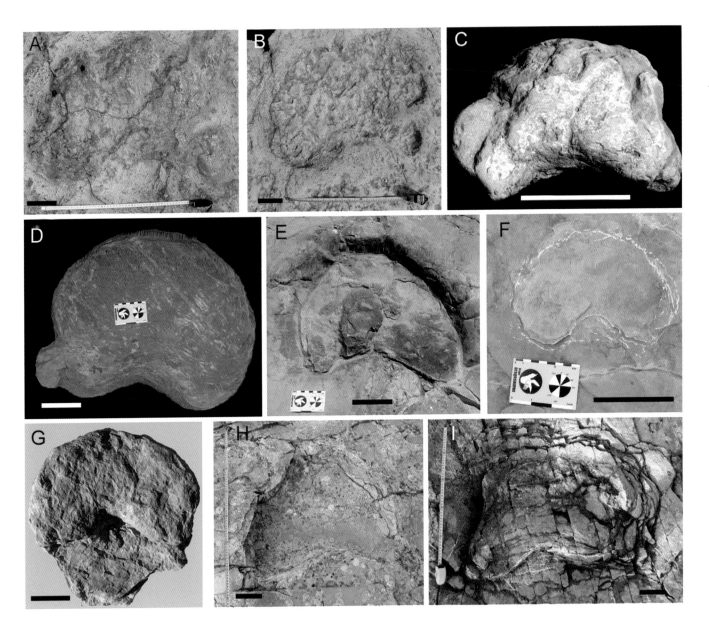

8.4. Photographs of some of the best-preserved sauropod manus tracks from the Jurassic and the Jurassic-Cretaceous interval of the Iberian Peninsula. (A) MNDPDSA-G5 (Monumento Natural das Pegadas de Dinossáurios da Serra de Aire, Galinha tracksite). (B) MNDPDSA-G1 (Galinha tracksite). (C) MUJA-1899 ([housed in Museo del Jurásico de Asturias], Tazones, Villaviciosa locality). (D) JMH1.7m (José María Herrero tracksite). (E) 1CT-2 (El Pozo tracksite). (F) LCR9.5m (Las Cerradicas tracksite, from Castanera et al., 2011). (G) LCU-I-37-12p (Las Cuestas I tracksite, modified from Pascual Arribas et al., 2008). (H) MI-VIII-B-1m (Miraflores I tracksite, from Castanera and Canudo, 2011). (I) SS1-R1-4 m (Salgar de Sillas tracksite, from Castanera et al., 2012). Scale bar = 10 cm.

Barco, et al., 2010; Castanera, Canudo, et al., 2010; Castanera and Canudo, 2011; Castanera et al., 2011, 2012; Cobos, 2011; Alcalá et al., 2014).

Level 5 at Pedra da Mua tracksite has some trackways that show well-preserved footprints (Meyer et al., 1994; Santos, 2003, 2008). Trackways numbers 1 and 2 (PM5-1 and PM5-2) have manus prints that are kidney-shaped (TLm/TWm ratio = 0.69) and without digit marks (track 9 in Fig. 8.3) and pes prints that are subtriangular in shape (TLp/TWp ratio = 1.37), with digit marks and claw marks oriented anterolaterally and decreasing in size from digit I to III (track 10 in Fig.

8.3; see also Figs. 8.5C, 8.5D). The claw marks also occupy a progressively more posterolateral position from digit I to digit III. The callosity marks of digits IV and V are lateral in orientation. The heel impression is rounded, and some of the tracks preserve a small lateral notch behind digit V (Fig. 8.5D).

Within the tracksites from the Iberian Range, the best-preserved manus prints are JMH1.7m (track 11 in Fig. 8.3; see also Fig. 8.4D), 1CT-2 (Fig. 8.4E), LCR9.5m (track 12 in Fig. 8.3; see also Fig. 8.4F), LCU-I-37-24m (track 13 in Fig. 8.3; see also Fig. 8.4G), MI-VIII-B-1m (track 14 in Fig. 8.3; see also

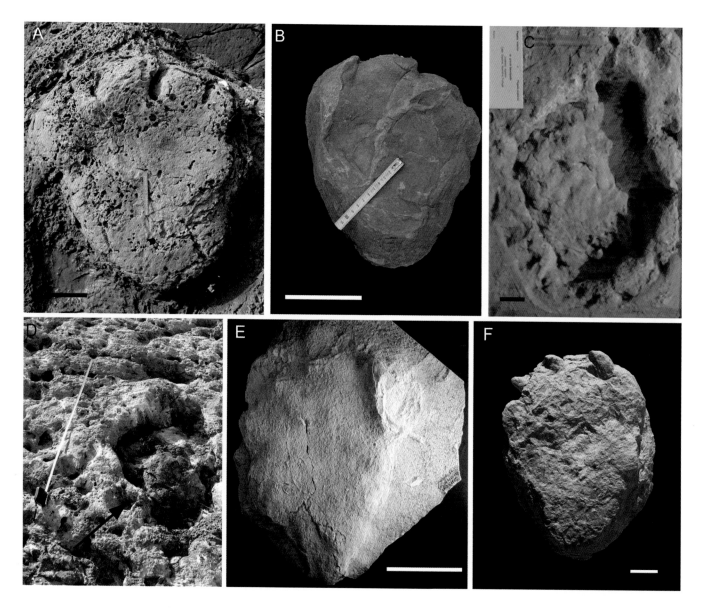

8.5. Photographs of some of the best-preserved sauropod pes tracks from the Jurassic and the Jurassic-Cretaceous interval of the Iberian Peninsula. (A) JVLCS-261 (Jurásico, Municipio de Villaviciosa, Formación Lastres, Cuadrúpedo, Suelto, Quintueles, Villaviciosa locality). (B) MUJA-1896 (Museo del Jurásico de Asturias, Luces, Colunga locality). (C) Plaster cast of MNHN-MG-P270 (Museu Nacional de História Natural e da Ciência da Universidade de Lisboa; PM5-2/E11, Pedra da Mua tracksite). (D) PM5-1 (Pedra da Mua tracksite). (E) LCR14.6p (Las Cerradicas tracksite, from Castanera et al., 2011). (F) LCU-I-37-12p (Las Cuestas I tracksite, from Castanera et al., 2012). Scale bar = 10 cm. Ruler in D = 1 m.

Fig. 8.4H), and SS1-R1–4m (track 15 in Fig. 8.3; see also Fig. 8.4I). All of these have a general kidney-shaped morphology, but further distinctions can be made. The manus prints JMH1.7m (TLm/TWm ratio = 0.72), LCU-I-37-24m (TLm/TWm ratio = 0.67), and MI-VIII-B-1m (TLm/TWm ratio = 0.7?) have pollex marks that are variable in size (Table 8.1) and position (medial to posterior). In the case of the 1CT-2 (TLm/TWm ratio = 0.76) and SS1-R1-4m (TLm/TWm ratio = 0.72), the position of digit V is well developed and oriented slightly laterally. In the former, the position of digit I is oriented posteriorly whereas in the latter it is poorly preserved. LCR9.5m has a perfect kidney-shaped morphology

(TLm/TWm ratio = 0.68) without evidence of a pollex mark. None of these manus tracks has marks of any individual digit.

The pes prints are poorly preserved, and only tracks from the Las Cerradicas and Las Cuestas I tracksites provide good examples. LCR14.6p (track 16 in Fig. 8.3; see also Fig. 8.5E) is subtriangular in shape (TLp/TWp ratio = 1.2), with a large anterior margin and marks of digits I to V. Digit I is the biggest and swells medially. Claw marks seem to be present for digits I–III, and these are lateral in orientation. The heel is subtriangular, narrow, and oriented slightly posterolaterally. A small lateral notch behind digit V can also be discerned. LCU-I-37-12p (track 17 in Fig. 8.3; see also Fig. 8.5F) is

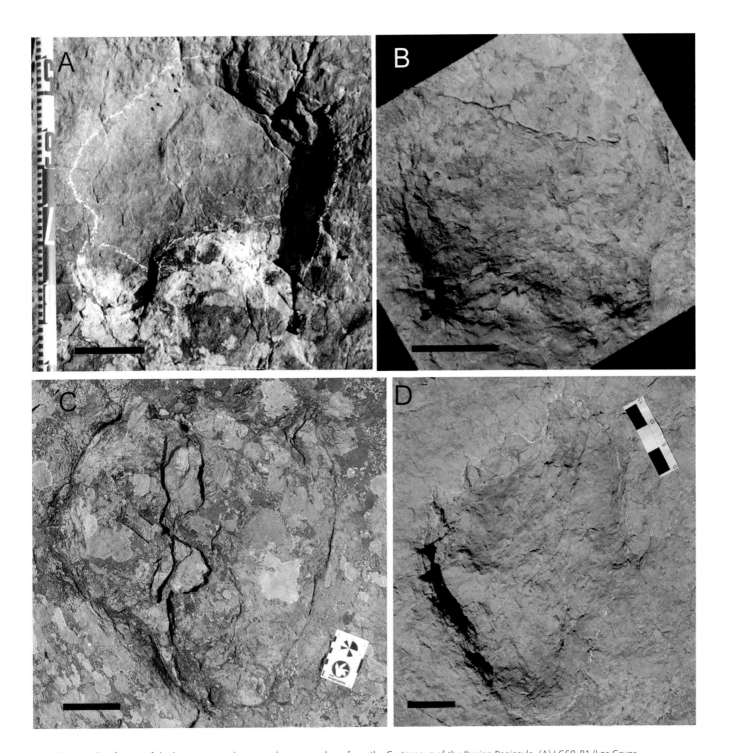

8.6. Photographs of some of the best-preserved sauropod manus and pes from the Cretaceous of the Iberian Peninsula. (A) LCS9-R1 (Los Cayos S tracksite, from Moratalla, Hernán, and Jiménez, 2003). (B) ME#42 (Mina Esquirol tracksite, Fumanya, from Vila, Oms, and Galobart, 2005). (C) EMajS2.1 (El Majadal tracksite, from Castanera et al., 2013). (D) FS#29 (Fumanya Sud tracksite, from Vila et al., 2008). Scale bar = 10 cm.

subtriangular (TLp/TWp ratio = 1.3), with three claw marks decreasing in size from digit I to digit III and rounded marks of digits IV and V. The heel impression is rounded. In both cases, the heteropody values are intermediate (1:3). In the Miraflores I tracksite, despite the poor preservation of the pes print, trackway MI-VIII-B-1 seems to have a low heteropody value (about 1:2).

Late Early Cretaceous Tracksites

Despite the great number of tracksites in the Jurassic-Cretaceous interval, there is a gap in information until the Barremian-Aptian; in the early Early Cretaceous, there are not so many sauropod tracksites reported (Santos, 2003, 2008; Moratalla, 2009). Among these, there are few tracksites that preserve tracks with good sauropod manus/pes morphologies. Moratalla, Hernán, and Jiménez (2003) and Moratalla and Hernán (2008) have described a trackway from Los Cayos S tracksite that shows a well-preserved manus print that exhibits digit traces. The morphology of the manus print (LCS9-R1, track 18 in Fig. 8.3; see also Fig. 8.6A) is horseshoe-shaped (TLm/TWm ratio = 0.94), with marks of digits I and V oriented posteriorly, digit V being slightly wider. There is no evidence of a claw mark on digit I. Digits II–IV are anteriorly located and are discernible in outline but not as individual digits. Recently, Castanera, Pascual, and Canudo (2013) have described an isolated pes print (EMajS2.1, track 19 in Fig. 8.3; see also Fig. 8.6C) in the El Majadal tracksite. This track is subtriangular in shape (TLp/TWp ratio = 1.08) and is characterized by the presence of five digit marks, preserving claw marks in the first three digits. These are oriented anteriorly (I) and anterolaterally (II–III). The callosity marks of digits IV–V are oriented laterally. The footprint is also characterized by a pronounced lateral notch behind digit V and a subtriangular rounded heel.

Late Cretaceous Tracksites

In the Late Cretaceous, there is also a gap in information regarding sauropod tracksites from the Cenomanian to the Late Campanian. The best-preserved sauropod print morphologies are from the Early Maastrichtian Fumanya tracksites (Vila, Oms, and Galobart, 2005; Vila et al., 2008). There, good examples of manus prints are horseshoe-shaped (TLm/TWm ratio = 0.86), with digits I and V oriented posteriorly (e.g., ME#42, track 20 in Fig. 8.3; see also Fig. 8.6B), whereas others are kidney- or crescent-shaped. The latter shape, however, is probably due to preservational effects. There are no claw marks in the manus prints. The best-preserved pes prints (e.g., FS#29, track 21 in Fig. 8.3; see also Fig 8.6D) are subtriangular (TLp/TWp ratio = 1.2), with four claw marks oriented anterolaterally, that of digit I being the largest and those of digits II–IV almost equal in size. The rounded marks of digit V are oriented laterally, and the heel impression is subtriangular and rounded. There is no evidence of a lateral notch. The heteropody is intermediate (1:3).

CHANGES IN TRACK MORPHOLOGY ASSOCIATED WITH THE OSTEOLOGICAL REMAINS

Manus Tracks

Analysis of the sauropod track morphology of Iberian sauropod tracksites reveals that the manus prints show four different general morphologies (Fig. 8.3.): (1) speech-bubble–shaped with prominent claw mark in digit I (Middle Jurassic);

(2) kidney-shaped with claw mark in digit I (Late Jurassic–Early Cretaceous); (3) kidney-shaped without claw mark in digit I (Late Jurassic–Early Cretaceous); (4) horseshoe-shaped with digits I and V posteriorly located (late Early Cretaceous and Late Cretaceous). Within the kidney-shaped morphologies, there are slight variations in TL/TW ratios (Table 8.1) that are associated with the anteroposterior development of the manus and the location of digits I and V more medially/laterally (e.g., LCR9.5m, PM5-1, LCU-1-37-24m) or posteromedially/posterolaterally (SS1-R1-4m; MI-VIII-B-1m; 1CT-2). There are also variations in the symmetry of the tracks, some of them being slightly asymmetrical (SS1-R1-4m, 1CT-2). These differences are considerable, but the absence of more tracks in the trackways prevents us from creating a fifth category for the moment. Furthermore, an intriguing question is whether the differences regarding the absence/presence of the pollex marks are a consequence of anatomy or are due to preservational factors (in relation to the depth to which the forelimb sank into the substrate and the composition of the substrate), or whether both factors may be involved.

The variations in the manus print morphologies reflect the evolution of the sauropod metacarpus, from the nontubular metacarpal arrangement observed in primitive sauropods to a tubular disposition (U-shaped, 270° metacarpal arch) in more derived sauropods (Bonnan, 2003; Apesteguía, 2005; Milàn, Christiansen, and Mateus, 2005; Wright, 2005; Santos, Moratalla, and Royo-Torres, 2009). In the Iberian Peninsula, the speech-bubble–shaped morphology seen in manus prints from trackways MNDPDSA-G1 and MNDPDSA-G5 (Galinha tracksite) suggests a manus intermediate between primitive and more derived sauropods. According to Santos, Moratalla, and Royo-Torres (2009:417), "the posterior position of the manus digit I impressions suggests that metacarpus I also occupies a posterior position, and consequently that the whole metacarpus was built as a semi-tubular structure."

From the Middle Jurassic onward, the manus morphologies suggest slightly different metacarpal arrangements. In the Late Jurassic and the Jurassic-Cretaceous transition, the kidney-shaped morphologies show an arrangement where digit I is positioned posteriorly to posteromedially and digit V posterolaterally. These morphologies may suggest the early stage of a tubular structure, but the metacarpal arch still does not reach the 270° angle present in derived sauropods (Apesteguía, 2005; Wright, 2005). In the late Early Cretaceous and Late Cretaceous, the horseshoe-shaped morphologies show digits I and V located in the posterior part (Moratalla, Hernán, and Jiménez, 2003; Vila, Oms, and Galobart, 2005; Vila et al., 2008); they are completely symmetrical rounded marks, suggesting a completely tubular structure for the metacarpals typical of derived sauropods (metacarpal arch of 270°).

A key question is the absence or presence of digit I manus claw impressions and whether or not the claw was used during locomotion (Thulborn, 1990; Upchurch, 1994). The sauropod predictive ichnology set forth by Wright (2005) proposed that except in titanosaurs the tracks would show substantial digit I manus claw impressions for all sauropods, so their absence in many tracks might be a preservational effect. The Iberian sauropod track record yields various tracks (MNDPDSA-G1, MNDPDSA-G5, JVLCS-262, JMH1.7m, LCU-I-37-24m, MI-VIII-B-1m, and ML 965 from Milàn, Christiansen, and Mateus, 2005) from the Jurassic and Jurassic-Cretaceous interval where manus claw impressions have been described (Fig. 8.3). Their position varies from posteromedial (MNDPDSA-G1, Galinha tracksite; MI-VIII-B-1m, Miraflores I tracksite) to medial (MNDPDSA-G5, Galinha; ML 965, Porto Dinheiro locality; JVLCS-262, Quintueles locality; JMH1.7m, José María Herrero tracksite) and seems to have significant taxonomic value. Considering the variations over time, a reduction in the size of the digit I claw impressions can be seen as well. This also reflects the osteological remains, namely, the reduction of the pollex during sauropod evolution, this being completely lost in the most derived sauropods (Salgado, Coria, and Calvo, 1997; Wilson, 2002; Upchurch, Barrett, and Dodson, 2004). The Iberian sauropod track record thus represents an exceptional sample of manus tracks showing pollex marks. It is noteworthy that this evidence of pollex marks is recorded in tracksites that are preserved in different sediments and where the tracks show different depths (shallow and deep tracks). Thus, MNDPDSA-G1 and MNDPDSA-G5 are preserved in limestone, and the tracks are shallow. MI-VIII-B-1m is preserved in sandstone, and the tracks are also shallow. The others (ML 965, JVLCS-262, JMH1.7m) are preserved as sandstone casts, and the tracks are quite deep. Analyzing these data, it should be noted that the Galinha tracksite at least, where MNDPDSA-G1 and MNDPDSA-G5 show shallow tracks with pollex marks that can be seen repeatedly along the whole trackway, provides evidence of a sauropod using its pollex during locomotion. The other tracks (ML 965, JVLCS-262, JMH1.7m) are isolated and are deep tracks, so it is not possible to ascertain how far the presence or absence of the pollex mark can be associated with the use of the pollex during locomotion or might depend on the substrate conditions.

Another interesting issue is the preservation of individual marks in the digits. The trackway MNDPDSA-G5 (Middle Jurassic, Galinha tracksite) is the only trackway in the whole Iberian Peninsula that preserves manus prints, which show individual marks in digits I–IV (Fig. 8.4A). This character is not usually reported in sauropod manus prints, which generally lack discrete digit traces (Thulborn, 1990; Wright, 2005), though blunt digit impressions have been described in some

tracksites (Lockley et al., 2002; Kim and Lockley, 2012). In ages subsequent to the Middle Jurassic, the manus prints lack any impression of clear individual digits. This could be a consequence of digits II to V being encapsulated in a tissue that formed a single unit, as was suggested by Milàn, Christiansen, and Mateus (2005), or it could be due to preservational effects.

Pes Tracks

By contrast with the manus prints, the pes prints are more conservative in general morphology throughout the Mesozoic (Fig. 8.3). They have a general subtriangular morphology, with a varying number and disposition of claw marks in the digits. There are also slight variations in the TLp/TWp ratio (Table 8.2). In the Middle Jurassic, four individual digits display claw marks, two of them oriented anteriorly (I–II) and two laterally (III–IV). In the Late Jurassic and Jurassic-Cretaceous interval, the pes prints generally have impressions of three claw marks (I–III), oriented laterally and decreasing in size from digits I to III. The impressions of blunt callosities at the distal ends of digits IV and V are also inferred in a lateral position. Some of the tracks show a small lateral notch behind digit V (tracks 6, 7, 10, 16, and 19 in Fig. 8.3; see also Figs. 8.5A, 8.5B, 8.5D, 8.5E, and 8.6C). In the late Early Cretaceous, specimen EMajS2.1 also has marks for the referred claws, but the orientation is slightly different and they seem to be oriented more anterolaterally (though it is an isolated track and we cannot rule out that the effect could be taphonomic). Moreover, a pronounced lateral notch (larger than those from the Late Jurassic and the Jurassic-Cretaceous interval) can also be discerned (Fig. 8.6C). Finally, in the Late Cretaceous, the claw marks of digits I–IV are again oriented in a completely lateral position. The blunt callosity of digit V can also be discerned. Other morphological characters such as the lateral notch or the heel do not show significant changes through the ages, and these small differences could be due to taxonomic or preservational effects.

The variations in pes print morphologies are not as great as are those of the manus tracks, though there seems to be a trend toward a more lateral orientation of the digits and claws (except in the case of the isolated track EMajS2.1 from El Majadal tracksite). Bonnan (2005) suggested that the diagnostic features of sauropod pedal anatomy were acquired early in sauropod evolution and remain relatively unmodified in the derived groups. As we have seen, in the Iberian sample the most noticeable changes are in the position and orientation and the number of claw marks. In terms of the number and orientation of claw marks, the most considerable change was produced between the Middle Jurassic and the Late Jurassic,

Table 8.3. Osteological characters preserved in sauropod trackways and their relationship with the main groups of sauropods

Type	Trackway gauge	Digit I Claw Mark	Group of sauropods
I	Narrow	Well developed	Nontitanosauriform
II	Intermediate	Well developed or reduced	Brachiosaurs and most basal titanosaurs
III	Wide	Reduced	Basal titanosaurs
IV	Wide	Absent	Advanced titanosaurs
V	Wide	Well developed	Nonneosauropod eusauropods

Source: Day et al. (2004); Santos, Moratalla, and Royo-Torres (2009)

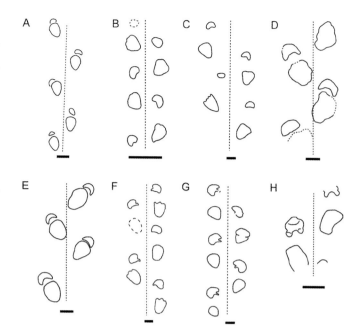

8.7. Trackways of the selected tracks. (A) PM5-4: wide-gauge, kidney-shaped manus print, no claw mark in digit I, titanosauriforms (Pedra da Mua, redrawn from Meyer et al., 1994). (B) LCR10: wide-gauge, kidney-shaped manus print, no claw mark in digit I, titanosauriforms (Las Cerradicas tracksite, redrawn from Castanera et al., 2011). (C) FS#29: wide-gauge, horseshoe-shaped manus print, no claw mark in digit I, titanosauria (Fumanya Sud tracksite, redrawn from Vila et al., 2008). (D) MI-VIII-B-1: narrow-gauge, kidney-shaped manus print, claw mark in digit I, nontitanosauriforms (Miraflores I tracksite, redrawn from Latorre-Macarrón et al., 2006). (E) LCU-I-37: narrow- to intermediate-gauge, kidney-shaped manus print, claw mark in digit I, nontitanosauriforms (Las Cuestas I tracksite, redrawn from Pascual Arribas et al., 2008). (F) MNDPDSA-G5: wide-gauge, speech-bubble–shaped manus print, claw mark in digit I, nonneosauropod eusauropods (Monumento Natural das Pegadas de Dinossáurios da Serra de Aire, Galinha tracksite, redrawn from Santos, Moratalla, and Royo-Torres, 2009). (G) MNDPDSA-G1: wide-gauge, speech-bubble–shaped manus print, claw mark in digit I, nonneosauropod eusauropods (Galinha tracksite, redrawn from Santos, Moratalla, and Royo-Torres, 2009). (H) LCS9-R1: narrow- to intermediate-gauge, horseshoe-shaped manus print, no claw mark in digit I, titanosauriforms (Los Cayos S tracksite, redrawn from Moratalla, Hernán, and Jiménez, 2003). Scale bar = 50 cm.

when the sauropod prints lose the claw mark in digit IV and the orientation changes to a more lateral position. The presence of four ungual phalanges with an anterior orientation is a primitive character within Sauropoda (Wilson, 2002; Bonnan, 2005; González-Riga, Calvo, and Porfiri, 2008). Nonetheless, during the Late Cretaceous, at least some footprints also seem to have four claw marks, though in this case laterally oriented (e.g., FS#29; Wright, 2005). Noteworthy is the presence of a claw mark in digit IV, because this is a character not usually reported with osteological remains (Nair and Salisbury, 2012).

IBERIAN SAUROPOD TRACKS AND TRACKMAKER INTERPRETATIONS

The sauropod footprints selected for study show enough characters to discern synapomorphies of some sauropod clades (Carrano and Wilson, 2001; Wright, 2005). The presence of the digit I claw mark throws useful light on the trackmaker. Day et al. (2004) predicted four types of sauropod trackways on the basis of osteological characters such as the presence and size of pollex marks and the trackway gauge (Table 8.3): "(1) narrow-gauge trackways with well-developed manus claw impressions, formed by non-titanosauriform sauropods; (2) intermediate-gauge trackways with well-developed or reduced manus claw impressions, produced by brachiosaurs and the most basal titanosaurs; (3) fully wide-gauge trackways, with reduction of the manus claw and other manual phalanges, formed by basal titanosaurs; and (4) fully wide-gauge trackways with no indication of the presence of any manual phalanges, produced by advanced titanosaurs" (333–334).

Some of the Iberian footprints can be classified into these groups, whereas others represent other categories. Thus, the trackways from the tracksites of Pedra da Mua level 5 (Tithonian), Las Cerradicas (Berriasian), and Fumanya (Maastrichtian) are wide gauge, without any claw mark in the digit I impression, and are related to titanosauriforms (Santos, 2003; Castanera et al., 2011) and titanosaurs (Vila et

al., 2008); these may thus represent types 3 and 4 described by Day et al. (2004), respectively (Figs. 8.7A–C, Fig. 8.8). These two groups of sauropods have been described on the basis of osteological remains in the Jurassic-Cretaceous interval and the Late Cretaceous of the Iberian Peninsula, respectively (Canudo et al., 2010; Vila et al., 2012).

Other trackways can be classified in the first category (nontitanosauriform) and the second category (brachiosaurs and the most basal titanosaurs). Examples are MI-VIII-B-1 from Miraflores I and LCU-I-37 from Las Cuestas I tracksite (Berriasian), which are narrow- to narrow-intermediate–gauge trackways and show digit I claw impressions (Figs. 8.7D, 8.7E). Nonetheless, these have been tentatively associated with diplodocids and nontitanosauriform macronarians, respectively, taking into account the gauge (narrow and intermediate) and the size of the digit I claw mark (Pascual-Arribas et al., 2008; Castanera and Canudo, 2011). Other Iberian

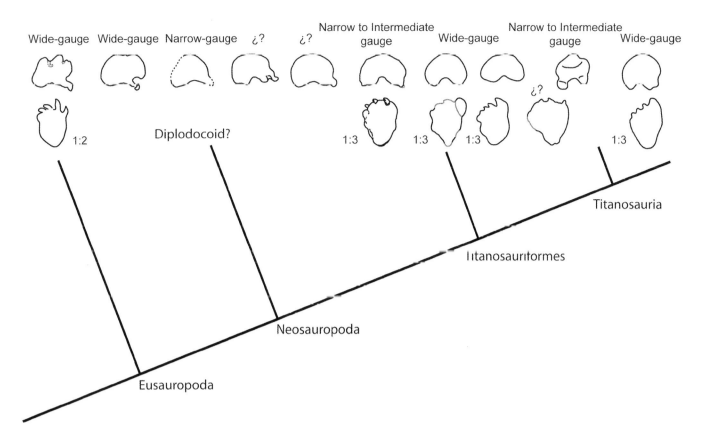

Wide-gauge Wide-gauge Narrow-gauge ¿? ¿? Narrow to Intermediate gauge Wide-gauge Narrow to Intermediate gauge Wide-gauge

1:2 Diplodocoid? 1:3 1:3 1:3 ¿? 1:3

Titanosauria

Titanosauriformes

Neosauropoda

Eusauropoda

8.8. Sauropod affinities of the studied sauropod tracks (based on synapomorphies proposed by Day et al. [2004], the cladogram and predicted ichnology proposed by Wright [2005], and the original papers). Drawings not to scale. The ¿? denotes uncertainty in the attribution of the tracks to a concrete group.

footprints either do not form part of a trackway or the gauge category is unknown, such as specimens ML965 (Kimmeridgian-Tithonian) and JMH1.7m (Tithonian-Berriasian), which have been associated with brachiosaurids (Milàn, Christiansen, and Mateus, 2005) and nontitanosauriform macronarians (Castanera, Canudo, et al., 2010; Castanera and Canudo, 2011) on the basis of the manus shape. These three groups of sauropods (diplodocids, nontitanosauriform macronarians, and basal titanosauriforms) have been described in the Jurassic-Cretaceous interval of the Iberian Peninsula, though manus and pedal remains are scarce at this time (Royo-Torres, 2009; Canudo et al., 2010; Barco and Canudo, 2012; Mannion et al., 2012) so these attributions are more complicated than in other cases due to the high levels of sauropod biodiversity in the Late Jurassic and the Jurassic-Cretaceous interval both worldwide and in the Iberian Peninsula (Royo-Torres et al., 2009; Mannion and Upchurch, 2010).

Trackways MNDPDSA-G1 and MNDPDSA-G5 from Galinha tracksite (Middle Jurassic) could represent a fifth type of sauropod trackway (Table 8.3) characterized by well-developed manus claw impressions and wide-gauge trackways (Figs. 8.7F, 8.7G); these footprints are thought to have been produced by nonneosauropod eusauropods (Santos, Moratalla, and Royo-Torres, 2009). This group was abundant in the Middle and Late Jurassic of Europe (Royo-Torres and

Upchurch, 2012). These trackways support the evidence that some groups of basal sauropods, as well as the titanosaurs, could have produced wide-gauge trackways (Wilson and Carrano, 1999; Santos, Moratalla, and Royo-Torres, 2009; D'Emic, 2012). Another challenging issue is the trackway from Los Cayos S, which does not show any evidence of a claw mark in the digit I impression in the manus tracks and is thought to have been produced by a titanosauriform even though it seems to be narrow/intermediate gauge (Fig. 8.7H) (Moratalla, Hernán, and Jiménez, 2003; Moratalla and Hernán, 2008). Titanosauriforms were abundant in the Lower Cretaceous faunas of the Iberian Peninsula (Canudo, Royo-Torres, and Cuenca-Bescós, 2008; Royo-Torres and Upchurch, 2012).

Within the subdivision of sauropod trackway types made by Day et al. (2004), the differences between the nontitanosauriforms, which are all represented by the first group, and the titanosauriforms, which are represented by the other three groups, suggest the difficulty of distinguishing between the various groups of nontitanosauriforms. In fact, Wright (2005) suggested that "the ideal tracks would also seem to indicate that the great majority of tracks were made by titanosauriforms or basal eusauropods and that nontitanosauriform neosauropods left very few tracks" (258). This author also proposed that this is possibly because "all neosauropods, possibly all eusauropods, had metacarpals arranged in a semicircular column, which would have produced horseshoe-shaped tracks" (258). Wright (2005) also proposed that the crescent-shaped morphology could be more indicative of neosauropods such as diplodocoids, which would have a wider metacarpal arch (210°), than other sauropods, which would have a horseshoe-shaped morphology (tight arch, 270°) and that this transition may occur in camarasauromorphs. Thus, in the light of Wright's proposal (2005), some of the Iberian kidney-shaped manus prints may represent the manus morphology of nontitanosauriform neosauropods (Fig. 8.8), though attribution to a concrete group of sauropods is difficult given the current state of knowledge.

Moreover, narrow-gauge trackways with high heteropody ratios have usually been associated with diplodocoids (Apesteguía, 2005; Wright, 2005). However, the narrow/intermediate trackways MI-VIII-B-1m and LCU-I-37 of the Iberian record have lower ratios (1:2 and 1:3, respectively), so according to these data, the trackmakers would not be diplodocoids. As yet, further work is needed in order to determine whether the distinction between different nontitanosauriform neosauropod groups is possible and to establish the role of parameters such as heteropody, trackway gauge, and size of the digit I claw mark in the different groups. As regards the pes print morphology, the most noticeable change occurred between the Middle and Late Jurassic, which seems to reflect a transition between eusauropods (claws anteriorly directed) and neosauropods (claws anterolaterally directed) and which may represent a functional change into a plantigrade posture (Bonnan, 2005; Wright, 2005). Attribution to or association with a distinct group of sauropods is uncertain due to the conservative morphology of the footprints. Nonetheless, it is noteworthy that the tracks that show the lateral notch have been associated with titanosauriforms (e.g., PM5-2/E11; LCU-I-37-12p; EMajS2.1). This lateral notch is also marked in the ichnogenus *Brontopodus birdi*, which has also usually been associated with titanosauriforms (Farlow, Pittman, and Hawthorne, 1989; Farlow, 1992; Farlow et al., 2012).

IMPLICATIONS FOR SAUROPOD ICHNOTAXONOMY

Since the initial subdivision of sauropod trackways on the basis of the trackway gauge (as narrow- and wide-gauge categories; Farlow, 1992), many trackways have been classified as belonging either to the ichnogenus *Parabrontopodus* or to *Brontopodus*, taking into account only this character (Lockley, Farlow, and Meyer, 1994). At present, this classification of sauropod trackways seems to be obsolete due to the description of new intermediate-/wide-gauge ichnogenera (Dalla Vecchia and Tarlao, 2000; Calvo and Mazzeta, 2004;

González-Riga and Calvo, 2009; Santos, Moratalla, and Royo-Torres, 2009). Recent papers have demonstrated that this parameter can change along the same trackway and that the subdivision is much more complicated than previously thought, being more like a continuum (Romano, Whyte, and Jackson, 2007; Marty, 2008; Moratalla, 2009; Santos, Moratalla, and Royo-Torres, 2009; Castanera et al., 2012).

The Iberian sauropod tracks from the studied trackways belong to various categories. In the Middle Jurassic (MND-PDSA-G1 and MNDPDSA-G5) and Late Cretaceous (ME#42, FS#29), they are truly wide-gauge trackways (Vila et al., 2008; Santos, Moratalla, and Royo-Torres, 2009), whereas in the Jurassic-Cretaceous interval and late Early Cretaceous, the limits are not so clear, the trackways being narrow in MI-VIII-B-1m (Castanera, Barco, et al., 2010), narrow-intermediate in LCU-I-37 and LCS9-R1 (Moratalla, Hernán, and Jiménez, 2003; Castanera et al., 2012), or intermediate to wide-gauge in LCR9-LCR14 (Castanera et al., 2011). In other trackways from these ages, the category cannot be established at all (SS1-R1, Meijide Fuentes, Fuentes Vidarte, and Meijide Calvo, 2001; Castanera et al., 2012; 1CT-2, Alcalá et al., 2014), whereas yet other studied tracks were found isolated, so the category is unknown. Lockley, Farlow, and Meyer (1994) proposed that heteropody is also a useful criterion for discriminating between sauropod tracks. In the case of the Iberian tracks, it ranges from 1:2 (MNDPDSA-G1, MNDPDSA-G5, and MI-VIII-B-1m) to 1:3 (JMH1, LCU-I-37, and LCR9–LCR14). This variation is not as great as that reported by Lockley, Farlow, and Meyer (1994), where the range is from 1:2 to 1:5, leading us to conclude that heteropody is not as useful for distinguishing between the Iberian sauropod tracks.

Despite the fact that heteropody, trackway gauge, and trackway configuration may play considerable roles in sauropod ichnotaxonomy (Lockley, Farlow, and Meyer, 1994; Marty et al., 2010), in the light of the Iberian sauropod track data, we agree with Wright (2005), who suggested that "sauropod tracks should be classified primarily on the basis of footprint morphology and only secondarily on the internal width of the trackway" (260). The variability in manus and pes morphology within the Iberian sauropod track record indicates that it may indeed be of major importance in sauropod ichnotaxonomy. Considering these data, we propose that:

1. The manus TLm/TWm ratio (Table 8.1) may be a valuable criterion for quantifying the metacarpal arrangement in order to discriminate between different types of manus.
2. Pollex size and orientation are also of considerable importance when clearly preserved.

3. The number and orientation of the digits and claw marks in the pes prints (cf. Lockley, Farlow, and Meyer, 1994) are also a significant element for discriminating between tracks.
4. The absence of a claw mark in digit IV is of major importance because it is not usually reported.
5. The presence/absence of a lateral notch might also be of considerable importance.

CONCLUSIONS

This selection of Iberian sauropod tracksites, which ranges in age from the Middle Jurassic to the Late Cretaceous, is unique worldwide as a sample of sauropod tracks showing anatomical details over a wide range of time. Belonging to four main types, the variety of sauropod manus tracks reflects the variation in the manus within sauropod groups through the ages. Nonneosauropod eusauropods (speech bubble–shaped, Middle Jurassic), nontitanosauriform neosauropods (kidney-shaped with claw mark, Late Jurassic–Early Cretaceous), and titanosauriforms (kidney-shaped without claw mark, Late Jurassic–Early Cretaceous) and titanosauriforms/titanosaurs (horseshoe-shaped, late Early Cretaceous–Late Cretaceous) are the groups probably represented. Among the pes prints, the presence of four claw marks indicates a primitive character (nonneosauropod eusauropods), whereas in the Late Cretaceous, the titanosaurs may have four claw marks as well, though in this case oriented laterally. The characters observed in the Iberian sauropod tracks indicate that the manus are varied enough to distinguish between different groups. Some characters in the pes tracks, such as claw marks (especially in digit IV), the lateral notch, or the morphology of the heel, are helpful as well. Other features such as heteropody and trackway gauge-type can help but are not so significant in the studied Iberian sample.

ACKNOWLEDGMENTS

This paper forms part of the projects CGL2010-16447 and CGL2011-30069-C02-01, subsidized by the Ministry of Economics and Competition, the European Regional Development Fund and the Government of Aragón, Spain (Grupos Consolidados and Dirección General de Patrimonio Cultural). This study was also funded by the PEst-OE/CTE/UI0611/2012 Project of the Centro de Geofísica da Universidade de Coimbra (Portugal). D.C. is the beneficiary of a grant from the Ministry of Education (AP2008-01340) and a grant from the Europa Cai-DGA program (CB 5/11). B.V. acknowledges support from the Ministry of Science and Innovation (Subprograma Juan de la Cierva [MICINN-JDC] 2011). V.F.S. acknowledges support from Centro de Investigação da Terra e do Espaço da UC-CITEUC, Universidade de Coimbra (Portugal). The authors want to acknowledge Alvaro Jalles, Celia Caciones, Gabriel Mendes, Marta Borges, Paulo Rodrigues, Sara Brito, and Vitor Amendoeira, members of the Federação Portuguesa de Espeleologia, for their help during the field trip and access to the Pedra da Mua tracksites. The authors also thank Alejandro Plaza from the Museo Numantino de Soria for the picture of the specimen (MNS2006/75/1) from Las Cuestas I tracksite and Rui Castanhinha and Alexandra Tomás from the Museu da Lourinhã for providing the abbreviation of the track (ML 1151). Rupert Glasgow revised the English grammar. The comments of the referees Matteo Belvedere and Sashima Läbe, and the editors greatly improved the manuscript and are also acknowledged.

REFERENCES

Alcalá, L., F. Pérez-Lorente, L. Luque, A. Cobos, R. Royo-Torres, and L. Mampel. 2014. Preservation of dinosaur footprints in shallow intertidal deposits of the Jurassic-Cretaceous transition in the Iberian Range (Teruel, Spain). Ichnos 21(1): 19–31.

Apesteguía, S. 2005. Evolution of the titanosaur metacarpus; pp. 321–345 in V. Tidwell and K. Carpenter (eds.), Thunder-Lizards: The Sauropodomorph Dinosaurs. Indiana University Press, Bloomington, Indiana.

Azeredo, A. C. 2007. Formalização da litostratigrafia do Jurássico Inferior e Médio do Maciço Calcário Estremenho (Bacia Lusitânica). Comunicações Geológicas 94: 29–51.

Barco, J. L., and J. I. Canudo. 2012. On the phylogenetic position of the sauropod *Galvesaurus*, and other reflections; pp.17–29 in P. Huerta and F. Torcida (eds.), Actas de V Jornadas Internacionales sobre Paleontología de Dinosaurios y su Entorno, Salas de los Infantes, Burgos.

Bonnan, M. F. 2003. The evolution of manus shape in sauropod dinosaurs: implications for functional morphology, forelimb orientation, and phylogeny. Journal of Vertebrate Paleontology 23 (3): 595–613.

Bonnan, M. F. 2005. Pes anatomy in sauropod dinosaurs: implications for functional morphology, evolution, and phylogeny; pp 346–380 in V. Tidwell and K. Carpenter (eds.), Thunder-Lizards: The Sauropodomorph Dinosaurs. Indiana University Press, Bloomington, Indiana.

Calvo, J., and G. V. Mazzetta. 2004. Nuevos hallazgos de huellas de dinosaurios en la Formación Candeleros (Albiano-Cenomaniano), Picún Leufú, Neuquén, Argentina. Ameghiniana 41(4): 545–554.

Canudo, J. I., R. Royo-Torres, and G. Cuenca-Bescós. 2008. A new sauropod: *Tastavinsaurus sanzi* gen. et sp. nov. from the Early Cretaceous (Aptian) of Spain. Journal of Vertebrate Paleontology 28(3): 712–731.

Canudo, J. I., J. L. Barco, D. Castanera, and F. Torcida Fernández Baldor. 2010. New record of a sauropod in the Jurassic-Cretaceous transition of the Iberian Peninsula (Spain): palaeobiogeographical implications. Paläontologische Zeitschrift 84: 427–435.

Carrano, M. T., and J. A. Wilson. 2001. Taxon distributions and the tetrapod track record. Paleobiology 27(3): 564–582.

Castanera, D., and J. I. Canudo. 2011. Los saurópodos del intervalo Jurásico-Cretácico de la Cordillera Ibérica: icnitas vs huesos; pp. 101–110 in A. Pérez-García, F. Gascó, J. M. Gasulla, and F. Escaso (eds.), Viajando a Mundos Pretéritos. Ayuntamiento de Morella, Morella, Castellón, Spain.

Castanera, D., C. Pascual, and J. I. Canudo. 2013 Primera evidencia de la morfología del pie en saurópodos en el grupo Urbión del Cretácico Inferior de la Cuenca de Cameros, (Soria). Geogaceta 53: 13–16.

Castanera, D., J. L. Barco, J. I. Canudo, and C. Pascual. 2010. Aproximación a la diversidad de morfotipos de icnitas de saurópodo en la aloformación Huérteles (Berriasiense) en Soria (España). Cidaris 30: 91–97.

Castanera, D., J. I. Canudo, I. Díaz-Martínez, J. Herrero Gascón, and F. Pérez-Lorente. 2010. Grandes contramoldes de icnitas de saurópodos en el Tithónico-Berriasiense de la Formación Villar del Arzobispo en Galve (Teruel); pp. 178–183 in J. L. Ruiz-Omeñaca, L. Piñuela, and J. C. García-Ramos (eds.), Comunicaciones del V Congreso del Jurásico de España. Museo del Jurásico de Asturias (MUJA), Colunga, Spain.

Castanera, D., C. Pascual, J. I. Canudo, N. Hernández, and J. L. Barco. 2012. Ethological variations in gauge in sauropod trackways from the Berriasian of Spain. Lethaia 45: 476–489.

Castanera, D., J. L. Barco, I. Díaz-Martínez, J. Herrero-Gascón, F. Pérez-Lorente, and J. I. Canudo. 2011. New evidence of a herd of titanosauriform sauropods from the Lower Berriasian of the Iberian range (Spain). Palaeogeography, Palaeoclimatology, Palaeoecology 310: 227–237.

Castanera, D., B. Vila, N. L. Razzolini, V. F. Santos, C. Pascual, and J. I. Canudo. 2014. Sauropod trackways of the Iberian Peninsula: palaeontological and palaeoenvironmental implications. Journal of Iberian Geology 40(1): 49–59.

Cobos, A. 2011. Los dinosaurios de Teruel como recurso para el desarrollo territorial. Ph.D. dissertation, Universidad del País Vasco, Bilbao, Spain, 560 pp.

Cobos, A., R. Royo-Torres, L. Luque, L. Alcalá, and L. Mampel. 2010. An Iberian stegosaurs paradise: the Villar del Arzobispo Formation (Tithonian-Berriasian) in Teruel (Spain). Palaeogeography, Palaeoclimatology, Palaeoecology 293: 223–236.

Dalla Vecchia, F. M., and A. Tarlao. 2000. New dinosaur tracks sites in the Albian (Early Cretaceous) of the Istrian peninsula (Croatia). Parte II–Paleontology. Memorie di Scienze Geologiche 52(2): 227–292.

Day, J. J., D. B. Norman, A. S. Gale, P. Upchurch, and P. Powell. 2004. A middle Jurassic dinosaur trackway site from Oxfordshire, UK. Palaeontology 47(2): 319–348.

D'Emic, M. D. 2012. The early evolution of titanosauriform sauropod dinosaurs. Zoological Journal of the Linnean Society 166: 624–671.

Farlow, J. O. 1992. Sauropod tracks and trackmarkers: integrating the ichnological and skeletal records. Zubia 10: 89–138.

Farlow, J. O., J. F. Pittman, and J. M. Hawthorne. 1989. Brontopodus birdi, Lower Cretaceous Sauropod footprints from the U.S. Gulf Coastal Plain; pp. 371–393 in D. D. Gillette and M. G. Lockley (eds.), Dinosaur Tracks and Traces. Cambridge University Press, Cambridge, UK.

Farlow, J. O., M. O'Brien, G. J. Kuban, B. F. Dattilo, K. T. Bates, P. L. Falkingham, L. Piñuela, A. Rose, A. Freels, C. Kumagai, C. Libben, J. Smith, and J. Whitcraft. 2012. Dinosaur tracksites of the Paluxy River Valley (Glen Rose Formation, Lower Cretaceous), Dinosaur Valley State Park, Somervell County, Texas; pp. 41–69 in P. Huerta and F. Torcida (eds.), Actas de V Jornadas Internacionales sobre Paleontología de Dinosaurios y su Entorno, Salas de los Infantes, Burgos.

Fernández Barrenechea J. 1993. Evolución de la mineralogía de arcillas en el tránsito diagénesis-metamorfismo de bajo grado en el Grupo Urbión (Cretácico inferior) de la cuenca de Los Cameros (Soria–La Rioja). Ph.D. dissertation, Universidad Complutense de Madrid, Facultad de Ciencias Geológicas, Madrid, Spain.

García-Ramos, J. C., J. Lires, and L. Piñuela. 2002. Dinosaurios: Rutas por el Jurásico de Asturias. La Voz de Asturias, Lugones, Spain, 204 pp.

García-Ramos, J. C., L. Piñuela, and J. Lires. 2006. Atlas del Jurásico de Asturias. Ediciones Nobel, Oviedo, Spain, 225 pp.

González-Riga, B., and J. O. Calvo. 2009. A new wide-gauge sauropod track site from the Late Cretaceous of Mendoza, Neuquén Basin, Argentina. Palaeontology 52(3): 631–640.

González-Riga, B. J., J. O. Calvo, and J. Porfiri. 2008. An articulated titanosaur from Patagonia (Argentina): new evidence of neosauropod pedal evolution. Palaeoworld 17: 33–40.

Hall, L. E., A. E. Fragomeni, and D. E. Fowler. 2016. The flexion of sauropod pedal unguals and testing the substrate grip hypothesis using the trackway fossil record; chap. 9 in P. L. Falkingham, D. Marty, and A. Richter (eds.), Dinosaur Tracks: The Next Steps. Indiana University Press, Bloomington, Indiana.

Hernández Medrano, N., C. Pascual-Arribas, P. Latorre Macarrón, and E. Sanz Pérez. 2008. Contribución de los yacimientos de icnitas sorianos al registro general de Cameros. Zubía 23–24: 79–120.

Kim, J. Y., and M. G. Lockley. 2012. New sauropod tracks (Brontopodus pentadactylus ichnosp. nov.) from the Early Cretaceous Haman Formation of Jinju Area, Korea: implications for sauropod manus morphology. Ichnos 19: 84–92.

Latorre Macarrón, P., C. Pascual-Arribas, E. Sanz Pérez, and N. Hernández Medrano. 2006. El yacimiento con huellas de saurópodos de Miraflores, Fuentes de Magaña (Soria, España); pp. 273–296 in Colectivo Arqueológico y Paleontológico de Salas (ed.), Actas de las III Jornadas Internacionales sobre Paleontología de Dinosaurios y su Entorno, Salas de los Infantes, Burgos.

Lockley, M. G., J. O. Farlow, and C. A. Meyer. 1994a. Brontopodus and Parabrontopodus ichnogen. nov. and the significance of wide- and narrow-gauge sauropod trackways. Gaia 10: 135–145.

Lockley, M. G., C. A. Meyer, and V. F. Santos. 1994b. Trackway evidence for a herd of juvenile sauropods from the Late Jurassic of Portugal. Gaia 10: 27–35.

Lockley, M., J. C. García-Ramos, L. Piñuela, and M. Avanzini. 2008. A review of vertebrate track assemblages from the Late Jurassic of Asturias, Spain with comparative notes on coeval ichnofaunas from the western U.S.A.: implications for faunal diversity in siliciclastic facies assemblages. Oryctos 8: 53–70.

Lockley, M. G., J. Lires, J. C. García-Ramos, L. Piñuela, and M. Avanzini. 2007. Shrinking the world's largest dinosaur tracks: observations of the Ichnotaxonomy of Gigantosauropus asturiensis and Hispanosauropus hauboldi from the Upper Jurassic of Asturias, Spain. Ichnos 14: 247–255.

Lockley, M., A. S. Schulp, C. A. Meyer, G. Leonardi, and D. K. Mamani. 2002. Titanosaurid trackways from the Upper Cretaceous of Bolivia: evidence for large manus, wide-gauge locomotion and gregarious behaviour. Cretaceous Research 23: 383–400.

Mannion, P. D., and P. Upchurch. 2010. A quantitative analysis of environmental associations in sauropod dinosaurs. Paleobiology 36(2): 253–282.

Mannion, P. D., P. Upchurch, O. Mateus, R. N. Barnes, and M. E. H. Jones. 2012. New information on the anatomy and systematic position of Dinheirosaurus lourinhanensis (Sauropoda: Diplodocoidea) from the Late Jurassic of Portugal, with a review of European diplodocoids. Journal of Systematic Palaeontology 10(3): 521–551.

Marmi, J., B. Vila, C. Martín-Closas, and S. Villalba-Breva. 2014. Reconstructing the foraging environment of the latest titanosaurs (Fumanya dinosaur tracksite, Catalonia). Palaeogeography, Palaeoclimatology, Palaeoecology 410: 380–389.

Marty, D. 2008. Sedimentology, taphonomy, and ichnology of Late Jurassic dinosaur tracks from the Jura carbonate platform (Chevenez-Combe Ronde tracksite, NW Switzerland): insights into the tidal-flat palaeoenvironment and dinosaur diversity, locomotion, and palaeoecology. GeoFocus 21. Fribourg University, Fribourg, Switzerland, 278 pp.

Marty, D., M. Belvedere, C. A. Meyer, P. Mietto, G. Paratte, C. Lovis, and B. Thüring. 2010. Comparative analysis of Late Jurassic sauropod trackways from the Jura Mountains (NW Switzerland) and the central High Atlas Mountains (Morocco): implications for sauropod ichnotaxonomy. Historical Biology 22(1–3): 109–133.

Mateus, O., and J. Milàn. 2010. A diverse Upper Jurassic dinosaur ichnofauna from central-west Portugal. Lethaia 43: 245–257.

Meijide Fuentes, F., C. Fuentes Vidarte, and M. Meijide Calvo. 2001. Primeras huellas de saurópodo en el Weald de Soria (España), Parabrontopodus distercii, nov. ichnoesp; pp. 407–415 in Colectivo Arqueológico y Paleontológico de Salas (ed.), Actas de las I Jornadas Internacionales sobre Paleontología de Dinosaurios y su Entorno, Salas de los Infantes, Burgos.

Meyer, C. A., M. G. Lockley, J. Robinson, and V. F. Santos. 1994. A comparison of well-preserved sauropod tracks from the Late Jurassic of Portugal and the western United States: evidence and implications. Gaia 10: 57–64.

Milàn, J., P. Christiansen, and O. Mateus. 2005. A three-dimensionally preserved sauropod manus impression from the Upper Jurassic of Portugal: implications for sauropod manus shape and locomotor mechanics. Kaupia. Darmstädter Beiträge zur Naturgeschichte 14: 47–52.

Moratalla, J. J. 2009. Sauropod tracks of the Cameros Basin (Spain): identification, trackway patterns and changes over the Jurassic-Cretaceous. Geobios 42: 797–811.

Moratalla, J. J., and J. Hernán. 2008. Los Cayos S y D: dos afloramientos con icnitas de saurópodos, terópodos y ornitópodos en el Cretácico inferior del área de Los Cayos (Cornago, La Rioja, España). Estudios Geológicos 64(2): 161–173.

Moratalla, J. J., J. Hernán, and S. Jiménez. 2003. Los Cayos dinosaur tracksite: an overview on the Lower Cretaceous ichno-diversity of the Cameros Basin (Cornago, La Rioja Province, Spain). Ichnos 10: 229–240.

Nair, J. P., and S. W. Sallisbury. 2012. New anatomical information on *Rhoetosaurus brownie* Longman, 1926, a gravisaurian sauropodomorph dinosaur from the Middle Jurassic of Queensland, Australia. Journal of Vertebrate Paleontology 32(2): 369–394.

Pascual-Arribas, C., and N. Hernández-Medrano. 2011. Nuevos datos sobre el yacimiento icnítico de Las Cuestas I (Santa Cruz de Yanguas, Soria. España). Studia Geologica Salmanticensia 46: 121–157.

Pascual-Arribas, C., N. Hernández-Medrano, P. Latorre Macarrón, and E. Sanz Pérez. 2008. Estudio de un rastro de huellas de saurópodo del yacimiento de Las Cuestas I (San Cruz de Yanguas, Soria, España): implicaciones taxonómicas. Studia Geologica Salmanticensia 44: 13–40.

Pérez-Lorente, F. 2003. Icnitas de dinosaurios del Cretácico en España; pp. 49–108 in F. Pérez-Lorente (coord.), Dinosaurios y Otros Reptiles Mesozoicos de España. Fundación Patrimonio Paleontológico de La Rioja. Instituto de Estudios Riojanos (IER) Ciencias de la Tierra, Logroño, Spain.

Quijada, I. E., P. Suarez-Gonzalez, M. I. Benito, and R. Mas. 2013. New insights on stratigraphy and sedimentology of the Oncala Group (eastern Cameros Basin): implications for the paleogeographic reconstruction of NE Iberia at Berriasian times. Journal of Iberian Geology 39(2): 313–334.

Romano, M., M. A. Whyte, and S. J. Jackson. 2007. Trackway ratio: a new look at trackway gauge in the analysis of quadrupedal dinosaur trackways and its implications for ichnotaxonomy. Ichnos 14: 257–270.

Royo-Torres, R. 2009. Los dinosaurios saurópodos en la Península Ibérica; pp. 139–166 in P. Huerta and F. Torcida (eds.), Actas de las IV Jornadas Internacionales sobre Paleontología de Dinosaurios y su Entorno. Salas de los Infantes, Burgos.

Royo-Torres, R., and P. Upchurch. 2012. The cranial anatomy of the sauropod *Turiasaurus riodevensis* and implications for its phylogenetic relationships. Journal of Systematic Palaeontology 10(3): 553–583.

Royo-Torres, R., A. Cobos, L. Luque, A. Aberasturi, E. Espílez, I. Fierro, A. González, L. Mampel, and L. Alcalá. 2009. High European sauropod dinosaur diversity during Jurassic-Cretaceous transition in Riodeva (Teruel, Spain). Palaeontology 5: 1009–1027.

Salgado, L., R. A. Coria, and J. O. Calvo. 1997. Evolution of titanosaurid sauropods. I: Phylogenetic analysis based on the postcranial evidence. Ameghiniana 34(1): 3–32.

Santos, V. F. 2003. Pistas de dinossáurio no Jurássico-Cretácico de Portugal. Considerações paleobiológicas e paleoecológicas. Ph.D. dissertation, Universidad Autónoma de Madrid, Madrid, Spain, 365 pp.

Santos, V. F. 2008. Pegadas de dinossáurios de Portugal. Museu Nacional de História Natural da Universidade de Lisboa, Lisboa, Spain, 124 pp.

Santos, V. F., J. F. Moratalla, and R. Royo-Torres. 2009. New sauropod trackways from the Middle Jurassic of Portugal. Acta Palaeontologica Polonica 54(3): 409–422.

Santos, V. F., M. G. Lockley, C. A. Meyer, J. Carvalho, A. M. Galopim de Carvalho, and J. J. Moratalla. 1994. A new sauropod tracksite from the Middle Jurassic of Portugal. Gaia 10: 5–13.

Schulp, A. S., and W. A. Brokx. 1999. Maastrichtian sauropod footprints from the Fumanya site, Berguedà, Spain. Ichnos 6(4): 239–250.

Thulborn, T. 1990. Dinosaur Tracks. Chapman and Hall, London, U.K., 410 pp.

Upchurch, P. 1994. Manus claw function in sauropod dinosaurs. Gaia 10: 161–173.

Upchurch, P., P. M. Barrett, and P. Dodson. 2004. Sauropoda; pp. 259–322 in D. B. Weishampel, P. Dodson, and H. Osmólska (eds.), The Dinosauria. University of California Press, Berkeley, California.

Vila, B., O. Oms, and À. Galobart. 2005. Manus-only titanosaurid trackway from Fumanya (Maastrichtian, Pyrenees): further evidence for an underprint origin. Lethaia 38: 211–218.

Vila, B., O. Oms, J. Marmi, and À. Galobart. 2008. Tracking Fumanya footprints (Maastrichtian, Pyrenees): historical and ichnological overview. Oryctos 8: 115–130.

Vila, B., À. Galobart, J. I. Canudo, J. Le Loeuff, J. Dinarès-Turell, V. Riera, O. Oms, T. Tortosa, and R. Gaete. 2012. The diversity of sauropod dinosaurs and their first taxonomic succession from the latest Cretaceous of southwestern Europe: clues to demise and extinction. Palaeogeography, Palaeoclimatology, Palaeoecology 350–352: 19–38.

Wilson, J. A. 2002. Sauropod dinosaur phylogeny: critique and cladistic analysis. Zoological Journal of the Linnean Society 136(2): 215–275.

Wilson, J. A., and M. T. Carrano. 1999. Titanosaurs and the origin of "wide-gauge" trackways: a biomechanical and systematic perspectives on sauropod locomotion. Paleobiology 25(2): 252–267.

Wright, J. L. 2005. Steps in understanding sauropod biology: the importance of sauropod tracks; pp. 252–284 in K. A. Curry Rogers and J. A. Wilson (eds.), The Sauropods: Evolution and Paleobiology. University of California Press, Berkeley, California.

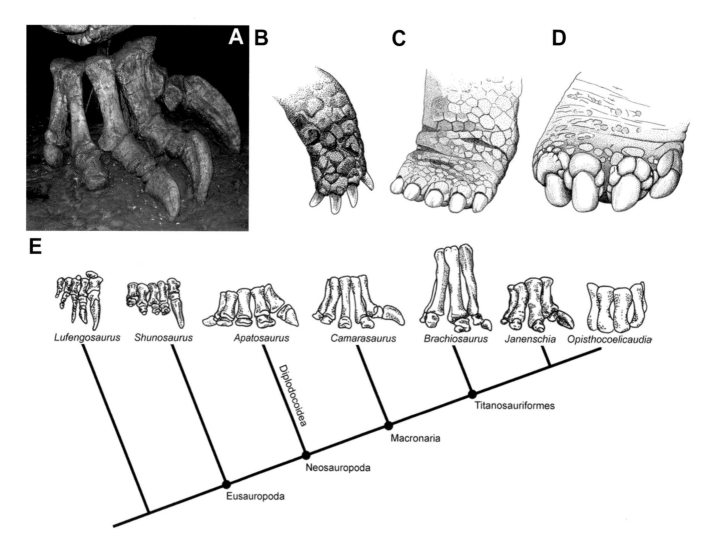

9.1. (A) Mounted diplodocid right pes. Note the deep, laterally compressed unguals and en-echelon arrangement. Morrison Formation, Late Jurassic. Currently on display at New Mexico Museum of Natural History and Science. (B–D) Feet of the specialized scratch-digging tortoise, *Gopherus*. Note the curvature of the flattened unguals and their similarly en-echelon arrangement. (B) Left manus of *Gopherus canyonensis*, from Bramble (1982); (C) left manus and (D) pes of *Gopherus polyphemus*, from Auffenberg (1976). (E) Phylogenetic distribution of sauropod manual morphology (right manus depicted). Manual phalanges exhibit a phylogenetic trend toward reduction and loss, retaining only digit I; derived titanosauriforms take this even further, losing all manual phalanges. *Reproduced from Figure 3, Fowler and Hall (2011).*

The Flexion of Sauropod Pedal Unguals and Testing the Substrate Grip Hypothesis Using the Trackway Fossil Record

9

Lee E. Hall, Ashley E. Fragomeni, and Denver W. Fowler

SAUROPOD PEDES EXHIBIT A UNIQUE, HIGHLY DERIVED pedal ungual morphology and articulation. During plantar flexion of the pes, the spade-like, laterally compressed unguals are rotated ventrally and deflected laterally across the front of the pes so the claws overlap in an en-echelon fashion; this positions the dorsal margins ventrally and the medial sides face posteriorly, creating a hoe-like structure oriented perpendicular to the plane of limb movement. Several functional hypotheses have been erected in an attempt to explain this feature. One, the miring-avoidance via substrate grip hypothesis, suggests orientation of the unguals during plantar flexion was utilized to generate traction and prevent miring while sauropods traversed muddy substrates. In this study, we test this hypothesis by examining the sauropod fossil trackway record for evidence of plantar flexion utilized during interaction between sauropod pedes and muddy substrates. A review of published sauropod trackways, various unpublished data, and examination of natural track casts (track infillings), found no evidence of plantar flexion being employed during locomotion on muddy substrates. In contrast, trackway evidence unequivocally shows that sauropods utilized plantar extension when walking in soft mud, directing the unguals anteriorly and orienting them vertically. Plantar extension may have prevented torsion or lateral sliding of the limb (much like sports cleats) and could have aided in miring prevention by carving channels for air to circulate around the pes. Detailed skin impressions show sauropod feet were also covered in a rugose, coarsely textured surface of scales and warty tubercles, which could have functioned like off-road tire tread. We maintain that plantar flexion was a feature adapted for scratch-digging behaviors, as similarly described in tortoises, and was utilized for excavating nesting structures.

INTRODUCTION

As in vivo records of substrate interaction, tracks have great potential to test paleobiological hypotheses of behavior. Published examples in the dinosaur literature include inferences of trophic relationships (Thulborn and Wade, 1984; Thomas

and Farlow, 1997), sociality (Lockley, Young, and Carpenter, 1983; Lockley, Farlow, and Meyer, 1994; Matsukawa, Matsui, and Lockley, 2001), and locomotor behaviors (Lockley and Hunt, 1995; Gatesy et al., 1999; Wilson and Carrano, 1999; Wright, 2005; Milner, Christiansen, and Mateus, 2006; Milner et al., 2009). Sauropod dinosaurs were the largest land animals of all time and exhibited a number of unusual morphological features potentially reflecting the interplay of large body size with the unique biology and life history of dinosaurs (Sander et al., 2010; Varricchio, 2011). One such feature is sauropod pedal unguals, which are large, laterally compressed, deep, and are oriented in an overlapping en-echelon arrangement (Fig. 9.1A; Fowler and Hall, 2011). Gallup (1989) noted that the short robust phalanges, strengthening of interphalangeal articulation, and mobility of the unguals of sauropod dinosaurs are similar to scratch-digging mammals (Hildebrand, 1985), and he suggested that sauropod pedal unguals' primary use was in scratch-digging (specifically for nest excavation). Later, Bonnan (2005) described the changes in ungual orientation that occurred during plantar flexion, where the asymmetric position of the flexor tubercle of the ungual causes lateral rotation of the claw. This brings the deep, flattened medial side of the unguals to face ventrally (down into the substrate; also noted in less detail by Langston, 1974; and Gallup, 1989) so that the claws would be at an angle to the flat base of the foot, thus forming a hoe-like shape (Fowler and Hall, 2011). However, Bonnan (2005) considered a scratch-digging function unlikely and instead preferred Gallup's alternative hypothesis: that unguals were adapted for substrate grip, providing traction and perhaps preventing miring (although it should be noted that this functional suggestion was only a very minor conclusion). In a 2011 study, we revived the scratch-digging function by comparing sauropod pedal unguals with those of extant tortoises (Fowler and Hall, 2011), which have a similarly unusual shape and orientation (Fig. 9.1B), and which are thought to be adapted for scratch-digging (Bramble, 1982; Ruby and Niblick, 1994). We further suggested that substrate grip was not a selection pressure acting on sauropod ungual morphology based on the observation that sauropod forelimbs underwent a rapid and drastic reduction of unguals throughout their evolution (Fig. 9.1C). Furthermore, sauropod eggs were laid into an elongate, trough-like hollow excavated into the substrate (Fowler and Hall, 2011:fig. 9), which is consistent with the hypothesis that the hind limbs were used during nest excavation (e.g., Vila et al., 2010).

Approaching function purely from a morphological perspective can be problematic as a single anatomical structure may be utilized for several purposes, which may require the same morphological characteristics. In our example, it is difficult to separate any potential morphological characteristics of a substrate grip function from that of scratch-digging, as scratch-digging requires initial gripping of the sediment before subsequent removal. As such, neither suggested function has been adequately tested.

Here we address this apparent impasse by investigating direct evidence of behavior, rather than inferring behavior or function from morphology alone. Despite the possibly similar physical requirements of scratch-digging and substrate grip, these behaviors would be employed at different moments in the animal's lifetime. This consequently permits formation of testable hypotheses dependent on the observation of specific behaviors being employed at the expected times. In our prior paper (Fowler and Hall, 2011), we reviewed evidence that clearly showed that the hind feet were used in nest excavation behavior; whether the pedal unguals were specifically adapted for nest digging remains an open question (we believe that they are), but it is incontrovertible that they were used to actually dig the nests: scratch-digging for nests passed from a hypothetical function (Gallup, 1989) into a factual function (Fowler and Hall, 2011).

In this study, we examine the fossil trackway record to assess sauropod pedal interaction with the substrate during walking, specifically to search for any evidence that may corroborate a hypothetical substrate grip function. Can this similarly be considered fact?

It is worth noting here the difference between function (or use) and adaptation, at least as used in this study. A function is any use for a structure, whereas an adaptation is a function that is being positively selected for through evolution (either to maintain a particular morphology or to further change morphology of the structure to be increasingly better suited to a particular function). The critical distinction is that it may not matter how many different functions a structure is used for if only one function is being selected for. Moreover, some functions may be being positively selected, whereas others (even common functions) may be negatively selected; a human example is that although we can use our arms and legs to climb trees, our ability to do so is markedly reduced from that of our ancestors. Human limb morphology has evolved (possibly adapted) away from arboreal functions,

despite our ability to still use our limbs for this function. This may imply that a different function is being selected for (although care must be taken to avoid straw-man arguments in this context). Thus, establishing adaptation is difficult, as you must be able to show evolution of a given structure and function through phylogeny. In the case of sauropod pedal unguals, adaptation is implied by the evolutionary development of unusual ungual shape and ungual orientation during plantar flexion of the hind limb, especially when contrasted with the reduction and loss of unguals (and in some cases, even all phalanges) on the forelimb. We would tentatively suggest that adaptation is demonstrable in sauropods. Regardless, any inference of adaptation first requires that a hypothetical function is actually employed in that manner by the animal, and this is something that we can easily look for in evidence of behavior. In the case of substrate grip during walking, tracks are that evidence.

TESTING HYPOTHESES

Fossil sauropod tracks and natural track casts often preserve impressions of individual digits and unguals, presenting an opportunity to reexamine the substrate grip hypothesis. Tracks are typically preserved in originally wet sediments; these are the conditions in which there is risk of the animal becoming mired, and hypothesized substrate grip behavior (Bonnan, 2005) might have been utilized. Bonnan (2005:358) reviewed sauropod tracks with ungual impressions and suggested that claws were "deeply impressed," but they did not note any tracks in which the pes digits were plantarflexed below the footpads into the substrate. This might be expected if the unusual ungual orientation that occurs during digital plantar flexion was an adaptation for substrate grip (Fig. 9.2). In contrast, Pittman and Gillette (1989; pers. comm., 2012) demonstrated that sauropod pedal digits were extended while walking in muddy substrates, and they suggested that they were "wrapped around the lateral margin of the foot" (331) on firmer substrates. A substrate grip function can therefore be broken down into two component functions: traction control (Gallup, 1989; Pittman, 1989; Bonnan, 2005) and miring-avoidance (Bonnan, 2005).

Traction control suggests that sauropod unguals functioned like the bladed-cleats mounted on the soles of many modern athletic shoes; being variably positioned along the long axis of the sole, bladed cleats present a broad surface that is oriented perpendicular to the direction of movement. This helps prevent lateral sliding, increases traction, and decreases torsion (helping avoid twisted ankles). The traction control hypothesis would be supported by observation of plantarflexed ungual impressions in sauropod trackways, especially so if they are oriented perpendicular to the direction of movement. Conversely, a lack of plantarflexed ungual impressions would fail to find support for the traction control hypothesis.

A miring-avoidance function is problematic because Bonnan (2005) did not explicitly explain the mechanics of how this works, and we could not find prior publications describing any man-made or animal morphological analogues. Some animals adapted for snow-covered areas (e.g., snowshoe hare) possess broad spreading feet to distribute weight over a larger area (preventing sinking), and some waterfowl (e.g., coots) possess lobed toes that increase surface area of the foot (aiding swimming) but no examples are known of animals possessing such feet to avoid miring in mud (Fowler and Hall, 2011; although see "post-holing" hypothesis in "Discussion").

This study further analyzes the track record to see whether support is found for either the traction control or miring-avoidance hypotheses. It is important to note that we are essentially querying the potential adaptive function of the unusual plantar flexion digit rotation that was described by Bonnan (2005); in other words, that if the unusual rotation and orientation evolved for either of the substrate-grip hypotheses, then pes tracks should record digital and/or ungual plantar flexion. Therefore, we expect to make one of three observations where claw impressions are visible in tracks made into muddy substrates: (1) plantar flexion; (2) plantar extension; or (3) "null" where the digits will be in a neutral or intermediate position.

MATERIALS AND METHODS

Track morphology was assessed by referencing photographs published in the literature and provided from researchers' personal data sets, and it was corroborated by physical examination of cast specimens. Specimens were selected using the following standards: (1) morphological landmarks of the pes preserved included details of the unguals to allow examination and interpretation of their positions and actions during track making; (2) specimens exhibited morphology consistent with Neosauropoda (sensu Bonaparte, 1986), wherein the anterolaterally directed, laterally compressed, shovel-shaped unguals are ventrolaterally deflected (Bonnan, 2005). We recorded the presence or absence of external morphological features of the pes, namely impressions of separate digits, including the unguals and/or skin and scales. Preservation of these features indicated little to no secondary deformation. Undertracks, tracks made in especially waterlogged sediments (e.g., the Blue Hole Ballroom large sauropod trackway of Farlow et al., 2012), or generally poor preservation may obscure the features of the pes and thus lack necessary detail. Natural track casts were included in this analysis provided

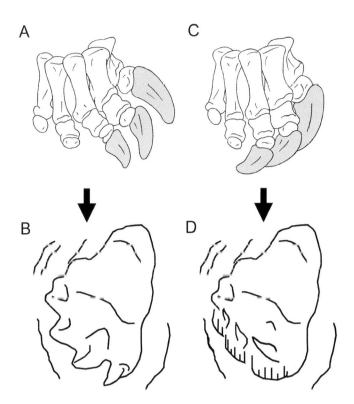

9.2. Extended and flexed sauropod unguals. (A) Foot skeleton and (B) track line drawing (from track 3 in the Blue Hole Ballroom "Small sauropod sequence") of a right sauropod pes in digital extension. The ungual of digit I is extended anteriorly and oriented nearly vertically, whereas digit II is deflected anterolaterally and slightly medially canted, and digit III is oriented laterally and strongly medially canted. Early Cretaceous Glen Rose Formation. (C) Foot skeleton and (D) hypothetical track line drawing of a right sauropod pes in a flexed position. The unguals of digits I–III are anterolaterally to laterally directed and strongly medially canted such that the broad medial surfaces of the unguals face nearly ventrally and overlap to form a "hoe-like" surface perpendicular to forward motion along the substrate plane. A hypothetical track should exhibit lateral deflection of the unguals along with gouge marks indicative of the anterior portion of the pes receiving an increased amount of force during the imprinting and push-off phases of the step, which also results in the anterior margin of the pes being most deeply impressed. However, these landmarks of a flexed pes were not observed in this study.

they satisfied the criteria set forth herein. Most trackways that we observed either consisted of undertracks (e.g., Moratalla et al., 1994; Bilbey et al., 2005) or were eroded or weathered such that recognition of morphological landmarks was not possible (e.g., Ferrusquía-Villafranca, Jiménez-Hidalgo, and Bravo-Cuevas, 1996; Ahmed, Lingham-Soliar, and T. Broderick, 2004; Moratalla, 2009). Out of the photos of tracks and trackways examined in the literature (including high-resolution images provided by researchers), 34 sauropod track specimens were found that exhibited especially well-preserved ungual impressions (Table 9.1).

Substrate type was noted for each specimen. Tracks tend to preserve in sediments such as wet mud, silt, or sand, which arguably may be considered a hindrance to locomotion. This satisfies the environmental conditions necessary for testing the original flexion-substrate-grip hypothesis as laid out by

Table 9.1. Sauropod pes tracks and natural track casts

Specimen	Formation	Age	Reference
'Morphotype 1' pes track from Agua del Choque track site level An-1	Anacleto	Late Cretaceous	González et al. (2014: fig. 3, 6)
'Brooklyn College' from West Bank Bird Site trackway S4	Glen Rose	Early Cretaceous	Farlow, Pittman, and Hawthorne (1989); Farlow et al. (2012: fig. 7A); Farlow (pers. comm., 2012)
Brontopodus, Blue Hole Ballroom 'Small sauropod sequence'. track 1	Glen Rose	Early Cretaceous	Farlow (et al. 2012: fig. 15B); Farlow (pers. comm., 2012)
Brontopodus, Blue Hole Ballroom 'Small sauropod sequence'. track 2	Glen Rose	Early Cretaceous	Farlow et al. (2012: fig. 15B); Farlow (pers. comm., 2012)
Brontopodus, Blue Hole Ballroom 'Small sauropod sequence'. track 3	Glen Rose	Early Cretaceous	Farlow et al. (2012: fig. 15B, 15C); Farlow (pers. August, 2012)
Brontopodus, 'Left manus-pes set, West Bank Bird Site'	Glen Rose	Early Cretaceous	Farlow et al. (2012: fig. 6B, 15E); Farlow (pers. comm., 2012)
'Right manus-pes set S2M (trackway S2)' and pes track cast of HMNH slab	Glen Rose	Early Cretaceous	Farlow, Pittman, and Hawthorne (1989: fig. 42.9 A–42.9C); Farlow et al. (2012: fig. 15F); Farlow (pers. comm., 2012)
'S2N' and pes track cast of AMNH slab 3065	Glen Rose	Early Cretaceous	Farlow (1989: fig. 42.9 D–42.9F); Farlow et al. (2012: fig. 20); Farlow (pers. comm., 2012)
'S2O' of AMNH slab 3065	Glen Rose	Early Cretaceous	Farlow, Pittman, and Hawthorne (1989: fig. 42.9G); Farlow et al. (2012: fig. 20); Farlow (pers. comm., 2012)
'S2P' of AMNH slab 3065	Glen Rose	Early Cretaceous	Farlow (1989: fig. 42.9H); Farlow et al. (2012: fig. 20); Farlow (pers. comm., 2012)
'S2Q' and pes track cast of AMNH slab 3065	Glen Rose	Early Cretaceous	Farlow, Pittman, and Hawthorne (1989: fig. 42.9I, 42.9J); Farlow et al. (2012: fig. 20); Farlow (pers. comm., 2012)
'S2R' and pes track cast of AMNH slab 3065	Glen Rose	Early Cretaceous	Farlow (1989: fig. 42.9K–42.9O); Farlow et al. (2012: fig. 20); Farlow (pers. comm., 2012)
'S2S' of TMM slab 40638–1	Glen Rose	Early Cretaceous	Farlow, Pittman, and Hawthorne (1989: fig. 42.4); Farlow (pers. comm., 2012)
'S2T' of TMM slab 40638–1	Glen Rose	Early Cretaceous	Farlow, Pittman, and Hawthorne (1989: fig. 42.4); Farlow (pers. comm., 2012)
'S2U' of TMM slab 40638–1	Glen Rose	Early Cretaceous	Farlow, Pittman, and Hawthorne (1989: fig. 42.4); Farlow (pers. comm., 2012)
'S2V' of TMM slab 40638–1	Glen Rose	Early Cretaceous	Farlow, Pittman, and Hawthorne (1989: fig. 42.4); Farlow (pers. comm., 2012)
'S2W' of TMM slab 40638–1	Glen Rose	Early Cretaceous	Farlow, Pittman, and Hawthorne (1989: fig. 42.4); Farlow (pers. comm., 2012)
LCU-I-37-12p pes track	Huerteles	Early Cretaceous	Castanera et al. (2012: fig. 4)
Dinosaur Farm Titanosaurid sauropod pes natural track cast	Wessex	Early Cretaceous	Fowler and Hall (2011: fig. 4)
'Praia das Gentias' pes track	Lourinhã	Late Jurassic	Coutinho (2012pers. comm.)
'Large sauropod pes natural track cast'	Lourinhã	Late Jurassic	Mateus and Milan (2009: fig. 3)
Large sauropod pes natural track cast with skin	Lourinhã	Late Jurassic	Mateus and Milan (2009: fig. 6)
CTD-TCH-1015-S4-LP4	Reuchenette	Late Jurassic	Marty (2013 pers. comm., 2013)
CTD-TCH-1000-S9-LP5	Reuchenette	Late Jurassic	Marty (2013 pers. comm., 2013)
CTD-BSY-1015-S2-LP3	Reuchenette	Late Jurassic	Marty (2013 pers. comm., 2013)
CTD-BSY-1015-S2-RP3	Reuchenette	Late Jurassic	Marty (2013 pers. comm., 2013)
'Morphotype 1' pes natural track cast	Morrison	Late Jurassic	Platt and Hasiotis (2006: fig. 4)
CU 194–2/LACM 154995 *Brontopodus*, pes track cast	Morrison	Late Jurassic	Lockley and Hunt (1995: fig. 4.55); (pers. obs.)
'Morphotype A' pes natural track cast near Whitby	Saltwick	Middle Jurassic	Romano, Whyte, and Manning (1999: fig. 4)
'Morphotype B?' pes natural track cast from Rail Hole Bight.	Saltwick	Middle Jurassic	Romano, Whyte, and Manning (1999: fig. 6A)
'Morphotype B' pes natural track cast from Maw Wyke	Saltwick	Middle Jurassic	Romano (1999: fig. 6B)
CU 122.88 pes natural track cast	Summerville	Middle Jurassic	Lockley (pers. comm., 2013)
'UCRC I 173' pes natural track cast	Kem Kem	Late Cretaceous	Ibrahim et al. (2014: fig. 15)
'CDUE Locality 100' Left pes	Ameskroud	Early-Middle Jurassic	Enniouar, Abdelouahed, and Habib (2014: fig. 6); Enniouar (pers. comm., 2014)

Note: AMNH, American Museum of Natural History; BSY, Bois de Sylleux; CDUE, l'Université Chouaib Doukkali; CTD, Courtedoux; CU, University of Colorado Denver, Dinosaur Tracks Museum; HMNH, Houston Museum of Natural History; LACM, Los Angeles County Museum of Natural History; LCU-I, Las Cuestas I; TCH, Tchâfouè; TMM, Texas Memorial Museum; UCRC, University of Chicago Research Collection.

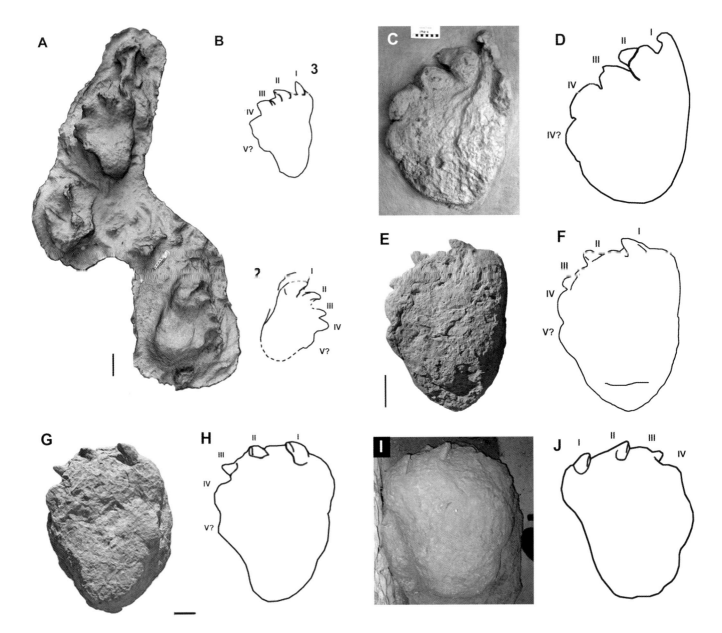

9.3. Well-preserved sauropod pes natural track casts. (A) Latex peel and (B) interpretive line drawing of the last two tracks in the Blue Hole Ballroom "Small sauropod sequence" of Farlow et al. (2012). The numbers 2 and 3 are labels used in this study as the terminal three tracks of the trackway were included in the data set. Note the anterior direction of the digit I ungual and its near-vertical orientation. Scale is 10 cm. Early Cretaceous. (C) Pes cast and (D) interpretive line drawing of *Brontopodus* specimen described in Lockley and Hunt (1995; CU cast 194-2 pictured). The shallow depth of the track and horizontal alignment of the digits indicate that this track was impressed into a relatively shallow substrate. The anterolateral orientation of digit I and the medial cant to the ungual suggests that this pes was in a neutral position when contacting the substrate as the digits neither extend nor overlap. Scale is 10 cm. Late Jurassic. (E) Pes cast and (F) interpretive line drawing of CU 122.88. Middle Jurassic. (G) Pes cast and (H) interpretive line drawing modified from Castanera et al. (2012), both exhibiting unflexed pedes of large sauropods that impressed fairly deeply into the substrate. Asymmetrical load distribution on the plantar surface is indicated by the staggered positions of the unguals. The anterolateral orientation of the unguals demonstrates that plantar flexion was not engaged. Scales are 10 cm. Early Cretaceous. (I) Pes cast and (J) interpretive line drawing of large *Brontopodus* type pes impressed moderately deep into the substrate. The anterolateral direction of the unguals, along with the weakly medially canted orientation of unguals I and II demonstrate that plantar flexion was not engaged during track making. Specimen length ~80 cm. Currently on display at the Dinosaur Farm Museum, Isle of Wight, United Kingdom. Early Cretaceous. *Modified from Fowler and Hall (2011).*

Bonnan (2005); as such, it was not required to preferentially select or omit tracks based on substrate type.

Tracks (and natural casts) were analyzed specifically for preservation of substrate interaction features left by the digits and unguals. This included examining and comparing relative impression depths of the unguals versus the rest of the pes, relative orientation of the unguals in relation to the pes, angle of entry of the unguals into the substrate and the orientation of the pes, amount of angulation or "spread" between unguals, and examination for any "subsurface" features (i.e., evidence of turning or twisting of an emplaced ungual). It was also necessary to form a hypothetical model

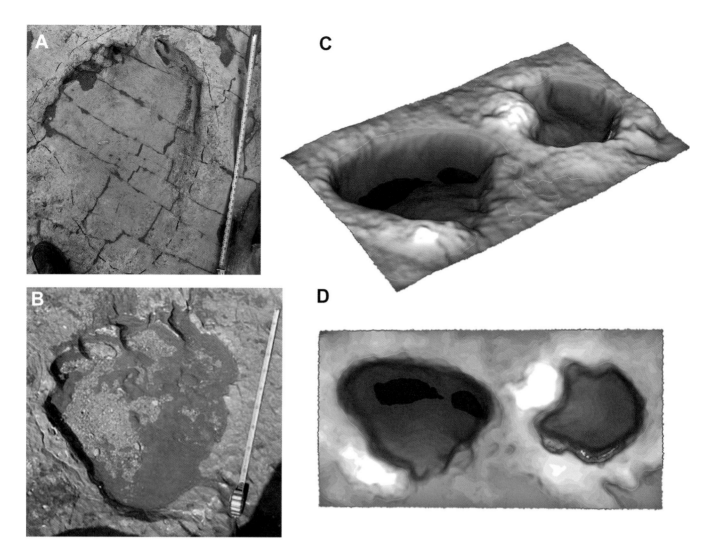

9.4. Sauropod pes tracks. (A) This *Brontopodus* specimen (Brooklyn College collection) from the West Bank Bird site, Glen Rose Formation displays morphology typical of the well-preserved pes tracks encountered in this study. Note that the impression of the digit I ungual is anterolaterally directed, oriented nearly vertically and medially removed from digit II. The condition of digit I suggests that the pes may have been partly extended at the time of impression into the substrate. A 1 m measurement tape is used for scale. Early Cretaceous. (B) Large sauropod pes track from the Lourinhã Formation of Portugal preserved in conditions similar to those of A. The nonoverlapping, anterolateral direction of the digits, and the nearly vertical orientations of the digits I and II unguals suggests that this track was not made by a plantar flexed pes. A 1-m measurement tape is used for scale. Late Jurassic. (C) Oblique and (D) planar views of a LiDAR topographic relief image of trackway S2M of the Glen Rose Formation. Lighter shades indicate higher relief whereas darker shades indicate lower relief. The plantar surface of the sauropod pes was asymmetrically loaded during the step. Pes length ~89 cm. Modified from Farlow et al. (2012).

depicting the track left by a plantarflexed pes in a muddy substrate (Fig. 9.2); to our knowledge, such a feature has not yet been identified from published sauropod trackways. Finally, mobility parameters for sauropod pedal joints were taken from Bonnan (2005) as discussed in our previous study (Fowler and Hall, 2011:4). Full descriptions of track and natural cast specimens are available in the supplementary information.

RESULTS

None of the tracks or natural track casts in our data set preserve any indication of plantar flexion being utilized for substrate grip. Furthermore, plantar flexion has not been explicitly noted or described in the literature, or by other track researchers (Farlow, pers. comm., 2012; Faria Santos, pers. comm., 2012; Manning, pers. comm., 2013). Many of the specimens we observed did appear to exhibit some degree of plantar extension. Evidence from the trackway record indicates that extension of the pes orients the unguals so that digit I is directed anteriorly or weakly anterolaterally while the claw is held near vertically or with a slight medial cant, digit II is directed anterolaterally or weakly laterally while the claw is canted medially or strongly medially, and digit III is oriented strongly laterally while the claw is canted fully medially (Fig. 9.3). See Table 9.1 for track citations.

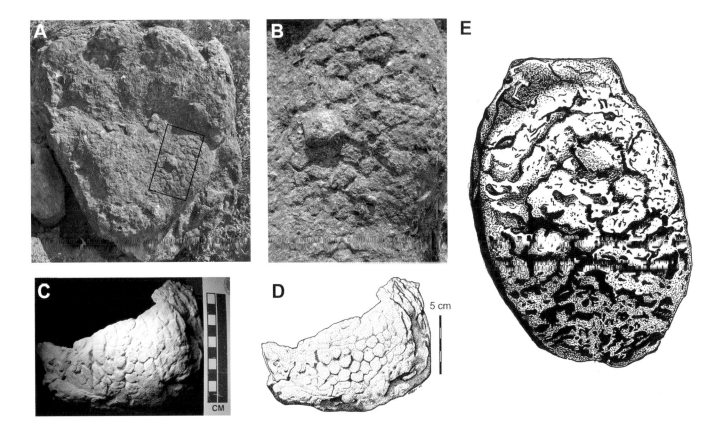

9.5. Sauropod natural track cast skin impressions. (A) Large sauropod pes cast from the Lourinhã Formation of Portugal exhibiting large (2–3 cm) tubercles. Box indicates position of details in B. (B) Close-up of plantar surface from A. Note the high relief between the large tubercle and the smaller "clusters" as well as the channels between them. (C) Natural cast and (D) shaded relief illustration of skin impressions on the plantar surface of a sauropod manus from the Morrison Formation. Modified from Platt and Hasiotis (2006). (E) Illustration of the plantar surface of a modern elephant pes; the rugose network of furrows, ridges, and pits enhance traction. Foot ~40 cm diameter.

DISCUSSION

Flexion-Traction-Control

The lack of any sauropod tracks that exhibit plantar flexion of claws into the substrate fails to provide support for the flexion-traction-control hypothesis. In tracks emplaced in quite wet sediments (described here and elsewhere in the literature), the digits are separated and extended anteriorly with the ungual of digit I oriented almost vertically, and digits II and III canted weakly to moderately medially (Figs. 9.4A, 9.4B). Furthermore, the most deeply emplaced portion of the pes is not the anterior end (where flexed unguals would be expected to have received force for flexion-traction-control), but the posteromedial margin, presumably where most of the force was actually focused during walking. This was observed consistently in nearly every track specimen examined for this study. There is no evidence to suggest that sauropods engaged plantar flexion in muddy sediment (as depicted in Figs. 9.4C, 9.4D); rather, the evidence shows they applied plantar extension, possibly to prevent lateral slipping, or torsion in the limb column. In some specimens, small ridges

of sediment appear to be pushed up between the extended unguals, apparently due to minor amounts of outward pedal rotation during stepping (e.g., González Riga et al., 2014; Fig. 9.3A). This corroborates the extension-traction-control hypothesis (Gallup, 1989; Pittman, 1989; Pittman and Gillette, 1989; Bonnan, 2005), and when considered alongside the phylogenetic trend of pedal ungual hypertrophy and manual ungual loss, suggests that the unusual orientation of the unguals during plantar flexion may be an adaptation for a purpose other than locomotion in wet, muddy substrates. Phrased another way, why put off-road tires on the rear axle and bald tires on the forward axle of a four-wheel drive? Sauropod limbs are clearly adapted toward weight support. The columnar and relatively inflexible nature of the limbs probably limited the ability of sauropods to excavate nests, thus requiring another adaptation to aid the behavior.

Substrate grip or traction in sauropods may have been enhanced by the rugose skin texture of both the palmar and plantar surfaces of the manus and pes (Figs. 9.5A–9.5D). Pedal and manual skin impressions have been reported as rugose, coarse, and composed of patterns of polygonal tubercles and scales from ~1 cm to 3 cm in diameter (Platt

A

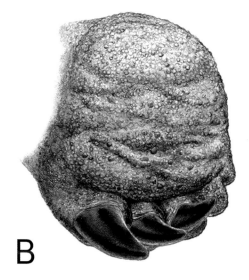

B

C

9.6. Comparison of pedal morphology in elephants, sauropods, and tortoises. (A) Ventral view of the plantar surface of a modern elephant pes from Figure 9.5E. (B) Ventral view life reconstruction of a sauropod right pes exhibiting plantar flexion; the laterally deflected, ventrally rotated unguals overlap, turning the broad lateral surfaces of the claws into a scraping blade; traction-enhancing scales and tubercles cover the plantar surface. Foot ~60 cm wide. (C) Oblique ventral view of the plantar surface of the right pes of the desert tortoise, *Gopherus agassizii*, in a neutral position. Under plantar flexion, the unguals would be drawn together to form a spade-like surface for removing sediment during nest excavation. Also note the rugged, tread-like texture created by the large, cylindrical scales on the plantar surface of the foot, separated by deep channels. Foot ~4 cm wide.

and Hasiotis, 2006; Mateus and Milàn, 2009; Marzola et al., 2014; Xing et al., 2015). Possible modern analogues occur on the plantar surfaces of elephant feet (Fig. 9.5E) whose thick keratinous pads are covered by a rugged network of furrows, pits, and ridges that aid in providing traction on an otherwise planar foot (Eltringham, 1982), as well as the feet of tortoises, which are covered and fringed with scales. However, we urge caution when drawing direct similarities between the feet of elephants and sauropods because they are only superficially similar (Fig. 9.6).

Miring-Avoidance

There is little evidence to suggest that sauropods' unique pedal morphology and position during plantar flexion was functional in miring-avoidance. However, incidences of miring are quite commonly reported for sauropodomorphs (Dodson et al., 1980; Sander, 1992; Hungerbühler, 1998; Montgomery and Fiorillo, 2001; Fowler et al., 2003; Kirkland, pers. comm., 2012), and extended unguals may have helped prevent "postholing" entrapment of the pes by vacuum-like forces in muddy substrate. As extended claws gouged the sediment during the downward motion of the pes, these gouges could have provided channels for airflow, preventing suction entrapment by allowing air to freely circulate around the pes (Kirkland, pers. comm., 2012). This presents a plausible mechanism by which miring may be avoided via ungual extension in a way that is corroborated by track evidence and is potentially testable by the use of models. Plantar flexion

during withdrawal of the foot may create a tapered profile, facilitating removal from sticky sediments in which miring is a potential hazard (Noto, pers. comm., 2014). However, the "tapering" scenario would likely require plantar flexion, thus leaving a distinct impression within a track, for which there is no evidence.

Flexion

The hypotheses presented so far do not provide a functional explanation for the unusual rotation and orientation of unguals during flexion (Bonnan, 2005). Pittman (1989; pers. comm., 2012) suggested that for tracks emplaced in "firmer substrates" the claws were "wrapped around the lateral margin of the foot" (especially so in firm, slippery substrates such as microbial mudflats [Marty, pers. comm., 2013]); this is closer to the null or neutral position, but it still does not involve plantar flexion. However, it is possible that plantar flexion was used for grip in conditions that did not preserve tracks, such as poorly indurated, coarse-grained substrates. This is indirectly testable by comparison to the similarly shaped unguals of extant tortoises (such as *Gopherus*; Fowler and Hall, 2011). Plantar flexion of the feet of tortoises positions the unguals to create a hoe-like shape, which is very similar to that hypothesized for sauropods (Bonnan, 2005). Thus, tortoises placed on a loose gravelly substrate could be observed to see whether flexion is used to grip the surface for locomotion. Although this potential behavior remains to be tested, tortoises have been observed using plantar-flexed hind (and fore-)limbs to excavate sediment for nests and burrows where flexion draws the claws together to aid in scraping up sediment (Bramble, 1982; Fowler and Hall, 2011). We maintain that the plantar-flexed pedal unguals of sauropods could have been utilized in a similar fashion, for scratch-digging nests as originally hypothesized by Gallup (1989). This is corroborated by fossil nest excavations attributed to titanosauriform sauropods (e.g., Chiappe et al., 2005; Vila et al., 2010) that clearly show elongated, trough-like margins that would have been excavated by the pedal unguals employing plantar flexion.

CONCLUSIONS

Evidence from trackways provides support for a suite of functions for sauropod unguals, similar to the original suggestion of Gallup (1989) and later Bonnan (2005). Although we failed to find support for the miring-avoidance hypothesis of Bonnan (2005), support is found for the extension-traction-control hypothesis of Pittman (1989; and others). Currently, there is no evidence of plantar flexion being employed during walking, so the suggestion that substrate grip was a potential function of the unusual unguals and plantar flexion posture remains unsupported. In contrast, plantar flexion of the digits is used by extant tortoises to make a hoe-shape of the foot for use in digging, and ichnological evidence suggests that at least titanosauriform sauropod feet were used to excavate nesting structures (e.g., Vila et al., 2010; Fowler and Hall, 2011). Therefore, we conclude that there is currently no evidence to support a traction control function, but there is evidence to support a scratch-digging nest-excavation function.

This study provides an example where the rich trackway fossil record can be used to test functional hypotheses. Although substrate features are part of this study, more detailed investigation into the specific traits of soft, muddy substrates (water content, thickness, sedimentology, and so on) and their relationship with the use of unguals during plantar extension (degree of "spread" between claws, consistency of cant angles, etc.) would be informative to better understanding sauropod locomotion behaviors. We encourage ichnologists to continue to record and publish in detail (especially clear, high-resolution photographs and photogrammetry or other three-dimensional data techniques) any evidence of ungual-substrate interaction as such information is indispensable and essential for conducting proper paleobiological analyses.

ACKNOWLEDGMENTS

We are tremendously grateful to the following people for their time and sharing of data, without whose generous contributions this research would not have been possible: A. Richter and T. van der Lubbe (Niedersächsisches Landesmuseum Hannover); J. Farlow (Indiana University, Fort Wayne); O. Mateus (Universidade Nova de Lisboa, Portugal); J. Pittman; V. F. Dos Santos (Museu Nacional de Historia Natural e da Ciência, Portugal); C. Coutinho (Portugal); D. Marty (Office de la culture: Paleontology A16, Switzerland); B. Platt (University of Kansas, Lawrence); J. Phillips and B. Phillips (Dinosaur Farm Museum, Isle of Wight); M. Green and K. Simmonds (Isle of Wight); M. Lockley (University of Colorado, Denver); L. Chiappe (Natural History Museum of Los Angeles County); P. Manning (The University of Manchester, England); J. Milàn (Geomuseum Faxe, Denmark); P. Mannion (Imperial College London); D. Loope (University of Nebraska, Lincoln); T. Martin (Emory University, Atlanta); B. Breithaupt (Bureau of Land Management, Wyoming), A. Enniouar (Université Chouaib Doukkali, Morocco), and D. Varricchio (Montana State University, Bozeman), and A. Farke (Raymond Alf Museum of Paleontology, California). Thanks to J. Horner and the Museum of the Rockies for funding and support. Support for the Museum of the Rockies graduate student fund was provided by D. Waggoner and D. Sands. Thanks to T. Martin and R. Boessenecker for helpful discussion. Reviews by C. Noto, S. Lucas, and D. Marty are appreciated and improved this manuscript. Elephant foot figure by A. Fragomeni. Sauropod and *Gopherus* pes restoration by L. Hall. Finally, we wish to express our thanks to the editors of this volume for their patience, enthusiasm, and no doubt countless hours spent wrangling and assembling manuscripts from researchers across the globe.

REFERENCES

Ahmed, A. A. K., T. Lingham-Soliar, and T. Broderick. 2004. Giant sauropod tracks from the Middle-Late Jurassic of Zimbabwe in close association of theropod tracks. Lethaia 37(4): 467–470.

Auffenberg, W. 1976. The genus Gopherus (Testudinidae): pt I. Osteology and relationships of extant species. Bulletin of the Florida State Museum Biological Sciences 20: 47–110.

Bilbey, S. A., D. L. Mickelson, J. E. Hall, K. Scott, C. Todd, and J. I. Kirkland. 2005. Vertebrate Ichnofossils from the Upper Jurassic Stump to Morrison Formational Transition-Flaming Gorge Reservoir, UT. Geological Society of America Abstracts with Programs 37(6): 12.

Bonaparte, J. F. 1986. The dinosaurs (Carnosaurs, Allosaurids, Sauropods, Cetiosaurids) of the middle Jurassic of Cerro Cóndor (Chubut, Argentina). Annales de Paléontologie (Vert.-Invert.) 72(4): 325–386.

Bonnan M. F. 2005. Pes anatomy in sauropod dinosaurs: implications for functional morphology, evolution, and phylogeny; pp. 346–380 in V. Tidwell and K. Carpenter (eds.), Thunder-Lizards: The Sauropodomorph Dinosaurs. Indiana University Press, Bloomington, Indiana.

Bramble, D. M. 1982. *Scaptochelys*: generic revision and evolution of gopher tortoises. Copeia 1982(4): 852–867.

Castanera, D., C. Pascual, J. I. Canudo, N. Hernandez, and J. L. Barco. 2012. Ethological variations in gauge in sauropod trackways from the Berriasian of Spain. Lethaia 45: 476–489.

Chiappe, L. M., F. Jackson, R. A. Coria, and L. Dingus. 2005. Nesting titanosaurs from Auca Mahuevo and adjacent sites; pp. 285–302 in K. C. Rogers and J. Wilson (eds.), The Sauropods: Evolution and Paleobiology. University of California Press, Berkeley, California.

Dodson, P., A. K. Behrensmeyer, R. T. Bakker, and J. S. Mcintosh. 1980. Taphonomy and paleoecology of the dinosaur beds of the Jurassic Morrison Formation. Paleobiology 6: 208–232.

Eltringham, S. K. 1982. Elephants. Blandford Press, London, U.K., 262 pp.

Enniouar, A., L. Abdelouahed, and A. Habib. 2014. A Middle Jurassic sauropod tracksite in the Argana Basin, Western High Atlas, Morocco: an example of paleoichnological heritage for sustainable geotourism. Proceedings of the Geologists' Association 125: 114–119.

Farlow, J. O., J. G. Pittman, and J. M. Hawthorne. 1989. *Brontopodus birdi*, Lower Cretaceous sauropod footprints from the U.S. Gulf Coastal Plain; pp. 372–394 in D. D. Gillette and M. G. Lockley (eds.), Dinosaur Tracks and Traces. Cambridge University Press, Cambridge, UK.

Farlow, J. O., M. O'Brien, G. J. Kuban, B. F. Dattilo, K. T. Bates, P. L. Falkingham, L. Pinuela, A. Rose, A. Freels, C. Kumagai, C. Libben, J. Smith, and J. Whitcraft. 2012. Dinosaur tracksites of the Paluxy River Valley (Glen Rose Formation, Lower Cretaceous), Dinosaur Valley State Park, Somervell County, Texas; pp. 41–69 in P. Huerta and F. Torcida (eds.), Actas de V Jornadas Internacionales sobre Paleontolgia de Dinosaurios y su Entorno, Salas de los Infantes, Burgos.

Ferrusquía-Villafranca, E., E. Jiménez-Hidalgo, and V. M. Bravo-Cuevas. 1996. Footprints of small sauropods from the Middle Jurassic of Oaxaca, southeastern Mexico. Continental Jurassic: Museum of Northern Arizona Bulletin 60: 119–126.

Fowler, D. W., and L. E. Hall. 2011. Scratch-digging sauropods, revisited. Historical Biology 23(1): 27–40.

Fowler, D. W., K. Simmonds, M. Green, and K. A. Stevens. 2003. The taphonomic setting of two mired sauropods (Wessex Fm, Isle of Wight, UK): palaeoecological implications and taxon preservation bias in a Lower Cretaceous wetland. Journal of Vertebrate Palaeontology 23(3, Abstracts volume): 51A.

Gallup M. R. 1989. Functional morphology of the hindfoot of the Texas sauropod *Pleurocoelus* sp. indet. in J. O. Farlow (ed.), Paleobiology of the Dinosaurs. Geological Society of America Special Paper 238: 71–74.

Gatesy, S. M., K. M. Middleton, F. A. Jenkins, and N. H. Shubin. 1999. Three-dimensional preservation of foot movements in Triassic theropod dinosaurs. Nature 399(6732): 141–143.

González Riga, B. J., D. Ortiz, L. D. Tomaselli, R. Candeiro, J. P. Coria, and M. Pramparo. 2014. Sauropod and theropod dinosaur tracks from the Upper Cretaceous of Mendoza (Argentina): trackmakers and anatomical evidences. Journal of South American Earth Sciences 61: 134–141. doi: 10.1016/j.jsames.2014.11.006.

Hildebrand M., 1985. Digging of quadrupeds; pp. 89–109 in M. Hildebrand, D. M. Bramble, K. F. Liem, and D. B. Wake (eds.), Functional Vertebrate Morphology. Harvard University Press, Cambridge, Massachusetts.

Hungerbühler, A. 1998. Taphonomy of the prosauropod dinosaur *Sellosaurus*, and its implications for carnivore faunas and feeding habits in the Late Triassic. Palaeogeography, Palaeoclimatology, Palaeoecology 143: 1–29.

Ibrahim, N., D. J. Varricchio, P. C. Sereno, J. A. Wilson, D. B. Dutheil, D. M. Martill, L. Baidder, and S. Zouhri. 2014. Dinosaur footprints and other ichnofauna from the Cretaceous Kem Kem beds of Morocco. PLoS One 9(3): e90751.

Langston Jr., W. 1974. Nonmammalian Comanchean tetrapods. Geoscience and Man (8): 77–102.

Lockley, M. G., and A. P. Hunt. 1995. Ceratopsid tracks and associated ichnofauna from the Laramie Formation (Upper Cretaceous: Maastrichtian) of Colorado. Journal of Vertebrate Paleontology 15(3): 592–614.

Lockley, M. G., J. O. Farlow, and C. A. Meyer. 1994. *Brontopodus* and *Parabrontopodus* ichnogen. nov. and the significance of wide- and narrow-gauge sauropod trackways. Gaia 10: 135–145.

Lockley, M. G., B. H. Young, and K. Carpenter. 1983. Hadrosaur locomotion and herding behavior: evidence from footprints in the Mesaverde Formation, Grand Mesa Coal Field, Colorado. Mountain Geologist 20: 5–14.

Marzola, M., O. Mateus, A. Schulp, L. Jacobs, M. Polcyn, and V. Pervov. 2014. Early Cretaceous tracks of a large mammaliamorph, a crocodylomorph, and dinosaurs from an Angolan diamond mine. Journal of Vertebrate Paleontology 36 (Abstracts volume): 181.

Mateus, O., and J. Milàn. 2009. A diverse Upper Jurassic dinosaur ichnofauna from central-west Portugal. Lethaia 43(2): 245–257.

Matsukawa, M., T. Matsui, and M. G. Lockley. 2001. Trackway evidence of herd structure among ornithopod dinosaurs from the Cretaceous Dakota Group of northeastern New Mexico, U.S.A. Ichnos 8(3–4): 197–206.

Milàn, J., P. Christiansen, and O. Mateus. 2005. A three-dimensionally preserved sauropod manus impression from the Upper Jurassic of Portugal: implications for sauropod manus shape and locomotor mechanics. Kaupia 14: 47–52.

Milner, A. R., J. D. Harris, M. G. Lockley, J. I. Kirkland, and N. A. Matthews. 2009. Bird-like anatomy, posture, and behavior revealed by an Early Jurassic theropod dinosaur resting trace. PloS One 4(3): e4591.

Montgomery, H., and A. Fiorillo. 2001. Depositional setting and paleoecological significance of a new sauropod bonebed in the Javelina Formation (Cretaceous) of Big Bend National Park, Texas. Geological Society of America, Abstracts with Programs 33: 196.

Moratalla, J. J. 2009. Sauropod tracks of the Cameros Basin (Spain): Identification, trackway patterns and changes over the Jurassic-Cretaceous. Geobios 42(6): 797–811.

Moratalla, J. J., J. García-Mondéjar, V. F. Santos, M. G. Lockley, J. L. Sanz, and S. Jiménez. 1994. Sauropod trackways from the Lower Cretaceous of Spain. Gaia 10: 75–84.

Pittman, J. G. 1989. Stratigraphy, lithography, depositional environment, and track type of dinosaur trackbearing beds of the Gulf Coastal plain; pp. 135–154 in D. D. Gillette and M. G. Lockley (eds.), Dinosaur Tracks and Traces. Cambridge University Press, Cambridge, UK.

Pittman, J. G., and D. D. Gillette. 1989. The Briar site: a new sauropod dinosaur tracksite in Lower Cretaceous beds of Arkansas, USA.; pp. 313–332 in D. D. Gillette and M. G. Lockley (eds.), Dinosaur Tracks and Traces. Cambridge University Press, Cambridge, UK.

Platt, B. F., and S. T. Hasiotis. 2006. Newly discovered sauropod dinosaur tracks with skin and foot-pad impressions from the Upper Jurassic Morrison Formation, Bighorn Basin, Wyoming, USA. Palaios 21(3): 249–261.

Romano, M., M. A. Whyte, and P. L. Manning. 1999. New sauropod dinosaur prints from the Saltwick Formation (Middle Jurassic) of the Cleveland basin, Yorkshire. Proceedings of the Yorkshire Geological Society 52(4): 361–369.

Ruby, D. E., and H. A. Niblick. 1994. A behavioral inventory of the desert tortoise: development of an ethogram. Herpetological Monograph 8: 88–102.

Sander, P. M. 1992. The Norian *Plateosaurus* bonebeds of central Europe and their taphonomy. Palaeogeography, Palaeoclimatology, Palaeoecology 93(3): 255–299.

Sander, P. M., A. Christian, M. Claus, R. Fechner, C. T. Gee, E. M. Griebeler, and U. Witzel. 2010. Biology of the sauropod dinosaurs: the evolution of gigantism. Biological Reviews 86(1): 177–155.

Thomas, D. A., and J. O. Farlow. 1997. Tracking a dinosaur attack. Scientific American 277(6): 74–79.

Thulborn, R. A., and M. Wade. 1984. Dinosaur trackways in the Winton Formation (mid-Cretaceous) of Queensland. Memoirs of the Queensland Museum 21(2): 413–517.

Varricchio, D. J. 2011. A distinct dinosaur life history? Historical Biology 23(1): 91–107.

Vila, B., F. D. Jackson, J. Fortuny, A. G. Sellés, and À. Galobart. 2010. 3-D modeling of megaloolithid clutches: insights about nest construction and dinosaur behaviour. PLoS One 5(5): e10362.

Wilson, J. A., and M. T. Carrano. 1999. Titanosaurs and the origin of 'wide-gauge' trackways: a biomechanical and systematic perspective on sauropod locomotion. Paleobiology 25(2): 252–267.

Wright, J. 2005. Steps in understanding sauropod biology; pp. 252–284 in K. C. Rogers and J. Wilson (eds.), The Sauropods: Evolution and Paleobiology. University of California Press, Berkeley, California.

Xing, L., D. Li, M. Lockley, D. Marty, J. Zhang, W. S. Persons, H. You, C. Peng, and S. B. Kummell. 2014. Dinosaur natural track casts from the Lower Cretaceous Hekou Group in the Lanzhou-Minhe Basin, Gansu, northwest China: Ichnology, track formation, and distribution. Cretaceous Research 52: 194–205.

Morphotype 1 pes track from Agua del Choque track site level An-1–Anacleto Formation, Late Cretaceous, Argentina

Description: Impressions of digits I–IV and the unguals of I–III are well preserved. The pes was shallowly emplaced in soft, silty mud. The unguals of digits I, II, and III are anterolaterally directed and are canted slightly medially. Shallow pressure ridges appear to have formed in the substrate on the lateral margins of the unguals. The plantar surface of the pes is asymmetrically loaded along the medial and posterior margins.

Brooklyn College from West Bank Bird Site Trackway S4–Glen Rose Formation, Early Cretaceous, United States

Description: Impressions of digits I–IV and the unguals of I–III are well preserved. The pes was deeply emplaced in soft mud. The ungual of digit I is anteriorly directed; the ungual of digit II is somewhat laterally directed; and digit III laterally directed. Digit I is nearly vertical; digit II is canted slightly medially; and digit III is canted strongly medially. The plantar surface is asymmetrically loaded along the medial and posterior margins.

Brontopodus Small sauropod sequence track 1–Glen Rose Formation, Early Cretaceous, United States

Description: Impressions of digits I–IV and the unguals of I–III are well preserved. The pes was moderately deeply emplaced in soft mud. The ungual of digit I is anterolaterally directed; the ungual of digit II is somewhat laterally directed; and digit III is laterally directed. Digit I is nearly vertical; digit II is canted slightly medially; and digit III is canted strongly medially. The plantar surface is asymmetrically loaded along the medial and posterior margins.

Brontopodus Small sauropod sequence track 2–Glen Rose Formation, Early Cretaceous, United States

See Figures 9.3A, 9.3B.

Description: Impressions of digits I–IV and the unguals of I–III are well preserved. The pes was moderately deeply emplaced in soft mud. The ungual of digit I is anterolaterally directed; the ungual of digit II is somewhat laterally directed; and digit III is laterally directed. Digit I is nearly vertical; digit II is canted slightly medially; and digit III is canted strongly medially. The plantar surface is asymmetrically loaded along the medial and posterior margins.

Brontopodus Small sauropod sequence track 3–Glen Rose Formation, Early Cretaceous, United States

See Figure 9.3A, 9.3B.

Description: Impressions of digits I–IV and the unguals of I–III are well preserved. The pes was moderately deeply emplaced in soft mud. The ungual of digit I is anteriorly directed; the ungual of digit II is anterolaterally directed; and digit III is laterally directed. Digit I is vertical; digit II is canted slightly medially; and digit III is canted strongly medially. The plantar surface is asymmetrically loaded along the medial and posterior margins.

Brontopodus Left manus-pes set–Glen Rose Formation, Early Cretaceous, United States

See Figure 9.4A.

Description: Impressions of digits I–IV and the unguals of I–III are well preserved. The pes was deeply emplaced in soft mud. The ungual of digit I is anterolaterally directed; the ungual of digit II is somewhat laterally directed; and digit III is laterally directed. Digit I is vertical; digit II is canted slightly medially; and digit III is canted strongly medially. The plantar surface is asymmetrically loaded along the medial and posterior margins.

S2M and pes cast: S2W–Glen Rose Formation, Early Cretaceous, United States

For S2M, see Figures 9.4C, 9.4D.

Description: Impressions of digits I–IV and the unguals of I–III are well preserved. The pes was deeply emplaced into soft mud. The unguals are anterolaterally directed with digits I and II canted slightly medially. The plantar surface is asymmetrically loaded along the medial and posterior margins.

S2N and pes cast; S2S; S2U–Glen Rose Formation, Early Cretaceous, United States

Description: Impressions of digits I–III and the unguals of I–II are well preserved. The pes was deeply emplaced into soft mud. The unguals are anterolaterally directed with digits I and II canted slightly medially. The plantar surface is asymmetrically loaded along the medial and posterior margins.

S2O; S2P- Glen Rose Formation, Early Cretaceous, United States

Description: Impressions of digits I–III and the unguals of I–III are well preserved. The pes was deeply emplaced into soft mud. The unguals are anterolaterally directed with digits I and II canted slightly medially and digit III more medially. The plantar surface is asymmetrically loaded along the medial and posterior margins.

S2T–Glen Rose Formation., Early Cretaceous, United States

Description: Impressions of digits I–III and the unguals of I–II are well preserved. The pes was deeply emplaced into soft mud. The unguals are anterolaterally directed with digits I and II canted slightly medially. The plantar surface is asymmetrically loaded along the medial and posterior margins. The medial displacement rim is overprinted by a tridactyl pes.

S2Q and pes cast; S2V–Glen Rose Formation, Early Cretaceous, United States

Description: Impressions of digits I–IV and the unguals of I–III are well preserved. The pes was deeply emplaced in soft mud. The ungual of digit I is anteriorly directed; the ungual of digit II is somewhat laterally directed; and digit III is laterally directed. Digit I is vertical; digit II is canted slightly medially; and digit III is canted strongly medially. The plantar surface is asymmetrically loaded along the medial and posterior margins.

S2R and pes cast–Glen Rose Formation, Early Cretaceous, United States

Description: The impression of digits I–IV and the ungual of digit I is well preserved. The plantar surface between digits II and III is overprinted by a theropod track, which distorts the sauropod ungual morphology. The pes was deeply emplaced in soft mud. The ungual of digit I is deeply impressed, anteriorly directed, and vertically oriented. The plantar surface is asymmetrically loaded along the medial and posterior margins.

LCU-I-37-12p–Huérteles Formation., Early Cretaceous, Soria, Spain (see Castanera et al., 2012:fig. 4A)

See Figures 9.3G, 9.3H.

Description: Digits I–IV and the unguals of I–III are well preserved. The pes was deeply emplaced in soft mud. The ungual of digit I is anterolaterally directed; the ungual of digit II is moderately laterally directed; and digit III is laterally directed. Digit I is canted medially, and digits II and III are canted strongly medially. The plantar surface is asymmetrically loaded along the medial and posterior margins.

Dinosaur Farm Titanosaurid sauropod pes natural track cast–Wessex Formation, Early Cretaceous, Isle of Wight, United Kingdom; Pes Cast 122.88–Summerville Formation; Middle Jurassic, Arizona, United States

For Titanosaurid, see Figures 9.3I, 9.3J; for 122.88, see Figures 9.3E, 9.3F.

Description: Digits I–III and the unguals of I–III are well preserved. The pes was moderately deeply emplaced into soft mud. The unguals of digits I and II are anterolaterally directed, and the ungual of digit III is laterally directed. Digits I and II are near vertical, and digit III is canted medially. The plantar surface is asymmetrically loaded along the medial and posterior margins.

Praia das Gentias pes track–Lourinhã Formation, Praia das Gentias, Late Jurassic, Portugal

See Figure 9.4B.

Description: Digits I–IV and the unguals of I–III are well preserved. The pes was moderately deeply emplaced in soft mud. The ungual of digit I is anterolaterally directed, and the unguals of digits II and III are laterally directed. Digit I is nearly vertical; digit II is canted medially; and digit III is canted strongly medially. The plantar surface is asymmetrically loaded along the medial and posterior margins.

Large sauropod pes natural track cast–Lourinhã Formation, Late Jurassic, Portugal

Description: Digits I–III and the unguals of I–III are well preserved. The pes was moderately deeply emplaced in soft mud. The ungual of digit I is anterolaterally directed; the ungual of digit II is somewhat laterally directed; and digit III laterally directed. Digit I is canted medially, and digits II and III are canted strongly medially. The plantar surface is asymmetrically loaded along the medial and posterior margins.

Large sauropod pes natural track cast with skin–Lourinhã Formation, Late Jurassic, Portugal

See Figures 9.5A, 9.5B.

Description: Digits I–IV and the unguals of I–III are well preserved. The pes was deeply emplaced in soft mud. The ungual of digit I is anteriorly directed; the ungual of digit II is anterolaterally directed; and digit III laterally directed. Digit I is vertical; digit II is canted medially; and digit III is canted strongly medially. Scale and tubercle impressions are present on the heel. The plantar surface is asymmetrically loaded along the medial and posterior margins.

CTD-TCH-1014-S4-LP4; CTD-TCH-1000-S9-LP5; CTD-BSY-1015-S2-LP3; CTD-BSY-1015-S2-RP3–Reuchenette Formation, Late Jurassic, Jura Canton, Switzerland

Description: Digits I–IV and the unguals of I–III are well preserved. The pes was shallowly emplaced in a microbial mud mat. The ungual of digit I is anterolaterally directed, and the unguals of digits II and III are laterally directed. Digit I is nearly vertical; digit II is canted medially; and digit III is canted strongly medially. The plantar surface appears evenly loaded.

Morphotype 1 pes natural track cast–Morrison Formation, Late Jurassic, United States

Description: Digits I–IV are preserved. Details of the unguals are not preserved. The pes was deeply emplaced in soft mud. Digit I is anterolaterally directed; digit II is laterally directed; and digit III posterolaterally directed. Long (~10 cm) vertical scour marks are associated with the tips of digits II–IV.

CU 194-2/LACM 15499 *Brontopodus* pes track cast–Morrison Formation, Late Jurassic, Utah, United States

See Figures 9.3C, 9.4D.

Description: Digits I–III and the unguals of I–III are well preserved. The pes was moderately deeply emplaced in soft mud. The ungual of digit I is anterolaterally directed; the ungual of digit II is moderately laterally directed; and digit III laterally directed. Digit I is canted slightly medially, and digits II and III are canted strongly medially. The plantar surface is asymmetrically loaded along the medial and posterior margins.

Morphotype A pes cast–Saltwick Formation, Late Jurassic, England

Description: Digits I–IV and the unguals of digits I–II are preserved. The pes was deeply emplaced in soft mud. The ungual of digit I is anterolaterally directed; the ungual of digit II is somewhat laterally directed; and digit III laterally directed. Digit I is nearly vertical; digit II is canted medially; and digit III is canted strongly medially. The plantar surface is asymmetrically loaded along the medial and posterior margins.

?Morphotype B pes cast 1–Saltwick Formation, Late Jurassic, England

Description: Digits I–II and the unguals of I–II are well preserved. The pes was deeply emplaced into soft mud. The ungual of digit I is anterolaterally directed, and the ungual of digit II is laterally directed. Digit I is canted medially, and digit II is canted strongly medially.

Morphotype B pes cast 2–Saltwick Formation, Late Jurassic, England

Description: Digits I–IV and the unguals of digits I–III are well preserved. The pes was moderately shallowly emplaced into soft mud. The unguals of digits I and II are anterolaterally directed, and the ungual of digit III is laterally directed. Digit I is canted medially, and digits II and III are canted strongly medially.

UCRC I 173 Natural pes track cast–Kem Kem Beds, Late Cretaceous, Er Remlia, Morocco

Description: Digits I–IV and the unguals of I and II are well preserved. The pes was deeply emplaced (~38 cm) in soft mud. Digit I is anterolaterally directed; digit II is strongly anterolaterally directed; and digit III appears laterally directed. Very long (~30 cm) subvertical scour marks are associated with the tips of digits I and II. Unguals of digits I and II are canted slightly medially, and digit III is incomplete. The plantar surface appears asymmetrically loaded along the medial margin.

CDUE Locality 100 Right pes track–Amiskoud Formation, Early Middle Jurassic, Tafaytour, Morocco

Description: Impressions of digits I–IV and the unguals of I–III are well preserved. The ungual of digit I is anterolaterally directed, the ungual of digit II is somewhat laterally directed and digit III is strongly laterally directed. The angular orientation of the unguals is not clearly discernable, though they do appear to have contacted the substrate in at least a subvertical position. The plantar surface of the pes is asymmetrically loaded along the medial and posterior margins, with a mediolaterally directed slide scour close to 20 cm long off the anteromedial portion of the track.

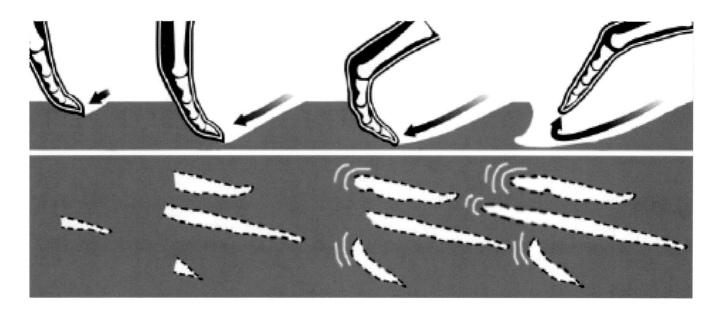

10.1. Pedal kinematics model of dinosaur swim track formation from left to right. Distal track to the right displays a posterior overhang. *Modified from Romilio, Tucker, and Salisbury. (2013).*

Dinosaur Swim Track Assemblages: Characteristics, Contexts, and Ichnofacies Implications

10

Andrew R. C. Milner and Martin G. Lockley

TRACES MADE BY SWIMMING TETRAPODS ARE SIMPLY known as "swim tracks." These trace fossils are of interest to paleontologists because they provide insight into the behavior of past vertebrates in aquatic environments. However, swim tracks have always been a controversial subject for several reasons. Often swim tracks show irregular morphologies and are incomplete, so interpretation of them can be problematic. Unlike tracks made by animals walking on firm ground, which supports most or all of their weight, swimming tetrapods are fully or partially buoyant, and if their feet or hands come into contact with the subaqueous substrate, they will register swim tracks, sometimes preserving elite swim tracks. It has been suggested that swim tracks rarely display regular step and stride patterns as observed in a walking trackway (Milner, Lockley, and Kirkland, 2006), although clear swim trackway patterns are sometimes distinguishable (McAllister, 1989a; Ezquerra et al., 2007; Romilio, Tucker, and Salisbury, 2013; Xing et al., 2013). Because swim tracks are sometimes incomplete and are often found to have irregular and confusing configurations, it is sometimes difficult to identify the trackmaker or to distinguish between manus and pes tracks if the producer was quadrupedal. Surprisingly, under closer examination of a variety of swim track types from different localities and of different ages, it is most often the case that a clear swim trackway pattern can be observed where there are large enough surfaces exposed and not too high a density of tracks, although these trackways can display considerable variation in overall morphology.

INTRODUCTION

Swim tracks raise interesting questions about dinosaur behavior that may challenge conventional wisdom. In this regard, history shows that many behavioral interpretations have been incorrect or at least very controversial when applied to swim tracks (e.g., manus-only sauropod swim tracks [Bird, 1944]; "dinosaur stampede" now interpreted as swimming dinosaurs [Romilio, Tucker, and Salisbury, 2013], although further debated by Thulborn [2013]). As the field of vertebrate ichnology continues to grow and develop new subdisciplines,

the subject of swim tracks has become a specialization in its own right with a rapidly growing literature base.

Reviewed herein are the distribution of dinosaur swim tracks and the claims and debates that have surrounded various sites and assemblages. In order to identify swim tracks correctly and place them in proper ichnological context, comparisons are made between dinosaur swim tracks and the swim tracks of other tetrapods, especially those of crocodilians, turtles, and pterosaurs, which often occur in the same deposits.

Some of the earliest evidence-based literature for swimming dinosaurs arose from the interpretation of manus-only sauropod tracks as having been made subaqueously, when they sought refuge from predators in the water (Bird, 1944; Coombs, 1975; Bird, 1985; Norman, 1985; Ishigaki, 1989; Czerkas and Czerkas, 1990; Thulborn, 1990; Lee and Huh, 2002; Henderson, 2004; Li et al., 2006). The assumption was that these predators (theropod dinosaurs) could not, or would not willingly, swim after sauropods (e.g., Colbert, 1945; Andrews, 1953; Hotton, 1963; Colbert, 1965; Paul, 1988). Ironically, we now know that most, if not all, purported sauropod swim tracks have been misinterpreted and were not made subaqueously (Lockley and Rice, 1990; Lockley, 1991; Lockley et al., 1994; Lockley and Meyer, 2000; Falkingham et al., 2011). In fact, most swimming dinosaur tracks are those of theropods. The questions remain whether ornithischians left swim tracks (Gierlinski and Potemska, 1987; Pienkowski and Gierlinski, 1987; McAllister, 1989a, 1989b; Fujita et al., 2012), and whether any convincing examples have been documented. New interpretations of swimming ornithischian dinosaurs from the early Late Cretaceous Winton Formation of Queensland, Australia (Romilio, Tucker, and Salisbury, 2013), are reviewed herein.

In order to place the dinosaur swim track record in its broader context, the record of swim tracks produced by crocodilians, turtles, pterosaurs, and other tetrapods are examined, especially where they occur in close association with dinosaur swimming and walking tracks (Milner, Lockley and Johnson, 2006; Milner, Lockley, and Kirkland, 2006; Lockley, Cart, et al., 2014). This approach is of considerable

10.2. Restoration of the coelophysoid dinosaur *Megapnosaurus* swimming in an Early Jurassic lake in southwestern Utah. © 2011 by H. Kyoht Luterman.

significance in understanding tetrapod ichnofacies that have already been defined on the basis of recurrent assemblages of walking and swimming tracks of dinosaurs, crocodilians, and pterosaurs (García Ramos, Lires, and Piñuela, 2002; Lockley, Cart, et al., 2014; Lockley, Kukihara, et al., 2014).

WHAT ARE SWIM TRACKS AND HOW ARE THEY MADE?

Swim tracks were made by swimming tetrapods that were partially or completely submerged in water, usually buoyed up, while their digits were able to strike the subaqueous substrate where they registered footmarks (Figs. 10.1 and 10.2). Sometimes not only digits register marks, but the entire foot can leave traces. Quadrupedal animals can leave both manus and pes swim tracks but these are often difficult to distinguish from one another (McAllister, 1989a, 1989b). In rare cases, tail traces and other areas on the ventral side of the body may leave marks on the substrate, but this is usually more often the case in shorter limbed tetrapods such as crocodilians, phytosaurs, and many kinds of amphibians. The possibility of feeding traces in association with swim tracks is not out of the question. Beak and/or toe prod marks and grazing traces could be registered by pterosaurs and birds as they floated and swam in the shallows. Although such feeding traces are very rare in the fossil record, some have been inferred and reported in association with pterosaur swim tracks (Lockley and Wright, 2003).

Most swim tracks appear as parallel to subparallel sets of distal phalangeal digit and/or claw marks (Figs. 10.1 and 10.3). Depending on the length and number of digits, one to five scrape marks can be registered (although many early tetrapods had more than five digits [Niedźwiedzki et al., 2010], no tetrapod swim tracks are known from the Devonian thus far). For example, bipedal, aerially produced theropod tracks such as *Grallator* and *Eubrontes* normally register digits II–IV (rarely digit I) with digit III being the longest (Figs. 10.1, 10.4A, 10.7C). If this same *Grallator* foot morphology is represented by a swim track, then digit III would normally have registered a longer, deeper groove than digits II and IV (Figs. 10.1 and 10.4B). Each of these digit marks would taper out at each end and occasionally sediment would have been piled up at the posterior border of each digit trace (Figs. 10.1 and 10.4C). Depth of digit marks may vary along the length of the traces. In a theropod, digit I would likely not register a mark while swimming and anywhere from one to three scratch marks could be represented. Slight variation in water depth, or thickness of soft, easily penetrated upper substrate layers, would account for differences in the number of digit marks, and sometimes digit mark lengths may vary along the course of a single trackway. If an animal's foot penetrated an upper layer of softer substrate, it could still register a swim trace because the overall movement of the body would still be consistent with some form of swimming behavior. This would presumably apply to all tetrapods capable of terrestrial locomotion (excluding plesiosaurs, mosasaurs, whales, etc.).

Until very recently, only a single ichnogenus attributed to swimming dinosaurs had been formally named as *Characichnos* (meaning "scratch mark") based on traces from the Middle Jurassic Saltwick Formation of England (Whyte and Romano, 2001) (Fig. 10.5). The majority of recorded dinosaur swim track specimens can easily be placed within this ichnotaxon, although this has never formally been done. Whyte and Romano (2001:232) described the ichnogenus *Characichnos* as having "two to four elongate, parallel hypichnial ridges (or epichnial grooves) which may be straight, gently curved or slightly sinuous. The termination of the ridges (or grooves) may be straight or sharply reflexed. Trackway consists of two rows of tracks; the long axes of the tracks are parallel to each other in a straight trackway, and either parallel or oblique to the midline of the trackway." This type of ichnospecies is *Characichnos tridactylus* (Whyte and Romano, 2001:232) and is described as having "tracks elongate, hypichnia consisting of three subparallel ridges with depression or scalloped margin at the posterior end. Left and right

10.3. Large theropod swim track (replica = SGDS 1447 [St. George Dinosaur Discovery Site]) preserved in fine-grained sandstone from the Lower Jurassic Kayenta Formation, Washington County, Utah. White arrows point to proximal ends of claw marks, and black arrows to distal ends of claw marks. *Photo of in situ specimen courtesy David Slauf.*

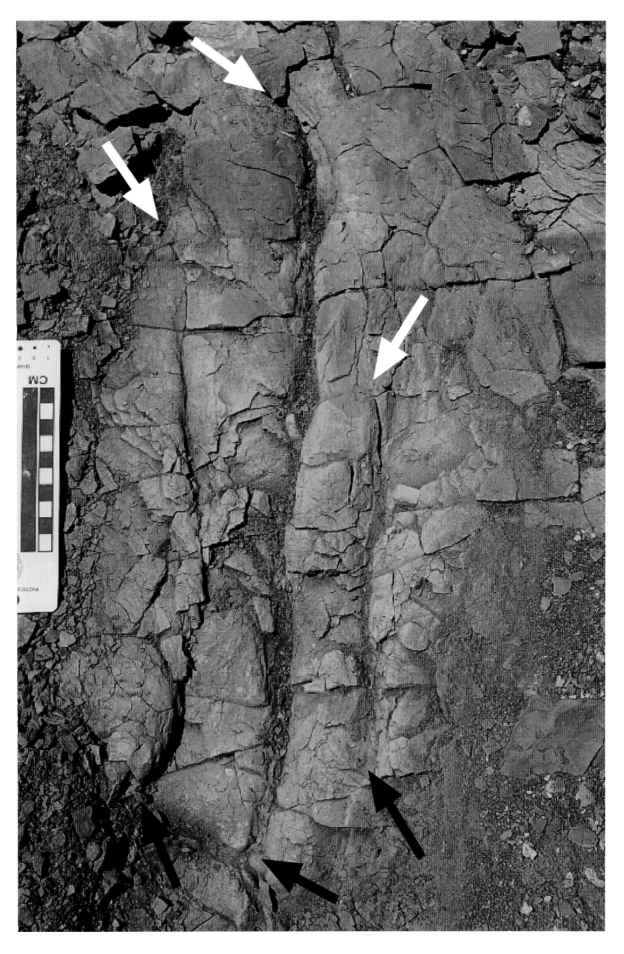

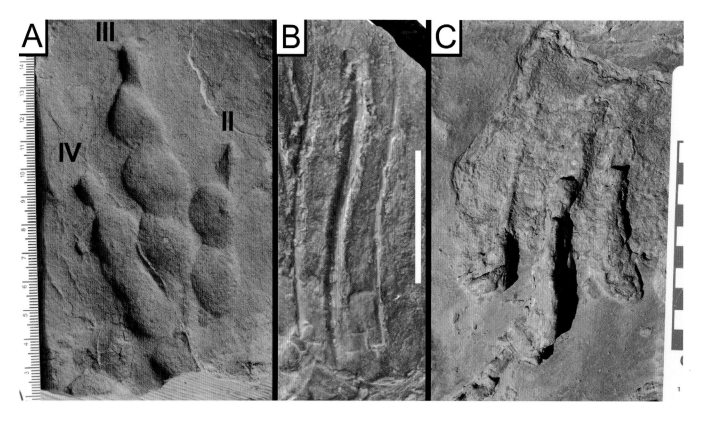

10.4. Different track preservation types from the Lower Jurassic St. George Dinosaur Discovery Site (SGDS), Moenave Formation, St. George, Utah. Color variations due to specimens coming from different stratigraphic levels. (A) Walking theropod tracks such as this *Grallator* (SGDS 1289) normally register digits II–IV (rarely digit I) with digit III being the longest. (B) *Grallator*-type swim track with a longer digit III trace than lateral digits II and IV (SGDS 167). Scale = 10 cm. (C) Each digit mark tapers out at each end and sediment is piled up at the posterior end (bottom of photo) of each claw trace (SGDS Field #SW 77). Scale in centimeters.

tracks only slightly offset from each other; long axes of tracks subparallel."

Based on recent discoveries and additional data about swim tracks in general, the ichnogenus *Characichnos* can be applied to more than just dinosaur swim tracks. Examples include swim tracks attributed to Late Triassic phytosaurs from the Chinle Formation of Utah (Lockley and Milner, 2006; Milner et al., unpubl. data) (Figs. 10.6A, 10.6B); trackways of *Batrachopus* produced by basal crocodylomorphs found transitioning from walk to swim and vice versa (Milner, Lockley, and Kirkland, 2006) (Fig. 10.6C); reptilian swim tracks from the Lower-Middle Triassic Moenkopi and Red Peak formations of Utah, Arizona, New Mexico, and Wyoming, respectively (Lammers, 1964; Boyd and Loope, 1984; Hunt, Lockley, and Lucas, 1993; McAllister and Kirby, 1998; Thomson and Lovelace, 2014) (Fig. 10.6D); and true crocodilian swim tracks (Whyte, Romano, and Elvidge, 2007; Lockley, Fanelli, et al., 2010; Lockley, Cart, et al., 2014) (Fig. 10.6E). However, whereas many tetradactyl trackmakers may produce tridactyl tracks resembling, or practically indistinguishable from,

Characichnos, in strictly morphological terms, the label *Characichnos* (implying *C. tridactylus*) may not be applied to swim tracks, which display four or more digit traces, as in the case of many examples of *Hatcherichnus* (McCrea, Pemberton, and Currie, 2004).

The overall morphology of theropod swim tracks has been reviewed in some detail by various authors (Milner, Lockley, and Kirkland, 2006; Ezquerra et al., 2007). Other ichnogenera now referred to by some as swim tracks include *Wintonopus latomorum* (with *Skartopus australis* now considered by some authors to be a junior synonym) from the famous early Late Cretaceous (Cenomanian) Lark Quarry in the Winton Formation of Queensland, Australia (Romilio, Tucker, and Salisbury, 2013). These purported swim tracks were originally interpreted as being created by stampeding ornithischian (*Wintonopus*) and theropods ("*Skartopus*") (Thulborn and Wade, 1979, 1984), a position still supported by Thulborn (2013) and currently the subject of active debate. Both theropod and ornithischian dinosaur swim tracks will be discussed in greater detail.

10.5. Holotype specimen of *Characichnos tridactylus* Whyte & Romano, 2001, from the Middle Jurassic Saltwick Formation of England preserved in a fine- to medium-grained sandstone as natural casts in siltstone. (A) Photo of natural cast specimen. (B) Interpretive drawing of walking trackways (labeled A and C) and swim trackways (labeled B, E, and F; track labeled D cannot be identified although it does resemble a partial walking track similar to those in trackway A). *Modified from Whyte and Romano (2001).*

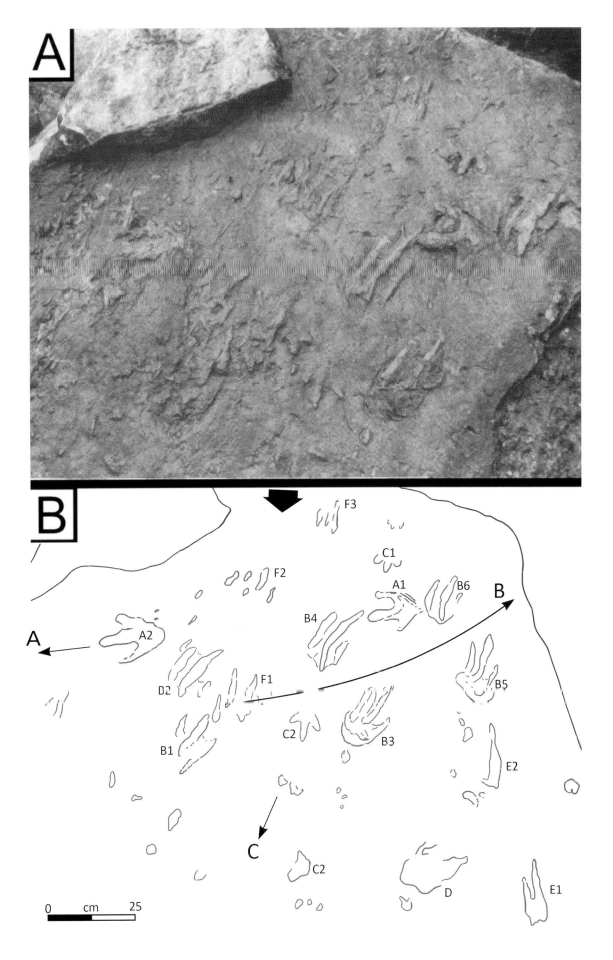

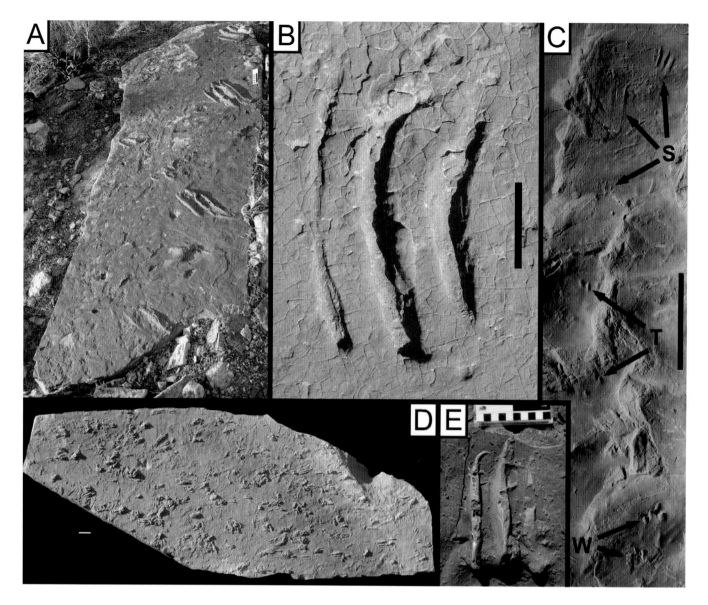

10.6. Examples of nondinosaurian swim tracks that could be placed within the ichnogenus, *Characichnos*. (A) Phytosaur swim trackway preserved in medium-grained sandstone from the Upper Triassic Shinarump Member, Chinle Formation, Harrisburg, Utah (SGDS 1210 [St. George Dinosaur Discovery Site]). Slab is about 2.5 m long and 1 m wide. (B) Large phytosaur swim track in coarse-grained sandstone from the Upper Triassic Church Rock Member, Chinle Formation, San Juan County, Utah (SGDS 1197). Scale = 10 cm. (C) *Batrachopus* crocodylomorph trackway that transitioned from walk to swim in fine-grained sandstone from the Lower Jurassic Whitmore Point Member, Moenave Formation, SGDS, Utah (SGDS 18.T5). Scale = 10 cm.

(D) Chirotheriid swim tracks in coarse-grained sandstone from the Lower Triassic Moenkopi Formation of southern Utah. Scale = 10 cm. (E) True crocodilian swim track from the Dakota Group of northwestern Colorado, from Lockley, Cart, et al. (2014). Abbreviations: S, swim tracks; T, transitional marks; W, walking tracks.

Swim tracks were once thought to be rare and controversial because of incomplete foot morphologies (Lockley and Meyer, 2000; Lockley and Wright, 2003). However, now that swim tracks are better understood, they appear to be quite common, and they were evidently overlooked because they were not previously recognized or adequately comprehended. The result is that the subdiscipline of vertebrate swim tracks has become a specialization in vertebrate ichnology with worldwide application.

Another controversial aspect of swim tracks is whether these traces are indeed true swim tracks or just undertracks with swim track–like morphology. A classic example of tracks misinterpreted as theropod swim tracks, later to be confirmed as undertracks are from the Early Jurassic East Berlin Formation at Dinosaur State Park in Rocky Hill, Connecticut (Coombs, 1980) (Fig. 10.7). According to Coombs (1980), the theropod trackmaker was probably the same kind of animal that made abundant *Eubrontes* tracks at the same locality. Others, however, interpret these to be undertracks for different reasons. Some authors (Farlow and Galton, 2003; Galton and Farlow, 2003; Rainforth and Howard, 2008) suggested these 'swim tracks' may have been made on a firmer substrate after they demonstrated that nearly identical tracks were produced by rhea walking on plaster of paris that was nearly hardened. Also, the Connecticut "swim tracks," for the most part, do not display typical swim track morphology (Milner, Lockley, and Kirkland, 2006).

So what are some of the differences and similarities between true swim tracks and undertracks with swim track–like appearance? To follow Coombs's (1980) example, the distal end of digit III touched down registering the distal phalangeal toe pad and a triangular claw impression having no associated caudally directed claw mark (Fig. 10.7B), which is typical in a swim track. However, digits II and IV have caudally directed claw marks with slight posterior mud rims that resemble swim tracks. When looking at the claw positions in a typical *Eubrontes* track (Fig. 10.7C), they match up closely to distal positions of claws in Coombs' illustration (1980:fig. 1) (Fig. 10.7B), although digit IV is situated closer to digit III. A very firm substrate as was already described (Farlow and Galton, 2003; Galton and Farlow, 2003), or a soft, easily penetrated layer above a firmer surface could produce partially penetrative undertracks as described by Rainforth and Howard (2008). Additionally the position of footfalls in the Rocky Hill trackways (Fig. 10.7A) match very well with typical *Eubrontes* walking trackways (Fig. 10.7D). Confusion and uncertainly can occur with some undertracks that can easily be mistaken for true swim tracks. For example, traces that look identical to swim tracks from the Carboniferous of eastern Canada were originally described as swim tracks (Dawson, 1868; Matthew, 1904; Sternberg, 1933), but now they are suggested to be undertracks of walking amphibians, yet no one has published this formally.

TERMINOLOGY AND MEASUREMENTS
SPECIFIC TO SWIM TRACKS

Considerable variations in tetrapod swim track morphologies are recognized in the fossil record due to both physical and biological influences. Swim tracks are produced subaqueously and are physically influenced by water depth, current flow directions (or lack of current flow), and substrate consistency. Biological influences on swim track formation include animal size, morphology of feet and limbs, buoyancy, and different swimming behaviors. The resulting traces show much variability in footmark lengths, digit reflectures, kickoff scours, and orientations of claw marks in relationship to one another (e.g., whether claw marks are parallel or subparallel, and how lengths and depths compare within a single swim track set). Claw marks in swim tracks often show posterior overhangs when preserved as natural casts or posterior undercutting of the sediment in natural molds (Fig. 10.8A).

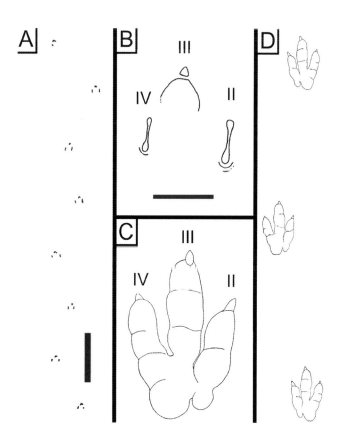

10.7. "Swim tracks" preserved in medium- to coarse-grained sandstone from the Lower Jurassic East Berlin Formation at Dinosaur State Park in Rocky Hill, Connecticut, adapted from Coombs (1980), compared with *Eubrontes* tracks. (A) Trackway. Scale = 1 m. (B) Close-up of individual "swim track" showing the positions of digits II–IV. Scale = 10 cm. (C) *Eubrontes* track showing the positions of digits II–IV (modified from Lull, 1953). (D) *Eubrontes* trackway. *Modified from Lull (1953)*.

In rare cases where elite swim tracks are preserved, cuticle details, skin impressions, scale scratch lines (= scale striations of some authors) can also be preserved (for example, Milner, Lockley and Johnson, 2006; Milner, Lockley, and Kirkland, 2006) (Figs 10.8A–10.8C, 10.8F). Again, in rare circumstances, the differences between left and right swim tracks can be identified even when clear trackways cannot be identified (this has only been applied to theropods, see Milner, Lockley, and Kirkland, 2006).

Posterior overhangs in casts, or undercuts in molds, refer to the caudally directed marks left by the distal ends of the digits as they were pushed into and beneath the substrate (McAllister, 1989b; Milner, Lockley, and Kirkland, 2006; Thomson and Lovelace, 2014) (Fig. 10.8A). Often associated with posterior overhangs and undercuts are sediment mounds and kickoff scours: for example, Lockley, Fanelli, et al. (2010:figs. 2–4). Sediment mounds are simply where the sediment has been pushed backward and up by the digits to form distinct caudally directed mud rim or mound, sometimes showing clear folds and distortion behind each digit mark (Figs. 10.6E and 10.8D). Kickoff scours as defined

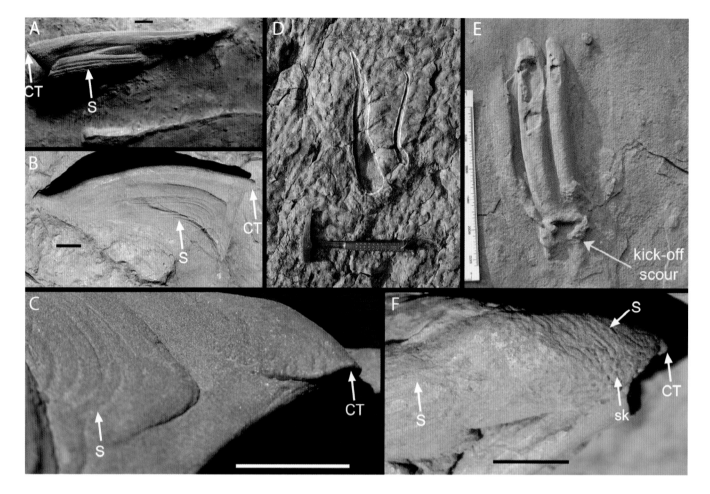

10.8. Details of swim track morphology. (A) Theropod swim track preserved as a natural cast showing posterior overhang and distal phalangeal toe pad with scale scratch lines. Specimen from the St. George Dinosaur Discovery Site (SGDS), Utah (SGDS Field #SW 104). (B, C) Cuticle detail and scale scratch lines on distal phalangeal toe pad preserved as a natural cast. C is a close-up of the claw tip of B showing fine details preserved. Specimen from the SGDS, Utah (SGDS 361). (D) Sediment mounded up along the posterior margin of swim track. Specimen from La Virgen del Campo Tracksite, Spain (courtesy Loic Costeur). (E) Kickoff scours behind well-preserved chirotheriid swim track. Specimen from the Moenkopi Formation, Utah (from Thomson and Lovelace, 2014) (see Fig. 10.5A). (F) Skin impressions associated with scale scratch lines in a theropod swim track preserved as a natural cast. Specimen from the SGDS, Utah (SGDS Field #SW 69). Abbreviations: CT, claw tip; S, scale scratch lines; sk, skin impressions. Scale in A–C, F = 1 cm.

by Thomson and Lovelace (2014:108) "can be identified by indistinct posterior margins of otherwise well-preserved footmarks where water and sediment was apparently swirled as the animal's foot left the substrate" (Fig. 10.8E).

Scale scratch lines are parallel striations that are usually only found on true track surfaces and indicate the direction of digit and foot motion as they entered and exited the sediment (Figs. 10.8B, 10.8C). These longitudinal striations are often better preserved in finer sediments, although they can appear in coarser sands. In rare situations, scale scratch lines and skin impressions can be found on swim tracks (Milner, Lockley, and Kirkland, 2006) (Fig. 10.8F). Variation in scale scratch line widths are recognized from different parts of the foot and from different ichnotaxa.

Most often each digit registered straight, parallel to subparallel scratch marks, but sometimes digit reflectures can occur. This means the foot changed direction one or more times during a single track registration (Thomson and Lovelace, 2014). Single reflectures have v- or c-shaped swim track marks, whereas double reflectures are produced when the digits push backward then slide forward, and then backward again to create an s- or z-shaped swim track (Thomson and Lovelace, 2014).

Measurements of individual swim tracks and swim trackways should include the following if preserved: total swim track length; total swim track width; maximum swim track depth (noting exact position of this depth in the trace); pace (measure from swim track midpoints); stride (measure from swim track midpoints); trackway width; pace angulation; individual digit trace lengths, depths, and widths; minimum and maximum distances between digit marks; average angle of swim track sets in relation to the midline of the trackway. Measuring all of these parameters along with good documentation of sedimentary structures, especially current flow directions, are critical in understanding swim track morphotypes.

Identifying whether swim tracks were made by the hands or feet has been addressed in detail by McAllister (1989b), although his interpretations are problematic, not least because he misidentified the trackmakers. Although McAllister (1989b) inferred differences between manual and pedal swim tracks from the Cretaceous Dakota Formation of Kansas and attributed them to swimming ornithischian dinosaurs, both hand and foot traces never appear to be in the same trackways. Kukihara and Lockley (2012) have reinterpreted very similar tracks from the same region as crocodilian in origin, and Lockley, Lucas, et al. (2010) have cited the Kansas material described by McAllister (1989b) as one of the better examples of a large assemblage of parallel swim tracks. This means it is difficult to identify smaller "manus" and larger "pes" tracks presumed to be from a single individual, from tracks of many different-sized individuals. Track size differences may even be attributable to differences in completeness, which in turn may be a function of buoyancy, water depth, and/or the effect of wave amplitude causing variation in water depths. Manus and pes swim tracks in the same trackway, however, have been documented in some cases, as in *Batrachopus* swim trackways from the Moenave Formation in Utah (Milner, Lockley, and Kirkland, 2006) (Fig. 10.6C).

HISTORICAL AND GEOLOGICAL RECORDS
OF TETRAPOD SWIM TRACKS

Some of the earliest reports of possible swimming tetrapod evidence, and likely the oldest recorded tetrapod swim tracks, include *Hylopus logani* Dawson, 1882 (type specimen is CMN 4622, housed at the Canadian Museum of Nature) from the Carboniferous of Horton Bluff, Nova Scotia. This specimen displays narrow digit marks that are parallel in sets of five (Sternberg, 1933, Harington et al., 2005) (Fig. 10.9A). In the classic book, *Acadian Geology* by Sir William Dawson (1868:355), he described this historic type specimen, then still in the collection of Sir William Logan, as "having a series of seven footprints in two rows, distant about three inches; the distance of the impressions in each row being three or four inches, and the individual impressions about one inch in length. They seem to have been made by the points of the toes, which must have been armed with strong and apparently blunt claws, and appear as if either the surface had been somewhat firm, or as if the body of the animal had been partly water-borne." Later, Sternberg (1933) also suggested that the body could have been partially supported by water. Matthew (1904) described a set of tracks from the

10.9. Historic "swim tracks." (A) "*Hylopus logani*" (Dawson, 1882), possible swim tracks in fine-grained sandstone from the Carboniferous of Horton Bluff, Nova Scotia (CMN 4622 [Canadian Museum of Nature]), although these could be undertracks. (B) "*Ornithoides (?) adamsi*" in fine-grained sandstone from the Pennsylvanian Joggins Formation of Nova Scotia (modified from Matthew, 1904). Scale = 0.5 cm.

Pennsylvanian Joggins Formation of Nova Scotia as having two and sometimes three sets of toe mark impressions, calling them *Ornithoides (?) adamsi* new species, and these resemble swim tracks as well (Fig. 10.9B). However, both of these examples could also be undertracks produced by a walking quadruped.

Latest Carboniferous (Stephanian) reptile swim tracks with associated tail drag marks were also reported from the Morteru area of France (Langiaux, 1980). These swim tracks show distal digit marks made by sharp claws in sets of two to four. They have posterior sediment mounds and most track sets have shorter manus swim tracks slightly anterior to and/ or overlapping the pes swim tracks (Langiaux, 1980:figs. 2–4). These small swim tracks were given the name *Amphisauropus acanthomorphus* Langiaux, 1980 attributed to captorhinomorphs. Similarly, Lucas et al. (2001:fig. 5) illustrate a single trackway specimen (NMMNH P-31674, housed at the New Mexico Museum of Natural History) from the Lower Permian Abo Formation of central New Mexico that they identify as *Amphisauropus latus*. They resemble swim tracks by having parallel sets of sinuous distal digit scrape marks that they say is "one of the extramorphologic traits associated with *A. latus*" (Lucas et al., 2001:74).

Permian *Serpentichnus* traces are suggested to have been made by snake-like amphibians (such as *Brachydectes*) swimming in shallow water and pushing off the bottom to create series of L-shaped marks produced by the body as it impacted the substrate (Braddy, Morrissey, and Yates, 2003). Braddy, Morrissey, and Yates (2003) also describe *Batrachichnus delicatulus* from the Permian of New Mexico. These traces are trackways with symmetrical, continuous, sinuous medial grooves with paired manus and pes tracks in alternating symmetry along the length of the medial grooves. Hunt and Lucas (2006) name the *Serpentichnus* ichnocoenosis for ichnofaunas characterized by the presence of small temnospondyl amphibian swim tracks such as *Serpentichnus* and *Batrachichnus* (e.g., Braddy, Morrissey, and Yates, 2003).

Other Permian swim tracks are reported from the Carapacha Formation in Argentina (Melchor and Sarjeant, 2004) in which they have been identified as *Characichnos*. This is the first time in the literature that swim tracks other than those made by dinosaurs have been identified as *Characichnos*. Swanson and Carlson (2002) illustrate a Lower Permian tracksite in the Wellington Formation of north-central Oklahoma on which approximately 1400 individual tetrapod tracks have been mapped. These include walking, wading, and typical swim tracks.

Mesozoic swim tracks have been reported from western North America from many formations. Triassic swim tracks include amphibians and reptiles, such as chirotheriid and phytosaurs (Boyd and Loope, 1984; Lockley, Milner, and Lucas, 2005; Lockley and Milner, 2006; Milner et al., unpubl. data). Cretaceous tracks originally assigned to ornithischian dinosaurs (McAllister, 1989a, 1989b) and pterosaurs (Gillette and Thomas, 1989), but these were later reassigned as produced by crocodylomorphs (Bennett, 1992; Lockley and Hunt, 1995; Kukihara, Lockley, and Houck, 2010; Kukihara and Lockley, 2012). Late Jurassic tracks from Cérin, France, reported as hopping dinosaurs (Bernier et al., 1982, 1984) were later interpreted as turtle swim tracks (Thulborn, 1989, 1990). These examples indicate how confusion can arise when attempting to attribute swim tracks to the actual track producers.

In the upper part of the Alemoa Member of the Late Triassic Santa Maria Formation in southern Brazil, swim tracks found in close association to *Rhynchosauroides* footprints were suggested to have been made by lacertoid reptiles and possibly sphenodonts (da Silva et al., 2008). These swim tracks (referred to as "drag marks" by da Silva et al., 2008) are in parallel sets of two to four, some are sinuous or only slightly curved in opposing directions when comparing right and left swim tracks in a single trackway. Some of these swim traces may also have been produced by manual digits, but most were certainly pedal in origin.

Walking trackways of *Gwyneddichnium* associated with purported swim tracks from the Late Triassic Chinle Formation in and around Dinosaur National Monument near Vernal, Utah, and northwestern Colorado were first reported by Lockley et al. (1992) and Lockley and Hunt (1995). A more detailed study by Lockley (2006) suggests that walking trackways of *Gwyneddichnium* have five-toed manus and pes

tracks with or without tail drag traces (Fig. 10.10). Trackways identified as being *Gwyneddichnium* swimming are unique, displaying pes-only tracks with wide divarication angles and webbing between the digits (Fig. 10.10). Right and left pes tracks are situated side by side, indicating the hind feet would have touched bottom at the same time in order to propel the completely or partially submerged animal forward. The hands may have been used in a swimming motion to keep the anterior part of the animal more buoyant, or the forelimbs were pulled to the side of the body providing a more streamline posture with the hind limbs and tail used to propel the animal forward through the water. A tanystropheid reptile is suggested as a possible producer of these interesting swim tracks (Lockley and Hunt, 1995; Lockley, 2006).

Purported dinosaur swim tracks have also produced debate over so-called swim tracks of sauropods, theropods, and ornithopods. Manus-dominated trackways of sauropods from the Early Cretaceous of Bandera County, Texas, were found and reported on by Bird (1944). He claimed these trackways were produced by swimming sauropods in which the longer forelimbs would touch bottom while the rear of the dinosaurs were floating, lifting the hind limbs above the substrate. Others followed Bird's (1944) interpretation (Coombs, 1975; Bird, 1985; Norman, 1985; Ishigaki, 1989; Thulborn, 1990; Czerkas and Czerkas, 1990; Lee and Huh, 2002; Henderson, 2004; Li et al., 2006) as these and other manus-dominated sauropod trackways represent evidence for swimming behavior in sauropod dinosaurs, thus creating a sauropod swim track paradigm. In contrast to this paradigm, others interpret these trackways as undertracks of walking, predominantly terrestrial sauropods (Lockley and Rice, 1990; Lockley, 1991; Lockley et al., 1994; Lockley and Meyer, 2000; Falkingham et al., 2011) rather than aquatic animals. For example, Falkingham et al. (2011) showed through experimentation that brachiosaurs shed more weight onto the forelimbs where the manus covered a much smaller surface area compared with the hind limbs and much larger pes. Depending on the kind of substrates, the front feet would sink deeper into the sediment while the hind feet would create minimal disturbance to the sediment (Falkingham et al., 2011).

Although later authors (Lockley and Rice, 1990; Lockley, 1991; Lockley et al., 1994; Lockley and Meyer, 2000; Falkingham et al., 2011) were of the opinion that there was no convincing evidence that manus-only sauropod tracks indicated swimming behavior, the debate was revived in reference to a set of purportedly manus-only sauropod swim tracks from a black mudstone unit in the Cretaceous of Korea (Lee and Huh, 2002; Lee and Lee, 2006). As noted by Lockley, Huh, and Kim (2012:13–14) in a review of research on these tracks, they differ from other manus-only or manus-dominated sauropod trackways in consisting of:

Trackways with deep (4–27 cm) rounded impressions partitioned by 2–6 conspicuous radial ridges, and in some cases a circular pocket in the center. . . . Thulborn (2004) argued that the circular impressions differ morphologically from the well-preserved prints of the sauropod manus, a position supported by Hwang et al. (2008). However, Thulborn (2004) also proposed a model in which the radial ridges were raised by suction, creating a crinkled morphology rather like that seen when hot milk forms a crinkled skin on the surface! . . . It is highly unlikely that the dinosaur tracks are true 'swim' tracks. . . . A much simpler interpretation has been proposed (Hwang et al., 2008). . . . The large tracks are undertracks and the radial ridges are the result of the black mud filling radial cracks that formed when large dinosaurs (ornithopods) walked on the sandy ash layer that had accumulated above the black lacustrine muds. . . . This is precisely the same mechanism and mode of preservation, as reported by Lockley, Matsukawa, and Obata (1989). . . . This model is supported by Song (2010).

In the past, speculative scenarios suggested that theropods were viewed as hydrophobic animals, unlike other kinds of dinosaurs such as sauropods that would have escaped theropods by simply entering the water (e.g., Colbert, 1945; Andrews, 1953; Hotton, 1963; Colbert, 1965; Paul, 1988). Coombs (1980) questioned these ideas while describing purported theropod swim tracks from the Lower Jurassic East Berlin Formation at Dinosaur State Park, Connecticut. Coombs (1980) suggested that these theropod swim tracks (Fig. 10.7) were probably made by the same kind of animal that was responsible for producing abundant *Eubrontes* tracks at the same locality. As discussed, these were later shown to be undertracks (Galton and Farlow, 2003; Milner, Lockley, and Kirkland, 2006). Suggested swim tracks from the Lower Jurassic of Lavini di Marco in northern Italy are reported (Avanzini, 2002), and they have similar track morphology to those described by Coombs (1980) (Fig. 10.7), so it is possible that the Italian traces may too represent undertracks (Farlow and Galton, 2003; Galton and Farlow, 2003; Rainforth and Howard, 2008). Additionally, traces reported as ornithopod swim tracks from the upper Hettangian Przysucha Formation in the Holy Cross Mountains of Poland (Gierlinski and Potemska, 1987; Pieńkowski and Gierlinski, 1987) are probably undertracks (Lockley, 1991).

Unlike typical terrestrial fossil tracks, paleontologists have made little attempt at naming tetrapod swim tracks with a few exceptions. As mentioned, Dawson (1882) named Carboniferous tracks resembling swim tracks (although probably undertracks) "*Hylopus logani*," and Bernier et al. (1984) named *Saltosauropus* (meaning "hopping saurian track") from Cérin, France, that are now considered turtle swim tracks referable to *Chelonichnium* (Thulborn, 1989; Lockley and Meyer, 2000). Of greater significance are swim tracks from the Middle Jurassic Saltwick Formation of England

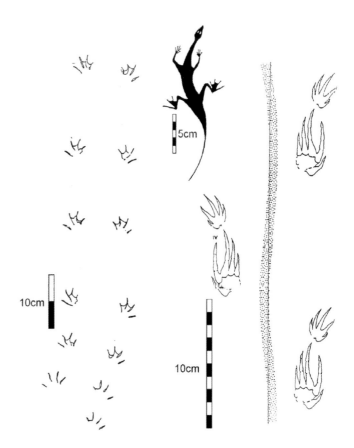

10.10. Comparison between inferred swim trackway of *Gwyneddichnium* on the left and walking trackway made by the same kind of animal on the right. *From Lockley (2006).*

named *Characichnos* by Whyte and Romano (2001). Additionally, *Characichnos* has been identified in the Lower Jurassic Zagaje Formation of Poland (Gierlinski, Niedźwiedzki, and Pieńkowski, 2004), and this name can certainly be applied to swim tracks from other sites, not only produced by dinosaurs but also made by other kinds of tetrapods as discussed herein.

A very important theropod and crocodylomorph swim track site in the Early Jurassic Whitmore Point Member of the Moenave Formation at the St. George Dinosaur Discovery Site at Johnson Farm (SGDS) in southwestern Utah (Milner et al., 2004, 2012; Milner, Kirkland, et al., 2005; Milner, Lockley, et al., 2005; Milner and Lockley, 2006; Milner, Lockley, and Johnson, 2006; Milner, Lockley, and Kirkland, 2006; Milner and Kirkland, 2007; Milner and Spears, 2007) contains thousands of well-preserved swim tracks that will be discussed in greater detail herein. Nearly all sites with dinosaur swim tracks only have small sample sizes, but this is not the case for the SGDS locality and several unpublished localities near St. George, Utah, in the lower part of the Silty Facies of the Kayenta Formation and the Kayenta-Navajo Transition zone (Milner et al., 2012). Additional theropod swim track sites are known from the Early Jurassic Moenave and Kayenta formations of southern Utah, many of which have received brief mention or have not yet been reported on in any detail (DeBlieux et al., 2003, 2005, 2006; Milner and Spears, 2007; Milner et al., 2009, 2012).

Another important locality includes large theropod swim tracks from the Early Cretaceous (late Barremian to Aptian Enciso Group) of La Virgen del Campo Tracksite (LVC) in the Cameros Basin, Spain (Ezquerra et al., 2004, 2007; Costeur et al., 2007; Ezquerra-Miguel, Costeur, and Pérez-Lorente, 2010). This spectacular swim trackway consists of 18 consecutive swim track sets preserved in lacustrine near-shore sediments. Current ripples on the swim track surface indicate a current flow direction from the west northwest, whereas the trackway direction is toward the northwest (Ezquerra et al., 2007) (Fig. 10.11). Note that left swim track set 16 is just over a meter in length, making the LVC swim tracks the largest ever recorded (Fig. 10.11). This then suggests that at the time of swim track formation, this large theropod dinosaur was swimming up-current and slightly oblique to it. This explains the inward-turned right foot-falls that are a typical configuration for swim tracks produced by animals swimming slightly against but oblique to current flow direction (Milner, Lockley, and Kirkland, 2006; Ezquerra et al., 2007).

Dinosaur swim tracks have also been reported by Kvale, Hasiotis, et al. (2001), Kvale, Johnson, et al. (2001), and Kvale et al. (2004) from the Middle Jurassic Gypsum Springs Formation of Wyoming. No attempt was made to determine whether these swim tracks are theropod or ornithischian and the authors (Kvale, Hasiotis, et al., 2001; Kvale et al., 2004) also suggested that they could be "crocodylian."

Modern debate over supposed theropod "didactyl" walking tracks from Niger (Mudroch et al., 2011) have these tracks assigned to a new ichnogenus and ichnospecies, *Paravipus didactyloides*, and the possibility of them being swim tracks was not discussed in the original report. However, we suggest later in this chapter, in detail, that they are swim tracks, an interpretation also suggested by Lockley et al. (2016). Finally, on the topic of contemporary debates, dinosaur tracks from the Winton Formation in Australia were originally interpreted as being produced by a dinosaur stampede (Thulborn and Wade, 1979, 1984); however, a recent study by Romilio, Tucker, and Salisbury (2013) refers to all of the stampeding dinosaur tracks as swim tracks. The debate continues with dissenting opinions by Thulborn (2013) (Fig. 10.13).

We find few published records of mammalian swim tracks (Rutherford and Russell, 1928; Russell, 1930; Bennett, Morse, and Falkingham, 2014). Rutherford and Russell (1928) described an unusual trackway from the Paleocene Paskapoo Formation in Alberta as being tracks of creodonts or condylarth mammals, but the trackmaker's identity was regarded as condylarth shortly afterward (Russell, 1930). Later, McCrea,

Pemberton, and Currie (2004) reexamined this same trackway, assigning it to a new ichnotaxon, *Albertasuchipes russellia*, declaring they were instead crocodilian swim tracks. Turtle swim tracks can be seen on the same slab (McCrea, Pemberton, and Currie, 2004). It should be noted that the *Albertasuchipes* trackway preserves elongate pes scrape marks showing evidence of possible webbing between digits, scale scratch lines, and posterior overhangs. Short distal digit impressions in sets of three and possibly four are located anterior to the pes traces (McCrea, Pemberton, and Currie, 2004). Additionally, Mikuláš (2010) described probable crocodilian swim tracks that he identifies as *Characichnos* from the Miocene of the Ceské Budejovice Basin in the Czech Republic. This is the second time that *Characichnos* has been used for nondinosaurian swim tracks.

Bennett, Morse, and Falkingham (2014) described the first genuine account of mammalian swim tracks from the Plio-Pleistocene Koobi Fora Formation in the Turkana Basin of Kenya. These swim tracks were produced by hippopotami that were probably completely submerged in water to produce traces typical of a hippo's bottom-walking or punting. This kind of swimming behavior is well known in modern hippopotami, which register tracks identical in morphology to those from the Koobi Fora Formation (Bennett, Morse, and Falkingham, 2014).

So why are swim tracks attributed to mammals so poorly known in the fossil record? The same can be said for bird swim traces, and this is certainly unusual since many modern species of birds and mammals are common swimmers. We really do not know the answer to this, but we can speculate that it may have something to do with collecting and/or reporting bias. Additionally, some could simply be mistaking swim tracks for sedimentary structures such as tool marks or similar structures. For example, Boyd and Loope (1984) demonstrated that structures from the Early Triassic Red Peak Formation were in fact swim tracks, not prod marks created by driftwood (Boyd, 1975). It is our opinion that mammalian and avian swim tracks will eventually come to light as people begin to recognized and report them as such.

NONTHEROPOD DINOSAUR SWIM TRACKS IN THE FOSSIL RECORD

Given that all purported sauropod swim track interpretations have been discredited or at least shown to be highly dubious, nearly all recorded dinosaur swim tracks can be attributed to theropod dinosaurs, possibly as a result of some behavioral and feeding characteristics (Milner and Kirkland, 2007; Milner et al., 2012). Although not proven to have been produced by ornithischian dinosaurs, some publications actually figure

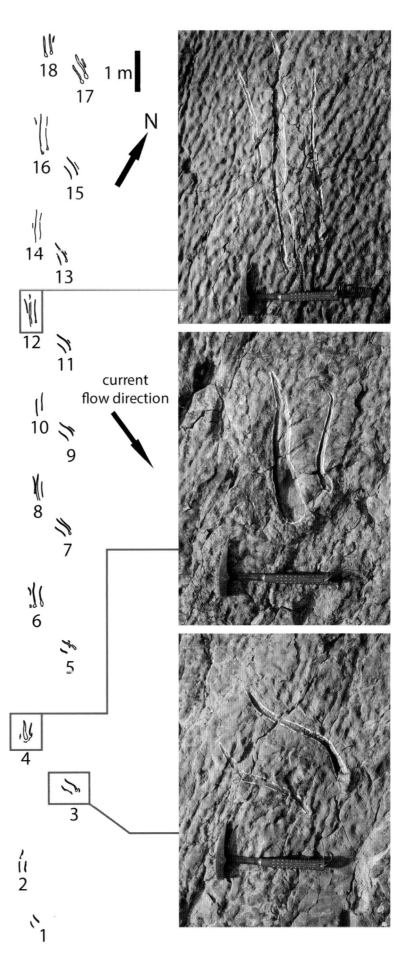

10.11. Large swim trackway from the Lower Cretaceous Enciso Group at the La Virgen del Campo Tracksite in the Cameros Basin, Spain. *Courtesy of Loic Costeur.*

1 m

N

current
flow direction

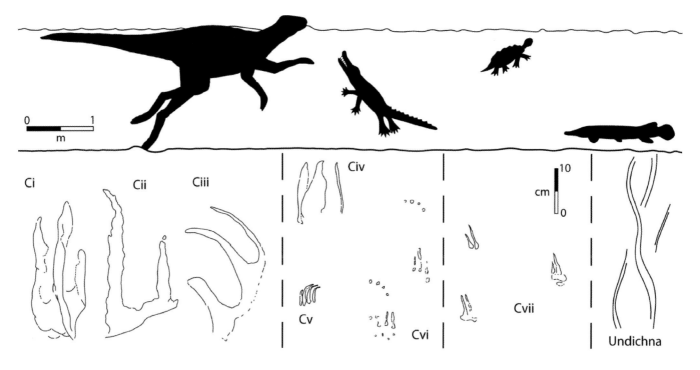

10.12. Diagram showing swim track types from the Middle Jurassic of England compared with possible producers. On the left is an ornithischian dinosaur swimming, which may have produced *Characichnos* swim tracks. *From Whyte, Romano, and Elvidge (2007).*

ornithischian dinosaurs as the animals producing dinosaur swim tracks (Pieñkowski and Gierlinski, 1987; Whyte and Romano, 2001; Whyte, Romano, and Elvidge, 2007) (Fig. 10.12). Also McAllister (1989a, 1989b) proposed swim tracks from the Early Cretaceous as being produced by swimming ornithischians, but these swim tracks were likely made by crocodilians (Kukihara, Lockley, and Houck, 2010; Kukihara and Lockley, 2012).

Currie (1983:66–67) reported on evidence of swimming ornithopod dinosaurs from the Lower Cretaceous Gething Formation in Peace River Canyon, British Columbia. "At Site 4, a trackway of *Amblydactylus* is found in ripple-marked, ferruginous sandstone. The pattern of the tracks at this site suggests that the animal was partially floating when the trackway was made, and that the tracks were made in the muddy bottom of a quiet body of water that was at least two meters in depth. Numerous other sites in the canyon appear to have been underwater when the footprints were made."

Proposed ornithopod swim tracks with associated tail drag marks are reported from the Lower Cretaceous Hekou Group, Gansu Province, China (Li et al., 2006; Fujita et al., 2012). Two trackways from the Yanguoxia site 2 each show sinuous grooves between left and right footprints with atypical track morphology. All of these tracks only show three distal digit impressions (digits II–IV) and never display the posterior portion of the feet. Digit III is always larger and deeper than digits II and IV are. All digit impressions are blunt,

wide, and rounded, and for this reason, Fujita et al. (2012) attribute the tracks to ornithopods. Ornithopod dinosaurs typically held their tails horizontal to the ground when they walked (Galton, 1970) and trackways often show evidence of them walking on all fours (e.g., Fujita et al., 2012; Castanera et al., 2013). Fujita et al. (2012) describe these two trackways as have being made by swimming ornithopods that were "tip-toe" walking on the firm subaqueous substrate with the tail angled downward on the substrate rather than horizontally as would be expected. This scenario seems unusual, but at this time, we cannot provide any alternative explanations for these unique trackways.

Purported ornithischian swim tracks from the early Late Cretaceous Winton Formation of Australia are described in detail by Romilio, Tucker, and Salisbury (2013). However, their interpretation radically changes previous interpretations (Thulborn and Wade, 1984) and has resulted in an on-going debate (Thulborn, 2013). Originally, the Lark Quarry site was interpreted as a mixed herd of small ornithischian dinosaurs represented by the track type *Wintonopus latomorum*, and small theropods represented by *Skartopus australis* (Thulborn and Wade, 1979, 1984). Romilio, Tucker, and Salisbury (2013) made *Skartopus* a junior synonym of *Wintonopus* after they reexamined trackways previously assigned to the *Skartopus*, finding that tracks assigned to *Skartopus* co-occur within individual trackways of the ornithopod-type tracks assigned to *Wintonopus* (Fig. 10.13A). *Wintonopus* are

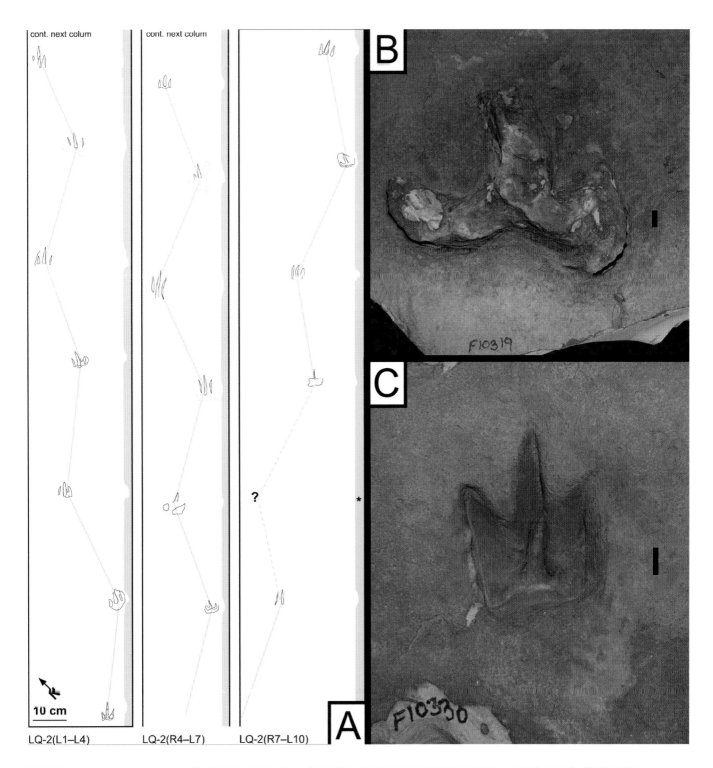

cont. next colum

cont. next colum

B

C

F10319

F10330

10 cm

A

LQ-2(L1–L4)   LQ-2(R4–L7)   LQ-2(R7–L10)

10.13. Purported swim tracks preserved in siltstone and sandstone from the Lower Cretaceous Lark Quarry in Queensland, Australia. (A) Example of a *Wintonopus* trackway (from Romilio, Tucker, and Salisbury, 2013). (B) The holotype track of *Wintonopus latomorum* (QMF10319 [Queensland Museum]). (C) The holotype track of *"Skartopus australis"* (QMF10330). Scale for B and C = 1 cm. *B and C courtesy of Anthony Romilio.*

short-toed, subunguligrade tracks (Fig. 10.13B), whereas *Skartopus* have more elongate digit impressions with short posterior undercuts typical of swim tracks (Fig. 10.13C). Additionally, the authors have shown that deep tracks that more closely resemble *Skartopus* on the substrate surface have a track base more like *Wintonopus*. Romilio, Tucker,

and Salisbury (2013) show no morphological differences between the two ichnotaxa and synonymize them. However, as we already mentioned, Thulborn (2013) does not support the data and conclusions provided by Romilio, Tucker, and Salisbury (2013).

THEROPOD SWIM TRACK MORPHOLOGY AND CURRENT FLOW INFLUENCES

Morphology of swim tracks is greatly influenced by current flow directions (Milner, Lockley, and Kirkland, 2006; Ezquerra et al., 2007). When observing a swim track site, it is critical to record any sedimentary structures that may reliably indicate current flow directions that correlate with the time of swim track registration. The overall morphology of individual swim track sets can be influenced by many factors, including water depth, substrate consistency, and, most importantly, swim direction in relationship to current flow directions. For example, McAllister (1989b) did not take into account the influences of current flow directions had on swim track morphology in his detail description. To reiterate, current flow influences, as well as substrate consistencies, are critical when interpreting swim tracks and swim trackways.

An excellent example of swim tracks influenced by current flow directions was originally described by Milner, Lockley, and Kirkland (2006). The world's largest and best-preserved collection of theropod dinosaur swim tracks comes from the "Main Track Layer" (MTL) at the base of the Johnson Farm Sandstone Bed (JFSB) (Milner et al., 2012) in the lowermost 5 meters of the Whitmore Point Member of the Moenave Formation at the SGDS in St. George, Utah (Milner, Lockley, and Kirkland, 2006, 2012; Milner and Kirkland, 2007; Milner and Spears, 2007; Milner et al., 2012). These swim tracks are summarized in detail based primarily on Milner, Lockley, and Kirkland (2006).

At the SGDS, important relationships between local paleogeography, trackway orientations, and sedimentary structures are evident, leading to important paleoenvironmental interpretations of the Moenave Formation in this region. It is possible to trace and partially map the MTL from the SGDS museum site ("onshore" location) toward the northwest, over to the nearby dinosaur swim track quarry sites, where abundant dinosaur swim tracks (*Characichnos*), subaqueous invertebrate traces (*Palaeophycus, Skolithos, Scoyenia, Kouphichnium*, etc.), and sedimentary structures represent an "offshore" location along the same MTL bedding surface. The dinosaur swim tracks are actually concentrated in a channel-like depression filled by the JFSB, and paralleled the paleoshoreline, situated approximately 80 m to the northwest (Milner, Lockley, and Kirkland, 2006).

This same pattern can be seen in sedimentary structures preserved on the MTL across the entire area, evidence of being formed onshore and offshore. Sedimentary structures, such as mudcracks, sulfate salt crystal casts, and rain drop impressions can typically be formed only when the sediment is subaerially exposed. Offshore, near-shore, and shoreline aqueous-subaqueous sedimentary structures include a variety of current and symmetrical ripples, tool marks, flute casts, and subaqueous scratch circles. In the offshore channel-like feature, current flow was from the north and paralleled the shoreline, based primarily on scratch circles, tool marks, flute casts, and other subaqueous sedimentary structures (Milner, Lockley, and Kirkland, 2006, 2012; Milner and Kirkland, 2007; Milner and Spears, 2007).

Dinosaur swim tracks at the SGDS are interpreted as being preserved along the western margin of freshwater Lake Dixie (Kirkland, Lockley, and Milner, 2002; Milner and Lockley, 2006). Natural casts of dinosaur tracks and swim tracks formed from infilling by fine-grained, well-sorted sand. The JFSB covers a 15 cm–thick horizon of purplish-gray, silty shale, and mudstone-claystone (Kirkland and Milner, 2006; Milner, Lockley, and Kirkland, 2006). Abundant invertebrate grazing traces and burrows indicate organic-rich, probably well-oxygenated sediments within the upper 1–2 cm of mudstone directly below the MTL (Milner, Lockley, and Kirkland, 2006).

The sand that filled the MTL in the area west of the museum that preserves natural casts of the swim tracks was laid down predominantly by unidirectional currents (toward the north). Thinner bedding planes near the base of the MTL at the SGDS are separated by clay-rich mudstone drapes that may represent fluctuations in sedimentation rate prior to a possible "main depositional event" influx of sediment that initially buried the track-bearing surface (Milner, Lockley, and Kirkland, 2006). Lateral variation in thickness (in some cases completely pinching out) of the JFSB reflects its subsequent erosion and indicates that the sediment was deposited with local thickness variation due to underlying topography. Fluvial depositional environments are typically associated with asymmetrical ripples produced by unidirectional currents. However, sedimentary features of the JFSB – thin, laterally extensive bed geometry and climbing-ripple cross-bedding – do not support a fluvial channel origin but instead indicate deposition in an offshore lacustrine setting (i.e., Lake Dixie), perhaps by longshore currents (Kirkland and Milner, 2006).

Areas located northwest of the paleoshoreline have a MTL that is bioturbated with abundant invertebrate feeding-grazing trails and burrows (Milner, Lockley, and Kirkland, 2006). The JFSB in this area ranges in thickness from 10–20 cm from the paleoshoreline to where it thickens in the channel-like trough previously described. The MTL at the base of the thinner JFSB has abundant scours, flute casts (with current orientations toward the north), and rare, scoured-out dinosaur tracks and scoured mud cracks. This surface indicates a shallow, near-shore environment. Vertebrate tracks and traces crossing the mudflat and beach indicate that the

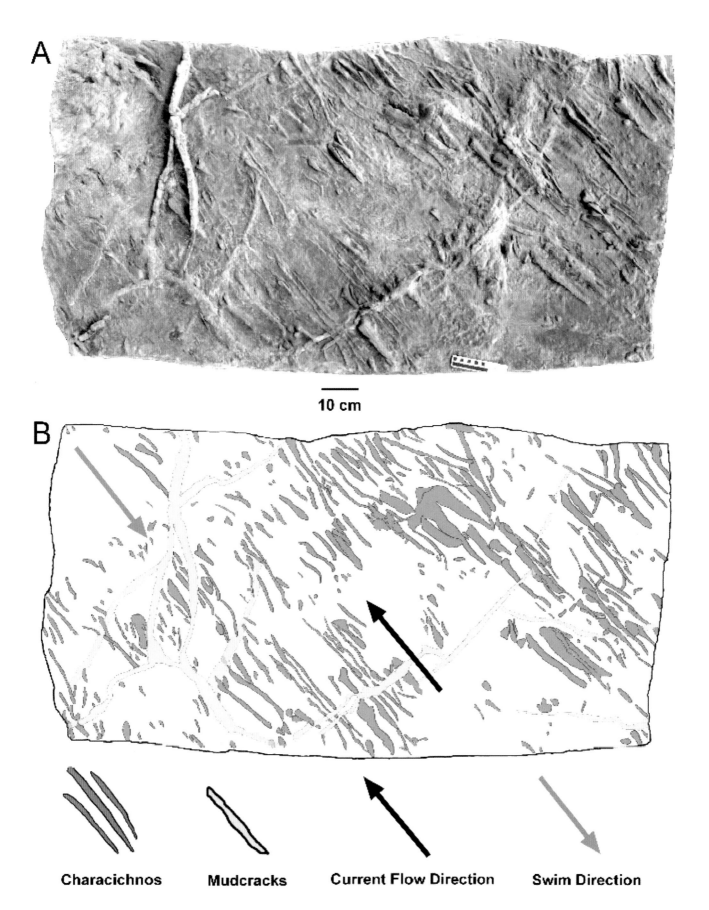

A

B

Characichnos **Mudcracks** **Current Flow Direction** **Swim Direction**

10 cm

10.14. Example of high density swim track block from the Early Jurassic Whitmore Point Member of the Moenave Formation from the St. George Dinosaur Discovery Site (SGDS) in southwestern Utah (SGDS Field #SW 22). Natural cast tracks that were formed in interbedded mudstones and siltstones, infilled with fine-grained sandstone.

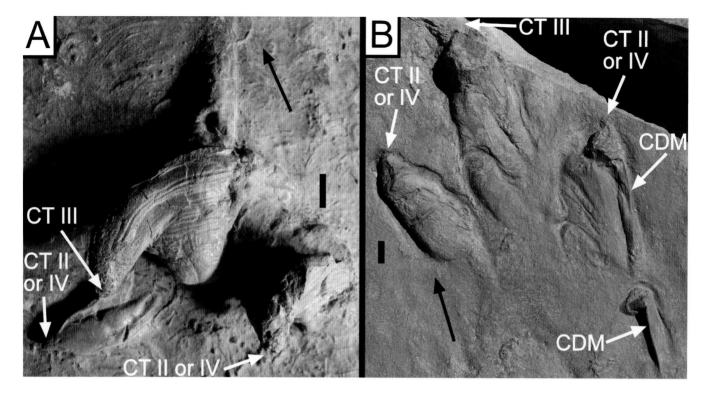

10.15. Morphology of down-current swim tracks from the Lower Jurassic Whitmore Point Member, Moenave Formation, St. George Dinosaur Discovery Site (SGDS), Utah. Traces are fine-grained sandstone natural casts preserved mostly in silty mudstone. (A) Down-current swim track with deepest and longest trace made by the mesaxonic digit (digit III in theropods) (SGDS Field #SW 104). Note arched fluvial scratch circles in upper half of photo. Scale = 1 cm. (B) Down-current *Grallator*-type swim track that resembles a "normal" walking track (SGDS Field #SW 90). Scale = 1 cm. Black arrow indicates current flow directions. Abbreviations: CDM, claw drag marks; CT, claw tip.

trackmakers possibly ventured out into shallow waters of the lake, and they could potentially leave footprints in a substrate showing varying states of cohesion and submergence. Mud cracks and some of the poorly preserved dinosaur tracks would have formed during a lake level regression and represent an extension of the same surface (MTL) now preserved in the museum to the east of the swim track quarry sites. Scouring of these mud cracks and tracks, and the formation of flute casts, probably occurred contemporaneously with the initial deposition of the basal portion of the JFSB in this area. This contemporaneous deposition may have been responsible for the simultaneous and rapid burial of swim tracks, other ichnites, and sedimentary structures as the sand swept across the exposed, underlying fine sediments. Changes in substrate and environment would likely give rise to a wide range of preservational track types (Whyte and Romano, 2001).

The SGDS reveals a variety of track types. These include typical walking to running theropod tracks of *Grallator* (Fig. 10.4A) and less common *Eubrontes*, slightly elongate tracks resembling scratch marks or slide marks, and highly elongate tracks that are identified as swim tracks (Figs. 10.4B–10.4C, 10.14, 10.15). It should be kept in mind that the vast majority (~95%) of dinosaur tracks at the site are those of *Grallator*, formed subaerially while walking, with a few *Eubrontes* tracks in which the typical foot morphology is clearly seen

(Milner, Kirkland, et al., 2005; Milner, Lockley, and Kirkland, 2006). This leads to the inference that the scratch/slide marks and swim tracks represent a different mode of preservation of traces made by the same trackmakers. Swim tracks at the SGDS generally consist of more elongate, parallel to subparallel "scrape marks" (as the name *Characichnos* implies) that occur in high densities (Fig. 10.14), almost invariably *without* associated walking tracks.

The majority of swim tracks at the SGDS were likely made by feet held almost perpendicular to bedding surfaces (Figs. 10.1 and 10.2). Most of the claw marks and thin swim tracks have been laterally distorted and vertically compressed over time due to weight of the overlying sediment. Because of this, most of the swim tracks are now preserved at an average angle of about 45° or greater due to sediment load compaction (Milner, Lockley, and Kirkland, 2006).

At least three categories of swim tracks at the SGDS appear to be present; all preserved as natural casts in high density assemblages (Fig. 10.14). It is important to note that down-, up-, and cross-current traces are all found on the same surfaces in the channel-like trough at the base of the JFSB (Milner et al., 2012). Down- and cross-current traces (*Grallator*) are uncommon in comparison with the up-current swimming traces at this site (Milner, Lockley, and Kirkland, 2006). These include:

1. Inferred down-current traces that:
   - have variable claw impression depths (typically 5–7 cm)
   - always have the deepest and longest trace made by the mesaxonic digit (presumably digit III) (Fig. 10.15A)
   - are accompanied by *Grallator* tracks that resemble normal walking footprints (Fig. 10.15B);
2. Traces oriented up-current that:
   - consist of ubiquitous, parallel scrape marks (*Characichnos*)
   - vary considerably in overall length (ranging in size from 5 to 40 cm long and 5–8 cm wide)
   - are usually in sets of three, although occasionally possess only two or one claw marks (digit III is assumed to always be present)
   - digit III marks are longer and deeper, whereas digits II and IV are shorter and shallower than digit III (Figs. 10.4B, 10.16); and
3. Cross-current swim tracks that:
   - are usually oriented more in an up-current than down-current direction
   - are usually situated at an angle of approximately 45° to current flow direction
   - have similar lengths and widths as the up-current swim tracks (Figs. 10.4C and 10.16).

All observations were made independently of those of Whyte and Romano (2001), who had made similar observations about Middle Jurassic, British swim tracks. Whyte and Romano (2001) concluded that the two styles of preservation they observed actually represented two different track-making episodes that related to changing water levels through time. This raises an important taxonomic consideration about the SGDS swim track assemblage: are *Characichnos* traces, therefore, merely extramorphological variants of *Grallator* and other tridactyl theropod track types made during similar or identical behaviors to those that produced "typical" *Grallator* tracks, or are they distinct "behavioral variants" (Milner, Lockley, and Kirkland, 2006)? The problem here is that it is impossible in most cases to differentiate between walking/running tracks of different ichnotaxon when only swim tracks can be observed. As we discussed, it can be very difficult to tell the difference among some examples of theropod, tridactyl ornithischian, and even some crocodilian swim tracks.

Additionally, Ezquerra et al. (2007) and others (Ezquerra et al., 2004; Costeur et al., 2007; Ezquerra-Miguel, Costeur, and Pérez-Lorente, 2010), described large theropod swim tracks from the Early Cretaceous (late Barremian to Aptian Enciso Group) of LVC in the Cameros Basin, Spain. This locality records a single swim trackway consisting of 18 consecutive swim track sets preserved in a near-shore lacustrine environment. Current ripples on the swim track surface indicate a current flow direction from the west-northwest, whereas the trackway direction is toward to northwest (Ezquerra et al., 2007) (Fig. 10.11). At the time of swim track formation, this large theropod dinosaur was swimming up-current and slightly oblique to it with the current pushing on the left side of the body. The inward-turned right foot-falls suggest a typical configuration for swim tracks produced by animals swimming slightly against but oblique to current flow direction where the right foot was kicking in a direction nearly parallel to current flow in an attempt to maintain a desired swim direction (Ezquerra et al., 2007). Oddly, this is a completely different swim trackway configuration from that produced by a smaller theropod from the SGDS in Utah (Fig. 10.16). Like the Spanish trackway, the SGDS swim trackway was swimming against but oblique to a current on the left anterior side of the body. Unlike the Spanish trackway, the left foot made two short kicks oriented perpendicular to current prior to the right foot making longer sweeps in the direction of travel that overlapped the left foot marks for the most part (Fig. 10.16).

Purported deinonychosaurian theropod tracks named *Paravipus didactyloides* (Fig. 10.17) from the Middle Jurassic Irhazer Group in the Republic of Niger, northwest Africa (Mudroch et al., 2011) are controversial (Lockley et al., 2016). They appear to have been produced by theropods, and we could accept that deinonychosaurs might have swum, although currently there is no evidence for this group in the southern hemisphere in the Middle Jurassic. This site preserves five individual swim trackways with 120 swim tracks that have a normal swimming gait uninfluenced by current flow of any kind (Mudroch et al., 2011). These swim tracks have many characteristics that suggest that the genus *Paravipus* should become a junior synonym of *Characichnos*; however, the ichnospecies could be considered valid because of the unique arrangement of digit marks, especially long digit III, slightly shorter digit IV, and a very short, somewhat oval-shaped digit II (Fig. 10.17A). We interpret these traces as swim tracks because all scrape marks were produced by the distal ends of the digits with stacked sediment at their posterior margins, which are undercut by the claws and distal phalanges. These tracks do not represent typical foot morphology of clear dromaeosaur tracks such as *Dromaeopodus* and *Velociraptorichnus*, which display metapodial pads and a clear connection between each digit trace (Li et al., 2006; Lockley et al., 2016). We agree that *Paravipus* could fit the morphology of a swimming dromaeosaur, or a trackmaker with dromaeosaur-like foot morphology but not a walking representative of this clade. *Paravipus* trackway morphology

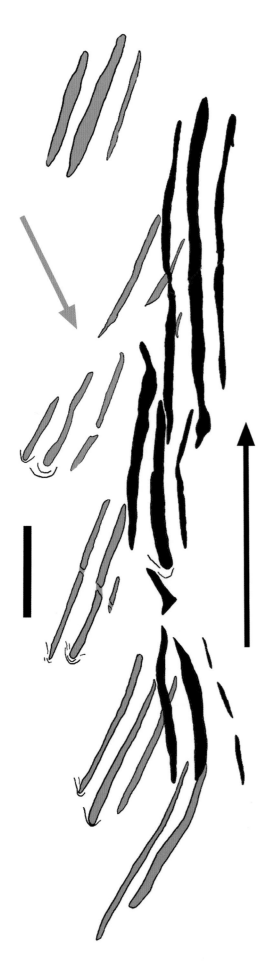

10.16. Theropod swim trackway from the Early Jurassic St. George Dinosaur Discovery Site (SGDS) in Utah (SGDS 167). This trackway represents an animal that was moving mostly up-current, but it does show a slight cross-current influence. Orange tracks = left foot strikes; brown tracks = right foot strikes; black arrow = swim direction; gray arrow = current flow direction. Scale = 10 cm.

(Fig. 10.17B) is also not consistent with obvious dromaeosaur trackways. The inward rotation of the tracks in relationship to the trackway midline, the very short pace and stride, and the wider gauge of the trackways do not compare well with typical dromaeosaur trackways from China and Utah. The *Paravipus* trackways suggest to us, theropods swimming in waters with a consistent depth. The shortened pace and stride lengths, consistent left and right track morphologies, and a lack of current flow indicators at the site would fit well with buoyant animals swimming in a calm lake or pond. Mudroch et al. (2011) describe the depositional environment of the Irhazer Group as fluvio-lacustrine, and the tracks were produced in a fine-grained sandstone covered with small mud cracks and microbial mats, although they do claim that mudcracks continued to form after the tracks were produced.

SUBSTRATE INFLUENCE ON SWIM TRACK MORPHOLOGY

Swim tracks can be formed in nearly any subaqueous substrate in many environments that include lacustrine, marginal marine, and fluvial settings. Swim tracks preserved in subaqueous marginal lacustrine deposits in the lower part of the Whitmore Point Member of the Moenave Formation at the SGDS in southwestern Utah were made in fine-grained clay-rich shales, mudstones, and siltstones (Kirkland and Milner, 2006; Milner, Lockley, and Kirkland, 2006). The high clay content was firm enough to hold its shape and preserve fine details but soft enough for the digits to easily penetrate the sediment. Geochemical studies of the sediments preserving swim tracks at the SGDS and other sites could be beneficial in order to estimate more accurately what the sediment consistency was like.

Coombs (1980) described what he interpreted to be swimming dinosaur tracks that may have been made by *Eubrontes* track producers preserved in medium- to coarse-grained poorly sorted sandstone within a cyclic lacustrine environment (Rainforth and Howard, 2008). Other authors (Farlow and Galton, 2003; Galton and Farlow, 2003; Milner, Lockley, and Kirkland, 2006; Rainforth and Howard, 2008) showed that these traces are controversial because they do not display typical swim track morphologies as described herein (Fig. 10.7). Experimentation with live rheas walking on hardening plaster of paris produce undertracks similar to the Rocky Hill tracks, although digits II and IV do not pull tighter into digit III with depth into the plaster and remain consistent with distances between these digits when compared with the true tracks (Farlow and Galton, 2003). Also, the experimental rhea tracks did not show posterior sediment mounds like those described from Rocky Hill (Farlow and Galton, 2003), although it is not inconceivable that mounds such as these could be made in partially unsealed penetrative tracks.

In experiments, typical tracks formed when an animal placed partial or full weight on the sediment deforming it downward and outward (Milàn and Bromley, 2006). Swim tracks usually were formed with minimal downward weight on the substrate, but instead most of the force was directed caudally as the digits cut through the substrate and compress it through backward motion rather than downward force as in a typical walking track.

Again, experimentation with modern flightless birds such as emus have provided valuable information about walking true track and undertrack morphologies in relationship to different kinds of sediment consistencies (Milàn, 2006; Milàn and Bromley, 2006). However, experimentation with swimming modern birds (and other kinds of tetrapods) and the traces they would produce in subaqueous substrates has not been attempted, although this kind of experimentation would likely prove very informative for comparative purposes to swim tracks preserved in the fossil record, and for biomechanical applications.

DIFFERENTIATING SWIM TRACKS OF THEROPODS FROM THOSE OF OTHER OFTEN-ASSOCIATED ARCHOSAURS

Recent studies of the "mid" Cretaceous (Albian-Cenomanian) Dakota Group in Colorado have shown that archosaurian swim tracks attributable to crocodilians and pterosaurs, as well as a few nonarchosaurian (turtle) swim tracks are abundant and co-occur with dinosaur tracks at multiple localities (Lockley, Cart, et al., 2014). However, to date, most evidence indicates that the dinosaur tracks were made by individuals walking, not swimming. Where swim and walking tracks of different trackmakers occur in close association, it is important to show the evidence for differentiating them in order to correctly interpret the track assemblage. This has been accomplished to some degree with swim track assemblages in the Moenave Formation of Utah (Milner, Lockley, and Kirkland, 2006) and has been well documented with tracks from Dakota Group sites in Colorado (Lockley, Cart, et al., 2014). This same association can also have implications for paleoecology and behavior. The fundamental criteria for differentiating tracks are diagnostic morphology, which in turn is useful for paleobiology and ichnotaxonomy.

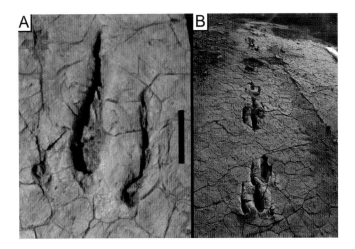

10.17. *Paravipus didactyloides* in fine-grained sandstone from the Middle Jurassic Irhazer Group in the Republic of Niger, northwest Africa. (A) Replica of holotype specimen (NMB-1887 in the Natural History Museum Braunschweig, Germany). Scale = 10 cm. (B) Overlapping in situ trackways. *Photos courtesy of Alexander Mudroch.*

The following analysis of Dakota Group swim tracks is summarized, with modification, from Lockley, Cart, et al. (2014).

Because swim tracks attributable to many vertebrate groups are inherently variable, diagnostic and nondiagnostic features of dinosaur, pterosaur, crocodilian, and turtle swim tracks may only be recognized in some cases, notably, where tracks are relatively complete and/or well preserved. We must bear in mind that:

1. Individual tracks may show traces of any number of the digits (i.e., from 1–5 depending on the foot [manus or pes] and the trackmaker involved). The trackmaker's foot, or the feet of several trackmakers, may have touched the substrate more than once in the same place, causing more than five digit traces to be registered (i.e., more digit traces than the maximum number of digits on the foot of any given individual) (Fig. 10.18A).

2. Only the distal part of the toes (claws) may be registered in some cases. In other cases, tracks may represent registration of the entire foot, including the "heel" (Fig. 10.15B).

3. Dragging of feet as they register may elongate the length of tracks and digit traces beyond dimensions that are diagnostic of foot length (Fig. 10.18B).

4. Swimming animals rarely produce regular trackways or trackway patterns over any distance, although bottom-walking turtles may produce regular patterns (Figs. 10.12 and 10.18C).

5. Interdigital web traces may or may not be present depending on the depth of tracks, the

trackmaker taxon involved, and the substrate conditions (Fig. 10.18D).

6. Size may be a useful indicator of trackmaker taxon, but there is considerable overlap between the sizes of the feet of various swim trackmakers, whether these represent different species or individuals of different sizes (e.g., juveniles and adults).

7. Inferences about the behavior of extant and extinct swim trackmakers may help discriminate the affinity of swim tracks that might otherwise be ambiguous.

For these reasons, we recognize the following diagnostic features of pterosaur swim tracks that help differentiate them from the theropod swim tracks described herein. Although, like theropods, almost all pterosaur swim tracks represent the distal end of pes digits, like walking traces they indicate four slender digits, of which II and III are equal in length and slightly longer than the traces of digits I and IV. In swim tracks, registered when the pes scraped or dragged over the substrate, it is typical to find four long parallel striations many times the length of normally registered pes tracks, which show narrow heel traces (Lockley and Wright, 2003; Lockley and Schumacher, 2014; Lockley, Kukihara, et al., 2014) (Fig. 10.18B). If individual pes footprints are registered in swim tracks assemblages (e.g., Kukihara and Lockley, 2012), they may help identify swim tracks unambiguously. Pterosaur pes swim tracks, like walking tracks, typically have two longer digit traces (length formula II = III > I = IV) at the touchdown point. Web traces, if present, appear attached to the distal ends of the digit traces just behind the claw traces (Fig. 10.18D): in other words, the pterosaur pes was fully, not partially webbed. The walking trackways of pterosaurs have been named *Pteraichnus* in most cases and may occur in association with swim tracks (Lockley and Wright, 2003). However, no different formal ichnotaxonomic names have been given to pterosaur swim tracks.

In contrast to the features of pterosaur swim tracks, diagnostic characteristics of crocodilian swim tracks have recently been described by Lockley, Fanelli, et al. (2010), Kukihara, Lockley, and Houck (2010) and Kukihara and Lockley (2012). Typically, assemblages consist of isolated manus and pes tracks with no discernible trackway patterns. Reports of fossil trackways suggestive of walking are rare (Mehl, 1931; Lockley, 2010; Houck et al., 2010) but are known from modern trackway studies (Farlow and Elsey, 2010). Trackways produced by the high walk, show a tetradactyl pes track with a posteriorly tapered heel. Pes digits I–III are robust with strong claws, but pes digit IV is more slender and less well developed (digit length ratio III>II = IV>I). The manus track is pentadactyl with the traces of digits II–IV much longer than the traces of digits I and V, and without a conspicuous posterior heel trace (digit length ratio III>II and IV>I and V). Neither the manus nor the pes has well-developed webbing. Rather the webbing only extends across the proximal part of the hypices between digits.

Inferred crocodilian swim tracks are known from many geologic formations and consist of tridactyl and tetradactyl, parallel, scrape marks that indicate trackmakers with robust digits up to ~2 cm wide (Figs. 10.12 and 10.19). These have been named *Hatcherichnus* based on an inferred manus-pes set (tetradactyl pes and incomplete manus) associated with sinuous tail traces (Foster and Lockley, 1997) and can be distinguished from type *Characichnos* on morphological grounds. However, such differentiation may be difficult in the case of scrape, scratch, or drag marks that are very abundant and elongated for distances of up to a meter or more, which shows many parallel traces in a Cretaceous Dakota assemblage from Colorado (Figs. 10.11 and 10.19) that has similarities to those reported from the Early Jurassic at the SGDS. In this case, the Dakota traces are all parallel to subparallel and indicate movement against the prevailing current as noted from other sites in Colorado and Kansas (Lockley, Fanelli, et al., 2010) (Fig. 10.19). Crocodilian swim tracks are often variable and incomplete and may consist of only three toe traces, representing the longer digits of the manus or pes (i.e., digits II–IV). This is because pes digit I and manus digits I and V are too short to touch the substrate until the other digits are quite deeply registered. Recently, *Hatcherichnus* have also been described from Late Jurassic (Kimmeridgian) of northern Spain (Avanzini et al., 2010).

Unlike pterosaurs and crocodilians, turtles produce rather similar manus and pes swim tracks (Figs. 10.12 and 10.18C). Diagnostic features reflect the fact that the manus and pes is typically short and pentadactyl with well-developed webbing. In comparison with theropod, crocodilian, or pterosaur swim tracks, which are usually tridactyl and tetradactyl and much larger, turtle tracks are typically short, wide, and often pentadactyl (Fig. 10.18C). Because turtles are capable of bottom-walking (punting), they may register more or less regular trackway patterns (Fig. 10.12) (Avanzini et al., 2005). Such trackways have been referred to as *Chelonipus* (Ruhle von Lilienstern, 1939; Lockley and Foster, 2006; Lovelace and Lovelace, 2012).

It is clear from the preceding descriptions of theropod, crocodilian, pterosaur, and turtle swim tracks that it is possible to differentiate them but only in cases where preservation is relatively complete and suitable for preserving diagnostic morphological features. Theropod and pterosaur swim tracks are only similar in being represented exclusively by pes tracks. Otherwise, they are different in the following diagnostic features: number of digit traces, relative length of digit traces, and presence of webbing traces only in pterosaur

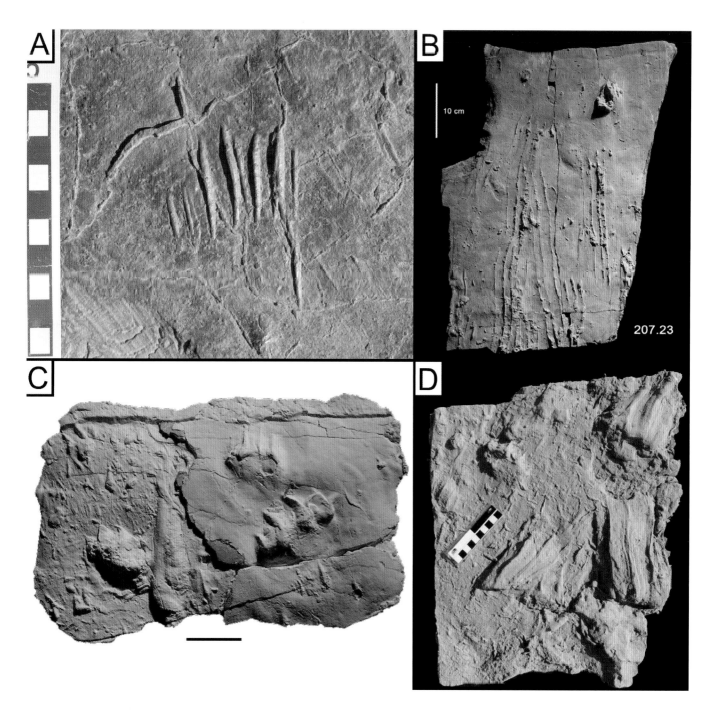

10.18. (A) Unknown swim trace type showing nine overlapping possible distal digit marks from the Moenave Formation of Utah (SGDS 421 [St. George Dinosaur Discovery Site]). (B) Set of pterosaur swim tracks from the Lower Cretaceous Dakota Group of northwestern Colorado showing characteristic claw traces that indicate diagnostic pes digit length ratios II = III > I = IV. (CU 207.23 [University of Colorado Denver, Dinosaur Tracks Museum collection] from Lockley, Cart, et al., 2014:fig. 7). (C) Turtle swim tracks from the Laramie Formation, Colorado. Scale bar = 5 cm (CU 223-3 from Wright and Lockley, 2001). (D) Pterosaur swim tracks with interdigital web traces from the Cretaceous Dakota Group of northwestern Colorado consisting of three digit traces. Scale bar in centimeters. (CU 207.57 from Lockley, Cart, et al., 2014:fig. 9.) All figured tracks are preserved fine-grained sandstone as natural casts.

tracks. Incidentally, both theropod and pterosaur swim tracks are good indicators of water depth.

Crocodilian and turtle swim tracks were made by animals that were quadrupedal, at least on land and when bottom-walking. However, it is not easy to differentiate between manus and pes tracks, which are often incomplete in swim tracks. Based on size, crocodilian tracks are larger than turtle tracks on average. However, as crocodilian swim tracks are often tridactyl, they may be confused with or convergent with theropod swim tracks. Thus, many tracks labeled as *Hatcherichnus* are similar to tracks that could also be labeled as *Characichnos*. As both crocodilians and turtles may register tracks on subaqueous substrates in water of any depth, their tracks are not useful water depth indicators.

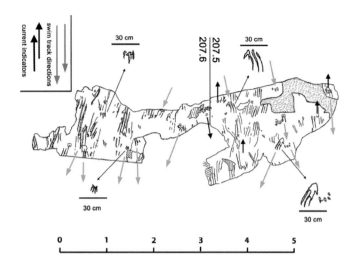

10.19. Large assemblage of elongate crocodilian tracks preserved as natural casts in fine-grained sandstone from the Lower Cretaceous Dakota Group of northwestern Colorado. The specimen is preserved in two parts as mold and replica (CU 207.5 and 207.6 [University of Colorado Denver, Dinosaur Tracks Museum collection]). Surface is ~ 6 m long and 1–2 m wide. Details of individual tracks are shown at twice map scale, with gray arrows indicating direction of travel. Black arrows indicate current flow directions, mostly opposite to the trackway orientations. *From Lockley, Cart, et al. (2014:fig. 11).*

Finally, no solid evidence for avian swim tracks has been reported in the literature, although Anfinson et al. (2009) report a single set of three parallel scratch marks resembling *Characichnos* from the Lower Cretaceous Dakota Formation near Vernal, Utah. This set of marks were found in close association with the bird ichnogenus *Koreanaornis* (Anfinson et al., 2009), but their true affinities cannot be determined with confidence.

SWIM TRACKS AND ICHNOFACIES

Hunt and Lucas (2007a) proposed the *Characichnos* ichnofacies in which most of the tracks preserved are swim tracks consisting of typical parallel scratch marks in association with fish swim traces such as *Undichna* and *Parundichna*, claiming this ichnofacies represents shallow lacustrine and tidal environments. Tetrapod swim tracks (e.g., *Characichnos* and *Hatcherichnus*) are also known from fluvial deposits (Foster and Lockley, 1997), and in the case of pterosaur swim tracks, often referred to as the *Pteraichnus* ichnofacies (Lockley et al., 1994; Lockley and Meyer, 2000), and marine turtle tracks, characteristic of the *Chelonichnium* ichnofacies (Lockley and Meyer, 2000); these swim tracks represent marine environments (see García Ramos, Lires, and Piñuela [2002] for further examples of pterosaur swim track assemblages). Thus, the "*Characichnos* ichnofacies" of Hunt and Lucas (2007a, 2007b) can be expanded to include certain fluvial and marine settings, and many previously defined ichnofacies, or what they prefer to call ichnocoenoses. For example, Hunt and Lucas (2007a) recognized the "*Hatcherichnus* ichnocoenosis" as a component of their *Characichnos* ichnofacies, and Lockley, Li, et al. (2010) emphasized that the *Hatcherichnus* ichnofacies interpenetrates the "*Caririchnium*" or "ornithopod ichnofacies" in the Dakota Group to such an extent that one may speak of crocodilian waterways interpenetrating the emergent parts of the dinosaur freeway. Moreover, it appears that a distinct lack of trackway evidence for large theropod predators in the Dakota Group indicates that the crocodilians were likely the top predators (Lockley, Fanelli, et al., 2010; Lockley and Lucas, 2011).

Hunt and Lucas (2007a:67) provide examples of the *Characichnos* ichnofacies including *Characichnos* assemblages from the Middle Jurassic Saltwick Formation in England (Whyte and Romano, 2001); *Hatcherichnus* assemblages from the Late Jurassic Morrison Formation of eastern Utah (Foster and Lockley, 1997); and the subaqueous, marginal lacustrine paleosurfaces at the SGDS in the Early Jurassic Whitmore Point Member of the Moenave Formation in southwestern Utah (Milner, Lockley, and Johnson, 2006; Milner, Lockley, and Kirkland, 2006; Milner and Kirkland, 2007; Milner and Spears, 2007; Milner et al., 2012). As indicated, the *Hatcherichnus* ichnofacies (sensu Lockley, Li, et al., 2010) is widespread throughout the Dakota Group and according to the definitions of Hunt and Lucas (2007a) represent a component of the *Characichnos* ichnofacies. Based on this rationale, the *Characichnos* ichnofacies of Hunt and Lucas (2007a) is widely represented by the *Hatcherichnus* ichnofacies of Lockley, Li, et al. (2010) in many additional depositional environments. However, while all assemblages assigned to the *Characichnos* ichnofacies represent swim tracks in the broadest sense, it now appears well established that the Moenave assemblages or ichnocoenoses represent theropods in terrestrial lacustrine settings, whereas the Morrison and Dakota assemblages represent crocodilians in fluvial and coastal plain settings. In both cases, the general paleoecology inferred from the totality of trace and body fossil evidence is quite different. As indicated, the *Pteraichnus* and *Chelonichnium* ichnofacies (Lockley et al., 1994; Lockley and Meyer, 2000; Lockley, 2007) must also be considered, as distinctive ichnofacies in many cases dominated by swim tracks.

CONCLUSIONS

Swim tracks were once considered to be controversial with little support or understanding of their true meanings. With advancements in research on these kinds of traces, swim tracks are now widely accepted as ichnites that provide valuable information on locomotory behavior of dinosaurs (and other tetrapods) in aquatic environments and how these animals responded to current influence. They are also valuable as paleoecological indicators as they provide evidence of which groups of aquatic and nonaquatic tetrapods co-occurred in different paleoenvironments, thereby creating various distinctive ichnofacies. While, it is not surprising to find swim tracks attributable to crocodilians, turtles, and pterosaurs in subaqueous deposits intercalated with evidence of dinosaur tracks on emergent surfaces, present evidence also suggests that theropod dinosaur swim tracks are the most abundant swim tracks attributable to dinosaurs. This evidence is contrary to traditional opinion, which held that theropods avoided swimming.

ACKNOWLEDGMENTS

We would like to thanks Martin Whyte (University of Sheffield), Mike Romano (University of Sheffield), Loïc Costeur (Natural History Museum Basel), C. Richard Harington (Canadian Museum of Nature), Anthony Romilio (University of Queensland), Alexander Mudroch (Initiative of Independent Palaeobiologists Deutschland), Tracy Thomson (University of California, Davis), Don DeBlieux (Utah Geological Survey), and David Slauf (St. George Dinosaur Discovery site at Johnson Farm) for providing some of their pictures and figures of swim tracks. Thank you to Ky Luterman for allowing us to use her wonderful painting in Figure 10.2. Kent Cart and Jason Martin (Grand Junction, Colorado) were particularly helpful in finding many of the Dakota Formation sites discussed here. The Dakota studies were funded, in part, by the Korean National Research Institute of Cultural Heritage. Thank you to Peter Falkingham (Liverpool John Moores), Annette Richter (Lower Saxony State Museum), and David Lovelace (University of Wisconsin-Madison) for their helpful reviews.

REFERENCES

Andrews, R. C. 1953. All about Dinosaurs. Random House, New York, New York, 146 pp.

Anfinson, O. A., M. G. Lockley, S. H. Kim, K. S. Kim, and J. Y. Kim. 2009. First report of the small bird track *Koreanaornis* from the Cretaceous of North America: implications for avian ichnotaxonomy and paleoecology. Cretaceous Research 30: 885–894.

Avanzini, M. 2002. Il Trentino dei dinosaur: I Lavini di Marco in Val Lagarina. Provincia Autonoma di Trento, Servizio Geologico–Museo Tridentino di Scienze Naturali, Edizioni Osiride, Rovereto, Italy, 37 pp.

Avanzini, M., L. Piñuela, J. I. Ruiz-Omeñaca, and J. C. Garcia-Ramos. 2010. The crocodile track *Hatcherichnus*, from the Upper Jurassic of Asturias (Spain); pp. 89–92 in J. Milàn, S. G. Lucas, M. G. Lockley, and J. A. Spielmann (eds.), Crocodyle Tracks and Traces. Bulletin 51. New Mexico Museum of Natural History and Science, Albuquerque, New Mexico.

Avanzini, M., J. García-Ramos, J. Lires, M. Menegon, L. Piñuela, and A. Fernández. 2005. Turtle tracks from the Late Jurassic of Asturias, Spain. Acta Palaeontologica Polonica 50(4): 743–755.

Bennett, S. C. 1992. Reinterpretation of problematical tracks at Clayton Lake State Park, New Mexico: not a pterosaur but several crocodiles. Ichnos 2: 37–42.

Bennett, M. R., S. A. Morse, and P. L. Falkingham. 2014. Tracks made by swimming Hippopotami: an example from Koobi Formation (Turkana Basin, Kenya). Palaeogeography, Palaeoclimatology, Palaeoecology 409: 9–23.

Bernier, P., G. Barale, J.-P. Bourseau, E. Buffetaut, G. Demathieu, C. Gaillard, and J.-C. Gall. 1982. Trace nouvelle de locomotion de chelonian et figures d'émersion associées dans les calcaires lithographiques de Cerin (Kimmeridgien superieur, Ain, France). Geobios 15: 447–467.

Bernier, P., G. Barale, J.-P. Bourseau, E. Buffetaut, G. Demathieu, C. Gaillard, and S.

Wenz. 1984. Découverte des pistes de dinosaurs sauteurs dans les calcaires lithographiques de Cerin (Kimmeridgien superieur, Ain, France): implications paléoecologiques. Geobios Mémoire Spécial 15: 177–185.

Bird, R. T. 1944. Did *Brontosaurus* ever walk on land? Natural History 53: 60–69.

Bird, R. T. 1985. Bones for Barnum Brown: Adventures of a Dinosaur Hunter. Texas Christian University Press, Fort Worth, Texas, 226 pp.

Boyd, D. W. 1975. False or misleading traces; pp. 65–83 in R. W. Frey (ed.), The Study of Trace Fossils. Springer-Verlag, New York, New York.

Boyd, D. W., and D. B. Loope. 1984. Probable vertebrate origin for certain sole marks in Triassic red beds of Wyoming. Journal of Paleontology 58: 467–476.

Braddy, S. J., L. B. Morrissey, and A. M. Yates. 2003. Amphibian swimming traces from the Lower Permian of southern New Mexico. Palaeontology 46: 671–683.

Castanera, D., B. Vila, N. L. Razzolini, P. L. Falkingham, J. I. Canudo, P. L. Manning, and À. Galobart. 2013. Manus track preservation bias as a key factor for assessing trackmaker identity and quadrupedalism in basal ornithopods. PLoS One 8(1): e54177. doi:10.1371/journal. pone.0054177.

Colbert, E. H. 1945. The Dinosaur Book. Man and Nature Publication Handbook 14. American Museum of Natural History, New York, New York, 156 pp.

Colbert, E. H. 1965. The Age of Reptiles. W.W. Norton & Company, New York, New York, 256 pp.

Coombs, W. P., Jr. 1975. Sauropod habits and habitats. Palaeogeography, Palaeoclimatology, Palaeoecology 17: 1–33.

Coombs, W. P., Jr. 1980. Swimming ability of carnivorous dinosaurs. Science 207: 1198–1200.

Costeur, L., R. Ezquerra, S. Doublet, P. M. Galton, and F. Perez-Lorente. 2007. Swimming

non-avian theropod trackway from the Early Cretaceous of Spain; p. 16 in Abstract volume of the Fifth meeting of the European Association of Vertebrate Palaeontologists, May 15–19, 2007, Carcassonne, France.

Currie, P. J. 1983. Hadrosaur trackways from the Lower Cretaceous of Canada. Acta Palaeontologica Polonica 28: 68–78.

Czerkas, S. J., and S. Czerkas. 1990. Dinosaurs: A Global View. Dragon's World Press, Limpsfield, U.K., 248 pp.

Da Silva, R. C., J. Ferigolo, I. D. S. Carvalho, and A. C. S. Fernandes. 2008. Lacertoid footprints from the Upper Triassic (Santa Maria Formation) of southern Brazil. Palaeogeography, Palaeoclimatology, Palaeoecology 262: 140–156.

Dawson, J. W. 1868. Acadian Geology. 2nd edition. MacMillan, Edinburgh, U.K., 422.

Dawson, J. W. 1882. On the results of recent explorations of erect trees containing animal remains in the Coal formation of Nova Scotia. Philosophical Transactions of the Royal Society of London 173(2): 621–659.

DeBlieux, D. D., J. I. Kirkland, J. A. Smith, J. McGuire, and V. L. Santucci. 2005. An overview of the vertebrate paleontology of Late Triassic and Early Jurassic rocks in Zion National Park, Utah; p. 2. in The Triassic/Jurassic Terrestrial Transition, St. George, Dixie State College, March 14–16, 2005.

DeBlieux, D. D., J. I. Kirkland, J. A. Smith, J. McGuire, and V. L. Santucci. 2006. An overview of the vertebrate paleontology of Upper Triassic and Lower Jurassic rocks in Zion National Park, Utah; pp. 490–501 in J. D. Harris, S. G. Lucas, J. A. Spielmann, M. G. Lockley, A. R. C. Milner, and J. I. Kirkland (eds.), The Triassic-Jurassic Terrestrial Transition. Bulletin 37. New Mexico Museum of Natural History and Science, Albuquerque, New Mexico.

DeBlieux, D. D., J. A. Smith, J. A. McGuire, J. I. Kirkland, V. L. Santucci, and M. Butler. 2003.

A paleontological inventory of Zion National Park, Utah and the use of GIS to create Paleontological Sensitivity Maps for use in resource management. Journal of Vertebrate Paleontology 23 (3, Supplement): 45A.

Ezquerra Miguel, R., L. Costeur, and F. Pérez-Lorente. 2010. Los dinosaurios también nadaban. Investigacion y Ciencia 411: 2–8.

Ezquerra, R., L. Costeur, S. Doublet, P. M. Galton, and F. Pérez-Lorente. 2004. Lower Cretaceous swimming theropod trackway from 'La Virgen del Campo' (La Rioja, Spain); p. 150 in Abstracts volume of the Palaeontological Association annual meeting, Lille, December 2004.

Ezquerra, R., S. Doublet, L. Costeur, P. M. Galton, and F. Pérez-Lorente. 2007. Were non-avian theropod dinosaurs able to swim? Supportive evidence from an Early Cretaceous trackway, Cameros Basin (La Rioja, Spain). Geology 35(6): 507–510.

Falkingham, P. L., K. T. Bates, L. Margetts, and P. L. Manning. 2011. Simulating sauropod manus-only trackway formation using finite-element analysis. Biology Letters 7: 142–145.

Farlow J. O., and R. M. Elsey. 2010. Footprints and trackways of the American alligator, Rockefeller Wildlife Refuge, Louisiana; pp. 31–40 in J. Milàn, S. G. Lucas, M. G. Lockley, and J. A. Spielmann (eds.), Crocodyle Tracks and Traces. Bulletin 51. New Mexico Museum of Natural History and Science, Albuquerque, New Mexico.

Farlow, J. O., and P. M. Galton. 2003. Dinosaur trackways of Dinosaur State Park, Rocky Hill, Connecticut; pp. 248–263 in P. M. Letourneau and P. E. Olsen (eds.), The Great Rift Valleys of Pangea in Eastern North America. Volume 2. Columbia University Press, New York, New York.

Foster, J. R., and M. G. Lockley. 1997. Probable crocodilian tracks and traces from the Morrison Formation (Upper Jurassic) of eastern Utah. Ichnos 5: 121–129.

Fujita, M., T.-N. Lee, Y. Azuma, and D. Li. 2012. Unusual tridactyl trackways with tail traces from the Lower Cretaceous Hekou Group, Gansu Province, China. Palaios 27(8): 560–570.

Galton, P. M. 1970. The posture of hadrosaurian dinosaurs. Journal of Paleontology 44: 464–473.

Galton, P. M., and J. O. Farlow. 2003. Dinosaur State Park, Connecticut, USA: history, footprints, trackways, exhibits. Zubia 21: 129–173.

García Ramos, J. C., J. Lires, and L. Piñuela. 2002. Dinosaurios: rutas por el Jurásico de Asturias. Illustrated by Arturo de Miguel and Raúl Martín. Group Zeta in conjunction with La Voz de Asturias, Polígono de Puente Nora, Lugones Asturias, Spain.

Gierlinski, G., and A. Potemska. 1987. Lower Jurassic dinosaur footprints from Gliniany Las, northern slope of the Holy Cross Mountains, Poland. Neues Jahrbuch für Geologie und Paläontologie Abhandlungen 175: 107–120.

Gierlinski, G., G. Niedźwiedzki, and G. Pieńkowski. 2004. Tetrapod track assemblage in the Hettangian of Sołtyków, Poland, and its paleoenvironmental background. Ichnos 11: 195–213.

Gillette, D. D., and D. A. Thomas. 1989. Problematical tracks and traces of Late Albian (Early Cretaceous) age, Clayton Lake State Park, New Mexico, USA; pp. 337–342 in D. D. Gillette and M. G. Lockley (eds.), Dinosaur Tracks and Traces. Cambridge University Press, Cambridge, UK.

Harington, C. R., H. Foster, R. Holmes, and P. J. Currie. 2005. Photographic catalogue of trackways in the Canadian Museum of Nature. Syllogeus 74: 1–151.

Henderson, D. M. 2004. Tipsy punters: sauropod dinosaur pneumaticity, buoyancy and aquatic habits. Proceedings of the Royal Society of London, Biology Letters 271: S180–S183.

Hotton, N., III. 1963. Dinosaurs. Pyramid Publications, New York, New York, 192 pp.

Houck, K., M. Lockley, M. Caldwell, and B. Clark. 2010. A well-preserved crocodylian trackway from the South Platte Formation (Lower Cretaceous), Golden, Colorado: Crocodyle tracks and traces. New Mexico Museum of Natural History and Science Bulletin 51: 115–120.

Hunt, A. P., and S. G. Lucas. 2007a. Tetrapod ichnofacies: a new paradigm. Ichnos 14: 59–68.

Hunt, A. P., and S. G. Lucas. 2007b. Late Triassic tetrapod tracks of western North America; pp. 215–230 in S. G. Lucas and J. A. Spielmann (eds.), Triassic of the American West. Bulletin 40. New Mexico Bulletin of Natural History and Science, Albuquerque, New Mexico.

Hunt, A. P., M. G. Lockley, and S. G. Lucas. 1993. Tetrapod footprints from the Middle Triassic Moenkopi Formation, west-central New Mexico; pp. G20–G23 in S. G. Lucas and M. Morales (eds.), The Nonmarine Triassic. Bulletin 3. New Mexico Bulletin of Natural History and Science, Albuquerque, New Mexico.

Hwang, K.-G., M. G. Lockley, M. Huh, and S. Paik. 2008. A reinterpretation of dinosaur footprints with internal ridges from the Upper Cretaceous Uhangri Formation, Korea. Palaeogeography, Palaeoclimatology, Palaeoecology 258: 59–70.

Ishigaki, S. 1989. Footprints of swimming sauropods from Morocco; pp. 83–86 in D. D. Gillette and M. G. Lockley (eds.), Dinosaur Tracks and Traces. Cambridge University Press, Cambridge, UK.

Kirkland, J. I., and A. R. C. Milner. 2006. The Moenave Formation at the St. George Dinosaur Discovery Site at Johnson Farm; pp. 289–309 in J. D. Harris, S. G. Lucas, J. A. Spielmann, M. G. Lockley, A. R. C. Milner, and J. I. Kirkland (eds.), The Triassic-Jurassic Terrestrial Transition. Bulletin 37. New Mexico Museum of Natural History and Science, Albuquerque, New Mexico.

Kirkland, J. I., M. G. Lockley, and A. R. C. Milner. 2002. The St. George dinosaur tracksite. Utah Geological Survey Notes 34: 4, 5, 12.

Kukihara, R., and M. G. Lockley. 2012. Fossil footprints from the Dakota Group (Cretaceous) John Martin Reservoir, Bent County, Colorado: new insights into the paleoecology of the dinosaur freeway. Cretaceous Research 33: 165–182.

Kukihara, R., M. G. Lockley, and K. Houck. 2010. Crocodile footprints from the Dakota Group (Cretaceous) John Martin Reservoir, Bent County, Colorado. New Mexico Museum of Natural History and Science, Bulletin 51: 121–136.

Kvale, E. P., S. T. Hasiotis, D. L. Mickelson, and G. D. Johnson. 2001. Middle and Late Jurassic dinosaur fossil-bearing horizons: implications for dinosaur paleoecology, northeastern Bighorn Basin, Wyoming; pp. 17–45 in C. L. Hill (ed.), Mesozoic and Cenozoic Paleontology in the Western Plains and Rocky Mountains: Guidebook for Field Trips, Society of Vertebrate Paleontology 61st Annual Meeting. Occasional Paper 3. Museum of the Rockies, Bozeman, Montana.

Kvale, E. P., D. L. Mickelson, S. T. Hasiotis, and G. D. Johnson. 2004. The history of dinosaur footprint discoveries in Wyoming with emphasis on the Bighorn Basin. Ichnos 11: 3–9.

Kvale, E. P., G. D. Johnson, D. L. Mickelson, K. Keller, L. C. Furer, and A. W. Archer. 2001. Middle Jurassic (Bajocian and Bathonian) dinosaur megatracksites, Bighorn Basin, Wyoming, USA. Palaios 16: 233–254.

Lammers, G. E. 1964. Reptile tracks and the paleoenvironment of the Triassic Moenkopi of Capitol Reef National Monument, Utah. Museum of Northern Arizona Bulletin 40: 49–55.

Langiaux, J. 1980. Premières observations a Morteru (Stéphanien de Blanzy-Montceau) ichnites, faune, flore. Revue Périodique de Vulgarisation des Sciences Naturelles et Historiques "La Physiophile" Montceau-le-Mines 93: 77–88.

Lee, Y.-N., and M. Huh. 2002. Manus-only sauropod tracks in the Uhangri Formation (Upper Cretaceous), Korea and their paleobiological implications. Journal of Paleontology 76: 558–564.

Lee, Y. N., and H. J. Lee. 2006. A sauropod trackway in Donghae-Myeon, Goseong County, South Gyeongsang Province, Korea and its paleobiological implications of Uhangri manus-only sauropod tracks. Journal of the Paleontological Society of Korea 22: 1–14.

Li, D., Y. Azuma, M. Fujita, Y.-N. Lee, and Y. Arakawa. 2006. A preliminary report on two new vertebrate track sites including dinosaurs from the Early Cretaceous Hekou Group, Gansu Province, China. Journal of the Paleontological Society of Korea 22(1): 29–49.

Lockley, M. G. 1991. Tracking Dinosaurs: A New Look at an Ancient World. Cambridge University Press, Cambridge, U.K., 238 pp.

Lockley, M. G. 2006. Observations on the ichnogenus Gwyneddichnium and Gwyneddichnium-like footprints and trackways from the Upper Triassic of the western United States; pp. 170–175 in J. D. Harris, S. G. Lucas, J. A. Spielmann, M. G. Lockley, A. R. C. Milner, and J. I. Kirkland (eds.), The Triassic-Jurassic Terrestrial Transition. Bulletin 37. New Mexico Museum of Natural History and Science, Albuquerque, New Mexico.

Lockley, M. G. 2007. A tale of two ichnologies: the different goals and missions of vertebrate and invertebrate ichnology and how they relate in ichnofacies analysis. Ichnos 14: 39–57.

Lockley, M. G. 2010. A solution to the Mehliella mystery: tracking, naming, identifying, and measuring the first crocodylian trackway reported from the Cretaceous (Dakota Group, Colorado); pp. 157–163 in J. Milàn, S. G. Lucas, M. G. Lockley, and J. A. Spielmann (eds.), Crocodyle Tracks and Traces. Bulletin 51. New Mexico Museum of Natural History and Science, Albuquerque, New Mexico.

Lockley, M. G., and A. P. Hunt. 1995. Dinosaur Tracks and Other Fossil Footprints of the Western United States. Columbia University Press, New York, New York, 360 pp.

Lockley, M. G., and A. R. C. Milner. 2006. Tetrapod tracksites from the Shinarump Formation (Chinle Group, Upper Triassic) of southwestern Utah; pp. 257–262 in J. D. Harris, S. G. Lucas, J. A. Spielmann, M. G. Lockley, A. R. C. Milner, and J. I. Kirkland (eds.), The Triassic-Jurassic Terrestrial Transition. Bulletin 37. New Mexico Museum of Natural History and Science, Albuquerque, New Mexico

Lockley, M. G., and A. Rice. 1990. Did "*Brontosaurus*" ever swim out to sea? Evidence from brontosaur and other dinosaur footprints. Ichnos 1: 81–90.

Lockley, M. G., and B. Schumacher. 2014. A new pterosaur swim tracks from the Cretaceous Dakota Group of eastern Colorado: implications for pterosaur swim track behavior, pp. 365–371 in M. G. Lockley and S. G. Lucas (eds.), Fossil Footprints of western North America. Bulletin 62. New Mexico Museum of Natural History and Science, Albuquerque, New Mexico.

Lockley, M. G., and C. A. Meyer. 2000. Dinosaur Tracks and Other Fossil Footprints of Europe. Columbia University Press, New York, New York, 342 pp.

Lockley, M. G., and J. L. Wright. 2003. Pterosaur swim tracks and other ichnological evidence of behaviour and ecology, pp. 297–313 in E. Buffetaut and J.-M. Mazin (eds.), Evolution and Palaeobiology of Pterosaurs. Special Publication 217. Geological Society of London, London, UK.

Lockley, M. G., and J. R. Foster. 2006. Dinosaur and turtle tracks from the Morrison Formation (Upper Jurassic) of Colorado National Monument, with observations on the taxonomy of vertebrate swim tracks; pp. 193–198 in J. R. Foster and S. G. Lucas (eds.), Paleontology of the Upper Jurassic Morrison Formation. Bulletin 36. New Mexico Museum of Natural History and Science, Albuquerque, New Mexico.

Lockley, M. G., and S. G. Lucas. 2011. Crocs not theropods were likely top predators on the Cretaceous dinosaur freeway: implications of a large track census. Journal of Vertebrate Paleontology 31(Program and abstracts supplement): 146.

Lockley, M. G., A. R. C. Milner, and S. G. Lucas. 2005. Archosaur tracks from the Chinle Group (Late Triassic), St. George area, southwestern Utah: tracking dinosaur origins; pp.13–14 in The Triassic/Jurassic Terrestrial Transition, St. George, Dixie State College, March 14–16, 2005.

Lockley, M. G., M. Huh, and J.-Y. Kim. 2012. Mesozoic Terrestrial ecosystems of the Korean Cretaceous Dinosaur Coast: A field guide to the excursions of the 11th Mesozoic Terrestrial Ecosystems Symposium (August 19–22). A publication supported by the Korean Federation of Science and Technology Societies, Seoul, Korea.

Lockley, M. G., M. Matsukawa, and I. Obata. 1989. Dinosaur tracks and radial cracks: unusual footprint features. Bulletin of the National Science Museum, Tokyo, Series C 15(4): 151–160.

Lockley, M. G., J. G. Pittman, C. A. Meyer, and V. F. Santos. 1994. On the common occurrence of manus-dominated sauropod trackways in Mesozoic carbonates. Gaia 10: 119–124.

Lockley, M. G., K. Conrad, M. Paquette, and A. Hamblin. 1992. Late Triassic vertebrate tracks in the Dinosaur National Monument area. Utah Geological Survey Miscellaneous Publications 92–93: 383–391.

Lockley, M. G., R. Kukihara, L. Pionek, and A. Delgalvis. 2014. A survey of new fossil footprint sites from Glen Canyon National Recreation Area (western USA), with special reference to the Kayenta Navajo transition zone (Glen Canyon Group, Lower Jurassic); pp. 157–180 in M. G. Lockley and S. G. Lucas (eds.), Fossil Footprints of Western North America. Bulletin 62. New Mexico Museum of Natural History and Science, Albuquerque, New Mexico.

Lockley, M. G., R. Li, M. Matsukawa, and J. Li. 2010. Tracking Chinese crocodilians: *Kuangyuanpus, Laiyangpus* and implications for naming crocodylian and crocodylian-like tracks and associated ichnofacies. New Mexico Museum of Natural History and Science Bulletin 51: 99–108.

Lockley, M. G., D. Fanelli, K. Honda, K. Houck, and N. A. Matthews. 2010. Crocodile waterways and dinosaur freeways: implications of multiple swim track assemblages from the Cretaceous Dakota Group, Golden area, Colorado; pp. 137–156 in J. Milàn, S. G. Lucas, M. G. Lockley, and J. A. Spielmann (eds.), Crocodyle Tracks and Traces. Bulletin 51. New Mexico Museum of Natural History and Science, Albuquerque, New Mexico.

Lockley, M. G., J. D. Harris, R. Li, L. Xing, and T. van der Lubbe. 2016. Two-toed tracks through time: on the trail of 'raptors' and their allies; chap. 11 in P. L. Falkingham, D. Marty, and A. Richter (eds.), Dinosaur Tracks: The Next Steps. Indiana University Press, Bloomington, Indiana.

Lockley, M. G., S. G. Lucas, J. Milàn, J. D. Harris, M. Avanzini, J. R. Foster, and J. A. Spielmann. 2010. The fossil record of crocodylian tracks and traces; pp. 1–13 in J. Milàn, S. G. Lucas, M. G., Lockley, and J. A. Spielmann (eds.), Crocodyle Tracks and Traces. Bulletin 51. New Mexico Museum of Natural History and Science, Albuquerque, New Mexico.

Lockley, M. G., K. Cart, J. Martin, R. Prunty, K. Hups, J.-D. Lim, K.-S. Kim, K. Houck, and G. Gierlinski. 2014. A bonanza of new tetrapod tracksites from the Cretaceous Dakota Group, western Colorado: implications for paleoecology; pp. 393–409 in M. G. Lockley and S. G. Lucas (eds.), Fossil Footprints of Western North America. Bulletin 62. New Mexico Museum of Natural History and Science, Albuquerque, New Mexico.

Lovelace, D. M., and S. D. Lovelace. 2012. Paleoenvironments and paleoecology of a Lower Triassic invertebrate and vertebrate ichnoassemblage from the Red Peak Formation (Chugwater Group), central Wyoming. Palaios 27(9): 636–657.

Lucas, S. G., A. J. Lerner, and H. Haubold. 2001. First record of Amphisauropus and Varanopus in the Lower Permian Abo Formation, central New Mexico. Hallesches Jahrbuch für Geowissenschaften B 23: 69–78.

Lull, R. S. 1953. Triassic life of the Connecticut Valley. Revised. State Geological and Natural History Survey of Connecticut Bulletin 81: 1–336.

Matthew, G. F. 1904. New species and a new genus of batrachian footprints of the Carboniferous system in eastern Canada. Transactions of the Royal Society of Canada 4: 77–119.

McAllister, J. A. 1989a. Dakota Formation tracks from Kansas: implications for the recognition of tetrapod subaqueous traces; pp. 343–348 in D. D. Gillette, and M. G. Lockley (eds.), Dinosaur Tracks and Traces. Cambridge University Press, Cambridge, UK.

McAllister, J. A. 1989b. Subaqueous vertebrate footmarks from the upper Dakota Formation (Cretaceous) of Kansas, U.S.A. Occasional Papers of the Museum of Natural History, University of Kansas 127: 1–22.

McAllister, J., and J. Kirby. 1998. An occurrence of reptile subaqueous traces in the Moenkopi Formation (Triassic) of Capitol Reef National Park, south central Utah, USA. Geologic Resources Division Technical Research NPS/NRGRD/GRDTR-98/01: 45–49.

McCrea, R. T., S. G. Pemberton, and P. J. Currie. 2004. New ichnotaxa of mammal and reptile tracks from the upper Paleocene of Alberta. Ichnos 11: 323–339.

Mehl, M. G. 1931. Additions to the vertebrate record of the Dakota Sandstone. American Journal of Science 21(125): 441–452.

Melchor, R. N., and W. A. S. Sarjeant. 2004. Small amphibian and reptile footprints from the Permian Carapacha Basin, Argentina. Ichnos 11: 57–78.

Mikuláš, R. 2010. *Characichnos* isp., probable crocodile swim traces from the Miocene of the Ceské Budejovice Basin, Czech Republic; pp. 179–181 in J. Milàn, S. G. Lucas, M. G. Lockley, and J. A. Spielmann (eds.), Crocodyle Tracks and Traces. Bulletin 51. New Mexico Museum of Natural History and Science,Albuquerque, New Mexico

Milàn, J. 2006. Variations in the morphology of Emu (*Dromaius novaehollandiae*) tracks reflecting differences in walking pattern and substrate consistency: ichnotaxonomic implications. Palaeontology 49(2): 405–420.

Milàn, J., and R. G. Bromley. 2006. True tracks, undertracks and eroded tracks, experimental work with tetrapod tracks in laboratory and field. Palaeogeography, Palaeoclimatology, Palaeoecology 231: 253–264.

Milner, A. R. C., and J. I. Kirkland. 2007. The case for fishing dinosaurs at the St. George Dinosaur Discovery Site at Johnson Farm. Utah Geological Survey Notes 39(1): 1–3.

Milner, A. R. C., and M. G. Lockley. 2006. History, geology, and paleontology: St. George Dinosaur Discovery Site at Johnson Farm, Utah; pp. 35–48 in R. E. Reynolds (ed.), Making Tracks across the Southwest: The 2006 Desert Symposium Field Guide and Abstracts from Proceedings. Desert Studies Consortium and LSA Associates, Zzyzx, California.

Milner, A. R. C., and S. Z. Spears. 2007. Mesozoic and Cenozoic paleoichnology of southwestern Utah. Geological Society of America, Rocky Mountain Section Annual Meeting, St. George, Utah. Utah Geological Association Publication 35: 1–85.

Milner, A. R. C., M. G. Lockley, and S. B. Johnson. 2006. The story of the St. George Dinosaur Discovery Site at Johnson Farm: an important new Lower Jurassic dinosaur tracksite from the Moenave Formation of southwestern Utah; pp. 329–345 in J. D. Harris, S. G. Lucas, J. A. Spielmann, M. G. Lockley, A. R. C. Milner, and J. I. Kirkland (eds.), The Triassic-Jurassic Terrestrial Transition. Bulletin 37. New Mexico Museum of Natural History and Science, Albuquerque, New Mexico.

Milner, A. R. C., M. G. Lockley, and J. I. Kirkland. 2006. A large collection of well-preserved theropod dinosaur swim tracks from the Lower Jurassic Moenave Formation, St. George, Utah; pp. 315–328 in J. D. Harris, S. G. Lucas, J. A. Spielmann, M. G. Lockley, A. R. C. Milner, and J. I. Kirkland (eds.), The Triassic-Jurassic Terrestrial Transition. Bulletin 37. New Mexico Museum of Natural History and Science, Albuquerque, New Mexico.

Milner, A. R. C., J. I. Kirkland, M. G. Lockley, and J. D. Harris. 2005a. Relative abundance of theropod dinosaur tracks in the Early Jurassic (Hettangian) Moenave Formation at a St. George dinosaur tracksite in southwestern Utah: bias produced by substrate consistency. Geological Society of America Abstracts with Programs 37: 5.

Milner, A. R. C., J. D. Harris, M. G. Lockley, J. I. Kirkland, and N. A. Matthews. 2009. Bird-like anatomy, posture, and behavior revealed by an Early Jurassic theropod dinosaur resting trace. PLoS One 4(3): e4591.

Milner, A. R. C., M. G. Lockley, J. I. Kirkland, D. L. Mickelson, and G. S. Vice. 2005. First reports of a large collection of well-preserved dinosaur swim tracks from the Lower Jurassic Moenave Formation, St. George, Utah: a preliminary evaluation; p. 18 in The Triassic/Jurassic Terrestrial Transition, St. George, Dixie State College, March 14–16, 2005.

Milner, A. R. C., M. G. Lockley, J. I. Kirkland, P. Bybee, and D. L. Mickelson. 2004. St. George tracksite, southwestern Utah: remarkable Early Jurassic (Hettangian) record of dinosaurs walking, swimming, and sitting provides a detailed view of the paleoecosystem along the shores of Lake Dixie. Journal of Vertebrate Paleontology 24(3, Supplement): 94A.

Milner, A. R. C., T. A. Birthisel, J. I. Kirkland, B. H. Breithaupt, N. A. Matthews, M. G. Lockley, V. L. Santucci, S. Z. Gibson, D. D. DeBlieux, M. Hurlbut, J. D. Harris, and P. E. Olsen. 2012. Tracking Early Jurassic dinosaurs across southwestern Utah and the Triassic-Jurassic transition; pp. 1–107 in J. W. Bonde and A. R. C. Milner (eds.), Field Trip Guide Book for the 71st Annual Meeting of the Society of Vertebrate Paleontology. Paleontological Papers 1. Nevada State Museum, Las Vegas, New Mexico.

Mudroch, A., U. Richter, U. Joger, R. Kosma, O. Idé, and A. Maga. 2011. Didactyl tracks of paravian theropods (Maniraptora) from the ?Middle Jurassic of Africa. PLoS One 6(2): e14642. doi:10.1371/journal.pone.0014642.

Niedźwiedzki, G., P. Szrek, K. Narkiewicz, M. Narkiewicz, and P. E. Ahlberg. 2010. Tetrapod trackways from the early Middle Devonian period of Poland. Nature 463: 43–48.

Norman, D. 1985. The Illustrated Encyclopedia of Dinosaurs. Crescent Books, New York, New York, 400 pp.

Paul, G. S. 1988. Predatory Dinosaurs of the World. Simon and Schuster, New York, New York, 464 pp.

Pieńkowski, G., and G. Gierlinski. 1987. New finds of dinosaur footprints in Liassic of the Holy Cross Mountains and its paleoenvironmental background. Prezegl'd Geologiczny 4: 199–205.

Rainforth, E. C., and M. Howard. 2008. Swimming theropod? A new investigation of unusual theropod footprints from Dinosaur State Park, Rocky Hill, CT (Newark Supergroup, eastern North America). Geological Society of America Abstracts and Programs 40(2): 79.

Romilio, A., R. T. Tucker, and S. W. Salisbury. 2013. Reevaluation of the Lark Quarry dinosaur tracksite (late Albian-Cenomanian Winton Formation, central-western Queensland, Australia): no longer a stampede? Journal of Vertebrate Paleontology 33: 102–120.

Ruhle von Lilienstern, H. 1939. Fährten und spüren im Chirotherium-Sandstein von Südthüringen. Forschritte der Geologie und Paläontologie 12: 293–387.

Russell, L. S. 1930. Early Tertiary mammal tracks from Alberta. Transactions of the Royal Canadian Institute 17: 217–221.

Rutherford, R. L., and L. S. Russell. 1928. Mammal tracks from the Paskapoo Beds of Alberta. American Journal of Science 15: 262–264.

Song, J. Y. 2010. A reinterpretation of unusual Uhangri dinosaur tracks from the view of functional morphology. Journal of the Paleontological Society of Korea 26: 95–105.

Sternberg, C. M. 1933. Carboniferous tracks from Nova Scotia. Geological Society of America Bulletin 44: 951–964.

Swanson, B. A., and K. J. Carlson. 2002. Walk, wade, or swim? Vertebrate traces on an Early Permian lakeshore. Palaios 17: 123–133.

Thomson, T. J., and D. M. Lovelace. 2014. Swim track morphotypes and new track localities from the Moenkopi and Red Peak formations (Early-Middle Triassic) with preliminary interpretations of aquatic behaviors; pp. 103–128 in M. G. Lockley and S. G. Lucas (eds.), Fossil Footprints of Western North America. Bulletin 62. New Mexico Museum of Natural History and Science, Albuquerque, New Mexico.

Thulborn, R. A. 1989, The gaits of dinosaurs; pp. 39–50 in D. D. Gillette and M. G. Lockley (eds.), Dinosaur Tracks and Traces. Cambridge University Press, Cambridge, UK.

Thulborn, R. A. 1990. Dinosaur Tracks. Chapman Hall, London, U.K., 410 pp.

Thulborn, R. A. 2004. Extramorphological features of sauropod dinosaur tracks in the Uhangri Formation (Cretaceous), Korea. Ichnos 11: 295–298.

Thulborn, R. A. 2013. Lark Quarry revisited: a critique of methods used to identify a large dinosaurian track-maker in the Winton Formation (Albian-Cenomanian), western Queensland, Australia. Alcheringa 37: 312–330. Available at http://dx.doi.org/10.1080/03115518.2013.748482. Accessed October 13, 2015.

Thulborn, R. A., and M. Wade. 1979. Dinosaur stampede in the Cretaceous of Queensland. Lethaia 12: 275–279.

Thulborn, R. A., and M. Wade. 1984. Dinosaur trackways in the Winton Formation (mid-Cretaceous) of Queensland. Memoirs Queensland Museum 21: 413–517.

Whyte, M. A., and M. Romano. 2001. A dinosaur ichnocoenosis from the Middle Jurassic of Yorkshire, UK. Ichnos 8: 233–234.

Whyte, M. A., M. Romano, and D. J. Elvidge. 2007. Reconstruction of Middle Jurassic dinosaur-dominated communities from the vertebrate ichnofauna of the Cleveland Basin of Yorkshire, UK. Ichnos 14: 117–129.

Wright, J. L., and M. G. Lockley. 2001. Dinosaur and turtle tracks from the Laramie/Arapahoe formations (Late Cretaceous), near Denver, Colorado, USA. Cretaceous Research 22: 365–376.

Xing, L.-D., M. G. Lockley, J.-P. Zhang, A. R. C. Milner, H. Klein, D.-Q. Li, W. S. Persons IV, and H.-F. Ebi. 2013. A new Early Cretaceous dinosaur track assemblage and the first definite non-avian theropod swim trackway from China. Chinese Science Bulletin 58(19): 2370–2378.

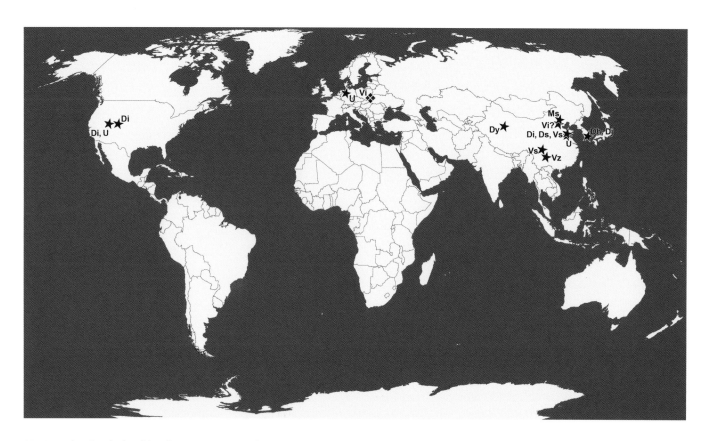

11.1. Map showing the localities of Cretaceous, didactyl tracks worldwide. The star indicates Early-"middle" Cretaceous and the diamond u, Late Cretaceous tracks. Abbreviations:

Dh, *Dromaeosauripus hamanensis;* Di, *Dromaeosauripus* isp.; Dj, *Dromaeosauripus jinjuensis*; Ds, *Dromaeopodus shandongensis;* Du, *Dromaeopodus* isp.; Dy, *Dromaeosauripus*

yongjingensis; Ms, *Menglongipus sinensis;* U, unnamed/unattributed; Vi, *Velociraptorichnus* isp.; Vs, *Velociraptorichnus sichuanensis;* Vz, *Velociraptorichnus zhangi.*

11

Two-Toed Tracks through Time: On the Trail of "Raptors" and Their Allies

Martin G. Lockley, Jerald D. Harris, Rihui Li, Lida Xing, and Torsten van der Lubbe

THE TWO-TOED, OR DIDACTYL, TRACKS OF DEINONYCHO-saurian dinosaurs, popularly known as "raptors," are among the most distinctive theropod tracks known. Including the first confirmed report from China in 1994, a total of 16 track-sites have been recognized, all from Cretaceous strata. These include nine Chinese, two Korean, three North American, and two European occurrences. Many of these tracks have been assigned to four ichnogenera: *Velociraptorichnus* (two ichnospecies), *Dromaeopodus*, *Menglongipus*, and *Dromaeosauripus* (three ichnospecies). Most of the tracks have been attributed to dromaeosaurid theropods, but in the case of the largest sample, from Germany, a troodontid trackmaker is inferred.

Here we review all 16 Cretaceous occurrences, indicating sample size, preservation quality, morphological character-istics, utility for ichnotaxonomy, and trackmaker identifica-tion. The size range (foot length) of known deinonychosau-rian trackmakers is between ~10 and 28 cm. Most reported instances were made by trackmakers inhabiting fluvio-la-custrine habitats in the later part of the Early Cretaceous (Barremian-Albian).

We also note two reports of pre-Cretaceous, didactyl–or pseudodidactyl–ichnogenera (*Evazoum* and *Paravipes*), both of probable saurischian affinity, with which Cretaceous dei-nonychosaurian tracks have been compared. In the case of *Evazoum*, we infer a morphological and behavioral conver-gence with deinonychosaurian morphology, related to the retraction of digit II. In the case of *Paravipes*, morphological similarities are less obvious, and we infer that this ichnoge-nus likely comprises swim tracks.

INTRODUCTION

Although the term "raptor" has broad, popular meaning for birds of prey, in popular depictions of nonavian dinosaurian paleontology it has come (thanks to *Jurassic Park*) to refer spe-cifically to deinonychosaurian (dromaeosaurid and troodon-tid) theropod dinosaurs. These theropods have a distinctive pedal morphology characterized by a retracted pedal digit II that bears a well-developed, hypertrophied 'sickle claw.'

Until quite recently, tracks of these Mesozoic 'raptors' were unknown, despite the clade being quite well-represented in the osteological record, especially in the Cretaceous (Turner, Makovicky, and Norell, 2012). In particular, their pedal spe-cialization, well preserved in many specimens (e.g., Ostrom, 1969; Norell and Makovicky, 1997; Xu and Norell, 2004; Gao et al., 2012), indicates that digit II was held raised during locomotion in life.

Because deinonychosaurians were known long before their tracks were identified, ichnologists speculated, based on known pedal skeletons, that deinonychosaurian tracks would be didactyl, comprising impressions of only digits III and IV. Tracks pertaining to these dinosaurs were not reported until 1994, when small, didactyl tracks from the Cretaceous of Sichuan Province, China, were described as *Velociraptorichnus* (Zhen et al., 1994). At present, 16 Creta-ceous deinonychosaurian track sites have been reported (Fig. 11.1, Table 11.1), including 11 from Asia (9 from China, 2 from Korea), 3 from North America, and 2 from Europe. These are in addition to reports of likely nondeinonychosaurian, didactyl track morphotypes from the Permian, Late Triassic–Early Jurassic, and Middle Jurassic. Here we give a detailed account of all these occurrences and their implications for distinguishing different deinonychosaurian track morphot-ypes. In this regard, some samples are instructive in pointing to differences between dromaeosaurid and troodontid tracks. We also take into consideration other essential, morphologi-cal factors, such as size and relative proportions of digit traces and trackway patterns, as well as nonmorphological factors, such as sample size, substrate preservation, geological age, and paleobiogeography.

PRE-CRETACEOUS DIDACTYL TRACKS

Some pre-Cretaceous, didactyl tracks clearly pertain to non-dinosaurian trackmakers. The best known example is *Dromo-pus didactylus* (Moodie, 1930), which comprises incompletely preserved tracks of a pentadactyl, lacertiform trackmaker from the Permian of Texas that show only the impressions of the longest digits (IV and III). Such tracks are small and

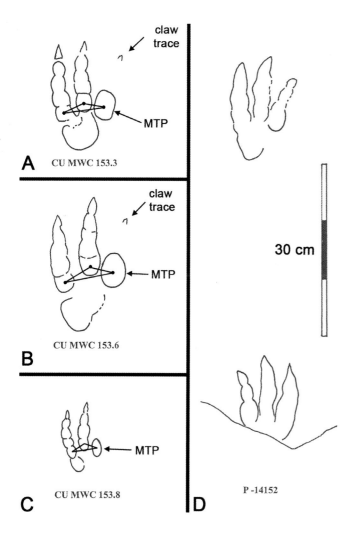

11.2. (A–C) *Evazoum gatewayensis* from the Chinle Group of western Colorado. University of Colorado Museum of Natural History specimens now part of the University of Colorado Denver, Dinosaur Tracks Museum collection (CU): CU MWC 153.3 (holotype, [Museum of Western Colorado collection]) (A), CU MWC 153.6 (B), and CU MWC 153.8 (C). Note the triangular configuration of the metatarsophalangeal pads (MTP) of digits II–IV. (D) An example of a more typical, tridactyl *Evazoum* track (P-14152) from the Late Triassic of New Mexico. *After Lockley and Lucas (2013).*

form part of a suite of trackways made by small, quadrupedal animals that cannot morphologically compare with the tracks of large, bipedal dinosaurs from the Mesozoic. However, they provide important evidence that animals with pentadactyl, tetradactyl, and tridactyl feet are capable of leaving didactyl tracks depending on how substrate conditions and/or locomotory behaviors affect footprint preservation, a subject discussed in some detail by Lockley and Lucas (2013).

Recently, Lockley, Lucas, and Hunt (2006) and Lockley and Lucas (2013) referred to four, Late Triassic–Early Jurassic ichnogenera (*Otozoum*, *Pseudotetrasauropus*, *Evazoum*, and *Kalosauropus*) as the "OPEK plexus." This group is accommodated in the previously named ichnofamily Otozoidae (Lull, 1904), which originally contained only *Otozoum*. Lockley et al. (2006) noted that type *Otozoum* and *Pseudotetrasauropus*

(O and P morphotypes) are large tracks, in some cases associated with manus tracks, indicating quadrupedal trackmakers. In contrast, *Evazoum* (Nicosia and Loi, 2003) and *Kalosauropus* (Ellenberger, 1972, 1974) (E and K morphotypes) are both small tracks made by bipeds. Of significance here are reports that *Evazoum* from North America includes a morphotype that is functionally didactyl, or pseudodidactyl, preserving for digit II only a swollen or enlarged, proximal pad trace. This morphotype has been illustrated on a number of occasions (Olsen, Schlische, and Gore, 1989; Gaston et al., 2003; Lockley et al., 2006) and recently formally described as the new ichnospecies *Evazoum gatewayensis* (Lockley and Lucas, 2013) (Fig. 11.2). All ichnotaxa in the OPEK plexus have been attributed to basal sauropodomorphs, if not unanimously then at least by a majority of authors (Ellenberger, 1972; Lockley and Hunt, 1995; Rainforth, 2003; Lockley, Lucas, and Hunt, 2006; D'Orazi Porchetti and Nicosia, 2007; Lockley and Lucas, 2013). Given this interpretation, some *Evazoum* tracks represent a foot morphology and/or locomotory behavior that appears to be a convergent precursor of the morphology of deinonychosaurian tracks that are now increasingly well known in the Cretaceous. The fact that both inferred trackmaking groups (Deinonychosauria and basal Sauropodomorpha) belong to the saurischian clade suggests the possibility of a reiterative developmental pattern or the generation of convergent morphology as a result of similar evolutionary adaptation.

The Middle Jurassic ichnogenus *Paravipus didactyloides*, of presumed theropod affinity, was recently named on the basis of unusual, "didactyl" tracks from Mali, Africa. These enigmatic tracks, described as having been made by a digitigrade, bipedal trackmaker, preserve paired, subparallel impressions assigned to digits III and IV, of which the purported digit III impressions were slightly longer than those of digit IV (Murdoch et al., 2011). Impressions of digit II were occasionally represented by small, round or oval "pad" impressions (Mudroch et al., 2011). These tracks show many similarities to theropod swim tracks from the Early Jurassic of Utah (Milner, Lockley, and Kirkland, 2006) (Fig. 11.3A) and discussed elsewhere in this volume (Milner and Lockley, 2016).

Another locality revealing large, apparently didactyl, tracks was reported by Ishigaki and Lockley (2010) from the Early Jurassic of Morocco. Although some of the individual tracks in this report have a striking didactyl appearance (Figs. 11.3B, 11.3C), these tracks were associated with, and in some cases parts of, trackways of tridactyl theropods. Their didactyl appearance is attributable to postregistration sediment collapse and/or irregular gaits associated with unusual preservational conditions. These tracks have not been named.

11.3. Large "didactyliform" tracks from the Jurassic of Africa. (A) Fiberglass replica of the holotype of *Paravipus didactyloides* from Middle Jurassic of Mali (CU 182.9 [University of Colorado Denver, Dinosaur Tracks Museum collection]). (B–C) Natural impressions and molds of apparently didactyl theropod tracks from the Early Jurassic of Morocco.

DEINONYCHOSAUR TRACK OCCURENCES
IN SPACE AND TIME

The First Discovery: Velociraptorichnus sichuanensis

The tracks named by Zhen et al. (1994) as *Velociraptorichnus sichuanensis* are part of an important assemblage of small tracks from the Cretaceous Jiaguan Formation of Sichuan Province, China. *Velociraptorichnus sichuanensis* was originally called "*Deinonychosaurichnus*" in an unillustrated abstract (Zhen et al., 1987) and a subsequent citation of this same abstract (Zhen et al., 1989). However, because abstracts are not valid venues in which to erect taxa, this name is not valid, and *V. sichuanensis*, as erected in Zhen et al. (1994), is the valid name. In Zhen et al. (1994), only one track, the holotype (CFEC-B-1 [Chongqing Natural History Museum collection]), was illustrated; two other paratype and topotype tracks (CFEC-B-2 and CFEC-B-3) were only reported (Fig. 11.4). However, measurements were given for all three tracks, which have lengths between 11.0 and 11.5 cm and widths between 6.0 and 6.5 cm.

The slab containing the type of *V. sichuanensis* (Figs. 11.4B–11.4C), also preserves a second, poorly preserved track that we attribute to *V. sichuanensis*. The two consecutive *V. sichuanensis* paratype tracks (Fig. 11.4D) occur on a different slab and have a step length of 55 cm. Neither slab was illustrated by Zhen et al. (1994), who provided only line drawings of the type. Although neither slab has previously been illustrated photographically, except for a small picture of the holotype in a popular book (Xing, 2010:26), the holotype slab was illustrated as a line drawing by Lockley et al. (2008:fig. 4) as part of a restudy of the ichnogenus *Minisauripus*, which is also preserved on the holotype slab (Fig. 11.4C). Likewise, the single step sequence on the holotype slab was illustrated in a line drawing by Lockley et al. (2004:fig. 7C) and Matsukawa, Lockley, and Li (2006:fig. 6). To rectify this situation, the type material is illustrated here photographically for the first time (Fig. 11.4). The main points given in the description of *V. sichuanensis* by Zhen et al. (1994) confirm that the ichnospecies can be unequivocally attributed to a deinonychosaurian theropod dinosaur: the trackmaker was clearly functionally didactyl, with digit II represented only by a small, proximal pad impression. However, whether the attribution by Zhen et al. (1994) of *V. sichuanensis* to a dromaeosaurid trackmaker, as opposed to a troodontid trackmaker, is correct is an open question.

The Perfect Trackway: Dromaeopodus
shandongensis, the Spoor of a Big Pack Hunter?

Li and Lockley (2005) first reported large (track length ~28 cm), didactyl, dromaeosaurid trackways from the Lower Cretaceous Tianjalou Formation of Shandong Province China found in association with a rich ichnofauna of bird and other dinosaur tracks (Li, Liu, and Lockley, 2005; Li

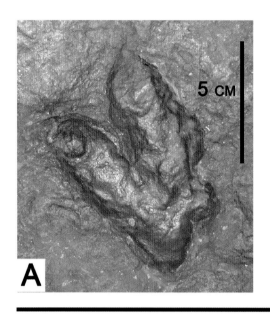

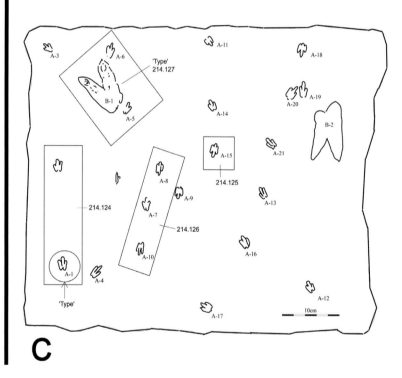

11.4. (A) Photograph of the holotype of *Velociraptorichnus sichuanensis* (CFEC-B-1 [Chongqing Natural History Museum collection]), located in the upper left corner of the type slab. (B) Photograph of the slab bearing the type of *V. sichuanensis*. (C) Map of the slab bearing the type of *V. sichuanensis* (after Lockley et al., 2008:fig. 4). The map also shows the positions of the type and paratypes of *Minisauripus chuanzhuensis*. Replicas of the *V. sichuanensis* holotype (CU 214.127 [University of Colorado Denver, Dinosaur Tracks Museum collection]) and the *M. chuanzhuensis* holotype (CU 214.124) and paratypes (CU 214.125 and 214.126) are reposited in the University of Colorado Museum of Natural History collections. (D) Paratype slab from *V. sichuanensis* type locality showing two consecutive tracks in trackway segment.

et al., 2008; Lockley et al., 2008). These were subsequently described in detail by Li et al. (2008), who erected for the didactyl tracks the ichnotaxon *Dromaeopodus shandongensis* (Fig. 11.5) based on well-preserved holotype (CU 214.111 [University of Colorado Denver, Dinosaur Tracks Museum collection]) and paratype (CU 214.112) trackways from two different, but closely associated, track-bearing horizons (Li et al., 2008:figs. 1–3). The holotype horizon yielded only one well-preserved *D. shandongensis* trackway, but the paratype horizon yielded six parallel trackways, strongly suggesting

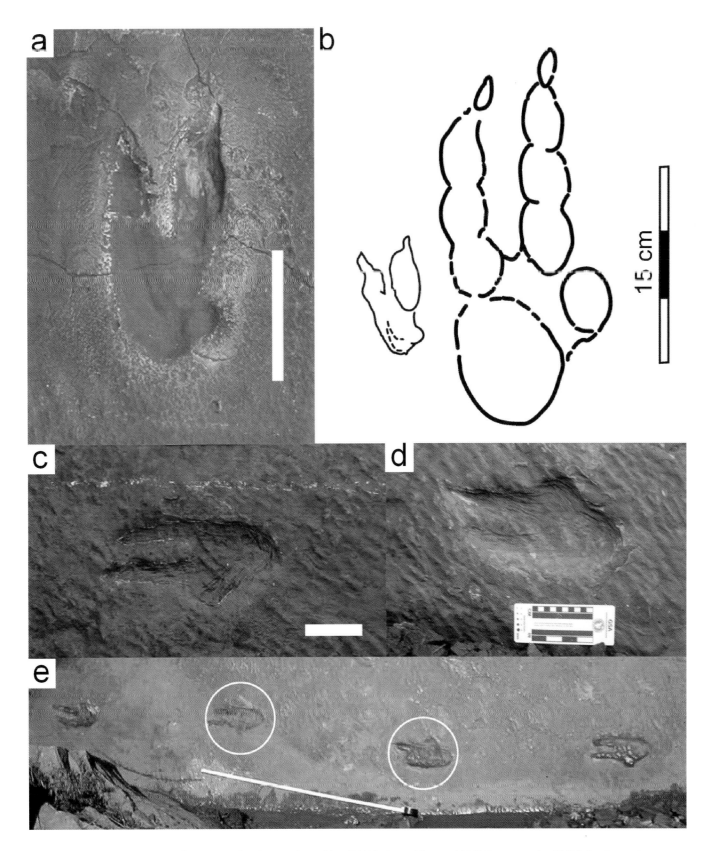

11.5. Didactyl dromaeosaurid tracks from Junan, Shandong Province, China: (A) *Velociraptorichnus* and (C–E) *Dromaeopodus*. (B) Relative sizes and anatomical differences between the smaller *Velociraptorichnus* (left) and larger *Dromaeopodus* (right). (C–D) Close-ups of the middle right (C) and left (D) prints, respectively, from the holotype *Dromaeopodus* trackway (E). Scale bars: A = 5 cm, C = 10 cm, and E = 1 m. *After Li et al. (2008).*

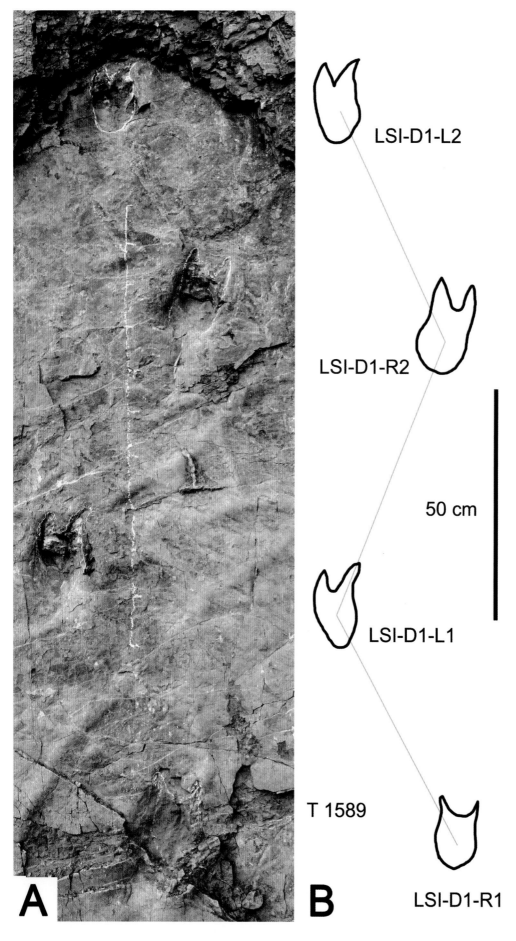

11.6. A newly discovered dromaeosaurid trackway from the Tianjialou Formation of Shandong China. (A) Photograph and (B) line drawing of four consecutive tracks of the Linshu site (LSI): LSI-D1-R1, LSI-D1-L1, LSI-D1-L2, LSI-D1-R2. T 1589 refers to the tracing of this trackway in the catalog in the University of Colorado Denver, Dinosaur Tracks Museum collection. *After Xing, Lockley, et al. (2013).*

LSI-D1-L2

LSI-D1-R2

50 cm

LSI-D1-L1

T 1589

LSI-D1-R1

A

B

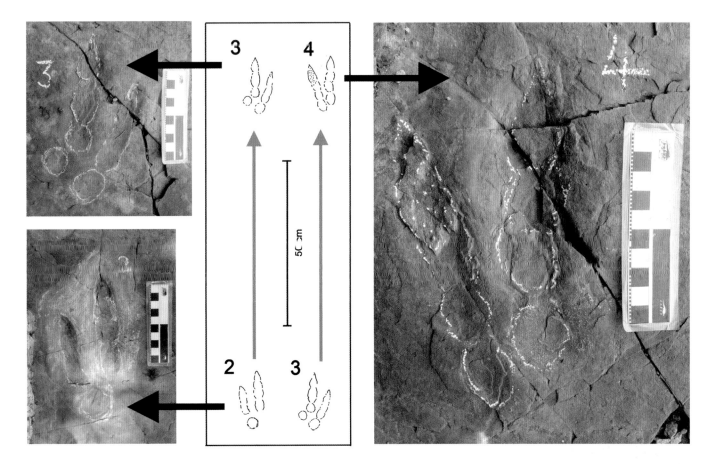

11.7. Line drawings and photographs of tracks 2–4 of the *Dromaeosauripus hamanensis* type trackway from the Lower Cretaceous Haman Formation, Korea (after Lockley, Huh, and Kim, 2012). Compare with Figures 11.4–11.6 and 11.8–11.9.

that they were made by a gregarious track-making taxon. This evidence is intriguing because prior to this discovery, dromaeosaurid theropods had speculatively been hypothesized as having engaged in gregarious behavior, possibly including wolf-like, pack-hunting behavior (Ostrom, 1969, 1990; Paul, 1988; Maxwell and Ostrom, 1995; q.v. Roach and Brinkman, 2007).

Li et al. (2008) compared *D. shandongensis* to *Velociraptorichnus* (Fig. 11.5), which also occurs at the Shandong site. This comparison emphasized a number of distinctive differences. In addition to the considerable size difference (28 cm vs. 11 cm), *Dromaeopodus* has slightly curved and more parallel digits III and IV traces that were more subequal in length than in the smaller *Velociraptorichnus. Dromaeopodus shandongensis* furthermore has sharp claw impressions, and its heel (metatarsophalangeal) pad and proximal pad of digit II both are well developed, in contrast to *V. sichuanensis*.

Another Dromaeosaurid Trackway from Shandong, China

In November 2012, two of the authors (M.G.L. and L.X.) discovered a new didactyl trackway in the Lower Cretaceous

(Aptian-Albian) Tianjialou Formation (Dasheng Group) of Shandong Province (Xing, Li, et al., 2013). The tracks (Fig. 11.6) are deep (~4 cm) and were made in thinly bedded lacustrine sediment. The trackway pertains to a trackmaker with feet about 18 cm long and 10 cm wide, although these measurements may be slightly enlarged as a result of preservational factors. The mean pace length is ~59 cm. This size is intermediate between type *Velociraptorichnus* and type *Dromaeopodus*, both of which, as noted, are known from the same region of China (Li et al., 2008). Although the didactyl morphology of the tracks is clear, details of pad impressions and the traces of the short, proximal portions of digit II seen in *Velociraptorichnus* and *Dromaeopodus* are not visible. It is therefore difficult to attribute this track to either of these ichnogenera. Based on size, they are similar to *Dromaeosauripus*. This new find establishes a considerable size range in dromaeosaurid tracks from Shandong.

More Asian Didactyl Tracks: Dromaeosauripus from the Cretaceous of Korea and China

Site 1: The Dromaeosauripus hamanensis type Kim et al. (2008) reported the first deinonychosaurian tracks from

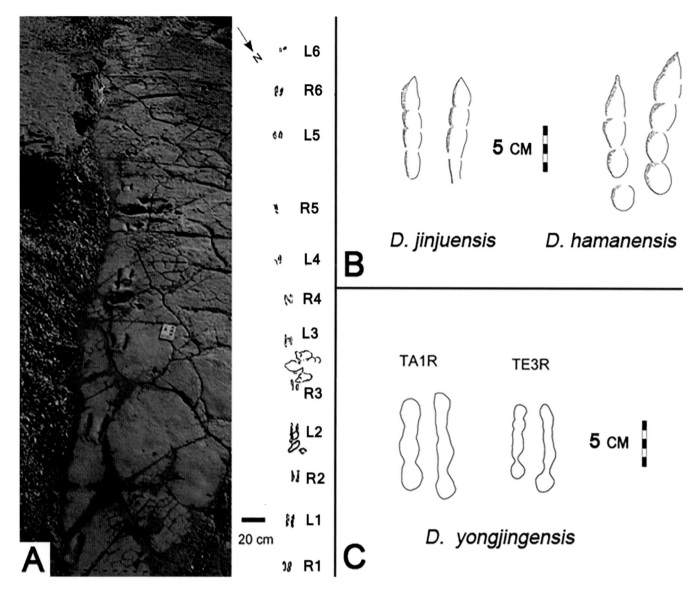

11.8. (A) Photograph and line drawing of the 12 tracks in the holotype *Dromaeosauripus jinjuensis* type trackway from the Lower Cretaceous Jinju Formation, Korea (after Kim et al., 2012).

(B) Comparison of individual tracks of *D. jinjuensis* and *D. hamanensis* (after Kim et al., 2012).
(C) Examples of *Dromaeosauripus yongjingensis* from the Lower Cretaceous Hekou Group of

Gansu Province, China, which are convergent with *D. jinjuensis* (after Xing, Li, et al., 2013). Compare with Figures 11.4–11.7 and 11.9.

Korea: a trackway consisting of four footprints from the Lower Cretaceous Haman Formation. Three of these tracks possess quite good morphological detail. The tracks in this trackway are clearly didactyl and have a mean length of ~15.5 cm, which is intermediate in size between *Velociraptorichnus* and *Dromaeopodus*. As initially reported, the tracks lack the proximal basal pad of digit II and have heel traces that are indistinct and whose precise locations relative to the digit traces are ambiguous. The fourth track in the sequence suggests a phalangeal pad formula for digits III and IV that differ from those of *Dromaeopodus*. These differences prompted Kim et al. (2008) to assign this ichnotaxon to the new ichnogenus and ichnospecies *Dromaeosauripus hamanensis* (Fig.

11.7). The individual digit traces show only slight inward curvature, which is less pronounced than in *Dromaeopodus*.

Subsequent examination of the type trackway (Lockley, Huh, and Kim, 2012) suggested that a faint trace of the basal pad of digit II in track number III can be discerned (Fig. 11.7). Furthermore, the proximal-most digit IV pad trace illustrated by Kim et al. (2008) for the fourth track in the holotype sequence could, in fact, be a faint heel trace. These observations suggest that *Dromaeosauripus* may be morphologically more similar to *Dromaeopodus* and *Velociraptorichnus* than previously supposed. However, this conclusion is tentative, and evidence suggests that the lack of posterior traces (heel and digit II) may reflect a distinctive, digitigrade

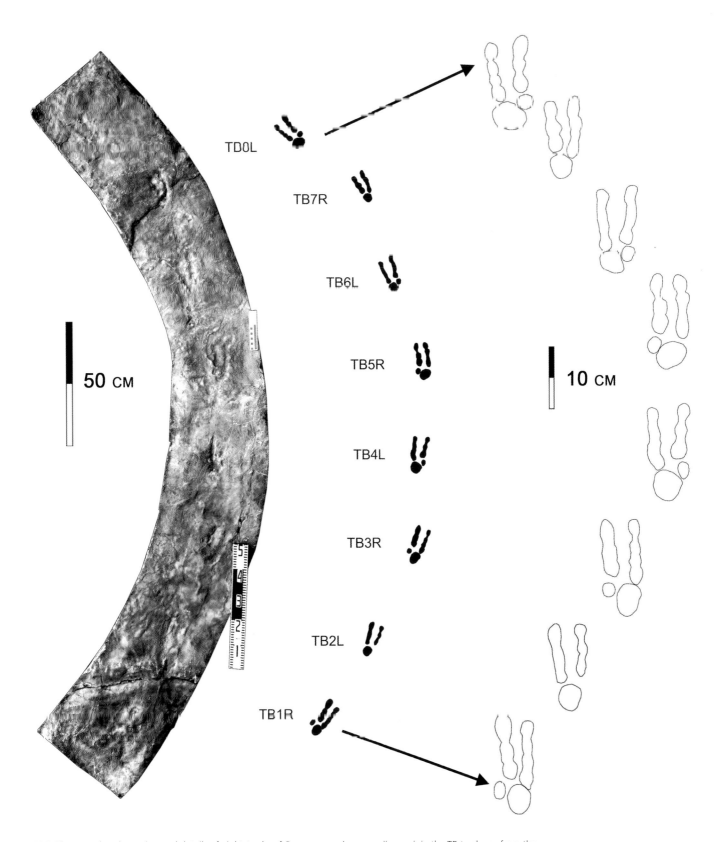

11.9. Photograph, schematics, and details of eight tracks of *Dromaeosauripus yongjingensis* in the TB trackway from the Lower Cretaceous Yanguoxia Formation (Hekou Group) of Gansu Province, China (modified after Xing, Lockley, et al., 2013). Compare with illustrations of *D. hamanensis* and *D. jinjuensis*, respectively, in Figures 11.7–11.8.

11.10. *Menglongipus* trackway. *After Xing et al. (2009).*

morphology. Replicas of *D. hamanensis* are reposited in the CU collections (CU 214.131 and 214.132).

Site 2: The *Dromaeosauripus jinjuensis* type Kim et al. (2012) described another trackway of a distinctive, didactyl biped from Korea, this time from the Lower Cretaceous Jinju Formation, which is older than the Haman Formation The new trackway comprises a continuous sequence of 12 tracks (Fig. 11.8). This trackway formed the basis for a new ichnospecies, *Dromaeosauripus jinjuensis*, which differs from *D. hamanensis* in a number of details. The tracks are shorter (mean length 9.3 cm) and have wider separations of the proximal digit traces. Thus, there is no sign that the digit traces converge proximally toward any trace of a heel, possibly indicating a distinctive, digitigrade posture. As noted herein, differential preservation may account for these differences. *D. jinjuensis* is represented in the CU collections by CU 214.243.

Site 3: The *Dromaeosauripus yongjingensis* type Xing, Lockley, et al. (2013) described trackways of another ichnospecies of *Dromaeosauripus*, *D. yongjingensis*, from the Lower Cretaceous Yanguoxia Formation (Hekou Group) at the Liujiaxia Dinosaur National Geopark in Gansu Province, China (see also Li et al., 2006). Six trackway segments were recognized, including a turning trackway from which eight consecutive tracks were illustrated (Fig. 11.9). They illustrated 19 individual tracks that show much the same range of variation seen in both Korean ichnospecies (*D. hamanensis* and *D. jinjuensis*). Among these the holotype (GSLTZP-S2-TE4L [Fossil Research and Development Center of the Third Geology and Mineral Resources Exploration Academy of Gansu Province collection]) is also represented by replica UCM 214.279 (University of Colorado Museum of Natural History collection).[1] The tracks are very similar in size (length 14.5–16.0 cm) to *D. hamanensis* and have very similar digit III and

IV lengths, but in contrast to the Korean ichnospecies, most of the Chinese tracks have well-developed heel pad traces. However, the basal pad of digit II is poorly impressed in most cases: at best, it is only faintly and inconspicuously illustrated in about half the tracks illustrated. Some tracks lack heel traces and therefore are short, similar to *D. jinjuensis*. These observations suggest that there is considerable variation in the morphology of the tracks so far attributed to ichnogenus *Dromaeosauripus*, and that the Chinese samples demonstrate that differences between the two Korean ichnospecies may, in part, be related to preservation. However, there appear to be genuine, morphological (not extramorphological) differences between *Dromaeosauripus* and the previously named ichnogenera (*Dromaeopodus* and *Velociraptorichnus*), especially in relation to the development of the heel pad and digit II trace.

Site 4: A problematic trackway: The *Menglongipus sinensis* type Xing et al. (2009) reported poorly preserved, ostensibly didactyl theropod tracks from a single trackway on a track surface that otherwise preserves only tridactyl *Grallator* tracks from the latest Jurassic or earliest Cretaceous Tuchengzi Formation of Hebei Province, China. They were named *Menglongipus sinensis* (Fig. 11.10). Based on the age of the Tuchengzi Formation, these tracks were proclaimed as the oldest known deinonychosaur tracks. Photogrammetric models of four tracks in a single trackway sequence show that the tracks are variably preserved and that, in all cases, the presumed digit IV is short compared to digit III. If this accurately reflects the pedal morphology of the trackmaker, then it differs from those of the trackmakers of *Velociraptorichnus*, *Dromaeopodus*, and *Dromaeosauripus*, in which digits III and IV are subequal in length. Xing et al. (2009) further illustrated a photogrammetric digital elevation model of one of the *Dromaeosauripus yongjingensis* tracks from

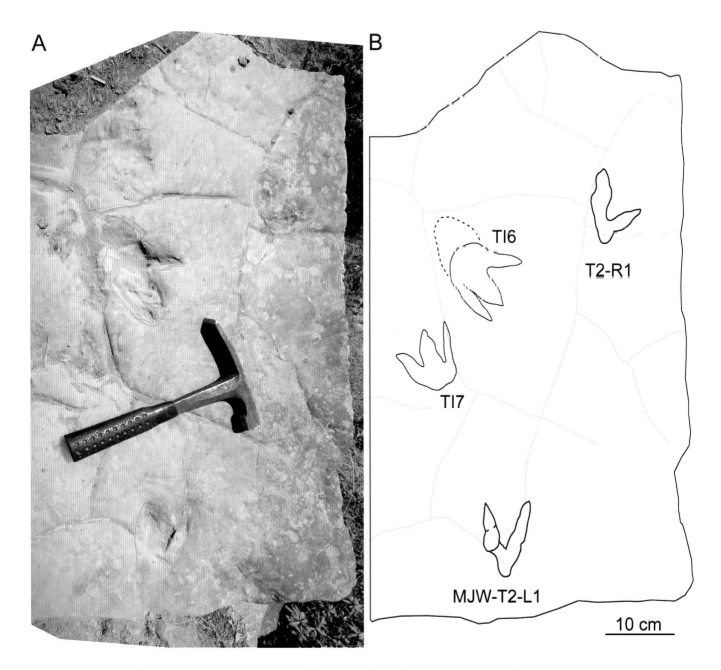

11.11. (A) Photograph and (B) an interpretative outline drawing of didactyl and tridactyl theropod tracks from the Mujiaowu (MJW) tracksite. The tridactyl morphotype has been named *Velociraptorichnus zhangi. Modified after Xing et al., 2015.*

the Gansu site (Xing, Lockley, et al., 2013), documenting the different morphology. Coarse-grained substrates in the Tuchengzi Formation preserve, at numerous sites, abundant, tridactyl theropod tracks attributed to *Grallator* (Matsukawa, Lockley, and Li, 2006). For this reason, as well as the poor preservation, *Menglongipus sinensis* could be merely a poorly preserved tridactyl track, which would render it a nomen dubium. However, although the tracks are difficult to distinguish from poorly preserved *Grallator* tracks, their occurrence in a single trackway that does not include tridactyl *Grallator* tracks led Lockley et al. (2013) to retain the validity of the ichnotaxon, at least provisionally. Despite this provisional acceptance of the ichnotaxon, the step is somewhat

irregular and the middle toe (digit III) impression of the first footprint in the sequence aligns poorly with the inferred trackway trend. Therefore, the case for the validity of *Menglongipus sinensis* is weaker than those for *Velociraptorichnus, Dromaeopodus,* and *Dromaeosauripus.* If it was truly made by a didactyl, presumably deinonychosaurian, trackmaker, the short digit IV suggests that the trackmaker was more likely a troodontid than a dromaeosaurid, though Xing et al. (2009) specifically stated that such a determination was not possible to make.

Site 5: Didactyl and tridactyl traces: *Velociraptorichnus zhangi* **ichnosp. nov. type** The most recent report of *Velociraptorichnus* comes from the Mujiawa tracksite in the

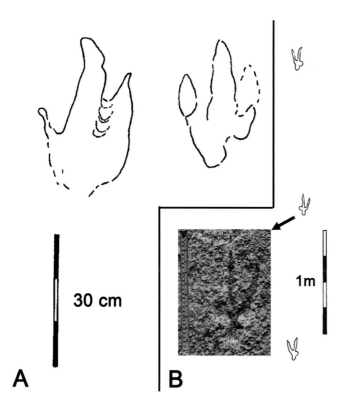

11.12. (A) Line drawings of two isolated tracks from the Lower Cretaceous Cedar Mountain Formation of Arches National Park, Utah (after Lockley et al., 2004). (B) Schematic diagram and photograph of a representative track from a second Cedar Mountain Formation deinonychosaurian trackway, attributed to *Dromaeosauripus* by Lockley, Gierlinski, Dubicka, et al. (2014).

Cretaceous Xiaoba Formation, Sichuan Province (Xing et al., 2015). The Mujiawa site has yielded both didactyl and tridactyl tracks that we interpret as different expressions of *Velociraptorichnus* caused by sporadic registration of the normally retracted digit II, either due to special substrate conditions or differences in behavior pertaining to degree of claw retraction in different individuals. The tridactyl morphotype has been named *Velociraptorichnus zhangi* (Xing et al., 2015); it possesses a very low divarication angle between the traces of digit II and III (Fig. 11.11).

At least one other Cretaceous tracksite from Sichuan Province that preserves a few didactyl tracks is currently under investigation by two of the authors (M.G.L. and X.L.). Descriptions of the tracks from this site will be published elsewhere.

Deinonychosaurian Tracks in North America

At present, only three North American dinosaur tracksites are known to have yielded deinonychosaurian tracks. Two occur in the Cedar Mountain Formation in eastern Utah, the third occurs in the Dakota Group of Colorado (Table 1). The first-discovered site yielded only isolated footprints, but the second-discovered site yieldes two well-preserved trackways.

Cedar Mountain Formation occurrences of *Dromaeopodus* and associated ichnites from Utah Isolated, possibly didactyl tracks reported from the Ruby Ranch Member of the Lower Cretaceous Cedar Mountain Formation within the boundaries of Arches National Park in eastern Utah were tentatively identified as the first deinonychosaur tracks from North America (White and Lockley, 2002; Lockley et al., 2004). Unequivocal confirmation of such tracks came from a five-track (four-step) trackway at another nearby site known as the Mill Canyon Dinosaur Tracksite, just west of Arches National Park (Lockley, Gierlinski, and Dubicka, et al., 2014). The two best-preserved tracks are represented in the cu collections by ucm 199.67 and 199.68. Further excavation at the site has since exposed another three-track (two-step) trackway (ucm 199.82) illustrated by Lockley, Gierlinski, and Houck, et al. (2014:fig. 8A). The Mill Canyon tracks were originally attributed to *Dromaeopodus* by Cowan, Lockley, and Gierlinski (2010) but have since been restudied and assigned to *Dromaeosauripus* (Lockley, Gierlinski, and Dubicka, et al., 2014; Lockley, Gierlinski, and Houck, et al., 2014). This trackway (Fig. 11.12) pertains to a smaller and more gracile trackmaker than that of the type *Dromaeopodus* from China. The Utah tracks have relatively small, though unequivocal, proximal pad traces on digit II. The tracks are smaller (footprint length 20–21 cm) than the isolated, apparently didactyl tracks reported in 2004 (Fig. 11.12A), which remain enigmatic. Recently, the Bureau of Land Management has installed interpretative signs at the Mill Canyon Dinosaur Tracksite, including one which uses the label *Dromaeosauripus* and shows a reconstructed "raptor" foot, with retracted claw, making a didactyl track.

Deinonychosaurian Tracks in Europe

At present, only two European dinosaur tracksites are reported to have yielded deinonychosaurian tracks. One is associated with the Early Cretaceous of Germany, the other with the late Cretaceous of Poland. The German site yielded a large assemblage of tracks, but it occurred as part of an extensively trampled surface. The Polish sites yielded a single trackway that was not very well preserved.

The Oberkirchen "Chicken Yard" Site Didactyl tracks from the Early Cretaceous (Berriasian) of northern Germany (van der Lubbe, Richter, and Böhme, 2009; van der Lubbe et al., 2012) occur amid a heavily 'trampled' tracksite on the company grounds of the Obernkirchener Sandsteinbrüche near the town of Obernkirchen (See also Richter and Böhme, 2016). The site, which has become popularly known as the Chicken Yard (Hühnerhof) is important not least because it preserves the largest assemblage of deinonychosaurian tracks known anywhere in the world. Their relative abundance

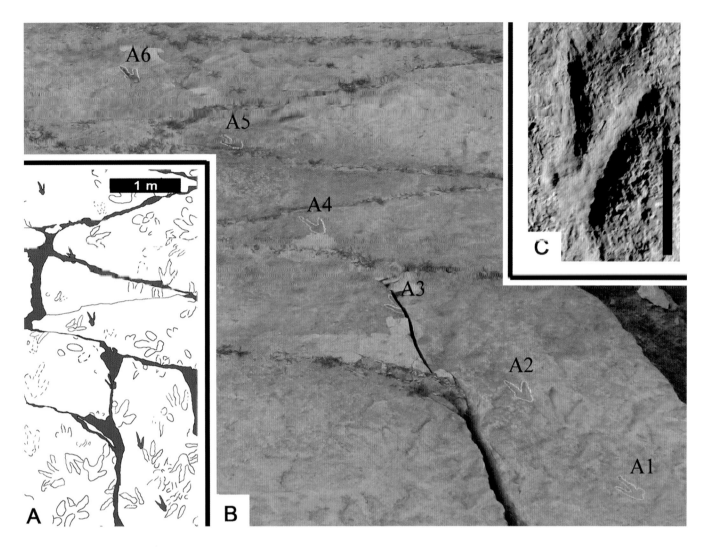

11.13. (A) Partial map and (B) photograph of deinonychosaurian trackway A from the Obernkirchen Chicken Yard site, Lower Cretaceous Obernkirchen Sandstone, Germany.

allows for a detailed description of general track morphology, as well as an opportunity to expand available data on didactyl tracks. The entire Obernkirchen ichnoassemblage will be described in detail elsewhere (e.g., Richter and Böhme, 2016).

A total of 86 didactyl tracks documented to date are preserved as shallow impressions (depths ~0.6–14 mm; average ~6 mm) in silty sandstone. Many of the tracks are well preserved, showing well-defined pad and sharp claw impressions. They generally are very slender, and the digit IV impressions are markedly shorter than those of digit III. A segment of 1 (trackway A) of the 17 didactyl trackways at the site (labeled A–R, omitting I) is illustrated herein (Fig. 11.13). Trackway width is narrow, and pace angulation is 170°–180°. Track sizes range from a total track length 13.0 cm to a maximum of 23.3 cm. Pace and stride lengths are long: the 105 cm average pace length of trackway F corresponds to a track length of 20.9 cm.

A slightly ovoid, almost circular heel impression constitutes the proximal-most part of each track. Digit II is represented by a small, subcircular to ovoid impression positioned slightly posterior to the impression of the first pad of digit III. Digit III impressions consist of three elongate, subrectangular to slightly ovoid pad impressions terminating in long, sharp claw imprints. Digit IV impressions consist of four pad impressions, but they are only well defined in a few specimens. Digit IV claw impressions are rare, and much shorter than those of digit III. Digit IV impressions are rather straight, and, in a single case (A 12), actually curves gently laterally. The angles of divarication between digit III and IV impressions range from 21° to 36° (average ~28°).

Digit III consistently left the deepest impressions (2.5–14 mm), followed by the metatarsal pad impressions (1.5–11.3 mm). The digit IV impressions are, on average, only slightly deeper than those of digit II (6 mm and 5.1 mm, respectively). The maximum value for a digit IV impression depth is 12.5 and the minimum value is 2.5 mm; the same values are 11.5 and 2.1 mm, respectively, for digit II. The morphology and measurements of the tracks are distinctive

Table 11.1. Summary data on 16 reported deinonychosaurian tracksites from the Cretaceous of Asia, North America, and Europe, with ichnotaxonomic designations, details of sample size, track size, quality of preservation, and sources of information

Site	Ichnotaxon	Number of tracks/ trackways	L and W (cm)	Preservation	Reference(s)
Emei County, Sichuan, China	*Velociraptorichnus sichuanensis*	4/3	L ~10 cm	Good, deep casts	Zhen et al. (1994)
Junan County, Shandong, China	*Dromaeopodus shandongensis*, *Velociraptorichnus*	14/7	L 28.0	Excellent molds in type; others variable	Li et al. (2008)
Linshu County, Shandong, China	dromaeosaurid track indet.	5/1	L ~19.0, W ~ 10.5	Fair, deep molds	Xing et al. (2013b)
Chicheng, Hebei, China	*Menglongipus sinensis*	4/1–2	Mean L = 6.3, mean W = 4.3	Poor molds	Xing et al. (2009)
Gansu, China	*Dromaeosauripus yongjingensis*	71/7	L 14.8, W 6.4	Fair to good molds	Xing et al. (2013a)
Chu Island, Korea	*Dromaeosauripus hamanensis*	4/1	mean L = 15.5	Fair, shallow molds	Kim et al. (2008)
Bito Island, Korea	*Dromaeosauripus jinjuensis*	12/1	mean L = 9.3	Deep incomplete molds	Kim et al. (2012)
Arches National Park, Utah	Unnamed, probable deinonychosaurian	2/2	L ~28–?~35	Fair, very deep molds	Lockley et al. (2004)
Mill Canyon site, Utah	*Dromaeosauripus*	4/1	L 21/	Good, shallow molds	Cowan et al. (2009)
Obernkirchen, Germany	Unnamed, probable troodontid	–/86	L = 13–23.3	Good molds	van der Lubbe et al. (2009), (2012)
Młynarka Mount, Poland	*Velociraptorichnus* isp.	3/1	L 17.0	Poor molds	Gierliński (2007, 2008, 2009)
Mujiaowu, Sichuan, China	*Velociraptorichnus zhangi*, *Velociraptorichnus* isp.	2/0 3/1	L ~10.4, W 9.4 L 11.3	Good molds Good molds	Xing et al. (2015)
Yanqing, Beijing, China	? *Velociraptorichnus*	1/1	L 10.8, W 5.3	Good mold	Xing et al. (2015)
Bajiu, Sichuan, China	cf. *Dromaeopodus*	2/2	L 17.4–24.8, W 5.9–14.5	Good mold	Xing et al. (2016)
Shimiaogou Sichuan, China	*Velociraptorichnus* isp.	5/1	L ~7.1 cm	Fair molds	Xing et al., in press
Dinosaur Ridge	*Dromaeosauripus*	2/2	L ~16.0-17.0	Natural casts	Lockley et al., 2016

Notes: Note that two distinct morphotypes are known from the Junan County site in Shandong province. L, length; W, width.

enough that they could support a new ichnotaxon when fully described.

A Late Cretaceous Track from Poland

Gierlinski (2007, 2008) illustrated two individual tracks attributed to *Velociraptorichnus* from the Late Cretaceous (late Maastrichtian) Młynarka Mount site in Poland and indicated that a specimen is preserved in the Polish Geological Institute (PIG 1704.11.6). According to the illustrations in these publications, the tracks are about 17 cm long. Later describing the same material in more detail, Gierlinski (2009) suggested that tracks might be more similar to *Dromaeopodus* from China (Li et al., 2008). The first two reports comprise illustrated abstracts that thus provide minimal description of the material. In the later paper, two tracks were shown in a purported trackway sequence, with the first (a right) followed, after a gap of about 1.87 m as measured from the scale of the photograph (Gierlinski, 2009:fig. 7), by a left. Gierlinski (2009) suggested that the tracks compose a portion of a single trackway with a footprint: pace length ratio of 1:12.30, indicating a remarkable speed of about 50 km/hr. Unfortunately, the trackway has since been destroyed (Gierlinski, pers. comm., 2012).

The illustration presented by Gierlinski (2007), which is of the same left footprint illustrated by Gierlinski (2009), does not show the proximal trace of digit II as well defined, but the illustration in Gierlinski (2008) indicates a different track, or differently illuminated view, of a left footprint with the proximal trace of digit II more clearly defined. However, the method of illustrating tracks that the author used in these papers involved isolating the tracks from photographs by cropping the matrix surrounding the tracks. This may have introduced some subjectivity in how the outline or margin of the track is defined.

DISCUSSION

Deinonychosaurian tracks have been reported from 16 Cretaceous tracksites in Asia, North America, and Europe (Fig. 11.1, Table 11.1), including the most recently reported sites from China (Xing et al. 2015a; 2015b; 2016, in press) and Colorado (Lockley et al., unpublished manuscript). With the exception of the German site, the sample sizes are small. The best preserved among these tracks are the types of *Velociraptorichnus* from Sichuan Province and *Dromaeopodus* from Shandong Province, China. The type of *Dromaeopodus* is particularly well preserved and is associated with a nearby assemblage of six other parallel trackways, suggestive of gregarious behavior (Li et al., 2008). The large, German sample consists of 86 tracks, and the preservation is sufficient both to differentiate the tracks from the other named ichnogenera and to suggest that they are of troodontid, rather than dromaeosaurid origin; they remain the only convincingly troodontid tracks known. As noted, the morphological differences between the as-yet unnamed German tracks and other, named tracks, such as *Dromaeopodus*, may well be sufficient to warrant the erecting of a new ichnotaxon. The only other track morphotype currently known that could be attributed to a troodontid, based on the differential lengths of the traces of digits III and IV, is *Menglongipus*. However, such attribution is only conjectural because the poor preservation of the type specimens renders such an inference uncertain.

Known deinonychosaurian tracks range widely in size (Table 11.1), varying from a mean length of only 6.3 cm in *Menglongipus sinensis* to 28 cm in *Dromaeopodus shandongensis*. One very deep track from the Arches National Park site in Utah is about 35 cm long, but its length may have been exaggerated during the registration and extraction of the foot.

Clearly, the footprint evidence complements the skeletal evidence that indicates that deinonychosaurian trackmakers were paleogeographically widespread. The ages of many of the track-bearing units, however, are less certain: not all are reliably dated to the level of epoch or age (Chen et al., 2006; Matsukawa, Lockley, and Li, 2006; Lockley et al., 2013). It is beyond the scope of this chapter to discuss in detail the ages of the formations from which these tracks originate or the resultant implications for the temporal distribution of deinonychosaurian tracksites. In general, most of these tracksites appear to be late Early Cretaceous (post-Neocomian; Barremian-Albian) in age. The Hebei and Yanqing tracksites are ostensibly the oldest, associated with the basal Cretaceous (or Jurassic-Cretaceous transition), and the Polish site is the youngest. The Polish site is also the only one associated with carbonate deposition in a marginal marine setting; all others are associated with fluvio-lacustrine settings.

Following the discovery of *Velociraptorichnus* in 1994, and the uncertain reports of tracks from Utah in 2004, all 14 subsequent discoveries have been reported since 2007. Why the bulk of discoveries have been made in such a short period of time may have to do with increased searching for and recognition of dinosaur tracks globally, but in any case, deinonychosaurian tracks are not as rare as might have previously been assumed. The reports of Li et al. (2008) and van der Lubbe, Richter, and Böhme (2009) and van der Lubbe et al. (2012), respectively, have indicated that it may be possible to differentiate dromaeosaurid from troodontid tracks, at least in some cases. Examinations of the foot skeletons of

members of both groups, to see how they might 'best fit' the footprints (Lockley, 1998), would facilitate such distinctions.

Ostrom (1969) interpreted the morphology of pedal digit II of *Deinonychus antirrhopus* as highly specialized for a predatory function such that it did not contact the ground at all under normal circumstances, an interpretation borne out by later discoveries and studies of other dromaeosaurids (Barsbold, 1974, 1998; Norell and Makovicky, 1997; Xu and Norell, 2004; Manning et al., 2005; Gao et al., 2012). The recent report of Xing et al. (2015), suggesting that in some cases traces of digit II (other than its proximal pad) may have registered, indicates that such cases are the exception, not the rule. As summarized in Table 11.1, more than 30 deinonychosaurian trackways have been described from the 16 sites included in this survey. Only two of these indicate occasional registration of the distal part of digit II. In terms of its effect on locomotion, the specialization of digit II shifted, albeit incompletely, the structural, weight-bearing axis of the foot from mesaxonic (along digit III) to ectaxonic (between digits III and IV); this shift is reflected by the subequal lengths of digits III and IV. As noted, digit II impressions have been reported in almost every well-preserved Cretaceous didactyl track documented to date, but with the exception of *V. zhangi* (Xing et al., 2015a), they consist solely of the proximal pad. These traces are not preserved in some examples of ichnogenus *Dromaeosauripus*, notably *D. jinjuensis* (Kim et al., 2012), probably due to preservational factors; the Gansu sample described by Xing, Lockley, et al. (2013) shows that the range of preservation of *Dromaeosauripus yongjingensis* is at least as wide as those of both the Korean ichnospecies (*D. hamanensis* and *D. jinjuensis*). Digit II, therefore, maintained a role, albeit a limited one, during normal locomotion. The depths of the Obernkirchen didactyl tracks suggest that the structural, weight-bearing axes of the feet of their trackmakers indeed had shifted from its normal position along the midline through digit III toward digit IV to a certain degree, confirming the prediction of Ostrom (1969). However, digit III constitutes the most deeply impressed part of the foot in all specimens documented so far and was undoubtedly the primary weight-bearing digit.

The primary character that distinguishes the Obernkirchen tracks from all the other didactyl tracks is the length difference between the impressions of digits III and IV (average ratio of digit IV to III impression length is 0.8). In all other didactyl tracks, except for *Menglongipus* (Xing et al., 2009), digit IV is either subequal to or slightly shorter than digit III, comparing favorably with dromaeosaurid pedal skeletons. The proportions of the pedal skeletons of *Troodon formosus* (Russell and Séguin, 1982), *Sinovenator changii* (Xu et al., 2002; van der Lubbe, pers. obs.), *Mei long* (Xu and Norell, 2004; van der Lubbe, pers. obs.), and *Talos sampsoni* (Zanno et al., 2011) match remarkably well the proportions of the didactyl tracks from Obernkirchen. However, the match is not universal among troodontids: digits III and IV are of subequal length in the pes of the holotype specimen of *Sinornithoides youngi* (Russell and Dong, 1993), which is a juvenile. Among dromaeosaurids, at least *Microraptor* (Xu, Zhou, and Wang, 2000) possesses a digit IV noticeably shorter than digit III (van der Lubbe, pers. obs.). Published data are only of limited use because measurements of unguals were usually made along their outer, and sometimes also inner, curvatures. During track formation, the claw will not imprint along its curvature but rather leave an imprint representing a line from its tip to the proximal end of the keratinous sheath; see Lockley and Lucas (2013) for similar traces in Triassic *Evazoum* tracks. An encompassing study of paravian pedal morphology with a special focus on characters (such as digit lengths) that are potentially relevant for ichnology (Sullivan, van der Lubbe, and Xing, 2012) may show whether the hypothesis briefly presented here can be further elaborated. Nevertheless, the didactyl tracks from Obernkirchen differ from all other published didactyl tracks in a character that is best explained as related to pedal morphology, and they are therefore probably of troodontid origin.

ACKNOWLEDGMENTS

The following all assisted in various ways. Li Jianjun, Beijing Musuem of Natural History, helped provide the senior author with access to replicas of the type material of *Velociraptorichnus*. The Shandong portion of this study was supported financially, in part, by the National Natural Science Foundation of China (grant 40972005) to Rihui Li, for study of the Junan site. The Qijiang International Dinosaur Tracks Symposium, Chongqing, 2012, and Linshu Land and Resources Bureau provided support to Lida Xing and Martin Lockley for access to the Linshu site. Jeong Yul Kim, Korea National University of Education, Cheongwon and Kyung Soo Kim, Korea National University of Education, Chinju, kindly helped the senior author with access to the Korean material. Study of the German tracksite was undertaken with the help of Annette Richter and Tom Hübner, Niedersächsisches Landesmuseum Hannover, Hannover, Germany; Annina Böhme, Department of Geobiology, Geoscience Centre of the University of Göttingen, Göttingen, Germany; and Corwin Sullivan, Institute of Vertebrate Paleontology and Paleoanthropology, Chinese Academy of Sciences, Beijing, Peoples Republic of China. Work at the Mill Canyon Dinosaur Tracksite, Grand County, Utah, was conducted under Bureau of Land Management permit UT 13 008E issued to the senior author (M.G.L.); funding for much of the work at this site was facilitated by the diligent work of Jong Deock Lim, National Research Institute of Cultural Heritage, Korea. Finally, we thank Jeong Yul Kim, Korea National University of Education Chungbuk, and Jesper Milàn, University of Copenhagen, for their helpful reviews of an earlier draft of this chapter.

REFERENCES

Barsbold, R. 1974. Dueling dinosaurs. Priroda 2: 81–83.

Barsbold, R. 1998. Fighting dinosaurs–they really fought; pp. 74–78 in A. M. G. de Carvalho, M. A. P. Cachão, A. M. P. S. A. F. Andrade, C. A. P. F. M. da Silva, and V. A. F. dos Santos (eds.), I Encontro Internacional sobre Paleobiologia dos Dinossáurios: Programa de Musealização para Pistas de Dinossáurios em Portugal. Museu Nacional de História Natural, Lisbon, Portugal.

Chen, P.-J., J. Li, M. Matsukawa, H. Zhang, Q. Wang, and M. G. Lockley. 2006. Geological ages of dinosaur-track-bearing formations in China. Cretaceous Research 27: 22–32.

Cowan, J., M. G. Lockley, and G. Gierlinski. 2010. First dromaeosaur trackways from North America: new evidence from a large site in the Cedar Mountain Formation (Early Cretaceous), eastern Utah. Journal of Vertebrate Paleontology 30(3, Supplement): 75A.

D'Orazi Porchetti, S., and U. Nicosia. 2007. Re-examination of some large early Mesozoic tetrapod footprints from the African collection of Paul Ellenberger. Ichnos 14: 219–245.

Ellenberger, P. 1972. Contribution à la classification des pistes de vertébrés du Trias: les types du Stormberg d'Afrique du Sud (I): Palaeovertebrata, Mémoire Extraordinaire 1972: 1–152.

Ellenberger, P. 1974. Contribution à la classification des pistes de vertébrés du Trias: Les types du Stromberg d'Afrique du Sud (II partie: Le Stromberg superieur–I. Le biome de la zone B/1 ou niveau de Moyeni: ses biocenoses). Palaeovertebrata, Mémoire Extraordinaire, 1–142.

Gao, C., E. M. Morschhauser, D. J. Varricchio, J. Liu, and B. Zhao. 2012. A second soundly sleeping dragon: new anatomical details of the Chinese troodontid *Mei long* with implications for phylogeny and taphonomy. PLoS One 7(9): e45203.

Gaston, R., M. G. Lockley, S. G. Lucas, and A. P. Hunt. 2003. *Grallator*-dominated fossil footprint assemblages and associated enigmatic footprints from the Chinle Group (Upper Triassic), Gateway area, Colorado. Ichnos 10: 153–163.

Gierlinski, G. D. 2007. New dinosaur tracks in the Triassic, Jurassic and Cretaceous of Poland; pp. 13–16 in P. Huerta and F. Torcida-Fernández-Baldor (eds.), IV Jornadas Internacionales sobre Paleontología de Dinosaurios y su Entorno Libros de Resúmenes. Salas de los Infantes, Burgos, Spain.

Gierlinski, G. D. 2008. Late Cretaceous dinosaur tracks from the Roztocze Hills of Poland; p. 44 in A. Uchman (ed.), Second International Congress on Ichnology Abstract Book. Polish Geological Institute, Warsaw, Poland.

Gierlinski, G. D. 2009. A preliminary report on new dinosaur tracks from the Triassic, Jurassic and Cretaceous of Poland; pp. 75–90 in Colectivo Arqueológico-Paleontológico de Salas (ed.), Actas de las IV Jornadas Internacionales sobre Paleontologia de Dinosaurios y su Entorno. Colectivo Arqueológico-Paleontológico de Salas de los Infantes, Burgos, Spain.

Ishigaki, S., and M. G. Lockley. 2010. Didactyl, tridactyl and tetradactyl theropod trackways from the Lower Jurassic of Morocco: evidence of limping, labouring and other irregular gaits. Historical Biology 22: 100–108.

Kim, S.-Y., M. G. Lockley, J. O. Woo, and S. H. Kim. 2012. Unusual didactyl traces from the Jinju Formation (Early Cretaceous, South Korea) indicate a new ichnospecies of *Dromaeosauripus*. Ichnos 19: 75–83.

Kim, J. Y., K. S. Kim, M. G. Lockley, S. Y. Yang, S. J. Seo, and H. I. Choi. 2008. New didactyl dinosaur footprints (*Dromaeosauripus hamanensis* ichnogen. et ichnosp. nov.) from the

Early Cretaceous Haman Formation, south coast of Korea. Palaeogeography, Palaeoclimatology, Palaeoecology 262: 72–78.

Li, R., and M. G. Lockley. 2005. Dromaeosaurid trackways from Shandong Province China. Journal of Vertebrate Paleontology 25 (3, Supplement): 84A.

Li, R., M. Liu, and M. G. Lockley. 2005. Early Cretaceous dinosaur tracks from the Houzuoshan Dinosaur Park in Junan County, Shandong Province, China. Geological Bulletin of China 24: 277–280.

Li, D., Y. Azuma, M. Fujita, Y. N. Lee, and T. Arakawa. 2006. A preliminary report on two new vertebrate track sites including dinosaurs from the Early Cretaceous Hekou Group, Gansu Province, China. Journal of the Paleontological Society of Korea 22: 29–49.

Li, R., M. G. Lockley, P. J. Makovicky, M. Matsukawa, M. A. Norell, J. D. Harris, and M. Liu. 2008. Behavioral and faunal implications of Early Cretaceous deinonychosaur trackways from China. Naturwissenschaften 95: 185–191.

Lockley, M. G. 1998. The vertebrate track record. Nature 396: 429–432.

Lockley, M. G., and A. P. Hunt. 1995. Dinosaur Tracks and Other Fossil Footprints of the Western United States. Columbia University Press, New York, New York, 360 pp.

Lockley, M. G., and S. G. Lucas. 2013. *Evazoum gatewayensis* a new Late Triassic archosaurian ichnospecies from Colorado: implications for footprints in the ichnofamily Otozoidae. New Mexico Museum of Natural History and Science Bulletin 61: 345–353.

Lockley, M. G., M. Huh, and J.-Y. Kim. 2012. Mesozoic Terrestrial Ecosystems of the Korean Cretaceous Dinosaur Coast: A Field Guide to the Excursions of the 11th Mesozoic Terrestrial Ecosystems Symposium (August 19–22). Korean Federation of Science and Technology Societies, Kwangju, Korea.

Lockley, M. G., S. G. Lucas, and A. P. Hunt. 2006. *Evazoum* and the renaming of Northern Hemisphere '*Pseudotetrasauropus*': implications for tetrapod ichnotaxonomy at the Triassic-Jurassic boundary. New Mexico Museum of Natural History and Science Bulletin 37: 199–206.

Lockley, M. G., D. White, J. Kirkland, and V. Santucci. 2004. Dinosaur tracks from the Cedar Mountain Formation (Lower Cretaceous), Arches National Park, Utah. Ichnos 11: 285–293.

Lockley, M. G., G. Gierlinski, Z. Dubicka, B. H. Breithaupt, and N. A. Matthews. 2014. A preliminary report on a new dinosaur tracksites in the Cedar Mountain Formation (Cretaceous) of eastern Utah. New Mexico Museum of Natural History and Science Bulletin 62: 279–286.

Lockley, M. G., K. Houck, S.-Y. Yang, M. Matsukawa. and S.-K. Lim. 2006. Dinosaur dominated footprint assemblages from the Cretaceous Jindong Formation, Hallayo Haesang National Park, Goseong County, South Korea: evidence and implications. Cretaceous Research 27: 70–101.

Lockley, M. G., J. Li, R. H. Li, M. Matsukawa, J. D. Harris, and L. Xing. 2013. A review of the tetrapod track record in China, with special reference to type ichnospecies: implications

for ichnotaxonomy and paleobiology. Acta Geologica Sinica 87: 1–20. [English edition]

Lockley, M. G., S. H. Kim, J.-Y. Kim, K. S. Kim, M. Matsukawa, R. Li, J. Li, and S.-Y. Yang. 2008. *Minisauripus*–the track of a diminutive dinosaur from the Cretaceous of China and Korea: implications for stratigraphic correlation and theropod foot morphodynamics. Cretaceous Research 29: 115–130.

Lockley, M. G., G. D. Gierlinski, K. Houck, J.-D. F. Lim, K.-S. Kim, D. Y. Kim, T. K. Kim, S. H. Kang, R. Hunt Foster, R. Li, C. Chesser, R. Gay, Z. Dubicka, K. Cart, and C. Wright. 2014. New excavations at the Mill Canyon dinosaur track site (Cedar Mountain Formation, Lower Cretaceous) of eastern Utah. New Mexico Museum of Natural History and Science Bulletin 62: 287–300.

Lockley, M. G., Xing, L., Matthews, N. A. and Breithaupt, B. 2016. Didactyl raptor tracks from the Cretaceous, Plainview Sandstone at Dinosaur Ridge, Colorado. Cretaceous Research 61: 161–168.

Lull, R. S. 1904. Fossil Footprints of the Jura-Trias of North America. Memoirs of the Boston Society of Natural History 5: 461–557.

Manning, P. L., D. Payne, J. Pennicott, P. M. Barrett, and R. A. Ennos. 2005. Dinosaur killer claws or climbing crampons? Biology Letters 2: 110–112.

Matsukawa, M., M. G. Lockley, and J. Li. 2006. Cretaceous terrestrial biotas of east and southeast Asia, with special reference to dinosaur dominated ichnofaunas: towards a synthesis. Cretaceous Research 27: 3–21.

Maxwell, W. D., and J. H. Ostrom. 1995. Taphonomy and paleobiological implications of *Tenontosaurus-Deinonychus* associations. Journal of Vertebrate Paleontology 15: 707–712.

Milner, A. R. C., and M. G. Lockley. 2016. Dinosaur swim track assemblages: characteristics, contexts, and ichnofacies implications; chap. 10 in P. L. Falkingham, D. Marty, and A. Richter (eds.), Dinosaur Tracks: The Next Steps. Indiana University Press, Bloomington, Indiana.

Milner, A. R. C., M. G. Lockley, and J. I. Kirkland. 2006. A large collection of well-preserved theropod dinosaur swim tracks from the Lower Jurassic Moenave Formation, St. George, Utah. New Mexico Museum of Natural History and Science Bulletin 37: 315–328.

Moodie, R. L. 1930. Vertebrate footprints from the red-beds of Texas. II. Journal of Geology 38: 548–565.

Mudroch, A., U. Richter, U. Joger, R. Kosma, O. Idé, and A. Maga. 2011. Didactyl tracks of paravian theropods (Maniraptora) from the ?Middle Jurassic of Africa. PLoS One 6(2). e14642.

Nicosia, U., and M. Loi. 2003. Triassic footprints from Lerici (La Spezia, northern Italy). Ichnos 10: 127–140.

Norell, M. A., and P. J. Makovicky. 1997. Important features of the dromaeosaur skeleton: information from a new specimen. American Museum Novitates 3215: 1–28.

Olsen, P. E., R. W. Schlische, and P. J. W. Gore, eds. 1989. Tectonic, Depositional, and Paleoecological History of Early Mesozoic Rift Basins, eastern North America, Volume T351. American Geophysical Union, Washington, D.C., 174 pp.

Ostrom, J. H. 1969. Osteology of *Deinonychus antirrhopus*, an unusual theropod from the Lower Cretaceous of Montana. Bulletin of the Peabody Museum of Natural History 30: 1–165.

Ostrom, J. H., 1990. Dromaeosauridae; pp. 269–279 in D. B. Weishampel, P. Dodson, and H. Osmálka (eds.), The Dinosauria. University of California Press, Berkeley, California, 733 pp.

Paul, G. S. 1988. Predatory Dinosaurs of the World. Simon and Schuster, New York, New York, 464 pp.

Rainforth, E. C. 2003. Revision and re-evaluation of the Early Jurassic dinosaurian ichnogenus *Otozoum*. Palaeontology 46: 803–838.

Richter, A., and A. Böhme. 2016. Too many tracks: preliminary description and interpretation of the diverse and heavily dinoturbated Early Cretaceous "Chicken Yard" ichnoassemblage (Obernkirchen tracksite, northern Germany); chap. 17 in P. L. Falkingham, D. Marty, and A. Richter (eds.), Dinosaur Tracks: The Next Steps. Indiana University Press, Bloomington, Indiana.

Roach, B. T., and D. B. Brinkman. 2007. A reevaluation of cooperative pack hunting and gregariousness in *Deinonychus antirrhopus* and other nonavian theropod dinosaurs. Bulletin of the Peabody Museum of Natural History 48: 103–138.

Russell, D. A., and Z. Dong. 1993. A nearly complete skeleton of a new troodontid dinosaur from the Early Cretaceous of the Ordos Basin, Inner Mongolia, People's Republic of China. Canadian Journal of Earth Sciences 30: 2163–2173.

Russell, D. A., and R. Séguin. 1982. Reconstruction of the small Cretaceous theropod *Stenonychosaurus inequalis* and a hypothetical dinosauroid. Syllogeus 37: 1–43.

Sullivan C., T. A. van der Lubbe, and X. Xing. 2012. Ichnological implications of the structural variation in the paravian foot; p. 55 in A. Richter and M. Reich (eds.), Dinosaur Tracks 2011: an International Symposium, Obernkirchen, April 14–17, 2011 Abstract Volume and Field Guide to Excursions. Universitätsverlag Göttingen, Göttingen, Germany.

Turner, A. H., P. J. Makovicky, and M. A. Norell. 2012. A review of dromaeosaurid systematics and paravian phylogeny. Bulletin of the American Museum of Natural History 371: 1–206.

van der Lubbe, T. A., A. Richter, and A. Böhme. 2009. *Velociraptor*'s sisters: first report of troodontid tracks from the Lower Cretaceous of northern Germany. Journal of Vertebrate Paleontology 29(3, Supplement): 194A–195A.

van der Lubbe, T. A., A. Richter, A. Böhme, C. Sullivan, and T. R. Hübner. 2012. Sorting out the sickle claws: how to distinguish between dromaeosaurid and troodontid tracks; p. 35 in A. Richter and M. Reich (eds.), Dinosaur Tracks 2011: An International Symposium, Obernkirchen, April 14–17, 2011 Abstract Volume and Field Guide to Excursions. Universitätsverlag Göttingen, Göttingen, Germany, 187 pp.

White, D., and M. G. Lockley. 2002. Probable dromaeosaur tracks and other dinosaur footprints from the Cedar Mountain Formation (Lower Cretaceous), Utah. Journal of Vertebrate Paleontology 22(3, Supplement): 119A.

Xing L.-D. 2010. Dinosaur Tracks. Shanghai Press of Scientific and Technological Education, Shanghai, China, 255 pp.

Xing, L., J. D. Harris, D.-H. Sun, and H.-Q. Zhao. 2009. The earliest known deinonychosaur tracks from the Jurassic-Cretaceous boundary in Hebei Province, China. Acta Palaeontologica Sinica 48: 662–671.

Xing, L., D. Li, J. D. Harris, P. R. Bell, Y. Azuma, M. Fujita, Y.-N. Lee, and P. J. Currie. 2013. A new *Dromaeosauripus* (Dinosauria: Theropoda) ichnospecies from the Lower Cretaceous Hekou Group, Gansu Province, China. Acta Palaeontologica Polonica 58: 723–730.

Xing, L. D., M. G. Lockley, G. Yang, X. Xu, J. Cao, H. Klein, W. S. Persons IV, H. J. Shen, and X. M. Zheng. 2015a. An unusual deinonychosaurian track morphology (*Velociraptorichnus zhangi* ichnosp. nov.) from the Lower Cretaceous Xiaoba Formation, Sichuan Province, China. Palaeoworld 24(3): 283–292.

Xing, L., J. Zhang, M.. G. Lockley, R. T. McCrea, H. Klein, L. Alcalá, L. G. Buckley, M. E. Burns, S. B. Kümmel, and Q. He. 2015b. Hints of the Early Jehol Biota: important dinosaur footprint assemblages from the Jurassic-Cretaceous Boundary Tuchengzi Formation in Beijing, China. PLoS ONE 10(4):e0122715, doi:10.1371/journal.pone.0122715.

Xing, L.D., M. G. Lockley, D. Marty, H. Klein, G. Yang, J. P. Zhang, G. Z. Peng, Y. Ye, W. S. Persons IV, X. Y. Yin, and T. Xu. In press. A diverse saurischian (theropod-sauropod) dominated footprint assemblage from the Lower Cretaceous Jiaguan Formation in the Sichuan Basin, southwestern China: a new ornithischian ichnotaxon, pterosaur tracks and an unusual sauropod walking pattern. Cretaceous Research.

Xing, L., M. G. Lockley, D. Marty, H. Klein, R. T. Buckley, R. T. McCrea, J. Zhang, G. G. Gierlinski, J. D. Divay, and Q. Wu. 2013. Diverse dinosaur ichnoassemblages from the Lower Cretaceous Dasheng Group in the Yishi fault zone, Shandong Province, China. Cretaceous Research 45: 114–134.

Xu, X., and M. A. Norell. 2004. A new troodontid dinosaur from China with avian-like sleeping posture. Nature 431: 838–841.

Xu, X., Z. Zhou, and X. Wang. 2000. The smallest known non-avian theropod dinosaur. Nature 408: 705–708.

Xu, X., M. A. Norell, X.-L. Wang, P. J. Makovicky, and X.-C. Wu. 2002. A basal troodontid from the Early Cretaceous of China. Nature 415: 780–784.

Zanno, L. E., D. J. Varricchio, P. M. O'Connor, A. L. Titus, and M. J. Knell. 2011. A new troodontid theropod, *Talos sampsoni* gen. et sp. nov., from the Upper Cretaceous Western Interior Basin of North America. PLoS One 6(9): e24487.

Zhen, S., B. Zhang, W. Chen, and S. Zhu. 1987. Bird and dinosaur footprints from the Lower Cretaceous of Emei County, Sichuan; pp. 37–38 in First International Symposium, Nonmarine Cretaceous Correlations. IGC, Urumqi, China.

Zhen, S., J. Li, B. Zhang, W. Chen, and S. Zhu. 1994. Dinosaur and bird footprints from the Lower Cretaceous of Emei County, Sichuan, China. Memoirs of Bejing Natural History Museum 54: 106–120.

Zhen, S., J. Li, C. Rao, N. Mateer, and M. G. Lockley. 1989. A review of dinosaur footprints in China; pp. 187197 in D. D. Gillette and M. G. Lockley (eds.), Dinosaurs Past and Present. Cambridge University Press, Cambridge, UK.

NOTE

1. Beginning in 2012, after the transfer of the CU collection to UCM, this prefix, if used, is synonymous with the CU prefixes cited in many of the publications referenced herein and elsewhere.

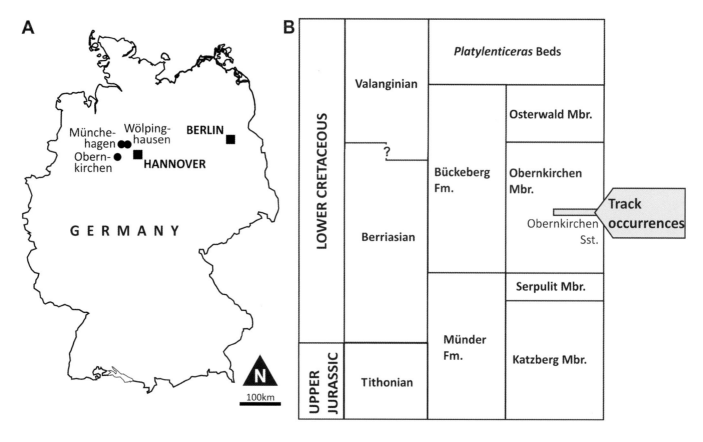

12.1. (A) Locality map and (B) stratigraphic position of the material studied herein.

Diversity, Ontogeny, or Both?: A Morphometric Approach to Iguanodontian Ornithopod (Dinosauria: Ornithischia) Track Assemblages from the Berriasian (Lower Cretaceous) of Northwestern Germany

12

Jahn J. Hornung, Annina Böhme, Nils Schlüter, and Mike Reich

IDENTIFYING THE CAUSES OF MORPHOLOGICAL VARIA tion (including taxonomic diversity, ontogeny, sexual dimorphism, and individual variation) observed in a set of vertebrate tracks–especially from different closely related trackmaker species–is difficult and often not straightforward due to imperfect knowledge of biological variation in the autopodia of the trackmakers, and a number of ethological, preservational, and taphonomical influences. Here we use multivariate data sets obtained from 14 homologous two-dimensional (2-D) landmarks to evaluate the range and potential causes of variation in iguanodontian ornithopod pes tracks from the Berriasian of northwestern Germany.

In order to minimize the nonanatomical-related effects associated with track morphology (extramorphological track modifications), the specimens included in the analysis were selected by application of a rigorous check of preservational quality. The data sets were normalized by full Procrustes fitting and reduced to one- or two-dimensional data by principal components analysis (PCA). Comparison of multitrack trackway segments with the entire data set shows that the variation found in the former only accounts for a maximum of 32.2% (median: 8.9%, arithmetic mean: 13.2%, n = 10) of total variation. The latter is therefore interpreted to be informative with respect to variation between individual trackmakers. Different approaches (nondiscriminatory PCA of the entire data set, group-wise PCA of a priori defined provenance groups, and discriminant analysis) found two overlapping clusters of scores, interpreted to represent two morphotypes (O and M), correlating to the geographical and stratigraphical provenance of the tracks. Morphological extremes are related to smaller size, whereas larger specimens occupy the zone of overlap between the morphotypes. In morphotype O, a clear morphological trend correlates with an increase in size, suggesting an ontogenetic signal and appendicular allometry of the trackmaker.

The data suggest that qualitatively quite different ornithopod tracks form a morphological continuum, in which variation is partly explainable by ontogeny. On the other hand, the data set is interpreted to contain at least two distinct morphotypes, which, however, cannot be sharply defined by qualitative characters due to a considerable morphological overlap. The method presented here is promising to evaluate variation in track data sets and to interpret it in terms of ichnotaxonomy, orthotaxonomy, and ontogenesis of the trackmaker.

INTRODUCTION

One outstanding character of vertebrate track fossils is the fact that they represent a mode of preservation of body parts of their producers, in contrast to most invertebrate tracks and traces, which mostly reflect the behavior and activity of the trackmaker but not necessarily its anatomy (e.g., Hunt and Lucas, 2007; Lockley, 2007). As such, their interpretation is subjugated largely to the same problems as that of skeletal remains, including the recognition of individual and ontogenetic variation, population structure and dynamics, and even paleopathologies. For example, ontogenetic allometry is often encountered in dinosaurs, commonly including the limbs (e.g., Foster and Chure, 2006; Brett Surman and Wagner, 2007; Kilbourne and Makovicky, 2010), and therefore potentially affecting the track morphology (e.g., Breithaupt and Matthews, 2012). However, these aspects are rarely addressed by ichnological studies because the influence of kinematics and substrate properties on track morphology cannot easily be examined. From this thematic complex, potential social behavior (which is affected by population dynamics) is most abundantly investigated in dinosaur track studies (e.g., Ostrom, 1972; Thulborn and Wade, 1979; Lockley, Houck, and Prince, 1986; Farlow, 1987; Leonardi, 1989; Pittman and Gillette, 1989; Hawthorne, 1990; Carpenter, 1992; Barnes and Lockley, 1994; Lockley, Meyer, and dos Santos, 1994; Matsukawa et al., 1997; Lockley and Matsukawa, 1999; Lockley, Schulp, et al., 2002; Lockley, Wright, et al., 2002; Day et

203

al., 2004; Barco, Canudo, and Ruiz-Omeñaca, 2006; Myers and Fiorillo, 2009; Castanera et al., 2011), whereas an ontogenetic or paleopathological impact on track morphologies is proposed sporadically (e.g., Abel, 1935; Dantas et al., 1994; Lockley et al., 1994; Matsukawa, Lockley, and Hunt, 1999; Avanzini, Piñuela, and García-Ramos, 2008; Ishigaki and Lockley, 2010). On the other hand, the limited range of morphological information on the trackmakers preserved in their tracks often hampers the assessment of the signal of biological variation and its potential causes (individual variation, ontogeny, taxonomy). This is further acerbated by imperfect knowledge of the degree, origins, and phenotypical expression of variation in the trackmakers. The second strong bias includes the impact of extramorphological and taphonomical modification frequently experienced by tracks (e.g., by differences in substrate, subsequent erosion), which can often only be assessed or estimated to a limited level of certainty. Current concepts in ichnotaxonomy, which are based mostly on isolated qualitative morphological traits (e.g., ichnotaxobases, Bromley, 1990; Bertling et al., 2006) rarely pay attention to these factors that may potentially prove to be problematic, such as when synapomorphy-based track/trackmaker relations are studied (Carrano and Wilson, 2001) or when ichnotaxonomic definitions are used to interpret ichnocoenoses in terms of biological diversity or ichnostratigraphy.

A potential approach to these problems is a multivariate, 2-D homologous landmark (MHL) analysis of large data sets of similar and taxonomically related tracks (for similar approaches and techniques in dinosaur tracks, see Rodrigues and Santos, 2004; Belvedere, 2008; Moratalla and Marugán-Lobón, 2009; Castanera et al., 2015; Buckley, McCrea, and Lockley, 2016; Wings, Lallensack, and Mallison, 2016). By categorizing the material in qualitatively defined groups, this method allows a quantification of morphological traits that can otherwise only be described in qualitative terms and has the potential to expose morphological trends, which would otherwise be obscured. On the other hand, unlike numerical analyses of continuous shape landmarks (based on an outline shape defined by a theoretically infinite number of [semi] landmarks, e.g., Rasskin-Gutmann et al., 1997; Zelditch, Swiderski, and Sheets, 2012), it allows the application of an a priori filter (using objective criteria) to eliminate random and other nonanatomy-related shape transformations from the data set. It also has the advantage of being applicable to isolated footprints (even those from old collections with limited data on their origin), independently from the preservation of trackways or their length (though their availability will strongly bolster the meaning and reliability of the data set). Even specimens represented only by (suitable) photographs or sketches can be used as data sources, provided that the

accuracy of both is properly assessed and included in the interpretation of the results (see also Falkingham, 2016). It does not require special hardware and employs only software that is freely available.

AIMS OF THIS STUDY

Here we apply MHL analysis to a data set obtained from iguanodontian ornithopod tracks occurring in the upper Berriasian Obernkirchen Sandstone (Bückeberg Formation) of northwestern Germany, which form a dominant component of the local dinosaur ichnofauna (e.g., Struckmann, 1880; Grabbe, 1881; Ballerstedt, 1905, 1914; Stechow, 1909; Dietrich, 1927; Lehmann, 1978; Diedrich, 2004; Lockley, Wright, and Thies, 2004; Böhme et al., 2009; Hornung and Reich, 2012). These tracks have been referred to as "tracks of *Iguanodon*" for most of their research history (e.g., Struckmann, 1880; Dollo, 1883; Kuhn, 1958; Lehmann, 1978). With increasing acceptance that there is no evidence to relate these tracks to the Barremian-Aptian orthogenus *Iguanodon* Mantell, 1825, and the advent of a more formalized ichnotaxonomic nomenclature for iguanodontian tracks (Sarjeant, Delair, and Lockley, 1998), these tracks were mostly collectively referred to the ichnogenus *Iguanodontipus* Sarjeant, Delair, and Lockley, 1998, with more or less certainty (e.g. Diedrich, 2004; Lockley, Wright, and Thies, 2004; Hornung and Reich, 2007; Böhme et al., 2009; Richter et al., 2012). However, there is growing evidence that the material encompasses a range of morphotypes: Böhme et al. (2009) additionally found a distinctly different morphotype, which they tentatively associated with the ichnogenus *Caririchnium* Leonardi, 1984, and Hornung and Reich (2012) stated that only a small part of the studied ornithopod tracks can be referred to *Iguanodontipus* when applying the qualitative definitions of this ichnogenus by Sarjeant, Delair, and Lockley (1998) and Meyer and Thüring (2003; but see also Díaz-Martínez et al., 2015). The relationships of very small, tridactyl ornithischian tracks from this stratum are evaluated by Hübner (2016).

At track localities in the Obernkirchen Sandstone, the ornithopod tracks and trackways often occur in large numbers on the same track horizon (Dietrich, 1927; Lehmann, 1978; Diedrich, 2004; Hornung and Reich, 2012; Richter et al., 2012; Wings et al., 2012). Though it may be open to discussion whether this indicates a level of true gregarious behavior (which seems probable at least in some places), these animals apparently occurred abundantly within very short time intervals. However, various sizes and variable morphologies of these tracks may indicate either the sympatric occurrence of different orthotaxa, of different ontogenetic stages, or of both. To test the proposed method and evaluate the preceding questions, we quantified the morphology of a subset of

Table 12.1. Overview of tracksites from which material was used for the MHL analyses

Locality	Abbreviation	Number of tracks	Number of trackway segments	Remarks	References
Obernkirchener Sandstone quarry, Bückeberg near Obernkirchen	OBK	11	2 (OBK-I, -M)	Several track horizons. Material examined herein are either from the "Chicken Yard" horizon (Richter, Böhme, and van der Lubbe, 2009; Richter et al., 2012; n = 1), from the "Upper Surface" (Böhme et al., 2009; Richter et al., 2012; n = 4), or from unknown levels (n = 6, partially described by Lehmann, 1978).	Lehmann (1978), Hornung et al. (2012: loc. 8), Richter et al. (2012)
F. Wesling KG quarry, Münchehagen, Rehburg-Loccum	MHG	38	9 (MHG -A, -B, -D, -E, -F, -I, -J, -K, -L)	Two main track horizons, each consisting of a few track layers (Wings et al. 2012). Most tracks are from the upper horizon (n = 30). Some isolated material cannot be assigned to a certain horizon (n = 8).	Hornung et al. (2012: loc. 1), Wings et al. (2012)
Uncertain, possibly near Wölpinghausen	XX?	?	0	Probably material from Struckmann (1880)	Wölpinghausen locality; Hornung et al. (2012: loc. 4); Hornung, Böhme, and Reich 2012b
Uncertain	XX1	4	0	—	—

Notes: Trackway segments refer to trackways (two or more footprints left by the same individual) used in this study. MHL, multivariate, two-dimensional homologous landmark.

these tracks, assembled by adhering to strict criteria of preservational quality, by obtaining a set of homologous landmarks from the pes tracks.

MATERIALS AND METHODS

Geological Setting

All material originates from a number of outcrops (Table 12.1, Figure 12.1) located in the Obernkirchen Sandstone, Wealden 3, *Cypridea alta formosa* ostracod subzone, Obernkirchen Member, lower Bückeberg Formation of Late Berriasian age. The tracksites are located in an area west of Hannover, Lower Saxony, northwestern Germany. The Obernkirchen Sandstone represents a relatively thin sandstone intercalation formed as a deltaic and sandy shore-face barrier system, prograding into predominantly pelitic lacustrine basin sediments deposited in the Lower Saxony Basin (Pelzer, 1998). Dinosaur tracks (including ornithopod, theropod, sauropod, and ankylosaur footprints) are abundant in a number of subenvironments, including barrier sands and shallow-water delta plains. For a detailed account on the geology, vertebrate ichnofauna, and paleoenvironment, as well as descriptions of tracksites from the Obernkirchen Sandstone, see Hornung et al. (2012) and Hornung, Böhme, and Reich (2012a).

A strict lateral correlation between track horizons in different outcrops and therefore proof of isochrony cannot be provided, and hence the Obernkirchen Sandstone track occurrences were not classified as a megatracksite by Hornung et al. (2012). However, according to the sedimentation rate model for the Bückeberg Formation by Berner (2011), and dating of the Obernkirchen Sandstone by Elstner and

Mutterlose (1996), the total time for deposition for this unit lasted 10^4 to 10^5 years, and its dinosaur ichnoassemblage is considered to represent a coherent ichnocoenosis (sensu Hunt and Lucas, 2007).

Though the paleoenvironmental interpretation differs between the track localities, the general lithofacies, type, and quality of track preservation are similar. The tracks are mostly preserved at bedding surfaces of heterolithic successions of thin- to thick-bedded, fine-grained, highly mature, mostly textureless but also planar laminated, cross-laminated, or cross-stratified, often oscillation ripple-marked sandstone with thin layers of mudstone, ranging in thickness from 0.1 to >5 cm. As an exception, tracks may also be preserved within sandstone successions at bedding planes between distinct beds. The most common type of preservation is that of true tracks or – in the case of heterolithic contacts – "nearly true" tracks in which the tracks may have been left in sand covered by a millimeter-thin layer of mud. Technically undertracks, or underprints, these tracks have the potential to preserve faithfully the autopodial morphology of the trackmaker because the very thin layer of mud only marginally influences the relief produced in the underlying sand. Deeper undertracks are rare and related to thinly bedded sandstone successions.

Structures (tracks, ripple-marks, etc.) on sandstone bedding planes are often preserved in pristine condition and were protected by the mudstone covers, suggesting rapid, episodic high-energy depositional events (storms, seasonal floods) for the sandstones, followed by calm suspension settling. Carbonaceous films on such well-preserved surfaces suggest the development of microbial mats, stabilizing the sediment, and favoring track preservation (see also Marty, Strasser, and Meyer, 2009). In some instances, especially

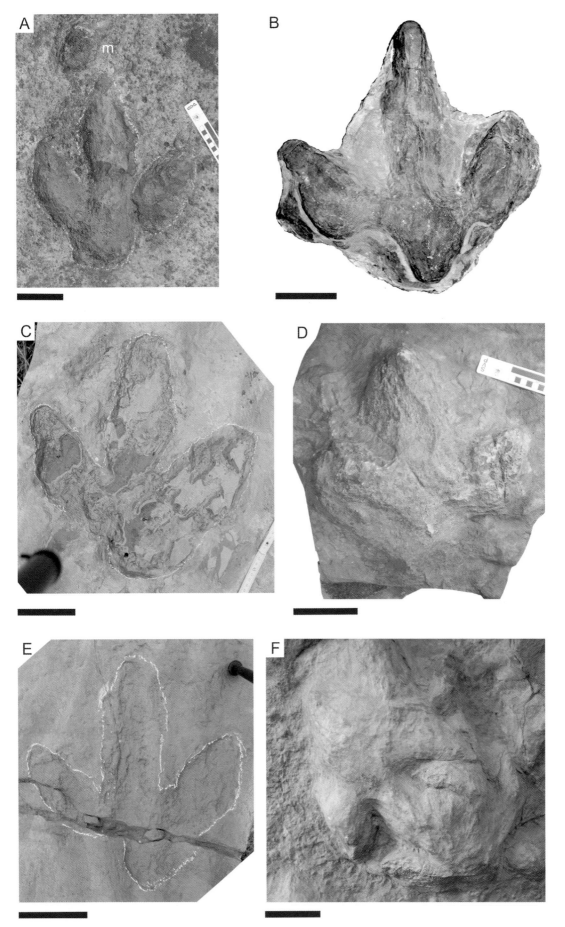

12.2. (A–D) Examples of track specimens included in the multivariate, two-dimensional homologous landmark (MHL) analysis. (A) Morphotype M, in situ specimen, data-ID OBK-N-1, epichnial relief, left pes impression (m, left manus impression), Obernkirchen quarry, Chicken Yard track horizon. (B) Morphotype ?O, NLMH 105.746 (in the Niedersächsisches Landesmuseum Hannover collection), data-ID XX2-X-2, natural hypichnial cast, left pes impression, uncertain locality, probably Wölpinghausen. (C) Morphotype M, NLMH unnumbered, data-ID MHG-L-2, artificial cast of natural epichnial relief, right pes impression, Münchehagen, Wesling quarry, upper track horizon. (D) Morphotype M, NLMH unnumbered, data-ID MHG-X-8, natural hypichnial cast, right pes impression, Münchehagen, Wesling Quarry, uncertain track horizon. (E) Tectonically distorted (clefted) specimen, not used for analysis, NLMH unnumbered, artificial cast of natural epichnial relief, right pes impression, Münchehagen, Wesling Quarry, upper track horizon. (F) Overstepping pes tracks, not used for analysis, GPMH unnumbered ([in the Geologisch-Paläontologisches Museum der Universität Hamburg collection]; see Lehmann, 1978) natural hypichnial casts, Obernkirchen Quarry, uncertain track horizon. (A, B) Preservational class I; (C, D) preservational class II; (E, F) preservational class III. Scale bars: 10 cm.

Table 12.2. Track data set

Data-ID	Pres. Class	Specimen-#	Morphotype	Site, track horizon	Auto-podium	Preservation	FL [cm]	FW [cm]
MHG-A-1	II	NLMH	M	Münchehagen, upper horizon	right	Epichnium	43	38
MHG-A-2	I	NLMH	M	Münchehagen, upper horizon	left	Epichnium	41	37
MHG-A-3	I	NLMH	M	Münchehagen, upper horizon	right	Epichnium	39	38
MHG-A-4	I	NLMH	M	Münchehagen, upper horizon	left	Epichnium	40	38
MHG-B-1	II	NLMH	M	Münchehagen, upper horizon	right	Epichnium	31.5	33
MHG-B-2	II	NLMH	M	Münchehagen, upper horizon	left	Epichnium	33	32
MHG-C-2	II	NLMH	M	Münchehagen, upper horizon	left	Epichnium	42	40.5
MHG-D-1	II	NLMH	M	Münchehagen, upper horizon	right	Epichnium	40	42
MHG-D-2	II	NLMH	M	Münchehagen, upper horizon	left	Epichnium	43.5	43
MHG-E-1	I	NLMH	M	Münchehagen, upper horizon	right	Epichnium	25	30
MHG-E-2	I	NLMH	M	Münchehagen, upper horizon	left	Epichnium	28	31
MHG-F-1	II	NLMH	M	Münchehagen, upper horizon	right	Epichnium	45	43.5
MHG-F-2	II	NLMH	M	Münchehagen, upper horizon	left	Epichnium	43	41
MHG-F-3	II	NLMH	M	Münchehagen, upper horizon	right	Epichnium	44	41.5
MHG-G-1	II	NLMH	M	Münchehagen, upper horizon	left	Epichnium	55	55
MHG-H-1	II	NLMH	M	Münchehagen, upper horizon	left	Epichnium	45	43
MHG-I-1	II	NLMH	M	Münchehagen, upper horizon	left	Epichnium	44	42
MHG-I-2	II	NLMH	M	Münchehagen, upper horizon	right	Epichnium	45	41
MHG-J-1	II	NLMH	M	Münchehagen, upper horizon	right	Epichnium	23	27
MHG-J-2	II	NLMH	M	Münchehagen, upper horizon	left	Epichnium	28	30
MHG-K-1	II	NLMH	M	Münchehagen, upper horizon	right	Epichnium	32	34
MHG-K-2	II	NLMH	M	Münchehagen, upper horizon	right	Epichnium	27	33
MHG-L-1	II	NLMH	M	Münchehagen, upper horizon	right	Epichnium	44.5	43
MHG-L-2	II	NLMH	M	Münchehagen, upper horizon	right	Epichnium	43	41.5
MHG-X-1	I	NLMH 103.173A	M	Münchehagen, n/a	left	Hypichnium	32.5	30
MHG-X-2	I	NLMH 103.173B	M	Münchehagen, n/a	left	Hypichnium	34	28.5
MHG-X-3	I	NLMH 103.173C	M	Münchehagen, n/a	right	Hypichnium	33	33.5
MHG-X-4	I	NLMH 103.173D	M	Münchehagen, n/a	right	Hypichnium	30	30.5
MHG-X-5	II	NLMH 103.267	M	Münchehagen, n/a	left	Hypichnium	41	42
MHG-X-6	I	NLMH 105.734	M	Münchehagen, n/a	left	Hypichnium	34	35
MHG-X-7	I	NLMH 105.735	M	Münchehagen, n/a	right	Hypichnium	45.5	43.5
MHG-X-8	II	NLMH	M	Münchehagen, n/a	right	Hypichnium	34.5	35
MHG-X-9	II	NLMH	M	Münchehagen, upper horizon	right	Epichnium	28	28,5
MHG-X-10	II	NLMH	M	Münchehagen, upper horizon	right	Epichnium	22.5	26.5
MHG-X-11	II	NLMH	M	Münchehagen, upper horizon	left	Epichnium	46.5	45
MHG-X-12	II	NLMH	M	Münchehagen, upper horizon	right	Epichnium	49	50
MHG-X-13	II	NLMH	M	Münchehagen, upper horizon	right	Epichnium	43	42
MHG-X-14	II	NLMH	M	Münchehagen, upper horizon	right	Epichnium	42	43
OBK-B-2	II	in situ	O	Obernkirchen, "Upper Surface"	right	Epichnium	45	45
OBK-I-1	I	in situ	O	Obernkirchen, "Upper Surface"	right	Epichnium	35	31,5
OBK-I-2	I	in situ	O	Obernkirchen, "Upper Surface"	left	Epichnium	35	31
OBK-I-3	I	in situ	O	Obernkirchen, "Upper Surface"	right	Epichnium	34	32,5
OBK-J-1	II	GPMH	O	Obernkirchen, n/a	right	Hypichnium	36.5	38
OBK-K-1	II	GPMH	O	Obernkirchen, n/a	right	Hypichnium	46	44.5
OBK-L-1	II	GPMH	O	Obernkirchen, n/a	right	Hypichnium	32	30
OBK-M-1	II	GPMH	O	Obernkirchen, n/a	left	Hypichnium	26	22
OBK-M-2	II	GPMH	O	Obernkirchen, n/a	right	Hypichnium	25.5	25
OBK-N-1	I	in situ	?M	Obernkirchen, "Chicken Yard"	left	Epichnium	39.5	33.5
OBK-X-1	I	NLMH	O	Obernkirchen, n/a	left	Hypichnium	47	46
XX1-X-1	I	NLMH	?M	?	right	Hypichnium	33	31
XX1-X-2	I	MSLB	?M	?	right	Hypichnium	45	45
XX1-X-3	II	MSLB	?M	?	left	Hypichnium	45	45
XX1-X-4	I	MSLB	?M	?	left	Hypichnium	49.5	49.5
XX2-X-1	I	NLMH 149.7	?O	?Wölpinghausen	left	Hypichnium	41	3
XX2-X-2	I	NLMH 105.746	?O	?Wölpinghausen	left	Hypichnium	44	47

Notes: Landmark coordinates and Procrustes fitting variance-covariance matrix are available online: http://dx.doi.org/10.6084/m9.figshare.1348327. FL, foot length; FW, foot width; GPMH, Geologisch-Paläontologisches Museum der Universität Hamburg; MSLB, Museum für Stadtgeschichte und Schaumburg-Lippische Landesgeschichte, Bückeburg; n/a, not available; NLMH, Niedersächsisches Landesmuseum Hannover; other abbreviations as in Table 12.1.

at Münchehagen, reticulated or knobby patterns of raised ridges and pits on sandstone surfaces are very reminiscent of similar structures in desiccating microbial mats (Bose and Chafetz, 2009), suggesting emersion after mat formation. Tracks on these surfaces often show poor preservation. At Obernkirchen, such surface structures are commonly associated with root-traces (Grupe, 1931), indicating longer phases of emersion. Other indicators of subaerial exposure (e.g., mud cracks) are extremely rare and the track horizons with well-preserved tracks were most probably permanently covered by shallow water until final burial. Omission surfaces are also very rare, though the densely track-covered, lower, "Chicken Yard" level at Obernkirchen (Richter et al., 2012) was probably exposed for some time, resulting in a slight surface degradation and destruction of subtle superficial sedimentary structures.

The track localities are located within a depositional system that encompassed the mouth of a river system draining the southern uplands and reaching the lake basin in the roughly funnel-shaped Hils Embayment, west and southwest of present-day Hannover (Pelzer, 1998; Hornung et al., 2012; Hornung, Böhme, and Reich, 2012a). The Obernkirchen track levels were located in a deltaic mouthbar complex (Hornung et al., 2012; Richter et al., 2012), with a slightly more distal location of the upper level that yielded the majority of the examples included in this work. The Münchehagen locality is interpreted as to be located in a sandy barrier complex, migrating landward over fine-grained back-barrier lagoonal deposits (Pelzer, 1998; Wings et al., 2012). The exact environmental setting of the Wölpinghausen locality is unclear, as the track-bearing strata are no longer exposed (Hornung, Böhme, and Reich, 2012b)

Track Material

According to the strict criteria outlined herein, we sampled data from 55 tracks from the Obernkirchen quarry (two stratigraphic levels, n = 11), Münchehagen Wesling quarry (n = 38), and from at least one or more unknown localities (insufficient locality data, n = 6), one of which is most probably Wölpinghausen, the classical site of Struckmann (1880). Details on the localities and material are provided in Tables 12.1–12.2; examples of the material are shown in Figure 12.2. The material was chosen in the order of importance according to (1) preservational quality, (2) range of size, (3) range of morphological variation, and (4) diversity of provenance. The iguanodontian ornithopod pedal track morphologies that are evaluated here have been qualitatively identified in the ichnocoenoses by the presence of three broad, bluntly terminating digits, no trace of digit I, presence of a posteromesial,

distally rounded metatarsal-phalangeal pad, confluent with the base of digits, subequal lengths of digits II and IV, subsymmetric arrangement of digits II and IV with respect to digit III, moderate to wide digital divarication (angle between digits II and IV: >45°), inward rotation of pes impressions in trackways and the association of small, ovate to triangular manual impressions in some specimens (see also Castanera et al., 2013). Examples exhibiting morphological differences, which can clearly be related to anatomical features of the trackmaker (e.g., ?*Caririchnium* isp. ornithopod tracks of Böhme et al., 2009; Richter et al., 2012; theropod tracks), were omitted from the data set.

Data Acquisition

Criteria of track data selection As MHL depends on the proper identification of anatomically homologous landmarks, the foremost objective of track selection was to ensure that the material reflects the pedal morphology of the trackmaker in the most unmodified way possible. Contraindicative to inclusion in the data set were tracks with strong intrinsic, extrinsic, or extramorphologic modifications (e.g., Diedrich, 2004; Manning, 2004; Jackson, Whyte, and Romano, 2009, 2010; Falkingham et al., 2009, 2011; Falkingham, 2014; Belvedere and Farlow, 2016) and undertracks. A catalog and classification of such modifications and modes of preservation are provided in Table 12.3 and Figure 12.3, leading to three preservational classes:

- Class I track data are considered to represent the trackmaker's autopodial morphology in an optimal quality and are regarded first-choice as for MHL.
- Class II track data represent suboptimally preserved specimens that are rated sufficient for MHL analysis, though special caution must be given to the interpretation of results. Contradictory results must be rated inferior in accuracy to results from class I tracks, and the use of class II data should be avoided if a sufficient database is available for the entire range of observed morphologies.
- Class III track data are unsuitable for MHL analysis as it inadequately represents the trackmaker's autopodial morphology.

According to the "Goldilocks" principle (for application in track formation and preservation see Falkingham et al., 2011), most tracks of an ichnocoenosis can be anticipated to rate as class III, with sequentially decreasing numbers of individual tracks usable (class II) and optimal (class I) for MHL

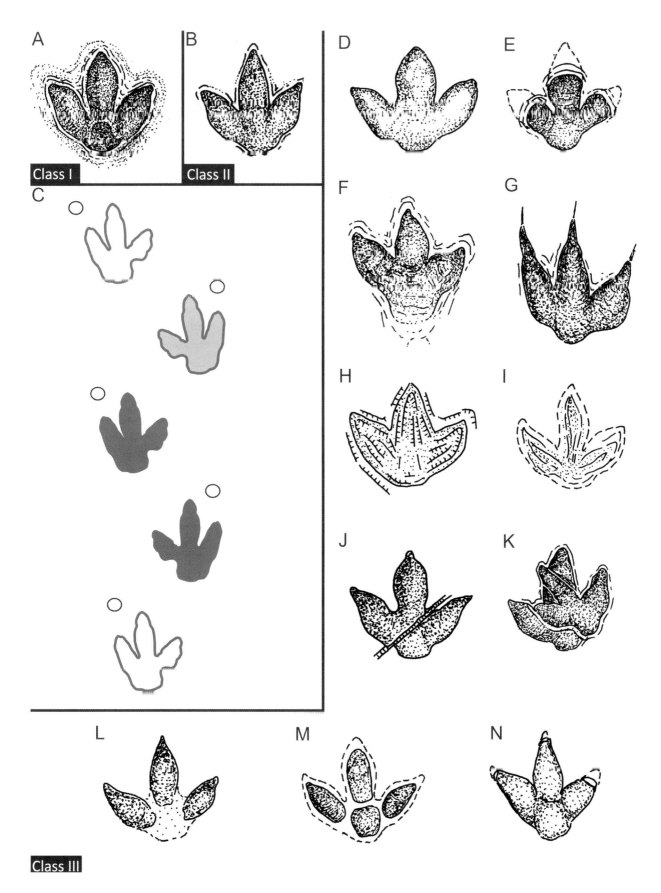

12.3. Track preservation quality and class rating. (A) Idealized example of class I track. (B) Idealized example of class II track. (C) Schematic example of trackway segment including tracks of class I (dark gray), class II (light gray), and class III (white), only the gray tracks will be used for multivariate, two-dimensional homologous landmark (MHL) analysis. White circles indicate manus impressions. (D–N) Idealized examples of class III tracks. For further explanations, see Table 12.3.

Table 12.3. Definition of track preservation classes

Preservational Class	Description
I	Pristine preservation of tracks with no to minimal extramorphological or epigenetic modifications (Figs. 12.2A, 12.2B, 12.3A). The tracks are deeply impressed (at least to a depth equaling half the width of the digits), with nearly vertical margins and well-defined outline. Preferrably used for MHL analysis.
II	Good preservation of tracks with well-preserved, undistorted outline but minor extramorphological or epigenetical modifications (Figs. 12.2C, 12.2D, 12.3B), including shallower pes impressions, minor modifications of the track floor, such as by adhesion spikes or microfaulting, local marginal collapse, or compression of interdigital sediment. Usable for MHL analysis.
III	Preservation severely affected by intrinsic or extrinsic, extramorphological or epigenetic modifications (Figs. 12.2E, 12.2F, 12.3D, 12.3N). These modifications may include but not be restricted to: too shallow impression (Fig. 12.3D), partial roofing of deeply impressed tracks (Fig. 12.3E), slipping and slip-faces (Fig. 12.3F), dragging and drag marks (Fig. 12.3G), outline distorted by intensive microfaulting (Fig. 3H), or high water content in sediment (Fig. 12.3I), tectonic deformation (Figs. 12.2E, 12.3J), overstepping (Figs. 12.2F, 12.3K), incomplete impression or preservation (Fig. 12.3L), partial preservation by firm substrate, undertracks (Fig. 12.3M), or overtracks or postexcavational damage to hypichnial casts (Fig. 12.3N). Unsuitable for MHL analysis.

Note: MHL, multivariate, two-dimensional homologous landmark.

analysis, respectively, resulting in a fastidious data selection (Figs. 12.2–12.3).

The entire outer perimeter of the track must be visible from above. Tracks in which the digits were inserted in the sediment surface subhorizontally (penetrated into the substrate) and the anterior part of the track is roofed (technically forming an endichnium rather than an epichnium, or a penetrative deep track, see Fig. 12.3E) are not suitable. Class I tracks are characterized by vertical or near-vertical track margins (preserving the unaltered, uncollapsed outline), and the foot was impressed and removed as vertically as possible, to at least half of its dorsoventral height into the substrate. It is assumed that a resulting track represents the most objective outline of the pes, including the most anatomical information possible, reduced to a 2-D plane. Though it may be argued that relinquishment of the third dimension in the data set is a reduction in information, the 2-D outline is considered to carry most of the anatomical information, because lateral sediment displacement during track formation is mostly controlled by the pes's volume and morphology. In contrast, the third dimension (vertical to the bedding plane) is often affected and modified by locomotion dynamics (e.g., Thulborn and Wade, 1989). Additionally, data acquisition for 2-D landmarks is much simpler, and equipment- and

time-efficient, and the loss of purpose-relevant information is considered negligible.

Critical areas are often the hypices between the digits, being affected by compression of the interdigital sediment (Falkingham et al., 2009), and the posterior margin of the foot (the metatarsal-phalangeal pad), being modified by the maximum angle of inclination of the metatarsus during the step cycle. Depending on the degree of such inaccuracies, tracks may still be rated as class II, as long as the defined homologous landmarks are unequivocally identifiable.

Optional missing data algorithms in data normalization or PCAs should not be used because the impact of such methods on result accuracy have not been established by meticulous analyses, which are beyond the scope of this work. For the time being, tracks in which the locations of all homologous landmarks cannot be unanimously identified are summarily rated class III, and excluded from MHL analysis.

The use of different preservational modes of tracks (epichnial relief, hypichnial casts, both natural as well as artificial) is not only possible but explicitly encouraged in order to avoid statistical effects potentially caused by different modes of preservation between two or more track populations. The validity of MHL analysis results is strengthened if they are found to be stable regardless of the preservational mode of the included tracks. Any results showing a correlation between morphospace definition and track preservation have to be treated with suspicion. Assessment of the superiority of a certain preservational mode in terms of data quality is not ever straightforward. For example, in this study, hypichnial casts were considered slightly advantageous in some cases, as preserved epichnial reliefs in sandstone layers were in fact undertracks created below a millimeter-thin veil of mud (Wings et al., 2012). In these cases, which are not always obvious, the casts better represent the original morphology of the pes.

These prerequisites reduce the amount of usable data. However, such a restriction is inevitable for a realistic judgment of the anatomical morphospace of the trackmaker's autopodium and all further expansion of the data set should be done with utmost care in adopting these principles. Naturally, the method applied requires the inclusion of data from more than one trackway; theoretically, any compared track can originate from a different trackway. However, to obtain a measure of objectivity, several (left and right) tracks from at least one trackway segment (Fig. 12.3C) should be included to compare the range of variation created by a single, evidently unique trackmaker to the entire data set.

The 2-D landmarks (x, y coordinates) were acquired with the free software ImageJ (Rasband, 1997–2012) supplemented with the open-source plugin Point Picker (Thévenaz, 2012) from vertical digital photographs of the tracks. In the case of extensive in situ track occurrences (e.g., the Obernkirchen quarry track surface), rectified orthophotos of the site may be a useful data source. Alternatively or additionally, data acquired by further imaging techniques, such as laser scanning or photogrammetry, can be employed.

Landmark definition The 2-D homologous landmarks are defined herein as representing locations of anatomically homologous points along the outline of tracks. An example are the apices of the digit impressions, provided that these are not modified by extramorphological processes (e.g., dragging of the foot) as previously defined. The specific choice of landmarks should be carefully considered with respect to their correlation to anatomical features of the trackmaker. For the purpose of this study, a set of 14 landmarks has been defined according to the scheme in Figure 12.4. In order to ensure strict adherence to homology in the data set, geometric effects of mirroring between the left and right foot and preservation as epichnium and hypichnium have to be considered. As a convention, the sequence of landmarks (1, 2, 3 . . .) has been defined based on the epirelief of the right pes, beginning at the apex of digit III and continuing clockwise, and adopted subsequently for other orientations (Fig. 12.4).

Landmarks 1, 5, 11: apices of digits III, IV, and II, respectively.

Landmarks 3, 13: hypices between digits III/IV, and II/III, respectively.

Landmarks 7, 9: locations of maximum lateral/medial indentation at the posterolateral/mediolateral base of digits IV and II, respectively.

Landmark 8: posterior margin of track, opposite to the apex of digit III.

Landmarks 2, 4, 6, 10, 12, 14: location of maximum convexity of the medial and lateral margins of the free digits.

Other parameters Foot length (FL) and foot width (FW) have been measured according to Figure 12.4.

Data grouping In order to carry out varying subtests (see study layout), data have been a priori grouped either by their locality provenance or by size. For practical use, the size has been classified in size classes based on FL (Fig. 12.5). These size classes are arbitrary (5-cm intervals), defined for the purpose of comparability, and not intended to represent individual age classes, which can (potentially) only be determined after testing.

Numerical Data Analysis

Following data acquisition, the landmark coordinates were imported to the Palaeontological Statistics (PAST) free

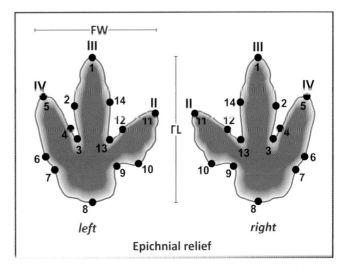

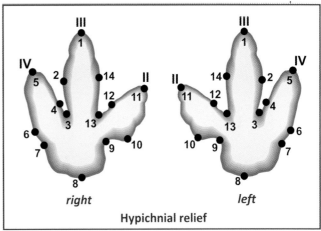

12.4. Two-dimensional landmark definition for epichnial reliefs and hypichnial casts.

software package (version 2.17, Hammer, Harper, and Ryan, 2001; Hammer and Harper, 2006; Hammer, 2012). In order to work with a size, position, and rotation normalized data set, the original data were fully standardized to 2-D Procrustes coordinates (algorithm by Rohlf and Slice, 1990) and aligned to their major axis.

The resulting multivariate data set was used as a variance-covariance matrix in subsequent PCAS (Davis, 1986; Harper, 1999). According to the purpose (see study layout), PCAS were either conducted for the entire unordered data set, or with the between-groups option of PAST for a priori defined data groups (subsets), calculating principal components (PCs) for group means (Hammer, 2012). In the latter case, a control PCA of ungrouped data was executed to compare the results. The significance of resulting PCs were evaluated by comparison to a "scree plot" of eigenvalues against a "broken stick" curve expected by a random model (Jackson, 1993). By default, the PCs with the highest significance obtained by this method were used for bivariate analysis and interpretation. The "meaning" of the PCs (the direction and degree of

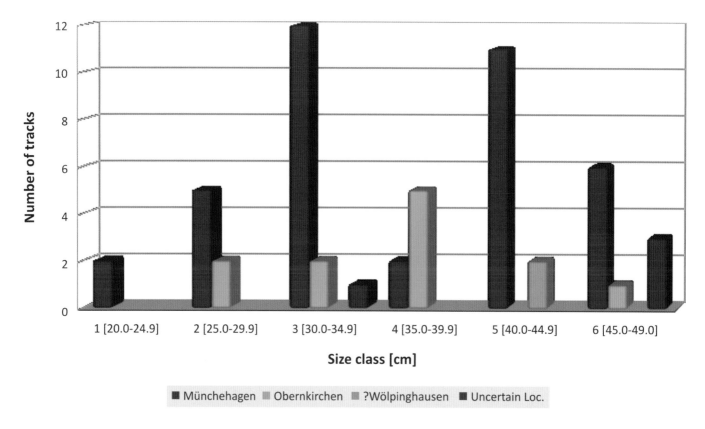

12.5. Size class distribution histogram.

morphological variation expressed by the eigenvector) are evaluated by (1) comparison of the PCA score correlation to the respective eigenvector, and (2) by simulating the principal deformation from the mean shape for each relevant PC, using the graphical shape deform (2-D) module (applying the thin plate spline deformation method) of PAST.

The clusters of PCA scores found and interpreted as morphotypes were tested with a discriminant analysis, including Hotelling's t-squared test, evaluating the probability of the same mean for both populations, with the number of variables set to 4 (Hammer, 2012).

In order to evaluate the morphological variation between tracks within distinct trackways compared to the whole data set, a PCA with an a priori grouped trackway data set versus a group comprising the remainder of the data is performed. This is reiterated for all individual trackway segments (n = 10) (Table 12.4). The range of PCA scores for the individual trackways (ΔsT) and the entire data set (ΔsN, equaling eigenvalue of the chosen PC) are calculated by subtracting the individual minimum scores from the individual maximum scores of both groups ($sT_{min} - sT_{max}$ and $sN_{min} - sN_{max}$, respectively). The quotient ΔsT/ΔsN produces the relative range of variance within the distinct trackway compared to the entire data set, which can be expressed as percentage value for convenience. It is proposed that the variance within an individual trackway

should be distinctly less than 50% of the whole range of variance in the data set for the method to be considered useful and we apply herein a threshold of 40% (ΔsT/ΔsN < 0.4).

To evaluate the potential correlation between individual size and the direction of the morphological variation, it is tested whether there is a linear relationship between size and morphological similarity. If there exists an allometric relationship between size (individual age) and autopodial morphology and proportions, it may be expected that the maximum and minimum scores of the most significant PCs each correlate with the largest and smallest size classes (or vice versa) and the intermediate size classes correlate to the numerical sequence of the intermediate PCA scores. It may be expected that the correlation is increasingly blurred by variation due to preservational and other effects with decreasing variance relative to the PC. Allometry cannot be detected by this method if the resulting plots show a chaotic arrangement of size classes across the scatter of scores.

Study Layout

In order to evaluate and interpret the various aspects of potential variation in the data, the data set and various subsets (groups) have been subjugated to a road map of analyses that are either independent from each other or may be dependent

Table 12.4. Level A MHL analysis

Trackway	Tracks	PC	PC_{var} [%]	sN_{max}	sN_{min}	ΔsN	sT_{max}	sT_{min}	ΔsT	$\Delta sT / \Delta sN$
OBK–I	3	1	95.092	0.21166	−0.19713	0.40879	0.21166	0.18714	0.02452	0.060
OBK–M	2	1	95.241	0.23591	−0.13925	0.37516	0.23591	0.17615	0.05976	0.159
MHG–A	4	1	87.470	0.14416	−0.17700	0.30921	0.15788	0.07875	0.05613	0.200
MHG–B	2	1	86.062	0.11743	−0.14542	0.26285	−0.12651	−0.14542	0.01891	0.072
MHG–D	2	1	87.415	0.16587	−0.18801	0.35388	0.16587	0.13916	0.02671	0.075
MHG–E	2	1	90.655	0.16047	−0.15219	0.31266	0.15728	0.15017	0.00711	0.023
MHG–F	3	1	85.351	0.14069	−0.14255	0.28324	0.14069	0.07060	0.07009	0.247
MHG–I	2	1	85.923	0.13284	−0.22529	0.35813	0.06231	0.04379	0.01852	0.052
MHG–K	2	1	87.651	0.15804	−0.24570	0.40374	−0.11557	−0.24570	0.13013	0.322
MHG–L	2	1	100.000	0.08822	−0.11447	0.20269	−0.09092	−0.11447	0.02355	0.116
									Median	0.096

Notes: Variation data for track segments versus the entire data set. PC, principal component used; PC_{var}, amount of variation represented by the principal component used; sN_{max}, maximum PCA score for entire data set; sN_{min}, minimum PCA score of entire data set; ΔsN, range of PCA scores for entire data set; sT_{max}, maximum PCA score for trackway; sT_{min}, minimum PCA score for trackway; ΔsT, range of PCA scores for trackway; other abbreviations as in Table 12.1.

in a logical cascade (Fig. 12.6). Key questions addressed include:

- Is there sufficient difference between the variation exhibited by tracks left by a single individual versus the variation among different individuals?
- How many morphometrically/morphologically definable types of tracks are present in the data set?
- How important is the morphometric variation within and between each of these types?
- How should this morphological variation be interpreted in terms of orthotaxonomy, ichnotaxonomy, sexual dimorphism, and ontogeny?

In order to solve these questions, we formulate a set of antagonistic hypotheses, inferences that result from these hypotheses, and anticipations of the results to be expected to either confirm or falsify these hypotheses. This set is arranged in a four-level sequence (A to D) of testing:

Level A: Significance of morphological variation between tracks of individuals (general test of method applicability) – Hypothesis A0: The variation observed in the entire data set is not more distinct than the variation found between tracks left by a single, biological individual. Logical prerequisites: None. All trackway segments must be tested individually against the entire data set.

Inferences: Theoretically, individual tracks from a single trackway, created by a single individual, should show no variation when compared to the overall data set. However, due to a broad range of reasons, including substrate conditions, walking speed, movement, preservation and other factors, no two tracks are identical, even if they were produced during subsequent steps. The applicability of the MHL analysis to the

study of variation between individual trackmakers depends upon the ability to differentiate between these obvious inaccuracies and a true anatomical signal. If hypothesis A0 is found probable, the method is unsuitable to evaluate the variation between groups of individual trackmakers, because the statistical signal obtained is not stronger than that from individual measurement errors and nonanatomy-related morphology of the tracks.

Expected PCA results (after a priori grouping for individual trackway segments and the remaining data): The spread of PCA scores across footprints from one or more trackway segments equals 60%–100% of the maximum spread (eigenvalue) along the most significant PC, encompassing the entire variation of the data set.

Hypothesis A1: The variation observed in the entire data set is more distinct than the variation found between single tracks left by a single, biological individual. Logical prerequisites: None. All trackways must be tested individually against the entire data set.

Inferences: The data set shows a wider range of variation than that associated to nonanatomy-related spread of PCA scores. The feasibility of MHL to evaluate the variation between groups of individual trackmakers is confirmed.

Expected PCA results (after a priori grouping for individual trackway segments and the remaining data): The spread of PCA scores across tracks from each trackway segments is distinctly smaller than the maximum spread of the entire data set along the most significant PC. Hypothesis A1 is considered confirmed if the spread within each of the individual trackway segments equals or is less than 40% of the eigenvalue of the entire data set.

Level B: Identification of number of track morphotypes in the data set – Hypothesis B0: The entire data set represents

Level A hypotheses:

- *A0: The variation observed in the whole dataset is not more distinct than the variation found between footprints left by a single, biological individual.*

- *A1: The variation observed in the whole dataset is more distinct than the variation found between single footprints left by a single, biological individual.*

Level B hypotheses:

- *B0: The whole dataset represents a single track morphotype.*

- *B1a: The whole dataset represents more than one track morphotype. The distribution of these morphotypes is independent from their distribution across different track localities.*

- *B1b: The whole dataset represents more than one track morphotype. The distribution of these morphotypes is dependent from their distribution across different track localities.*

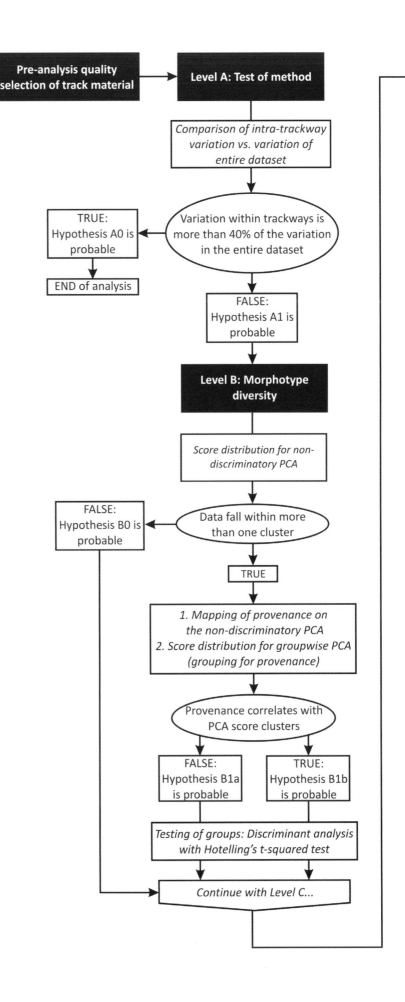

12.6. Flow-diagram of study layout. Abbreviation: PCA, principal components analysis.

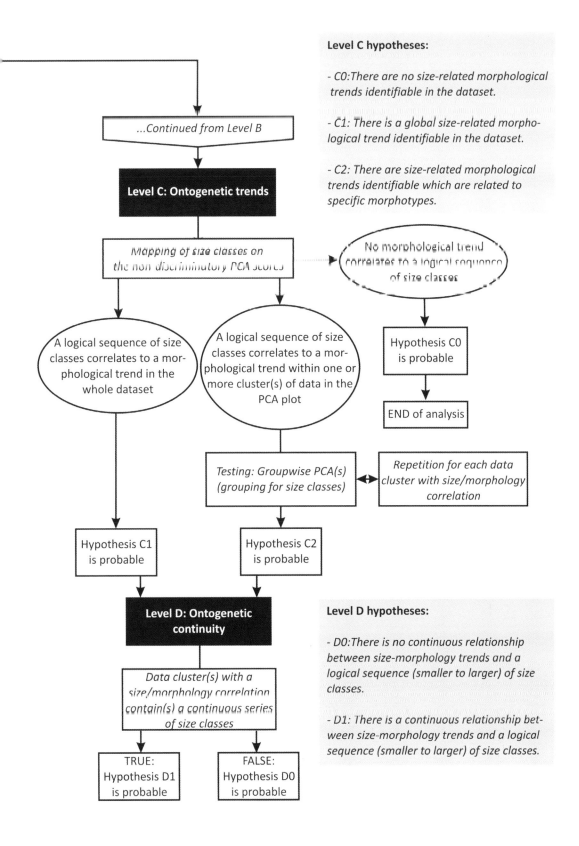

a single track morphotype. Logical prerequisites: Depends on confirmation of hypothesis A1.

Inferences: A single track morphotype is present in the data set. Observed variation resulted from preservational, individual, or ontogenetic variation. Potential morphological differences of trackmakers (if present) are not expressed by the track record.

Expected PCA results: Hypothesis B0 is considered probable if scores from the entire, nondiscriminatory data set do not group into more than one distinct morphospace along the significant PCs.

Hypothesis B1a: The entire data set represents more than one track morphotype. The distribution of these morphotypes is independent from their distribution across different track localities. Logical prerequisites: Depends on confirmation of hypothesis A1.

Inferences: Two or more different morphotypes of trackmakers were present sympatrically in at least some of the track localities. These could be interpreted to represent two or more morphotypes of trackmakers, which in turn can be interpreted as either proxies for the presence of several taxa of trackmakers, or (in the case of exactly two morphotypes) as indicators of sexual dimorphism.

Expected PCA results: Hypothesis B1a is considered to be confirmed if scores of the entire, nondiscriminatory data set group into two or more distinct clusters along the significant PCs, and no relationships are found by mapping of provenance on a whole data set PCA.

Hypothesis B1b: The entire data set represents more than one track morphotype. The distribution of these morphotypes is dependent on their distribution across different track localities. Logical prerequisites: Depends upon confirmation of hypothesis A1.

Inferences: Two or more track morphotypes occur not or only partially sympatrically in different localities. This can be interpreted as the occurrence of two or more different morphs/taxa of trackmakers, which occurred spatially or spatially and temporally separated from each other. In the case of exactly two different morphotypes, a spatial separation of dimorphic genders is theoretically possible.

Expected PCA results: Hypothesis B1b is considered to be confirmed if scores of the entire, nondiscriminatory data set group into two or more distinct clusters along the significant PCs and relationships between these clusters are found by mapping of provenance on entire data set PCAs, and the distribution of these clusters is repeated by a PCA with a priori grouping of data according to their locality provenance.

Level C: Identification of size-related morphological trends (allometry) within the data set – Hypothesis C0:

There are no size-related morphological trends identifiable in the data set. Logical prerequisites: Depends upon confirmation of hypothesis A1.

Inferences: Allometric effects do not affect anatomical features of the trackmakers that were involved in trackmaking. The degree of variation observed does not correlate with body size and must be explained by alternative means.

Expected PCA results: Mapping of size classes on an entire data set PCA, as well as PCAs with data grouped a priori for size classes, fail to find any relationship of a logical sequence of size classes to score distributions along the significant PCs.

Hypothesis C1: There is a global size-related morphological trend identifiable in the data set. Logical prerequisites: Depends upon confirmation of hypothesis A1.

Inferences: Allometric effects equally affect anatomical features of all trackmaker morphotypes involved. The degree of variation observed does at least partially correlate to body size and can be interpreted as a proxy of ontogenetic variation. The proxy is independent from morphotype discrimination.

Expected PCA results: Mapping of size classes on an entire data set PCA, as well as PCAs with data grouped a priori for size, shows a relationship of a logical sequence of size classes to score distributions along the significant PCs.

Hypothesis C2: There are size-related morphological trends identifiable that are related to specific morphotypes. Logical prerequisites: Depends upon confirmation of hypotheses A1 and B1a or A1 and B1b. The test must be carried out for all morphotypes.

Inferences: Allometric effects affect anatomical features of some trackmaker morphotypes involved, possibly in various degrees and variants. The degree of variation observed does at least partially correlate to body size of the trackmakers and can be interpreted as a proxy of ontogenetic variation. Applicability of the proxy depends on track morphotype.

Expected PCA results: Mapping of size classes on morphotype subset PCA, as well as subset PCAs with data grouped a priori for size, find a relationship of a logical sequence of size classes to score distributions along the significant PCs.

Level D: Identification of continuous or discontinuous size-morphology trends – Hypothesis D0: There is no continuous relationship between size-morphology trends and a logical sequence (smaller to larger) of size classes. Logical prerequisites: Depends upon confirmation of hypotheses A1 and C1 or C2. The test must be carried out for any recognized morphotype and ontogenetic trend.

Inferences: The continuous presence of various ontogenetic stages of a trackmaker in the data set is not evident. The data may be interpreted as showing an episodic presence or recruitments of distinct ontogenetic stages (age classes) of the same trackmaker taxon, or as a hint toward a presence of one or several "cryptic" additional morph(s) of trackmakers, which were not identified at Level B.

Expected PCA results. A mapping of size-class distribution on significant PCs does not reveal a continuous, logical sequence.

Hypothesis D1: There is a continuous relationship between size-morphology trends and a logical sequence (smaller to larger) of size classes. Logical prerequisites: Depends upon confirmation of hypotheses A0 and C1. The test must be carried out for all recognized morphotypes.

Inferences: The continuous presence of various ontogenetic stages of a trackmaker in the data set is evident. The data may be interpreted as showing a continuous representation of a trackmaker over its ontogeny in the data set. This may have happened either in the form of ontogenetically mixed social groups, or by time-averaging of the recorded passage of single individuals.

Expected PCA results: A mapping of size class distribution on significant PCs does reveal a continuous, logical sequence.

RESULTS

Level A: Applicability of Method

The proportion of variance of individual trackway segments compared to the entire data set (Table 12.4) ranges from 1.9% to 32.2%, with a median at 9.6% and an arithmetic mean of 13.2%. The high value of 32.2% (Münchehagen trackway MHG-K) may be an exception due to data quality as the two footprint specimens (MHG-K-1 and MHG-K-2) are widely separated in all PCA results. The second highest value scores at 24.7% (MHG-F).

As a result, the variance within the entire data set is distinctly higher than that from tracks of a single individual and the method is considered informative with respect to variation between individuals (hypothesis A1 confirmed).

Level B: Number of Morphotypes in the Data Set

The bivariate plot of the most significant PC1 and PC2 for the entire, nondiscriminatory data set shows no obvious clustering of scores. However, mapping the provenance of the tracks on the same plot reveals a bimodal eigenvalue distribution with a concentration of scores for Münchehagen specimens in the upper left quadrant and for Obernkirchen specimens in the lower right quadrant (Fig. 12.7A). There is a considerable zone of overlap between both populations.

This picture is sharpened when the PCA is carried out with the between-groups option for an a priori grouping by provenance. Fitting to the broken stick curve finds PC2 and PC3 most significant, and a bivariate scatter plot of these shows a similar result to the nondiscriminatory PCA (Fig. 12.7B). However, the groups are slightly more separated, though an area of overlap remains. In the group-wise PCA, most tracks from unknown localities group with the Münchehagen tracks, whereas those from Wölpinghausen fall closer to the Obernkirchen population, though still in the zone of overlap.

Interestingly, OBK-N-1, from the theropod-dominated Chicken Yard track level (Richter, Böhme, and van der Lubbe, 2009; Richter et al., 2012; Richter and Böhme, 2016) at Obernkirchen, falls outside the cloud of other Obernkirchen specimens, which were obtained either from the stratigraphically younger "Upper Surface" track level (Richter et al., 2012) or from unknown horizons at this location. Furthermore, it groups more closely with the specimens from Münchehagen (Figs. 12.7A, 12.7B).

A discriminant analysis with Hotelling's t-squared test (p-value threshold p = 0.05 for null-hypothesis: both populations have the same mean, Fig. 12.8) finds the populations from Obernkirchen and Münchehagen separated when OBK-N-1 is grouped before the analysis with the Münchehagen tracks (p = 0.09334) or omitted from the test (p = 0.07912), whereas distinction of both populations is blurred by inclusion with the Obernkirchen tracks (p = 0.04596). When OBK-N-1 is grouped a priori with the Münchehagen tracks, this position is confirmed by the discriminant analysis (score: 1.5181), when it is not included in the predefined groups, the discriminant function results with classification in the Obernkirchen group (score: −2.89168). Together with good correspondence of OBK-N-1 with the Münchehagen morphotype in the PCAs, these results may be interpreted tentatively as the occurrence of the Münchehagen morphotype trackmaker in the lower stratigraphic level at Obernkirchen. The shift in classification when classified retrospectively may be the effect of overall relatively small sample, overlap of morphospaces, and noise. Probably due to the same reason, one of the four tracks in trackway MHG-A (MHG-A-1) and an isolated track (MHG-X-9) group with the Obernkirchen morphotype. These aberrations are independent from the grouping of OBK-N-1. The otherwise good distinction of both morphotypes in the discriminant analysis contrasting with their broad overlap in PCAs result from unimodal abundance distributions with well-separated maxima but overlapping spreads. These spreads may be the result of individual variation, allometry, the presence of unrecognized, "cryptic" morphotypes (blending with the morphological spread of the recognized morphotypes), preservation, and other data noise.

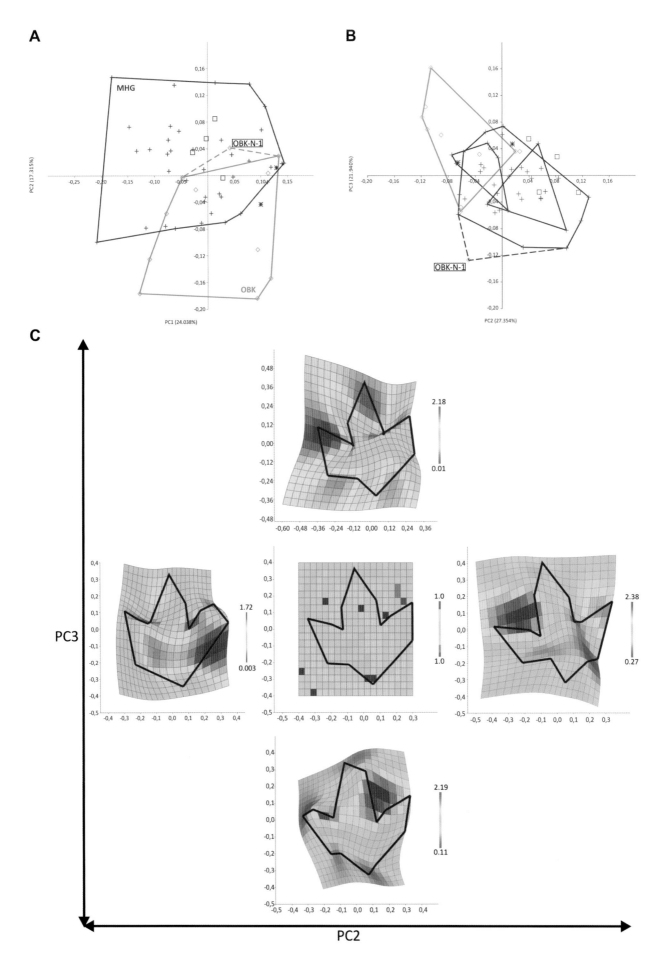

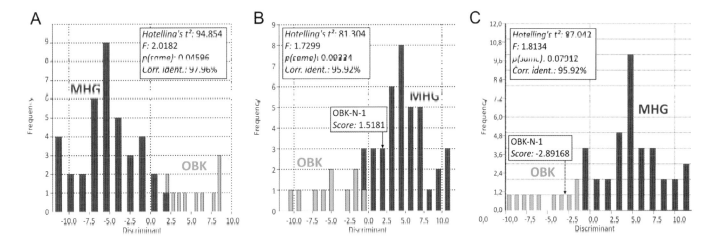

12.8. Level B, discriminant analysis with Hotelling's t-squared test for tracks from Obernkirchen ([OBK], green, morphotype O) and Münchehagen ([MHG], red, morphotype M) (A) OBK-N-1 grouped with morphotype O (B) OBK-N-1 grouped with morphotype M. (C) OBK-N-1 omitted. The discriminant score calculated for OBK-N-1 is indicated in B and C.

Simulating principal deformation from the mean shape for PC2 from the group-wise PCA (Fig. 12.7C) reveals that this component embraces proportional lengthening and narrowing of the digits compared to FL and a relative transformation from a uniformly isosceles-triangular toward a more cylindrical (digit III) or spindle-shaped (digits II and IV) outline of digits along its eigenvector. PC3 mostly describes a proportional shortening of free digit III in relation to digits II, IV, and the FL along its eigenvector (Fig. 12.7C). Both PCs include increasing digital divarication as a component of shape transformation.

As a result, the group formed by tracks from Obernkirchen (morphotype O) is characterized by a proportionally elongate and slender digit III in extreme examples; toward the data set average, this distinction vanishes. The group formed by tracks from Münchehagen (morphotype M) is characterized by a proportionally short digit III in extreme examples, toward the data set average, this distinction vanishes. The opposing morphological trends and the broad zone of overlap between both morphotypes would itself suggest that the data represent a morphological continuum. However, the extremes of both groups are confined to their respective localities. They are independent from size distribution, and show different ontogenetic signals and are therefore considered different morphotypes (hypothesis B1b confirmed). At present, though, there is a relatively broad overlap between both morphotypes, and specimens falling in this overlap zone may not be properly determinable.

Level C: Ontogenetic Trends

A mapping of size classes on an entire data set, nondiscriminatory PCA (PC1/PC2) bivariate plot, as well as a PCA carried out on a priori–defined size class groups do not yield a global logical distribution of size classes (hypothesis C1 falsified) (Fig. 12.9A).

A mapping of size classes on a morphotype M data subset PCA (PC1/PC2) bivariate plot, as well as a morphotype M data subset PCA after a priori size class grouping do not reveal any logical distribution of size classes (hypothesis C2 falsified for morphotype M) (Fig. 12.9B, 12.9C).

A mapping of size classes on a morphotype O data subset PCA finds a logical, sequential size class distribution across the PCA scores, which is confirmed after a priori size class grouping (Fig. 12.9D). The group-wise PCA results in seven PCs, of which PC2 and PC3 are significant and account for 29.364% and 17.565% of variance, respectively. However, a logical sequence of size classes correlates best with PC2, though there is a broad overlap in the intermediate sizes (Fig. 12.10). This is interpreted as documenting an allometric trend toward proportionally longer, narrower, and more spindle-shaped digits during ontogeny (hypothesis C2 confirmed for morphotype O), though caution must be given to the small sample of morphotype O. When compared to morphotype M, the largest specimens approach the zone of overlap of both morphotypes, indicating that the most distinctive morphology of this morphotype is

12.7. Level B, multivariate, two-dimensional homologous landmark (MHL) analysis. (A) Principal components analysis (PCA), bivariate scatter plot of PC1/PC2 of entire data set with the resulting groups of provenance indicated (the groups were not predefined for data analysis). Dashed green outline: expansion of Obernkirchen (OBK) group to include OBK-N-1. (B) PCA, bivariate scatter plot of PC2/PC3. PCA with between-groups option for entire data set after a priori grouping for provenance. Dashed red line: expansion of Münchehagen (MHG) group to include OBK-N-1. (C) Principal deformation simulation on the mean shape of entire data set for PC2 and PC3 from the analysis in B. Green diamonds: Obernkirchen specimens; red crosses: Münchehagen specimens; navy blue asterisks: Wölpinghausen specimens; blue squares: other specimens from uncertain locality.

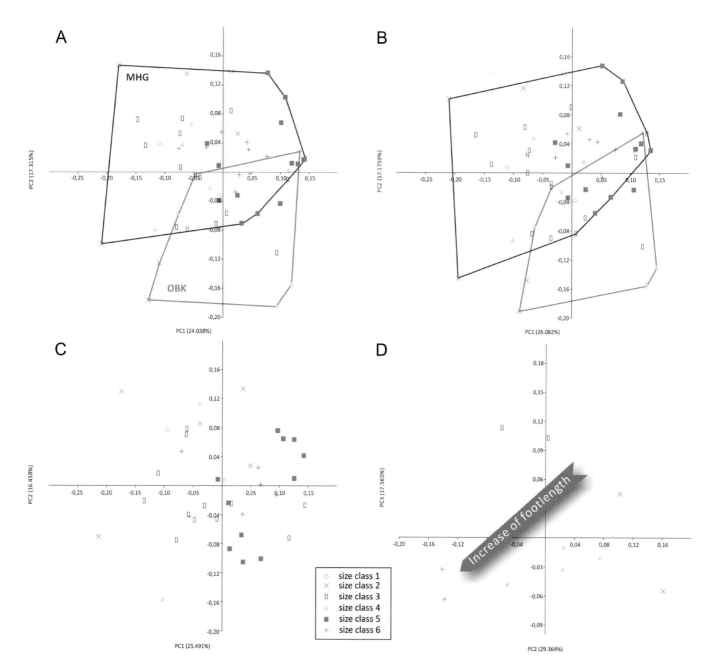

12.9. Level C and D, MHL analyses. (A) PCA, bivariate scatter plot for PC1/PC2 of entire data set with the distribution of size classes mapped by symbols (the groups were not predefined for data analysis). (B) PCA, bivariate scatter plot for PC1/PC2. PCA with between-groups option for entire data set after a priori grouping for class sizes. (C) PCA, bivariate scatter plot for PC1/PC2, calculated with between-groups option for morphotype M tracks, after a priori grouping for class sizes. (D) PCA, bivariate scatter plot for PC2/PC3, calculated with between-groups option for morphotype O tracks, after a priori grouping for class sizes (OBK-N-1 omitted). Abbreviations as in Figure 12.7.

exhibited by specimens from earlier ontogenetic stages (Fig. 12.11).

Level D: Ontogenetic Continuity

Size classes 1 and 5 are missing from the sample of morphotype O. However, it has to be kept in mind, that the subsample, upon which this test applies, is relatively small (n = 10) and the absence of these size classes do not permit a reasonable conclusion on ontogenetic continuity in the data set.

DISCUSSION

The multivariate homologous 2-D-landmark eigenanalyses of the data set reveal quantitatively defined morphological groups that surprisingly do not reflect intuitive qualitatively assessed groupings. From a qualitative point of view, some track specimens seem to be quite different from others, raising the question whether they were made by different taxa of trackmakers or by different ontogenetic stages of the same trackmaker taxon (e.g., Hornung and Reich, 2012). This also

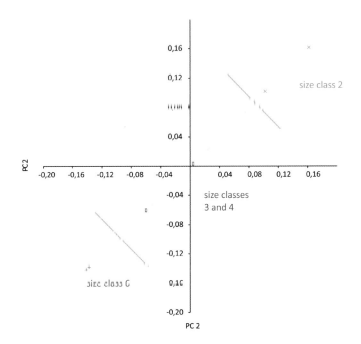

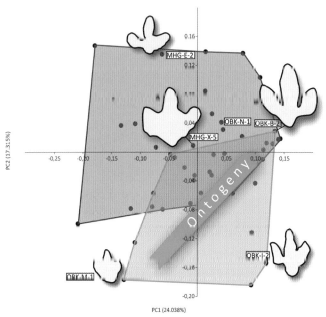

12.10. Level C and D, MHL analysis. Scatter plot for PC2/PC2 score distribution, calculated with between-groups option for morphotype O tracks, after a priori grouping for size classes (OBK-N-1 omitted). Note the linear arrangement of size class distribution with the extremes (classes 2 and 6) on opposite poles of variance. Size class symbols as in Figure 12.9. Abbreviations as in Figure 12.7.

12.11. Interpretation of morphotype grouping, ontogenetic trend, and outlines of example specimens representing morphological extremes for each quadrant and near-average morphologies of tracks for the entire data set PCA (compare with Fig. 12.7A). All tracks mirrored to represent right-side impressions. Abbreviations as in Figure 12.7.

holds true for specimens of preservational classes I and II, in which the autopodial morphology is well represented and substrate differences or other nonanatomy-related modifications do not play an important role. However, the numerical analysis shows that a broad range of morphologies form two wide morphometric clusters, which further overlap in a considerable area. According to the facts that specimens forming the two clusters are found separated at two different localities and that one of the clusters shows an ontogenetic, allometric signal while the other does not, they are considered to represent two distinct morphotypes (M and O, Fig. 12.11) and probably two different orthotaxa of trackmakers. Both morphotypes encompass short toed and broadly triangular-toed morphologies as used to define the ichnogenus *Iguanodontipus* (Sarjeant, Delair, and Lockley, 1998), as well as the far more abundant long- and slender-toed type, which was most commonly described from Germany (e.g., Struckmann, 1880; Ballerstedt, 1905, 1914; Dietrich, 1927; Lehmann, 1978; Diedrich, 2004; Lockley, Wright, and Thies, 2004), England (e.g., Beckles, 1852; Woodhams and Hines, 1989; Parkes, 1993; Radley, Barker, and Harding, 1998; Pollard and Radley, 2011), Spain (e.g., Pascual-Arribas et al., 2009; Cobos and Gascó, 2012; Castanera et al., 2013), and Portugal (Santos, Callapaz, and Rodrigues, 2012). The most important morphological differences are found here to be related to ontogeny and allometry, at least for part of the tracks (morphotype O) (Fig. 12.11). Tracks exhibiting a morphology similar to the

ichnoholotype of *Iguanodontipus burreyi* Sarjeant, Delair, and Lockley, 1998, are located near one end of an ontogenetic continuum between both morphologies, representing young individuals (Figs. 12.10 and 12.11). Interestingly, the morphometrical analysis of 43 iguanodontian tracks from the Cameros Basin (Spain) by Moratalla and Marugán-Lobón (2009) also resulted in a potential allometric trend in the sample, but, contrary to morphotype O, the tracks became relatively broader with increasing size. Even in the morphotype M, with no recognizable ontogenetic pattern in the data set, this morphology is scattered unordered (when considering size) among the morphological cluster and does not form a separate group. This may be a hint that further "cryptic" morphotypes may be hidden in the morphotype M cluster. This, however, cannot be resolved with the applied method. The continuity of the presence of morphotype O trackmakers of all ontogenetic stages (above a minimum size not yet recorded by tracks less than 23 cm FL) cannot be evaluated, considering the small data sample. The close association of tracks from several ontogenetic stages (GPMH specimen from Geologisch-Paläontologisches Museum der Universität Hamburg) on the same track horizon with the same general direction of locomotion (Lehmann, 1978) may suggest that a continuum of several ontogenetic stages formed social groups, though this aspect must be studied further.

From analyzing FL, FW, and the ratio between both of a much larger track data set (n = 598), Matsukawa, Lockley, and

Hunt (1999) identified three age classes (juvenile, subadult, and adult) of iguanodontian ornithopods in the "mid"-Cretaceous Dakota Group of the western United States. Mapping these age classes across their occurrences showed that all three co-occurred in most localities with large data samples. However, the presence of distinctive age classes in the sample suggest that the tracksites were not frequented continuously by individuals of all ontogenetic stages but probably represent a periodic (yearly?) passing of the population(s). It should be noted that Matsukawa, Lockley, and Hunt (1999) treated the entire track population as having been created by a single orthotaxon (or at least as representing the same ontogenetic pattern of several closely related taxa), an assumption that has to be considered with caution regarding the osteological evidence for relatively high taxonomic diversity of sympatric iguanodontian communities (e.g., Horner, Weishampel, and Foster, 2004; Norman, 2011).

From an ichnotaxonomic point of view, the results bring up a number of critical aspects. On the one hand, it seems that the data show that (potential) distinctions of vertebrate ichnotaxa based upon qualitative morphological criteria may easily underestimate the continuous morphological bandwidth of such a morphological group that is revealed when it is evaluated by quantitative analysis. This is especially true for aspects of ontogenetic variation, which is nearly impossible to determine based upon qualitative analysis of specimens, with exception in the case in which it may be postulated based upon independent evidence, for example, for gregarious behavior. On the other hand, the numerical analysis may expose a higher diversity of trackmakers than is obvious from qualitative analysis of the track record. Given that the data quality and quantity is in many cases not sufficient for a numerical analysis—a fate shared with orthotaxonomy—a qualitative approach to ichnotaxonomy is inevitable. However, utmost caution is due if ichnotaxonomic decisions are based solely or mostly upon morphometric criteria, such as length ratios, toe divarication angles, and so on. Ichnotaxa should be based primarily on qualitative characters that can be described as present or absent, or being of a qualitatively definable shape, rather than by size relations. An example are the ornithopod tracks from Obernkirchen mentioned by Böhme et al. (2009) and Richter et al. (2012) as ?*Caririchnium* isp., which clearly differ from the material used in this study by characters such as the presence of additional pad-like structures in the metatarsal region, which are regarded as primary anatomical features of its producer. Optimally such characters are identifiable on a single autopodial impression, with trackway characters adding a secondary level of characters if available (and not related to trackmaker behavior rather than anatomy). It can be argued that, as a result of the analysis presented herein, the ichnogenus *Iguanodontipus* is

suitable to contain all of the studied material (and therefore the bulk of large ornithopod tracks from the early Cretaceous of Lower Saxony). This view is qualitatively supported by the latest review of iguanodontian ornithopod ichnotaxonomy by Díaz-Martínez et al. (2015). These authors strongly reduced the vast amount of proposed ichnotaxa and modified the diagnosis of *Iguanodontipus* as previously given by Sarjeant, Delair, and Lockley (1998) and Meyer and Thüring (2003) to include a wider range of morphologies accordingly (see also Castanera et al., 2013). However, a thorough evaluation of this matter is beyond the scope of this work.

The proposed method to evaluate questions of trackmaker diversity, paleobiology, and ichnotaxonomic concepts and their meaning seems to be promising. It can, however, only be a first attempt and further work should concentrate on expanding the database and including a wider range of morphotypes to elucidate their relationships. It should also be a line of further investigation to improve the chosen system of landmarks for better results and the handling of missing and suboptimal data.

CONCLUSIONS

Multivariate homologous 2-D-landmark analyses are carried out to evaluate their usefulness and results for a set of iguanodontian ornithopod pes tracks from the Berriasian of northwestern Germany. Comparison of multitrack trackway segments with the entire data set shows that the range of variance found in the former only accounts for a maximum of 32.2% of the entire variation. The latter is therefore interpreted to represent variation between individuals. A nondiscriminatory PCA of the entire data set found two populations of scores, which form two overlapping clusters, interpreted to represent two morphotypes (O and M). The distribution of morphotypes correlates with the provenance of the tracks, further supporting the existence of two different populations of trackmakers. The two morphotypes were also found in a discriminatory PCA after an a priori grouping for provenances, as well as in a discriminant analysis with Hotelling's t-squared test. Morphotype O is less abundant than morphotype M and occurs with certainty only at the Upper Surface track level in Obernkirchen. Morphotype M mainly occurs at Münchehagen, a single occurrence may be at the lower Chicken Yard track level at Obernkirchen. Other material from uncertain provenance falls more or less in the zone of morphological overlap between both morphotypes and is ambiguous in its determination.

Mapping size classes on the nondiscriminatory PCA, as well as a discriminatory PCA after a priori grouping for size classes show that morphological extremes are related to smaller size, whereas larger specimens occupy the zone of overlap between the morphotypes. In the case of morphotype O, a clear morphological trend correlates to an increase of size, suggesting an ontogenetic signal and allometry in the trackmaker. Ontogenetic continuity cannot be assessed for morphotype O because of the small sample size. For morphotype M, size does not follow a recognizable trend in morphology, and there was no indication for an ontogenetic signal.

The data suggest that qualitatively quite different ornithopod tracks form a morphological continuum, in which variation is partly explainable by ontogeny. On the other hand, the data set may contain more than one morphotype that cannot be defined by qualitative characters. This raises the warning in base-scale ichnotaxonomic separation of morphotypes on qualitative characters rather than on morphometric differences of otherwise very similar morphologies. The method presented here is promising to evaluate variation in track data sets and to interpret it in terms of ichnotaxonomy, orthotaxonomy, and ontogenesis of the trackmaker. This is an interesting perspective regarding the relative abundance of tracks when compared to skeletal material. However, the method requires further refinement and confirmation by adding larger data sets and studying alternative landmark sets and analysis methods.

ACKNOWLEDGMENTS

We thank D. Castanera, J. Moratalla, and the editor P. Falkingham for thoughtful reviews of the manuscript. U. Kotthoff, Geologisch-Paläontologisches Museum der Universität Hamburg (GPMH), and A. Twachtmann-Schlichter, Museum Bückeburg (Museum für Stadtgeschichte und Schaumburg-Lippische Landesgeschichte, Bückeburg), kindly granted access to specimens under their care, and T. Sachs and K. Sachs provided additional information on the GPMH specimens. We additionally thank A. Richter, "E." Widmann, Niedersächsisches Landesmuseum Hannover (NLMH), and S. Guichard (student apprentice at NLMH) for support with the track material at the NLMH, as well as V. J. Roden for a linguistic review of the manuscript.

APPENDIX

Supplemental files can be found at http://dx.doi.org/10.6084/m9.figshare.1348327.

REFERENCES

Abel, O. 1935. Vorzeitliche Lebensspuren. G. Fischer, Jena, Germany, 644 pp.

Avanzini, M., L. Piñuela, and J. C. García-Ramos. 2008. Theropod palaeopathology inferred from a Late Jurassic trackway, Asturias (N. Spain). Oryctos 8: 71–75.

Ballerstedt, M. 1905. Über Saurierfährten der Wealdenformation Bückeburgs. Naturwissenschaftliche Wochenschrift (Neue Folge) 4(31): 481–485.

Ballerstedt, M. 1914. Bemerkungen zu den älteren Berichten über Saurierfährten im Wealdensandstein und Behandlung einer neuen, aus 5 Fußabdrücken bestehenden Spur. Centralblatt für Mineralogie, Geologie und Paläontologie [1914](2): 48–64.

Barco, J. L., J. I. Canudo, and J. I. Ruiz-Omeñaca 2006. New data on *Therangospodus oncalensis* from the Berriasian Fuentesalvo tracksite (Villar del Río, Soria, Spain): an example of gregarious behavior in theropod dinosaurs. Ichnos 13: 237–248.

Barnes, F. A., and M. G. Lockley. 1994. Trackway evidence for social sauropods from the Morrison Formation, eastern Utah (USA). Gaia 10: 37–42.

Beckles, S. H. 1852. On the *Ornithoidichnites* of the Wealden. Quarterly Journal of the Geological Society 8: 396–397.

Belvedere, M. 2008. Ichnological researches on the Upper Jurassic dinosaur tracks in the Iouaridène area (Demnat, central High-Atlas, Morocco). Ph. D. dissertation, Università degli Studi di Padova, Padua, Italy, 121 pp.

Belvedere, M. and J. O. Farlow. 2016. A numerical scale for quantifying the quality of preservation of vertebrate tracks; chap. 6 in P. L. Falkingham, D. Marty, and A. Richter (eds.), Dinosaur Tracks: The Next Steps. Indiana University Press, Bloomington, Indiana.

Berner, U. 2011. The German Wealden, an unconventional hydrocarbon play? Erdöl, Erdgas, Kohle 127: 303–306.

Bertling, M., S. J. Braddy, R. G. Bromley, G. R. Demathieu, J. Genise, R. Mikulás, J. K. Nielsen, K. S. S. Nielsen, A. K. Rindsberg, M. Schlirf, and A. Uchman. 2006. Names for trace fossils: a uniform approach. Lethaia 39: 265–286.

Böhme, A., U. Stratmann, M. Wiggenhagen, T. van der Lubbe, and A. Richter. 2009. New tracks on the rock: parallel trackways of a new type of *Iguanodontipus-Caririchnium*-like morphology from the Lower Cretaceous sandstones of Obernkirchen, northern Germany. Journal of Vertebrate Paleontology 29(3): 66A.

Bose, S., and H. S. Chafetz. 2009. Topographic control on distribution of modern microbially induced sedimentary structures (MISS): a case study from Texas coast. Sedimentary Geology 213: 136–149.

Breithaupt, B. H., and N. A. Matthews. 2012. Neoichnology and photogrammetric ichnology to interpret theropod community dynamics; p. 17 in A. Richter and M. Reich (eds.), Dinosaur Tracks 2011. An International Symposium, Obernkirchen, April 14–17, 2011. Abstract Volume and Field Guide to Excursions. Universitäts-Verlag, Göttingen, Germany.

Brett-Surman, M. K., and J. R. Wagner. 2007. Discussion of character analysis of the appendicular anatomy in Campanian and Maastrichtian North American hadrosaurids–variation and ontogeny; pp. 135–170 in K. Carpenter (ed.), Horns and Beaks. Ceratopsian and Ornithopod Dinosaurs. Indiana University Press, Bloomington, Indiana.

Bromley, R. G. 1990. Trace Fossils: Biology and Taphonomy. Unwin Hyman, London, U.K. 280 pp.

Buckley, L. G., R. T. McCrea, and M. G. Lockley. 2016. Analysing and resolving Cretaceous avian ichnotaxonomy using multivariate statistical analyses: approaches and results; chap. 15 in P. L. Falkingham, D. Marty, and A. Richter (eds.), Dinosaur Tracks. The Next Steps. Indiana University Press, Bloomington, Indiana.

Carpenter, K. 1992. Behavior of hadrosaurs as interpreted from footprints in the 'Mesaverde' Group (Campanian) of Colorado, Utah, and Wyoming. Rocky Mountain Geology 29: 81–96.

Carrano, M. T., and J. A. Wilson. 2001. Taxon distributions and the tetrapod track record. Paleobiology 27: 563–581.

Castanera, D., J. Colmenar, V. Sauqué, and J. I. Canudo. 2015. Geometric morphometric analysis applied to theropod tracks from the Lower Cretaceous (Berriasian) of Spain. Palaeontology 58(1): 183–200.

Castanera, D., C. Pascual, N. L. Razzolini, B. Vila, J. L. Barco, and J. I. Sanz. 2013. Discriminating between medium-sized tridactyl trackmakers: tracking Ornithopod tracks in the base of the Cretaceous (Berriasian, Spain). PLoS One 8(11): e81830. doi: 10.1371/journal.pone.0081830.

Castanera, D., J. L. Barco, I. Díaz-Martínez, J. Herrero Gascón, F. Pérez-Lorente, and J. I. Canudo. 2011. New evidence of a herd of titanosauriform sauropods from the lower Berriasian of the Iberian range (Spain). Palaeogeography, Palaeoclimatology, Palaeoecology 310: 227–237.

Cobos, A., and F. Gascó. 2012. Presencia del icnogénero *Iguanodontipus* en el Cretácico Inferior de la provincia de Teruel (España). Geogaceta 52: 185–188.

Dantas, P., V. F. dos Santos, M. G. Lockley, and C. A. Meyer. 1994. Footprint evidence for limping dinosaurs from the Upper Jurassic of Portugal. Gaia 10: 43–48.

Davis, J. C. 1986. Statistics and Data Analysis in Geology. John Wiley & Sons, New York, New York, 550 pp.

Day, J. J., D. B. Norman, A. S. Gale, P. Upchurch, and H. P. Powell. 2004. A Middle Jurassic dinosaur trackway site from Oxfordshire, UK. Palaeontology 47: 319–348.

Díaz-Martínez, I., X. Pereda-Suberbiola, F. Pérez-Lorente, and J. I. Canudo. 2015. Ichnotaxonomic review of large ornithopod dinosaur tracks: temporal and geographic implications. PLoS One 10(2): e0115477.

Diedrich, C. 2004. New important iguanodontid and theropod trackways of the tracksite Obernkirchen in the Berriasian of NW Germany and megatracksite concept of Central Europe. Ichnos 11: 215–228.

Dietrich, O. W. 1927. Über Fährten ornithopodider Saurier im Oberkirchner Sandstein. Zeitschrift der Deutschen Geologischen Gesellschaft 78: 614–621.

Dollo, L. 1883. Troisième note sur les dinosauriens de Bernissart. Bulletin du Musée Royal d'Histoire Naturelle de Belgique 2: 85–126.

Elstner, F., and J. Mutterlose. 1996. The Lower Cretaceous (Berriasian and Valanginian) in NW Germany. Cretaceous Research 17: 119–133.

Falkingham, P. L. 2014. Interpreting ecology and behaviour from the vertebrate fossil track record. Journal of Zoology 292: 222–228. doi: 10.1111/jzo.12110.

Falkingham, P. L. 2016. Applying objective methods to subjective track outlines; chap. 4 in P. L. Falkingham, D. Marty, and A. Richter (eds.), Dinosaur Tracks: The Next Steps. Indiana University Press, Bloomington, Indiana.

Falkingham, P. L., K. T. Bates, L. Margetts, and P. L. Manning. 2011. The 'Goldilocks' effect: preservation bias in vertebrate track assemblages. Journal of the Royal Society, Interface 8(61): 1142–1154.

Falkingham, P. L., L. Margetts, I. M. Smith, and P. L. Manning. 2009. Reinterpretation of palmate and semi-palmate (webbed) fossil tracks: insights from finite element modelling. Palaeogeography, Palaeoclimatology, Palaeoecology 271: 69–76.

Farlow, J. O. 1987. Lower Cretaceous dinosaur tracks, Paluxy River Valley, Texas; pp. 1–50 in Field Trip Guidebook, South-Central Section, Geological Society of America, Baylor University, Waco. Geological Society of America, Boulder, Colorado.

Foster, J. R., and D. J. Chure. 2006. Hindlimb allometry in the Late Jurassic theropod dinosaur *Allosaurus* with comments on its abundance and distribution; pp. 119–122 in J. R. Foster and S. G. Lucas (eds.), Paleontology and Geology of the Upper Jurassic Morrison Formation. Bulletin 36. New Mexico Museum of Natural History and Science, Albuquerque, New Mexico.

Grabbe, H. 1881. Neue Funde von Saurier-Fährten im Wealdensandstein der Bückeberges. Verhandlungen des naturhistorischen Vereines der preussischen Rheinlande und Westfalens, Correspondenzblatt 38 (=Vierte Folge, 8): 161–164.

Grupe, O. 1931. Über Wurzelböden im Wealdensandstein der Bückeberge um ihre Bedeutung für den Rhythmus dynamischer Vorgänge. Zeitschrift der Deutschen Geologischen Gesellschaft 83: 224–234.

Hammer, Ø. 2012. PAST, PALeontological STatistics, Version 2.17. Reference Manual. Natural History Museum, University of Oslo, Oslo, Norway, 229 pp.

Hammer, Ø., and D. A. T. Harper. 2006. Paleontological Data Analysis. Blackwell Publishing, New York, New York, 351 pp.

Hammer, Ø., D. A. T. Harper, and P. D. Ryan. 2001. PAST: Paleontological Statistics software package for education and data analysis. Palaeontologia Electronica 4(1): 1–9.

Harper, D. A. T. (ed.) 1999. Numerical Palaeobiology. New York: John Wiley & Sons, New York, New York, 478 pp.

Hawthorne, J. M. 1990. Dinosaur track-bearing strata of the Lampasas Cut Plain and Edwards Plateau, Texas. Baylor Geological Studies Bulletin 49: 1–45.

Horner, J. R., D. B. Weishampel, and C. A. Foster. 2004. Hadrosauridae; pp. 438–463 in D. B. Weishampel, P. Dodson, and H. Osmolska (eds.), The Dinosauria. 2nd edition. University of California Press, Berkeley, California.

Hornung, J. J., and M. Reich. 2007. Krokodile, Schildkröten & Dinosaurier. Die 'Wealden'-Sammlung der Universität Göttingen. Fossilien 24(1): 32–36.

Hornung J. J., and M. Reich. 2012. Excursion Guide C2: Dinosaur tracks from the Berriasian Obernkirchen Sandstone on exhibit at the Göttingen University Geopark; pp. 169–187 in A. Richter and M. Reich (eds.), Dinosaur Tracks 2011. An International Symposium, Obernkirchen, April 14–17, 2011. Abstract Volume and Field Guide to Excursions. Universitäts-Verlag, Göttingen, Germany.

Hornung, J. J., A. Böhme, and M. Reich. 2012a. Excursion Guides A–C: the 'German Wealden' and the Obernkirchen Sandstone: an introduction; pp. 62–72 in Richter, A. and M. Reich (eds.), Dinosaur Tracks 2011. An International Symposium, Obernkirchen, April 14–17, 2011. Abstract Volume and Field Guide to Excursions. Universitäts-Verlag, Göttingen, Germany.

Hornung, J. J., A. Böhme, and M. Reich. 2012b. Excursion Guide B2: Wölpinghausen; pp. 143–149 in A. Richter and M. Reich (eds.), Dinosaur Tracks 2011. An International Symposium, Obernkirchen, April 14–17, 2011. Abstract Volume and Field Guide to Excursions. Universitäts-Verlag, Göttingen, Germany.

Hornung, J. J., A. Böhme, T. van der Lubbe, M. Reich, and A. Richter. 2012. Vertebrate tracksites in the Obernkirchen Sandstone (late Berriasian, Early Cretaceous) of northwest Germany – their stratigraphical, palaeogeographical, palaeoecological, and historical context. Paläontologische Zeitschrift 86: 231–267.

Hübner, T. 2016. Elusive ornithischian tracks in the famous Berriasian (Lower Cretaceous) Chicken Yard-tracksite of northern Germany: quantitative differentiation between small tridactyl track makers; chap. 16 in P. L. Falkingham, D. Marty, and A. Richter (eds.), Dinosaur Tracks: The Next Steps. Indiana University Press, Bloomington, Indiana.

Hunt, A. P., and S. G. Lucas. 2007. Tetrapod ichnofacies: a new paradigm. Ichnos 14: 59–68.

Ishigaki, S., and M. G. Lockley. 2010. Didactyl, tridactyl and tetradactyl theropod trackways from the Lower Jurassic of Morocco: evidence of limping, labouring and other irregular gaits. Historical Biology 22(1–3): 100–108.

Jackson, D. A. 1993. Stopping rules in principal components analysis: a comparison of heuristical and statistical approaches. Ecology 74: 2204–2214.

Jackson, S. J., M. A. Whyte, and M. Romano. 2009. Laboratory-controlled simulations of dinosaur footprints in sand: a key to understanding vertebrate track formation and preservation. Palaios 24: 222–238.

Jackson, S. J., M. A. Whyte, and M. Romano. 2010. Range of experimental dinosaur (*Hypsilophodon foxii*) footprints due to variation in sand consistency: how wet was the track? Ichnos 17: 197–214.

Kilbourne, B. M., and P. J. Makovicky. 2010. Limb bone allometry during postnatal ontogeny in non-avian dinosaurs. Journal of Anatomy 217: 135–152.

Kuhn, O. 1958. Die Fährten der vorzeitlichen Amphibien und Reptilien. Meisenbach, Bamberg, Germany, 64 pp.

Lehmann, U. 1978. Eine Platte mit Fährten von *Iguanodon* aus dem Obernkirchener Sandstein (Wealden). Mitteilungen aus dem Geologisch-Paläontologischen Institut der Universität Hamburg 48: 101–114.

Leonardi, G. 1984. Le impronte fossili di dinosauri; pp. 165–186 in J. F. Bonaparte, E. H. Colbert, P. J. Currie, A. de Ricqles, Z. Kielan-Jaworowska, G. Leonardi, N. Morello, and P. Taquet (eds.), Sulle orme dei dinosauri. Editio Editrice, Venice, Italy.

Leonardi, G. 1989. Inventory and statistics of the South American dinosaurian ichnofauna and its paleobiological interpretation; pp. 333–336 in D. D. Gillette and M. G. Lockley (eds.), Dinosaur Tracks and Traces. Cambridge University Press, Cambridge, UK.

Lockley, M. G. 2007. A tale of two ichnologies: the different goals and potentials of invertebrate and vertebrate (tetrapod) ichnotaxonomy and how they relate to ichnofacies analysis. Ichnos 14: 39–57.

Lockley, M. G., and M. Matsukawa. 1999. Some observations on trackway evidence for gregarious behavior among small bipedal dinosaurs. Palaeogeography, Palaeoclimatology, Palaeoecology 150: 25–31.

Lockley, M., C. Meyer, and V. F. dos Santos. 1994. Trackway evidence for a herd of juvenile sauropods from the Late Jurassic of Portugal. Gaia 10: 27–35.

Lockley, M. G., J. L. Wright, and D. Thies. 2004. Some observations on the dinosaur tracks at Münchehagen (Lower Cretaceous), Germany. Ichnos 11: 261–274.

Lockley, M. G., K. J. Houck, and N. K. Prince. 1986. North America's largest dinosaur trackway site: implications for Morrison Formation paleoecology. Geological Society of America Bulletin 97: 1163–1176.

Lockley, M. G., A. P. Hunt, J. Moratalla, and M. Matsukawa. 1994. Limping dinosaurs? Trackway evidence for abnormal gaits. Ichnos 3: 193–202.

Lockley, M., A. S. Schulp, C. A. Meyer, G. Leonardi, and M. Maimani. 2002. Titanosaurid trackways from the Upper Cretaceous of Bolivia: evidence for large manus, wide-gauge locomotion and gregarious behaviour. Cretaceous Research 23: 383–400.

Lockley, M., J. Wright, D. White, M. Matsukawa, L. Jianjun, F. Lu, and L. Hong. 2002. The first sauropod trackways from China. Cretaceous Research 23: 363–381.

Manning, P. L. 2004. A new approach to the analysis and interpretation of tracks: examples from the Dinosauria; pp. 93–123 in D. McIllroy (ed.), The Application of Ichnology to Palaeoenvironmental and Stratigraphic Analysis. Special Publications 228. Geological Society London, London, UK.

Mantell, G. A. 1825. Notice on the *Iguanodon*, a newly discovered fossil reptile, from the sandstone of Tilgate Forest, in Sussex. Philosophical Transactions of the Royal Society of London 115: 179–186.

Matsukawa, M., M. G. Lockley, and A. P. Hunt. 1999. Three age groups of ornithopods inferred from footprints in the mid-Cretaceous Dakota Group, eastern Colorado, North America. Palaeogeography, Palaeoclimatology, Palaeoecology 147: 39–51.

Matsukawa, M., T. Hamuro, T. Mizukami, and S. Fujii. 1997. First trackway evidence of gregarious dinosaurs from the Lower Cretaceous Tetori Group of eastern Toyama prefecture, central Japan. Cretaceous Research 18: 603–619.

Marty, D., A. Strasser, and C. A. Meyer. 2009. Formation and taphonomy of human footprints in microbial mats of present day tidal flat environments: implications for the study of fossil footprints. Ichnos 16: 127–142.

Meyer, C. A., and B. Thüring. 2003. The first iguanodontid dinosaur tracks from the Swiss Alps (Schrattenkalk Formation, Aptian). Ichnos 10: 221–228.

Moratalla, J. J., and J. Marugán-Lobón. 2009. Assessing dinosaur ichno-variability with geometric morphometrics: the ornithopod tracks from the Cameros Basin (Lower Cretaceous, Spain) as a case study. Journal of Vertebrate Paleontology 29 (3, Supplement): 151A.

Myers, T. S., and A. R. Fiorillo. 2009. Evidence for gregarious behavior and age segregation in sauropod dinosaurs. Palaeogeography, Palaeoclimatology, Palaeoecology 274: 96–104.

Norman, D. B. 2011. Ornithopod dinosaurs; pp. 407–475 in D. J. Batten (ed.), English Wealden Fossils. Palaeontological Association Field Guide to Fossils 14. The Palaeontological Association. London, UK.

Ostrom, J. H. 1972. Were some dinosaurs gregarious? Palaeogeography, Palaeoclimatology, Palaeoecology 11: 287–301.

Parkes, A. S. 1993. Dinosaur footprints in the Wealden at Fairlight, East Sussex. Proceedings of the Geologists' Association 104: 15–21.

Pascual-Arribas, C., N. Hernández-Medrano, P. Latorre-Macarrón, and E. Sanz-Pérez. 2009. El icnogénero Iguanodontipus en el yacimiento de 'Las Cuestas I' (Santa Cruz de Yanguas, Soria, España). Studia Geologica Salamanticensia 45(2): 105–128.

Pelzer, G. 1998. Sedimentologie und Palynologie der Wealden-Fazies im Hannoverschen Bergland. Courier Forschungsinstitut Senckenberg 207: 1–211.

Pittman, J. G., and D. D. Gillette. 1989. The Briar Site: a new sauropod dinosaur tracksite in Lower Cretaceous beds of Arkansas, USA; pp. 313–332 in D. D. Gillette and M. G. Lockley (eds.), Dinosaur Tracks and Traces. Cambridge University Press, Cambridge, UK.

Pollard, J. E., and J. D. Radley. 2011. Trace fossils; pp. 652–676 in D. J. Batten (ed.), English Wealden Fossils. Palaeontological Association Field Guide to Fossils 14. The Palaeontological Association, London, UK.

Radley, J. D., M. J. Barker, and I. C. Harding. 1998. Palaeoenvironment and taphonomy of dinosaur tracks in the Vectis Formation (Lower Cretaceous) of the Wessex Sub-basin, southern England. Cretaceous Research 19: 471–487.

Rasband, W. S. 1997–2012. ImageJ. National Institutes of Health, Bethesda, Maryland. Available at http://imagej.nih.gov/ij/. Accessed November 6, 2015.

Rasskin-Gutmann, D., G. Hunt, R. E. Chapman, J. L. Sanz, and J. J. Moratalla. 1997. The shapes of tridactyl dinosaur footprints: procedures, problems, and potentials. Dinofest International Proceedings 1997: 377–383.

Richter, A., and A. Böhme. 2016. Too many tracks – preliminary description and interpretation of the diverse and heavily dinoturbated Early Cretaceous 'Chicken Yard' ichnoassemblage (Obernkirchen tracksite, northern Germany, chap. 17 in P. L. Falkingham, D. Marty, and A. Richter (eds.), Dinosaur Tracks: The Next Steps. Indiana University Press, Bloomington, Indiana.

Richter, A., A. Böhme, and T. van der Lubbe. 2009. 'Chicken Run': a new unusual, heavily dino-turbated tracksite from the Lower Cretaceous sandstones of Obernkirchen, northern Germany. Journal of Vertebrate Paleontology 29(3): 171A.

Richter, A., J. J. Hornung, A. Böhme, and U. Stratmann. 2012. Excursion Guide A1: Obernkirchen sandstone quarries: a natural workstone lagerstaette and a dinosaur tracksite; pp. 73–99 in A. Richter, T. and M. Reich (eds.), Dinosaur Tracks 2011. An International Symposium, Obernkirchen, April 14–17, 2011. Abstract Volume and Field Guide to Excursions. Universitäts-Verlag, Göttingen, Germany.

Rodrigues, L. A., and V. F. dos Santos. 2004. Sauropod tracks: a geometric morphometric study; pp. 129–142 in A. M. T. Elewa (ed.), Morphometrics. Applications in Biology and Paleontology. Springer Verlag, Berlin, Germany.

Rohlf, F. J., and D. Slice. 1990. Extensions of the Procrustes method for the optimal superimposition of landmarks. Systematic Zoology 39: 40–59.

Santos, V. F. dos, P. M. Callapaz, and N. P. C. Rodrigues. 2012. Dinosaur footprints from the Lower Cretaceous of the Algarve Basin (Portugal): new data on the ornithopod palaeoecology and palaeobiogeography of the Iberian Peninsula. Cretaceous Research 40: 158–169.

Sarjeant, W. A. S., J. B. Delair, and M. G. Lockley. 1998. The footprints of *Iguanodon*: a history and taxonomic study. Ichnos 6: 183–202.

Stechow, E. 1909. Neue Funde von *Iguanodon*-Fährten. Centralblatt für Mineralogie, Geologie und Paläontologie [1909]: 700–705.

Struckmann, C. 1880. Vorläufige Nachricht über das Vorkommen großer, vogelähnlicher Thierfährten (*Ornithoidichnites*) im Hastingssandsteine von Bad Rehburg bei Hannover. Neues Jahrbuch für Mineralogie, Geologie und Paläontologie [1880](1): 125–128.

Thévenaz, P. 2012. Point Picker. École Polytechnique Fédérale de Lausanne, Lausanne, Switzerland. Available at http://bigwww.epfl.ch/thevenaz/pointpicker/. Accessed November 6, 2015.

Thulborn, R. A., and M. Wade. 1979. Dinosaur stampede in the Cretaceous of Queensland. Lethaia 12: 275–279.

Thulborn R. A., and M. Wade. 1989. A footprint as a history of movement; pp. 51–56 in D. D. Gillette and M. G. Lockley (eds.), Dinosaur Tracks and Traces. Cambridge University Press, Cambridge, U.K.

Wings, O., J. N. Lallensack, and H. Mallison. 2016. The Early Cretaceous dinosaur trackways in Münchehagen (Lower Saxony, Germany): 3D photogrammetry as basis for geometric morphometric analysis of shape variation and evaluation of material loss during excavation; chap. 3 in P. L. Falkingham, D. Marty, and A. Richter (eds.), Dinosaur Tracks: The Next Steps. Indiana University Press, Bloomington, Indiana.

Wings, O., D. Falk, N. Knötschke, and A. Richter. 2012. Excursion Guide B1: The Early Cretaceous dinosaur trackways in Münchehagen (Lower Saxony, Germany): the Natural Monument 'Saurierfährten Münchehagen' and the adjacent Wesling Quarry; pp. 113–142 in A. Richter and M. Reich (eds.), Dinosaur Tracks 2011. An International Symposium, Obernkirchen, April 14–17, 2011. Abstract Volume and Field Guide to Excursions. Universitäts-Verlag, Göttingen, Germany.

Woodhams, K. E., and J. S. Hines. 1989. Dinosaur footprints from the Lower Cretaceous of East Sussex, England; pp. 301–307 in D. D. Gillette and M. G. Lockley (eds.), Dinosaur Tracks and Traces. Cambridge University Press, Cambridge, UK.

Zelditch, M. L., D. L. Swiderski, and H. D. Sheets. 2012. Geometric Morphometrics for Biologists: A primer. 2nd edition. Elsevier, Amsterdam, the Netherlands, 437 pp.

13.1. This small sample of three distinct track-ways shows the wealth of information in the placement of the steps (and seeming missteps) of the trackmakers at the A16 tracksites. Derived from the Courtedoux–Béchat Bovais track-site (level 515). Note that the elliptical manus (light gray) and pes (dark gray) track markers schematically represent the dimensions and orientation of each track. The triangle on each indicates the anterior direction of the manus or pes. Blue disks indicate tracks that were not definitely associated with any trackway.

Uncertainty and Ambiguity in the Interpretation of Sauropod Trackways

13

Kent A. Stevens, Scott Ernst, and Daniel Marty

TRACKWAY INTERPRETATION, THE DRAWING OF INFER- ences about a trackmaker and its movements from a pattern of trace impressions, is examined from the perspective of the information in the pattern of individual tracks along a trackway, with emphasis here on sauropod trackways. Al- though trackways are commonly regarded as direct records of locomotion behavior, their interpretation is in fact less straightforward than is often expected. Even the basic es- timation of trackmaker size (e.g., glenoacetabular distance, a common proxy for trackmaker size) is not generally valid. Moreover, without knowledge of trackmaker size, any ob- served pattern of manus and pes tracks has arbitrarily many possible solutions in terms of limb phase and duty factor, the primary components of gait. An analysis of the relationship between trackmaker size, stride length, and limb phase (i.e., gait) reveals a previously unappreciated interdependence among these parameters. A new approach is introduced to address the problem of supporting inferences in the presence of ambiguity and uncertainty.

INTRODUCTION

Whereas fossil trackways are often regarded as providing a direct record of the behavior of a trackmaker during its passage across a substrate (e.g., Lockley, 1998), trackway interpretation presents more ambiguity and requires more assumptions than is commonly expected. A trackway is mean- ingful only with regard to what may be directly inferred about the trackmaker (such as its taxonomic identity, size, and ontogenetic stage of development) but also its move- ment (such as its specific gait and speed). To make headway with any one such aspect of trackway interpretation, many of the other unknown factors are presumed but not indepen- dently verified. In interpreting a trackway, these unknown various factors are mutually interrelated and confounded to an extent that is likely unrecognized in most interpretive studies.

This study considers the information provided by the geometric pattern in the placement of tracks that compose a trackway, with emphasis on the interpretation of sauropod trackways. Trackway descriptions typically classify quadrupe- dal trackways in terms of the degree of overstepping between adjacent manus and pes tracks (e.g., so-called pes-dominated and pes-only trackways), the placement of individual tracks relative to the trackway (e.g., whether the pes or manus tracks are closer to the trackway midline), their orientation, and the trackway width or gauge (terminology according to Marty, Meyer, and Billon-Bruyat, 2006; Marty, 2008; Mar- ty, Belvedere, et al., 2010). Explaining how these patterns arise in terms of the trackmakers' physical characteristics, movements, and behaviors is in fact more difficult than has been recognized. The following first discusses the sources of ambiguity and uncertainty that prevent tracks from being straightforwardly interpreted by trackmaker simulation. An intractably large space of possible interpretations is present- ed as the number of unknowns far exceeds the number of knowns or safe presumptions required to propose a concrete trackmaker. A new method for incrementally constraining the space of possible trackway interpretations is introduced as an alternative to trackmaker simulation. Software named Cadence is under development to facilitate the interpretation of dinosaur trackways discovered on several large tracksites prior to the construction of Highway A16 near Porrentruy, Switzerland (Marty et al., 2003, 2004; Marty, 2008; Marty, Paratte, et al., 2010).

Abbreviations and Definitions

Acetabular height (H_A): The hip height of a trackmaker.
Duty factor (DF): That fraction of a walk cycle during which a limb is in contact with the ground and bearing weight.
Gleno-acetabular distance (D_{GA}): A conventional proxy for trackmaker size, as measured from the glenohumeral (shoulder joint) to acetabulum (hip joint).
Glenoid height (H_G): The shoulder height of a trackmaker.
Limb phase (LP): The fraction ($0.0 \leq LP \leq 1.0$) of the walk cycle when a given limb strikes the ground, relative to the left hind limb (LP = 0.0).
Locators: In a trackmaker model, the location of the left and right manus (L_{LM}, L_{RM}) and the left and right pes (LLP and LRP).

227

Manus-pes distance (D_{MP}): The distance from the center of a pes track to the center of the next manus track as measured in the direction of travel D_{MP} is 0.0 in the case of complete pes overprinting.

Pes length (PL) and pes width (PW): The length and width of an individual hind foot track.

Reference points: A mathematical model of a quadrupedal trackmaker is located in space by two points associated with the axial skeleton: the anterior reference point (RP_A) and the posterior reference point (RP_P). These may be regarded to correspond to the pectoral and pelvic girdles, respectively, whereupon $|RP_A - RP_P| = D_{GA}$.

Stride length (SL): Center-to-center distance between successive ipsilateral manus or pes tracks.

Track phase (TP): The placement of a manus track relative to the previous and subsequent ipsilateral pes tracks ($0.0 \leq TP \leq 1.0$). A manus track placed midway has $TP = 0.5$; complete pes overprinting corresponds to $TP = 0.0$. $TP = D_{MP}/SL$.

Trackway ratio (TR): The ratio of track width to the total width of the trackway (both measured perpendicular to the long axis of the trackway).

Sources of Uncertainty, Ambiguity, and Constraint

A fossil trackway is expected to preserve information about the trackmaker, including the size and proportions of the trackmaker, its taxonomic identification and developmental stage, and information about its locomotion during its progression across the substrate including its gait and speed. As will be reviewed, empirical relationships between trackway measurements and these various trackmaker properties have been proposed, such as, using measured manus-pes separations to estimate glenoacetabular distance, and from estimated glenoacetabular distance to estimate hip height and from estimated hip height to estimated walking speed, or from measured pes track dimensions to estimated hip height to estimated walking speed. Track measurements have also been used to infer the trackmaker taxonomy when there is the potential for subadults of a larger taxon to be confused for adults of a smaller taxon or when differences in stance (from wide to narrow) might be confused for taxonomic differences. The following discussion reviews the uncertainty and ambiguity in the process of such trackway interpretations. Later it will be suggested that rather than use these uncertain and ambiguous sources of information separately, they are more usefully regarded as constraints, with some interdependent and some independent sources of information.

Trackway Measurement Uncertainty

Various quantitative heuristics have been proposed to derive estimates based on trackway measurements, such as trackmaker size, speed, and other properties (e.g., Alexander, 1976, 1989; Thulborn, 1981, 1982, 1989, 1990). Beyond the uncertainty presented by such empirical relationships, measurement uncertainty is also unavoidably present as it is in other empirical sciences. Some physical sciences are more accustomed to explicitly dealing with measurement uncertainty, wherein data is often accompanied by an explicit estimate of the uncertainty associated with the measurement. The concern in ichnology goes beyond conventional measurement error (as when using vernier calipers), for tracks often simply do not support much measurement precision. A trackway may preserve some elite tracks where anatomical details of the foot such as pad, digit, and claw impressions and the surrounding displacement rims are all crisply defined and permit precise measurements of the foot dimensions and orientation, and yet a few steps later, the track preservation may be poor (e.g., Padian and Olsen, 1984b; Milàn, 2006; Marty, Strasser, and Meyer, 2009). Some physical sciences routinely deal with data of variable measurement uncertainty and explicitly associate a measure of confidence with the measurements so that low confidence data have relatively less influence on the computation of correlations and other statistical analyses (Taylor, 1997; Bevington and Robinson, 2002). This would likewise be important in ichnology, where the data may be highly variable within a trackway and the statistical power is often weak due to small sample sizes.

When setting out to measure a trackway, there are many well-recognized sources of uncertainty, such as whether the tracks are true tracks and the extent to which they have been modified or altered (e.g., Marty, Strasser, and Meyer, 2009) or whether the tracks are undertracks (Milàn and Bromley, 2006, 2008) and the type and conditions of the substrate at the time of track formation (Manning, 2004). These sources of uncertainty are compounded by the task of measurement itself, either in the field or from collected images or scan data, which ultimately rely on arbitrary criteria. Such measurement uncertainty is unavoidable and exacerbated by ambiguity in the assignment of tracks to trackways where multiple trackways intermingle or cross (e.g., Fig. 13.1). Even

13.2. Two extraordinary long and parallel trackways (S18 and S19, each with more than 200 tracks and each over 100 m long) from the Courtedoux–Béchat Bovais tracksite (level 515). They show several small turns and multiple abrupt changes in trackway pattern. The mean ratios between the width of the angulation pattern and the pes length (trackway width sensu Marty, 2008) vary along the course of these trackways, between narrow gauge and wide gauge. These variations even occur over a couple of steps and demonstrate that these two locomotor styles could have been used by one and the same sauropod trackmaker (Marty, Belvedere, et al., 2010; Marty, Paratte, et al., 2010).

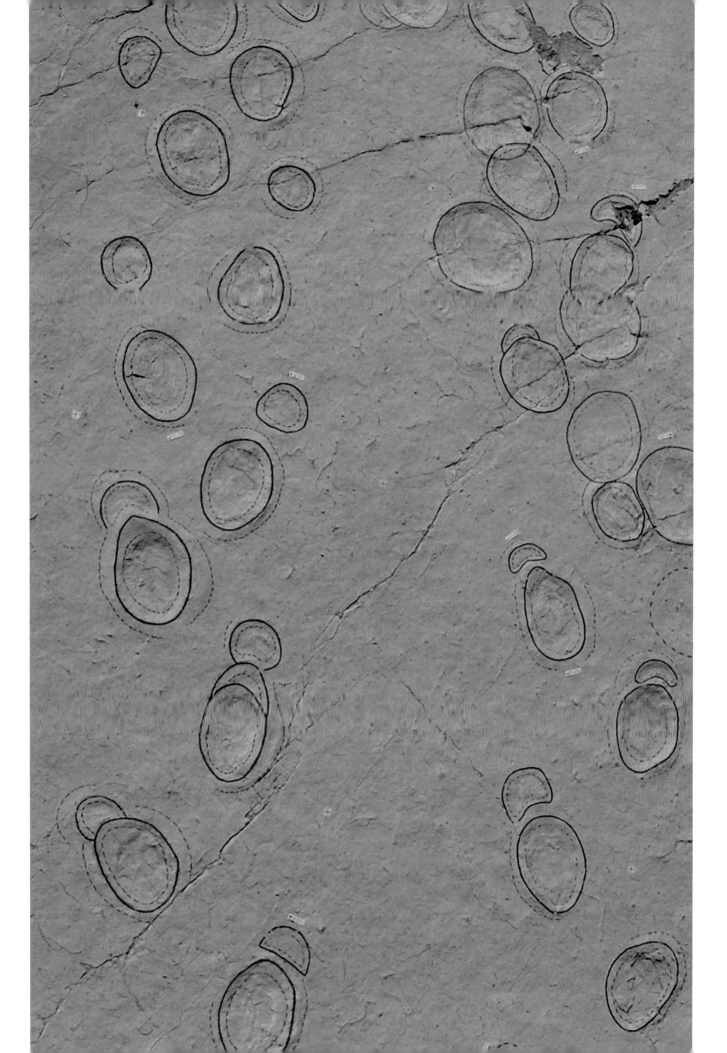

in an isolated trackway, some individual tracks may also be missing at a location where expected, due to either some failure of preservation or unusual behavior of the trackmaker. Occasionally, additional individual tracks may be found in association with a given trackway that are difficult to assign to another trackway. They should not be dismissed as spurious because they may indicate significant transient behavior of the trackmaker, such as additional steps due to hesitation or idle shifting of weight while standing stationary or pausing. Although such deviations from a regular progression of tracks add to the uncertainty of a simple "steady state" interpretation, they may nonetheless provide information about how the trackmaker balanced, especially at low speeds.

Measurement limitations, preservation artifacts, and ambiguous attribution of tracks to trackways are not the only contributions to the uncertainty in their interpretation. Neither the physical characteristics of the trackmaker nor its pattern of locomotion are known with certainty. Whereas these factors taken together might seem to reduce the confidence in any one trackway interpretation, taken together they may provide weak but additive constraint on such interpretation.

Taxonomic Uncertainty

Small or juvenile members of one species may leave tracks the size and shape of a smaller, unrelated species (Thompson, White, and Morgan, 2007). The Highway A16 trackways present a large range of quadrupedal trackway size classes with pes length ranging from 0.1 to 1.2 m, presumably reflecting a range of juvenile and subadult individuals from multiple sauropod taxa that attained differing sizes in their adult forms (Marty, 2008, 2009). Trackways created by subadults of a larger taxon might well be confused with those of older individuals of a smaller taxon. The potential for confusion is even greater for tridactyl trackways of varying sizes, given the greater diversity of bipedal taxa among saurischians and ornithischians, and the great range of adult sizes (Thulborn, 1990; Weishampel, Dodson, and Osmólska, 2004).

Fortunately, some sauropod trackways preserve anatomical foot details such as pad, digit, and claw impressions (e.g., Farlow, Pittman, and Hawthorne, 1989; Farlow, 1992; Santos, Moratalla, and Royo-Torres, 2009; Marty, Belvedere, et al., 2010) that correlate well with osteology, and as the number of both trackways and skeletal specimens increases, the case for inferring the taxonomic identity of the trackmakers based on track morphology is steadily improving, at least for some taxa, and some trackways (Farlow, 1992; Wright, 2005), even though most foot skeletal features are not typically considered to be synapomorphies (e.g., Salgado, Coria, and Calvo, 1997; Wilson and Sereno, 1998; Wilson and Carrano, 1999). Such taxonomic information, even if imprecise, could translate to

bounds on trackmaker model parameters, constraints that are graded according to the specificity of the taxonomic information derived from track morphology. In the absence of diagnostic morphological characters in individual tracks to distinguish taxa of trackmakers, hopefully their trackways are statistically distinguishable by other measures.

Sauropod ichnotaxa vary in terms of relative trackway width or "trackway ratio," the ratio of track width to total trackway width (Romano, Whyte, and Jackson, 2007). Although the measurement of gauge is straightforward, the interpretation is not. A broad distinction is made, where "narrow-gauge" trackways are believed to be associated with diplodocids and "wide-gauge" trackways with brachiosaurids or titanosaurids (Farlow, Pittman, and Hawthorne, 1989; Farlow, 1992; Lockley et al., 1994; Wilson and Carrano, 1999; Henderson, 2006; Santos, Moratalla, and Royo-Torres, 2009; Marty, Paratte, et al., 2010). Wilson and Carrano (1999) regard the wide-gauge trackways associated with titanosaurs to be consequences of their broad skeletal plan. Whereas wide-gauge trackmakers may not have been capable of creating narrow-gauge trackways, narrow-gauge trackmakers seemed capable of creating, under some circumstances, wide-gauge trackways (as we can when walking carefully on a slippery substrate?). Lockley et al. (2001, 2002) suggest that some observed variations in trackway gauge may be partly ontogenetic. To add to the uncertainty, some variations in gauge may also be behavioral. On level 515 of the Highway A16 Béchat Bovais tracksite, some sauropod trackways varied repeatedly from narrow to intermediate to wide gauge, while traveling an extensive path (>100 m), as if these individuals readily shifted between these two locomotor styles over a period of only a few steps (Fig. 13.2). These trackways present a particularly rich source of information about the potential relationships between gauge and variability in stride length (and presumably speed) and even activity, as some of these trackways are parallel, with correlated turns, divergences, and convergences. On the other hand, trackways with similar track morphology were found to be either narrow or wide gauge, even on single levels and in spatial proximity. Given that trackway width is only a weak source of information toward taxonomic identification of the trackmaker, there remains considerable taxonomic uncertainty presented by some sauropod trackways.

Manus pronation was apparently highly variable in some sauropods based on the large rotational differences observed in trackways (Farlow, Pittman, and Hawthorne, 1989; Lockley and Hunt, 1995; Marty, 2008). Although the mechanism underlying the rotation is not well understood, manus pronation was likely passive (Bonnan, 2003). The variability in manus track orientation is greater than that of the pes (Marty, 2008) In addition to manus placement, heteropody, track

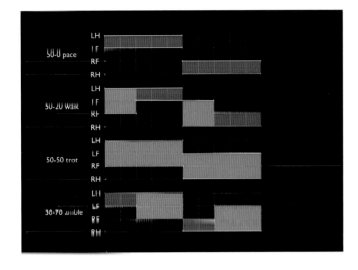

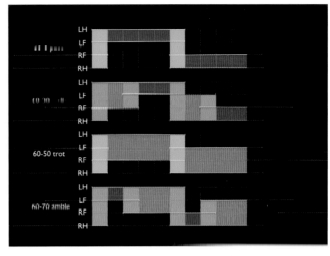

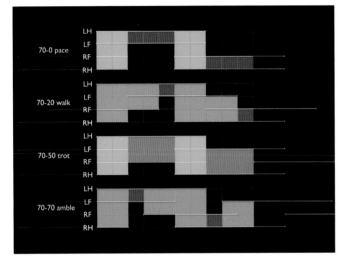

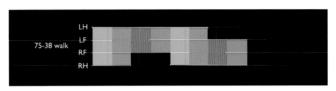

13.3. (*top left*) Gait diagrams for pace, walk, trot, and amble showing static stability for different limb phases, holding duty factor constant at 0.5 (50%). See also Figures 13.4–13.6 for other duty factors. Red indicates where the center of mass is not straddled by supporting limbs; hence, it is not statically stable. Pink indicates where the center of mass is straddled, but only two limbs are supporting. On the left, the first number is duty factor as a percentage (50% in this figure) and the second number is limb phase as a percentage (0.0 = pace; 0.2 = walk; 0.5 = trot; 0.7 = amble). For this duty factor, only two legs are in support phase, and manus track morphology, pronation variability

phase at a time, with dark red indicating they are ipsilateral and pink indicating they are diagonal. Longer duty factors permit periods of static stability during the walk cycle (Figures 13.3–13.4). Abbreviations: LF, left forelimb; LH, left hind limb; RF, right forelimb; RH, right hind limb.

13.4. (*top right*) Gait diagrams for a duty factor of 0.6 (60%) permits moments of static stability (compare with Fig. 13.3.) The walk cycle has periods of support from three legs (dark green) and all four legs (light green). Abbreviations as in Figure 13.3.

13.5. (*bottom left*) Gait diagrams that show static stability is still greater for a duty factor of 0.7, but at the expense of stride length, which must be greatly reduced for the limbs to remain in contact with the ground for 70% of the walk cycle. Abbreviations as in Figure 13.3.

13.6. (*bottom right*) Gait diagram for turtle walk cycle with limb phase 0.38 (intermediate between walk 0.25 and trot 0.5 and with a duty factor of 0.75 (from Walter, 2003).

may be diagnostic.

GAIT AMBIGUITY

The relative timing of a walk cycle for a given quadruped can be characterized by two parameters: limb phase (LP), the relative timing of the left forelimb making ground contact relative to when the left hind limb has just touched ground, and duty factor (DF), the fraction of the stride during which a given limb is in support phase, that is, ground contact

(Alexander and Jayes, 1983; Biewener, 1983). The term "gait" will be used here to refer to a specific combination of LP and DF, which corresponds to a point in a conventional gait diagram or "gait space" (Hildebrand, 1965; Biknevicius and Reilly, 2006).

Figures 13.3 to 13.5 show the timing of foot placement for combinations of limb phase (LP = 0.0, 0.25, 0.5, and 0.75) and duty factor (DF = 0.5 in Fig. 13.3, 0.6 in Fig. 13.4, and 0.7 in Fig. 13.5). Each gait diagram shows a complete step cycle (comprising left and right steps) divided into 10 equal intervals. The color code indicates the degree of stability (dark red

13.7. These two unusual sauropod trackways from the Porrentruy-CPP tracksite (level 525/530) follow parallel, tightly curved paths comprising very short steps. Besides perhaps recording an intimate interaction between the two sauropod trackmakers, these pes-only trackways represent very short stride lengths, wherein glenoacetabular distance (D_{GA}) clearly cannot be straightforwardly computed by the conventional trackway measurement techniques.

for the case where the animal is statically unstable with only two ipsilateral limbs in support, pink where two diagonal limbs are in support phase, and dark and bright green for three and four limbs in support phase, respectively. As shown in these figures, for duty factors less than about 0.7 all gaits show significant periods of static instability in which only two limbs are in support phase, and especially when those two are ipsilateral. Nonetheless, most quadrupedal mammals and reptiles frequently engage in walking gaits that are less statically stable but for which they can maintain dynamic stability by alternating triangles of support, regaining balance as the animals tip alternately to the left and right. This is true even in very slowly moving quadrupeds (e.g., turtles) where the stride length is sufficiently short to permit very long duty factors (DF ≈ 0.8), whereupon it remains in static stability for about 75% of the walk cycle (Walker, 1971; Hildebrand, 1980, 1985); see Figure 13.6.

In modern quadrupeds, the body plan (limb length, stance, range of motion, and relative body proportions) influences the preferred gait. Those with short limbs and slow limb movements are restricted in their locomotory repertoire and static stability becomes particularly important. Those with longer legs have the opportunity to adapt their gaits to satisfy a range of goals (Hildebrand, 1980, 1985). Extant mammals vary their gait with speed (Alexander and Jayes, 1983); the slow gait typically is a walk (LP ≈ 0.25), although a pace (LP ≈ 0) is also used in slow locomotion by some very long-legged quadrupeds to reduce the likelihood of interference between the hind and forelimbs (Hildebrand, 1980).

What gait might be attributed to the giant sauropods? It could be argued a priori that because lateral sequence gaits are more stable than diagonal-sequence gaits are (Grey, 1944), one might presume sauropods were engaged in a walk gait with LP ≈ 0.2 and a conservative, stable, duty factor, DF ≈ 0.7 (as in Fig. 13.5). Alternatively, a reptilian trot (LP ≈ 0.5) might be thought the primitive saurischian condition, and whereas it has substantial periods where all four limbs are in support given a long duty factor (Fig. 13.3), there is a long period of support by only two diagonal limbs. Intriguingly, the turtle (Fig. 13.5) has a single-foot lateral gait that is actually intermediate between walk and trot (LP ≈ 0.38) but with a very long 0.75 duty factor permitted by its short stride lengths (Walker, 1971). Vila et al. (2013) conclude that the titanosaur trackways they studied were most likely created by an amble gait, and Sellers et al. (2013) found that their simulations tended toward a pace for the titanosaur they modeled. But as will be shown, the relationship between trackway pattern and trackmaker gait is more ambiguous than is often appreciated, and sauropods might have engaged in any of a wide range of gaits from pace to amble.

Perhaps their gigantism influenced their choice of gait in order to maximize stability, as well as to minimize the cost of locomotion. As speed diminishes, as indicated by short stride lengths, the requirement for static stability increases. Some sauropod trackways from the Late Jurassic Highway A16 tracksites also have remarkably short relative stride lengths (Fig. 13.7). Their steps were so short, however, that the duty factor could have been very high, allowing the sauropod to walk with static stability in virtually all gaits (except perhaps a pace). But as stride length increases toward the maximum achievable by the trackmaker, duty factor necessarily reduces, resulting in unacceptably long periods of static instability for most gaits (as is apparent in the gait diagrams shown in Figs. 13.3–13.5). Moreover, as shown by Henderson (2006), having three limbs in support phase does not ensure stability, because the center of mass may either be at the margin of the triangle of support or approaching it.

Pes-only trackways are common among the sauropod trackways of the Highway A16 tracksites (Marty, Meyer, and Billon-Bruyat, 2006; Marty, 2008), and other tracksites (e.g., Meyer, 1990, Li et al., 2006; Lockley, Huh, and Kim, 2012). Does complete manus overprinting arise as a consequence of a particular combination of parameters (DF, LP, D_{GA}, and SL)? This hypothesis was examined by systematically choosing variations in D_{GA} and limb phase for a trackway of given stride length SL. Rather than revealing that pes-only trackways occur with a specific gait, complete manus overprinting can be achieved with any gait, for specific choices of D_{GA} relative to SL. Indeed, without knowing the relative scale of the trackmaker, gait is indeterminate based on trackway pattern alone.

Figure 13.8 shows synthetic trackways for three sizes of a given trackmaker (D_{GA} = 200, 250, and 300 cm). For each trackmaker size, four trackways were generated, corresponding to pace (LP = 0.00), walk (LP = 0.25), trot (LP = 0.50), and amble (LP = 0.75). Stride length was constant across all trackways (SL = 100 cm). The direction of travel is from left to right, with the pes tracks (larger tan-colored disks) and manus tracks (smaller light-blue disks) forming the corresponding trackway pattern according to each combination of D_{GA} and LP.

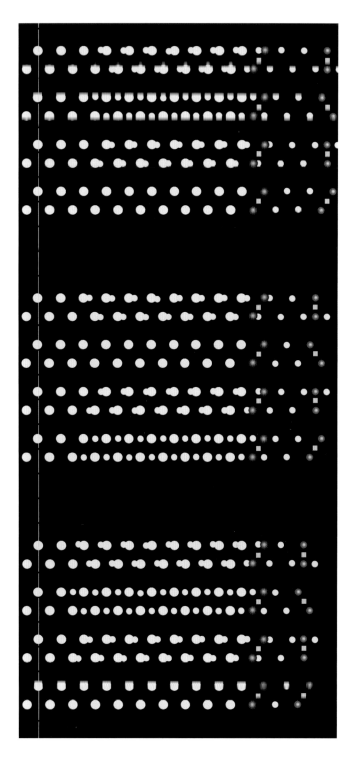

13.8. Quadrupedal trackways for combinations of trackmakers' size (D_{GA}) and limb phase (LP), for a constant stride length SL = 100 cm. The direction of travel is from left to right, with manus and pes tracks indicated by smaller and larger disks, respectively. Trackways are shown for combinations of three values of glenoacetabular distance D_{GA} (from 200, 250, and 300 cm) and limb phase LP (0.0, 0.25, 0.5, and 0.75). Note that each track phase can arise for each size of trackmaker but at different values of LP. For example, for D_{GA} = 200 cm, the pes oversteps the manus in a pace (LP = 0.0), whereas for a trackmaker 50 cm longer, the same overstepping occurs for a trot (LP = 0.50). The trackmaker is represented to the right of each trackway, indicating the position of the anterior and posterior reference points (separated by D_{GA}). The manus pair can be several step cycles ahead of the pes pair during the generation of these trackways (depending upon LP and the ratio D_{GA} to SL).

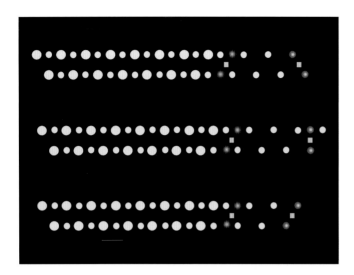

13.9. Three synthetic quadrupedal trackways created with constant stride length (SL = 100 cm). The three trackways are identical, with manus tracks precisely midway between pes tracks (track phase TP = 0.5), but they result from different combinations of LP and D_{GA}: top trackway LP = 0.5 (trot) and DGA = 300 cm. The middle trackway replicates this pattern with LP = 0.25 (walk) and D_{GA} = 325 cm, and the bottom trackway is for LP = 0.0 (pace) and D_{GA} = 250 cm. As in Figure 13.6, the trackmaker proceeds from left to right and is seen at the right extreme of each trackway, with pes and manus locators and their associated with posterior and anterior reference points, respectively. Note that the forelimb pair can be several complete step cycles ahead of the hind limb pair.

To generate these synthetic trackways, an abstract trackmaker is modeled as a mathematical abstraction consisting of two manus and two pes 'locators,' plus two additional points, the anterior and posterior 'reference points' (RP_A and RP_P) that correspond to the pectoral and pelvic girdles, respectively. The manus and pes locators are shown at the far right of each trackway in Figure 13.8, depicted as spheres (red for the pes locators, and blue for the manus locators). Each trackway records the progress of the four locators as they progress from left to right in accordance with the parameters of a phase generator (with specified D_{GA}, LP, and SL). Each locator alternates between static placement on the substrate during the period in which the corresponding limb is in support phase, and movement during the suspended phase as it protracts to take the next step. Upon contact with the substrate, each locator leaves a circular disk representing the corresponding manus or pes track. The timing of the locators is governed by the gait (LP and DF, and the initial positions are determined by the specified D_{GA} and SL). The anterior and posterior reference points are depicted by pink rectangles (the larger denoting RP_P). In a simple approximation to the trackmaker, the reference points RP_P and RP_A can be computed as the midpoint of the pes locators and the midpoint of the manus locators, respectively. The glenoacetabular length D_{GA} corresponds to the distance between the small and large squares at the end of each trackway, or $|RP_A - RP_P|$.

Even though the trackways in Figure 13.8 show different patterns of track placement, all patterns are readily described by a simple phase relationship between any given manus track and the two ipsilateral pes tracks that straddle it. Note that the pace, walk, trot, and amble are symmetrical gaits (uniform DF), with identical patterns of alternating manus and pes tracks on each side but with those on the left side simply shifted by SL/2 relative to those on the right. Consider the alternating pattern of manus and pes tracks along just the left side of one of the trackways in Figure 13.8. Note that the manus tracks in a given trackway are located between successive pes tracks at a constant "track phase" TP = D_{MP}/SL, where the distance D_{MP} from a pes track to the next manus track is a constant proportion of the stride length for that trackway. Track phase varies from 0.0 ≤ TP ≤ 1.0, where TP = 0.0 corresponds to the case of pes overprinting; hence, the manus tracks are obliterated and TP = 0.5 corresponds to placement of the manus tracks halfway between successive pes tracks.

For a given stride length SL, it is possible to create a trackway with any track phase by suitable combinations of a given trackmaker size D_{GA} and limb phase LP. There is thus a basic ambiguity in the interpretation of gait from track phase: the trackways created by two trackmakers differing in size by an integral multiple of SL will be identical. For example, in Figure 13.8, two sizes of trackmaker (D_{GA} = 200 and 300 cm) result in the same trackway patterns for each of the four limb phases plotted; however, the forelimbs in the latter case are one stride length (100 cm) farther ahead of the hind limbs than in the former case.

The case that originally caught our interest, the pes-only trackways, can likewise occur for different combinations of D_{GA} and LP. The underlying situation, more generally, is that any track phase can be achieved by any gait, for some value of D_{GA}. Specifically, assuming linear velocity, track phase is approximated by:

$$TP \approx ((LP \times SL + D_{GA}) \bmod SL)/SL$$

To demonstrate the importance of accurately reconstructing trackmaker size (D_{GA}), note that in Figure 13.9, the same trackway is created by three distinct gaits. In this case, the manus tracks are centered midway between "preceding" and "subsequent" pes tracks (TP = 0.5). Although the resultant three trackways are identical, the top trackway was created with a limb phase of 0.5 (i.e., a trot gait) and D_{GA} of 300 cm, the middle trackway replicates this pattern for LP = 0.25 (a walk) and D_{GA} of 325 cm, and the bottom trackway shows that this pattern also occurs for limb phase of 0.0 (a pace) and a D_{GA} of 250 cm. Trackmakers with D_{GA} of 300 cm versus 325 cm differ by less than 8%, in other words, within the expected uncertainty with which D_{GA} would be estimated from trackway measurements by known methods (e.g., Alexander, 1976; González Riga, 2011, as discussed further herein).

Gait (LP and DF) cannot be determined from a trackway pattern (TD and SL) without knowing the size (D_{GA}) of the trackmaker, and as will be discussed next, trackmaker size cannot be sufficiently precisely estimated, hence the trackmaker's gait will remain indeterminate until additional constraints can be introduced. First, consider the problems of estimating trackmaker size.

TRACKMAKER SIZE

Body length measured from shoulder to hip (glenoacetabular distance, or D_{GA}) is often estimated from a trackway as the distance measured between the midpoint of a pair of left and right pes tracks and the midpoint of the subsequent pair of left and right manus tracks (e.g., Soergel, 1925; Padian and Olsen, 1984a; Leonardi, 1987; Schult and Farlow, 1992; Mazzetta and Blanco, 2001; Belvedere, 2008). Some formulae are dependent upon the presumed gait: $D_{GA} = D_{MP} + SL$ (for amble) or $D_{MP} + SL/2$ (alternate gait) or $D_{MP} + SL/3$ (asynchronous gait) as described in (Leonardi, 1987). It seems intuitive to match a left-right pair of pes tracks with the subsequent left-right pair of manus tracks and to conclude that their separation corresponds to D_{GA}. But as was just discussed, a given trackway may be created by different-sized trackmakers, with multiple intervening steps separating those pes and manus tracks that were laid down at a given time.

As discussed with regard to Figures 13.8 and 13.9, when creating the various trackway patterns, the pes tracks are laid down one or two complete walk cycles behind the manus tracks at any point in time. With a fossil trackway, it follows that the pair of manus tracks corresponding to a given pair of pes tracks (in order to correctly derive glenoacetabular distance) could have been two or three strides ahead of the pes tracks. For walk cycles with sufficiently short strides, it is possible for the manus to be fully four steps ahead of the pes. Henderson (2006) interpreted the forelimb pair as two complete step cycles ahead of the hind limb pair; however, this factor does not hold for all trackways independently of SL.

Estimating glenoacetabular distance based on the distance between manus and pes tracks might seem particularly challenging in the case of complete pes overprinting, whereupon the manus tracks are obliterated. As illustrated in Figure 13.6, complete overprinting could occur for SL = 100 with (LP = 0.5, D_{GA} = 250), whereupon the manus on one side is one track ahead of its ipsilateral counterpart pes and the manus on the other side is two tracks ahead of its counterpart.

The near absence of manus tracks in some sauropod trackways has led to the speculation that the sauropods might have engaged in bipedality in those instances (Mazzetta and Blanco, 2001; Li et al., 2006). At the Highway A16 tracksites, some very long trackways are almost entirely composed of pes tracks (e.g., at the Courtedoux-Béchat Bovais tracksite, level 515, trackway S23 has but 1 manus track out of 44, and S24 has only 2 out of 66). Those very few intact manus tracks may constitute significant evidence for an interruption in an otherwise tightly regulated gait that tends to create pes-only trackways.

When both manus and pes tracks are distinctly preserved, estimation of glenoacetabular distance requires identifying which pair of manus tracks were associated with a given pair of pes tracks when the latter tracks were being laid down. But without knowledge of the specific gait, estimating D_{GA} is underconstrained and especially difficult when SL is small. At the Highway A16 tracksites, many sauropod trackways exhibit very short stride lengths relative to the size of their individual tracks, and often with pes-only tracks (Fig. 13.7).

An independent method for estimating trackmaker size derives from manus or pes track measurements, scaling from the size of the track to an estimate of hip height, then from hip height to body length. For example, the hip height would be estimated by a rough proportionality on pes track length $H_A \approx 4PL$ (Alexander, 1976), or using pes track width $H_A \approx 4PW$ (Lockley, 1986), or $H_A \approx 5.9PL$ (Thulborn, 1990), or to three decimal places $H_A \approx 4.586PL$ (González Riga, 2011) for a titanosaur. Given an estimate of hip height H_A, body length D_{GA} can then be estimated using an empirically derived proportionality of hip height to glenoacetabular distance D_{GA}/H_A, which was estimated to vary from 0.97 to 1.26 for sauropods (e.g., Mazzetta and Blanco, 2001; Mazzetta, Christiansen, and Fariña, 2004). But in fact numerical estimation of trackmaker dimensions such as D_{GA} and H_A from track measurement such as PL or PW are necessarily highly uncertain given that (1) track dimensions reflect the conditions of the substrate, and not just the dimensions of the trackmaker (Manning, 2004), (2) sauropod articular skeleton proportions vary considerably across taxa, (3) skeletal proportions also vary with ontogenetic stage, and (4) determining the taxonomic identity of a trackmaker and its developmental state is difficult to do in other than the broadest terms.

TRACKMAKER SPEED AND LIMB PROPORTIONS

Estimates of speed are generally based on the trackmaker's inferred hip (acetabular) height $H_A^{-1.17}$ (Alexander, 1976, 2006):

$$V = 0.25 \, G^{0.5} \, SL^{1.67} \, H_A^{-1.17}$$

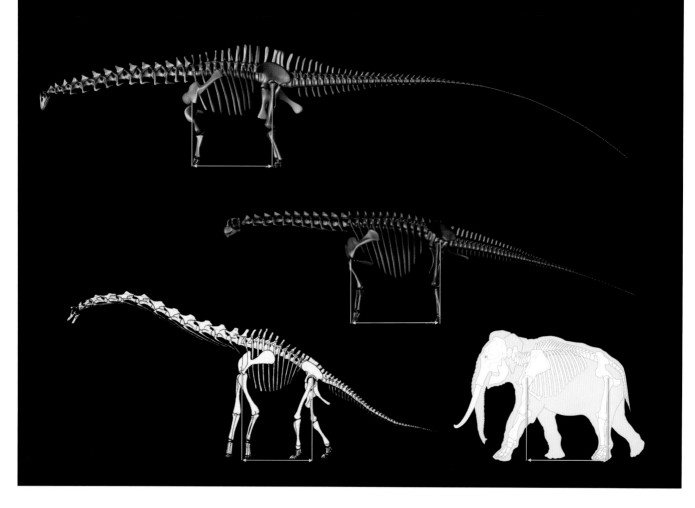

13.10. Forelimb and hind limb height (H$_G$ and H$_A$) relative to D$_{GA}$ for *Apatosaurus* (top), *Camarasaurus* (middle), and *Giraffatitan* (lower left) and African elephant. See text for details. *Giraffatitan* image after (Czerkas and Czerkas, 1991:132), elephant image from (Beauval, 2012, (C) ArchéoZoo.org). Not to scale.

Hip height, in turn, can be estimated by alternative methods, as was just discussed. Estimation of H$_A$ from D$_{GA}$ would be subject to the uncertainty in estimated D$_{GA}$ compounded with the uncertainty associated with the proportionality D$_{GA}$/H$_A$ for an unknown sauropod trackmaker. If H$_A$ were estimated from pes width PW or length PL, that estimate would be subject to uncertainty in the proportionality between those measurements and H$_A$.

Sauropod appendicular skeletons varied considerably in limb lengths (H$_G$ and H$_A$) and limb proportions (H$_G$/D$_{GA}$, H$_A$/D$_{GA}$, and H$_G$/H$_A$). At one extreme, the diplodocid sauropods had substantially shorter forelimbs than hind limbs (H$_G$/H$_A$ ≈ 0.6–0.7), as shown in Figure 13.10 (see also Table 13.1). Given this disparity, the limiting factor in determining maximum stride length in diplodocids may have been the forelimb height H$_G$, not hip height H$_A$, as generally has been expected for quadrupeds.

Sauropods varied considerably in the ratio of forelimb to hind limb height (0.6 ≤ H$_G$/H$_A$ ≤ 1.1). This ratio is fundamental to its feeding height, as the acetabular axis can be regarded as a fulcrum about which the vertebral column pivots, and varying H$_G$/H$_A$ varies the slope of the column,

and hence the height of the head that is suspended at the end of the presacral vertebra some distance from the sacrum. Because of the ventral placement of the pectoral girdles in sauropods (Schwarz, Frey, and Meyer, 2007; Stevens, 2013), the forelimbs need not be as long as the hind limbs in order that the dorsal vertebral column be horizontal. The relatively short diplodocid forelimbs (H$_G$/H$_A$ ≈ 0.7) result in an approximately horizontal anterior dorsals column (Fig. 13.10). The cervicodorsal column emerged from the shoulders as straight and horizontal and thus was well-adapted for low browsing, especially for sweeping laterally and downward. In camarasaurids, the ratio of forelimb to hind limb length was somewhat higher (H$_G$/H$_A$ ≈ 0.8), which resulted in an upward slope to the back as it passed through the shoulders. This slope was further accentuated by the very tall brachiosaurid forelimbs (H$_G$/H$_A$ ≈ 1.1). Whereas a given increase in forelimb length relative to the hind limbs elevates the pectoral girdles accordingly, more significantly, it creates an inclined vertebral column as if the presacral vertebral column pivoted about a fulcrum, the acetabulum, and a modest increase in forelimb length translates trigonometrically into a substantial increase in head height at the far end of the sloping neck.

Increasing both forelimb and hind limb length in equal measure raised both the head and heart equally. Contrary to the popular depiction of some sauropods (such as brachiosaurids and camarasaurids) with near-vertical, swan-like or giraffe-like necks, in all known sauropod taxa, the cervicodorsal vertebrae were in fact osteologically straight extensions of the anterior dorsal column. Whereas sauropod necks were straight at the base, they continued the slope of the back with a slope determined by the sauropod's limb proportions (Stevens, 2013).

The two extremes of limb proportions (H_C/H_A) and dimensions in sauropods could be expected to be reflected in their trackways patterns. The appendicular specializations that facilitated high browsing in brachiosaurs may also have conferred them an advantage for locomotion, because the very long limbs were not only capable of longer absolute stride lengths but also of longer duty factors given their height relative to body length (H_C/D_{GA}). The brachiosaur ratio of limb height to body length was close to that of an elephant ($H_C/D_{GA} = 1.0$) but much less than that of a giraffe ($H_C/D_{GA} = 1.7$). Brachiosaurs may have had the opportunity to engage in a greater repertoire of gaits than diplodocids did. These differences might be more apparent for higher speed locomotion involving longer stride lengths and less so when the sauropods were taking short steps.

Speed estimates using the preceding formula presume the trackmaker is engaged in an efficient steady-state locomotion. Some sauropod trackways undoubtedly represent the traces of individuals making determined process on their way, but some of the Highway A16 trackways may also represent cases of very slow, even hesitant and interrupted, locomotion. In addition to very short stride lengths, apparent "missteps" and additional tracks may constitute evidence for slow trackmaker speeds that may derive from coercing a trackmaker model to follow such highly variable, erratic trackways.

TRACKMAKER KINEMATICS

The proportions, stance or posture, and range of motion of a quadruped's appendicular skeleton are reflected in its characteristic locomotion patterns, which in turn leave identifiable traces of that motion in its trackways. This is especially apparent in modern vertebrates with sprawling gaits, in which the limbs achieve protraction and retraction through the enlistment of lateral flexion of the trunk (Ritter, 1992; Farley and Ko, 1997; Reilly and Elias, 1998; Kubo and Osaka, 2008; Carpenter, 2009; Kubo, 2010a, 2010b). Likewise the semisprawling forelimb posture of ceratopsians (Johnson and Ostrom, 1995; Dodson and Farlow, 1997; Paul and Christiansen,

Table 13.1. Estimates of ratios of forelimb to hind limb height and relative height of forelimb (H_C) and hind limb (H_A) compared to glenoacetabular distance (D_{GA})

	H_C/H_A	H_C/D_{GA}	H_A/D_{GA}
Dicraeosaurus	0.6	0.6	1.0
Mamenchisaurus	0.9	0.7	0.8
Diplodocus	0.7	0.7	1.0
Camarasaurus	0.8	0.7	0.9
Supersaurus	0.7	0.8	1.1
Apatosaurus	0.7	0.8	1.1
Camel	1.0	0.9	0.9
Dog	1.0	0.9	0.9
Horse	0.9	0.9	1.0
Boar	1.0	1.0	1.0
Elephant	1.0	1.0	1.0
Giraffatitan	1.1	1.0	1.0
Llama	0.9	1.1	1.3
Giraffe	1.0	1.7	1.7

Notes: Table sorted by increasing relative forelimb height H_C/D_{GA}, see text. Measurements averaged from archival and recent skeletal illustrations in lateral view, within 10%. See also Mazzetta and Blanco (2001: table 2).

2000; Thompson and Holmes, 2007) and the broad hind limb stance of titanosaurs (Wilson and Carrano, 1999) are expected to be reflected in trackway gauge or relative width.

Considerable attention has been given to explaining trackway gauge but less so to the kinematics of creating forward progress. Sauropod limbs, though fundamentally columnar, do not articulate in a parasagittal plane as do those of elephants and other graviportal mammals, due to the posteroventral orientation of the glenohumeral joint and the anteromedial path of the distal forelimb during both humeral rotation and elbow flexion (Bonnan, 2003; Wills, 2008; Stevens and Wills, 2009). The forelimbs, though more columnar than the semisprawling ceratopsian configuration, provide important constraint on the interpretation of sauropod trackways. Digital reconstructions of the forelimb and pectoral girdle of *Apatosaurus* with articulation as in Bonnan (2003) and pectoral girdle articulation as in Schwarz, Frey, and Meyer (2007), were explored kinematically (Wills, 2008; Stevens and Wills, 2009). A genetic algorithm search strategy examined a very large "configuration space" of possible limb poses to find a step cycle that achieves a smooth path (which minimizes lateral, vertical, and angular deviations of the anterior dorsals). Each limb comprises a kinematic chain from a reference point on the anterior axial skeleton (RP_A) down to the manus locator (L_{LM} or L_{RM}), with degrees of freedom at the scapula (rotation and elevation), shoulder (flexion/extension, abduction/adduction, and humeral rotation), elbow (flexion/extension), antebrachium (pronation/supination), and wrist (flexion/extension). These eight degrees of freedom were sampled, and combinations thereof constituted a

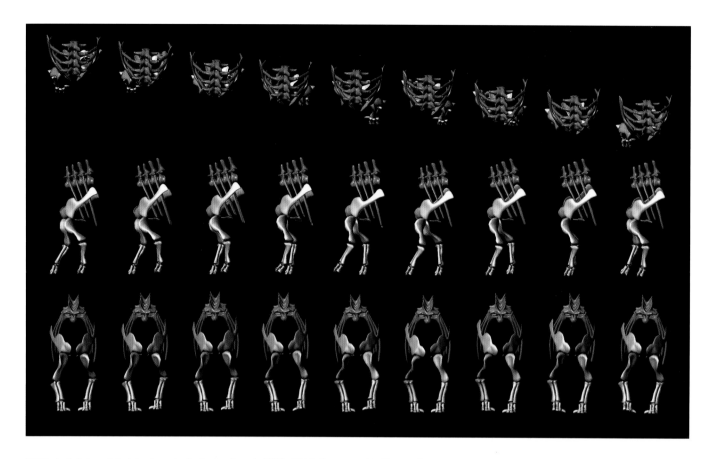

13.11. A digital model of *Apatosaurus louisae* walk cycle (Wills, 2008; Stevens and Wills, 2009).

large configuration space of possible limb/girdle poses. The left and right forelimbs were then exercised out of phase to solve for a symmetrical step cycle for specified parameters DF, H_{GA}, and SL that minimized deviations in RP_A from a smooth forward path. It was found that whereas each forelimb indeed flexes in an anteromedial rather than parasagittal plane individually, as part of a module comprising the two limbs, their girdles and the anterior dorsal column, the RP_A could be propelled down a reasonably smooth forward path, for sufficiently short SL (of roughly 1–1.5 m); see the Muybridge-style motion sequence in Figure 13.11. As stride length increases, the nonparasagittal limb movements no longer permit a smooth step cycle (Wills, 2008). The pattern of joint movements shown in the animation stills represents the near maximum stride length achievable without the manus breaking contact and twisting on the substrate, or the anterior dorsal column pitching and rolling excessively. The kinematics of sauropod forelimbs are critical in constraining estimates of maximum stride lengths, as they are more restrictive than hind limb kinematics are and show relatively greater variability across sauropod taxa.

Reflecting upon the preceding discussion, there are multiple disjoint lines of evidence, or in other words constraints, upon the interpretation of trackways. Some lines are diagnostic, and, if true, rule out some possibilities categorically and support others. Others are at least roughly quantitative and permit placing limits or bounds on some numerical estimates. Although such formulae may capture a rough empirical proportionality (e.g., between pes length and hip height), they can also be used as constraints rather than computing one term from the other (as discussed in the following). Trackway interpretation would be stymied if these various weak, ambiguous, and uncertain lines of evidence cannot be somehow combined.

A RESEARCH PERSPECTIVE: ACTIVE VERSUS PASSIVE SOLVERS

A trackway records position but not timing, whereas the gait only determines timing, but not position. To interpret a specific trackway in terms of the motion of a trackmaker across the substrate requires somehow determining what trackmaker was involved (to some specificity) and how it moved in order to have created that trackway. Of the two broad approaches to trackway interpretation considered here, one is through simulation: creating an animated, parametric model of the trackmaker, then finding a set of values for those parameters that results in an animation sequence that successfully replicates some aspects of the trackway, if not reproducing the trackway pattern itself. This will be termed an "active solver."

Active Solvers

To illustrate an application of an active solver, suppose the trackmaker is known but its movement pattern is unknown. The basic method involves specifying all the parameters of the trackmaker model, setting the model into motion, and observing the consequences (e.g., Sellers and Manning, 2007; Sellers et al., 2009; Sellers et al., 2013). The trackmaker model incorporates large numbers of parameters to encode the contribution of many major muscle groups, each with parameters to describe their origin and insertion points and activation patterns. Additionally, the model is initialized with constants specifying skeletal dimensions, joint locations, insertions and origins of muscle groups, and the body's mass distribution (Sellers et al., 2013). The trackmaker is thus well determined, but its movement is not. In this form of active solver, the trackmaker is specified by initial conditions, but its movements are left to be discovered by genetic algorithm. Rather than presume specific values for gait parameters (DF, LP, and so forth), a huge space of possible patterns of muscle activations is searched to find an efficient combination that permits efficient locomotion. When a suitably stable and efficient step cycle is eventually found, the observables include the parameters of the resultant trackway pattern, the trackmaker's gait parameters, and speed.

As discussed earlier, a study of limb kinematics (Wills, 2008; Stevens and Wills, 2009) also used genetic algorithms to search for a kinematic solution to efficiently protract/retract a limb during its walk cycle, as discussed regarding Figure 13.11. This study, although kinematic and not dynamic, shared the goal of efficiently exploring a huge space of possible combinations of unknowns, after carefully preloading the algorithm with sufficiently many known (constants specific to the simulation). That is, part of the problem space is strongly constrained (e.g., to model a particular trackmaker in adequate detail) and the the rest is allowed to grow combinatorially (within practical computational limits). Not all active solvers involve searching a large space: a strongly constrained trackmaker is constrained to a specific motion so that the result can be tested (Henderson, 2006). The trackmaker model, a parametric representation that could be set to the overall dimensions and mass distribution of either of two sauropods (brachiosaur and diplodocid) was constrained to move with either a narrow or wide gauge, with fixed gait parameters (LP, DF, and SL). When set in motion, the behavior of the model was observed, in this case to determine how well the different trackmakers could have maintained balance when following the two alternative trackways.

But how to proceed if one is presented with only the specific trackway and seeks an interpretation in terms of both an unknown trackmaker and unknown movements? A straightforward search of the combined space of possible trackmakers and their possible movements by an active solver? With that many unknowns, it would require searching over an intractably large space of combinations of possible candidate trackmakers and candidate movements.

Passive Solvers

Trackway interpretation is thus characterized by its many sources of uncertainty: the basic measurement process itself is subject to uncertainty, compounded by sources of ambiguity in the interpretation of gait, size, speed, and taxonomic identity (these constituting a representative, but by no means exhaustive, list of confounding factors). The computational goal, then, is to introduce independent constraints to reveal underlying, interpretable patterns in the behavior of the trackmaker. The result of this process is a constrained interpretation, not a unique interpretation.

The various formulae discussed herein can be regarded as simple models, each describing a functional relationship between some trackmaker property and measurable track properties, with empirically determined constants of proportionality and coefficients. As mentioned, these formulae are sometimes used "in series" to estimate terms in other expressions (such as estimating speed based on an estimate of H_A based on an estimate of D_{GA} based on a presumed relationship between D_{GA} and trackway measurements), but the uncertainties are compounded by such cascading. Trackway interpretation likely cannot advance without attempting more sophisticated models that are capable of simultaneously expressing more functional relationships and constraints than an simple formulae.

In conceptual terms, a quantitative model of the trackmaker is created and another of the trackway, and the latter is used as input to constrain the former. Consider a highly simplified model of a trackmaker that embodies minimal parameters for body size and limb proportions and gait parameters, positional information in terms of locators and reference points: {D_{GA}, H_A, LP, DF, L_{LM} L_{RM}, LLP, LRP, RP_A, RP_P}. A specific trackway would consist minimally of a set of track descriptors {LP_i, RP_i, LM_i, RM_i}, one for each track (left or right, manus or pes; LP_i describes the i-th left pes track, and so forth). An individual track is described by a set of attributes including its dimensions (e.g., PL, PW for the pes), position, and orientation at a minimum. With all track positions and orientation measured relative to a common coordinate frame (such as a geographic marker), other trackway measurements such as stride length SL, track phase TP, and pace angulation are readily computed.

In this framework, a trackway interpretation is a set of model parameter values more or less consistent with the specific trackway data. Each parameter can be assigned a specific value (such as assigning a particular length for $D_{GA} = 3.25$) or a range of values (such as $3.0 \leq D_{GA} \leq 3.5$). The interpretation process is one of finding a fit between the trackmaker model and the trackway data. The ambiguities and uncertainties discussed earlier result in there being many such trackway interpretations consistent with any set of trackway data. Returning to Figure 13.9, for example, the trackway {SL = 100, TP = 0.5} has three distinct interpretations (of many): {LP = 0.5, D_{GA} = 300 cm}, {LP = 0.25, D_{GA} = 325 cm}, and {LP = 0.0, D_{GA} = 250 cm}. Note that while the trackways in Figures 13.8 and 13.9 were generated from a specific assignment of values to the model parameters; trackway interpretation would proceed in the opposite direction, from a specific set of trackway data to a set of model parameter values.

Given any set of trackway data, there is an enormous space of possible combinations of trackway parameter values (or configuration space) to explore in order to select possible solutions. Measurement uncertainties would prevent determining a unique solution (set of model parameter values). Instead of selecting a specific point in this configuration space, trackway interpretation would, at best, define a small region in this configuration space, or set of such regions if very different interpretations can be derived.

Cadence, a Passive Solver

A passive-solver approach is introduced to combine two seemingly disjointed forms of information by the addition of internal constraints involving both position and timing. Our approach is to add such constraints conservatively and to explore what they yield incrementally, rather than attempt to create a highly elaborate model that entangles many such factors. We go from the ground up, the pes and manus locators, and build above them a scaffolding, a model. Because the speed of locomotion cannot be assumed to be uniform, we adopt the notion of a timeline from animation so that we might visualize the model as moving along that timeline, but with specifying the rate (in fact halts in forward progress are not ruled out). Each of the four feet are then in one of two states: either stationary on the ground bearing weight, or lifted wherein they are transported forward until they again are placed on the ground by the timing dictated by the gait.

Software named Cadence is under development to facilitate the interpretation of dinosaur trackways discovered during the excavation for Highway A16 near Porrentruy, Switzerland. Trackway data, compiled into a relational database, permit generation of a statistical model of the trackway to better understand covariances and recurrent characteristics of the trackways—relationships to be explored in Cadence. Although dealing with a generalized, statistically averaged, trackway avoids the idiosyncrasies in any specific trackway, later those outliers and transient behaviors (possible indicators of starting, stopping, and steps in preparation of turning) will be of greater importance.

The trackmaker model builds on a set of four locators, each of which alternates between support and suspended states, in other words, between stationary contact with the substrate aligned with one track and protraction as it lifts from one track to take the next step to land upon the next track, as controlled by a universal time variable. Time is measured such that 0.0 corresponds to the instant the left pes just touches ground and 1.0 to the completion of a full walk cycle. The position and state of all locators are thus controlled by this time variable, and together they reconstruct a pattern of steps as each foot comes into contact with the substrate and then lifts again. The visual impression is of four locators stepping in according with a given gait to be placed in registration with a trackway pattern. The animation plays out at any speed and can be paused or "scrubbed" as in conventional digital animation. The resulting spatiotemporal pattern of foot placements is constrained spatially by the trackway and temporally by the gait parameters (limb phase and duty factor). These constraints vary as a function of time, therefore, and propagate up from the locators into the model whereupon their effects can be observed. By analogy, the feet of a traditional marionette are suspended from strings; pull the strings and the feet will follow, and in the given sequence, they give the illusion of walking. The marionette is passive, driven externally, and yet those inputs propagate inward to the elbows and knees, shoulders, hips, and body, which adjust accordingly, in passive response. The marionette—being subject to, and controlled by external forces—is passively solving a reaction to those external forces subject to internal constraints; it is a "passive solver." Likewise, our virtual trackmaker is controlled externally yet subject to internal constraints. We recognize that the greatest potential in the A16 trackway data is in providing insights that are revealed only statistically. The variability along a typical trackway is often too high to permit drawing well-supported conclusions from just individual trackway samples in isolation.

Forcing the locators to follow a specific trackway constitutes an external "forcing function" to be reconciled with the internal constraints (such as the specified gait parameters, physical dimensions, and kinematic constraints). Each locator alternates between static support (in stationary contact with the substrate) and protraction to take the next step. Cadence permits visualizing a sequence of footsteps by animating the four manus and pes locators as they traverse the successive track positions along a trackway, as constrained by timing and other parameters. Combinations of internal constraints that cannot be reconciled with the external constraints are excluded, permitting a narrowing of the search space. Observables such as accelerations, decelerations, hesitations, and pauses in the model through the step cycle or along a trackway then guide an analysis of the goodness of fit between a given set of parameter values (i.e., model) and the trackway. It is here where the passive solver aspect of Cadence comes into play, allowing for an understanding of the role of those model parameters either singly or in combination. For example, it is presumptuous to assign a specific value to duty factor (the period of time when each limb is in support phase, that is, in contact with the substrate, as a fraction of the full step cycle). Duty factor is not an independent variable; it derives from, and trades off with, other constraints.

NEXT STEPS

The space of possible combinations is far too great to explore exhaustively without understanding more about the nature of these interdependencies. Insight will come from understanding how trackways are constrained, and although some of the more dramatic constraints will come from transient non-steady-state situations, such as turns, starts and stops, and hesitations—which should have predictably different consequences for different types of trackmakers—the most basic results will come from exploring a large number of configurations (i.e., potential combinations of free parameters) to search out patterns that will allow us to reduce the space of possible solutions. As this space of possible solutions shrinks, the model can be refined and the search process

repeated on the increasingly confined solution space. The expectation is that much can be learned from a very simple kinematic model that builds up from the locators, prior to introducing physics (sensu Cavagna, Thys, and Zamboni, 1976; Alexander and Jayes, 1978a, 1978b; Kokkevis, Metaxas, and Balder, 1995; Griffin, Main, and Farley, 2004).

With a passive solver, the goal is not mimicking behavior but understanding the influences of variations of the model on its resultant behavior. Passive solvers permit exploring the relationship between trackmaker and trackway with a gradual, general-to-specific approach. Importantly, rather than the resultant trackway being observable, it may be part of the initial conditions imposed on the model, such that the model is "coerced" to replicate a given pattern of tracks such as one derived from actual trackway data. Searching a space of parametric trackway models allows for exploration of tradeoffs between specified initial conditions and observables. The approach is incremental and conservative, and it follows a scientific method adapted from physics, which is metaphorically referred to in terms of a "spherical cow."[1] Toward this end, the model begins simply, incorporating very weak constraints, then the model is progressively refined to incorporate more specific assumptions about gait, limb range of motion, inertial and gravitational influences, and so forth, all adding to the complexity to the trackmaker model.

ACKNOWLEDGMENTS

We wish to thank the editors for inviting us to publish in this volume. The Swiss Federal Roads Authority and Canton Jura funded the excavation and documentation of the Highway A16 tracksites and part of this research. These contributions are gratefully acknowledged. The authors thank Drs. Karl Bates, Peter Falkingham, and Donald Henderson for their very useful suggestions in their review of the manuscript.

NOTE

1. The term is metaphorical and refers to the strategy wherein to model some complex phenomenon (a cow for instance), one starts with the simplest representation (a sphere in the case of a cow), then incrementally elaborates that model to capture increasingly more specific aspects of the object of study.

REFERENCES

Alexander, R. McN. 1976. Estimates of the speed of dinosaurs. Nature 261: 129–130.

Alexander, R. McN. 1989. Dynamics of Dinosaurs and Other Extinct Giants. Columbia University Press, New York, New York, 167 pp.

Alexander, R. McN. 2006. Dinosaur biomechanics. Proceedings of the Royal Society 273: 1849–1855

Alexander, R. McN., and A. S. Jayes. 1978a. Vertical movements in walking and running. Journal of Zoology 185: 27–40.

Alexander, R. McN., and A. S. Jayes. 1978b. Optimum walking techniques for idealized animals. Journal of Zoology 186: 61–81.

Alexander, R.M., and A. S. Jayes. 1983. A dynamic similarity hypothesis for the gaits of quadrupedal mammals. Journal of Zoology 201(1): 135 152.

Beauval, C. (artist). 2012. Straight-tusked elephant (Elephas [pallaeoloxodon] antiquus. [Web Graphic]. Available at http://photos .archeozoo.org/picture/2594-elephas_antiquus /category/86-proboscidiens_langen _proboscidea_lang_langes_proboscidios _lang_. Accessed November 6, 2015.

Belvedere, M. 2008. Ichnological researches on the Upper Jurassic dinosaur tracks in the Iouaridène area (Demnat, Central High-Atlas, Morocco). Ph.D. dissertation, Università degli Studi di Padova, Padua, Italy. 121 pp.

Bevington, P., and D. K. Robinson. 2002. Data Reduction and Error Analysis for the Physical

Sciences. 3rd edition. McGraw-Hill Higher Education, New York, New York, 336 pp.

Biewener, A. A. 1983. Allometry of quadrupedal locomotion: the scaling of duty factor, bone curvature and limb orientation to body size. Journal of Experimental Biology 105: 147–171.

Biknevicius, A. R., and S. M. Reilly. 2006. Correlation of symmetrical gaits and whole body mechanics: debunking myths in loco-motor biodynamics. Journal of Experimental Zoology 305A: 923–934.

Bonnan, M. F. 2003. The evolution of manus shape in sauropod dinosaurs: implications for functional morphology, forelimb orientation, and phylogeny. Journal of Vertebrate Paleontology 23(3): 595–613.

Carpenter, K. 2009. Role of lateral body bending in crocodylian track making. Ichnos 16(3): 202–207.

Cavagna, G.A., H. Thys, and A. Zamboni. 1976. The sources of external work in level walking and running. Journal of Physiology 262: 639–657.

Czerkas, S. A., and S. Czerkas. 1991. Dinosaurs: A Global View. Mallard Press, New York, New York, 247 pp.

Dodson, P., and J. O. Farlow. 1997. The fore-limb carriage of ceratopsid dinosaurs; pp. 393–398 in D. L. Wolberg, E. Stump, and G. D. Rosenberg (eds.), Dinofest International: Proceedings of a Symposium Held at Arizona State University. Academy of Natural Sciences, Philadelphia, Pennsylvania.

Farley, C. T., and T. C. Ko. 1997. Mechanics of locomotion in lizards. Journal of Experimental Biology 200: 2177–2188.

Farlow, J. O. 1992. Sauropod tracks and trackmakers: integrating the ichnological and skeletal record. Zubia 10: 89–138.

Farlow, J. O., J. G. Pittmann, and J. M. Hawthorne. 1989. Brontopodus birdi, Lower Cretaceous sauropod footprints from the U.S. Gulf Coastal plain; pp. 371–394 in D. D. Gillette and G. M. and Lockley (eds.), Dinosaur Tracks and Traces. Cambridge University Press, Cambridge, UK.

González Riga, B.J. 2011. Speeds and stance of titanosaur sauropods: analysis of Titanopodus tracks from the Late Cretaceous of Mendoza, Argentina. Annals of the Brazilian Academy of Science 83(1): 279–290.

Grey, J. 1944. Studies in the mechanics of the tetrapod skeleton. Journal of Experimental Biology 20: 88–116.

Griffin, T. M., R. P. Main, and C. T. Farley. 2004. Biomechanics of quadrupedal walking: how do four-legged animals achieve inverted pendulum-like movements? Journal of Experimental Biology 207: 3545–3558.

Henderson, D. M. 2006. Burly gaits: centers of mass, stability, and the trackways of sauropod dinosaurs. Journal of Vertebrate Paleontology 26(4): 907–921.

Hildebrand, M. 1965. Symmetrical gaits of horses. Science 150: 701–708.

Hildebrand, M. 1980. The adaptive significance of tetrapod gait selection. Integrative and Comparative Biology 20: 255–267.

Hildebrand, M. 1985. Walking and running; pp. 38–57 in M. Hildebrand, D. M. Bramble, K. F. Liem, and D. B. Wake (eds.), Functional Vertebrate Morphology. Harvard University Press, Cambridge, Massachusetts.

Johnson, R. E., and J. H. Ostrom. 1995. The forelimb of Torosaurus and an analysis of the posture and gait of ceratopsian dinosaurs; pp. 205–218 in J. J. Thomason (ed.), Functional Morphology in Vertebrate Palaeontology. Cambridge University Press, New York, New York.

Kokkevis, E., D. Metaxas, and N. I. Balder. 1995. Autonomous animation and control of four-legged animals; pp. 10–17 in Proceedings of Graphics Interface '95.

Kubo, T. 2010a. Extant lizard tracks: variation and implications for paleoichnology. Ichnos 17(3): 187–196.

Kubo, T. 2010b. Variation in modern crocodylian limb kinematics and its effect on trackways; pp. 51–53 in J. Milàn, J., S. G. Lucas, M. G. Lockley, and J. A. Spielmann (eds.), Crocodile Tracks and Traces. New Mexico Museum of Natural History and Science, Albuquerque, New Mexico.

Kubo, T., and M. Osaka. 2009. Does pace angulation correlate with limb posture? Palaeogeography, Palaeoclimatology, Palaeocology 275: 54–58.

Leonardi, G. 1987. Glossary and Manual of Tetrapod Footprint Palaeoichnology. Publicação do Departemento Nacional da Produção Mineral Brasil, Brasília, Brazil.

Li, D., Y. Azuma, M. Fugita, Y.-N. Lee, and Y. Arakawa. 2006. A preliminary report on two new vertebrate track sites including dinosaurs from the Early Cretaceous Hekou Group, Gansu Province, China. Proceedings of the Goseong International Dinosaur Symposium. Journal of the Paleontological Society of Korea 22(1): 29–49.

Lockley, M. G. 1986. A guide to dinosaur track-sites of the Colorado Plateau and American Southwest. Geology Department Magazine, University of Colorado at Denver, Special Issue 1: 56.

Lockley, M. G. 1998. The vertebrate track record. Nature 396: 429–432.

Lockley, M. G., and A. P. Hunt. 1995. Ceratopsid tracks and associated ichnofauna from the Laramie Formation (Upper Cretaceous: Maastrichtian) of Colorado. Journal of Vertebrate Paleontology 15(3): 592–614.

Lockley, M. G., M. Huh, and B. S. Kim. 2012. Ornithopodichnus and pes-only sauropod track-ways from the Hwasun tracksite, Cretaceous of Korea. Ichnos 19(1–2): 93–100.

Lockley, M. G., C. A. Meyer, A. P. Hunt, and S. G. Lucas. 1994. The distribution of sauropod tracks and trackmakers. Gaia 10: 233–248.

Lockley, M. G., J. L Wright, S. G. Lucas, and A. P. Hunt. 2001. The Late Triassic sauropod track record comes into focus: old legacies and new paradigms; pp. 181–190 in S. G. Lucas and D. Ulmer-Scholle (eds.), Geology of Llano Estacado. New Mexico Geological Society Fall Field Conference Guidebook 52.. New Mexico Geological Society, Socorro, New Mexico.

Lockley, M. G., A. S. Schulp, C. A. Meyer, G. Leonardi, and D. K. Mamani. 2002. Titanosaurid trackways from the Upper Cretaceous of Bolivia: evidence for large manus, wide-gauge locomotion and gregarious behaviour. Cretaceous Research 23: 383–400.

Manning, P. L. 2004. A new approach to the analysis and interpretation of tracks: examples from the Dinosauria; pp. 93–123 in D.

McIlroy (ed.), The Application of Ichnology to Palaeoenvironmental and Stratigraphic Analysis. Special Publications 228. Geological Society London, London, UK.

Marty, D. 2008. Sedimentology, taphonomy, and ichnology of Late Jurassic dinosaur tracks from the Jura carbonate platform (Chevenez–Combe Ronde tracksite, NW Switzerland): insights into the tidal-flat palaeoenvironment and dinosaur diversity, locomotion, and palaeoecology. Ph.D. dissertation, University of Fribourg, Fribourg, Switzerland; in GeoFocus 21. University of Fribourg, Fribourg, Switzerland, 278 pp.

Marty, D. 2009. Sedimentology, taphonomy, and ichnology of Late Jurassic dinosaur tracks from the Jura carbonate platform (NW Switzerland): insights into the tidal-flat palaeoenvironment and dinosaur diversity, locomotion, and palaeo-ecology. Journal of Vertebrate Paleontology 29 (3, Supplement A): 144A.

Marty, D., C. A. Meyer, and J.-P. Billon-Bruyat. 2006. Sauropod trackway patterns expres-sion of special behaviour related to substrate consistency? An example from the Late Jurassic of northwestern Switzerland. Hantkeniana 5: 38–41.

Marty, D., A. Strasser, and C. Meyer. 2009. Formation and taphonomy of human footprints in microbial mats of present-day tidal-flat envi-ronments: implications for the study of fossil footprints. Ichnos 16: 127–142.

Marty, D., G. Paratte, C. Lovis, M. Jacquemet, and C. A. Meyer. 2010. Extraordinary sauropod trackways from the Late Jurassic Béchat Bovais tracksite (Canton Jura, NW Switzerland): implications for sauropod locomotor styles. 8th Annual Meeting of the European Association of Vertebrate Palaeontologists, June 7–12, Aix-en-Provence, France, 56 pp.

Marty, D., L. Cavin, W. A. Hug, C. A. Meyer, M. G. Lockley, and A. Iberg. 2003. Preliminary report on the Courtedoux dinosaur tracksite from the Kimmeridgian of Switzerland. Ichnos 10: 209–219.

Marty, D., L. Cavin, W. A. Hug, P. Jordan, M. G. Lockley, and C. A. Meyer. 2004. The protec-tion, conservation and sustainable use of the Courtedoux dinosaur tracksite, Canton Jura, Switzerland. Revue de Paléobiologie Volume Spécial 9: 39–49.

Marty, D., M. Belvedere, C. A. Meyer, P. Mietto, G. Paratte, C. Lovis, and B. Thüring. 2010. Comparative analysis of Late Jurassic sauropod trackways from the Jura Mountains (NW Switzerland) and the central High Atlas Mountains (Morocco): implications for sauropod ichnotaxonomy. Historical Biology 22(1–3): 109–133.

Mazzetta, G. V., and R. E. Blanco. 2001. Speeds of dinosaurs from the Albian-Cenomanian of Patagonia and sauropod stance and gait. Acta Palaeontologica Polonica 46: 235–246.

Mazzetta, G. V., P. Christiansen, and R. A. Fariña. 2004. Giants and bizarres: body size of some southern South American Cretaceous dino-saurs. Historical Biology 16(2–4): 71–83.

Meyer, C.A. 1990. Sauropod tracks from the Upper Jurassic Reuchenette Formation (Kimmeridgian, Lommiswil, Kt. Solothurn) of northern Switzerland. Eclogae Geologicae Helvetiae 82(2): 389–397.

Milàn, J. 2006. Variations in the morphology of emu (Dromaius novaehollandiae) tracks reflecting differences in walking pattern and substrate consistency: ichnotaxonomic implications. Palaeontology 49(2): 405–420.

Milàn, J., and R. G. Bromley. 2006. True tracks, undertracks and eroded tracks, experimental work with tetrapod tracks in laboratory and field. Palaeogeography, Palaeoclimatology, Palaeoecology 231: 253–264.

Milàn, J., and R. G. Bromley. 2008. The impact of sediment consistency on track and undertrack morphology: experiments with emu tracks in layered cement. Ichnos 15(1): 18–24.

Padian, K., and P. E. Olsen. 1984a. Footprints of the Komodo monitor and the trackways of fossil reptiles. Copeia 1984: 662–671.

Padian, K., and P. E. Olsen. 1984b. The fossil trackways Pteraichnus: not pterosaurian, but crocodilian. Journal of Paleontology 58: 178–184.

Paul, G. S., and P. Christiansen. 2000. Forelimb posture in neoceratopsian dinosaurs: implications for gait and locomotion. Paleobiology 26(3): 450–465.

Reilly, S. M., and J. A. Elias. 1998. Locomotion in Alligator mississippiensis: kinematic effects of speed and posture and their relevance to the sprawling-to-erect paradigm. Journal of Experimental Biology 201: 2559–2574.

Ritter, D. 1992. Lateral bending during lizard locomotion. Journal of Experimental Biology 173: 1–10.

Romano, M., M. A. Whyte, and S. J. Jackson. 2007 Trackway ratio: a new look at trackway gauge in the analysis of quadrupedal dinosaur trackways and its implications for ichnotaxonomy. Ichnos 14(3–4): 257–270.

Salgado, L., R. A. Coria, and J. O. Calvo. 1997. Evolution of titanosaurid dinosaurs: I. phylogenetic analysis based on the postcranial evidence. Ameghiniana 34: 3–32.

Santos, V. F., J. J. Moratalla, and R. Royo-Torres. 2009. New sauropod trackways from the Middle Jurassic of Portugal. Acta Palaeontologica Polonica 54(3): 409–422.

Schult, M. F., and J. O. Farlow. 1992. Vertebrate trace fossils; pp. 34–66 in C. G. Maples and R. R. West (eds.), Trace Fossils: Paleontological Society Short Courses in Paleontology 5. University of Tennessee, Knoxville, Tennessee.

Schwarz, D., E. Frey, and C. A. Meyer. 2007. Novel reconstruction of the orientation of the pectoral girdle in sauropods. Anatomical Record 290: 32–47.

Sellers, W. I., and P. L. Manning. 2007. Estimating dinosaur maximum running speeds using evolutionary robotics. Proceedings of the Royal Society, Series B 274: 2711–2716.

Sellers, W. I., L. Margetts, R. A. Coria, and P. L. Manning. 2013. March of the titans: the locomotor capabilities of sauropod dinosaurs. PLoS One 8(10): e78733. doi:10.1371/journal.pone.0078733.

Sellers, W. I., R. L. Manning, T. Lyson, K. Stevens, and L. Margetts. 2009. Virtual palaeontology: gait reconstruction of extinct vertebrates using high performance computing. Palaeontologia Electronica 12(3): 26 pp.

Soergel, W. 1925. Die Fahrten der Chirotheria. Eine paläobiologische Studie. G. Fischer, Berlin, Germany, 92 pp.

Stevens, K. A. 2013. The articulation of sauropod necks: methodology and mythology. PLoS One 8(10): e78572. doi:10.1371/journal.pone.0078572.

Stevens, K. A., and E. D. Wills. 2009. Non-parasagittal yet efficient: the role of the pectoral girdles and trunk in the walk of Triceratops and Apatosaurus. Journal of Vertebrate Paleontology 29 (3, Supplement A): 198.

Taylor, J. R. 1997. An Introduction to Error Analysis: The Study of Uncertainties in Physical Measurements. 2nd edition. University Science Books, Sausalito, California, 327 pp.

Thompson, M. E., R. S. White, and G. S. Morgan. 2007. Pace versus trot: can medium speed gait be determined from fossil trackways?; pp. 309–314 in S. G. Lucas et al. (eds.), Cenozoic Tracks. Bulletin 42. New Mexico Museum of Natural History and Science, Albuquerque, New Mexico.

Thompson, S., and R. Holmes. 2007. Forelimb stance and step cycle in Chasmosaurus irvinenesis (Dinosauria: Neoceratopsia). Palaeontologia Electronica 10(1): 5A:17 pp. Available at http://palaeo-electronica.org/2007_1/step/index.html. Accessed November 6, 2015.

Thulborn, R. A. 1981. Estimated speed of a giant bipedal dinosaur. Nature 292: 273–274.

Thulborn, R. A. 1982. Speeds and gaits of dinosaurs. Palaeogeography, Palaeoclimatology, Palaeoecology 38: 227–256.

Thulborn, R. A. 1989. The gaits of dinosaurs; pp. 39–50 in D. D. Gillette and M. G. Lockley (eds.), Dinosaur Tracks and Traces. Cambridge University Press, Cambridge, UK.

Thulborn, R. A. 1990. Dinosaur Tracks. Chapman and Hall, London, U.K., 410 pp.

Vila, B., O. Oms, À. Galobart, K. T. Bates, V. M. Egerton, and P. L. Manning. 2013. Dynamic similarity in titanosaur sauropods: ichnological evidence from the Fumanya dinosaur tracksite (Southern Pyrenees). PLoS One 8(2): e57408. doi:10.1371/journal.pone.0057408.

Walker, W. F., II. 1971. A structural and functional analysis of walking in the turtle, Chrysemys picta marginata. Journal of Morphology 134: 195–213.

Weishampel, D. B., P. Dodson, and H. Osmólska. 2004. The Dinosauria. 2nd edition. University of California Press, Berkeley, California, 880 pp.

Wills, E. D. 2008. Gait animation and analysis for biomechanically-articulated skeletons. Ph.D. dissertation, University of Oregon, Department of Computer and Information Science, Eugene, Oregon, 287 pp.

Wilson, J. A., and M. T. Carrano. 1999. Titanosaurs and the origin of 'wide-gauge' trackways: a biomechanical and systematic perspective on sauropod locomotion. Paleobiology 25: 252–267.

Wilson, J. A., and P. C. Sereno. 1998. Early evolution and higher-level phylogeny of sauropod dinosaurs. Journal of Vertebrate Paleontology Memoir 5, 68 pp.

Wright, J. L. 2005. Steps in understanding sauropod biology: the importance of sauropod tracks; pp. 252–280 in K. A. Curry Rogers and J. A. Wilson (eds.), The Sauropods. University of California Press, Berkeley, California.

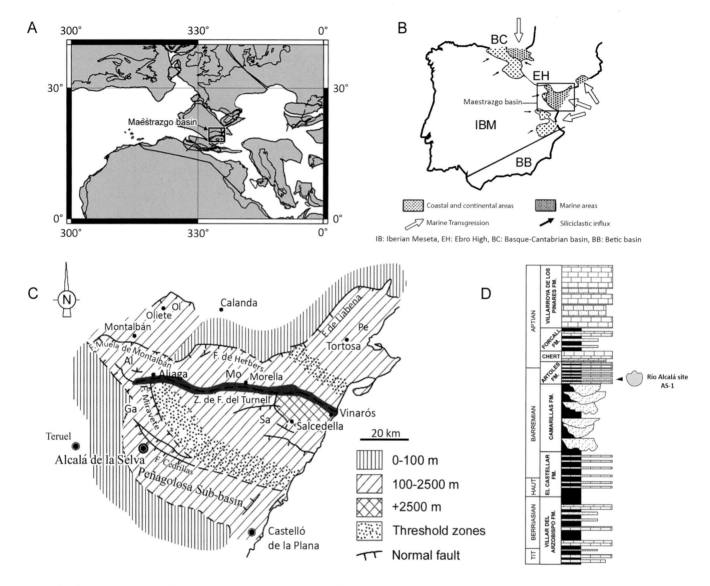

14.1. (A) Tethyan paleogeography for southwestern Europe during the Early Aptian (modified from Skelton, Granier, and Moullade, 2013). (B) Paleogeographic map showing the evolution of Iberia during the Late Barremian to earliest Aptian (modified from Salas et al., 2001). (C) Simplified structural map of the Maestrazgo Basin during the Early Cretaceous (modified from Salas et al., 2001). (D) General stratigraphic column in the area of Río Alcalá tracksite.

Abbreviations: AS-1, Rio Alcalá tracksite; BB, Betic Basin; BC, Basque-Cantabrian Basin; EH, Ebro High; FM, formation; IB, Iberian Meseta.

244

Dinosaur Tracks as "Four-Dimensional Phenomena" Reveal How Different Species Moved

14

Alberto Cobos, Francisco Gascó, Rafael Royo-Torres, Martin G. Lockley, and Luis Alcalá

ALTHOUGH THOUSANDS OF DINOSAUR TRACKS HAVE been found worldwide, three-dimensional (3-D) natural track casts are still relatively poorly documented. Those few that have been published, however, sometimes show impressions of reticulated skin, toe pads, and scratch marks made by scales and may even record how the sole of the foot bore the trackmaker's weight. In very exceptional circumstances, such casts can even preserve evidence of distal limb kinematics of the trackmaker by recording the movement of the feet during track-making: in other words, footfall or footfall registration dynamics. Here we present a description of natural track casts that show all the features just outlined, allowing the contemplation of a new concept: "four-dimensional (4-D) tracks." Highly informative and representative examples of such tracks casts come from a new and exceptional Early Cretaceous tracksite in the Province of Teruel (Spain), as well as from selected North American sites. 4-D tracks are defined as true tracks or their infillings showing slide marks or grooves that reveal the trajectory of the trackmaker's foot within the sediment more completely than do most tracks. Each one of these tracks therefore reflects the time and motion involved in their registration: that is, they more obviously fossilize the fourth dimension of motion and time than typical 3-D tracks do. The Teruel track casts were made by large theropods and ornithopods, possibly spinosaurids and basal hadrosauriforms, walking on deep, firm mud. The infilling sediment responsible for this exceptional preservation consists of fine-grained sandstones with a high proportion of quartz. These were deposited in an estuarine shallow-water carbonate platform with freshwater discharges during the beginning of the Barremian-Aptian transgression of the Tethys Sea in southwestern Europe. Tracks with 4-D characteristics are probably more common than previously thought. Thus, the 4-D track concept has great potential to shed light on foot and distal limb kinematics.

INTRODUCTION

Thousands of dinosaur tracks have been found worldwide, but very few provide information regarding within-substrate limb excursion (Gatesy et al., 1999). Such information can be observed in some natural track casts (or infillings) (Lockley, 1991), although it can also be acquired exceptionally by cutting sections of well-preserved true tracks (Gatesy et al., 1999; Milàn, 2006). However, such methods are laborious and generally destructive. In contrast, foot movements can be traced in infillings (casts) when these preserve the general morphology of the foot along with scratch marks created by the skin as the foot slid within the substrate (Gatesy, 2001; Diffley and Ekdale, 2002). The preservation of such details is quite exceptional and has been described in the tracks of theropods (Gatesy, 2001; Currie, Badamgarav, and Kopphelhus, 2003; Avanzini, Piñuela, and García-Ramos, 2011; Huerta et al., 2012), sauropods (Currie, Badamgarav, and Kippelhus, 2003; Milàn, Christiansen, and Mateus, 2005; Platt and Hatiosis, 2006; Lockley et al., 2008; Lockley, 2014; Xing et al., 2015), ornithopods (Currie, Nadon, and Lockley, 1991; Xing et al., 2012, 2015), stegosaurs (Lockley et al., 2008; Mateus, Milàn, et al., 2011), and ankylosaurs (McCrea, Lockley, and Meyer, 2001). Tracks that preserve the scale impressions responsible for scratch marks are even more exceptional. For example, sensu Davis (2012) only 57 tracks with scale impressions have been recorded in the literature between 1841 and 2010 (see Noe, Lockley, and Hadden [2014], for an additional example). Nevertheless, many natural casts show slide marks on the track wall, indicating that they are true tracks.

Against this background, this work examines taphonomic, paleobiologic, and paleoecologic issues regarding the ichnological material from a new tracksite known as Río Alcalá (AS-1), located in the municipality of Alcalá de la Selva (Teruel Province, Spain), and compares these specimens with a few others from the United States that demonstrate evidence of dynamic movement. The Río Alcalá tracksite was discovered during prospecting work carried out by the Fundación Conjunto Paleontológico de Teruel-Dinópolis in 2012. The exquisite preservation of the AS-1 tracks reveals new anatomical features of ornithopod feet. In addition, details of the footfall dynamics of these ornithopods, and of a large theropod, are preserved. This Spanish tracksite, and the additional North American examples, significantly increase the number of known tracks containing skin impressions of dinosaurs in general and of ornithopods in particular. We

14.2. General overview (main sector) of the Cretaceous Río Alcalá site (AS-1; Universal Transverse Mercator coordinates UTM ED50 Huso 30: E688509, N4467317; Alcalá de la Selva, Teruel Province, Spain) with the natural casts on the underside of a sandstone layer (2AS-1) overlaying mudstone. Scale bar = 200 cm.

propose the new paleoichnological term "four-dimensional tracks" to describe tracks that reveal the trajectory (dynamic movement) of the trackmaker's foot within the sediment more completely than do most tracks.

GEOLOGICAL SETTING FOR RÍO ALCALÁ SITE (TERUEL, SPAIN)

The Río Alcalá tracksite (AS-1) is found in the Peñagolosa sub-basin (the southeastern sector of the Maestrazgo Basin) in the Iberian Chain (Spain). This basin was a significant embayment opened to the Tethyan Alpine area, which was invaded by an epeiric sea during the Barremian/Early Aptian. In this context, the marine sediments started in the transition between the fluvial deposits of the Camarillas Formation (Early Barremian) (Díaz-Molina and Yébenes, 1987; Martín-Closas and Salas, 1994; Soria et al., 1995) and the Artoles Formation. The basal part of this latter unit, 63 m thick, shows an alternation of massive-looking fine-grained (where the AS-1 site occurs) to medium-grained sandstones and purple and gray clays deposited in estuarine shallow-water carbonate platforms with freshwater discharges. Some bioclastic carbonate levels with bivalves, ostreids, and gastropods predominate at the upper part of this formation. Its age

has been established as terminal Barremian–basal Aptian (Salas et al., 1995) based on sequence stratigraphy. The location of the Río Alcalá tracksite on the most basal section of the unit points to a Late Barremian age (Fig. 14.1).

DINOSAUR TRACKS AND TRACKMAKERS FROM RÍO ALCALÁ TRACKSITE

AS-1 contains natural casts, formed by the infilling of deep, true tracks, produced in mud of a consistency sufficient to prevent the collapse of the track walls following the registration of the track. These casts appear on the underside of a sandstone layer (2AS-1) overlaying mudstone (Fig. 14.2). The fine-grained sandstones show mainly quartz (45%), iron oxide (18%), and feldspars (15%). The high proportion of quartz and iron oxide (hard components) was decisive in the taphonomic (biostratinomic) processes, allowing this exquisite preservation (Fig. 14.3).

Twenty casts can be seen in situ in the section in the same stratum—for this reason, it is impossible to identify striding (trackway) sequences—in the exposed 29.3 m length of the track-bearing layer. Another four have been recovered *ex situ* (2AS-1-2; 2AS-1-21; 2AS-1-22; 2AS-1-23) (Figs. 14.2, 14.4, and 14.5, Table 14.1).

14.3. (Left) Plane polarized light and (right) cross polarized light microphotographs of the sandstone infill of the tracks. It shows a high proportion of quartz (45%) and iron oxide (18%).

Theropod Morphotype

Two different morphotypes can be distinguished among these infillings. One is represented by a single tridactyl track, 2AS-1-13 (Figs. 14.4A, 14.4B, and 14.5A), with a maximum depth of 20 cm. Digits II and IV (their tips separated by 40 cm) are preserved, but digit III is missing as the result of recent breakage. Erosion has exposed the upper faces of digits II and IV. Well-preserved traces of a large lateromedially compressed claw measuring 50 mm across and 100 mm in length, belonging to digit II or IV (Fig. 14.4B), are visible. Sigmoidal slide marks or grooves produced by the skin (Gatesy, 2001) of the metatarsophalangeal pad during its registration within the muddy substrate can also be seen (Fig. 14.5A). This track is assigned to a large theropod trackmaker (Thulborn, 1990; Lockley, 1991), perhaps a spinosaurid. With an estimated foot length of 50 cm (theropods' tracks are longer than they are wide), the estimated hip height of the trackmaker would be 245 cm (h = 4.9 × foot length sensu Thulborn, 1990). The lithostratigraphic units of the Barremian–Early Aptian of the Maestrazgo Basin, which represent similar fluvial and estuarine environments, contain sites in which a few teeth and bones belonging to members of the Spinosauridae (Sánchez-Hernández, Benton, and Naish, 2007; Gasulla et al., 2011) have been found. Little is known about the pes of these theropods, but an almost complete pedal spinosaurid ungual phalanx (from Portugal) assigned to *Baryonyx walkeri* has an anteroposterior length of 78 mm and lateromedial width of 44 mm (Mateus, Araújo, et al., 2011). These measurements, and their ratio, are different to those of the pes of other tetanurans of the same period, but quite similar to those of 2AS-1-13.

Ornithopod Morphotype

The second morphotype is more abundant. In lateral view, some tracks have an elongated, semicylindrical shape (Figs. 14.5B–14.5C), whereas others are bell- or boot-shaped (Figs. 14.4C, 14.4F, and 14.5B–14.5E). Their different sizes reveal that several individuals made them. These animals, with foot lengths of 33–42 cm, would have had hip heights ranging from 194 to 247 cm (where h = 5.9 × foot length sensu Thulborn, 1990). In the pair of tracks 2AS-1-4/5 (Fig. 14.5B), one can see how in the semicylindrical track (on the left), the digits are placed almost vertically with respect to the horizontal plane. Thus, the trackmaker's foot penetrated deeper into the substrate than the boot-shaped track did (right). Probably 2AS-1-4 corresponds to a manus track and 2AS-1-5 to a pes track. It would be an example of the "Goldilocks effect" in the sense used by Falkingham et al. (2011). This association indicates the quadrupedal locomotion of the trackmaker over this substrate. The pes tracks (e.g., 2AS-1-21, which is 37 cm deep or 2AS-1-22, 42 cm deep) are deeper than the 2AS-1-13 theropod track. Among the factors that may have contributed to the ornithopod tracks being deeper than the theropod track are body weight, foot shape, type of locomotion and local variations in sediment firmness. Specimens 2AS-1-2 (Figs. 14.4C, 14.4D), 2AS-1-21 (Figs. 14.4E–14.4G), and 2AS-1-22 (Fig. 14.5E) show three short, wide toes with a wide interdigital angle, and a slightly bilobed "heel," which are characteristics of some ornithopod tracks (Thulborn, 1990; Lockley, 1991; Lockley et al., 2014). At least 15 of these tracks show abundant sliding grooves representing the movement of the foot within the substrate. Six also show scales' impressions and the scratch marks (Gatesy, 2001) they produced

Table 14.1. Some variables, measurements, and features of the best-preserved tracks from the AS-1 site

Track Number	Trackmaker	Maximum Depth into the mud (cm)	Maximum Length (cm)	Maximum Width (cm)	L/W	Hip Height (cm)	Presence of scale impressions	Presence of grooves/scratch marks
2A S-1-13	Theropod	20	?	40	?	If we estimate L = 50 cm (theropod footprint longer than wide: h = 245 (h = 4.9 × foot length sensu Thulborn 1990)	No	Yes. Parallel to each other at the "heel"
2A S-1-2	Ornithopod	40	42	41	1.02	247 (h = 5.9 × foot length sensu Thulborn 1990)	Yes. Between the "heel" and the plantar surface	Yes. Parallel, with a distance of 0.7–1 cm. They are 3 mm wide and 2 mm deep
2A S-1-21	Ornithopod	37	33	33	1	194 (h = 5.9 × foot length sensu Thulborn 1990)	Yes. In some places of the plantar surface (digits II and III), they measure 1–1.3 cm in diameter; 0.2–0.4 cm for interdigital zones	Yes. Parallel, with a distance of 0.3–0.8 cm
2A S-1-22	Ornithopod	42	38	36	1.05	224 (h = 5.9 × foot length sensu Thulborn 1990)	Yes, Between the "heel" and the plantar surface; 1–1.2 cm in diameter	Yes, Parallel, with a distance of 0.7–1 cm

Note: AS-1, Río Alcalá tracksite; L, length; W, width.

(Figs. 14.4C–14.4I and 14.5B–14.5G). These six tracks together represent approximately 45% of all known ornithopod tracks (and approximately 10% of all dinosaur tracks) bearing skin impressions (Avanzini, Piñuela, and García-Ramos, 2011; Davis, 2012). The skin pattern consists of hexagons and pentagons with a diameter of 2–4 mm in the interdigital area in 2AS-1-21, and 10–13 mm for the rest of the sole. The best preserved reticulated pattern is found at the interface between the sole and the lateral borders of the foot, a consequence of the pressure exerted in the weight-bearing phase (Thulborn and Wade, 1989; Gatesy, 2001) compressing the fleshy pads on the sole and the following lateral expansion (particularly noticeable in the "heel" area). However, the bottom of 2AS-1-21 also shows reticules, or small tubercles on the distal-most lateral part of the toe traces (Figs. 14.4E–14.4G and 14.5F), revealing the absence of a pointed claw (unlike that seen in primitive ornithopods). Instead, this reticulated pattern indicates that these trackmakers had distal phalanges with blunt truncated tips or a hoof-like shape, suggesting the trackmaker was a member of Hadrosauriformes (McDonald, 2012). This also suggests that the ungual phalanges were ventrally, laterally, and anteriorly surrounded by reticulate, soft tissue; a keratinous cover was, therefore, probably only present on the dorsal side (Figs. 14.6D–14.6F).

An abundance of bones and teeth belonging to large ornithopods has been described for the Early Cretaceous of the Maestrazgo Basin, including specimens belonging to the iguanodontians *Delapparentia turolensis* and *Iguanodon bernissartensis* (for which an especially complete record exists) (Sánchez-Hernández, Benton, and Naish, 2007; Gasulla

et al., 2011; Ruiz-Omeñaca, 2011). The known pedal ungual phalanges of hadrosauriforms, such as *Iguanodon*, are dorsoventrally flattened and elongate, with blunt truncated tips (McDonald, 2012). These features, which match the description of the ornithopod tracks from AS-1, suggest the trackmaker of the second morphotype of tracks was a related basal hadrosauriform.

SIMILAR TRACKS FROM THE WESTERN UNITED STATES: GEOLOGICAL SETTING

Although only representing single tracks, in contrast to the multiple tracks reported for the Spanish assemblages described, the two examples here reported from the western United States nevertheless provide useful material for comparison. The older track, a natural sandstone cast (UCM 188.25 [in the University of Colorado Museum of Natural History collection]), from the Middle-Late Jurassic Summerville Formation of northern Arizona was illustrated by Lockley, Meyer, and dos Santos (2000) as an example of *Megalosauripus* (Figure 14.7), derived from an assemblage of well-preserved theropod, sauropod, and associated pterosaur tracks (Lockley and Mickelson, 1997) from marginal marine siltstones and sandstones deposited on the southern shores of a large shallow marine embayment.

The Upper Cretaceous track (MOR 541) illustrated below (Fig. 14.8) derives from the Two Medicine Formation of Montana (Lockley, 2014). Although it also represents a natural sandstone cast, it was not found in situ. Thus, it is not known whether it formed part of a larger assemblage, and it

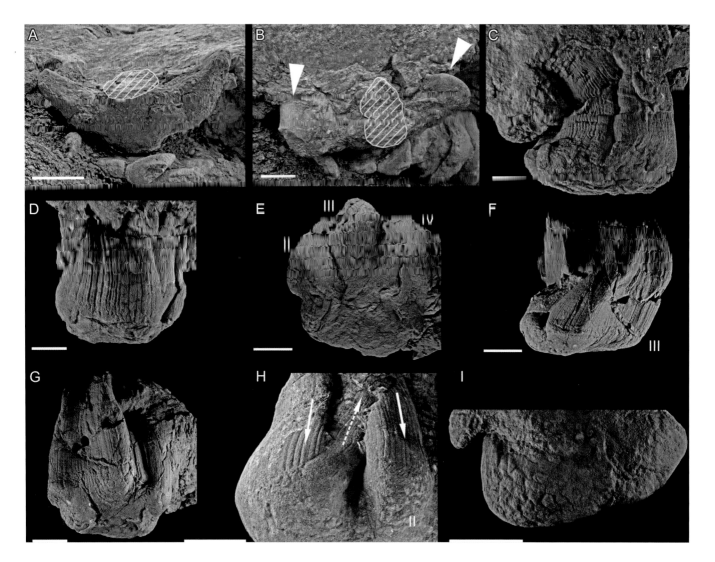

14.4. Tracks preserved at the Cretaceous Río Alcalá site (AS-1). They are assigned to a large theropod trackmaker (A, B) and to several large ornithopod trackmakers (C–I). (A) Pes track 2AS-1-13 in ventral and (B) lateral or medial anterior views. Notice the claws (arrows) on digits II and IV (digit III is broken). (C) Left pes track 2AS-1-2 in lateral and (D) posterior views. In 2AS-1-2 the vertical and parallel sliding grooves on the "heel" show the vertical trajectory of the foot as it penetrated the mud. The lateral, curved trajectories of these slide marks or grooves indicate that during the kickoff phase, the ornithopod lifted the foot upward to release it from the mud (D). The animal only changed the trajectory of the foot toward the direction of locomotion (forward) when the resistance offered by the mud was very reduced (C). (E) Left pes track 2AS-1-21 in plantar, (F) medial, and (G) anteromedial views. The general morphology and recorded trajectories of the foot show a backward-directed penetration of the substrate; later the foot was pulled out forward (F, G). Detail of slide marks indicating the dynamics of digit II and the "heel" during entry into the mud (solid arrow) and withdrawal from it (dashed arrow) in 2AS-1-22 (H). Scale impressions of the plantar surface of 2AS-1-23 (I). Tracks 2AS-1-2, 2AS-1-21, 2AS-1-22, and 2AS-1-23 are deposited at the Museo Aragonés de Paleontología (Teruel, Spain). Scale bar = 10 cm.

is not possible to infer precise details of the local depositional environment. Nevertheless, the Two Medicine Formation represents a fluvio-lacustrine floodplain paleoenvironment.

"FOUR-DIMENSIONAL TRACKS" AND FOOTFALL DYNAMICS

All dinosaur tracks and their infillings are inherently 3-D structures (because all the points that define them can be defined using three coordinates: x, y, and z), although generally they are described as two-dimensional (2-D), and illustrated on 2-D surfaces. The importance of these hyper-3-D tracks and others previously reported in the literature (e.g., Avanzini, Piñuela, and García-Ramos, 2011; Lockley, Meyer, and dos Santos, 2000) is that they record the motion of the trackmakers' feet via abundant slide marks or grooves that were formed over a short time period as the animals' tracks registered in the sediment (Gatesy, 2001). This allows us the contemplation of a new concept: four-dimensional (4-D) tracks. Four-dimensional tracks are defined as true tracks or their infillings showing slide traces that reveal the trajectory of the trackmaker's foot within the sediment more completely than do most 3-D tracks. Each one of these tracks therefore reflects the time and motion involved in their registration.

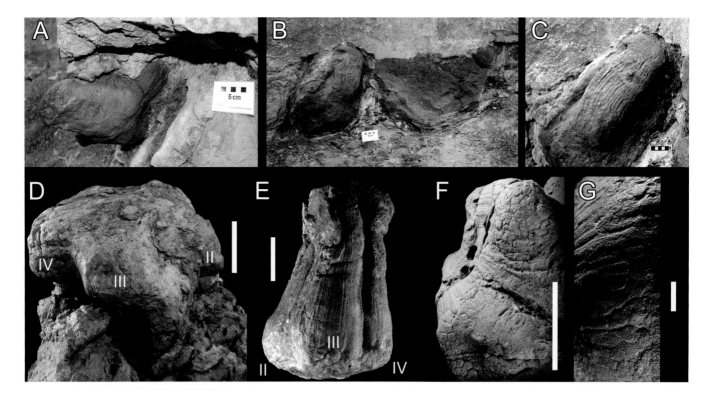

14.5. (A) Lateroplantar or medioplantar view of the theropod track (2AS-1-13) with sigmoidal slide marks produced by the skin. (B) Left manus track (2AS-1-4), showing a semicylindrical elongated shape (on the left), and a left pes track (2AS-1-5) (on the right) of an ornithopod trackmaker. (C) Notice the slide marks produced by the scales of the reticulated integument of (probably) digit V of the manus 2AS-1-4. (D) Anteroplantar view of the left pes track (inverted position) 2AS-1-2. (Note the record of lateral displacement as the foot penetrated the mud). (E) Anterior view of the 2AS-1-22 bell-shaped ornithopod track. (F) Plantar surface of the left pes (2AS-1-21) with scale impressions on digits II and III highlighting the absence of a pointed claw in these ornithopods. (G) Scale impressions and their parallel slide marks at the lateral border of 2AS-1-22. Scale bars in A, B, C = 5 cm; in D, E, F = 10 cm; and in G = 2 cm.

Thus, in effect evidence of the fourth temporal dimension becomes fossilized (Minkowski, 1909). This essential dimension occurs in all track formation although it is not usually considered in any detail. These 4-D tracks therefore permit the study of the interaction of the trackmakers' limbs with the substrate to determine the kind of locomotion represented in their step cycles and the motion of their toes in response to the conditions imposed by the mud (Milàn, 2006; Xing et al., 2012, 2015). Sometimes they are an unusually complete picture of the foot, recording anatomical details and showing the sides and top as well as the usual bottom, or sole, features.

The general morphology of Spanish tracks 2AS-1-13 shows that, on this muddy substrate, the theropod inserted its foot obliquely into the mud during the touchdown phase (Thulborn and Wade, 1989), with an entry angle of approximately 45° with respect to the horizontal plane, until reaching the weight-bearing phase (Thulborn and Wade, 1989) (Figs. 14.6A, 14.6B). The fact that the impression of the morphology of the digits (including the claw) was not destroyed during the kickoff phase (Thulborn and Wade, 1989) confirms that the foot was withdrawn backward, with the progressive retraction of digits II–IV. The animal then removed its foot posteriorly, in the direction of the "heel," possibly due in part to the constraints of the soft mud, making it impossible to perform the normal lift that would occur when walking on a more solid substrate (Avanzini, Piñuela, and García-Ramos, 2011; Huerta et al., 2012) (Figs. 14.6B, 14.6C). This differs from the forward movement of the foot inferred from Triassic theropod tracks (Gatesy et al., 1999) and the forward motion evident in specimen CU 188.25 (University of Colorado Denver, Dinosaur Tracks Museum collection) from the Late Jurassic (Lockley, Meyer, and dos Santos, 2000) from

14.6. Reconstruction of the footfall dynamics recorded in the 'four-dimensional tracks' of the Río Alcalá (AS-1) tracksite. (A–C) Locomotion of the large theropod trackmaker and (D–F) of the large ornithopod trackmakers. Some differences can be observed between these dinosaur groups in each walking phase. On touchdown (first phase), the theropod inserted its foot with an entry angle of approximately 45° with respect to the horizontal (A), whereas the ornithopod foot may have entered almost perpendicularly into the substrate (D). During the weight-bearing (second phase) (B, E) the foot transmitted pressure to the substrate, causing the expansion of the plantar pad. This expansion was a lot larger in the ornithopod than the theropod footfalls. Finally, during kickoff (third and final phase), the theropod moved the foot backward to remove the heel first, allowing the exceptional preservation of the claw impressions (C). However, the ornithopod lifted the foot in the direction of locomotion (F). The arrows show the trajectory of the feet during the step cycle for both dinosaurs.

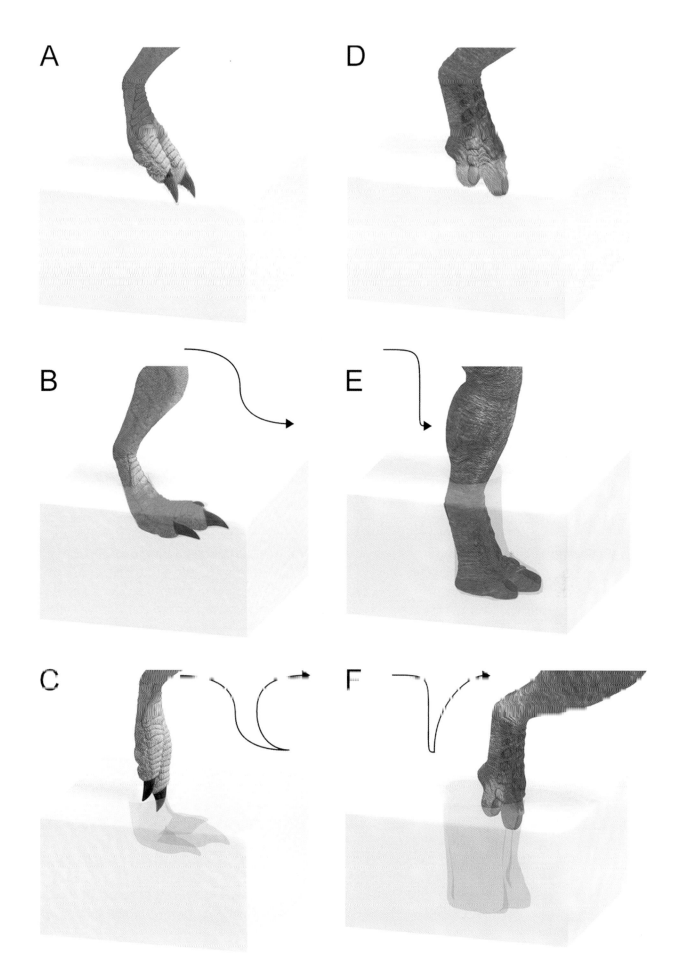

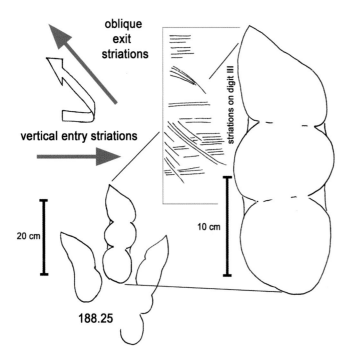

14.7. Drawings of a Jurassic theropod natural track cast (UCM 188.25 [in the University of Colorado Museum of Natural History collection]) from the western United States showing vertical entry striations (slide marks) and anteriorly oriented exit striations associated with the medial side of the two distal pads of digit III.

the United States (Fig. 14.7). This latter fossil from the Summerville Formation is a very well-preserved sandstone cast and reveals entry striations crosscut by exit striations. In this case, the entry striations are oriented vertically downward, and the exit striations obliquely upward and forward toward the anterior (Fig. 14.7). These European and American examples lack large metatarsal impressions, in contrast to the examples noted by Kuban (1989), which indicate that some theropods occasionally used a plantigrade-like locomotion.

The footfall dynamics of the basal hadrosauriforms, inferred from tracks in the Spanish sample, indicate registration in a mainly vertical plane, as shown by the vertically oriented traces of the metatarsophalangeal region. The foot entered almost perpendicular into the substrate, the flexed digits entering first (as reflected by the perimeter of the entry opening, which is significantly smaller than the track perimeter at the end of the weight-bearing phase (Figs. 14.4C, 14.4F, and 14.6D), bearing in mind that in a mud with modeling clay consistency, a wide reduction of the perimeter of entrance due to collapse is highly unlikely and inconsistent with the general typology of these tracks. While penetrating the substrate, the toes extended before the "heel" was brought down. This indicates a voluntary movement of the toes beyond the normal extension required for load transfer in the weight-bearing phase (Fig. 14.6E). During this phase, the elastic properties of the fleshy cushion permitted the pedal surface area to expand, acting as an energy absorber

(Miller et al., 2008). Track 2AS-1-21 even shows the marks of cushion flattening due to the load on digit III during the beginning of the kickoff phase (Figs. 14.4E and 14.5F). Flexing of the digits occurred during this phase, when the foot was lifted in the direction of overall locomotion (Fig. 14.6F); this is manifested by striations created by the toes and "heel" while lifting transversally, cutting the previous entry marks (especially easy to see in 2AS-1-21 and 2AS-1-22) (Fig. 14.4H). This implies that the toes of these not quite digitigrade feet were remarkably mobile. Such mobility is greater than that previously described for hadrosauriforms in the mid-Cretaceous of China (Xing et al., 2012), and quite different to the relatively inflexible subunguligrade foot recorded for hadrosaurids (Moreno, Carrano, and Snyder, 2007). The digits of the manus show less mobility than those of the pes during walking (except for digit V), which suggests a subunguligrade posture for the forelimb.

Moreover, the striations on a sandstone cast of a single toe of presumed ornithopod affinity (Lockley, 2014), from the Upper Cretaceous Two Medicine Formation at Landslide Butte Montana (MOR 541) (Fig. 14.8), indicate moderate forward motion of the foot during track registration. Sensu Lockley (2014), MOR 541 is probably an ornithopod track toe cast because the trace suggests a blunt ungual, the fusiform shape of the toe as seen in profile, from both left and lateral sides, the distal part of the toe cast is deepest, and the proximal part of the toe cast evidently represents an arched area toward the posterior end of the toe. MOR 541 is quite similar to the Jurassic theropod tracks illustrated by Avanzini, Piñuela, and García-Ramos (2011:figs. 6A–6D), suggesting some similarities in the dynamics of footprint registration. However, extracting the foot evidently resulted in a sideways (right lateral) and posterior motion of at least one toe of the foot. In other words, the foot was not extracted in a simple anterior, upward, or backward motion in the plane of the toe's long axis. This supports the suggestion that the foot was flexible, and that the toe moved out of the plane of symmetry represented by the toe trace at the point of touchdown.

Although infillings indicative of the 4-D dynamics of track registration have not been described in much detail until recently (e.g., Avanzini, Piñuela, and García-Ramos, 2011; Xing et al., 2015), they are probably more common than was previously thought. The examples described herein from Europe and North America indicate that individual foot kinematics during footprint registration can be reconstructed in considerable detail in cases where skin traces and striations are well preserved (Gatesy, 2001). It is also evident that there is considerable variation in the vectors of motion during what Thulborn and Wade (1989) called the three dynamic phases of touchdown, weight-bearing, and push off, during the "history of movement" of the foot while registering a footprint.

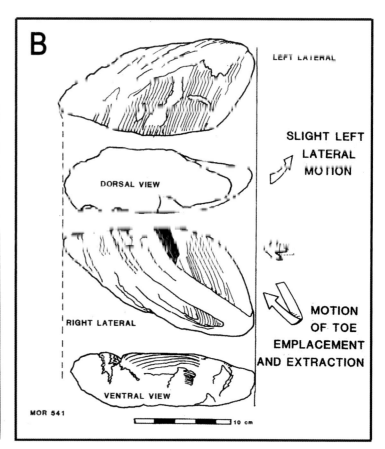

14.8. (A) Photographs of MOR 541 (Two Medicine Formation of Montana) from the western United States, also represented in the University of Colorado Museum of Natural History collections as specimen UCM 212.16 show a sandstone cast of a partial dinosaur track (toe trace) in left lateral (top) and right lateral views. (B) Drawings of the MOR 541 specimen, in left lateral, dorsal, right lateral, and ventral views. Arrow suggests anterior motion during foot entry is followed by posterior motion during exit. Tridactyl track outline shown to suggest that the toe cast may represent digit III (after Lockley, 2014).

However, this three-phase model is rather too simplified because it relies on a general theoretical biomechanical model that does not take into consideration the evidence of skin traces and striations, which are available in the samples cited here. Thus, the Thulborn and Wade (1989) model, though useful, implies linear forward motion in the parasagittal plane and in the general direction of motion of the animal over relatively uniform substrates. The examples cited here show that the foot may move downward, or forward, or both, during touchdown, and that during the weight-bearing phase, the fleshy pads may expand and move dynamically, whereas during kickoff (or extraction), the foot may move forward, backward, or sideways. Further study is needed to show whether the striking differences in touchdown and extraction dynamics recorded here for individual tracks (i.e., individual trackmakers) have the potential to reveal consistently different patterns of movement that are attributed to taxonomic differences and/or foot mobility characteristics that help distinguish among track-making groups, substrate conditions, or other factors.

ACKNOWLEDGMENTS

We thank the Ayuntamiento de Alcalá de la Selva (especially Jesús Edo and José Luis Tena), F. Berrón, and J. Canales for their support during field works, and D. Ayala for technical assistance. This contribution is part of the Departamento de Educación, Universidad, Cultura y Deporte (Dirección General de Patrimonio Cultural, Exp. 164/2012) and Departamento de Industria e Innovación (Aragón Government and Fondo Social Europeo); Spanish R&D project CGL2009-07792 DINOSARAGÓN and CGL2013-41295-P DINOTUR (Ministerio de Econonomía y Competitividad, Spanish Government); Research Group E-62 FOCONTUR project; Instituto Aragonés de Fomento; and Instituto de Estudios Turolenses (Diputación Provincial de Teruel). The second author is the holder of an Formación del Profesorado Universitario grant from the Ministerio de Educación (Ref. AP2008-00846). Thanks to Ray Rogers, Macalester College, St. Paul, Minnesota, for drawing attention to the Two Medicine Formation track cast described herein (Fig. 14.8).

Avanzini, M., L. Piñuela, and J. C. García-Ramos. 2011. Late Jurassic footprints reveal walking kinematics of theropod dinosaurs. Lethaia 45: 238–252.

Currie, P. J., D. Badamgarav, and E. B. Koppelhus. 2003. The first Late Cretaceous footprints from the Nemegt locality in the Gobi of Mongolia. Ichnos 10: 1–13.

Currie, P. J., G. Nadon, and M. G. Lockley. 1991. Dinosaur footprints with skin impressions from the Cretaceous of Alberta and Colorado. Canadian Journal of Earth Sciences 28: 102–115.

Davis, M. 2012. Census of dinosaur skin reveals lithology may not be the most important factor in increased preservation of hadrosaurid skin. Acta Palaeontologica Polonica 59: 601–605.

Díaz-Molina, M., and A. Yébenes. 1987. La sedimentación litoral y continental durante el Cretácico Inferior. Sinclinal de Galve, Teruel. Estudios Geológicos, Volumen Extraordinario Galve-Tremp: 3–21.

Diffley, R. L., and A. A. Ekdale. 2002. Footprints of Utah's last dinosaurs; tracks beds in the Upper Cretaceous (Maastrichtian) North Horn Formation of the Wasatch Plateau, central Utah. Palaios 17: 327–346.

Falkingham, P. L., K. T. Bates, L. Margetts, and P. L. Manning. 2011. The 'Goldilocks' effect: preservation bias in vertebrate track assemblages. Journal of the Royal Society: Interface 8(61): 1142–1154.

Gasulla, J. M., F. Ortega, F. Escaso, and A. Pérez-García. 2011. Los yacimientos de vertebrados de la Formación Arcillas de Morella (Aptiense Inferior); pp. 157–173 in A. Pérez-García, F. Gascó, J. M. Gasulla, and F. Escaso (eds), Viajando a Mundos Pretéritos. Ayuntamiento de Morella: Castellón, Spain.

Gatesy, S. M. 2001. Skin impressions of Triassic theropods as records of foot movement. Bulletin of the Museum of Comparative Zoology 156: 137–149.

Gatesy, S. M., K. M. Middelton, F. A. Jenkins, and N. M. Shubin. 1999. Three dimensional preservation of foot movement in the Late Triassic theropod dinosaur. Nature 399: 141–144.

Huerta, P., F. Torcida Fernández-Baldor, J. O. Farlow, and D. Montero, D. 2012. Exceptional preservation processes of 3D dinosaur footprint casts in Costalomo (Lower Cretaceous, Cameros Basin, Spain). Terra Nova 24: 136–141.

Kuban, G. J. 1989. Elongate dinosaur tracks; pp. 57–72 in D. D. Gillette and M. G. Lockley (eds.), Dinosaur Tracks and Traces. Cambridge University Press, Cambridge, UK.

Lockley, M. G. 1991. Tracking Dinosaurs: A New Look at an Ancient World. Cambridge University Press, Cambridge, UK, 238 pp.

Lockley, M. G. 2014. Tracking dinosaurs in a tepee: rare traces of Cretaceous tetrapods from Montana. New Mexico Museum of Natural History and Science Bulletin 62: 421–427.

Lockley, M. G., and D. Mickelson. 1997. Dinosaur and pterosaur tracks in the Summerville and Bluff (Jurassic) beds of eastern Utah and northeastern Arizona; pp. 133–138 in O. Anderson, B. Kues, and S. Lucas (eds.), Mesozoic Geology

and Paleontology of the Four Corners Area. New Mexico Geological Society 48th Annual Fall Field Conference Guidebook.New Mexico Geological Society, Socorro, New Mexico.

Lockley, M. G., C. A. Meyer, and V. F. dos Santos. 2000. Megalosauripus, and the problematic concept of Megalosaur footprints. Gaia 15: 313–337.

Lockley, M. G., L. Xing, J. Lockwood, and S. Pond. 2014. A review of large ornithopod footprints with special reference to their Ichnotaxonomy. Biological Journal of the Linnaean Society 113(3): 721–736.

Lockley, M. G., J. C. García-Ramos, J. Lires, L. Pinuela, and M. Avanzini. 2008. A review of vertebrate track assemblages from the Late Jurassic of Asturias, Spain with comparative notes on coeval ichnofaunas from the western USA: implications for faunal diversity in association with siliciclastic facies assemblages. Oryctos 8: 53–70.

Martín-Closas, C., and R. Salas. 1994. Lower Cretaceous Charophytes: Biostratigraphy and Evolution in the Maestrat Basin (Eastern Iberian Ranges). Excursion Guide for the VIII Meeting of the European Group of Charophyte Specialists, University of Barcelona, Spain, 89 pp.

Mateus, O., J. Milàn, M. Romano, and M. A. Whyte. 2011. New finds of stegosaur tracks from the Upper Jurassic Lourinhã Formation, Portugal. Acta Palaeontologica Polonica 56: 651–658.

Mateus, O., R. Araújo, C. Natário, and R. Castanhinha. 2011. A new specimen of the theropod dinosaur Baryonyx from the Early Cretaceous of Portugal and taxonomic validity of Suchosaurus. Zootaxa 2827: 54–68.

McCrea, R. T., M. G. Lockley, and C. A. Meyer. 2001. Global distribution of purported Ankylosaur track occurrences; pp. 413–454 in K. Carpenter (ed.), The Armored Dinosaurs. Indiana University Press, Bloomington, Indiana.

McDonald, A. T. 2012. Phylogeny of basal iguanodonts (Dinosauria: Ornithischia): an update. PLoS One 7–5: e36745.

Milàn, J. 2006. Variations in the morphology of emu (Dromaius novaehollandiae) tracks, reflecting differences in walking pattern and substrate consistency: ichnotaxonomical implications. Palaeontology 49: 405–420.

Milàn, J., P. Christiansen, and O. Mateus. 2005. A three-dimensionally preserved sauropod manus impression from the Upper Jurassic of Portugal: implications for sauropod manus shape and locomotor mechanics. Kaupia 14: 47–52.

Miller, C. M., C. Basu, G. Fritsch, T. Hildebrandt, and J. R. Hutchinson. 2008. Ontogenetic scaling of foot musculoskeletal anatomy in elephants. Journal of the Royal Society: Interface 5: 465–476.

Minkowski, H. 1909. Raum und Zeit. Jahresbericht der deutschen Mathematikervereinigung 18: 75–88.

Moreno, K., M. T. Carrano, and R. Snyder. 2007. Morphological changes in pedal phalanges through ornithopod dinosaur evolution: a biomechanical approach. Journal of Morphology 268: 50–63.

Noe, D., M. G. Lockley, and G. Hadden. 2014. Vertebrate tracks from the Cretaceous Dakota Group, Gunnison Gorge National Conservation Area, Delta County, Colorado. New Mexico Museum of Natural History and Science Bulletin 62: 385–392.

Platt, B., and S. Hasiotis. 2006. Newly discovered sauropod dinosaur tracks with skin and foot pad impressions from the Upper Jurassic Morrison Formation, Bighorn Basin Wyoming, USA. Palaios 21: 249–261.

Ruiz-Omeñaca, J. I. 2011. Delapparentia turolensis nov. gen et sp., un nuevo dinosaurio iguanodontoideo (Ornithischia: Ornithopoda) en el Cretácico Inferior de Galve. Estudios Geológicos 67: 83–110.

Salas, R., C. Martín-Closas, X. Querol, J. Guimerà, and E. Roca. 1995. Evolución tectosedimentaria de las cuencas del Maestrazgo y Aliaga-Penyagolosa durante el Cretácico inferior; pp. 13–94 in R. Salas and C. Martín-Closas (eds.), El Cretácico Inferior del Nordeste de Iberia. Publicacions de la Universitat de Barcelona, Barcelona, Spain.

Salas, R., J. Guimerà, R. Mas, C. Martín-Closas, A. Meléndez, and A. Alonso. 2001. Evolution of the Mesozoic Central Iberian Rift System and its Cainozoic inversion (Iberian chain); pp. 145–185 in P. A. Ziegler, W. Cavazza, A. H. F. Robertson, and S. Crasquin-Soleau (eds.), Peri-Tethys Memoir 6. Mémorian Muséum National d´Histoire Naturelle 186. Muséum National d´Histoire Naturelle, Paris, France.

Sánchez-Hernández, B., M. J. Benton, and D. Naish. 2007. Dinosaurs and other fossil vertebrates from the Late Jurassic and Early Cretaceous of the Galve area, NE Spain. Palaeogeography, Palaeoclimatology, Palaeoecology 249: 180–215.

Skelton, P. W., B. Granier, and M. Moullade. 2013. Introduction to thematic issue, 'Spatial patterns of change in Aptian carbonate platforms and related events.' Cretaceous Research 39: 1–5.

Soria, A. R., C. Martín-Closas, A. Meléndez, M. N. Meléndez, and M. Aurell. 1995. Estratigrafía del Cretácico Inferior continental de la Cordillera Ibérica Central. Estudios Geológicos 511: 141–152.

Thulborn, T. 1990. Dinosaur Tracks. Chapman Hall, London, UK, 410 pp.

Thulborn, R. A., and M. Wade. 1989. A footprint as a history of movement; pp. 51–56 in D. D. Gillette and M. G. Lockley (eds.), Dinosaur Tracks and Traces. Cambridge University Press, Cambridge, UK.

Xing, L., P. R. Bell, J. D. Harris, and P. J. Currie. 2012. An unusual, three-dimensionally preserved, large hadrosauriform pes track from 'mid'-Cretaceous Jiaguan Formation of Chongqing, China. Acta Geológica Sinica 86: 304–312.

Xing, L., D. Li, M. G. Lockley, D. Marty, J. Zhang, W. S. Persons IV, H.-L. You, C. Peng, and S. Kümmell. 2015. Dinosaur natural track casts from the Lower Cretaceous Hekou Group in the Lanzhou-Minhe Basin, Gansu, Northwest China: Ichnology, track formation, and distribution. Cretaceous Research 52: 194–205.

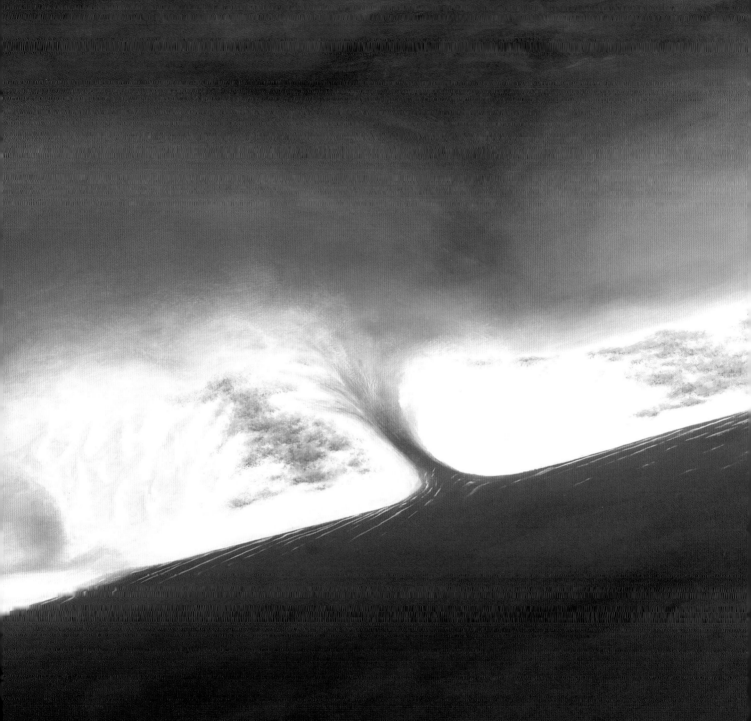

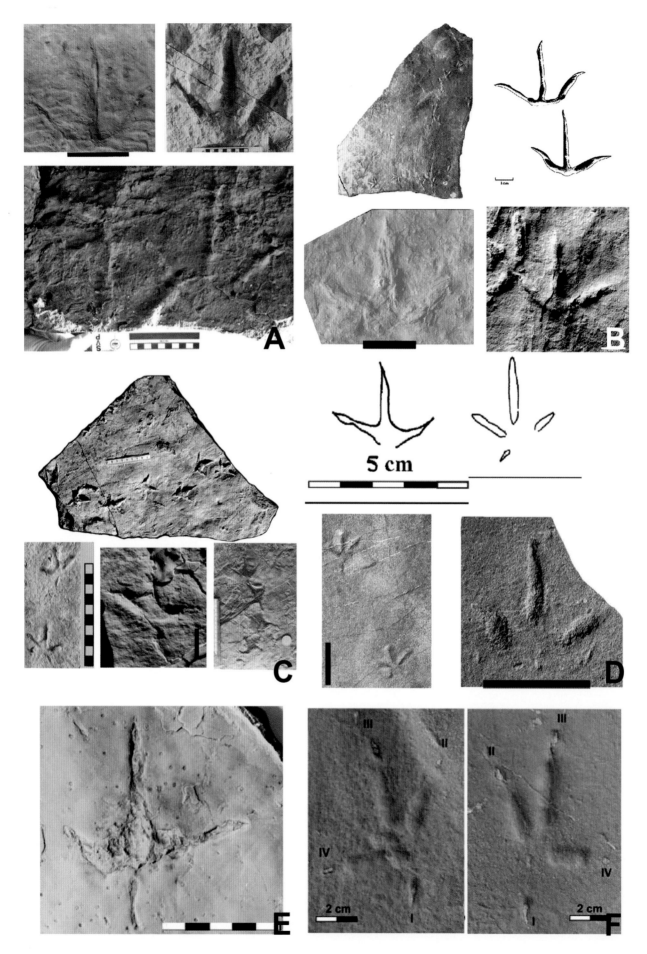

Analyzing and Resolving Cretaceous Avian Ichnotaxonomy Using Multivariate Statistical Analyses: Approaches and Results

15

Lisa G. Buckley, Richard T. McCrea, and Martin G. Lockley

SEVERAL NEW ICHNOTAXA OF AVIAN TRACKS HAVE BEEN described in recent years, adding to the known ichnodiversity of Cretaceous avians. The naming of new avian ichnospecies and ichnogenera has resulted in the creation of several avian ichnofamilies, but due to the challenges of documenting bird tracks, there are several ichnogenera that to date remain unassigned to any ichnofamily. Multivariate statistical analyses can be used to quantitatively test avian ichnotaxonomic assignments. Ichnospecies within the ichnofamilies Avipedidae (*Aquatilavipes swiboldae, A. izumiensis,* and *A. curriei*), Ignotornidae (*Ignotornis mcconnelli, I. yangi, I. gajinensis, Hwangsanipes choughi, Goseongornipes* ichnosp., *G. markjonesi*), Koreanaornipodidae (*Koreanaornis hamanensis, K. dodsoni, Pullornipes aureus*), Shandongornipodidae (*Shandongornipes muxiai*), and Jindongornipodidae (*Jindongornipes kimi*), were analyzed. Also included are the ichnotaxa *Magnoavipes* (*M. lowei, M. caneeri, M. denaliensis*), *Uhangrichnus chuni, Dongyangornipes sinensis, Morguiornipes robusta, Barrosopus slobodai,* and *Tatarornipes chabuensis,* and data from 59 tracks of small theropods from the Early Cretaceous of western Canada.

The results show strong statistical support for all described ichnogenera and ichnofamilies. The ichnogenus *Magnoavipes* does not group with any other avian ichnotaxon in this analysis, suggesting that interpretations of a nonavian affinity for the ichnogenus are accurate. Also, several unassigned ichnogenera occupy the same morphospaces as those of described ichnofamilies. However, the statistically well-differentiated ichnofamilies Koreanaornipodidae and

Avipedidae occupy a similar morphospace despite large morphologic differences. The data also show that both avian and theropod tracks have overlapping morphologies that make a 100% separation improbable. Total digit divarication is a reliable discriminatory value only on a population level. Trackway data (pace and stride lengths, track rotation, pace angulation) provide the most accurate data in separating bird trackways from those of theropods.

Multivariate statistical analyses have the potential to both test ichnotaxonomic assignments and show morphospace groupings that reveal previously undocumented patterns. However, a multivariate data set is only as strong as the most incomplete data table and is not immune to a priori assumptions regarding ichnotaxonomic assignment, data collection techniques, and/or preservational condition of the specimens in question. Increasing the number of measured parameters when reporting new avian ichnotaxa will serve to strengthen the utility of multivariate analyses in vertebrate ichnology.

INTRODUCTION

Avian track types (or morphotypes) can be generally assigned to ecological niches using overall track size and pace and stride data (i.e., long-legged wading birds vs. short-legged shorebirds). Other track features, such as the extent of interdigital webbing and the degree of rotation of individual tracks in a trackway, may also provide information that allows ichnologists to propose a well-supported modern analogue for the Cretaceous avian trackmaker. In recent years, many

15.1. Ichnofamilies considered in this study. (A) *Magnoavipes,* a previously contentious ichnogenus originally described as the trace of a large avian but now considered to be that of a nonavian theropod (Matsukawa et al., 2014). (B) Avipedidae and Limiavipedidae: (upper left) *Avipeda* (modified from Vialov, 1965); (upper right) *Aquatilavipes swiboldae,* scale = 1.0 cm (modified from Currie, 1981); (lower right) *Aquatilavipes izumiensis,* scale = 1.0 cm (modified from Azuma et al., 2002); (lower left)

Limiavipes curriei, scale = 5.0 cm (reassigned from *Aquatilavipes curriei,* McCrea and Sarjeant, 2001; McCrea et al., 2014). *L. curriei* is much larger than any Mesozoic ichnospecies of *Aquatilavipes.* (C) Ignotornidae: (top) *Ignotornis mcconnelli,* holotype (Lockley et al., 2009); (lower left) *Goseongornipes markjonesi* (Lockley, Houck, et al., 2006); (bottom center) *Ignotornis yangi* (Kim et al., 2006); (lower right) *Hwangsanipes choughi* (Yang et al., 1995). Scale divisions in centimeters. (D) Koreanaornipodidae: (top) *Koreanaornis*

hamanensis (Kim, 1969); (lower left) *Pullornipes aureus* (Lockley, Matsukawa, et al., 2006); (lower right) *Koreanaornis dodsoni* (Xing et al., 2011). Scale bar = 5.0 cm. (E) Jindongornipodidae. *Jindongornipes kimi* (Lockley and Rainforth, 2002). Scale = 5.0 cm. (F) Shandongornipodidae. *Shandongornipes muxiai* (Lockley et al., 2007): (left) left track LRH-DZ70 and (right) right track LRH-DZ67 (from the Qingdao Institute of Marine Geology). Both tracks from *S. muxiai* holotype trackway LRH-DH01. Scale = 2.0 cm.

259

novel traces attributed to Mesozoic birds have been described, as well as higher order classifications, such as avian ichnofamilies (Lockley et al., 1992; Lockley, Matsukawa, et al., 2006; Lockley and Harris, 2010). Avian ichnotaxonomy exists to provide a means by which to both document and discuss discrete patterns in the variation of footprint shapes that are attributable to now extinct birds. Ichnotaxonomic groupings for traces identified as avian in origin have the potential to not only identify ecological partitions among extinct avians (Falk, Martin, and Hasiotis, 2011), but also to potentially describe the biologic diversity of known Mesozoic avians whose traces are preserved within the ecological niches of "shorebird" and "wading bird."

Avian ichnites are assumed to avoid many of the pitfalls ascribed to large vertebrate tracks due to their nature; however, many of the features that are considered synapomorphy-based (presence of halluces) and phenetic-based (e.g., size, digit slenderness, divarication, morphology of unguals) characters are influenced by both anatomical and preservational variation (Buckley, McCrea, and Lockley, 2015, and references within). Avian tracks are subject to metatarsal or "heel" drag and digit drag marks, sediment displacement bulges, track margin collapse, and so on, as seen with features interpreted as toe drag marks in trackways of *Pullornipes aureus* (Lockley, Matsukawa, et al., 2006). Close examination of multiple qualitative (visual) differences among groups of avian ichnites is more likely to reveal anatomical features of the trackmaker.

The study of avian ichnites, however, is not without challenges. First, the challenge of identifying the potential trackmaker of "large bird" versus "small nonavian theropod" for certain ichnotaxa remains despite many efforts to clarify the issue (Lockley, Wright, and Matsukawa, 2001; Wright, 2004; Fiorillo et al., 2011; Xing et al., 2015): this is not surprising given that Aves occur within Theropoda. Wright (2004) observed many bird-like features in dinosaur tracks and noted many of the issues distinguishing the tracks of bipedal, tridactyl dinosaurs and avians that may be subject to similar functional constraints in locomotion. However, theropod locomotion is more hip-driven (Gatsey, 1990; Farlow et al., 2000) than knee-driven, as it is in birds (Rubenson et al., 2007), but the extent to which these factors contribute to qualitative and/or quantitative ichnotaxonomic differences is unknown. Whereas a high average total divarication (~110°) has been used to tentatively identify avian tracks (e.g., Lockley, Wright, and Matsukawa, 2001), using an average calculated from a large sample to determine the identity of one track has the potential to be misleading. The tracks of many extant shorebirds (Scolopacidae) exhibit divarications as low as 75° (i.e., *Tringa solitaria*, PRPRC NI2012.01 [Peace Region

Palaeontology Research Centre]), which is in the range of potential total divarications seen in the tracks of small theropods. Also, divarication can be highly variable within an individual trackway. The total divarication within one extant shorebird trackway (PRPRC NI2012.01) ranges between 75° and 116°, which is an observation that can also be made of fossil bird tracks: tracks from Korea assigned to *Aquatilavipes* ichnosp. (Huh et al., 2012) range in total divarication from 70° to 134°.

One ichnotaxon in particular is problematic in identifying the affinity of the trackmaker. *Magnoavipes* (Fig. 15.1A) was first described by Lee (1997) as a trace attributable to a large avian from the Woodbine Formation (Cenomanian) based on the slender digits and wide total divarication. A reanalysis of the ichnogenus led Lockley, Wright, and Matsukawa (2001) to attribute *Magnoavipes* to a theropod, rather than avian, trackmaker based on large size, lack of a hallux (as seen in extant large wading birds), and narrow trackway. However, Fiorillo et al. (2011) describe a new ichnospecies of *Magnoavipes* (*M. denaliensis*) and attribute it to an avian trackmaker based on the high total divarication. The strengths of the opposing diagnostic criteria (high total divarication vs. trackway characteristics) have yet to be statistically tested.

Second, while bird tracks are generally immune to common extramorphologic features sometimes preserved in the tracks of much larger animals, they do exhibit variability in preservation of certain morphologic features, specifically of digit I (hallux) and of webbing. There are examples in the literature where, within an individual trackway, the hallux (i.e., *Goseongornipes markjonesi*, Kim et al., 2012:fig. 9A) is inconsistent. The presence of semipalmate webbing (webbing restricted to the proximal part of the digits) may be difficult to detect due to sediment consistency and preservation.

A Brief Review of Avian Ichnofamilies from the Mesozoic

There are six ichnofamilies currently attributed to avian traces from the Mesozoic. Refer to the references cited within this section for more detailed information on the systematics of these ichnofamilies and the assignments of the ichnogenera therein.

Avipedidae

Avipedidae (Fig. 15.1B) was originally described by Sarjeant and Langston (1994:12) as containing "avian footprints showing three digits, all directed forward. Digits united or separate[d] proximally. Webbing lacking or limited to the most proximal part of the interdigital angles." The type ichnogenus for Avipedidae is *Avipeda* (Vialov, 1965), whose original diagnosis and description was emended by Sarjeant and Langston (1994) to remove the "wastebasket" nature of both *Avipeda* and the ichnofamily (or morphofamily of Sarjeant and Langston, 1994) for which *Avipeda* is the type ichnogenus. Sarjeant and Langston (1994) also assign *Aquatilavipes swiboldae* (Currie, 1981) to Avipedidae. Avipedidae, to date, includes *A. swiboldae* (Currie, 1981), and *A. izumiensis* (Azuma et al., 2002).

Limiavipedidae

It has been known for some time (Lockley and Harris, 2010) that *Aquatilavipes curriei* (McCrea and Sarjeant, 2001) did not match the ichnomorphology of other ichnospecies of *Aquatilavipes* (*A. swiboldae*, *A. izumiensis*). *A. curriei* prints are attributed to "large, long-legged avian track-maker[s] . . . [with a] functionally tridactyl pes tracks with no obvious webbing" (McCrea et al., 2014:86). Tracks attributed to Limiavipedidae also possess no hallux impressions and have a pace and stride that are relatively short compared to similarly sized theropod ichnotaxa but that are longer when compared to other avian ichnotaxa. Tracks of the Limiavipedidae are also strongly rotated toward the midline of the trackway. McCrea et al. (2014) erected a novel ichnofamily, Limiavipedidae, for avian ichnotaxa that are attributed to those tracks previously described as *A. curriei*. Currently, the only two ichnospecies ascribed to Limiavipedidae are *Limiavipes curriei* (McCrea and Sarjeant, 2001; McCrea et al., 2014) and *Wupus agilis* (Xing et al., 2015).

Ignotornidae

The ichnofamily Ignotornidae was first erected by Lockley et al. (1992:121) and diagnosed as "tetradactyl, slightly asymmetric bird tracks with variably preserved, posteriorly directed hallux impressions typically showing digit I and medial rotation towards trackway midline." Originally, Ignotornidae included both *Koreanaornis hamanensis* and *Jindongornipes kimi*, which were originally assigned to the ignotornids based on the presence of a well-defined, posteriorly oriented hallux. However, as more tetradactyl bird tracks were discovered with halluces of varying lengths and orientations, Ignotornidae was revised (by the emendation to *Ignotornis* by Kim et al., 2006) to include only those ichnotaxa that possess a slight proximal webbing prominent posteromedially oriented hallux impression, including *Hwangsanipes choughi* (Kim et al., 2006), and *Goseongornipes markjonesi* (Lockley, Houck, et al., 2006) (Fig. 15.1C).

Koreanaornipodidae

Koreanaornipodidae (Fig. 15.1D) was erected by Lockley, Houck, et al. (2006b95) to include all "small, wide sub symmetric functionally tridactyl tracks with slender digit impressions, and wide divarication between digits II and IV . . . small hallux occasionally present and posteromedially directed about 180° away from digit IV." This differs in the diagnosis of Avipedidae in that a hallux is not usually preserved in avipedids. Koreanaornipodidae includes not only the type ichnogenus *K. hamanensis* but also *K. dodsoni* (Xing et al., 2011) and *Pullornipes aureus* (Lockley, Matsukawa, et al., 2006).

Jindongornipodidae

Jindongornipodidae (Fig. 15.1E) was named by Lockley, Houck, et al. (2006) in their emendation of Ignotornidae. Jindongornipodidae is diagnosed by medium sized tetradactyl tracks with digit II shorter than digit IV, and a moderately long, posteriorly directed hallux (Lockley, Houck, et al., 2006). To date, Jindongornipodidae is a monoichnospecific ichnofamily, whose type ichnospecies is *J. kimi* (Lockley et al., 1992).

Shandongornipodidae

Shandongornipodidae (Fig. 15.1F) represents traces attributable to tetradactyl, paraxonic, zygodactyl avians (Lockley et al., 2007), and is currently represented by only one ichnospecies, *Shandongornipes muxiai* (Li et al., 2005).

Rationale for Study

The goals of this study are (1) to examine the current ichnotaxonomic assignments of bird tracks using multivariate statistical analyses for additional statistical support and (2) to examine data of both bird and small theropod tracks and trackways to test other morphologic criteria by which to distinguish between tracks of birds and small theropods. The working hypothesis to be tested is that footprint proportions, as expressed by ratios will reveal significant differences between the tracks of theropods and birds.

MATERIALS AND METHODS

Data Used

The linear and angular data (App. 15.1) for 584 tracks was collected from ichnotaxonomic descriptions and data tables

Table 15.1. Avian ichnotaxa used in this study

Ichnotaxa	Geographic Location	Formation, age	Reference(s)
Aquatilavipes swiboldae	Peace Region, British Columbia	Gething Formation, Lower Cretaceous (Aptian)	Currie (1981)
Aquatilavipes izumiensis	Fukui Prefecture, Japan	Tetori Group, Lower Cretaceous (Berriasian-Valanginian)	Azuma et al. (2002)
Barrosopus slobodai	Neuquén, Argentina	Anacleto Formation, Upper Cretaceous (Campanian)	Coria et al. (2002)
Dongyangornipes sinensis	Zhejiang Province, China	Jinhua Formation, Lower Cretaceous	Azuma et al. (2012)
Goseongornipes markjonesi	Kosong County, South Korea	Jindong Formation, Lower Cretaceous	Lockley, Houck, et al. (2006)
*Gruipeda vegrandiunus**	Denali National Park, Alaska	Cantwell Formation, Upper Cretaceous (upper Campanian–lower Maastrichtian)	Fiorillo et al. (2011)
Hwangsanipes choughi	Hwangsan Basin, South Korea	Uhangri Formation, Upper Cretaceous	Yang et al. (1995)
Ignotornis mcconnelli	Golden, Colorado	Dakota Group Lower Cretaceous (Aptian-Cenomanian)	Mehl (1931); Lockley et al. (2009)
Ignotornis yangi	Changseon and Sinsu Islands, South Korea	Haman Formation, Lower Cretaceous (Aptian-Cenomanian)	Kim et al. (2006)
Ignotornis gajinensis	Gajin, Korea	Haman Formation, Lower Cretaceous (Aptian–middle Albian)	Kim et al. (2012)
Jindongornipes kimi	Kyeongsangnam Province, Korea	Jindong Formation, Lower Cretaceous	Lockley et al. (1992)
Koreanaornis hamanensis	Gyeongsang Province, Korea.	Haman Formation, Lower Cretaceous (Aptian-Cenomanian)	Kim (1969)
Koreanaornis dodsoni	Xinjiang Uyghur Autonomous Region, China	Lower Layer, Tugulu Group, Lower Cretaceous	Xing et al. (2011)
Limiavipes† curriei	Grande Cache, Alberta	Grande Cache Member, Gates Formation, Lower Cretaceous (lower Albian)	McCrea and Sarjeant (2001); McCrea et al. (2014)
Magnoavipes lowei‡	Denton County, Texas	Woodbine Formation, Upper Cretaceous (Cenomanian)	Lee (1997)
Magnoavipes caneeri‡	Dinosaur Ridge, Colorado	South Platte Formation, Dakota Group, Upper Cretaceous (Cenomanian)	Lockley, Wright, and Matsukawa (2001)
Magnoavipes denaliensis‡	Denali National Park, Alaska	Cantwell Formation, Upper Cretaceous (upper Campanian–lower Maastrichtian)	Fiorillo et al. (2011)
Morguiornipes robusta	Wuerhe District, China	Lower Layer, Tugulu Group, Lower Cretaceous	Xing et al. (2011)
Pullornipes aureus	Laoning Province, China	Tuchengzi Formation, Lower Cretaceous (Tithonian-Berriasian)	Lockley, Matsukawa et al. (2006)
*Sarjeantopodus semipalmatus**	Niobrara County, Wyoming	Lance Formation, Upper Cretaceous (Maastrichtian)	Lockley, Nadon, and Currie (2004)
Shandongornipes muxiai	Shandong Province, China	Tianjialou Formation, Lower Cretaceous (Aptian-Albian)	Li et al. (2005); Lockley et al. (2007)
Tatarornipes chabuensis	Chabu Area, Inner Mongolia	Jingchuan Formation, Lower Cretaceous (Barremian–lower Aptian)	Lockley et al. (2012)
Uhangrichnus chuni	Uhangri, China	Uhangri Formation, Upper Cretaceous	Yang et al. (1995)

Source: Amended from Lockley and Harris (2010)

Notes: * Denotes ichnotaxa for which there is too little data reported to be used in the multivariate statistical analyses. † The ichnospecies *Aquatilavipes curriei* (McCrea and Sargeant, 2001) has been recently redescribed as not belonging to the ichnogenus *Aquatilavipes* but reassigned to the ichnofamily Limiavipedidae and redescribed as *Limiavipes curriei* (McCrea et al., 2014). ‡ Denotes ichnotaxa whose assignment to Aves is contentious: Lockley, Wright, and Matsukawa (2001) convincingly demonstrate that *Magnoavipes* is an ichnogenus attributed to theropods.

made available in the avian ichnology literature (see references and App. 15.1 for a complete list). The avian ichnotaxa examined in this study are listed in Table 15.1. The amount and types of data collected varied considerably: some data reported only maximum values and averages for the ichnotaxa were reported, whereas the data for all measured tracks were presented for other ichnotaxa (App. 15.1).

Standardization of Data

Data also varied in the number of variables that were reported. Some ichnotaxa were reported only with their length, width, and total divarication, whereas other ichnotaxa were presented with more comprehensive data. All data were converted to millimeters. Data were entered to fit into a standard table (App. 15.1). Data that were not available, either due to exclusion in the published data sets or due to incompleteness of the measured tracks, were entered as a question mark. The data set is set up to accommodate tetradactyl tracks (digits I–IV), and for tridactyl tracks digit I was coded as 0 mm (rather than a question mark for missing) to indicate the absence of that digit for the ichnotaxon in question. For ichnotaxa with trackways where digit I is inconsistently preserved, however, digit I was coded as a question mark to indicate missing data. This allows for tridactyl and tetradactyl tracks to be analyzed together, and the structural presence (or absence) of digit I to be recognized in the total analysis. If digit I were coded as a question mark for functionally tridactyl footprints, the multivariate program Paleontological Statistics (PAST) version 2.17 would treat functionally tridactyl tracks as having missing data and use pair-wise substitution (Hammer, Harper, and Ryan, 2001) for the value digit I instead of recognizing that digit I was not present.

Linear measurements

The following linear measurements are used in this study.

- Digit lengths (DL): the lengths of the individual digits (DL-I, digit I length; DL-II, digit II length; etc.) as measured from either the footprint center or the entire preserved proximodistal length of the footprint, whichever is best preserved;
- Digit widths (DW): the mediolateral widths of the individual digits (DW-I, digit I width; DW-II, digit II width; etc.) as measured at the most proximal point of the free digit;
- Footprint center: not a linear measurement but a point of reference, as measured from straight lines drawn from the distal ends of digits II–IV to their proximal point of convergence;
- Footprint length (FL): the distance as measured from the caudal (posterior) margin of the metatarsal pad (or, lacking the preservation of the metatarsal pad, the caudal points of digits II and IV) to the most distal end of digit III. In some cases, the length of digit III is used as FL;
- Footprint length with hallux (FLwH): the distance as measured from the distal tip of digit III to the distal tip of digit I. If no hallux is present, FLwH is essentially the same as the same as FL;[?]
- Footprint width (FW): the distance as measured from the most distal point of digit II to the most distal point of digit IV;
- Pace length (PL): the distance measured from a distinct point (usually footprint center) on one footprint to the next footprint in the sequence of the trackway (left footprint to right footprint, or right footprint to left footprint);
- Stride length (SL): the distance measured from a distinct point (usually footprint center) from left footprint to left footprint, or from right footprint to right footprint.

Angular measurements

The following angular measurements are used in this study.

- Divarication of digits I and II (DIV I–II): the angle formed between the distal points of digits I and II, with the footprint center as the origin of the angle;
- Divarication of digits II and III (DIV II–III): the angle formed between the distal points of digits II and III, with the footprint center as the origin of the angle;
- Divarication of digits III and IV (DIV III–IV): the angle formed between the distal points of digits III and IV, with the footprint center as the origin of the angle;
- Footprint rotation (FR): the angle at which the individual footprint is rotated toward (positive) or away (negative) from a straight line running through the center of the footprint that is parallel to the midline of the trackway;
- Pace angulation (PA): the angle formed by three consecutive footprints in a trackway, with the angle formed at the footprint center of the middle footprint of the consecutive sequence;
- Total divarication (DIVTOT): the angle formed between the distal points of digits II and IV, with the footprint center as the origin of the angle. DIVTOT is also known in the literature as DIV II–IV.

Ratios

Ratios are useful in comparing ichnologic data as they remove the size component from the analyses.

- Footprint length to footprint width ratio (FL/FW): the ratio obtained by dividing FL by FW. This value can also be described as the splay of the digits of a footprint. FL/FW is the only ratio that is consistently calculated in vertebrate ichnology.

Digit length ratios: the ratios of the different digit lengths provide the relative lengths of one digit as compared to the other digit in the ratio. The following digit length ratios are used in this analysis.

- DL-I/DL-II: the ratio of digit I length and digit II length;
- DL-I/DL-III: the ratio of digit I length and digit III length;
- DL-I/DL-IV: the ratio of digit I length and digit IV length;

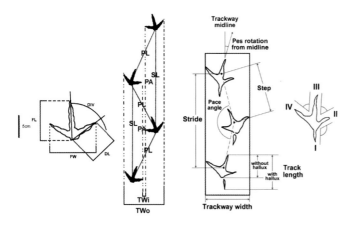

15.2. Measurements collected for avian ichnites. (Left) Complete data that is recommended to be collected (where possible) for both previously described and novel avian ichnotaxa; (right) an example of thorough data collection in the reassessment of *Ignotornis mcconnelli* (Lockley et al., 2009). Abbreviations: DIV, digit divarication; DL, digit length; FL, footprint length; FW, footprint width; PA, pace angulation; PL, pace length; SL, stride length; TWi, inner trackway width; TWo, outer trackway width.

Table 15.2. Comparative total divarication data for Mesozoic tridactyl semipalmate avian tracks, the tracks of the ichnogenus *Magnoavipes*, and Early Cretaceous theropod tracks

Total Divarication	Number	Mean	Minimum	Maximum
Theropod	26	87.8°	58°	120°
Semipalmate Avian	202	111.0°	58°	170°
Magnoavipes	23	93.4°	65°	118°

Note: The data show that, although there are large differences among the mean total divarications of the three groups, the minimum and maximum value range has a great deal of overlap. This amount of variation between the total divarication of theropod and bird tracks makes using the average total divarication value (or the total divarication of an isolated footprint) problematic.

- DL-II/DL-III: the ratio of digit II length and digit III length;
- DL-IV/DL-III: the ratio of digit III length and digit IV length.

Divarication ratios: the ratios of divarication data describe the percentage of splay on each lateral side of the footprint of the total divarication. The following digit divarication ratios are used in this study.

- DIV II–III/DIVTOT: the percentage of divarication between digits II–III of the total divarication;
- DIV III–IV/DIVTOT: the percentage of divarication between digits III–IV of the total divarication;
- DIV II–III/DIV III–IV: the ratio of DIV II–III as compared to DIV III–IV.

Statistical Analyses

Bivariate and multivariate analyses on the avian and theropod data set were performed using PAST version 2.17 (Hammer, Harper, and Ryan, 2001). Analyses performed were the t-test (bivariate), and principle component, discriminant, and canonical variate analyses (i.e., multivariate; Hammer and Harper, 2006).

Principal components analysis (PCA) is the two-dimensinoal (2-D) projection of multivariate data to identify the components that account for the maximum amount of variance in the data (Hammer and Harper, 2006). It reveals the relative variation contributed to the data set by each measured variable, produces principal component ordinance plots that visually project three-dimensinoal (3-D) plots of specimens in 2-D, and may reveal discrete groupings among specimens. PCA ordinance plots are often displayed with variance vectors that show the relative amount of variation that each measured variable contributes to the overall variation in the data set (Hammer and Harper, 2006). The first principal component (PC1) represents variation based on size (even in log-adjusted data) and is usually the largest principal component in terms of percentage of total variance within the sample (Hammer and Harper, 2006). However, careful examination of the variance vectors is required to determine the exact nature of PC1. PCA was used to find the percentage of total variation (variance) that each measured variable, or combination of variables, contributed to the total variation in the data set. It replaces missing data using pairwise substitution (Hammer, Harper, and Ryan, 2001). The strength of PCA is not in determining the significance of the differences among qualitative groupings – PCA is not statistics (Hammer and Harper, 2006) – but in revealing which variables contribute to distinguishing among ichnotaxonomic groups.

Discriminant analysis (DA) projects a multivariate data set down to one dimension in a way that maximizes separation between two a priori separated groups: in this case, the a priori groups are ichnotaxonomic assignments of Mesozoic bird tracks. This is a useful tool for testing hypotheses of morphologic similarity or difference between two groups. A 90% or greater separation between two groups is considered sufficient support for the presence of two taxonomically distinct morphotypes (Hammer and Harper, 2006); however, 100% is ideal. Canonical variate analysis (CVA) compares specimens a priori categorized in three or more groups using the same principles as DA does. The p_{same} between two a priori groups was determined using Hotelling's t-squared test (the multivariate version of the t-test; Hammer and Harper, 2006) to determine significance at $p \geq .05$.

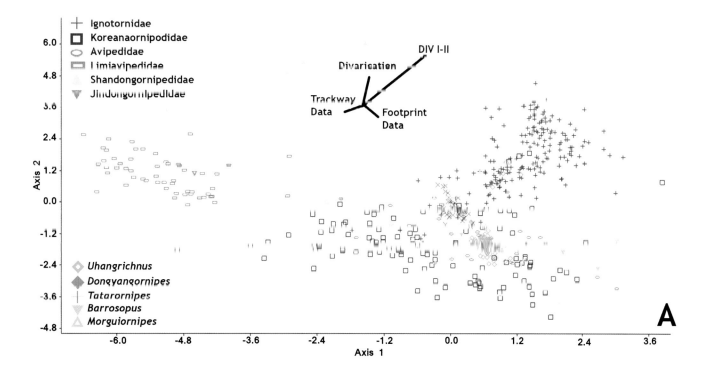

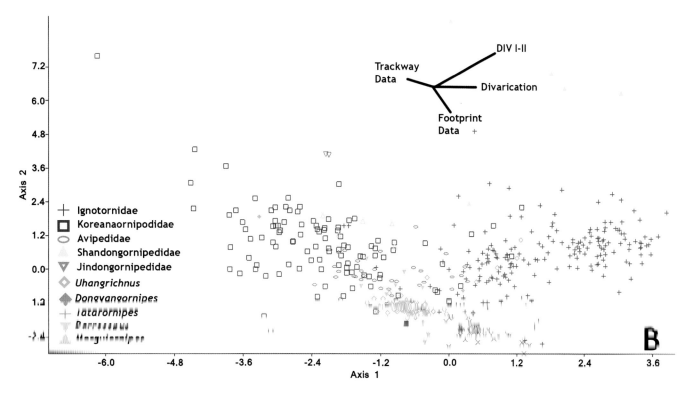

15.3. (A) Canonical variate analysis plot of all Mesozoic avian ichnofamilies (except for *Sarjeantopodus* and *Gruipeda*, see Table 15.1). The inclusion of the Limiavipedidae (*Limiavipes curriei*), whose tracks are much larger than the rest of the Mesozoic avian ichnotaxa, restricts the other ichnotaxa data points into one section of morphospace. All avian ichnogenera as categorized as ichnofamilies are shown to be significantly different from one another, despite their apparent overlap in morphospace (Table 15.3). (B) Canonical variate analysis plot of Mesozoic avian ichnofamilies excluding *Limiavipes curriei*. The removal of the much larger *L. curriei* allows for a more accurate analysis of the smaller avian ichnotaxa. There is a general grouping between tridactyl (lower left) and tetradatyl (upper right) avian tracks. *Shandongornipes muxiai* and *Dongyangornipes sinensis* cluster discretely from the rest of the analyzed Mesozoic avian tracks. Vectors show the relative amounts of variation each measured variable contributes to the total data set. Abbreviation: DIV, digit divarication.

Challenges and Assumptions of Performing
Multivariate Statistical Analyses on Ichnologic Data

Limited data and small sample size

In the cases where only maximum values and averages were presented in ichnotaxonomic descriptions, those ichnotaxa provided limited data to the overall analysis. Sample sizes of less than three will cause a multivariate analysis of a specimen to fail. Despite their limited contribution of variation data, *J. kimi*, *D. sinensis*, and *M. denaliensis* were left in most of the analyses to determine where in morphospace these ichnotaxa were likely (but not conclusively) to group if more data were available. However, only sparse data (or the mean values of data) for *Sarjeantopodus semipalmatus* (Lockley, Nadon, and Currie, 2004) and *Gruipeda vegrandiunus* (Fiorillo et al., 2011) were provided with the ichnotaxonomic descriptions, and they were not included in the analyses.

Assumption of consistency in data collection techniques

Unless specifically described in the methodology section of each published data source (i.e., Lockley et al., 2009) (Fig. 15.2), there is an assumption of a general standardized method for collecting both track and trackway data for the analyses herein.

The importance of size in multivariate statistical analyses of avian tracks

Given the large size range of the ichnotaxa in this study, analyses were conducted on $\log_{10}$-transformed data. However, there is justification for performing both nontransformed and $\log_{10}$-transformed data. Extant shorebirds reach adult size quickly, decreasing the likelihood that a significant difference in size between two ichnomorphotypes is ontogenetic. Although shorebird young are precocial and their young

mature at a slower rate relative to birds with altricial young (Gill, 2007), the young of shorebirds reach maturity quickly: young of *Charadrius vociferus* (killdeer) leave the nest within 24 hours of hatching, and at day 17 the growth curve of chicks asymptotes (Bunni, 1959; Jackson and Jackson, 2000), and the young of *Actitis macularia* (spotted sandpiper) reach 82% of their adult wing-tip to wing-tip length at day 15 (Oring, Graym, and Reed, 1997). In terms of hind limb development, long bones of the hind limb are roughly 30% to 35% of adult length at the time of hatching in *Larus californicus* (California gull) and increase isometrically (along with foot surface area) with body mass (Carrier and Leon, 1990). Unless two different track size classes are documented within the separate ichnotaxon in question, it is parsimonious to assume that the trackmakers of each ichnotaxon are of or approaching adult size. Two avian ichnotaxa that exhibit a significant difference in size may reflect two separate track-making species. The quick attainment of adult body size by many extant shorebirds allows us to assume that, given their equal investment in both cursorial and aerial locomotory modes (Dial, 2003), shorebird hind limbs become functional shortly after hatching (Carrier and Leon, 1990) and foot length of shorebirds reaches adult body size within days to weeks (Bunni, 1959; Carrier and Leon, 1990; Oring, Graym, and Reed, 1997; Jackson and Jackson, 2000).

A priori groupings and contentious avian ichnotaxa

One must proceed with caution when comparing data from ichnotaxa that have different potential trackmakers with a potential for large morphological similarities. Only specimens for which the affinity of the trackmaker is unambiguous should be used in a multivariate statistical analysis, as many of the comparisons rely on accurate a priori ichnotaxonomic assignments, especially when testing the strength of ichnotaxonomic assignments.[1] As such, ichnologic specimens from the Jurassic Period are not considered, because there is not

15.4. (A) Canonical variate analysis plot of the ichnospecies of Avipedidae (*Aquatilavipes izumiensis*, *A. swiboldae*), the Koreanaornipodidae (*Koreanaornis hamanensis*, *K. dodsoni*, *Pullornipes aureus*) and those ichnotaxa not currently assigned to an ichnofamily (*Uhangrichnus*, *Dongyangornipes*, *Barrosopus*, *Morguiornipes*, *Tatarornipes*). Despite being assigned to the same ichnogenus, *K. hamanensis* and *K. dodsoni* are significantly different and have 100% separation in morphospace. Due to the overall similarity in shape, there is a great deal of morphospace overlap between the avipedids and the koreanaornipedids. *Tatarornipes* forms a discrete cluster, suggesting that *Tatarornipes* does not belong within either the avipedids

or the koreanaornipedids. (B) Principal component (PC) analysis graphical results of PC1 (footprint size) and PC2 (divarication/digit splay ratio) axes of ichnospecies of Avipedidae (*Aquatilavipes izumiensis*, *A. swiboldae*) and Koreanaornipodidae (*Koreanaornis hamanensis*, *K. dodsoni*, *Pullornipes aureus*), with ichnospecies unassigned to an ichnofamily. (C) Principal component analysis graphical results of PC2 (digit divarication/splay ratio) and PC3 (digit I length/DIV III–IV ratio) axes of ichnospecies of Avipedidae (*Aquatilavipes izumiensis*, *A. swiboldae*) and Koreanaornipodidae (*Koreanaornis hamanensis*, *K. dodsoni*, *Pullornipes aureus*), with ichnospecies unassigned to an ichnofamily. *Pullornipes* has a relatively larger pace and stride, whereas *Barrosopus* has

relatively larger digit lengths. (D) Discriminant analysis graphical results of a comparison between the ichnogenera *Aquatilavipes* and *Koreanaornis*. The similarity between the two ichnogenera makes the results of the discriminant analysis contradictory: the two groups are significantly different ($p_{same} = 9.80 \times 10^{-04}$), yet the percentage of individual prints correctly assigned to their a priori groups was only 72.3%. (E) Discriminant analysis on *Koreanaornis dodsoni* and *Morguiornipes*. Despite the large amount of overlap between the two ichnotaxa in Figures 15.5A–15.5C, these two taxa are not significantly different ($p_{same} = .0569$), with 90.5% of the footprints correctly assigned to their a priori determined groups. Abbreviations: DIV, digit divarication; DL, digit length.

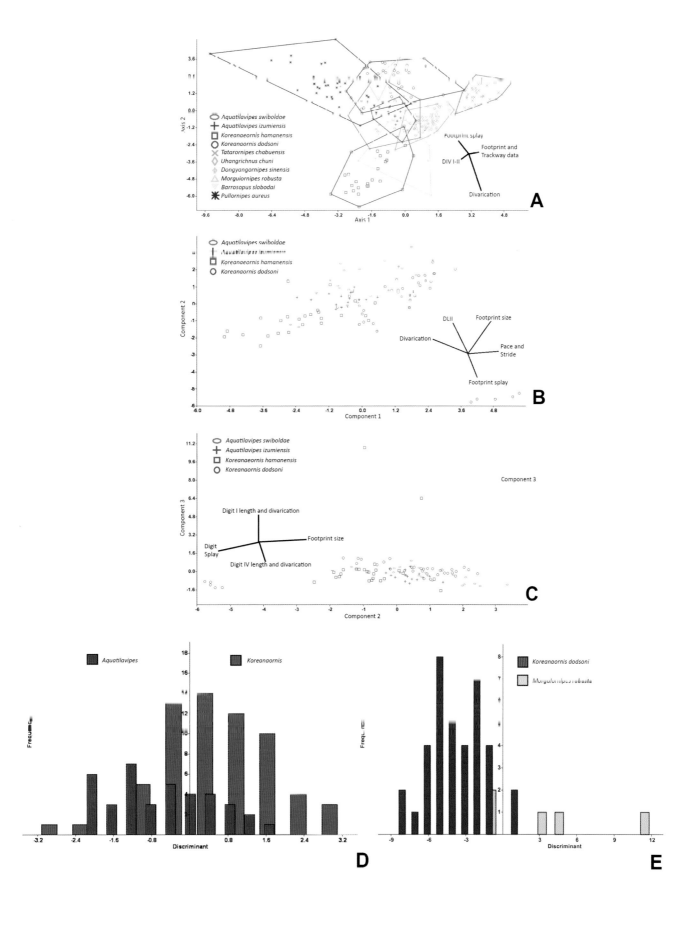

Table 15.3. Canonical variate analysis results of $\log_{10}$-transformed linear and unadjusted angle data (App. 15.1) of Mesozoic avian ichnotaxa a priori assigned to current ichnofamilies and those Mesozoic avian ichnotaxa currently unattributed to ichnofamilies

Icnofamily/taxon	Ignotornidae	Koreanaornipodidae	Tatarornipes	Morguiornipes	Uhangrichnus
Ignotornidae	0	7.88×10^{-84}	7.51×10^{-44}	6.26×10^{-09}	2.14×10^{-61}
Koreanapodidiae	7.88×10^{-84}	0	6.25×10^{-36}	0.100	5.46×10^{-26}
Tatarornipes	7.51×10^{-44}	6.25×10^{-36}	0	2.12×10^{-03}	1.06×10^{-15}
Morguiornipes	6.26×10^{-09}	0.100	2.12×10^{-03}	0	1.19×10^{-03}
Uhangrichnus	2.14×10^{-61}	5.46×10^{-26}	1.06×10^{-15}	1.19×10^{-03}	0
Limiavipedidae	9.67×10^{-96}	5.66×10^{-60}	5.60×10^{-33}	4.88×10^{-05}	1.56×10^{-55}
Avipedidae	2.12×10^{-34}	1.63×10^{-10}	1.46×10^{-09}	0.148	7.89×10^{-03}
Shandongornipodidae	6.32×10^{-32}	1.38×10^{-24}	1.14×10^{-10}	Fail	3.18×10^{-22}
Jindongornipodidae	4.03×10^{-14}	3.96×10^{-09}	6.68×10^{-05}	Fail	1.26×10^{-09}
Barrosopus	1.16×10^{-12}	3.39×10^{-05}	5.16×10^{-04}	Fail	0.589
Dongyangornipes	0.132	0.912	0.373	Fail	0.635

Note: "Fail" indicates a priori groups for which too few data were available for statistical analysis

enough data available to determine whether the trackmakers were true Avialae, nonavian theropods or whether they belonged to an intermediate branch of the theropod–avian phylogeny.

Ichnotaxa for which the trackmaker is of debatable affinity are also problematic in a multivariate statistical analysis. For example, the trackmaker for the ichnogenus *Magnoavipes* has been described by different authors as either avian (Lee, 1997; Fiorillo et al., 2011) or theropod (Lockley, Wright, and Matsukawa, 2001) (Table 15.2). In such cases, the strength of the a priori assignment of either avian or theropod for the ichnotaxon in question may be tested in separately run analyses.

Also, PAST version 2.17 does not offer more than 16 separate symbols for a priori categorization of specimens. This required that the ichnotaxa be a priori grouped by ichnogenera when analyzing all specimens. Specimens were coded as ichnospecies when in smaller analyses that involve fewer a priori groupings.

RESULTS

Support for Current Avian Ichnotaxonomic Assignments

Avian ichnofamilial assignments

Canonical variate analyses show that there is strong statistical support for all ichnofamilies analyzed (Hotelling's t-squared: $p_{same} \lll .01$) (Table 15.3). There is overlap in the graphical projection of the 3-D data onto a 2-D xy graph (Fig. 15.3A). The inclusion of *L. curriei* results in the data points of the much smaller ichnotaxa to concentrate on one side of the morphospace plot due to the larger overall footprint size and trackway dimensions, despite their significant differences. For example, *Dongyangornipes* is not significantly different

from the Ignotornidae (p_{same} = .132), Koreanaornipodidae (p_{same} = .912), *Tatarornipes* (p_{same} = .373), *Uhangrichnus* (p_{same} = .635), and Avipedidae (p_{same} = .900), despite the obvious visual differences between *Dongyangornipes* and the members of these ichnofamilies. Removing *L. curriei* from the ichnofamily analyses allows for a more accurate interpretation of the morphospace groupings among the remaining avian ichnotaxa. Canonical variate analyses with *L. curriei* removed still show strong statistical support for all analyzed ichnofamilies (Hotelling's t-squared: $p_{same} \lll .01$) (Fig. 15.3B); however, *Morguiornipes* (p_{same} = .187) and *Barrosopus* (p_{same} = .112) are not significantly different from the Avipedidae in this specific analysis.

Tridactyl and functionally tridactyl tracks

Avipedidae Due to the close association of Avipedidae and Koreanaornipodidae, the results of the analyses for both ichnofamilies will be discussed together (Fig. 15.4A, canonical variate graph; 15.4B, 15.4C, principle component analyses; and 15.4D, 15.4E discriminant analyses). The initial CVA of all of the ichnotaxa currently assigned to Avipedidae (*Aquatilavipes izumiensis*, *A. swiboldae*) with the ichnotaxa assigned to Koreanaornipodidae and the unassigned ichnogenera reveals the same results as the analyses that included all avian ichnotaxa: the inclusion of *Limiavipes curriei* (McCrea et al., 2014) skews the results by forcing the ichnogenera with smaller overall footprint size into a concentrated area of morphospace.

The CVA of the tridactyl tracks reveals interesting results. *Tatarornipes chabuensis* forms a discrete cluster in morphospace, save for one footprint of *Koreanaornis dodsoni* (Fig. 15.4A). *T. chabuensis* has a larger footprint width and digit III length relative to the other ichnotaxa in the analysis.[2]

Uhangrichnus chuni and *Dongyangornipes sinensis* are not significantly different (p_{same} = .282), which is not surprising

Limiavipedidae	Avipedidae	Shandongornipodidae	Jindongornipodidae	Barrosopus	Dongyangornipes
9.67×10^{-96}	2.12×10^{-34}	6.32×10^{-32}	4.03×10^{-14}	1.16×10^{-12}	0.132
4.88×10^{-00}	1.05×10^{-1}	1.05×10^{-11}	5.30×10^{-44}	3.53×10^{-30}	0.012
5.60×10^{-33}	1.46×10^{-09}	1.14×10^{-10}	6.68×10^{-05}	5.16×10^{-04}	0.373
4.88×10^{-05}	0.148	Fail	Fail	Fail	Fail
1.56×10^{-55}	7.89×10^{-03}	3.18×10^{-22}	1.26×10^{-09}	0.589	0.635
0	1.65×10^{-31}	2.17×10^{-13}	5.67×10^{-04}	3.40×10^{-15}	1.12×10^{-03}
1.65×10^{-31}	0	5.20×10^{-09}	8.33×10^{-04}	0.323	0.900
2.17×10^{-13}	5.20×10^{-09}	0	Fail	Fail	Fail
5.67×10^{-04}	8.33×10^{-04}	Fail	0	Fail	Fail
3.40×10^{-15}	0.323	Fail	Fail	0	Fail
1.12×10^{-03}	0.900	Fail	Fail	Fail	0

given their overall visual similarities (see "Ichnotaxonomic revision of *Dongyangornipes sinensis*"). However, the small sample size of *D. sinensis* prints in the data set may be also resulting in the nonsignificant differences seen both in the analysis of ichnofamilies (Fig. 15.3) and in the analysis of tridactyl prints only (Fig. 15.4A–15.4C). *Aquatilavipes swiboldae* and *A. izumiensis* show 100% overlap in morphospace despite a significant difference ($p_{same} = 9.87 \times 10^{-03}$), whereas *Koreanaornis hamanensis* and *K. dodsoni* show 100% separation and are significantly different ($p_{same} = 6.26 \times 10^{-19}$). However, *A. swiboldae* shows considerable overlap with *Pullornipes*, *Morguiornipes*, and *K. dodsoni*.

It is not unexpected that *Aquatilavipes* and *Koreanaornis* should occupy a similar morphospace: tracks of *Aquatilavipes* and *Koreanaornis* are not significantly different in footprint length (FL, t-test: $p_{same} = .86$), or footprint splay (F/W ratio, t-test: $p_{same} = .56$). However, tracks of *Aquatilavipes* and *Koreanaornis* do differ significantly in their total digit divarication (DIVTOT, t-test: $p_{same} = .02$). Also, *Aquatilavipes* has not been reported with a preserved digit I (Currie, 1981; Azuma et al., 2002), whereas *Koreanaornis hamanensis* has been illustrated with a small posteriomedially oriented hallux (Lockley et al., 1992) (Figs. 15.1B, 15.1D). The combination of both qualitative and quantitative differences between *Koreanaornis* (small hallux and higher total divarication, respectively) and *Aquatilavipes* are enough to justify their separation as discrete ichnomorphotaxa, and the high degree of morphospace overlap between these two ichnogenera can easily be explained by their similarities in footprint length and digit splay. Discriminant analyses on *Aquatilavipes* and *Koreanaornis* reveal contradictory results. While the two groups are significantly different ($p_{same} = 9.80 \times 10^{-04}$), the percentage of individual prints correctly assigned to their a priori groups was only 72.3% (Fig. 15.4D).

Principal component analysis on only Avipedidae and Koreanaornipodidae also provides equivocal results. The first principal component (size, 27.5%) does not reveal much information, except to group the ichnotaxa by size, with *K. hamanensis* being smaller (mean FL = 26.3 mm, n = 24) and *K. dodsoni* being larger (mean FL = 45.5 mm, n = 37) (Fig. 15.4B). Principal components two (PC2, divarication/digit splay ratio) and three (PC3, digit I length/DIV III–IV ratio) do little to differentiate the ichnotaxa in morphospace (Fig. 15.4C).

Avipedidae shows considerable overlap with the Koreanaornipodidae along PC1–PC2 in a PCA that includes all tridactyl ichnotaxa. There is a small separation seen between *A. izumiensis* and *A. swiboldae* along PC3 (PL and SL – digit lengths ratio): *A. izumiensis* has a slightly smaller mean footprint length (x = 37.7 mm) than *A. swiboldae* (x = 39.0 mm), although the difference is not significant (t-test: $p_{same} = .69$). There is also a slight, although not significant (t-test: $p_{same} = .27$), difference in total divarication between *A. izumiensis* (x = 120°, n = 17) and *A. swiboldae* (x = 114°, n = 20), which results in a separation along PC3 (digit splay – digit divarications ratio) (Figs. 15.4B, 15.4C).

Morguiornipes robusta is not shown to be significantly different from either ichnospecies of *Aquatilavipes* (*A. izumiensis*, $p_{same} = .544$; *A. swiboldae*, $p_{same} = .744$). Discriminant analyses comparing *M. robusta* to *Aquatilavipes* show that they are significantly different and have above 90% correct assignment to a priori categories (*A. izumiensis*: $p_{same} = 5.91 \times 10^{-03}$, 100%; *A. swiboldae*: $p_{same} = .0241$, 90.1%).

Koreanaornipodidae Canonical variate analysis on the ichnospecies within Koreanaornipodidae (*Koreanaornis hamanensis*, *K. dodsoni*, and *Pullornipes aureus*) and other tridactyl ichnospecies shows similar results to that of the CVA results of the Avipedidae. There is no overlap between *K. hamanensis* and *K. dodsoni* (Fig. 15.4A). *Pullornipes aureus* shows a small amount of morphospace overlap with *K. dodsoni* ($p_{same} = 3.11 \times 10^{-13}$), and no overlap with *K. hamanensis* ($p_{same} = 3.78 \times 10^{-22}$). *Morguiornipes* occupies almost the exact

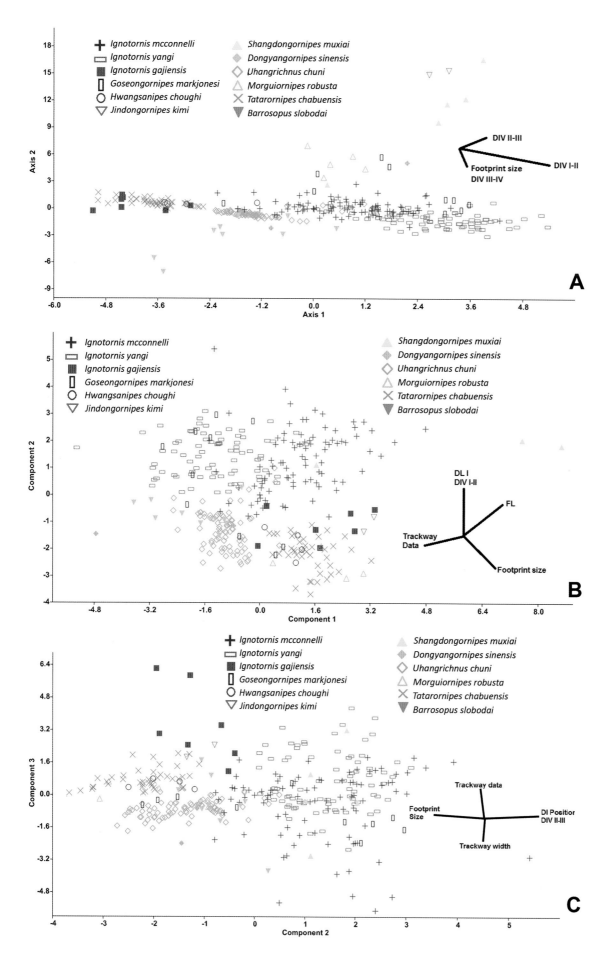

same morphospace as *K. dodsoni* does (p_{same} = .610). Although there is neither statistical nor morphospace support for the separation of *K. dodsoni* and *Morguiornipes*, they are visually distinct. In discriminant analyses, *Morguiornipes robusta* is not significantly different from *Koreanaornis dodsoni* (p_{same} = .0569), although 90.5% of prints were correctly identified (Fig. 15.4E). Also, *Barrosopus* and *K. hamanensis* occupy a similar morphospace and are not significantly different (p_{same} = .092) (Fig. 15.4A) despite visual differences. (See "Avipedidae" for detailed results of the comparison between Avipedidae and Koreanaornipodidae.)

Tetradactyl avian tracks

Ignotornidae The individual ichnogenera within the Ignotornidae (*Ignotornis*, *Hwangsanipes*, *Goseongornipes*) occupy the same morphospace, although *Hwangsanipes* and *Goseongornipes* occupy the outer edges of the *Ignotornis* morphospace (Figs. 15.1C and 15.5; Table 15.4). Canonical variate analysis of the ichnospecies within the ichnofamily Ignotornidae analyzed with avian ichnotaxa currently not assigned to an ichnofamily reveals that, despite the strong statistical support for all of the individual ichnofamilies, there is a strong grouping in morphospace for *Ignotornis mcconnelli*, *I. yangi*, and *Goseongornipes markjonesi* (Fig. 15.5A). However, *I. gajinensis* does not occupy a similar morphospace to that of *I. mcconnelli*, *I. yangi*, and *G. markjonesi* (and is significantly different from these ichnospecies) but occupies a similar morphospace for *Hwangsanipes* (Ignotornidae) and *Tatarornipes* (ichnofamily unassigned) (Fig. 15.5A). A principal component analysis performed on these ichnotaxa reveals that the main quantitative difference between the *I. gajinensis*–*Hwangsanipes* group and the *Ignotornis-Goseongornipes* group is the size of digit I: the *Ignotornis-Goseongornipes* group has a longer digit I, as presented in the PCA morphospace plot (Fig. 15.5B, 15.5C).

Given the morphologic similarity between *Ignotornis* and *Jindongornipes*, it was expected that the two ichnogenera would occupy a similar morphospace; both ichnogenera have a well-defined digit I, digits II and IV have wide angles of divarication from digit III, and some indication of webbing

between digits III and IV. One issue that affects the analysis of *Jindongornipes* is that there are only two data points made available from the systematic description of *J. kimi*. However, based on the available data, the Ignotornidae and *Jindongornipes* are significantly different based on DA (p_{same} = 2.97 × 10^{-17}). The Ignotornidae are smaller, have a longer digit I relative to footprint length and a higher digit divarication between digits I and II, whereas *Jindongornipes* has a relatively higher divarication between digits II and III (Fig. 15.5). The results of this analysis could be greatly altered by the inclusion of more data.

Shandongornipodidae There is both strong qualitative and quantitative support for the ichnofamily Shandongornipodidae. *Shandongornipes muxiai* is the only described trackway of a zygodactyl trackmaker from the Mesozoic: the unique positioning and splay of the digits makes this ichnotaxon easily distinguishable from other described Mesozoic avian ichnotaxa. In CVA, *S. muxiai* forms a discrete cluster in morphospace (Fig. 15.5A, with the exception of one track that contained a large amount of missing data, which groups with *I. mcconnelli*). Principal component analysis reveals that *S. muxiai* is distinguished from the other tetradactyl avian ichnotaxa by the footprint length including the length of digit I along PC1 (footprint size–trackway dimensions ratio) (Fig. 15.5B). When size (PC1, 19.5% of total variation) is removed, *S. muxiai* groups with the tetradactyl ichnotaxa possessing a relatively large digit I and a high digit divarication DIV III–IV (PC2, 17.4% of total data set variation) (Fig. 15.5C).

Jindongornipodidae The avian ichnofamily Jindongornipodidae is represented only by one ichnospecies, *Jindongornipes kimi*. Other than the observations made when compared to the Ignotornidae, the amount of footprint and trackway data available in the literature for these analyses (n = 2) caused many of the analyses to fail. However, with the data available, *J. kimi* does form a discrete group in the CVA.

Avian ichnospecies currently unassigned to ichnofamilies

Tatarornipes chabuensis In the analyses of tridactyl ichnotaxa, *Tatarornipes chabuensis* forms a discrete cluster from

15.5. (A) Canonical variate analysis of tetradactyl Mesozoic avian ichnites and the unassigned avian ichnotaxa (*Uhangrichnus*, *Dongyangornipes*, *Barrosopus*, *Morguiornipes*, *Tatarornipes*). *Shandongornipes muxiai* and *Jindongornipes kimi* form discrete groups, whereas the ignotornids separate into the *I. mcconnelli–I. yangi–Goseongornipes*, and the *Hwangsanipes–I. gajinensis* groups. (B) Principal component (PC) analysis graphical results of the tetradactyl Mesozoic avian ichnotaxa and the unassigned

avian ichnotaxa. PC1 represents data related to digit I, whereas PC2 represents the footprint and trackway size data/divarication ratio. As in Figure 15.6A, *Ignotornis mcconnelli* and *I. yangi* cluster together. This grouping is due to similarities in divarications and footprint size, specifically having a relatively larger DIV I–II with a smaller foot and pace and stride. *Shandongornipes muxiai* forms a discrete cluster due to the relatively larger divarications between digits I and II and total divarication. (C) Principal component analysis graphical

results of the tetradactyl Mesozoic avian ichnotaxa and the unassigned avian ichnotaxa with PC1 (relative size of digit I) removed. The PC2 axis represents digit I placement-/footprint size ratio, and PC3 represents trackway size/trackway width ratio. The groupings show more overlap, but as in Figures 15.6A and 15.6B, *Ignotornis mcconnelli* and *I. yangi* cluster together. Abbreviations: DI, digit I; other abbreviations as in Figure 15.2.

Table 15.4. Canonical variate analysis results of $\log_{10}$-transformed linear and unadjusted angle data (App. 15.1) of tetradactyl Mesozoic avian ichnotaxa a priori assigned to current ichnospecies and those Mesozoic avian ichnotaxa currently unattributed to ichnofamilies.

Ichnotaxon	Hwangsanipes	Uhangrichnus	Ignotornis yangi	Ignotornis gajinensis	Goseongornipes
Hwangsanipes	0	7.18×10^{-03}	1.15×10^{-08}	Fail	Fail
Uhangrichnus	7.18×10^{-03}	0	2.51×10^{-55}	1.11×10^{-17}	1.89×10^{-12}
Ignotornis yangi	1.15×10^{-08}	2.51×10^{-55}	0	4.07×10^{-22}	1.64×10^{-13}
Ignotornis gajinensis	Fail	1.11×10^{-17}	4.07×10^{-22}	0	Fail
Goseongornipes	Fail	1.89×10^{-12}	1.64×10^{-13}	Fail	0
Morguiornipes	Fail	6.82×10^{-14}	2.27×10^{-16}	Fail	Fail
Jindongornipes	Fail	2.63×10^{-26}	3.01×10^{-24}	Fail	Fail
Shandongornipes	Fail	5.74×10^{-30}	1.02×10^{-27}	Fail	Fail
Dongyangornipes	Fail	1.23×10^{-05}	5.00×10^{-06}	Fail	Fail
Ignotornis mcconnelli	1.60×10^{-05}	1.84×10^{-51}	5.24×10^{-44}	6.08×10^{-18}	2.31×10^{-15}
Barrosopus	Fail	3.08×10^{-06}	3.17×10^{-17}	Fail	0.731
Tatarornipes	0.987	1.97×10^{-34}	2.41×10^{-49}	7.21×10^{-06}	2.32×10^{-12}

Note: "Fail" indicates a priori groups for which too few data were available for statistical analysis.

the Avipedidae, the Koreanaornipodidae, and the unassigned ichnospecies. The qualitative assignment to a discrete ichnospecies is justified statistically ($p_{same} \lll .01$). *Tatarornipes* is separated from the rest of the tridactyl ichnotaxa by its relatively large footprint width and digit III length (Fig. 15.4).

Morguiornipes robusta As described in the results of the Avipedidae and Koreanaornipodidae analyses, *Morguiornipes robusta* is not statistically different from *Aquatilavipes izumiensis*, *A. swiboldae*, and *Koreanaornis dodsoni*. Principal component analysis shows that *M. robusta* groups with the individual tracks of *A. swiboldae*, *K. hamanensis*, *K. dodsoni*, and *Barrosopus slobodai* with similar digit lengths (App. 15.1). However, this grouping is due to similarity in variables: *M. robusta* is qualitatively distinct from *Koreanaornis* and *Aquatilavipes* in that it has much thicker (or wider) digits than the two aforementioned ichnogenera. The thickness of the digits is reminiscent of that observed in *Tatarornipes chabuensis*, but the digits of *M. robusta* do not taper sharply distally as the digits do in *T. chabuensis* (Fig. 15.6).

Barrosopus slobodai Prints of *Barrosopus slobodai* are not significantly different from *Aquatilavipes izumiensis*, *A. swiboldae*, or *Koreanaornis hamanensis* (Fig. 15.4A, Table 15.3). Discriminant analysis confirms the quantitative similarities of *B. slobodai* and *A. swiboldae* ($p_{same} = .831$, 76.7% correctly identified), *A. izumiensis* ($p_{same} = .403$, 73.1% correctly identified), and *K. hamanensis* ($p_{same} = .755$, 82.9% correctly identified). *Barrosopus* and *K. hamanensis* occupy a similar morphospace due to shared characteristics in overall footprint and trackway dimensions: they do not significantly differ in total divarication (t-test: $p_{same} = .50$), pace length (t-test: $p_{same} = .12$), or digit splay (FL/FW, t-test: $p_{same} = .087$).

Avian Ichnites versus Small Nonavian Theropod Ichnites (Magnoavipes)

Magnoavipes is a contentious ichnogenus: the relatively large size, with footprint lengths approaching 200 mm or more (App. 15.1) is indicative of a small theropod trackmaker, whereas its high divarication (*M. lowi* = 110°, *M. caneeli* = 85.1°, *M. denaliensis* = 107°) suggests a trackmaker of avian affinity (Table 15.2). Divarication and trackway data were analyzed separately on Mesozoic avians, *Magnoavipes*, and Early Cretaceous theropods (n_{total} = 59; McCrea, 2000) to determine the diagnostic strength of total divarication and trackway data for distinguishing between bird and theropod ichnites.

First, using total divarication as a discriminatory tool between theropod and avian tracks holds when using t-tests of means: comparing the means of the total divarication of theropods (*Irenichnites*, *Irenisauripus*, *Columbosauripus*, and an unidentified small theropod; mean = 87.8°, n = 26) and tracks of semipalmate avians (*Aquatilavipes*, *Koreanaornis*, *Tatarornipes*, *Morguiornipes*; mean = 111°, n = 202) does show a significant difference ($p_{same} = 5.40 \times 10^{-08}$). However, the range of data shows a considerable amount of overlap (Table 15.2). The total divarication of *Magnoavipes* is not significantly different than that of the theropod sample (t-test: $p_{same} = .245$), but the total divarication of *Magnoavipes* is significantly different than that of the semipalmate avians (t-test: $p_{same} = 7.23 \times 10^{-05}$).

Discriminant analyses on DIV II–III, DIV III–IV, and DIV-TOT confirm the univariate statistical results. *Magnoavipes* is not significantly different than the sampled theropods (p_{same}

Morguliornipes	Uhangnornipes	Shandongornipes	Donyangornipes	Ignotornis mcconnelli	Barrosopus	Tatarornipes
Fail	Fail	Fail	Fail	1.60×10^{-05}	Fail	0.987
6.02×10^{-14}	2.63×10^{-26}	5.74×10^{-30}	1.37×10^{-05}	1.94×10^{-51}	3.08×10^{-06}	1.97×10^{-34}
2.27×10^{-16}	3.01×10^{-24}	1.02×10^{-27}	5.00×10^{-06}	5.24×10^{-44}	3.17×10^{-17}	2.41×10^{-49}
Fail	Fail	Fail	Fail	6.08×10^{-18}	Fail	7.21×10^{-06}
Fail	Fail	Fail	Fail	2.31×10^{-15}	0.731	2.32×10^{-12}
0	Fail	Fail	Fail	4.07×10^{-17}	Fail	5.35×10^{-09}
Fail	0	Fail	Fail	3.70×10^{-27}	Fail	8.29×10^{-12}
Fail	Fail	0	Fail	2.30×10^{-29}	Fail	2.96×10^{-14}
Fail	Fail	Fail	0	2.51×10^{-07}	Fail	3.32×10^{-04}
4.07×10^{-17}	3.70×10^{-27}	2.30×10^{-29}	2.51×10^{-07}	0	7.48×10^{-18}	5.80×10^{-42}
Fail	Fail	Fail	Fail	7.48×10^{-18}	0	5.44×10^{-10}
5.35×10^{-09}	8.29×10^{-12}	2.96×10^{-14}	3.32×10^{-04}	5.80×10^{-42}	5.44×10^{-10}	0

= .179, 67.8% correctly identified). *Magnoavipes* is significantly different than the avian sample (p_{same} = 1.37 × 10^{-03}, 71.3% correctly identified), and the theropod sample was significantly different than the avian sample (p_{same} = 2.79 × 10^{-04}, 67.8% correctly identified). In all three DA, there was difficulty in correctly assigning each print to its a priori ichnotaxon. This is likely due to the large amount of variation in digit divarication of both birds and theropods: theropod tracks can reach total divarications of 120°, and bird tracks can exhibit total divarications as low as 60° (Table 15.2; App. 15.1). There is no one divarication value that clearly separates bird from theropod tracks.

Because of the great disparity in size between the tracks attributed to birds and those attributed to theropods and the ichnogenus *Magnoavipes*, the CVA comparing these ichnotaxonomic groups was performed on log$_{10}$-transformed linear data and unaltered angle data. The results show that all three groups are significantly different, but there is a great deal of overlap among the three groups (Fig. 15.7A). Although the largest amount of relative variation occurs along the footprint data. Footprint data occur in morphospace. There is very little separation among the three ichnotaxonomic groups along this axis.

In order to determine whether footprint ratios could be used to discriminate between the tracks of birds (including *Limiavipes curriei*) and theropods (and to remove size as a factor in the analyses), the footprint ratios were analyzed using CVA. Even though birds, *Magnoavipes*, and theropods were all significantly different from one another, they all occupy the same morphospace. The only indication of any ichnotaxonomic grouping was that tracks of *Magnoavipes*

occupy almost the same morphospace as that occupied by tracks of the theropod sample (Fig. 15.7B). It is evident that the footprint data alone (save for size) is not sufficient to separate the tracks of theropods from those of birds in a multivariate analysis.

Trackway data (pace and stride length, pace angulation, and footprint rotation) were analyzed using CVA, and the results show that both theropods (p_{same} = 1.66 × 10^{-31}) and *Magnoavipes* (p_{same} = 1.94 × 10^{-27}) are significantly different from birds, whereas the prints of *Magnoavipes* are not significantly different from prints of theropods (p_{same} = .987). However, these differences are largely size-based, so the same analysis was run using log$_{10}$-transformed pace and stride lengths. The results are similar: both theropods (p_{same} = 9.84 × 10^{-35}) and *Magnoavipes* (p_{same} = 7.27 × 10^{-20}) (Fig. 15.7C) are significantly different from the trackways of birds. These results are supported by the bivariate statistics: log$_{10}$-pace lengths are significantly different between theropod and bird trackways (p_{same} = 1.68 × 10^{-6}), and between bird trackways and the trackways of *Magnoavipes* (p_{same} = 2.00 × 10^{-03}), whereas the difference between the log$_{10}$-pace lengths of *Magnoavipes* and theropod trackways is significant at p > .05 but not at p > .01 (p_{same} = .014).[3]

ICHNOTAXONOMIC REVISION OF DONGYANGORNIPES SINENSIS

Uhangrichnus chuni was established by Yang et al. (1995) and was the first described functionally tridactyl, palmate avian track. The description of *U. chuni* was emended by Lockley, Lim, et al. (2012) based on the description by Lockley and

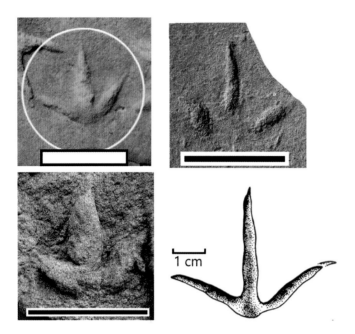

15.6. Functionally tridactyl Mesozoic avian footprints: (upper left) *Tatarornipes chabuensis* (Lockley, Li, et al., 2012); (upper right) *Koreanaornis dodsoni* (Xing et al., 2011); (lower left) *Morguiornipes robusta* (Xing et al., 2011); (lower right) *Aquatilavipes swiboldae* (Currie, 1981). Despite their morphologic differences, these four ichnotaxa tend to group together in multivariate statistical analyses due to the similarity in their measured variables. Collecting the variable of digit width might cause these groups to separate in morphospace. Scale = 5.0 cm.

impression between II and III is connected from the apex of digit II to the posterior third of digit III, whereas the web impression between digits III and IV is linked from the apex of digit IV to the middle of digit III."

We now consider these differences in webbing to be more parsimoniously explained by preservational variation: it is not uncommon for webbed tracks of one trackmaker to present a variety of webbing conditions. Several of the tracks of *Uhangrichnus chuni* pictured in figure 5 of Yang et al. (1995) display the webbing conditions (connected at the apices of digits II and IV and connecting on the posterior third and middle of digit III, respectively; asymmetrical web impressions) described as unique to *Dongyangornipes sinensis* and that separate *D. sinensis* from *U. chuni*. Also, the tracks of *U. chuni* and *D. sinensis* are similar in size, as noted by Azuma et al. (2013): the average (presumably) *D. sinensis* is described as footprint length of 3.64 cm and footprint width of 3.96 cm. This falls within the size range reported for *U. chuni* (FL: 3.30–4.62 cm, average 3.70 cm; FW: 3.8–5.4 cm, average 4.58 cm). The only quantitative difference between *U. chuni* and *D. sinensis* is that of footprint splay, as represented by total (II–IV) divarication and FL/FW. *U. chuni* has a wider splay (L/W: 0.81, total divarication 110°) than *D. sinensis* does (L/W: 0.87, total divarication 100°); however, given the small sample size of *D. sinensis* compared to that of *U. chuni*, there are prints of *U. chuni* that also exhibit the same footprint splay seen in *D. sinensis*.

Given that the unique webbing characteristic of *Dongyangornipes* reported by Azuma et al. (2013) to separate *Uhangrichnus* from *Dongyangornipes* is also observed (if not reported) in the original description of *Uhangrichnus*, and that the sample size of *Dongyangornipes* is reasonably too small to preserve a sporadically preserved hallux, we feel that there is very little to visually differentiate *Uhangrichnus* and *Dongyangornipes*. We consider *Dongyangornipes sinensis* a subjective junior synonym of *Uhangrichnus chuni*.

Harris (2010) of a topotype specimen of *U. chuni* with a short, posteromedially directed hallux. However, the emended diagnosis states that "web configuration palmate (i.e., well-developed) and equally developed in hypices between digits II and III and III and IV" (Lockley, Lim, et al., 2012:20). This is different than the webbing reported for *Dongyangornipes sinensis* (Azuma et al., 2013:5), stating: "although similar to *U. chuni* in size and general morphological characteristics, *Dongyangornipes sinensis* differs in that the anterior margin of the web impression is connected from the apex of digit III to the apex of digits II and IV in *U. chuni*," and "the web

15.7. (A) Canonical variate analysis on log₁₀-transformed footprint and trackway data of a priori separated groups of *Magnoavipes*, Mesozoic tridactyl semipalmate avians, and Early Cretaceous theropods (*Irenisauripus, Irenichnites, Columbosauripus*, and one unidentified small theropod ichnite). There is a large amount of overlap among the three groups, and a slight separation of the *Magnoavipes* and theropod groups along the footprint size axis. Both *Magnoavipes* and the theropods occupy a close, but not overlapping, section of morphospace. (B) Canonical variate analysis on footprint measurement ratios (FL/FW, DL2/DL3, DL4/DL3) of *Magnoavipes*, Mesozoic tridactyl semipalmate avians, and Early Cretaceous theropods (*Irenisauripus, Irenichnites, Columbosauripus*, and one unidentified small theropod ichnite). The footprint ratios used in this analysis did not provide any for separating theropod ichnites from bird ichnites, but they did show that theropods, because of the narrower splay on their digits, have a relatively larger FL/FW. All a priori groups are significantly different from one another. (C) Canonical variate analysis on log₁₀-tranformed pace (PL) and stride (SL) data, footprint rotation (FR), pace angulation (PA), and footprint length/footprint width ratios (FL/FW) of *Magnoavipes*, Mesozoic tridactyl semipalmate avians, and Early Cretaceous theropods (*Irenisauripus, Irenichnites, Columbosauripus*, and one unidentified small theropod ichnite). Birds have a relatively larger pace angulation and footprint rotation, whereas *Magnoavipes* and theropods have a relatively larger FL/FW. *Magnoavipes* and bird trackway data are significant (p_same = 2.00 × 10⁻⁰³), whereas *Magnoavipes* and theropod trackway data are significant at p ≥ .05 but not at p ≥ .01 (p_same = .014).

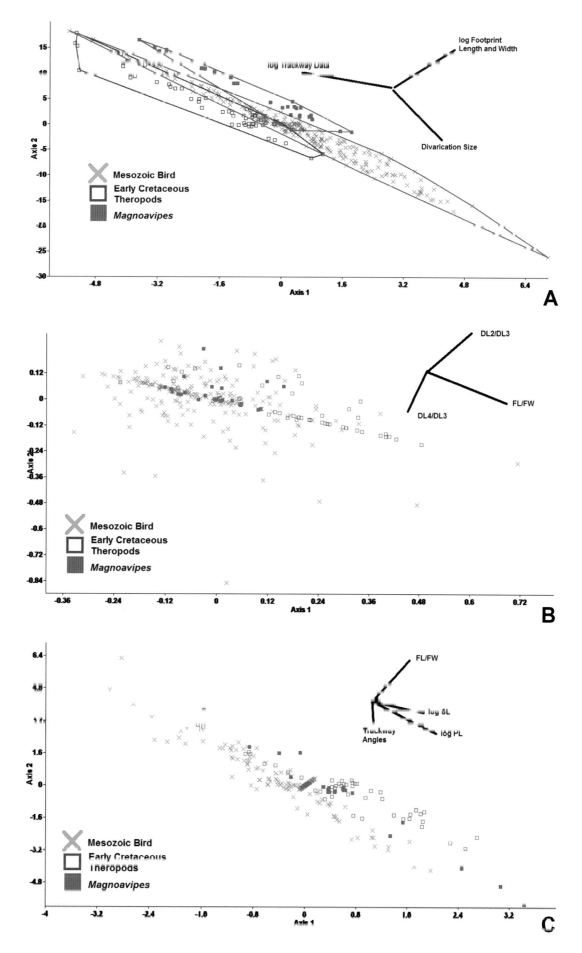

Systematics

Class Aves
Ichnofamily indet.
Uhangrichnus chuni (Yang et al., 1995;
emend. Lockley, Lim, et al., 2012)
Referred material
Chun, 1990:10a
"tracks of a bird with webbed feet"
Lockley et al., 1992:figure 9
Uhangrichnus chuni Yang et al., 1995:figure 5
Uhangrichnus chuni Yang et al., 1997:figures 3–5
Uhangrichnus chuni Lockley &
Rainforth, 2002:figure 17.12B
Uhangrichnus chuni Kim et al., 2006: figure 3D
Uhangrichnus chuni Lockley, 2007:figure 1D
Uhangrichnus chuni Lockley &
Matsukawa, 2009:figure 15D
Uhangrichnus Lockley & Harris, 2010:figure 9C
Dongyangornipes sinensis Azuma et al., 2013

Emended diagnosis Small, functionally tridactyl tracks of a web-footed bird with small, posteromedially directed hallux trace are sporadically preserved. Web configuration palmate, with the posterior margins of the webbing ranging from equally developed between digits II and III and digits III and IV, to connecting at the middle to the posterior third of digit III. Footprint, excluding hallux, wider (W) than long (L), averaging 3.70 and 4.58 cm, respectively (L/W = 0.81), but footprint length with hallux slightly longer than wide (L/W = 1.1). Trackway narrow with short step and stride (7.8 and 15.7 cm, respectively) and strong inward rotation (mean 20°) of digit III relative to trackway midline.

DISCUSSION

Qualitative Ichnotaxonomic Assignments of Avian Ichnotaxa Have Statistical Support

Multivariate statistical analyses have been shown to be a useful tool in testing the qualitative ichnotaxonomic assignments of avian ichnotaxa. These qualitative assignments, for the most part, are well-supported statistically. Although taxonomic assignments of ichnology specimens have received some criticism for being too subject to preservation and substrate consistency, the strong statistical support for the current standing avian ichnotaxa demonstrates that the assignment of vertebrate ichnotaxa is not random: the qualitative and simple quantitative differences observed in Mesozoic avian ichnotaxa have strong quantitative support.

*Separating Tracks of Large Avians
from Small Theropods*

Save for size, there is a great deal of overlap in the track morphology data of small theropods and large avians. The multivariate statistical analyses support the assignment of the ichnogenus *Magnoavipes* to a theropod, rather than an avian, trackmaker by Lockley, Wright, and Matsukawa (2001). The strongest support came from the trackway data: for all of the avian characteristics of the tracks of *Magnoavipes* (high total divarication, narrow digit widths; Lee, 1997), the data suggest that the trackmaker of *Magnoavipes* walks more like a theropod than a bird. The difference in pace angulation and footprint rotation between theropods and avians has long been observed: trackways attributed to theropods have higher angles in pace angulation (footprints placed closer to the midline of the trackway) and lower footprint rotation (footprints closer to parallel with the midline of the trackway) (Lockley, Wright, and Matsukawa, 2001).

Also, total divarication (DIV II–IV, as used in the original description of *Magnoavipes*, Lee, 1997) is unreliable for separating bird tracks from theropod tracks, and there is a large amount of overlap in the range of total divarications exhibited by Mesozoic avian ichnites and those of Early Cretaceous theropod ichnites (Table 15.2). Also, digit divarication can vary considerably within a single trackway in both avians and theropods (App. 15.1). Determining the affinity of a trackmaker based largely on the average total divarication is arbitrary and ignores both extant avian track data and fossil data for tracks of both theropods and avians alike. Given that both extant and extinct shorebirds are capable of such relative extremes in digit divarication, ichnologists must accept that Mesozoic analogues to extant shorebirds were likely capable of such relative extremes in digit divarication as well.

Trackways ratios, using the most common data collected for both avian and theropod tracks (footprint length; footprint width; digit lengths II, III, and IV; digit divarications) have the benefit of removing absolute size from the analyses but do not contain enough data to discriminate between avian and theropod tracks. Analyzed together, the number of measured footprint parameters outnumbers the trackway parameters, which may account for the lack of separation in morphospace between birds and theropods.

*Multivariate Statistical Analyses Are Not a
Primary Tool for Ichnotaxonomic Assignment*

While multivariate statistical analyses are a useful tool for testing previously established ichnotaxonomic assignments, and are also useful to test the quantitative support of the systematic assignment of future avian ichnotaxa, multivariate statistical analyses should not be used as the sole tool in either assigning an ichnite to an ichnotaxon or as the sole means of identifying the potential trackmaker of one footprint. Statistics, whether they be bivariate or multivariate, are best in a supporting role in vertebrate ichnology rather than as the primary source of interpretations.

Criticisms of using multivariate analyses as the sole means of identifying the trackmaker of an ichnite have been discussed in length by Thulborn (2013) in his analysis of the reinterpretation of the large tridactyl trackmaker in the famous Lark Quarry track site (Romilio and Salisbury, 2011). Many of these criticisms are valid: one cannot use a multivariate statistically determined "cut-off" value between two groups with any morphologic similarity to determine the accurate placement of single ichnites, or even single trackways (Thulborn, 2013). Statistically determined threshold values used to either assign previously unassigned ichnites to existing ichnotaxa or to determine trackmaker affinity are only as accurate as the degree of separation between the samples used to create the threshold value. The analyses are subject to the initial a priori assignments of the researchers. Any threshold value established between two groups with less than 100% separation will be inaccurate depending on the amount of morphologic overlap between the two groups: the more overlap, the greater the inaccuracy of the identification of the ichnite using the threshold.

Recommendations for Future Data Collection and Data Reporting

Multivariate statistical analyses are only as accurate as the data used within the analyses. The analyses herein would have been greatly improved had data been reported for all of the track and trackway variables. This is not referring to the missing data that is inherent in so many vertebrate ichnology datasets: data cannot be reported for prints that are incompletely preserved. Data should, however, be collected for all footprint and trackway variables that can be measured, even if they do not immediately aid in the systematic classification of the ichnite in question. Depending on the nature of the avian ichnites, all data variables may not be available to collect. For example, *Aquatilavipes swiboldae* (Currie, 1981) and *Uhangrichnus chuni* (Yang et al., 1995) were described from specimens where individual trackways were difficult to discern, making the accurate collection of trackway data

unfeasible. However, where feasible, thorough footprint and trackway data should be both collected and reported.

Reexaminations of avian ichnotaxa for which the original descriptions did not supply a large amount of data (either due to the original paucity of specimens at the time of description or lack of publishing the originally collected data), such as the reexamination of *Ignotornis mcconnelli* by Lockley et al. (2009), are extremely important to the study of avian ichnotaxonomy. As more specimens of existing avian ichnotaxa are recovered, such reexaminations of existing avian ichnotaxa will be beneficial for further resolving avian diversity in the Mesozoic.

CONCLUSIONS

This is not the final statistical review of Mesozoic avian ichnotaxonomy. As more specimens of existing avian ichnotaxa are recovered, and as novel avian ichnotaxa are erected, the results herein will undoubtedly be refined. The analyses herein do reveal useful information in demonstrating that the valid Mesozoic avian ichnotaxa have strong statistical support. These analyses also demonstrate that ichnotaxa erected using qualitative observations on size, shape, and basic statistical data are not arbitrarily named shapes. Ichnotaxonomy is a discipline that will continue to draw heavily on qualitative information, and thus far qualitative information has proved reliable. However, with the development of new data collection technologies, useful quantitative data can be used to support the qualitative observations.

To aid in the future quantitative analyses of vertebrate ichnites, we offer the following recommendations:

1. A reanalysis of the existing avian ichnotaxa, in the manner of Lockley et al. (2009), to amend the previously published data sets;
2. Collect and publish as many track and trackway variables as is feasible for each specimen. It is not enough to merely report and discuss a novel ichnotaxa; all data should be made available for more accurate comparisons and future statistical analyses;
3. As potentially useful as multivariate statistical analyses are, the utility of such analyses in resolving contentious ichnotaxonomic assignments or providing information to support a novel ichnotaxonomic assignment will only be as accurate as the input data.

ACKNOWLEDGMENTS

We wish to first thank the researchers who collected and published the data used in this statistical review. We also wish to thank J. Milàn, an anonymous reviewer, and A. Richter for their helpful reviews of the manuscript.

NOTES

1. Even though ichnotaxonomy is the focus of this study, this caveat is applicable to all morpho-taxa used in multivariate statistical analyses.
2. One of the diagnostic characteristics of *Tatarornipes chabuensis* is the robust width of the proximal digits (Lockley, Li, et al., 2012). Digit width is not a variable that is often reported in avian tracks: one exception is Azuma et al. (2013), who report digit width in their description of *Dongyangornipes sinensis*. Due to the large amount of missing data that digit width would have introduced to the analyses, digit width was not included.
3. Note in proof: since the production of this manuscript, two studies have been submitted for publication (Xing et al., 2015; Buckley et al., 2015) which detail new metrics and considerations for distinguishing between the tracks of large avians and those of small nonavian theropods.

APPENDIX 15.1

Data used in multivariate analyses of Mesozoic avian ichno-taxa. Data were compiled from references cited within the table. All data were standardized to fit the table presented here. Missing data are indicated by a question mark. Data that are not present due to morphology (i.e., no digit I) are indicated by a zero. All linear measurements are in millimeters. All angle measurements are in degrees. All data are unadjusted.

DATA ABBREVIATIONS

DIV, digit divarication; DIVTOT, digit divarication II–IV; DL, digit length; DW, digit width; FL, footprint length; FLwH, footprint length including hallux; FR, footprint rotation; FW, footprint width; (H), holotype; I, digit I; II, digit II; III, digit III; IV, digit IV; KoHa, *Koreanaornis hamanensis*; l/w, footprint length to footprint width ratio; (P), paratype; PA, pace angulation; PL, pace length; SL, stride length; (T), topotype; TW, trackway width.

INSTITUTIONAL ABBREVIATIONS

CU, University of Colorado Denver, Dinosaur Tracks Museum; DMNH, Denver Museum of Natural History, Denver, Colorado; FPDM, Fukui Prefectural Dinosaur Museum, Japan; KNUE, Korea National University of Education, Cheongwon, Chungbuk, Korea; KPE, Kyungpook National University, Earth Science Education Department, Taegu, South Korea; LRH, Qingdao Institute of Marine Geology; MGCM, Moguicheng Dinosaur and Bizarre Stone Museum, Xinjiang, China; MWC, Museum of Western Colorado; PVPH, Paleontolga de Vertebrados, Museo del Neuquen, Argentina ; TMP, Tyrell Museum of Palaeontology; UCM; University of Colorado Museum of Natural History at Boulder, Colorado; ZMNH, Zhejiang Natural History Museum, Zheijiang, China.

Ichnotaxon	Ichnofamily	Track #	FL	FLwH	FW	L/W	Digit lengths I	II	III	IV	Digit widths I	II	III	IV	Reference(s)
Aquatilavipes izumiensis	Aviipedidae	FPDM-V43.1	37.9	37.9	47	0.80	0	18	29.6	23.2	0	?	?	?	Azuma et al. (2002)
		FPDM-V43.2	43.8	43.8	36	1.22	0	12	35.1	22.5	0	?	?	?	
		FPDM-V43.3	37.2	37.2	47	0.80	0	19	31.5	22.6	0	?	?	?	
		FPDM-V43.4	39	39	45	0.86	0	18	30.6	24.6	0	?	?	?	
		FPDM-V43.5	40.8	40.8	47	0.86	0	20	33.8	23.5	0	?	?	?	
		FPDM-V43.6	34.3	34.3	50	0.69	0	20	28.9	24.2	0	?	?	?	
		FPDM-V43.7	35.3	35.3	42	0.84	0	16	29.7	22.9	0	?	?	?	
		FPDM-V43.8	33.4	33.4	44	0.76	0	16	27.7	22.5	0	?	?	?	
		FPDM-V43.9	39.2	39.2	36	1.08	0	16	27.9	18.7	0	?	?	?	
		FPDM-V43.10	43.7	43.7	41	1.07	0	16	32.7	22.8	0	?	?	?	
		FPDM-V43.12	31.9	31.9	46	0.70	0	16	26.7	21.9	0	?	?	?	
		FPDM-V43.13	40.5	40.5	49	0.83	0	20	31.3	24.2	0	?	?	?	
		FPDM-V43.14	36.5	36.5	45	0.82	0	18	28.9	21.9	0	?	?	?	
		FPDM-V43.15	34	34	40	0.85	0	18	27.2	23	0	?	?	?	
		FPDM-V44.B1	40.5	40.5	52	0.79	0	20	30.8	22.6	0	?	?	?	
		FPDM-V44.B2	34.3	34.3	47	0.74	0	20	27.8	24.2	0	?	?	?	
Aquatilavipes swiboldae	Aviipedidae	TMP 98.89.21-A	47	47	55	0.85	0	34	47	32	0	?	?	?	Currie (1981); McCrea and Sarjeant (2001)
		TMP 98.89.21-B	44	44	57	0.77	0	42	44	38	0	?	?	?	
		TMP 98.89.21-C	42	42	33	1.27	0	32	42	30	0	?	?	?	
		TMP 98.89.21-D	32	32	37	0.86	0	25	32	27	0	?	?	?	
		TMP 98.89.21-E	31	31	43	0.72	0	23	31	24	0	?	?	?	
		TMP 98.89.21-F	35	35	35	1.00	0	26	35	26	0	?	?	?	
		TMP 98.89.20-B	45	45	63	0.71	0	34	45	38	0	?	?	?	
		TMP 98.89.20-C	33	33	45	0.73	0	24	33	33	0	?	?	?	
		TMP 98.89.20-D	?	?	49	?	0	28	?	28	0	?	?	?	
		TMP 98.89.20-E	25	25	31	0.81	0	19	25	20	0	?	?	?	
		TMP 98.89.20-F	?	?	40	?	0	?	?	?	0	?	?	?	
		TMP 98.89.20-G	?	?	?	?	0	40	?	?	0	?	?	?	
Hwangsanipes choughi	Ignotornidae	TMP 90.30.1-A	40	40	55	0.73	0	33	40	27	0	?	?	?	Yang et al. (1995)
		TMP 90.30.1-B	37	37	57	0.65	0	26	37	35	0	?	?	?	
		TMP 90.30.1-C	41	41	55	0.75	0	29	41	32	0	?	?	?	
		TMP 79.23.3-	38	38	47	0.81	0	22	34	24	0	?	?	?	
		KPE50101-A38	50	?	?	?	?	?	50	?	?	?	?	?	
		KPE50101-A39	?	?	61	?	?	?	?	?	?	?	?	?	
		KPE50101-A40	48	?	66	0.73	?	?	48	?	?	?	?	?	
		KPE50101-A41	48	?	62	0.77	?	?	48	?	?	?	?	?	
Ignotornis mcconnelli	Ignotornidae	Average	40	?	50	?	11	40	?	?	?	?	?	?	Lockley et al. (2006)
		(-)-UCM17614-1.1	39	59	51	0.76	20	?	39	?	?	?	?	?	
		(-)-UCM17614-1.2	40	57	50	0.80	17	?	40	?	?	?	?	?	

Table 15A.1. *continued*

Ichnotaxon	Ichnofamily	Track #	FL	FLwH	FW	L/W	Digit lengths				Digit widths				Reference(s)
							I	II	III	IV	I	II	III	IV	
		(H)-UCM17614-1.3	37	51	52	0.71	14	?	37	?	?	?	?	?	
		(H)-UCM17614-1.4	40	56	47	0.85	16	?	40	?	?	?	?	?	
		(H)-UCM17614-1.5	40	53	49	0.82	13	?	40	?	?	?	?	?	
		(H)-UCM17614-1.6	?	?	46	?	?	?	?	?	?	?	?	?	
		(P1)-UCM98120-2.1	45	63	52	0.87	18	?	45	?	?	?	?	?	
		(P1)-UCM98120-2.2	45	66	52	0.87	21	?	45	?	?	?	?	?	
		(P1)-UCM98120-2.3	44	64	52	0.85	20	?	44	?	?	?	?	?	
		(P1)-UCM98120-3.1	41	54	46	0.89	13	?	41	?	?	?	?	?	
		(P1)-UCM98120-3.2	44	59	43	1.02	15	?	44	?	?	?	?	?	
		(P1)-UCM98120-3.3	42	53	46	0.91	11	?	42	?	?	?	?	?	
		(P1)-UCM98120-3.4	44	56	44	1.00	12	?	44	?	?	?	?	?	
		(P1)-UCM98120-3.5	?	?	44	?	?	?	?	?	?	?	?	?	
		(P1)-UCM98120-4.1	44	61	46	0.96	17	?	44	?	?	?	?	?	
		(P1)-UCM98120-4.2	42	61	42	1.00	19	?	42	?	?	?	?	?	
		(P1)-UCM98120-4.3	41	56	41	1.00	15	?	41	?	?	?	?	?	
		(P1)-UCM98120-5.1	34	52	57	0.60	18	?	34	?	?	?	?	?	
		(P1)-UCM98120-5.2	?	?	51	?	?	?	?	?	?	?	?	?	
		(P1)-UCM98120-5.3	37	53	55	0.67	16	?	37	?	?	?	?	?	
		(P1)-UCM98120-5.4	37	53	51	0.73	16	?	37	?	?	?	?	?	
		(P2)-UCM98121-6.1	39	53	50	0.78	14	?	39	?	?	?	?	?	
		(P2)-UCM98121-6.2	39	49	50	0.78	10	?	39	?	?	?	?	?	
		(P2)-UCM98121-6.3	?	?	52	?	?	?	?	?	?	?	?	?	
		(P2)-UCM98121-7	42	54	50	0.84	12	?	42	?	?	?	?	?	
		(P2)-UCM98121-8	40	51	46	0.87	11	?	40	?	?	?	?	?	
		(P2)-UCM98121-9	?	?	?	?	?	?	?	?	?	?	?	?	
		(P3)-UCM98122-10	?	54	51	?	?	?	?	?	?	?	?	?	
		(P3)-UCM98122-11	40	52	52	0.77	12	?	40	?	?	?	?	?	
		(P3)-UCM98122-12	35	?	?	?	?	?	35	?	?	?	?	?	
		(P3)-UCM98122-13.1	35	?	45	0.78	?	?	35	?	?	?	?	?	
		(P3)-UCM98122-13.2	39	53	50	0.78	14	?	39	?	?	?	?	?	
		(P3)-UCM98122-13.3	35	51	45	0.78	16	?	35	?	?	?	?	?	
		(P3)-UCM98122-14	?	?	52	?	?	?	?	?	?	?	?	?	
		(P3)-UCM98122-15	38	?	45	0.84	?	?	38	?	?	?	?	?	
		(T)-CU203.28-G1.1	35	?	44	0.80	?	?	35	?	?	?	?	?	
		(T)-CU203.28-G1.2	43	?	40	1.08	?	?	43	?	?	?	?	?	
		(T)-CU203.28-G1.3	40	?	49	0.82	?	?	40	?	?	?	?	?	
		(T)-CU203.28-G2.1	?	?	?	?	?	?	?	?	?	?	?	?	
		(T)-CU203.28-G3.1	40	?	50	0.80	?	?	40	?	?	?	?	?	
		(T)-CU203.28-G4.1	45	?	50	0.90	?	?	45	?	?	?	?	?	
		(T)-CU203.28-G4.2	44	55	50	0.88	11	?	44	?	?	?	?	?	
		(T)-CU203.28-G4.3	50	60	49	1.02	10	?	50	?	?	?	?	?	

Specimen											
(?)-CU203.28-G5.1	?	?	?	?	?	44	?	16	0.86	51	60	44
(?)-CU203.28-G5.2	?	?	?	?	?	44	?	9	0.83	53	53	44
(?)-CU203.28-G5.3	?	?	?	?	?	42	?	8	0.79	53	50	42
(?)-CU203.28-G6.1	?	?	?	?	?	40	?	14	1.00	40	54	40
(?)-CU203.28-G6.2	?	?	?	?	?	41	?	9	1.05	39	50	41
(?)-CU203.28-G6.3	?	?	?	?	?	38	?	12	0.97	39	50	38
(?)-CU203.28-G6.4	?	?	?	?	?	41	?	8	1.11	37	49	41
(?)-CU203.28-G6.5	?	?	?	?	?	40	?	11	1.00	40	51	40
(?)-CU203.28-G6.6	?	?	?	?	?	39	?	10	0.98	40	49	39
(?)-CU203.28-G6.7	?	?	?	?	?	40	?	?	1.00	40	?	40
(?)-CU203.28-G6.8	?	?	?	?	?	47	?	8	?	?	55	47
(?)-CU203.28-G7.1	?	?	?	?	?	?	?	?	?	?	?	?
(?)-CU203.28-G7.2	?	?	?	?	?	43	?	14	0.96	45	57	43
(?)-CU203.28-G7.3	?	?	?	?	?	42	?	16	0.93	45	58	42
(?)-CU203.28-G8.1	?	?	?	?	?	40	?	?	0.85	47	?	40
(?)-CU203.28-G8.2	?	?	?	?	?	40	?	?	?	?	?	40
(?)-CU203.28-G8.3	?	?	?	?	?	43	?	?	0.96	45	?	43
(?)-CU203.28-G9.1	?	?	?	?	?	42	?	?	0.93	45	?	42
(?)-CU203.28-G9.2	?	?	?	?	?	39	?	?	?	?	?	39
(?)-CU203.28-G9.3	?	?	?	?	?	39	?	?	?	?	?	39
(?)-CU203.28-G9.4	?	?	?	?	?	41	?	10	0.89	46	51	41
(?)-CU203.28-G9.5	?	?	?	?	?	48	?	11	1.02	47	59	48
(?)-CU203.28-G9.6	?	?	?	?	?	40	?	12	0.83	48	52	40
(?)-CU203.28-G10.1	?	?	?	?	?	31	?	15	0.84	37	46	31
(?)-CU203.28-G10.2	?	?	?	?	?	35	?	?	1.17	30	?	35
(?)-CU203.28-G10.3	?	?	?	?	?	?	?	?	?	?	?	?
(?)-CU203.28-G11.1	?	?	?	?	?	45	?	17	0.90	50	62	45
(?)-CU203.28-G11.2	?	?	?	?	?	45	?	?	0.90	50	?	45
(?)-CU203.28-G11.3	?	?	?	?	?	35	?	30	0.78	45	70	35
(?)-CU203.28-G12.1	?	?	?	?	?	40	?	?	0.78	51	?	40
(?)-CU203.28-G12.2	?	?	?	?	?	45	?	10	0.90	50	50	45
(?)-CU203.28-G12.3	?	?	?	?	?	40	?	15	0.82	49	60	40
(?)-CU203.28-G13.1	?	?	?	?	?	45	?	16	1.10	41	46	45
(?)-CU203.28-G13.2	?	?	?	?	?	30	?	?	0.75	40	45	30
(?)-CU203.28-G13.3	?	?	?	?	?	?	?	17	?	36	61	?
(?)-CU203.28-G14	?	?	?	?	?	44	?	15	0.90	49	53	44
(?)-CU203.28-G15.1	?	?	?	?	?	?	?	?	?	?	57	?
(?)-CU203.28-G15.2	?	?	?	?	?	42	?	?	?	48	?	42
(?)-CU203.28-G16	?	?	?	?	?	45	?	?	0.83	52	?	45
(?)-CU203.28-G17.1	?	?	?	?	?	40	?	13	0.77	53	52	40
(?)-CU203.28-G17.2	?	?	?	?	?	40	?	15	0.87	58	59	40
(?)-CU203.28-G17.3	?	?	?	?	?	46	?	?	0.67	54	?	46
(?)-CU203.28-G18.1	?	?	?	?	?	39	?	13	0.67	58	52	39
(?)-CU203.28-G18.2	?	?	?	?	?	44	?	15	0.81	54	59	44

Table 15A.1. *continued*

Ichnotaxon	Ichnofamily	Track #	FL	FLwH	FW	L/W	Digit lengths				Digit widths				Reference(s)
							I	II	III	IV	I	II	III	IV	
		(T)-CU203.28-G18.3	43	67	57	0.75	24	?	43	?	?	?	?	?	
		(T)-CU203.28-G18.4	44	52	55	0.80	8	?	44	?	?	?	?	?	
		(T)-CU203.28-G18.5	44	58	54	0.81	14	?	44	?	?	?	?	?	
		(T)-CU203.28-G18.6	45	58	56	0.80	13	?	45	?	?	?	?	?	
		(T)-CU203.28-G18.7	38	53	51	0.75	15	?	38	?	?	?	?	?	
		(T)-CU203.28-G18.8	42	61	52	0.81	19	?	42	?	?	?	?	?	
		Average	40	?	50	?	11	40	?	?	?	?	?	?	
Ignotornis yangi	Ignotornidae	KNUE040417-T1-01	34	56	51	0.67	16	28	34	33	?	?	?	?	Kim et al. (2006)
		KNUE040417-T1-02	34	?	?	?	26	?	34	?	?	?	?	?	
		KNUE040417-T1-03	35	55	48	0.73	10	22	35	32	?	?	?	?	
		KNUE040417-T1-04	34	56	46	0.74	18	24	34	30	?	?	?	?	
		KNUE040417-T1-05	33	55	47	0.70	12	23	33	30	?	?	?	?	
		KNUE040417-T1-06	32	52	46	0.70	11	22	32	30	?	?	?	?	
		KNUE040417-T1-07	33	58	48	0.69	13	23	33	31	?	?	?	?	
		KNUE040417-T1-08	35	59	48	0.73	19	26	35	29	?	?	?	?	
		KNUE040417-T2-09	37	58	49	0.76	17	24	37	32	?	?	?	?	
		KNUE040417-T2-10	35	?	45	0.78	?	23	35	30	?	?	?	?	
		KNUE040417-T2-11	36	54	51	0.71	17	23	36	31	?	?	?	?	
		KNUE040417-T2-12	36	60	47	0.77	15	24	36	30	?	?	?	?	
		KNUE040417-T2-13	28	?	38	0.74	?	22	28	22	?	?	?	?	
		KNUE040417-T2-14	29	?	38	0.76	?	23	29	20	?	?	?	?	
		KNUE040417-T2-15	29	?	40	0.73	?	24	29	25	?	?	?	?	
		KNUE040417-T2-16	31	?	40	0.78	?	24	31	23	?	?	?	?	
		KNUE040417-T2-17	28	?	40	0.70	?	22	28	25	?	?	?	?	
		KNUE040417-T3-18	31	38	43	0.72	10	23	31	28	?	?	?	?	
		KNUE040417-T3-19	34	44	45	0.76	4	25	34	28	?	?	?	?	
		KNUE040417-T3-20	34	47	47	0.72	8	25	34	32	?	?	?	?	
		KNUE040417-T3-21	34	?	45	0.76	?	26	34	29	?	?	?	?	
		KNUE040417-T3-22	35	45	48	0.73	13	27	35	27	?	?	?	?	
		KNUE040417-T3-23	34	?	45	0.76	?	26	34	30	?	?	?	?	
		KNUE040417-T3-24	35	?	45	0.78	?	23	35	32	?	?	?	?	
		KNUE040417-T3-25	32	?	42	0.76	?	27	32	20	?	?	?	?	
		KNUE040417-T3-26	31	?	43	0.72	?	25	31	26	?	?	?	?	
		KNUE040417-T3-27	30	?	41	0.73	?	25	30	22	?	?	?	?	
		KNUE040417-T3-28	31	?	42	0.74	?	22	31	27	?	?	?	?	
		KNUE040417-T3-29	31	48	43	0.72	8	24	31	30	?	?	?	?	
		KNUE040417-T4-30	35	45	46	0.76	9	22	35	30	?	?	?	?	
		KNUE040417-T4-31	32	?	47	0.68	?	25	32	32	?	?	?	?	
		KNUE040417-T4-32	35	56	47	0.74	15	26	35	28	?	?	?	?	
		KNUE040417-T4-33	35	?	45	0.78	?	23	35	30	?	?	?	?	
		KNUE040417-T4-34	?	63	45	?	16	24	?	28	?	?	?	?	

Specimen												
KNUE040417-T5-35	?	?	?	?	33	35	23	13	0.74	47	48	35
KNUE040417-T5-36	?	?	?	?	30	30	20	13	0.64	47	48	30
KNUE040417-T5-37	?	?	?	?	29	35	21	12	0.76	46	55	35
KNUE040417-T5-38	?	?	?	?	30	32	22	?	0.73	44	?	32
KNUE040417-T5-39	?	?	?	?	34	33	23	16	0.70	47	63	33
KNUE040417-T5-40	?	?	?	?	?	?	?	13	?	?	?	?
KNUE040417-T5-41	?	?	?	?	23	30	22	13	0.73	41	48	30
KNUE040417-T5-42	?	?	?	?	31	35	22	18	0.78	45	62	35
KNUE040417-T5-43	?	?	?	?	25	30	23	9	0.67	45	48	30
KNUE040417-T6-44	?	?	?	?	29	35	25	8	0.74	47	45	35
KNUE040417-T6-45	?	?	?	?	30	35	23	?	0.76	46	?	35
KNUE040417-T6-46	?	?	?	?	25	32	25	15	0.71	45	48	32
KNUE040417-T6-47	?	?	?	?	27	36	30	?	0.77	47	?	36
KNUE040417-T7-48	?	?	?	?	28	30	27	?	0.73	41	?	30
KNUE040417-T7-49	?	?	?	?	27	31	23	5	0.74	42	47	31
KNUE040417-T7-50	?	?	?	?	22	32	25	?	0.78	41	?	32
KNUE040417-T7-51	?	?	?	?	30	30	22	6	0.71	42	48	30
KNUE040417-T7-53	?	?	?	?	25	30	23	10	0.71	42	47	30
KNUE040417-T8-55	?	?	?	?	29	34	26	12	0.72	47	53	34
KNUE040417-T8-56	?	?	?	?	27	30	20	?	0.70	43	?	30
KNUE040417-T8-57	?	?	?	?	30	33	22	6	0.70	47	46	33
KNUE040417-T8-58	?	?	?	?	23	34	25	?	0.76	45	?	34
KNUE040417-T8-59	?	?	?	?	30	34	23	?	0.76	45	?	34
KNUE040417-T8-60	?	?	?	?	30	30	26	8	0.73	41	52	30
KNUE040417-T8-61	?	?	?	?	?	35	?	12	0.78	45	46	35
KNUE040417-T8-62	?	?	?	?	30	33	23	?	0.73	45	?	33
KNUE040417-T9-63	?	?	?	?	30	35	22	7	0.74	47	?	35
KNUE040417-T9-64	?	?	?	?	28	29	24	9	0.66	44	46	29
KNUE040417-T9-65	?	?	?	?	30	33	23	12	0.69	48	62	33
KNUE040417-T9-66	?	?	?	?	27	32	23	?	0.71	45	48	32
KNUE040417-T10-67	?	?	?	?	25	31	25	?	0.70	44	?	31
KNUE040417-T10-68	?	?	?	?	24	30	30	8	?	?	?	30
KNUE040417-T10-69	?	?	?	?	33	33	23	?	0.69	48	51	33
KNUE040417-T10-70	?	?	?	?	23	31	25	8	0.74	42	?	31
KNUE040417-T10-71	?	?	?	?	25	32	26	?	0.74	43	?	32
KNUE040417-T10-72	?	?	?	?	25	31	27	?	0.72	43	?	31
KNUE040417-T11-73	?	?	?	?	26	30	26	8	0.70	43	43	30
KNUE040417-T11-74	?	?	?	?	29	31	23	?	0.78	40	?	31
KNUE040417-75	?	?	?	?	30	35	25	8	0.73	48	47	35
KNUE040417-76	?	?	?	?	30	33	23	?	0.70	47	?	33
KNUE040417-77	?	?	?	?	34	35	24	8	0.74	47	47	35
KNUE040417-78	?	?	?	?	32	33	22	?	0.69	48	?	33
KNUE040417-79	?	?	?	?	27	30	23	12	0.75	40	48	30
KNUE040417-80	?	?	?	?	25	27	18	?	0.71	38	?	27

Table 15A.1. continued

Ichnotaxon	Ichnofamily	Track #	FL	FLwH	FW	L/W	Digit lengths				Digit widths				Reference(s)
							I	II	III	IV	I	II	III	IV	
Ignotornis gajinensis	Ignotornidae	KNUE081001-1	63	?	53	1.19	?	?	63	?	?	?	?	?	Kim et al. (2012)
		KNUE081001-2	43	?	54	0.80	?	?	43	?	?	?	?	?	
		KNUE081001-3	58	?	51	1.14	?	?	58	?	?	?	?	?	
		KNUE081001-4	48	?	53	0.91	?	?	48	?	?	?	?	?	
		KNUE081001-5	63	?	53	1.19	?	?	63	?	?	?	?	?	
		KNUE081001-6	68	?	55	1.24	?	?	68	?	?	?	?	?	
		KNUE081001-7	64	?	58	1.10	?	?	64	?	?	?	?	?	
Goseongornipes markjonesi	Ignotornidae	KPE50005	30.6	46	43	0.71	15	?	30.6	?	?	?	?	?	Lockley, Houck, et al. (2006)
		KPE5005.041	26	38	37	0.70	12	?	26	?	?	?	?	?	
		KPE5005.042	27	36	36	0.75	9	?	27	?	?	?	?	?	
		KPE5005.043	26	35	33	0.79	9	?	26	?	?	?	?	?	
		KPE5005.044	27	38	38	0.71	11	?	27	?	?	?	?	?	
Goseongornipes isp.	Ignotornidae	MGCM.H23a	44	44	56	0.79	?	24	35	28	?	?	?	?	Xing et al. (2011)
		MGCM.H23b	38	38	58	0.66	?	23	38	39	?	?	?	?	
		MGCM.H23c	43	43	60	0.72	?	31	32	29	?	?	?	?	
		MGCM.H23d	38	38	55	0.69	?	24	32	22	?	?	?	?	
		MGCM.H23e	38	38	48	0.79	?	24	27	23	?	?	?	?	
Koreanaornis hamanensis	Koreanaornipodidae	KoHaAverage	?	?	25	?	?	?	?	?	?	?	?	?	Lockley et al (1992)
		KoHaAverage	?	?	44	?	?	?	?	?	?	?	?	?	
		ZMNH-M5010 Average	26	?	36	0.72	?	?	?	?	?	?	?	?	Azuma et al. (2013)
		ZMNH-M8774 Average(4)	33	49	27	1.22	8	18	23	19	?	?	?	?	
		ZMNH-M8774 Average(5)	35	54	34	1.03	19	?	35	?	?	?	?	?	
		ZMNH-M8772 Average	24	?	34	0.71	?	?	24	?	?	?	?	?	Kim et al. (2012)
		KoHa-T1-L1	22.6	?	30	0.75	?	?	22.6	?	?	?	?	?	
		KoHa-T1-R1	24.5	?	39	0.63	?	?	24.5	?	?	?	?	?	
		KoHa-T1-L2	26.4	?	37	0.71	?	?	26.4	?	?	?	?	?	
		KoHa-T1-R2	34.7	?	38	0.92	?	?	34.7	?	?	?	?	?	
		KoHa-T1-L3	26.4	?	38	0.70	?	?	26.4	?	?	?	?	?	
		KoHa-T1-R3	22.6	?	38	0.60	?	?	22.6	?	?	?	?	?	
		KoHa-T1-L4	18.9	?	33	0.57	?	?	18.9	?	?	?	?	?	
		KoHa-T2-L1	38	?	38	1.00	?	?	38	?	?	?	?	?	
		KoHa-T2-R1	27.2	?	38	0.71	?	?	27.2	?	?	?	?	?	
		KoHa-T2-L2	27.2	?	35	0.77	?	?	27.2	?	?	?	?	?	
		KoHa-T2-R2	32.6	?	41	0.80	?	?	32.6	?	?	?	?	?	
		KoHa-T3-L1	20.4	?	30	0.69	?	?	20.4	?	?	?	?	?	
		KoHa-T3-R1	20.4	?	26	0.80	?	?	20.4	?	?	?	?	?	
		KoHa-T3-L2	20.4	?	31	0.67	?	?	20.4	?	?	?	?	?	
		KoHa-T3-R2	20.4	?	31	0.67	?	?	20.4	?	?	?	?	?	
		KoHa-T4-R1	28.5	?	33	0.86	?	?	28.5	?	?	?	?	?	
		KoHa-T4-L1	29.1	?	32	0.89	?	?	29.1	?	?	?	?	?	

Ichnotaxon	Ichnofamily	Specimen	1	2	3	4	5	6	7	8	9	10	11	12	13	Reference
Koreanaornis dodsoni		KSHa-T4-R2	25.1	?	34	0.73	?	?	25.1	?	?	?	?	?	?	
		KSHa-T4-L2	25.1	?	34	0.75	?	?	25.1	?	?	?	?	?	?	
		KSHa-T4-R3	22.8	?	34	0.67	?	?	22.8	?	?	?	?	?	?	
	Koreanaornipodidae	MGCM.H10a	42	42	63	0.67	0	13	36	30	?	?	?	?	?	Xing et al. (2011)
		MGCM.H10b	40	40	53	0.75	0	15	29	12	?	?	?	?	?	
		MGCM.H11a	30	30	37	0.81	0	22	27	25	?	?	?	?	?	
		MGCM.H11a	59	59	55	1.07	0	26	41	31	?	?	?	?	?	
		MGCM.H11b	52	52	52	1.00	0	31	40	30	?	?	?	?	?	
		MGCM.H11c	52	52	53	0.98	0	28	40	33	?	?	?	?	?	
		MGCM.H11c	34	34	32	1.06	0	17	26	25	?	?	?	?	?	
		MGCM.H11d	35	35	34	1.03	0	21	27	22	?	?	?	?	?	
		MGCM.H11e	35	35	38	0.92	0	20	32	19	?	?	?	?	?	
		MGCM.H12	32	32	32	1.00	0	23	30	23	?	?	?	?	?	
		MGCM.H14(H)	44	44	47	0.94	0	23	37	24	?	?	?	?	?	
		MGCM.H15a	45	45	50	0.90	0	24	38	28	?	?	?	?	?	
		MGCM.H15b	47	47	38	1.24	0	19	32	22	?	?	?	?	?	
		MGCM.H15c	42	42	55	0.76	0	28	32	37	?	?	?	?	?	
		MGCM.H15d	51	51	49	1.04	0	30	45	39	?	?	?	?	?	
		MGCM.H15e	42	42	52	0.81	0	24	32	33	?	?	?	?	?	
		MGCM.H16a	63	63	55	1.15	0	23	49	29	?	?	?	?	?	
		MGCM.H16b	48	48	55	0.87	0	29	41	31	?	?	?	?	?	
		MGCM.H16c	42	42	57	0.74	0	21	35	31	?	?	?	?	?	
		MGCM.H16d	45	45	54	0.83	0	18	31	24	?	?	?	?	?	
		MGCM.H17	45	45	54	0.83	0	20	42	26	?	?	?	?	?	
		MGCM.H18a	43	43	46	0.93	0	26	38	29	?	?	?	?	?	
		MGCM.H18b	34	34	40	0.85	0	22	27	21	?	?	?	?	?	
		MGCM.H19a	44	44	54	0.81	0	21	37	30	?	?	?	?	?	
		MGCM.H19b	49	49	50	0.98	0	21	41	32	?	?	?	?	?	
		MGCM.H19c	44	44	50	0.88	0	22	34	20	?	?	?	?	?	
		MGCM.H19d	34	34	39	0.87	0	21	28	24	?	?	?	?	?	
		MGCM.H19e	45	45	48	0.94	0	23	39	25	?	?	?	?	?	
		MGCM.H20a	52	52	68	0.76	0	30	43	39	?	?	?	?	?	
		MGCM.H20b	54	54	56	0.96	0	22	40	28	?	?	?	?	?	
		MGCM.H20c	46	46	59	0.78	0	31	41	31	?	?	?	?	?	
		MGCM.H20d	61	61	60	1.02	0	28	46	29	?	?	?	?	?	
		MGCM.H20e	46	46	61	0.75	0	29	36	36	?	?	?	?	?	
		MGCM.H20f	55	55	65	0.85	0	30	48	36	?	?	?	?	?	
		MGCM.H20g	51	51	64	0.80	0	28	42	38	?	?	?	?	?	
		MGCM.H20h	45	45	54	0.83	0	30	39	31	?	?	?	?	?	
		MGCM.H20i	56	56	54	1.04	0	28	41	19	?	?	?	?	?	
Jindongornipes kimi	Jindongornipedidae	KPE50006	80	80	65	1.23	?	?	80	?	?	?	?	?	?	Lockley et al. (1992)
		KPE50006	80	80	75	1.07	?	?	80	?	?	?	?	?	?	Li, Lockley, and Liu (2005);
Shandongornipes muxiai	Shandongornipodidae	LRH-DZ66	87	87	45	1.93	15	40	46	40	?	?	?	?	?	Lockley et al. (200?)

Table 15A.1. *continued*

Ichnotaxon	Ichnofamily	Track #	FL	FLwH	FW	L/W	Digit lengths I	II	III	IV	Digit widths I	II	III	IV	Reference(s)
		LRH-DZ67(H)	61	90	64	0.95	14	30	46	44	?	?	?	?	Lockley, Matsukawa, et al. (2006)
		LRH-DZ68(P)	54	88	57	0.95	13	30	45	40	?	?	?	?	
		LRH-DZ69	?	?	?	?	15	?	?	?	?	?	?	?	
		LRH-DZ70	60	82	67	0.90	14	40	45	41	?	?	?	?	
Pullornipes aureus	Koreanornipodidae	CU 212.21/22-A01	33	33	47	0.70	?	15	19	23	?	?	?	?	
		CU 212.21/22-A02	39	39	43	0.91	?	19	23	21	?	?	?	?	
		CU 212.21/22-A03	46	46	48	0.96	?	21	24	23	?	?	?	?	
		CU 212.21/22-A04	40	40	47	0.85	?	31	25	24	?	?	?	?	
		CU 212.21/22-A05	?	?	47	?	?	20	?	23	?	?	?	?	
		CU 212.21/22-A06	42	42	44	0.95	?	23	25	22	?	?	?	?	
		CU 212.21/22-A07	47	47	44	1.07	?	20	20	23	?	?	?	?	
		CU 212.21/22-A08	39	39	40	0.98	?	18	12	25	?	?	?	?	
		CU 212.21/22-A12	41	41	42	0.98	?	16	28	23	?	?	?	?	
		CU 212.21/22-A13	?	?	?	?	?	?	?	20	?	?	?	?	
		CU 212.21/22-A14	39	39	33	1.18	?	16	23	21	?	?	?	?	
		CU 212.21/22-A15	44	44	44	1.00	?	20	33	30	?	?	?	?	
		CU 212.21/22-A27	?	?	47	?	?	13	25	29	?	?	?	?	
		CU 212.21/22-A28	40	40	?	?	?	25	25	?	?	?	?	?	
		CU 212.21/22-A29	45	45	?	?	?	22	26	25	?	?	?	?	
		CU 212.21/22-A30	37	37	35	1.06	?	23	26	25	?	?	?	?	
		CU 212.21/22-A31	47	47	42	1.12	?	14	26	23	?	?	?	?	
		CU 212.21/22-A32	43	43	35	1.23	?	24	28	25	?	?	?	?	
		CU 212.21/22-A33	45	45	49	0.92	?	23	27	25	?	?	?	?	
		CU 212.21/22-A34	37	37	45	0.82	?	25	23	19	?	?	?	?	
		CU 212.21/22-A35	39	39	46	0.85	?	19	22	25	?	?	?	?	
		CU 212.21/22-A36	?	?	45	?	?	20	24	22	?	?	?	?	
		CU 212.21/22-A37	40	40	51	0.78	?	20	18	24	?	?	?	?	
		CU 212.21/22-A38	36	36	51	0.71	?	20	21	24	?	?	?	?	
		CU 212.21/22-A39	?	?	38	?	?	16	19	15	?	?	?	?	
		CU 212.21/22-A40	38	38	41	0.93	?	21	22	16	?	?	?	?	
		CU 212.21/22-A41	44	44	45	0.98	?	18	20	25	?	?	?	?	
		CU 212.21/22-A42	41	41	44	0.93	?	31	31	23	?	?	?	?	
		CU 212.21/22-A43	44	44	49	0.90	?	19	22	26	?	?	?	?	
		CU 212.21/22-A44	46	46	47	0.98	?	22	26	20	?	?	?	?	
		CU 212.21/22-A45	43	43	48	0.90	?	24	26	24	?	?	?	?	
Limiavipes curriei	Limiavipedidae	aa1R	77	77	112	0.69	0	?	?	?	?	?	?	?	McCrea and Sarjeant (2001); McCrea et al. (2014)
		aa2L	85	85	115	0.74	0	?	?	?	?	?	?	?	
		aa3R	80	80	105	0.76	0	?	?	?	?	?	?	?	
		aa4L	65	65	110	0.59	0	?	?	?	?	?	?	?	

Taxon								0	ratio			
a_R	?	?	?	?	?	?	?	0	0.74	95	70	70
a_5L	?	?	?	?	62	65	55	0	0.62	105	65	65
a_R	?	?	?	?	?	?	?	0	0.72	95	68	68
a_3L	?	?	?	?	55	70	60	0	0.83	87	72	72
c_R	?	?	?	?	?	?	?	0	0.69	101	70	70
c_L	?	?	?	?	?	?	?	0	0.64	107	68	68
c_3R	?	?	?	?	68	90	77	0	0.80	95	76	76
c_L	?	?	?	?	56	64	52	0	0.62	108	67	67
[?]	?	?	?	?	?	?	?	0	0.82	110	90	90
[?]	?	?	?	?	?	?	?	0	0.74	108	80	80
[?]	?	?	?	?	?	?	?	0	0.68	94	64	64
[?]	?	?	?	?	?	?	?	0	0.78	94	73	73
[?]	?	?	?	?	67	90	71	0	0.94	77	72	72
[?]	?	?	?	?	?	?	?	0	0.81	86	70	70
[?]	?	?	?	?	?	?	?	0	0.89	107	95	95
[?]	?	?	?	?	55	75	58	0	0.85	100	85	85
F_R	?	?	?	?	?	?	?	0	0.82	110	90	90
C1L	?	?	?	?	?	?	?	0	0.76	115	87	87
C2R	?	?	?	?	65	75	68	0	0.79	99	78	78
C4R	?	?	?	?	?	?	?	0	0.81	100	81	81
C5L	?	?	?	?	?	?	?	0	0.72	104	75	75
C6R	?	?	?	?	?	?	?	0	0.83	96	80	80
F1L	?	?	?	?	?	?	?	0	0.75	110	82	82
F2R	?	?	?	?	65	70	60	0	0.68	103	70	70
F3L	?	?	?	?	67	90	73	0	0.73	103	75	75
F4R	?	?	?	?	60	83	79	0	0.71	102	72	72
F5L	?	?	?	?	62	95	55	0	?	?	74	74
F6R	?	?	?	?	71	101	62	0	0.70	106	74	74
Z7L	?	?	?	?	67	89	68	0	?	?	72	72
[?]L	?	?	?	?	74	88	65	0	?	?	76	76
[?]R	?	?	?	?	69	88	68	0	?	?	63	63
[?]L	?	?	?	?	?	?	?	0	0.61	115	75	75
[?]R	?	?	?	?	65	70	60	0	?	?	70	70
H_R	?	?	?	?	67	90	73	0	0.75	120	80	80
RA1R	?	?	?	?	60	83	79	0	0.66	126	90	90
RA2L	?	?	?	?	62	95	55	0	0.86	110	83	83
RA3R	?	?	?	?	71	101	62	0	0.87	116	95	95
RA4L	?	?	?	?	67	89	68	0	0.77	116	101	101
RA5R	?	?	?	?	74	88	65	0	0.74	119	89	89
RA6L	?	?	?	?	69	88	68	0	0.75	117	88	88
RA7R	?	?	?	?	?	?	?	?	?	?	88	88
RA8L	?	?	?	?	69	80	67	0	0.65	123	80	?
RA9R	?	?	?	?	60	89	?	0	?	?	89	80
RA10L	?	?	?	?	?	?	?	?	?	?	?	89

Table 15A.1. *continued*

Ichnotaxon	Ichnofamily	Track #	FL	FLwH	FW	L/W	Digit lengths I	II	III	IV	Digit widths I	II	III	IV	Reference(s)
		PARB1R	85	85	107	0.79	0	54	85	60	?	?	?	?	
		PARB2L	?	?	?	?	?	?	?	?	?	?	?	?	
		PARB3R	83	83	108	0.77	0	60	83	69	?	?	?	?	
		PARB4L	82	82	99	0.83	0	55	82	63	?	?	?	?	
		PARB5R	85	85	111	0.77	0	61	85	66	?	?	?	?	
		PARB6L	?	?	?	?	?	?	?	?	?	?	?	?	
		PARB7R	88	88	95	0.93	0	56	88	64	?	?	?	?	
		PARB8L	80	80	110	0.73	0	71	80	67	?	?	?	?	
		PARB9R	88	88	116	0.76	0	71	80	67	?	?	?	?	
Barrosopus slobodai	?	PVPH-SB-415-17C-86	?	?	?	?	0	?	?	?	0	?	?	?	Coria et al. (2002)
		PVPH-SB-415-17C-87	29	29	35	0.83	0	21	29	29.7	0	5.6	6.8	5	
		PVPH-SB-415-17C-88	29	29	31	0.94	0	26	29	28	0	5	5.6	5	
		PVPH-SB-415-17C-89	34.5	34.5	37	0.93	0	20	34.5	26.8	0	6.6	6.7	6	
		PVPH-SB-415-17C-90	30	30	?	?	0	?	?	?	0	?	?	?	
		PVPH-SB-415-17C-91	40	40	40	1.00	0	?	?	?	0	?	?	?	
		PVPH-SB-415-17C-92	30	30	36	0.83	0	?	?	?	0	?	?	?	
		PVPH-SB-415-17C-93	35.5	35.5	44	0.81	0	?	?	?	0	?	?	?	
		PVPH-SB-415-17C-94	28	28	38	0.75	0	?	?	?	0	?	?	?	
Dongyangornipes sinensis	?	ZMNH-M8774 Average	34.8	?	40	?	0	?	34.8	?	?	?	?	?	Azuma et al. (2013)
		ZMNH-M8774 Holotype	36.4	36.4	40	0.92	0	17	22.8	21.3	?	3.7	6.4	4	
Moguiornipes robusta	?	MGCM.H25a	45	45	50	0.90	0	25	33	33	?	?	?	?	Xing et al. (2011)
		MGCM.H25b	46	46	60	0.77	0	36	38	32	?	?	?	?	
		MGCM.H25c	49	49	?	?	0	34	43	?	?	?	?	?	
		MGCM.H25d	58	58	63	0.92	0	32	43	?	?	?	?	?	
		MGCM.H27a	40	40	?	?	0	?	30	26	?	?	?	?	
Tatarornipes chabuensis	?	1-CU 214	50	50	54	0.93	0	?	50	?	?	?	?	?	Lockley, Li, et al. (2012)
		2-CU 214.184	48	48	56	0.86	0	?	48	?	?	?	?	?	
		3-CU 214.185	45	45	60	0.75	0	?	45	?	?	?	?	?	
		4-CU 214.184	46	46	62	0.74	0	?	46	?	?	?	?	?	
		5-CU 214.187	58	58	62	0.94	0	?	58	?	?	?	?	?	
		6-CU 214.187	50	50	60	0.83	0	?	50	?	?	?	?	?	
		7-CU 214.188	50	50	64	0.78	0	?	50	?	?	?	?	?	
		8-CU 214.186	50	50	54	0.93	0	?	50	?	?	?	?	?	
		9-CU 214.184	58	58	54	1.07	0	?	58	?	?	?	?	?	
		10-CU 214.189	64	64	72	0.89	0	?	64	?	?	?	?	?	
		11-CU 214.14	60	60	53	1.13	0	?	60	?	?	?	?	?	
		12-CU 214.14	56	56	60	0.93	0	?	56	?	?	?	?	?	
		13-CU 214.11	51	51	57	0.89	0	?	51	?	?	?	?	?	
		14-CU 214.12	47	47	63	0.75	0	?	47	?	?	?	?	?	
		15-CU 214.10	52	52	64	0.81	0	?	52	?	?	?	?	?	

Yang et al. (1995)

Specimen													
CU 214.9	?	?	?	?	45	?	0	?	0.79	57	45	45	?
CU 214.6	?	?	?	?	58	?	0	?	0.82	71	58	58	?
CU 214.7	?	?	?	?	42	?	0	?	0.70	60	42	42	?
CU 214.8	?	?	?	?	56	?	0	?	0.79	71	56	56	?
CU 214.8	?	?	?	?	47	?	0	?	0.90	52	47	47	?
CU 214.3	?	?	?	?	52	?	0	?	0.95	55	52	52	?
CU 214.3	?	?	?	?	45	?	0	?	0.88	51	45	45	?
CU 214.3	?	?	?	?	44	?	0	?	0.83	53	44	44	?
CU 214.5	?	?	?	?	46	?	0	?	0.74	62	46	46	?
CU 214.5	?	?	?	?	50	?	0	?	0.88	57	50	50	?
CU 214.18	?	?	?	?	56	?	0	?	0.70	80	56	56	?
CU 214.18	?	?	?	?	48	?	0	?	0.67	72	48	48	?
CU 214.160	?	?	?	?	50	?	0	?	0.69	72	50	50	?
CU 214.19	?	?	?	?	53	?	0	?	0.76	70	53	53	?
CU 214.19	?	?	?	?	43	?	0	?	0.69	62	43	43	?
CU 214.148	?	?	?	?	52	?	0	?	0.80	65	52	52	?
CU 214.149	?	?	?	?	55	?	0	?	0.96	57	55	55	?
CU 214.150	?	?	?	?	54	?	0	?	0.90	60	54	54	?
CU 214.151	?	?	?	?	54	?	0	?	0.95	57	54	54	?
CU 214.152	?	?	?	?	49	?	0	?	0.80	61	49	49	?
CU 214.152	?	?	?	?	54	?	0	?	0.77	70	54	54	?
CU 214.153	?	?	?	?	47	?	0	?	0.78	60	47	47	?
CU 214.155	?	?	?	?	54	?	0	?	0.90	60	54	54	?
CU 214.155	?	?	?	?	48	?	0	?	0.80	60	48	48	?
CU.214.156	?	?	?	?	55	?	0	?	1.00	55	55	55	?
CU 214.154	?	?	?	?	46	?	0	?	0.72	64	46	46	?
CU 214.154	?	?	?	?	47	?	0	?	0.70	67	47	47	?
CU 214.154	?	?	?	?	52	?	0	?	0.81	64	52	52	?
E50101-A01	?	?	?	?	35	?	?	?	0.90	39	35	35	?
E50101-A02	?	?	?	?	35	?	?	?	0.74	47	35	35	?
E50101-A03	?	?	?	?	34.6	?	?	?	0.82	42	34.6	34.6	?
E50101-A04	?	?	?	?	33	?	?	?	?	?	33	33	?
E50101-A05	?	?	?	?	41.6	?	?	?	0.80	52	41.6	41.6	?
E50101-A06	?	?	?	?	42.2	?	?	?	?	?	42.2	42.2	?
E50101-A07	?	?	?	?	38	?	?	?	0.83	46	38	38	?
E50101-A08	?	?	?	?	36.3	?	?	?	0.75	49	36.3	36.3	?
E50101-A09	?	?	?	?	32.5	?	?	?	0.71	46	32.5	32.5	?
E50101-A10	?	?	?	?	40.3	?	?	?	0.88	46	40.3	40.3	?
E50101-A11	?	?	?	?	37.8	?	?	?	0.78	49	37.8	37.8	?
E50101-A12	?	?	?	?	37.6	?	?	?	0.90	42	37.6	37.6	?
E50101-A13	?	?	?	?	39.3	?	?	?	0.80	49	39.3	39.3	?
E50101-A14	?	?	?	?	36.1	?	?	?	0.80	45	36.1	36.1	?
E50101-A15	?	?	?	?	37	?	?	?	0.84	44	37	37	?
E50101-A16	?	?	?	?	37	?	?	?	0.93	40	37	37	?

Table 15A.1. *continued*

Ichnotaxon	Ichnofamily	Track #	FL	FLwH	FW	L/W	Digit lengths I	II	III	IV	Digit widths I	II	III	IV	Reference(s)
		KPE50101-A17	37	37	40	0.93	?	?	37	?	?	?	?	?	?
		KPE50101-A18	35	35	50	0.70	?	?	35	?	?	?	?	?	?
		KPE50101-A19	36	36	40	0.90	?	?	36	?	?	?	?	?	?
		KPE50101-A20	37.9	37.9	45	0.85	?	?	37.9	?	?	?	?	?	?
		KPE50101-A22	36.1	36.1	44	0.82	?	?	36.1	?	?	?	?	?	?
		KPE50101-A24	38.5	38.5	45	0.86	?	?	38.5	?	?	?	?	?	?
		KPE50101-A25	37.4	37.4	42	0.89	?	?	37.4	?	?	?	?	?	?
		KPE50101-A26	39.2	39.2	48	0.82	?	?	39.2	?	?	?	?	?	?
		KPE50101-A27	40.2	40.2	47	0.86	?	?	40.2	?	?	?	?	?	?
		KPE50101-A28	35.3	35.3	41	0.86	?	?	35.3	?	?	?	?	?	?
		KPE50101-A29	50	50	?	?	?	?	50	?	?	?	?	?	?
		KPE50101-A30	37	37	?	?	?	?	37	?	?	?	?	?	?
		KPE50101-A31	37.3	37.3	45	0.83	?	?	37.3	?	?	?	?	?	?
		KPE50101-A32	36.5	36.5	43	0.85	?	?	36.5	?	?	?	?	?	?
		KPE50101-A33	37	37	?	?	?	?	37	?	?	?	?	?	?
		KPE50101-A34	36.9	36.9	45	0.82	?	?	36.9	?	?	?	?	?	?
		KPE50101-A34	36.9	36.9	45	0.82	?	?	36.9	?	?	?	?	?	?
		KPE50101-A36	37	37	47	0.79	?	?	37	?	?	?	?	?	?
		KPE50101-A37	37.9	37.9	?	?	?	?	37.9	?	?	?	?	?	?
		KPE50101-A42	37.1	37.1	51	0.73	?	?	37.1	?	?	?	?	?	?
		KPE50101-A43	33	33	46	0.72	?	?	33	?	?	?	?	?	?
		KPE50101-A44	40.5	40.5	46	0.88	?	?	40.5	?	?	?	?	?	?
		KPE50101-A45	37	37	48	0.78	?	?	37	?	?	?	?	?	?
		KPE50101-A46	39	39	49	0.79	?	?	39	?	?	?	?	?	?
		KPE50101-A47	40	40	48	0.83	?	?	40	?	?	?	?	?	?
		KPE50101-A48	34	34	38	0.89	?	?	34	?	?	?	?	?	?
		KPE50101-A49	36.9	36.9	43	0.86	?	?	36.9	?	?	?	?	?	?
		KPE50101-A50	33	33	44	0.75	?	?	33	?	?	?	?	?	?
		KPE50101-A51	38	38	?	?	?	?	38	?	?	?	?	?	?
		KPE50101-A52	33.8	33.8	43	0.79	?	?	33.8	?	?	?	?	?	?
		KPE50101-A54	41.7	41.7	?	?	?	?	41.7	?	?	?	?	?	?
		KPE50101-A55	34.1	34.1	40	0.85	?	?	34.1	?	?	?	?	?	?
		KPE50101-A56	30	30	38	0.79	?	?	30	?	?	?	?	?	?
		KPE50101-A57	32.5	32.5	42	0.78	?	?	32.5	?	?	?	?	?	?
		KPE50101-A58	29.7	29.7	42	0.71	?	?	29.7	?	?	?	?	?	?
		KPE50101-A59	42	42	47	0.89	?	?	42	?	?	?	?	?	?
		KPE50101-A60	37	37	48	0.77	?	?	37	?	?	?	?	?	?
		KPE50101-B01	37	37	49	0.76	?	?	37	?	?	?	?	?	?
		KPE50101-B02	34.2	34.2	?	?	?	?	34.2	?	?	?	?	?	?
		KPE50101-B03	39.6	39.6	53	0.74	?	?	39.6	?	?	?	?	?	?
		KPE50101-B04	36	36	43	0.84	?	?	36	?	?	?	?	?	?

Specimen	Ichnotaxon	Trackmaker											Reference
KPE50101-B05			37.3	37.3	48	0.78	?	?	37.3	?	?	?	?	
KPE50101-B06			41.3	41.3	48	0.87	?	?	41.3	?	?	?	?	
KPE50101-B07			40	40	47	0.85	?	?	40	?	?	?	?	
KPE50101-B08			36.2	36.2	46	0.78	?	?	36.2	?	?	?	?	
KPE50101-B09			40.7	40.7	43	0.95	?	?	40.7	?	?	?	?	
KPE50101-B10			39.4	39.4	42	0.93	?	?	39.4	?	?	?	?	
KPE50101-B11			37.1	37	48	0.77	?	?	37.1	?	?	?	?	
KPE50101-B12			37	37	45	0.82	?	?	37	?	?	?	?	
KPE50101-B13			32.8	32.8	?	?	?	?	32.8	?	?	?	?	
KPE50101-B14			40.9	40.9	48	0.85	?	?	40.9	?	?	?	?	
KPE50101-B15			37	37	45	0.82	?	?	37	?	?	?	?	
KPE50101-B16			35.8	35.8	49	0.72	?	?	35.8	?	?	?	?	
KPE50101-B17			38	38	42	0.90	?	?	38	?	?	?	?	
KPE50101-B18			?	?	46	?	?	?	?	?	?	?	?	
KPE50101-B19			35.1	35.1	?	?	?	?	35.1	?	?	?	?	
KPE50101-B20			39	39	47	0.82	?	?	39	?	?	?	?	
KPE50101-C01			35.6	35.6	49	0.73	?	?	35.6	?	?	?	?	
KPE50101-C02			33.7	33.7	?	?	?	?	33.7	?	?	?	?	
KPE50101-C03			37	37	45	0.82	?	?	37	?	?	?	?	
KPE50101-C04			46.2	46.2	50	0.92	?	?	46.2	?	?	?	?	
KPE50101-C05			35	35	52	0.68	?	?	35	?	?	?	?	
KPE50101-C06			36	36	52	0.70	?	?	36	?	?	?	?	
KPE50101-C07			35	35	52	0.67	?	?	35	?	?	?	?	
KPE50101-C08			35.1	35.1	45	0.78	?	?	35.1	?	?	?	?	
KPE50101-C09			34	34	?	?	?	?	34	?	?	?	?	
KPE50101-C10			37	37	48	0.77	?	?	37	?	?	?	?	
KPE50101-T1	*Uhangrichnus* topotype	?	36	?	?	?	?	?	36	?	?	?	?	Lockley, Lim, et al. (2012)
KPE50101-T2			35	?	43	0.81	?	?	35	?	?	?	?	
KPE50101-T4			40	49	48	0.83	9	?	40	?	?	?	?	
KPE50101-T3			38	51	45	0.84	13	?	38	?	?	?	?	
DMNH918.1	*Magnoavipes lowei*	Theropod	210	210	250	0.84	0	175	210	180	?	?	?	Lee (1997)
DMNH918.2			190	190	250	0.76	0	160	190	160	?	?	?	
DMNH918.3			210	210	250	0.84	0	167	210	145	?	?	?	
DMNH918.4			210	210	240	0.88	0	155	210	160	?	?	?	
DMNH918.5			200	200	270	0.74	0	198	200	160	?	?	?	
DMNH918.6			210	210	240	0.88	0	160	210	180	?	?	?	
CU MWC200-1.1	*Magnoavipes caneeri*	Theropod	?	?	?	?	0	?	?	?	?	?	?	Lockley, Wright, and Matsukawa (2001)
CU MWC200-1.2			200	200	250	0.80	0	?	200	?	?	?	?	
CU MWC200-1.3			?	?	?	?	0	?	?	?	?	?	?	
CU MWC200-1.4			?	?	?	?	0	?	?	?	?	?	?	
CU MWC200-1.5			210	210	260	0.81	0	?	210	?	?	?	?	
CU MWC200-1.6			?	?	?	?	0	?	?	?	?	?	?	
CU MWC200-1.7			210	210	260	0.81	0	?	210	?	?	?	?	

Table 15A.1. *continued*

Ichnotaxon	Ichnofamily	Track #	FL	FLwH	FW	L/W	Digit lengths				Digit widths				Reference(s)
							I	II	III	IV	I	II	III	IV	
		CU MWC200-1.8	230	230	270	0.85	0	?	230	?	?	?	?	?	
		CU MWC200-1.9	?	?	?	?	0	?	?	?	?	?	?	?	
		CU MWC200-1.10	220	220	240	0.92	0	?	220	?	?	?	?	?	
		CU MWC200-2.01	210	210	210	1.00	0	?	210	?	?	?	?	?	
		CU MWC200-2.02	180	180	210	0.86	0	?	180	?	?	?	?	?	
		CU MWC200-2.03	180	180	200	0.90	0	?	180	?	?	?	?	?	
		CU MWC200-2.04	180	180	?	?	0	?	180	?	?	?	?	?	
		CU MWC200-3.01	190	190	200	0.95	0	?	190	?	?	?	?	?	
		CU MWC200-3.02	180	180	190	0.95	0	?	180	?	?	?	?	?	
		CU MWC200-3.03	190	190	190	1.00	0	?	190	?	?	?	?	?	
		CU MWC200-3.04	170	170	180	0.94	0	?	170	?	?	?	?	?	
		CU MWC200-4.01	?	?	?	?	0	?	?	?	?	?	?	?	
		CU MWC200-4.02	200	200	260	0.77	0	?	200	?	?	?	?	?	
		CU MWC200-4.03	180	180	240	0.75	0	?	180	?	?	?	?	?	
		CU MWC200-5.01	180	180	200	0.90	0	?	180	?	?	?	?	?	
		CU MWC200-5.02	180	180	230	0.78	0	?	180	?	?	?	?	?	
Magnoavipes denaliensis	Theropod	Smallest	195	195	200	0.98	0	150	195	130	?	?	?	?	Fiorillo et al. (2011)
		Largest	205	205	200	1.03	0	150	205	130	?	?	?	?	

Table 5A.2. Ratios

Ichnotaxon	Ichnofamily	Track #	Divarication				PL	SL	PA	FR	TW	Reference(s)
			I	II–III	III–IV	TOT						
Acuatavipes aumiensis	Avipedidae	FPDM-V43.1	0	48	68.5	117	?	?	?	?	?	Azuma et al. (2002)
		FPDM-V43.2	0	48	60	108	?	?	?	?	?	
		FPDM-V43.3	0	45	80.5	126	?	?	?	?	?	
		FPDM-V43.4	0	45	65	110	?	?	?	?	?	
		FPDM-V43.5	0	56	66	122	?	?	?	?	?	
		FPDM-V43.6	0	65	66.5	131	?	?	?	?	?	
		FPDM-V43.7	0	44	77	121	?	?	?	?	?	
		FPDM-V43.8	0	62	80	142	?	?	?	?	?	
		FPDM-V43.9	0	47	62	109	?	?	?	?	?	
		FPDM-V43.10	0	55	40	95	?	?	?	?	?	
		FPDM-V43.11	0	46	69.5	115	?	?	?	?	?	
		FPDM-V43.12	0	71	80.5	152	?	?	?	?	?	
		FPDM-V43.13	0	54	68	122	?	?	?	?	?	
		FPDM-V43.14	0	52	57.5	109	?	?	?	?	?	
		FPDM-V43.15	0	49	62	111	?	?	?	?	?	
		FPDM-V44.B1	0	71	75	146	?	?	?	?	?	
		FPDM-V44.B2	0	42	65.5	108	?	?	?	?	?	
Aquatilavipes svitoldae	Avipedidae	TMP 98.89.21-A	0	46	62	108	?	?	?	?	?	Currie (1981); McCrea and Sarjeant (2001)
		TMP 98.89.21-B	0	42	48	90	?	?	?	?	?	
		TMP 98.89.21-C	0	37	34	71	?	?	?	?	?	
		TMP 98.89.21-D	0	55	43	98	?	?	?	?	?	
		TMP 98.89.21-E	0	77	53	130	?	?	?	?	?	
		TMP 98.89.21-F	0	47	48	95	?	?	?	?	?	
		TMP 98.89.20-A	0	58	68	126	?	?	?	?	?	
		TMP 98.89.20-B	0	61	71	132	?	?	?	?	?	
		TMP 98.89.20-C	0	53	52	105	?	?	?	?	?	
		TMP 98.89.20-D	0	68	66	134	?	?	?	?	?	
		TMP 98.89.20-E	0	53	52	105	?	?	?	?	?	
		TMP 98.89.20-F	0	?	?	?	?	?	?	?	?	
		TMP 98.89.20-G	0	61	62	123	?	?	?	?	?	
		TMP 90.30.1-A	0	75	73	148	?	?	?	?	?	
		TMP 90.30.1-B	0	84	61	145	?	?	?	?	?	
		TMP 90.30.1-C	0	64	74	138	?	?	?	?	?	
		TMP 79.23.3-	0	48	70	118	?	?	?	?	?	
Hwangsanipes choughi	Ignotornidae	KP50101-A38	?	60	?	?	?	?	?	?	?	Yang et al. (1995)
		KP50101-A39	?	?	?	?	?	?	?	?	?	
		KP50101-A40	?	57	59	116	?	?	?	?	?	
		KP50101-A41	?	51	55	106	?	?	?	?	?	
Ignotornis mcconnelli	Ignotornidae	Average	?	?	?	90	?	?	?	?	?	Lockley et al. (2009)
		(H)-UCM17614-1.1	85	62	63	125	?	?	?	18	?	

Table 15A.2. *continued*

Ichnotaxon	Ichnofamily	Track #	Divarication				PL	SL	PA	FR	TW	Reference(s)
			I	II–III	III–IV	TOT						
		(H)-UCM17614-1.2	115	60	65	125	72	?	143	18	70	
		(H)-UCM17614-1.3	95	60	63	123	73	140	150	21	70	
		(H)-UCM17614-1.4	120	50	55	105	80	147	118	17	84	
		(H)-UCM17614-1.5	90	60	65	125	60	121	120	5	76	
		(H)-UCM17614-1.6	100	57	50	107	49	94	?	15	?	
		(P1)-UCM98120-2.1	90	60	46	106	?	?	?	40	?	
		(P1)-UCM98120-2.2	95	57	50	107	87	?	?	25	?	
		(P1)-UCM98120-2.3	90	60	50	110	84	168	160	25	56	
		(P1)-UCM98120-3.1	95	80	60	140	?	?	?	15	?	
		(P1)-UCM98120-3.2	115	80	60	140	103	?	152	5	?	
		(P1)-UCM98120-3.3	90	80	55	135	100	197	160	8	60	
		(P1)-UCM98120-3.4	105	82	52	134	85	183	158	8	60	
		(P1)-UCM98120-3.5	110	67	53	120	87	169	?	10	60	
		(P1)-UCM98120-4.1	120	60	60	120	?	?	?	18	?	
		(P1)-UCM98120-4.2	100	52	68	120	98	?	140	0	67	
		(P1)-UCM98120-4.3	100	70	65	135	70	158	?	15	?	
		(P1)-UCM98120-5.1	65	90	72	162	?	?	?	18	?	
		(P1)-UCM98120-5.2	80	?	?	?	74	?	135	?	80	
		(P1)-UCM98120-5.3	80	80	65	145	86	146	135	11	85	
		(P1)-UCM98120-5.4	90	72	63	135	86	155	?	11	?	
		(P2)-UCM98121-6.1	95	60	50	110	?	?	?	−16	?	
		(P2)-UCM98121-6.2	70	50	75	125	62	?	128	38	75	
		(P2)-UCM98121-6.3	110	75	55	130	48	97	?	?	?	
		(P2)-UCM98121-7	85	60	65	125	?	?	?	?	?	
		(P2)-UCM98121-8	90	80	75	155	?	?	?	?	?	
		(P2)-UCM98121-9	?	?	?	?	?	?	?	?	?	
		(P3)-UCM98122-10	92	73	73	146	?	?	?	?	?	
		(P3)-UCM98122-11	?	80	60	140	?	?	?	?	?	
		(P3)-UCM98122-12	?	75	55	130	?	?	?	?	?	
		(P3)-UCM98122-13.1	?	37	68	105	?	?	?	16	?	
		(P3)-UCM98122-13.2	90	64	51	115	49	?	65	10	81	
		(P3)-UCM98122-13.3	122	65	47	112	42	49	?	6	?	
		(P3)-UCM98122-14	105	70	55	125	?	?	?	?	?	
		(P3)-UCM98122-15	?	65	55	120	?	?	?	10	?	
		(T)-CU203.28-G1.1	?	65	105	170	?	?	?	?	?	
		(T)-CU203.28-G1.2	?	?	87	?	85	?	150	28	59	
		(T)-CU203.28-G1.3	?	57	60	117	73	152	?	30	?	
		(T)-CU203.28-G2.1	?	?	?	?	?	?	?	?	?	
		(T)-CU203.28-G3.1	?	48	62	110	?	?	?	?	?	
		(T)-CU203.28-G4.1	?	?	?	?	?	?	?	?	?	
		(T)-CU203.28-G4.2	107	63	65	128	72	?	138	25	70	

Specimen									
(T)-CU203.28-G4.3	80	75	75	150	60	123	?	25	?
(T)-CU203.28-G5.1	95	80	53	133	?	?	?	10	?
(T)-CU203.28-G5.2	?	50	77	127	75	?	148	30	68
(T)-CU203.28-G5.3	100	62	58	120	70	138	?	24	?
(T)-CU203.28-G6.1	112	68	57	125	?	?	?	42	?
(T)-CU203.28-G6.2	112	43	62	105	65	120	155	34	50
(T)-CU203.28-G6.3	128	35	60	95	60	116	155	23	49
(T)-CU203.28-G6.4	92	57	47	104	57	114	167	33	41
(T)-CU203.28-G6.5	130	45	70	115	62	53	50	20	75
(T)-CU203.28-G6.6	100	62	58	120	32	65	?	10	70
(T)-CU203.28-G6.7	?	?	?	?	63	142	70	34	55
(T)-CU203.28-G6.8	?	?	?	?	78	?	?	5	?
(T)-CU203.28-G7.1	63	?	50	?	?	?	?	18	?
(T)-CU203.28-G7.2	93	77	55	132	90	142	120	19	76
(T)-CU203.28-G7.3	?	62	56	118	70	?	?	-11	?
(T)-CU203.28-G8.1	?	71	71	142	?	?	?	12	?
(T)-CU203.28-G8.2	?	?	57	?	95	170	170	30	52
(T)-CU203.28-G8.3	?	45	70	115	82	?	?	37	?
(T)-CU203.28-G9.1	?	50	63	113	?	?	?	12	?
(T)-CU203.28-G9.2	?	?	60	?	90	170	158	20	59
(T)-CU203.28-G9.3	133	65	57	?	85	162	161	25	56
(T)-CU203.28-G9.4	92	58	45	110	78	147	144	10	66
(T)-CU203.28-G9.5	95	80	42	100	74	?	?	14	?
(T)-CU203.28-G9.6	100	52	45	125	?	?	?	12	?
(T)-CU203.28-G10.1	?	?	64	116	105	?	162	35	50
(T)-CU203.28-G10.2	?	?	?	?	120	225	?	10	?
(T)-CU203.28-G10.3	?	?	?	?	?	?	?	28	?
(T)-CU203.28-G11.1	?	?	?	?	?	?	?	32	?
(T)-CU203.28-G11.2	?	?	?	?	?	?	?	?	?
(T)-CU203.28-G11.3	?	?	?	?	?	?	?	24	?
(T)-CU203.28-G12.1	103	60	75	?	97	?	172	?	52
(T)-CU203.28-G12.2	?	?	?	135	99	195	?	?	?
(T)-CU203.28-G12.3	110	46	50	?	71	120	116	?	72
(T)-CU203.28-G13.1	?	?	?	96	65	90	101	?	70
(T)-CU203.28-G13.2	?	?	?	?	52	?	?	?	?
(T)-CU203.28-G13.3	?	?	?	?	?	97	?	?	?
(T)-CU203.28-G14	110	55	67	122	?	?	?	?	?
(T)-CU203.28-G15.1	?	?	?	?	?	?	?	?	?
(T)-CU203.28-G15.2	?	?	?	?	?	?	?	9	?
(T)-CU203.28-G16	?	60	70	130	85	?	90	23	?
(T)-CU203.28-G17.1	?	?	?	?	52	?	?	20	?
(T)-CU203.28-G17.2	?	55	55	110	?	97	?	35	96
(T)-CU203.28-G17.3	?	?	70	140	?	?	?	?	?
(T)-CU203.28-G18.1	78	70	70	?	?	?	?	?	?

Ichnotaxon	Ichnofamily	Track #	Divarication				PL	SL	PA	FR	TW	References(s)
			I	II–III	III–IV	TOT						
		(T)-CU203.28-G18.2	78	70	58	128	57	?	80	0	90	
		(T)-CU203.28-G18.3	85	68	56	124	72	85	92	–15	103	
		(T)-CU203.28-G18.4	?	65	61	126	65	98	135	0	82	
		(T)-CU203.28-G18.5	94	60	58	118	80	131	89	0	107	
		(T)-CU203.28-G18.6	90	85	61	146	94	120	139	15	75	
		(T)-CU203.28-G18.7	95	58	67	125	66	150	97	12	87	
		(T)-CU203.28-G18.8	90	68	56	124	56	93	?	13	?	
		Average?	?	?	115	?	?	?	?	?	?	
Ignotornis yangi	Ignotornidae	KNUE040417-T1-01	83	63	59	122	?	?	?	?	?	Kim et al. (2006)
		KNUE040417-T1-02	?	60	?	?	127	?	?	?	?	
		KNUE040417-T1-03	97	60	59	119	117	238	149	?	29	
		KNUE040417-T1-04	90	66	56	122	122	230	142	?	32	
		KNUE040417-T1-05	93	60	60	120	101	219	142	?	37	
		KNUE040417-T1-06	82	61	59	120	102	200	145	?	31	
		KNUE040417-T1-07	92	58	60	118	98	172	118	?	50	
		KNUE040417-T1-08	76	72	54	126	?	?	?	?	?	
		KNUE040417-T2-09	80	52	77	129	112	?	?	?	?	
		KNUE040417-T2-10	?	67	65	132	107	217	163	?	17	
		KNUE040417-T2-11	80	53	75	128	104	214	180	?	4	
		KNUE040417-T2-12	112	62	60	122	115	213	161	?	19	
		KNUE040417-T2-13	?	70	60	130	119	232	180	?	6	
		KNUE040417-T2-14	?	67	60	127	75	180	137	?	34	
		KNUE040417-T2-15	?	65	68	133	93	150	125	?	36	
		KNUE040417-T2-16	?	60	65	125	75	155	135	?	25	
		KNUE040417-T2-17	?	65	60	125	92	152	130	?	25	
		KNUE040417-T3-18	50	56	66	122	?	?	?	?	?	
		KNUE040417-T3-19	92	57	63	120	47	?	?	?	?	
		KNUE040417-T3-20	65	55	73	128	77	94	102	?	36	
		KNUE040417-T3-21	?	53	62	115	85	157	150	?	22	
		KNUE040417-T3-22	73	68	57	125	109	193	172	?	8	
		KNUE040417-T3-23	?	60	57	117	94	200	170	?	7	
		KNUE040417-T3-24	?	55	67	122	110	185	132	?	43	
		KNUE040417-T3-25	?	60	58	118	125	230	155	?	24	
		KNUE040417-T3-26	?	55	65	120	85	183	120	?	51	
		KNUE040417-T3-27	?	61	70	131	?	?	?	?	?	
		KNUE040417-T3-28	?	60	55	115	91	94	?	?	?	
		KNUE040417-T3-29	77	55	50	105	98	187	173	?	3	
		KNUE040417-T4-30	90	59	60	119	?	?	?	?	?	
		KNUE040417-T4-31	?	60	58	118	106	?	?	?	?	
		KNUE040417-T4-32	85	56	64	120	124	210	133	?	40	
		KNUE040417-T4-33	?	63	52	115	106	220	151	?	30	

Specimen									
KNUE040417-T4-34	85	76	54	130	115	220	168	?	13
KNUE040417-T5-35	80	55	62	117	?	?	?	?	?
KNUE040417-T5-36	79	76	61	137	142	?	?	?	?
KNUE040417-T5-37	87	56	79	135	138	270	150	?	37
KNUE040417-T5-38	?	59	60	119	79	205	140	?	35
KNUE040417-T5-39	120	62	59	121	98	165	145	?	27
KNUE040417-T5-40	?	?	?	?	95	165	120	?	48
KNUE040417-T5-41	52	68	60	128	64	156	167	?	9
KNUE040417-T5-42	100	50	64	114	113	175	159	?	15
KNUE040417-T5-43	85	55	75	130	97	193	132	?	42
KNUE040417-T6-44	78	62	58	120	?	?	?	?	?
KNUE040417-T6-45	?	70	56	126	74	?	?	?	?
KNUE040417-T6-46	78	66	73	139	115	182	148	?	25
KNUE040417-T6-47	?	70	55	125	81	184	138	?	33
KNUE040417-T7-48	?	63	55	118	?	?	?	?	?
KNUE040417-T7-49	95	57	68	125	90	?	?	?	?
KNUE040417-T7-50	?	58	66	124	55	136	137	?	27
KNUE040417-T7-51	113	55	68	123	93	142	147	?	19
KNUE040417-T7-52	90	59	70	129	93	176	155	?	19
KNUE040417-T7-53	79	57	63	120	94	176	148	?	25
KNUE040417-T8-54	?	68	63	131	?	?	?	?	?
KNUE040417-T8-55	90	45	77	122	120	?	?	?	?
KNUE040417-T8-56	?	70	54	124	130	243	150	?	32
KNUE040417-T8-57	110	47	60	107	99	234	180	?	13
KNUE040417-T8-58	?	68	56	124	126	225	175	?	12
KNUE040417-T8-59	?	56	59	115	100	223	164	?	14
KNUE040417-T8-60	110	64	56	120	111	203	150	?	23
KNUE040417-T8-61	92	62	67	129	104	215	150	?	28
KNUE040417-T8-62	?	68	59	127	127	225	157	?	23
KNUE040417-T9-63	?	55	65	120	?	?	?	?	?
KNUE040417-T9-64	60	64	62	126	96	?	?	?	?
KNUE040417-T9-65	91	52	69	121	103	195	156	?	20
KNUE040417-T9-66	88	63	63	126	94	190	151	?	24
KNUE040417-T10-67	?	60	65	125	?	?	?	?	?
KNUE040417-T10-68	90	65	65	?	100	?	?	?	?
KNUE040417-T10-69	?	55	68	123	110	204	115	?	23
KNUE040417-T10-70	?	61	59	120	122	225	156	?	25
KNUE040417-T10-71	?	58	68	126	130	240	148	?	33
KNUE040417-T10-72	?	69	54	123	119	245	164	?	19
KNUE040417-T11-73	92	68	60	120	?	?	?	?	?
KNUE040417-T11-74	56	64	58	126	75	?	?	?	?
KNUE040417-75	73	51	?	120	?	?	?	?	?
KNUE040417-76	72	50	90	124	?	?	?	?	?
KNUE040417-77	?	?	?	122	?	?	?	?	?

Table 15A.2. *continued*

Ichnotaxon	Ichnofamily	Track #	Divarication				PL	SL	PA	FR	TW	Reference(s)
			I	II–III	III–IV	TOT						
		KNUE040417-78	74	51	?	125	?	?	?	?	?	
		KNUE040417-79	52	68	85	120	?	?	?	?	?	
		KNUE040417-80	70	60	?	130	?	?	?	?	?	
Ignotornis gajinensis	Ignotornidae	KNUE081001-1	?	70	60	130	106	?	?	?	?	Kim et al. (2012)
		KNUE081001-2	?	73	67	140	104	207	160	?	72	
		KNUE081001-3	?	80	60	140	87	173	130	?	92	
		KNUE081001-4	?	?	?	?	140	226	172	?	63	
		KNUE081001-5	?	75	70	145	104	226	145	?	58	
		KNUE081001-6	?	70	60	130	207	311	165	?	58	
		KNUE081001-7	?	75	65	140	144	342	155	?	?	
Goseongornipes markjonesi	Ignotornidae	KPE5005	?	71	65.1	137	64	129	150	8.9	?	Lockley, Houck, et al. (2006)
		KPE5005.041	?	70	55	125	67	133	170	15	?	
		KPE5005.042	?	72	54	126	71	128	130	28	?	
		KPE5005.043	?	61	59	120	74	142	174	22	?	
		KPE5005.044	?	75	63	138	60	111	166	16	?	
Goseongornipes isp.	Ignotornidae	MGCM.H23a	?	45	70	115	?	?	?	?	?	Xing et al. (2011)
		MGCM.H23b	?	53	48	101	?	?	?	?	?	
		MGCM.H23c	?	52	57	109	?	?	140	?	?	
		MGCM.H23d	?	67	85	152	?	?	?	?	?	
		MGCM.H23e	?	63	59	122	?	?	?	?	?	
Koreanaornis hamanensis	Koreanaornipodidae	KoHaAverage	?	?	?	105	?	?	?	?	?	Lockley et al. (1992)
		KoHaAverage	?	?	?	125	?	?	?	?	?	
		ZMNH-M5010 Average	?	?	?	128	127	251	171	?	?	Azuma et al. (2013)
		ZMNH-M8774 Average(4)	92	34	41	74	?	?	?	?	?	
		ZMNH-M8774 Average(5)	22	?	?	138	?	?	?	?	?	
		ZMNH-M8772 Average	?	?	?	118	?	?	?	?	?	
		KoHa-T1-L1	?	50	75	125	82.9	?	?	?	?	Kim et al. (2012)
		KoHa-T1-R1	?	65	70	135	?	?	?	?	?	
		KoHa-T1-L2	?	?	?	?	56.55	139.5	160	?	52.78	
		KoHa-T1-R2	?	80	90	170	84.83	156.5	145	?	60.32	
		KoHa-T1-L3	?	50	80	130	77.29	165.9	177	?	45.24	
		KoHa-T1-R3	?	70	80	150	71.63	147	177	?	43.36	
		KoHa-T1-L4	?	40	85	125	82.94	152.7	160	?	52.78	
		KoHa-T2-L1	?	68	74	142	?	?	?	?	?	
		KoHa-T2-R1	?	75	60	135	130.3	?	?	?	?	
		KoHa-T2-L2	?	65	50	115	86.88	211.8	146	?	65.16	
		KoHa-T2-R2	?	65	45	110	114	200.9	180	?	38.01	
		KoHa-T3-L1	?	70	50	120	?	?	?	?	?	
		KoHa-T3-R1	?	70	60	130	45.9	?	?	?	?	
		KoHa-T3-L2	?	70	50	120	51	91.8	145	?	43.35	
		KoHa-T3-R2	?	70	60	130	76.5	122.4	145	?	51	

Ichnotaxon	Specimen	1	2	3	4	5	6	7	8	9	10	Reference
Foreanornis oodsoni (Koreanaornipodidae)	KoHa-T4-R1	?	50	60	110	?	?	?	?	?	?	Xing et al. (2011)
	KoHa-T4-L1	?	50	60	110	62.7	?	?	?	?	?	
	KoHa-T4-R2	?	60	60	120	79.8	142.5	180	?	?	34.2	
	KoHa-T4-L2	?	55	60	115	51.3	142.5	170	?	?	45.6	
	KoHa-T4-R3	?	50	65	115	76.95	142.5	167	?	?	45.6	
	MGCM.H10a	?	48	54	102	?	?	?	?	?	?	
	MGCM.H10b	?	49	55	104	?	?	?	?	?	?	
	MGCM.H11a	?	36	36	72	?	?	?	?	?	?	
	MGCM.H111a	?	41	47	88	?	?	?	?	?	?	
	MGCM.H111b	?	48	40	88	?	?	?	?	?	?	
	MGCM.H111c	?	36	44	80	?	?	?	?	?	?	
	MGCM.H11c	?	32	26	58	?	?	?	?	?	?	
	MGCM.H11d	?	32	31	63	?	?	?	?	?	?	
	MGCM.H11e	?	39	41	80	?	?	?	?	?	?	
	MGCM.H12	?	34	34	68	?	?	?	?	?	?	
	MGCM.H14(H)	?	40	51	91	?	?	?	?	?	?	
	MGCM.H15a	?	34	44	78	?	?	?	?	?	?	
	MGCM.H15b	?	30	32	62	?	?	?	?	?	?	
	MGCM.H15c	?	41	34	75	?	?	?	?	?	?	
	MGCM.H15d	?	45	46	91	?	?	?	?	?	?	
	MGCM.H15e	?	34	42	76	?	?	?	?	?	?	
	MGCM.H16a	?	38	38	76	?	?	?	?	?	?	
	MGCM.H16b	?	43	50	93	?	?	?	?	?	?	
	MGCM.H16c	?	47	62	109	?	?	?	?	?	?	
	MGCM.H16d	?	53	43	96	?	?	?	?	?	?	
	MGCM.H17	?	51	50	101	?	?	?	?	?	?	
	MGCM.H18a	?	32	43	75	?	?	?	?	?	?	
	MGCM.H18b	?	53	51	104	?	?	?	?	?	?	
	MGCM.H19a	?	42	51	93	?	?	?	?	?	?	
	MGCM.H19b	?	36	39	75	?	?	?	?	?	?	
	MGCM.H19c	?	56	44	100	?	?	?	?	?	?	
	MGCM.H19d	?	45	37	82	?	?	?	?	?	?	
	MGCM.H19e	?	57	46	103	?	?	?	?	?	?	
Koreanornis dodsoni (Koreanaornipodidae)	MGCM.H20a	?	46	49	95	?	?	?	?	?	?	
	MGCM.H20b	?	57	47	104	?	?	?	?	?	?	
	MGCM.H20c	?	40	37	77	?	?	?	?	?	?	
	MGCM.H20d	?	46	52	98	?	?	?	?	?	?	
	MGCM.H20e	?	44	43	87	?	?	?	?	?	?	
	MGCM.H20f	?	45	55	100	?	?	?	?	?	?	
	MGCM.H20g	?	43	46	89	?	?	?	?	?	?	
	MGCM.H20h	?	40	36	76	?	?	?	?	?	?	
	MGCM.H20i	?	50	55	105	?	?	?	?	?	?	
Jindongornipes kimi (Jindongornipedidae)	KPE50006	?	?	?	125	?	?	140	?	?	?	Lockey et al. (1992)
	KPE50006	?	?	?	150	?	?	140	?	?	?	

Table 15A.2. *continued*

Ichnotaxon	Ichnofamily	Track #	Divarication				PL	SL	PA	FR	TW	Reference(s)
			I	II–III	III–IV	TOT						
Shandongornipes muxiai	Shandongornipodidae	LRH-DZ66	?	118	24	142	?	?	?	?	?	Li et al. (2005); Lockley et al. (2007)
		LRH-DZ67(H)	?	91	37	128	41	?	?	?	?	
		LRH-DZ68(P)	?	100	35	135	46	86	?	?	?	
		LRH-DZ69	?	?	?	?	44	90	?	?	?	
		LRH-DZ70	?	92	38	130	45	88	?	?	?	
Pullornipes aureus	Koreanornipodidae	CU 212.21/22-A01	?	82	50	132	?	?	?	2	?	Lockley, Matsukawa, et al. (2006)
		CU 212.21/22-A02	?	49	56	105	151	?	?	10	?	
		CU 212.21/22-A03	?	45	42	87	177	327	?	25	?	
		CU 212.21/22-A04	?	39	80	119	148	325	?	15	?	
		CU 212.21/22-A05	?	?	?	?	?	?	?	?	?	
		CU 212.21/22-A06	?	44	75	119	?	315	?	20	?	
		CU 212.21/22-A07	?	50	51	101	163	?	?	2	?	
		CU 212.21/22-A08	?	60	50	110	127	288	?	?	?	
		CU 212.21/22-A12	?	40	40	80	?	?	?	1	?	
		CU 212.21/22-A13	?	?	?	?	?	?	?	?	?	
		CU 212.21/22-A14	?	36	52	88	?	280	?	10	?	
		CU 212.21/22-A15	?	39	55	94	155	?	?	7	?	
		CU 212.21/22-A27	?	53	63	116	?	?	?	9	?	
		CU 212.21/22-A28	?	28	?	?	152	?	?	?	?	
		CU 212.21/22-A29	?	49	?	?	220	352	?	?	?	
		CU 212.21/22-A30	?	42	62	104	147	347	?	11	?	
		CU 212.21/22-A31	?	61	56	117	177	320	?	1	?	
		CU 212.21/22-A32	?	48	55	103	166	340	?	?	?	
		CU 212.21/22-A33	?	77	64	141	170	335	?	1	?	
		CU 212.21/22-A34	?	56	84	140	150	320	?	2	?	
		CU 212.21/22-A35	?	53	70	123	172	322	?	?	?	
		CU 212.21/22-A36	?	51	68	119	142	313	?	?	?	
		CU 212.21/22-A37	?	70	56	126	182	323	?	5	?	
		CU 212.21/22-A38	?	64	61	125	112	299	?	?	?	
		CU 212.21/22-A39	?	74	67	141	184	300	?	2	?	
		CU 212.21/22-A40	?	50	50	100	130	314	?	2	?	
		CU 212.21/22-A41	?	40	59	99	170	298	?	5	?	
		CU 212.21/22-A42	?	50	64	114	128	295	?	1	?	
		CU 212.21/22-A43	?	53	85	138	162	288	?	12	?	
		CU 212.21/22-A44	?	90	42	132	122	286	?	10	?	
		CU 212.21/22-A45	?	49	90	139	136	260	?	15	?	
Limiavipes curriei	Limiavipedidae	aa1R	0	?	?	?	?	?	?	?	?	McCrea and Sarjeant (2001); McCrea et al. (2014)
		aa2L	0	?	?	?	240	?	150	?	?	
		aa3R	0	?	?	?	219	445	153	?	?	
		aa4L	0	?	?	?	248	450	157	?	?	

Taxon										
aa5R	0	?	?	?	230	462	158	?	?	?
aa6L	0	68	54	112	235	455	159	?	?	?
aa7R	0	?	?	?	252	479	151	?	?	?
aa8L	0	?	?	?	240	473	?	?	?	?
cd1R	0	50	70	120	?	?	?	?	?	?
cd2L	0	?	?	?	220	?	135	?	?	?
cd3R	0	?	?	?	180	365	150	?	?	?
cd4L	0	?	?	?	235	405	?	?	?	?
A1L	0	50	60	110	230	?	?	?	?	?
A2R	0	?	?	135	?	?	?	?	?	?
AqCu-B1R	0	72	63	?	?	?	163	?	?	?
AqCu-B2L	0	?	?	?	261	514	178	?	?	?
AqCu-B3R	0	?	?	?	261	514	?	?	?	?
AqCu-B4L	0	?	?	?	254	?	?	?	?	?
AqCu-F1L	0	?	?	?	?	?	140	?	?	?
AqCu-F2R	0	?	?	?	220	410	132	?	?	?
AqCu-F3L	0	55	69	124	210	420	?	?	?	?
AqCu-F4R	0	?	?	?	250	?	127	?	?	?
AqCu-GF1L	0	?	?	?	?	396	144	?	?	?
AqCu-GF2R	0	?	?	?	218	423	149	?	?	?
AqCu-GF3L	0	?	?	?	225	432	153	?	?	?
AqCu-GF4R	0	67	65	132	220	442	?	?	?	?
AqCu-GF5L	0	?	?	?	225	?	165	?	?	?
AqCu-GF6R	0	?	?	?	229	460	140	?	?	?
AqCu-FG1L	0	?	?	?	?	390	147	?	?	?
AqCu-FG2R	0	60	65	125	230	430	151	?	?	?
AqCu-FG3L	0	?	?	?	235	480	151	?	?	?
AqCu-FG4R	0	?	?	?	190	530	?	?	?	?
AqCu-FG5L	0	?	?	?	270	?	?	?	?	?
AqCu-FG6R	0	?	?	?	225	?	?	?	?	?
AqCu-FG7L	0	?	?	?	315	?	?	?	?	?
AqCu-H1L	0	?	?	?	?	600	?	?	?	?
AqCu-H2R	0	?	?	?	300	570	166	?	?	?
AqCu-H2R	0	?	?	?	300	?	166	?	?	?
AqCu-H3L	0	62	59	131	305	?	172	?	?	?
AqCu-H4R	0	?	?	?	260	?	?	?	?	?
AqCu-PARA1R	0	50	69	119	208	476	156	?	30	?
AqCu-PARA2L	0	52	82	134	273	489	156	?	-4	?
AqCu-PARA3R	0	63	74	137	227	433	159	?	22	?
AqCu-PARA4L	0	58	68	126	215	?	163	?	35	?
AqCu-PARA5R	0	54	70	124	215	452	?	?	14	?
AqCu-PARA6L	0	61	58	119	221	?	?	?	23	?
AqCu-PARA7R	0	58	61	119	?	?	?	?	25	?
AqCu-PARA8L	0	?	?	?	?	?	?	?	19	?

Table 15A.2. *continued*

Ichnotaxon	Ichnofamily	Track #	Divarication				PL	SL	PA	FR	TW	Reference(s)
			I	II–III	III–IV	TOT						
		AqCu-PARA9R	0	61	67	128	?	?	?	?	?	
		AqCu-PARA10L	0	55	76	131	234	?	?	41	?	
		AqCu-PARB1R	0	71	79	150	?	?	?	?	?	
		AqCu-PARB2L	0	?	?	?	?	?	?	?	?	
		AqCu-PARB3R	0	58	55	113	?	510	?	30	?	
		AqCu-PARB4L	0	61	53	114	248	?	?	16	?	
		AqCu-PARB5R	0	64	55	119	237	476	158	24	?	
		AqCu-PARB6L	0	?	?	?	?	?	?	-30	?	
		AqCu-PARB7R	0	50	57	107	298	525	?	16	?	
		AqCu-PARB8L	0	44	52	96	?	?	?	0	?	
		AqCu-PARB9R	0	84	53	137	214	510	173	27	?	
Barrosopus slobodai	?	PVPH-SB-415-17C-86	0	?	?	120	120	230	?	?	?	Coria et al. (2002)
		PVPH-SB-415-17C-87	0	63	52	115	110	210	?	9	?	
		PVPH-SB-415-17C-88	0	66	54	120	100	235	?	19	?	
		PVPH-SB-415-17C-89	0	68	42	110	135	225	155	6	?	
		PVPH-SB-415-17C-90	0	?	?	?	90	165	?	?	?	
		PVPH-SB-415-17C-91	0	?	?	100	110	185	?	?	?	
		PVPH-SB-415-17C-92	0	?	?	140	30	135	?	?	?	
		PVPH-SB-415-17C-93	0	?	?	120	?	?	?	?	?	
		PVPH-SB-415-17C-94	0	?	?	120	?	?	?	?	?	
Dongyangornipes sinensis	?	DoSi-Average	?	44	44	98	?	?	?	?	?	Azuma et al. (2013)
		DoSi-Holotype	?	?	?	89	?	?	?	?	?	
Moguiornipes robusta	?	MGCM.H25a	?	48	42	90	?	?	?	?	?	Xing et al. (2011)
		MGCM.H25b	?	54	39	93	?	?	?	?	?	
		MGCM.H25c	?	50	?	?	?	?	?	?	?	
		MGCM.H25d	?	57	42	99	?	?	?	?	?	
		MGCM.H27a	?	?	50	?	?	?	?	?	?	
Tatarornipes chabuensis	?	1-CU 214	?	?	?	119	?	?	?	?	?	Lockley, Li, et al. (2012)
		2-CU 214.184	?	?	?	105	?	?	?	?	?	
		3-CU 214.185	?	?	?	106	?	?	?	?	?	
		4-CU 214.184	?	?	?	125	?	?	?	?	?	
		5-CU 214.187	?	41	61	102	?	?	?	?	?	
		6-CU 214.187	?	?	?	108	?	?	?	?	?	
		7-CU 214.188	?	55	65	120	?	?	?	?	?	
		8-CU 214.186	?	?	?	105	?	?	?	?	?	
		9-CU 214.184	?	48	52	100	?	?	?	?	?	
		10-CU 214.189	?	?	?	118	?	?	?	?	?	
		11-CU 214.14	?	?	?	87	?	?	?	?	?	
		12-CU 214.14	?	?	?	97	?	?	?	?	?	
		13-CU 214.11	?	56	54	110	?	?	?	?	?	
		14-CU 214.12	?	?	?	107	?	?	?	?	?	

				Yang et al. (1995)
15-CU 214.10	?	44	53	97	?	?	?	?
16-CU 214.9	?	?	?	98	?	?	?	?
17-CU 214.6	?	61	72	133	?	?	?	?
18-CU 214.7	?	?	?	122	?	?	?	?
19-CU 214.8	?	55	83	138	?	?	?	?
20-CU 214.8	?	61	53	105	?	?	?	?
1-CU 214.3	?	?	?	112	?	?	?	?
2-CU 214.3	?	?	?	118	?	?	?	?
3-CU 214.3	?	?	?	118	?	?	?	?
4-CU 214.5	?	?	?	101	?	?	?	?
5-CU 214.5	?	?	?	109	?	?	?	?
6-CU 214.18	?	65	59	124	?	?	?	?
7-CU 214.18	?	68	66	134	?	?	?	?
8-CU 214.160	?	?	?	135	?	?	?	?
9-CU 214.19	?	?	?	111	?	?	?	?
10-CU 214.19	?	?	?	98	?	?	?	?
11-CU 214.148	?	53	54	107	?	?	?	?
12-CU 214.149	?	40	56	81	?	?	?	?
13-CU 214.150	?	40	56	98	?	?	?	?
14-CU 214.151	?	46	48	94	?	?	?	?
15-CU 214.152	?	?	?	102	?	?	?	?
16-CU 214.152	?	?	?	126	?	?	?	?
17-CU 214.153	?	59	68	127	?	?	?	?
18-CU 214.155	?	?	?	115	?	?	?	?
19-CU 214.155	?	?	?	102	?	?	?	?
20:214.156	?	?	?	82	?	?	?	?
21-CU 214.154	?	?	?	129	?	?	?	?
22-CU 214.154	?	?	?	122	?	?	?	?
23-CU 214.154	?	32	63	95	?	?	?	?
KPE50101-A01	?	45	50	95	?	?	?	?
KPE50101-A02	?	54	55	109	?	?	?	?
KPE50101-A03	?	?	?	?	?	?	?	?
KPE50101-A04	?	52	60	112	?	?	?	?
KPE50101-A05	?	45	59	104	?	?	?	?
KPE50101-A06	?	42	56	98	?	?	?	?
KPE50101-A07	?	49	54	103	?	?	?	?
KPE50101-A08	?	58	59	117	?	?	?	?
KPE50101-A09	?	?	?	?	?	?	?	?
KPE50101-A10	?	51	52	103	?	?	?	?
KPE50101-A11	?	55	61	116	?	?	?	?
KPE50101-A12	?	46	51	97	?	?	?	?
KPE50101-A13	?	52	60	112	?	?	?	?
KPE50101-A14	?	46	60	106	?	?	?	?
KPE50101-A15	?	48	52	100	?	?	?	?

Uhangrichnus chuni

Table 15A.2. *continued*

Ichnotaxon	Ichnofamily	Track #	Divarication				PL	SL	PA	FR	TW	Reference(s)
			I	II–III	III–IV	TOT						
		KPE50101-A16	?	45	55	100	?	?	?	?	?	
		KPE50101-A17	?	?	?	?	?	?	?	?	?	
		KPE50101-A18	?	52	55	107	?	?	?	?	?	
		KPE50101-A19	?	?	?	?	?	?	?	?	?	
		KPE50101-A20	?	54	58	112	?	?	?	?	?	
		KPE50101-A22	?	46	58	104	?	?	?	?	?	
		KPE50101-A24	?	47	53	100	?	?	?	?	?	
		KPE50101-A25	?	54	56	110	?	?	?	?	?	
		KPE50101-A26	?	48	54	102	?	?	?	?	?	
		KPE50101-A27	?	52	60	112	?	?	?	?	?	
		KPE50101-A28	?	40	48	88	?	?	?	?	?	
		KPE50101-A29	?	60	?	?	?	?	?	?	?	
		KPE50101-A30	?	55	62	117	?	?	?	?	?	
		KPE50101-A31	?	58	59	117	?	?	?	?	?	
		KPE50101-A32	?	54	56	110	?	?	?	?	?	
		KPE50101-A33	?	60	60	120	?	?	?	?	?	
		KPE50101-A34	?	50	50	100	?	?	?	?	?	
		KPE50101-A34	?	50	50	100	?	?	?	?	?	
		KPE50101-A36	?	50	57	107	?	?	?	?	?	
		KPE50101-A37	?	56	66	122	?	?	?	?	?	
		KPE50101-A42	?	57	63	120	?	?	?	?	?	
		KPE50101-A43	?	56	66	122	?	?	?	?	?	
		KPE50101-A44	?	61	62	123	?	?	?	?	?	
		KPE50101-A45	?	60	66	126	?	?	?	?	?	
		KPE50101-A46	?	47	53	100	?	?	?	?	?	
		KPE50101-A47	?	53	57	110	?	?	?	?	?	
		KPE50101-A48	?	58	60	118	?	?	?	?	?	
		KPE50101-A49	?	55	65	120	?	?	?	?	?	
		KPE50101-A50	?	63	65	128	?	?	?	?	?	
		KPE50101-A51	?	67	67	134	?	?	?	?	?	
		KPE50101-A52	?	58	62	120	?	?	?	?	?	
		KPE50101-A54	?	42	63	105	?	?	?	?	?	
		KPE50101-A55	?	56	64	120	?	?	?	?	?	
		KPE50101-A56	?	61	63	124	?	?	?	?	?	
		KPE50101-A57	?	42	68	110	?	?	?	?	?	
		KPE50101-A58	?	57	61	118	?	?	?	?	?	
		KPE50101-A59	?	42	68	110	?	?	?	?	?	
		KPE50101-A60	?	57	61	118	?	?	?	?	?	
		KPE50101-B01	?	47	50	97	?	?	?	?	?	
		KPE50101-B02	?	?	?	?	?	?	?	?	?	
		KPE50101-B03	?	48	52	100	?	?	?	?	?	

Taxon	Class.	Specimen											Reference
		KPE50101-B04	?	43	53	96	?	?	?	?	?	?	
		KPE50101-B05	?	57	58	115	?	?	?	?	?	?	
		KPE50101-B06	?	46	51	97	?	?	?	?	?	?	
		KPE50101-B07	?	45	48	93	?	?	?	?	?	?	
		KPE50101-B08	?	53	61	114	?	?	?	?	?	?	
		KPE50101-B09	?	54	66	120	?	?	?	?	?	?	
		KPE50101-B10	?	46	50	96	?	?	?	?	?	?	
		KPE50101-B11	?	51	59	110	?	?	?	?	?	?	
		KPE50101-B12	?	42	45	87	?	?	?	?	?	?	
		KPE50101-B13	?	?	?	?	?	?	?	?	?	?	
		KPE50101-B14	?	41	56	97	?	?	?	?	?	?	
		KPE50101-B15	?	48	60	108	?	?	?	?	?	?	
		KPE50101-B16	?	54	63	117	?	?	?	?	?	?	
		KPE50101-B17	?	58	62	120	?	?	?	?	?	?	
		KPE50101-B18	?	?	?	?	?	?	?	?	?	?	
		KPE50101-B19	?	42	63	105	?	?	?	?	?	?	
		KPE50101-B20	?	53	55	108	?	?	?	?	?	?	
		KPE50101-C01	?	59	66	125	?	?	?	?	?	?	
		KPE50101-C02	?	?	?	?	?	?	?	?	?	?	
		KPE50101-C03	?	51	68	119	?	?	?	?	?	?	
		KPE50101-C04	?	54	57	111	?	?	?	?	?	?	
		KPE50101-C05	?	61	63	124	?	?	?	?	?	?	
		KPE50101-C06	?	66	69	135	?	?	?	?	?	?	
		KPE50101-C07	?	55	58	113	?	?	?	?	?	?	
		KPE50101-C08	?	59	61	120	?	?	?	?	?	?	
		KPE50101-C09	?	?	?	?	?	?	?	?	?	?	
		KPE50101-C10	?	53	60	113	?	?	?	?	?	?	
Uhangrichnus topotype	?	KPE50101-T1	?	?	?	?	?	?	?	?	?	?	Lockley, Lim, et al. (2012)
		KPE50101-T2	?	?	?	?	?	77	?	?	30	?	
		KPE50101-T4	?	?	?	?	?	81	159	?	18	?	
		KPE50101-T3	?	?	?	?	?	77	155	?	12	?	
Magnoavipes lowei	Theropod	DMNH918.1	?	53	47	100	?	?	?	?	10	?	Lee (1997)
		DMNH918.2	?	64	54	118	?	1070	?	?	28	?	
		DMNH918.3	?	60	50	110	?	1080	2170	?	14	?	
		DMNH918.4	?	62	48	110	?	1040	2080	?	35	?	
		DMNH918.5	?	62	47	109	?	980	2000	?	25	?	
		DMNH918.6	?	58	54	112	?	1100	2080	?	40	?	
Magnoavipes caneeri	Theropod	CU MWC200-1.1	?	40	60	100	?	1400	?	?	0	?	Lockley et al. (2001)
		CU MWC200-1.2	?	40	60	100	?	1400	?	?	0	?	
		CU MWC200-1.3	?	?	?	?	?	2800	?	?	?	?	
		CU MWC200-1.4	?	?	?	?	?	?	?	?	?	?	
		CU MWC200-1.5	?	35	52	87	?	1370	2780	?	?	?	
		CU MWC200-1.6	?	?	41	?	?	1400	2830	?	0	?	
		CU MWC200-1.7	?	37	43	80	?	1420	?	?	?	?	

Table 15A.2. *continued*

Ichnotaxon	Ichnofamily	Track #	Divarication				PL	SL	PA	FR	TW	Reference(s)
			I	II–III	III–IV	TOT						
		CU MWC200-1.8	?	51	50	101	1380	?	?	?	?	
		CU MWC200-1.9	?	?	?	?	?	2680	?	?	?	
		CU MWC200-1.10	?	45	40	85	?	?	?	?	?	
		CU MWC200-2.01	?	35	45	80	?	?	?	?	?	
		CU MWC200-2.02	?	35	60	95	1000	?	?	?	?	
		CU MWC200-2.03	?	48	57	105	1050	2050	?	?	?	
		CU MWC200-2.04	?	?	50	?	990	2040	?	?	?	
		CU MWC200-3.01	?	32	37	69	?	?	?	?	?	
		CU MWC200-3.02	?	32	33	65	930	?	?	?	?	
		CU MWC200-3.03	?	32	37	69	920	1840	?	?	?	
		CU MWC200-3.04	?	34	37	71	930	1840	?	?	?	
		CU MWC200-4.01	?	?	?	?	?	?	?	?	?	
		CU MWC200-4.02	?	?	?	?	1170	?	?	?	?	
		CU MWC200-4.03	?	40	40	80	1130	?	?	?	?	
		CU MWC200-5.01	?	42	45	87	?	?	?	?	?	
		CU MWC200-5.02	?	43	60	103	?	?	?	?	?	
Magnoavipes denaliensis	Theropod	Smallest	?	?	?	97	?	?	?	?	?	Fiorillo et al. (2011)
		Largest	?	?	?	116	?	?	?	?	?	

REFERENCES

Azuma, Y., Y. Arakawa, Y. Tomida, and P. J. Currie. 2002. Early Cretaceous bird tracks from the Tetori Group, Fukui Prefecture, Japan. Memoir of the Fukui Prefectural Dinosaur Museum 1: 1–6.

Azuma, Y., J. Lü, X. Jin, Y. Noda, M. Shibata, R. Chen, and W. Zheng. 2013. A bird footprint assemblage of early Late Cretaceous age, Dongyang City, Zhejiang Province, China. Cretaceous Research 40: 3–9.

Bunni, M. K. 1959. The Killdeer (Charadrius vociferus), Linnaeus, in the breeding season: ecology, behavior, and the development of homoiothermism. Ph.D. dissertation, University of Michigan, Ann Arbor, Michigan, 348 pp.

Buckley, L. G., R. T. McCrea, and M. G. Lockley. 2015. Birding by foot: a critical look at the synapomorphy- and phenetic-based approaches to trackmaker identification of enigmatic tridactyl Mesozoic traces. Ichnos 22(3–4): 192–207.

Carrier, D., and L. R. Leon. 1990. Skeletal growth and function in the California gull (Larus californicus). Journal of Zoology 222: 375–389.

Chun, S. S. 1990. Sedimentary processes depositional environments and tectonic setting of the Cretaceous Uhangri Formation. Ph.D. dissertation, Seoul National University, Department of Oceanography, Seoul, Korea, 328 pp.

Coria, R. A., P. J. Currie, D. Eberth, and A. Garrido. 2002. Bird footprints from the Anacleto Formation (Late Cretaceous) in Neuquén Province, Argentina. Ameghiniana 39: 1–11.

Currie, P. J. 1981. Bird footprints from the Gething Formation (Aptian, Lower Cretaceous) of northeastern British Columbia, Canada. Journal of Vertebrate Paleontology 1(3–4): 257–264.

Dial, K. P. 2003. Evolution of avian locomotion: correlates of flight style, locomotor modules, nesting biology, body size, development, and the origin of flapping flight. Auk 120(4): 941–952.

Falk, A. R., L. D. Martin, and S. T. Hasiotis. 2011. A morphologic criterion to distinguish bird tracks. Journal of Ornithology 152: 701–716.

Farlow, J. O., S. M. Gatesy, T. R. Holtz, J. R. Hutchinson, and J. M. Robinson. 2000. Theropod locomotion. American Zoologist 40: 640–663.

Fiorillo, A. R., S. T. Hasiotis, Y. Kobayashi, B. H. Breithaupt, and P. J. McCarthy. 2011. Bird tracks from the Upper Cretaceous Cantwell Formation of Denali National Park, Alaska, USA: a new perspective on ancient northern polar diversity. Journal of Systematic Palaeontology 9(1): 33–49.

Gatesy, S. M. 1990. Caudofemoral musculature and the evolution of theropod locomotion. Paleobiology, 170–186.

Gill, F. B. 2007. Ornithology. 3rd edition. New York: W. H. Freeman and Company, 758 pp.

Hammer, Ø., and D. A. T. Harper. 2006. Paleontological data analysis. Wiley-Blackwell, Malden, Massachusetts, 351 pp.

Hammer, Ø., D. A. T. Harper, and P. D. Ryan. 2001. PAST: paleontological statistics software package for education and data analysis. Palaeontologia Electronica 4: 9 pp.

Huh, M., M. G. Lockley, K. S. Kim, J. Y. Kim, and S. G. Gwak. 2012. First report of Aquatilavipes from Korea: new finds from Cretaceous strata in the Yeosu Islands Archipelago. Ichnos 19(1–2): 43–49.

Jackson, B. J., and J. A. Jackson. 2000. Killdeer (Charadrius vociferus); in A. Poole (ed.), The Birds of North America Online. Cornell Lab of Ornithology, Ithaca, New York. Available at http://bna.birds.cornell.edu.login.ezproxy.library.ualberta.ca/bna/species/517, doi:10.2173/bna.517. Accessed November 9, 2015.

Kim, B. K. 1969. A study of several sole marks in the Haman Formation. Journal of the Geological Society of Korea 5(4): 243–258.

Kim, J. Y., S. H. Kim, K. S. Kim, and M. G. Lockley. 2006. The oldest record of webbed bird and pterosaur tracks from South Korea (Cretaceous Haman Formation, Changseon and Sinsu Islands): more evidence of high avian diversity in East Asia. Cretaceous Research 27(1): 56–69.

Kim, J. Y., M. G. Lockley, S. J. Seo, K. S. Kim, S. H. Kim, and K. S. Baek. 2012. A paradise of Mesozoic birds: the world's richest and most diverse Cretaceous bird track assemblage from the Early Cretaceous Haman Formation of the Gajin tracksite, Jinju, Korea. Ichnos 19(1–2): 28–42.

Lee, Y. -N. 1997. Bird and dinosaur footprints in the Woodbine Formation (Cenomanian), Texas. Cretaceous Research 18(6): 849–864.

Li, R., M. G. Lockley, & M. Liu. 2005. A new ichnotaxon of fossil bird track from the Early Cretaceous Tianjialou Formation (Barremian-Albian), Shandong Province, China. Chinese Science Bulletin 50(11): 1149–1154.

Lockley, M. G. 2007. A 25-year anniversary celebration of the discovery of fossil footprints of South Korea. Haenam-gun, Jeollanam-do, southwestern Korea; pp. 41–62 in Proceedings of the Haenam Uhangri International Dinosaur Symposium, February 22–23.

Lockley, M. G., and E. Rainforth. 2002. The tracks record of Mesozoic birds and pterosaurs: an ichnological and paleoecological perspective; pp. 405–418 in L. Chiappe and L. M. Witmer (eds.), Mesozoic Birds above the Heads of Dinosaurs. University of California Press, Berkeley, California.

Lockley, M. G., and J. D. Harris. 2010. On the trail of early birds: a review of the fossil footprint record of avian morphological and behavioral evolution; pp. 1–63 in P. K. Ulrich and J. H. Willett (eds.), Trends in Ornithology Research. Nova Science Publishers, Hauppauge, New York.

Lockley, M. G., and M. Matsukawa. 2009. A review of vertebrate track distributions in East and Southeast Asia. Journal Paleontological Society of Korea 25(1): 17–42.

Lockley, M. G., G. Nadon, and P. J. Currie. 2004. A diverse dinosaur-bird footprint assemblage from the Lance Formation, Upper Cretaceous, eastern Wyoming: implications for ichnotaxonomy. Ichnos 11(3–4): 229–249.

Lockley, M. G., J. L. Wright, and M. Matsukawa. 2001. A new look at Magnoavipes and so-called big bird tracks from Dinosaur Ridge (Cretaceous, Colorado). Mountain Geologist 38(3): 137–146.

Lockley, M. G., J. Li, M. Matsukawa, and R. Li. 2012. A new avian ichnotaxon from the Cretaceous of Nei Mongol, China. Cretaceous Research 34: 84–93.

Lockley, M., K. Chin, K. Houck, M. Matsukawa, and R. Kukihara. 2009. New interpretations of Ignotornis, the first-reported Mesozoic avian footprints: implications for the paleoecology and behavior of an enigmatic Cretaceous bird. Cretaceous Research 30(4): 1041–1061.

Lockley, M. G., K. Houck, S. Y. Yang, M. Matsukawa, and S. K. Lim. 2006. Dinosaur-dominated footprint assemblages from the Cretaceous Jindong Formation, Hallyo Haesang national park area, Goseong County, South Korea: evidence and implications. Cretaceous Research 27(1): 70–101.

Lockley, M. G., R. Li, J. D. Harris, M. Matsukawa, and M. Liu. 2007. Earliest zygodactyl bird feet: evidence from Early Cretaceous roadrunner-like tracks. Naturwissenschaften 94(8): 657–665.

Lockley, M. G., S. Y. Yang, M. Matsukawa, F. Fleming, and S. K. Lim. 1992. The track record of Mesozoic birds: evidence and implications. Philosophical Transactions of the Royal Society B: Biological Sciences 336(1277): 113–134.

Lockley, M. G., J. D. Lim, J. Y. Kim, K. S. Kim, M. Huh, and K. G. Hwang. 2012. Tracking Korea's early birds: a review of Cretaceous avian ichnology and its implications for evolution and behavior. Ichnos 19(1–2): 17–27.

Lockley, M. G., M. Matsukawa, H. Ohira, J. Li, J. Wright, D. White, and P. Chen. 2006. Bird tracks from Liaoning Province, China: new insights into avian evolution during the Jurassic-Cretaceous transition. Cretaceous Research 27(1): 33–43.

Matsukawa, M., M. G. Lockley, K. Hayashi, K. Korai, C. Peiji, and Z. Haichun. 2014. First report of the ichnogenus Magnoavipes from China: new discovery from the Lower Cretaceous inter-mountain basin of Shangzhou, Shaanxi Province, central China. Cretaceous Research 47: 131–139.

McCrea, R. T. 2000. Vertebrate palaeoichnology of the Lower Cretaceous (lower Albian) Gates Formation of Alberta. Master's thesis, University of Saskatchewan, Saskatoon, Saskatchewan, Canada, 133 pp.

McCrea, R. T., and W. A. S. Sarjeant. 2001. New ichnotaxa of bird and mammal footprints from the Lower Cretaceous (Albian) Gates Formation of Alberta; pp. 453–478 in D. Tanke and K. Carpenter (eds.), Mesozoic Vertebrate Life. Indiana University Press, Bloomington, Indiana.

McCrea, R. T., L. G. Buckley, A. G. Plint, P. J. Currie, J. W. Haggart, C. W. Helm, and S. G. Pemberton. 2014. A review of vertebrate track bearing formations from the Mesozoic and earliest Cenozoic of western Canada with a description of a new theropod ichnospecies and reassignment of an avian ichnogenus: fossil footprints of western North America. New Mexico Museum of Natural History Science Bulletin 62: 5–93.

Mehl, M. G. 1931. Additions to the vertebrate record of the Dakota Sandstone. American Journal of Science 21: 441–452.

Oring, L. W., E. M. Graym, and J. M. Reed. 1997. Spotted Sandpiper (*Actitis macularius*); in A. Poole (ed.), The Birds of North America Online. Cornell Lab of Ornithology, Ithaca, New York. Available at http//bna.birds.cornell.edu.login.ezproxy.library.ualberta.ca/bna/species/289doi:10.2173/bna.289. Accessed November 9, 2015.

Romilio, A., and S. W. Salisbury. 2011. A reassessment of large theropod dinosaur tracks from the mid-Cretaceous (late Albian-Cenomanian) Winton Formation of Lark Quarry, central-western Queensland, Australia: a case for mistaken identity. Cretaceous Research 32(2): 135–142.

Rubenson, J., D. G. Lloyd, T. F. Besier, D. B. Heliams, and P. A. Fournier. 2007. Running in ostriches (Struthio camelus): three-dimensional joint axes alignment and joint kinematics. Journal of Experimental Biology 210(14): 2548–2562.

Sarjeant, W. A. S., and W. Langston. 1994. Vertebrate Footprints and Invertebrate Traces from the Chadronian (Late Eocene) of Trans-Pecos Texas. Texas Memorial Museum, University of Texas at Austin, Austin, Texas.

Thulborn, R. A. 2013. Lark Quarry revisited: a critique of methods used to identify a large dinosaurian track-maker in the Winton Formation (Albian-Cenomanian), western Queensland, Australia. Alcheringa: An Australasian Journal of Palaeontology 37(3): 312–330.

Vialov, O. S. 1965. Stratigrafiya neogenovix molass Predcarpatskogo progiba. Naukova Dumka, Kiev, Ukraine, part K, 191 pp. [Russian]

Wright, J. L. 2004. Bird-like features of dinosaur footprints; pp. 167–181 in P. J. Currie, E. B. Koppelus, M. A. Shugar, and J. L. Wright (eds.), Feathered Dragons: Studies on the Transition from Dinosaurs to Birds. Indiana University Press, Bloomington, Indiana.

Xing, L. D., J. D. Harris, C. K. Jia, Z. J. Luo, S. N. Wang, and J. F. An. 2011. Early Cretaceous bird-dominated and dinosaur footprint assemblages from the northwestern margin of the Junggar Basin, Xinjiang, China. Palaeoworld 20(4): 308–321.

Xing, L. D., L. G. Buckley, R. T. McCrea, M. G. Lockley, J. Zhang, L. Piñuela, H. Klein, and F. Wang. 2015. Reanalysis of *Wupus agilis* (Early Cretaceous) of Chongqing, China as a large avian trace: differentiating between large bird and small non-avian theropod tracks. PLoS One, 10(5): e0124039.

Yang, S. Y., M. G. Lockley, S. K. Lim, and S. S. Chun. 1997. Cretaceous bird tracks in Korea. Journal of the Paleontological Society of Korea Special Publication 2: 33–42.

Yang, S. Y., M. G. Lockley, R. Greben, B. R. Erickson, and S. K. Lim. 1995. Flamingo and duck-like bird tracks from the Late Cretaceous and early Tertiary: evidence and implications. Ichnos 4(1): 21–34.

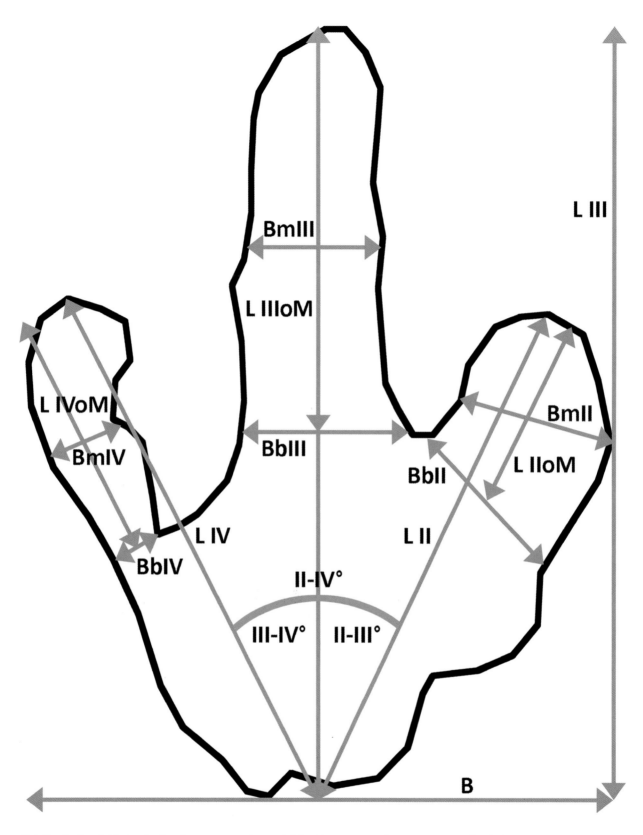

16.1. Sketch of track 1/3 showing locations of the measured distances taken for quantitative analyses. Abbreviations: LII, length along second digit; LIII, length along third digit, the same as total track length; LIV, length along fourth digit; B, total track width; LIIoM, length of second digit without metatarsal; LIIIoM, length of third digit without metatarsal; LIVoM, length of fourth digit without metatarsal; BbII, width at proximal third or base of second digit; BbIII, width at proximal third or base of third digit; BbIV, width at proximal third or base of fourth digit; BmII, width at mid length of second digit; BmIII, width at mid length of third digit; BmIV, width at mid length of fourth digit; II–III°, divarication angle between digits II and III; III–IV°, divarication angle between digits III and IV; II–IV°, divarication angle between digits II and IV.

Elusive Ornithischian Tracks in the Famous Berriasian (Lower Cretaceous) "Chicken Yard" Tracksite of Northern Germany: Quantitative Differentiation between Small Tridactyl Trackmakers

16

Tom Hübner

THE "CHICKEN YARD" TRACKSITE IN NORTHERN Germany, stratigraphically located in the Berriasian (lowermost Cretaceous) Bückeberg Formation, is famous for the extraordinary abundance of typical tridactyl theropod dinosaur tracks and the first didactyl footprints of deinonychosaurian theropods from Europe. Among the vast number of theropod tracks, there are a few small tridactyl tracks, which usually differ from the former by weaker mesaxony, wider divarication angles, stronger symmetry, broader digits, and a rounder and more consistent metatarsal imprint. These features indicate ornithischian track makers, but the strong morphological variation often hampers unambiguous differentiation from less well-preserved theropod tracks. Thus, measurements of the possible ornithischian tracks were quantitatively analyzed to identify potential common features among the ornithischian tracks as well as differing features setting them apart from small theropod tracks. The weaker mesaxony and broader digits are therefore interpreted as diagnostic features of the ornithischian tracks, but other dimensions have no significance in this respect. However, these results imply that the standardization and combination of data sets from a larger number of tracksites around the world will lead to a much better secured discrimination between small tridactyl trackmakers.

INTRODUCTION

The differentiation between ornithopod/ornithischian and theropod tridactyl tracks is a long debated and still not satisfactorily resolved problem in the field of paleoichnology (e.g., Moratalla, Sanz, and Jiménez, 1988; Thulborn, 1990, 2013; Lockley, 1991, 2009; Farlow and Lockley, 1993; Foster and Lockley, 1995; Farlow and Chapman, 1997; Azuma et al., 2006; Mateus and Milàn, 2008; Gierlinski, Niedzwiedzki, and Nowacki, 2009; Belvedere, Mietto, and Ishigaki, 2010; Martin et al., 2011; Romilio and Salisbury, 2011; Castanera et al., 2013; Romilio, Tucker, and Salisbury, 2013; Xing, Lin,

et al., 2014). The problem is on the one hand affected by the similar pes anatomy of both groups and on the other hand by significant variation in shape and dimensions of tracks, even within a single trackway, due to preservation (e.g., Thulborn and Wade, 1984; Manning, 2004; Henderson, 2006; Milàn, 2006; Milàn and Loope, 2007; Romilio and Salisbury, 2011; Romilio, Tucker, and Salisbury, 2013; Thulborn, 2013). The tracks of small ornithischians are especially difficult to identify when preservation is poor. However, in the case of good preservation, some unambiguous features do exist, including: the existence of manus imprints; an anteriorly directed and long first digit (in theropods, it is relatively short and often directed medially or posteriorly–the hallux); the absence of sharp and offset claw marks; strong symmetry of the track; wide digits with only a small number of weak digital pads; and a rounded, wide metatarsal "heel" imprint (e.g., Moratalla, Sanz, and Jiménez, 1988; Thulborn, 1990; Stanford, Weems, and Lockley, 2004; Lockley, McCrea, and Matsukawa, 2009; Castanera et al., 2013). Unfortunately, good preservation is usually not the case, so that some authors have attempted to discriminate theropod from ornithopod (not ornithischians in general) tracks statistically. Moratalla, Sanz, and Jiménez (1988) tested how significant previous subjective assignments of tracks to either theropods or ornithopods actually were (see also a review of that method in Thulborn [2013] and comments by Lockley [2009]), which was also applied by later studies (Mateus and Milàn, 2008; Romilio and Salisbury, 2011).

The Chicken Yard tracksite is located in the Berriasian Obernkirchen Sandstone within the Bückeberg Formation of northern Germany, which is covered by hundreds of tracks comprising didactyl and tridactyl tracks attributed to theropods, iguanodontid ornithopods, and small bipedal ornithischians of uncertain affinity (Richter, Böhme, and van der Lubbe, 2009; van der Lubbe, Richter, and Böhme, 2009; Hornung et al., 2012; Richter and Böhme, 2016). The latter are the rarest track type on the Chicken Yard tracksite,

16

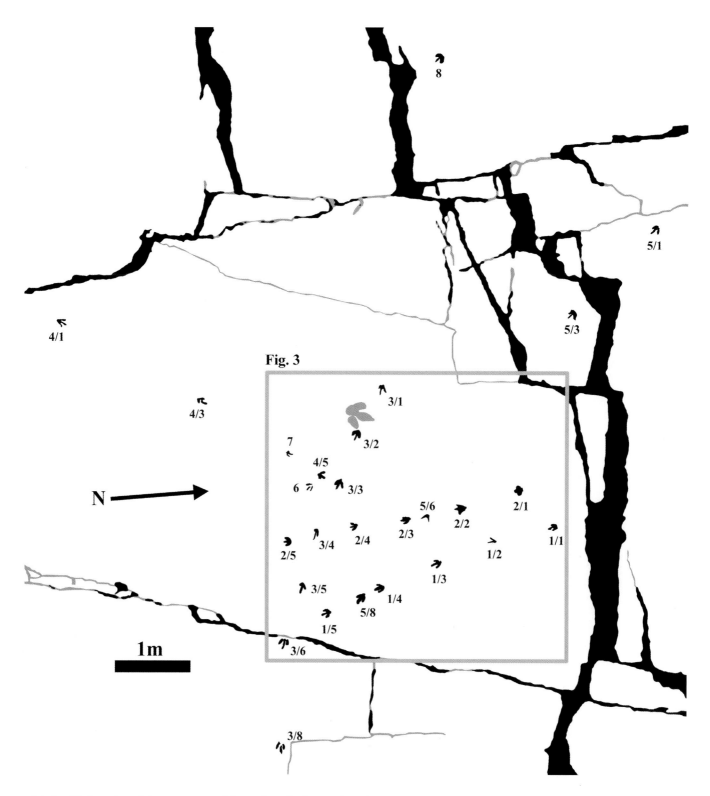

16.2. Simplified overview of the arrangement of the small ornithischian tracks and trackways on the central part of the Chicken Yard Tracksite (see Richter and Böhme, 2016, for an overview of the whole tracksite). The labeling is the same as in the text (see Appendix). The large *Iguanodontipus*-like track in gray was used together with some published tracks for comparison of ratios. Inset corresponds to Figure 16.3.

although up to five individual trackways can be recognized. The track morphology is immensely variable even within trackways. Easily identifiable theropod tracks of similar size are also present on the Chicken Yard tracksite and some possible ornithischian tracks would have been identified as a theropod, if found isolated. Thus, the purpose of this study is to clarify the most likely affinity of the tracks after considering preservational effects and using quantitative analyses slightly different from the methods of Moratalla, Sanz, and Jiménez (1988).

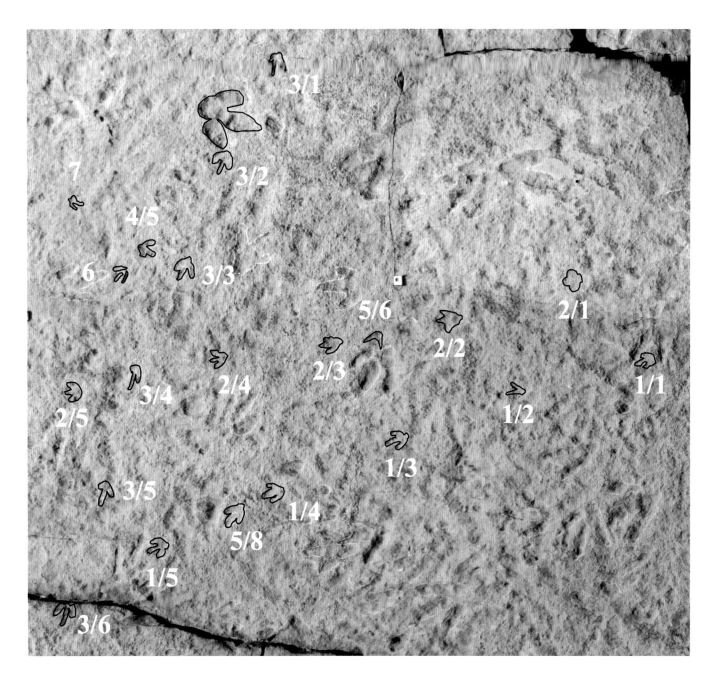

16.3. Close-up of Figure 16.2 with the highest density of possible ornithischian tracks showing strong dinoturbation on the Chicken Yard Tracksite.

MATERIAL AND METHODS

The tracks were measured in the field with a caliper and a goniometer. The measured distances of a track are mainly configured as in former studies (e.g., Thulborn, 1990; Romilio and Salisbury, 2011), but the width of the digits is slightly differently defined due to preservation bias (Fig. 16.1). One parameter is the width at digital mid length (BmII, BmIII, or BmIV for individual digits) and the other is the width at the proximal third or base of individual digits (BbII, BbIII, or BbIV). The hypex distances K and M (Moratalla, Sanz, and Jiménez, 1988; Romilio and Salisbury, 2011) were not measured, because most tracks had no consistent outline between the digits, where the distal endpoints of these distances could be defined. Divarication angles were measured between the central posterior border of the metatarsal imprint and the utmost distal tip of each digit. The ratios of the sample of published tracks were obtained directly from the respective figures in these papers.

The gathered data were exported into Microsoft Excel for bivariate regression analyses of simple values as well as ratios. Complete data sets were also exported into PAST (Palaeontological Statistics, version 1.38; Hammer, Harper, and Ryan, 2001) for conducting principal component analyses (PCA) and discriminant analyses. The latter are automatically combined with the Hotelling's t-squared test comparing

Table 16.1. Measurements obtained from the small ornithischian tracks of the Chicken Yard Tracksite

Track	LII	LIII	LIV	B	LIIoM	LIIIoM	LIVoM	BbII	BbIII
1/1sin	——	12.5		8.6+	—	7.7	—	—	—
1/2dex	—	10.4+	11.2	10.4	—	—	—	—	2.4
1/3sin	10.4	13.6	10.2	10	6.5	8.8	7.2	2.6	2.4
1/4dex	8.7	12.2	9.5	10.2	—	7.8	6.4	2.5	—
1/5sin	10.5	12	9	10.8	8.1	8.4	6.8	3.4	2.9
2/1sin	10.5	13	10.3	12.3	—	—	—	—	—
2/2dex	13	15.8	13.2	11.6	7.4	8.8	7	—	—
2/3sin	10.5	13	11.4	7.6+	7.2	6.9	7.7	2.8	2.7
2/4dex	9.8	13	9.9	10.3	6.2	9	—	2.4	—
2/5sin	8.4	12.5	7.8	12	—	—	—	—	—
3/1sin	10	15.2	12	—	7.6	11.6	9.6	3	3
3/2dex	12	13.4	13.5	13.5	8.5	10.5	8.1	3.8	3.8
3/3sin	11.1	14.5	13.3	12.5	7.3	9.1	9.4	3.5	2.9
3/4dex	11.6	14.5	—	—	7.6	10.1	—	2.8	2.6
3/5sin	—	13.7	10	—	—	9.9	9	—	2.6
3/6dex	—	12.6	10.4	13	—	7.9	7.9	—	—
3/8dex	11.5	13	11	12.6	7.2	8	7.9	2	2.3
4/1sin	10.2	14.1	12.2	11	—	8.2	7.1	—	—
4/3sin	8.2	13.9	9.6	9.9	5.3	9.5	6.8	2.8	3.3
4/5sin	—	11.3	—	—	—	7	5.3	—	—
5/1sin	9	13.9	10.9	11.4	6	9.3	8.3	2.5	3
5/3sin	—	12.5	8.8	—	—	—	—	—	—
5/6dex	—	—	12	10.4	—	—	—	—	—
5/8dex	11.2	16.2	11.7	11	—	10.3	—	—	3
5/12dex	10.4	13.2	10.5	11.6	8.2	8.8	7	2.4	2.5
6dex	7	11.7	9	11.2	—	—	—	—	—
7dex	—	9.3	7.4	7.6+	—	6.6	5.4	—	1.7
8sin	10	12.7	10.3	12	—	8.1	7.5	3.6	2.7

Notes: The track labels are in accordance with Figure 16.2. Values with a plus sign are minimum values due to incompleteness of the respective distance. These are treated as missing values in all quantitative analyses. Abbreviations: LII, length along second digit; LIII, length along third digit, the same as total track length; LIV, length along fourth digit; B, total track width; LIIoM, length of second digit without metatarsal; LIIIoM, length of third digit without metatarsal; LIVoM, length of fourth digit without metatarsal; BbII, width at proximal third or base of second digit; BbIII, width at proximal third or base of third digit; BbIV, width at proximal third or base of fourth digit; BmII, width at mid length of second digit; BmIII, width at mid length of third digit; BmIV, width at mid length of fourth digit; II–III°, divarication angle between digits II and III; III–IV°, divarication angle between digits III and IV; II–IV°, divarication angle between digits II and IV; sin, sinistral; dex, dextral.

the means of multivariate data sets (Hammer and Harper, 2006). Three-dimensional images of some of the tracks were provided by van der Lubbe, from which the gray-shaded contour images of Figure 16.4 are derived. He used the software PhotoModeler from Eos Systems.

RESULTS

General Observations of Tracks and Trackways

Twenty-eight tracks of possible ornithischian affinity were found. Three tracks are isolated and 25 tracks belong to five trackways. All tracks and trackways are located in the central third of the Chicken Yard tracksite, comprising approximately 80 m², and only trackways 1 and 2 have subparallel directions (Figs. 16.2 and 16.3).

The central area shows a high degree of dinoturbation, so that single tracks within three trackways are missing.

There are strong differences in track preservation ranging from shallow but well-imprinted tracks (~1.5 cm), with all three toes clearly visible, to very flat and incomplete tracks, where only one or two toes are recognizable and the track outline is very poorly preserved. These latter tracks were only identified by following and observing along the sequence of the better preserved tracks of a given trackway. With one possible exception, none of the tracks show unambiguous signs of separate claw marks. Pace angulations are generally close to 180°. All measurable distances and angles obtained from the tracks are presented in Table 16.1. A detailed description of individual tracks and trackways can be found in Appendix 16.1.

Analyses of Track Measurements

The goal was to look for quantitative differences between the possible ornithischian tracks and the theropod tracks of

RhIV	RmII	RmIII	RmIV	II–III°	II–IV°	III–IV°	Pace	Stride
2.4	—	2.8	—	28	62	34	—	—
—	—	2.9	3	26	59	32	82	—
2.8	—	—	—	31	64	30	78	153.5
—	—	3	—	15	45	24.5	79.1	155.4
3.3	2.5	—	2	29.5	61	28	75.9	155.1
—	3.2	—	—	29.5	56	28	—	—
—	2.7	3.4	3.3	23.7	47.5	22.2	76	—
2.4	—	2.5	—	20	37.2	12	70.9	146.5
—	3	2.8	3.3	21.6	46	17	69	139.5
—	3.3	—	2.8	37.9	82.7	42.1	89.5	157.3
2.2	—	—	2.9	26.4	58	29.2	—	—
3.5	2.9	2.6	2.6	18.4	53.8	29.2	67.3	—
3	3.4	2.8	3.1	28.5	55	31.4	64.5	131.5
—	2.9	2.3	—	27.5	41.5	22.5	68.5	133.6
3	—	2.8	3.9	30.5	75	35.5	72.3	140
—	—	2.8	—	38.7	69.5	34	71.5	144
2.4	2.1	2.4	2.5	19.5	51	30.5	—	127
2	—	2.2	2.4	26.5	51	31	—	—
2.3	—	2.7	2.7	21	48	23	—	205
—	—	3.4	1.6	—	—	31	—	183
2.4	2.2	2.5	3	21.8	57	29	—	—
—	—	—	—	—	—	—	—	197.5
—	—	—	—	—	—	—	—	—
3.2	—	3.2	3.6	16	43.4	30.3	—	132
2.6	2.9	3	3	26	53	31	—	—
—	—	—	—	—	62	—	—	—
1.6	—	1.9	1.6	28	60	29	—	—
3.7	2	3.5	3.1	33	68	36	—	—

the Chicken Yard tracksite, which could then be used to test the previous trackmaker identification hypotheses originally based only on general qualitative morphology.

First, simple bivariate regressions were applied to look for the degree of comparability among tracks and trackways of assumed ornithischian affinity and selected theropod tracks (see Tables 16.1 and 16.2 for measurements) of the Chicken Yard tracksite, respectively. Correlations between variables of the ornithischian tracks were generally poor to negligible. The highest correlations reached only a coefficient R^2 of between 0.5 and 0.7 in 10 of 120 regressions, which comprise mostly relationships of length to width or between the lateral divarication angles. In contrast, correlations between variables of the measured theropod tracks reached coefficients between 0.7 and 1 in 53 of 120 regressions. This demonstrates much better comparability among the chosen, better preserved theropod tracks (Fig. 16.6) than among possible ornithischian tracks (Figs 16.4, 16.5).

A PCA of most ornithischian tracks (excluding all tracks with 50% or more missing values) was performed together with the measured theropod tracks to find variables, which are responsible for most of the variance in the data set. The first two principal components (PC) explain only 64% of the variance. Nearly all metric variables are directed along PC1 (47% of variance), which therefore accounts for size differences in the data set. PC2 (17% of variance) is explained by the differences in divarication angles among the tracks. However, no clear separation between the ornithischian tracks and the theropod tracks are observed. Thus, to exclude size as the major cause for variance, a data set based on ratios of variables was also analyzed by PCA (Fig. 16.7). It becomes apparent that at least five PCs are necessary to explain 95% of the variance, mainly represented by variable ratios of lengths versus digit widths, within the data set. The position of several points within the scatter plot is additionally strongly influenced by missing values, and there is still a strong overlap

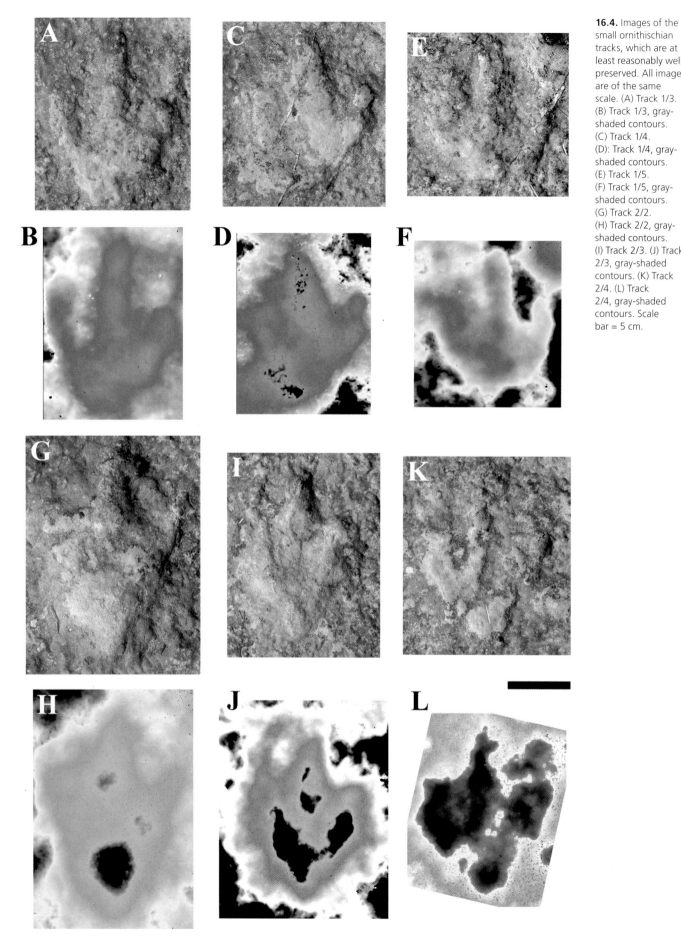

16.4. Images of the small ornithischian tracks, which are at least reasonably well preserved. All images are of the same scale. (A) Track 1/3. (B) Track 1/3, gray-shaded contours. (C) Track 1/4. (D): Track 1/4, gray-shaded contours. (E) Track 1/5. (F) Track 1/5, gray-shaded contours. (G) Track 2/2. (H) Track 2/2, gray-shaded contours. (I) Track 2/3. (J) Track 2/3, gray-shaded contours. (K) Track 2/4. (L) Track 2/4, gray-shaded contours. Scale bar = 5 cm.

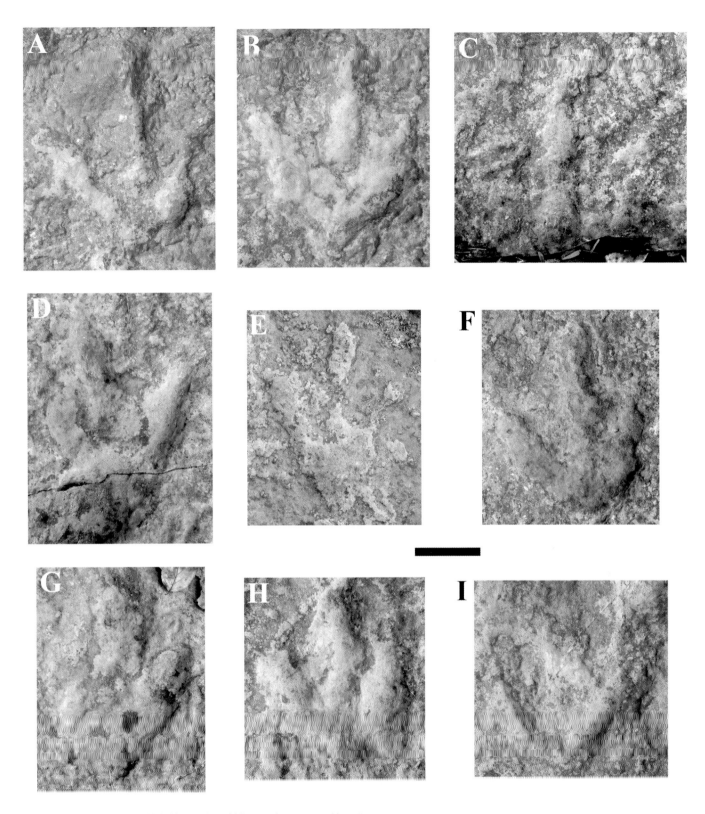

16.5. Images of the small ornithischian tracks, which are at least reasonably well preserved. All images are of the same scale. (A) Track 3/1. (B) Track 3/3. (C) Track 3/6. (D) Track 3/8. (E) Track 4/1. (F) Track 5/1. (G) Track 5/8. (H) Track 5/12. (I) Track 6. Scale bar = 5 cm.

Table 16.2. Measurements obtained from the chosen theropod tracks of the Chicken Yard Tracksite

Track	LII	LIII	LIV	B	LIIoM	LIIIoM	LIVoM	BbII
Tw1/1dex	9.2	13.9	12.6	12.4	7.6	9.6	9.3	2.7
Tw1/2sin	—	—	—	14.8	9	10.7	9.1	2.8
T1dex-gracile	18.2	28.8	20.7	21.6	12.8	19.8	14	5
T2sin-claws	32	42	28.6	28	11	27.7	—	5.6
T3sin-pads	28.6	33.4	27.9	29.4	18.6	21	17.5	5.4
T4 sin	15.9	19.9+	16.2	14.9	11.4	11.8+	13	2.4
T5dex?	8	11.2	8.1	9.6	5.1	8.1	7.2	2.4
T6 sin	10.2	13.8	9.8	10	6.8	8.4	—	2.2
T7dex-typical	25.4	34	24	29.2	16.6	23.8	15	4.4
T9sin	9.5	11	8	10.5	6.5	8	6.8	—
T10sin	10.8	14.5	10.4	9.2	6.5	9.5	—	—

Notes: Values with a plus sign are minimum values due to incompleteness of the respective distance. These are treated as missing values in all quantitative analyses. Abbreviations: Tw, trackway; other abbreviations as in Table 16.1.

and, thus, no secured separation between possible ornithischian and theropod tracks.

In the next step, bivariate regressions of ratios were performed to find indications as to whether the possible ornithischian and theropod tracks of the Chicken Yard tracksite can be distinguished in this way. Only 11 of 87 created combinations showed a detectable separation of the two groups (single outliers can still be involved). These cases were then chosen to test whether clearly identified ornithischian and theropod tracks chosen from the literature (Table 16.3) also plot within the respective areas.

Apart from still-visible restricted overlap and isolated outliers, there are mainly two morphological characters that separate both groups of tracks (Figs. 16.8 and 16.9). First, theropod tracks have generally shorter lateral digits compared to the third digit (strong mesaxony), whereas the ornithischian tracks (including two ichnospecies of *Anomoepus*) tend to have smaller differences in length between these digits (weaker mesaxony; see Lockley, 2009). Even clearer is the tendency of theropod tracks to have more slender digits compared to length than the ornithischian tracks, especially regarding the ratio of total track length (LIII) relative to width of the third digit at its mid length (BmIII). However, cases of covariance of both track groups are also present (Fig. 16.9), reflecting the highly similar bauplan of their feet. For instance, more slender lateral digits are correlated with a more slender third digit and a relatively longer second digit compared to the third digit means also increased slenderness of the former.

The final test for disparity between the two track groups was the discriminant analysis of the multivariate data set (Fig. 16.10). This was again performed without the tracks, which provided 50% or less of the necessary values, but the published tracks of unambiguous ornithischian and theropod tracks were included. This analysis resulted in a significant (Hotelling's t-squared with p = .0657) separation of both track groups, where 92.31% of the tracks could be grouped correctly.

DISCUSSION

Traditional Discrimination of the Small Tridactyls of the Chicken Yard Tracksite

Only a minority of the smaller tracks on the Chicken Yard tracksite are of possible ornithischian affinity. They were initially recognized by observation, because they differed in several features from most of the well-recognizable theropod tracks. These comprise the lack of claw marks, blunt ends of the digits, thicker digits with only two digital pads at maximum even in the third and fourth digit, the overall length-width ratio, the close proximity or partial fusion of digital imprints posteriorly, and especially the rear of the track with an often continuous metatarsal imprint and symmetrical posterior portion of the track. In contrast, the metatarsal imprint of typical theropod tracks on the Chicken Yard tracksite is often a kind of posterior extension of the fourth digit resulting in an asymmetrical posterior track shape; the individual digital imprints of a track are in many cases even posteriorly well-separated; the third and fourth digits have often more than two digital pads; less splayed tracks are longer than wide; and claw marks are quite common. These subjective criteria are similar to the features, which have been discussed for decades, as those utilized for potentially discriminating ornithopod from theropod tracks (e.g., Thulborn and Wade, 1984; Moratalla, Sanz, and Jiménez, 1988; Thulborn, 1990; Foster and Lockley, 1995; Lockley, 2009; Belvedere, Mietto, and Ishigaki, 2010; Martin et al., 2011; Romilio and Salisbury, 2011; Castanera et al., 2013; Romilio, Tucker, and Salisbury, 2013).

BbIII	BbIV	BmII	BmIII	BmIV	II–III°	II–IV°	III–IV°
1.9	3	2.8	2.6	2.7	22	57	33
2.5	2.6	2.9	2.5	1.9	38	73	37
4	3.6	—	3.8	—	29	61	32
5.8	6.4	6.6	7.6	7.3	25	55	31
6	2	5	5	2.9	24	59	28
3.8	2.5	3	3.5	2.8	22	47	19
2.4	2.4	1.8	2	2.4	33	70	36.5
3.2	2	2.2	—	2.2	22	59	34
6	4.8	4.4	5.5	4.7	33	72	43
2.5	2.1	1.4	2	2.3	37	63.5	34
—	—	2.5	2.6	2.9	23.4	45	28

The main reason for the difficulties in tridactyl track differentiation is the conservative bauplan of the foot skeleton of theropods and many small and/or less-derived ornithischians, including small ornithopods (Thulborn, 1990; Farlow and Lockley, 1993; Farlow and Chapman, 1997; Farlow et al., 2012). Both have the same numbers of phalanges in the weight bearing digits (II = 3, III = 4, IV = 5) and a similar degree of digitigrady due to their mostly longer than wide phalanges. Even the shape of the claws is similar because these are slightly curved and pointed at their tip in many small ornithischians including basal ornithopods and iguanodontians. Exactly this claw morphology can also be observed in some theropods, such as ornithomimosaurs (see Choiniere, Forster, and de Klerk, 2012:fig. 14F). It is therefore not surprising to find a strong graphical overlap of both groups in the bivariate regression and PCA (Figs. 16.7, 16.8, and 16.9; see also Farlow and Chapman, 1997; Farlow et al., 2012) of simple variables and ratios, respectively.

There are some observed features of the possible ornithischian tracks discussed here that do not match the traditional scheme, such as the weak or absent inward rotation of individual tracks, the high pace angulation, or the large pace and stride lengths. The stride distances of the trackways (see Table 16.1 for additional hip heights, relative stride lengths, and speeds) are indeed longer than noted for ornithischian trackways often recorded elsewhere (e.g., compare with Lockley et al., 1998; Azuma et al., 2006), but it has been shown that these calculations and associated measurements can vary. Furthermore, it is unlikely that these dinosaurs moved slowly all the time, so that short steps and strides including pace angulation cannot serve as ornithischian-only track features and are more related to the trackmakers' gait at the time of track formation (Romilio and Salisbury, 2014). The inward rotation of individual tracks is also not universally applicable to all identified ornithischian trackways (e.g., see *Delatorrichnus* in de Valais, 2011; sitting traces of *Anomoepus* in Olsen and Rainforth, 2003), although many do show this feature (e.g., Lockley et al., 1998; Moratalla, Hernan, and Jiménez, 2003; Gierlinski, 2009; Kim et al., 2009; Lockley, McCrea, and Matsukawa, 2009; Li et al., 2011; Castanera et al., 2013; Xing and Lockley, 2014). However, the ichnogenus *Siamopodus*, identified as made by a theropod due to claw marks and deep hypices, shows inward rotation of individual tracks (Xing, Lockley, et al., 2014), as in some other theropod tracks, so that this feature cannot universally serve as typical for ornithischian tracks as well.

Some other features of the tracks described here, such as the absence of hallux imprints or the stronger mesaxony in some of the tracks, naturally more resemble theropod tracks. These potential flaws in the assignment to ornithischian trackmakers are also contradictable. On the one hand, hallux imprints are found in ornithischians with a rather long hallux (most basal ornithischians, basal marginocephalians, basal ornithopods, and some basal iguanodontids) only preserved in very deep imprints or in sitting/resting traces (e.g., *Anomoepus*). On the other hand, the degree of mesaxony is also not an unambiguous feature for either ornithischians or theropods, because there are strong indications of convergence between the morphometrics II and III as well as differences within each of the clades (Lockley, 2009). Ichnogroup doubt also has to be kept in mind, because the length of the digits is negatively allometric in growth (Hübner, 2011), which is also observable in tracks of different sizes within single ichnogenera (Lockley, 2009). There are, for instance, ornithopod tracks known with a comparatively strong mesaxony (Castanera et al., 2013) and theropod tracks with weak mesaxony (Xing, Lockley, et al., 2014).

Overall, the described small tracks of the Chicken Yard tracksite are qualitatively different to the often much easier identifiable theropod tracks, so that different trackmakers

16.6. Images of some of the theropod tracks from the Chicken Yard Tracksite, which were measured for this study (see Table 16.2 for measurements). (A) Track TW1/2sin. (B) T1dex-gracile. (C) T2sin-claws. (D) T3sin-pads. (E) T9sin. (F) T10sin. Scale bars = 5 cm. Abbreviations as in Figure 16.1.

are very likely. The best alternative are small ornithischians, because of their foot morphology and because body fossils are well known in Europe around this time interval.

Preservation Complicates Matters

The assignment of some tracks to either of the two tridactyl groups is qualitatively equivocal. The poor preservation of several ornithischian tracks naturally obscures many features, but even better preserved tracks within a trackway show remarkable morphological variability including the shape of the metatarsal imprint, the degree of separation of the digit imprints, the degree of mesaxony, the relative widths of the digits as well as of the whole track, either blunt or acute distal ends of the digits, and the interdigital angles. Thus, the tracks could not unambiguously be assigned to a single known ichnospecies, and it is also not advisable to name a new ichnospecies. The combination of poor preservation and strong variability is thus the main reason for the poor correlation of measured distances and angles among the ornithischian tracks. The theropod tracks measured from the Chicken Yard tracksite show much higher correlation because only well-preserved examples were chosen for this study.

Richter and Böhme (2016) have demonstrated that the Chicken Yard tracksite is in most parts heavily altered by dinoturbation representing an unknown number of trampling events. The ornithischian tracks are definitely not the last of these events, because some of the tracks are partially overprinted or at least altered in shape by tracks made later close by. It is also clear that the potential trackmakers were not moving together over this area at exactly the same time, because the trackways are not parallel to each other nor do they show other signs of gregarious behavior (Cotton, Cotton, and Hunt, 1998; Lockley et al., 1998; Lockley and Matsukawa, 1999). However, as is demonstrated by trackway 3, intense trampling has not led to an evenly consolidated surface. The first track (3/1) was made on a slippery surface because the third digit slipped sideways before making the final imprint, which probably does not reflect the true (intended digit length) and was narrowed after withdrawing of the digit by the collapsing right wall (Fig. 16.5A). Three other flat tracks (3/2, 3/5, 3/6) (Fig. 16.5C) were made on firm, well-consolidated ground with lower water content. The third track (3/3) has well-defined steeper track walls (Fig. 16.5B), indicating firm but more moist ground. Finally, the last track (3/8) has shallowly inclined track walls and no details preserved on the very smooth track bottom (Fig. 16.5D), which might indicate that the sandy substrate was highly saturated at the time of track formation (Milàn, 2006) or that this spot was covered by shallow standing water, such as a puddle. Dry

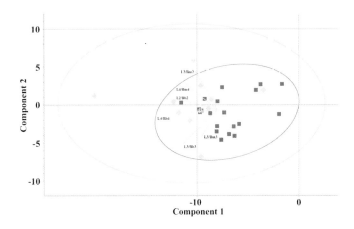

16.7. Plot of the first versus second principal component of variable ratios. Gray diamonds and respective 95% ellipse represent the Chicken Yard Tracksite theropod tracks. Black squares and respective small 95% ellipse represent the small ornithischian tracks. The lines in the center (biplot graphs) show which ratios best describe the first and/or second principal component (PC). Although together these two PCs explain only 57% of the variation, the biplots of the third through fifth PCs consist of the same ratios with only slightly different relative lengths. Therefore, 95% of the variation (represented by five PCs) is explained by the plotted ratios. Abbreviations as in Figure 16.1.

conditions are unlikely due to the absence of contouring fractures (Jackson, Whyte, and Romano, 2009). Interestingly, the nearby track 5/12 shows exactly the same preservational mode (Fig.e 16.5H). Saturated substrate at the time of track formation is also evident in track 4/1, where the track wall of the second digit has partially collapsed (Manning, 2004) (Fig. 16.5E). Overgrowth by microbial mats (Marty, Strasser, and Meyer, 2009), although difficult to prove directly (Richter and Böhme, 2016), had also taken place in variable intensity in several of the ornithischian tracks. Some even show strong overgrowth intensity leading to the formation of internal overtracks (compare Figs. 16.4K and 16.5E with Marty, Strasser, and Meyer, 2009:fig. 8). All these influences contributed to the strong preservational impact on the morphology of the small ornithischian tracks, which can therefore be classified as modified true tracks or internal overtracks (sensu Marty, Strasser, and Meyer, 2009).

The Solution Is in Sight

As there are no associated manus imprints visible, as the only unambiguous evidence for small ornithischians (e.g., Thulborn, 1990; Olsen and Rainforth, 2003; Stanford, Weems, and Lockley, 2004; Castanera et al., 2013), other less obvious features, which may be present despite the variable track preservation, had to be found in this study. Such features are found here and comprise mainly the average degree of mesaxony (Lockley, 2009) and the length-width ratio of the single digits. Thus, many theropod track types show a stronger mesaxony and also have more slender digits than the average

Table 16.3. Ratios of measured distances of some published tracks with generally accepted track affinity, including theropods, ornithopods, and small ornithischians

Ratios	Lockley et al. (1998); *Dinehichnus*		Castanera et al. (2013)	Olsen and Rainforth (2003); *Anomoepus scambus*	Olsen and Rainforth (2003); *Anomoepus intermedius* without claw impressions	Stanford, Weems, and Lockley (2004); *Hypsiloichnus*	*Iguanodontipus*-like track between 3/1 and 3/2 on "Chicken Yard"	Milner, Lockley, and Johnson (2006); *Eubrontes* without claw impressions	
	Fig. 3; Holotype (left center)	Fig. 3; Topotype (lower right)	Fig. 5B; photo of LCR8.7	Fig. 19.15D	Fig. 19.24C	Fig. 6 (right)		Fig. 17A (right); SGDS.59	Fig. 17C (center)
LIII/B	0.95	1.07	1.43	1.29	1.11	1.08	1.11	1.23	1.44
LII/LIV	1.15	1.11	0.78	1.10	1.02	1.15	1.00	1.12	0.85
LII/LIII	0.89	0.76	0.53	0.83	0.81	0.86	0.71	0.70	0.63
LIV/LIII	0.77	0.69	0.68	0.75	0.79	0.74	0.71	0.63	0.74
LII/LIIoM	1.72	1.78	1.50	1.61	1.64	1.71	1.53	1.58	1.38
LIII/LIIIoM	1.67	1.63	1.60	1.54	1.84	1.56	2.05	1.69	1.53
LIV/LIVoM	1.80	1.68	1.93	1.43	1.45	1.49	1.93	1.36	1.42
LII/BbII	3.44	3.96	2.63	3.47	3.83	4.29	2.64	3.80	3.63
LIII/BbIII	4.38	4.38	6.15	5.71	4.75	4.12	3.73	5.40	5.11
LIV/BbIV	3.60	4.80	4.50	4.29	4.50	3.06	2.90	3.40	4.25
LII/BmII	4.43	4.46	3.00	5.50	3.83	4.00	2.90	6.33	4.83
LIII/BmIII	4.38	5.83	6.15	5.71	4.75	4.67	3.73	6.75	5.75
LIV/BmIV	3.00	4.17	3.86	4.62	4.50	2.89	2.90	4.25	5.67
II–III°/II–IV°	1.74	2.34	2.57	2.87	2.74	1.69	1.90	2.68	4.10
III–IV°/II–IV°	0.97	0.91	1.07	0.96	1.12	0.91	1.34	1.07	1.11
II–III°/III–IV°	1.80	2.56	2.41	3.00	2.45	1.86	1.41	2.52	3.70

Note: LCR, Las Cerradicas tracksite; SGDS, St. George Dinosaur Discovery Site; other abbreviations as in Table 16.1

ornithischian track does. These features were formerly used for discriminating the two groups (e.g., Moratalla, Sanz, and Jiménez, 1988; Thulborn, 1990; Gierlinski, 2009; Kim et al., 2009; Li et al., 2012; Castanera et al., 2013) and their significance can now be supported, although the strong overlap in the plots as well as exceptional ichnogenera from elsewhere (e.g., *Corpulentapus*, Li et al., 2011; Lockley et al., 2015) have to be kept in mind. Divarication angles for instance are not at all usable for this purpose, as was already pointed out by Milàn (2006). Other, less clearly identifiable features may also contribute to the surprisingly distinct discrimination of the two track types. However, the plotting of some published tracks of well-accepted ornithischian affinity (e.g., *Dinehichnus*; see Figs. 16.8 and 16.9) within the scatter plot of the theropod tracks additionally demonstrates the anatomical similarities between ornithischian and theropod feet. Much more comprehensive data sets should be created in future studies incorporating all well-defined and supported ichnospecies of ornithischian affinity as well as key ichnotaxa of theropods of all Mesozoic time intervals, respectively. By building on and by combining the results of Moratalla, Sanz, and Jiménez (1988) with the current study, it should then be possible to clarify, more definitively, which unambiguous features are most informative for distinguishing between well-preserved ornithischian and theropod tracks of all sizes. Further experimental studies on the relationship between gait and track formation would also be eligible to clarify for instance, whether and how gait influences important features such as rotation of individual tracks.

Ornithischian Trackmaker Candidates

The true biodiversity of small-bodied species is often much larger than indicated by both the fossil body and track record (Brown et al., 2013). Despite this, up to four different groups of ornithischians are represented by body fossils in close spatial and/or temporal proximity to the Chicken Yard tracksite, so that these at least should to be taken into account as possible small trackmakers. These groups comprise heterodontosaurids, basal ornithopods, dryosaurid ornithopods, and basal marginocephalians.

The first candidate is *Echinodon becklesii* from the Berriasian Purbeck Limestone Formation of southern England, which is only known from skull fragments and teeth (Norman

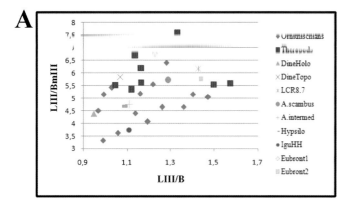

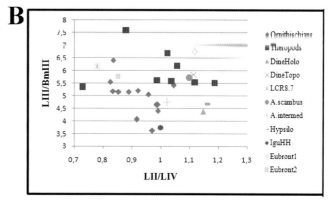

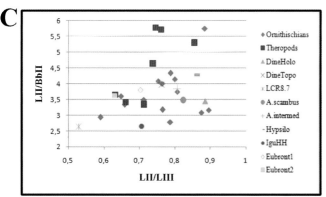

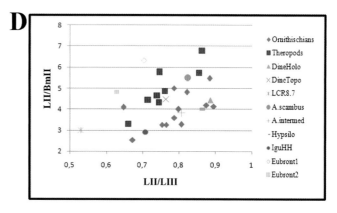

16.8. Bivariate plots of ratios of tracks of small ornithischians (blue diamonds) and theropods (red squares) from the Chicken Yard Tracksite. Note simple variance (e.g., top right) of track ratios. Plots of published tracks as well as of the large *Iguanodontipus*-like track from the Chicken Yard Tracksite (see Table 16.3 for detailed references) are added subsequently. Abbreviations: A. intermed, *Anomoepus intermedius*; A. scambus, *Anomoepus scambus*; DineHolo, Holotype track of *Dinehichnus socialis*; DineTopo, Topotype of the latter; Eubront1, *Eubrontes* 1; Eubront 2, *Eubrontes* 2; Hypsilo, *Hypsiloichnus marylandicus*; IguHH, *Iguanodontipus*-like track from the Chicken Yard Tracksite; LCR8.7, Respective ornithopod track from the Las Cerradicas tracksite; other abbreviations as in Figure 16.1.

and Barrett, 2002; Galton, 2012). It is commonly classified as a basal heterodontosaurid (Norman and Barrett, 2002; Butler et al., 2010; Butler et al., 2012), although Pol, Rauhut, and Becerra (2011) found it to be placed as a basal ornithischian. The only related taxa with preserved feet are *Heterodontosaurus tucki* (Santa Luca, 1980) and *Tianyulong confuciusi* (Zheng et al., 2009), which have three well-developed functional toes (II to VI) and a reduced first digit, respectively (Fig. 16.11A). Although this pedal anatomy would fit nicely to the small ornithischian tracks of the Chicken Yard Tracksite, the diminutive size of all known late heterodontosaurids of significantly less than 1 m in total body length (Zheng et al., 2009; Butler et al., 2010, 2012; Galton, 2012) makes heterodontosaurids less likely trackmaker candidates.

More likely in terms of body size are basal ornithopods similar to the Barremian taxa *Hypsilophodon foxii* (United Kingdom; Galton, 1974, 2009, 2012) and *Gideonmantellia amosanjuanae* (Spain; Ruiz-Omeñaca et al., 2012). There are no body fossil records known from the Berriasian of Europe, but the morphology of the tracks of the Chicken Yard tracksite fit nicely to the pedes of basal ornithopods (Galton, 1974)

with only three functional digits (Fig. 16.11B). The first digit is still relatively long, but the tracks are not deep enough to produce an imprint, and the assumed fast gait implies the imprint of the hallux unlikely. The North American ichnotaxon *Hypsiloichnus* was also referred to a basal ornithopod similar to *Hypsilophodon* or *Zephyrosaurus schaffi* (Stanford, Weems, and Lockley, 2004), although the tetradactyl pes imprints do not look like deep tracks sensu Marty, Strasser, and Meyer (2009). However, a close relative of *Zephyrosaurus*, *Orodromeus makelai* (Scheetz, 1999), has a slightly longer first digit than *Hypsilophodon* does; the fourth finger of the manus is also relatively longer, and the arms were stoutly built with possibly hyperextendable fingers as a sign for at least facultative quadrupedality (Scheetz, pers. comm., April 2013). It is therefore indeed possible that *Hypsiloichnus* was produced by a yet unknown basal ornithopod, although the known taxa with preserved hands and feet (Gilmore, 1915; Galton, 1974; Scheetz, 1999) all have slightly differing proportions with relatively shorter fourth fingers of the manus and first digits of the pedes. Other ornithischians, especially basal neoceratopsians, also have much smaller forelimbs

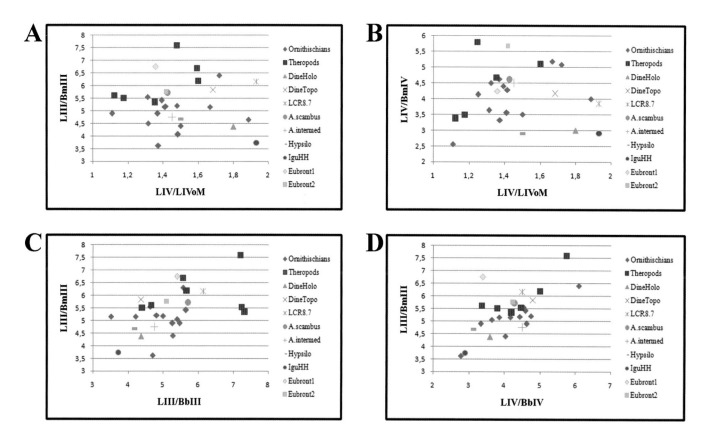

16.9. Bivariate plots of ratios of tracks of small ornithischians (blue diamonds) and theropods (red squares) from the Chicken Yard Tracksite. Note simple covariance (e.g., bottom right) of track ratios. Plots of published tracks as well as of the large *Iguanodontipus*-like track from the Chicken Yard Tracksite (see Table 16.3 for detailed references) are added subsequently. Abbreviations as in Figures 16.1 and 16.8.

than hind limbs as well as manus and pes morphologies very similar to basal ornithopods (Hailu and Dodson, 2004). They were even present in North America during the late Early Cretaceous (Chinnery-Allgeier and Kirkland, 2010), so that they should equally be considered as possible trackmakers of *Hypsiloichnus*.

European taxa of the third group of potential trackmakers, dryosaurid ornithopods, are known from the Middle Jurassic of England (*Callovosaurus leedsi*; Ruiz-Omeñaca, Pereda Suberbiola, and Galton, 2007), the Upper Jurassic of Portugal (*Eousdryosaurus nanohallucis*; Escaso et al., 2014), and the Early Cretaceous of England (*Valdosaurus canaliculatus*; e.g., Galton, 2009; Barrett et al., 2011). The English specimens of *Valdosaurus* are from the Valanginian and Barremian and therefore of younger age than the Chicken Yard tracksite, whereas the Berriasian specimens from Romania formerly assigned to *Valdosaurus* or Dryosauridae indet. (Posmoşanu, 2003) are now all considered to be Euornithopoda indet. (Galton, 2009). Nevertheless, the Chicken Yard tracksite is bracketed in time by the secured dryosaurid occurrences, so that they were most likely also present here. Dryosaurs have typical tridactyl feet with a strongly reduced first digit without nonungual phalanges

(Norman, 2004; Escaso et al., 2014) (Fig. 16.11C). The unguals are more hoof-like than in basal ornithopods, but they still taper acutely distally (Galton, 1981). Dryosaurs are further treated as obligate bipeds (Hübner, 2011; contra. Heinrich, Ruff, and Weishampel, 1993) better resembling with their highly cursorial pes morphology the basal ornithopods such as *Hypsilophodon* than most of the less-derived basal iguanodontians (Norman, 2004) so that their tracks would be hardly distinguishable from the former. *Dinehichnus socialis* is the only ichnotaxon, which is currently tentatively assigned to dryosaurs (Lockley et al., 1998), although a camptosaur affinity is also likely (Gierlinski and Sabath, 2008). It is known from the Upper Jurassic of western North America, Portugal, and Poland (Lockley et al., 1998; Gierlinski, Niedzwiedzki, and Nowacki, 2009) and probably from the Berriasian of Spain and Germany (Hornung and Reich, 2012; Castanera et al., 2013). The average size of the tracks is slightly larger than the small tracks from the Chicken Yard tracksite and only tracks 1/3 and 3/6 look roughly similar to *Dinehichnus* with an indicated rounded and separated metatarsal imprint (Figs. 16.4A and 16.5C). Despite these exceptions, the tracks cannot securely be assigned to *Dinehichnus* due to their often continuous posterior metatarsal border and less well-separated and more slender digit imprints. However, small dryosaurs are still potential trackmakers of the ornithischian tracks of the Chicken Yard tracksite, because it is not clear yet whether all dryosaurus are bound to produce *Dinehichnus* track types in every possible substrate.

The associated oval manus imprints of the Las Cerradicas locality may also indicate a basal ankylopollexian ornithopod as their potential trackmaker (Castanera et al., 2013), because the reduced manus of dryosaurids is unlikely to produce such an oval-shaped imprint. In contrast, basal ankylopollexians had still relatively slender feet but the manus already showed significant adaptations toward weight-bearing and quadruped locomotion (Carpenter and Wilson, 2008). Members of the latter clade were definitely present in Europe during the Late Jurassic and Early Cretaceous (Galton and Powell, 1980; Norman and Barrett, 2002) and may well be the more probable trackmaker at the Las Cerradicas tracksite.

The possible smoking gun for the identity of the small trackmakers on the Chicken Yard tracksite might be the partial skeleton of *Stenopelix valdensis* (Richter et al., 2012). It was found in a now historical quarry of the Harrl Hill near Bückeburg in the same formation (Obernkirchen Sandstone) as the Chicken Yard tracksite (Hornung, Böhme, and Reich, 2012; Richter et al., 2012). Its phylogenetic position is close to the base of Marginocephalia (Butler and Sullivan, 2009) or even at the base of Ceratopsia (Butler et al., 2011). *Stenopelix* possesses a slender pes with three functional digits with pointed but only weakly recurved unguals, whereas the first digit is reduced in length (Fig. 16.11D). This is very similar to the configuration seen in basal ornithopods, and it would therefore have been able to produce similar tracks. The pes morphology, the body size, and the secured occurrence in the Obernkirchen Sandstone make this small dinosaur a good candidate as producer of the small ornithischian tracks on the Chicken Yard tracksite, although the three other options are also likely.

CONCLUSIONS

The finding of eight potential ornithischian tracks identified on the Chicken Yard tracksite comprises five trackways and three isolated tracks. They differ from typical theropod tracks of this site mainly by the observations of having a more symmetrical outline, an often consistent posterior metatarsal border and the absence of claw marks. Other traditionally used discriminating features are more ambiguous, especially due to the strong and variable preservational influence on the track morphology. The tracks are in general preserved as modified true tracks or internal overtracks. Bivariate

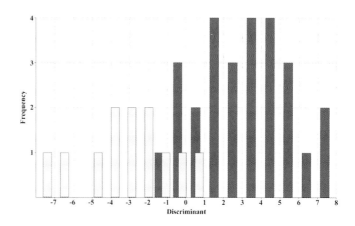

16.10. Graphical result of the discriminant analysis of all ratios of ornithischian and theropod tracks, except tracks with 50% or more missing values, and also including the chosen published tracks (see Table 16.3 for identity). Light gray indicates theropod tracks; dark gray indicates ornithischian tracks.

regressions analyses and PCA yielded no clear results, but a minority of bivariate regressions of ratios of distances as well as a multivariate discriminant analysis resulted in a distinct separation of the two track types. Subsequently included examples of well-defined published ichnospecies fitted well into this scheme. Theropod tracks can accordingly be distinguished from ornithischian tracks by a larger length difference between the third and the lateral digits (mesaxony) and by generally more slender digits, but the morphological variation within the latter hampers a sharp differentiation. These results are preliminary due to the comparatively small data set, but in combination with similar studies on large ornithopod tracks and a more comprehensive data set, they may be a good starting point for future studies finally leading to secured discriminating features for theropod and ornithischian tridactyl tracks. The small ornithischian tracks of the Chicken Yard tracksite may have been produced either by heterodontosaurids, basal ornithopods, dryosaurid ornithopods, and/or basal marginocephalians. The latter taxon is represented by a partial skeleton of *Stenopelix* found nearby and in the same formation, which makes marginocephalian ornithischians currently a plausible trackmaker

ACKNOWLEDGMENTS

I would like to thank Annette Richter for her unmatched support. I also thank Annina Böhme for her helpful suggestions I especially thank Torsten van der Lubbe, who introduced me to the Chicken Yard tracksite and the basics of paleoichnology. I also thank Jahn Hornung, Martin Lockley, and Brent Breithaupt for their very helpful comments on this manuscript. Finally, I am very grateful to my wife for her patience and encouragement.

Table 16.4. The average total track length (Ø LIII) of each of the five ornithischian trackways and the average stride lengths were used to calculate hip heights and speed of the trackmaker

Trackway	Ø LIII (cm)	Ø Pace (cm)	Ø Stride (cm)	Hip Height 4	Hip Height 4.6	SL/h 4	SL/h 4.6	Ø v (km/h) 4	Ø v (km/h) 4.6
1	12.575	78.75	155.67	0.50	0.58	3.09	2.69	14.60	12.78
2	13.46	76.35	147.77	0.54	0.62	2.74	2.39	12.67	11.11
3	13.84	68.82	135.22	0.55	0.64	2.44	2.12	10.85	9.52
4	13.1	—	194	0.52	0.60	3.70	3.22	19.35	16.92
5	13.95	—	164.75	0.56	0.64	2.95	2.57	14.37	12.57

Notes: The conversion factor 4 was taken from Alexander (1976) and the conversion factor 4.6 was taken from Thulborn (1990) for small biped dinosaurs. The relative stride lengths (SL/h) show that the trackmakers have all been trotting (>2.0) or running (>2.9).

APPENDIX 16.1: DESCRIPTION OF THE POTENTIAL SMALL ORNITHISCHIAN TRACKS OF THE CHICKEN YARD TRACKSITE

Trackway 1 consists of five consecutive tracks. Overall dimensions of individual tracks as well as pace and stride are almost identical (Table 16.1). The first track (1/1) is partially overprinted by larger tracks, so that the distal ends of both lateral digits are missing. A low displacement rim is present around the third digit and the slightly asymmetrical and wide metatarsal imprint, respectively. The distal third of the third digit imprint is the deepest part of the track. The second track (1/2) is of very poor preservation, because it is disturbed by a large track on its left side, it is very shallow, and its outline is very weakly visible. According to its position, 1/2 is a right track and the only preserved outlines belong to the distal ends of the weakly separated third and fourth digits. The third track (1/3) is reasonably well preserved in comparison to the former two tracks, because all three digits are visible and contain two individual pad imprints each (Figs. 16.4A, 16.4B). Digits II and IV are of roughly the same length and divarication angle. The second digit is distinctly separated from the metatarsal imprint, whereas the fourth digit merges to a slightly concave outline with the latter. The fourth track (1/4) is slightly disturbed by another track on its left side but it is otherwise of good preservation (Figs. 16.4C, 16.4D). The relative lengths of the digits, the divarication angles, and the total length/width ratio (slightly longer than wide; see Table 16.1) are similar to the third track. Differences consist of the much larger width of the third digit imprint and the weak appearance of the fourth digit. The fifth track (1/5) is of equal quality to those of the third and fourth digits and all relevant features are also comparable (Figs. 16.4E, 16.4F). One difference is the posteriorly acute metatarsal imprint.

Trackway 2 consists of five consecutive tracks that are of generally poorer preservation compared to trackway 1. Overall dimensions as well as pace and stride are also slightly more variable (Table 16.1). The first track (2/1) is only recognizable by its three-toed outline. No other internal details are preserved. The anterior separation of the digits is very short and the third digit is thinner than the lateral digits are. Track 2/1 is nearly as wide as long and the lateral digits are almost symmetrical in length and divarication angle. The metatarsal imprint of the second track (2/2) is probably overprinted by a partial other track (no visible outline) because it is deeper and well separated by an internal ridge from the digit imprints. The overall length of the track was still measurable by extrapolating the well-visible course of the lateral outline. The lateral digits are again similar in length and divarication angle (Figs. 16.4G, 16.4H). The imprint of the fourth digit is distinctly tapering, although a separate claw mark is absent. The outline of the other digits is blurred and irregular. The third track (2/3) looks more slender than the other tracks of trackway 2 (Figs. 16.4I, 16.4J). This is in part caused by remaining sediment infill at the rear of the fourth digit imprint and by the smaller divarication angles (Table 16.1). The digit imprints are otherwise well separated from each other by smooth, transitional swellings. The fourth digit is the only imprint, which is fused with the acute but rounded metatarsal imprint. The fourth track (2/4) is similar in symmetry and divarication angles to the former track, but it is generally of inferior preservation (Figs. 16.4K, 16.4L). It is shallower and the outline is more irregular than that of track 2/3, but the main features (long fourth digit united with slender but rounded metatarsal imprint, and second and third digit imprints well separated) are still recognizable. The fifth track (2/5) is even less well preserved because of its shallowness and the very weak outlines of the third digit and of the metatarsal imprint. The best preserved parts are the imprints of the lateral digits, which are of similar length and divarication angle. The latter is significantly wider than in the other tracks of trackway 2 (Table 16.1).

Trackway 3 consists of seven tracks (3/1–3/8), in which the first six tracks are in succession and one track is missing between the sixth and eighth tracks. Overall dimensions and pace are again roughly similar between the individual tracks. The first track (3/1) has well-separated digital imprints with the second digit slightly shorter than the fourth one. The former shows in addition a bulgy proximal portion, whereas the distal part is very slender (Fig. 16.5A). The third digit is unusually long in comparison, but the left track wall is much less inclined than the right wall, which is almost perpendicular to the ground. The divarication angles of both sides are comparable. The deepest parts are within the second and third digital imprint, whereas the shallower fourth digit and the metatarsal imprint are generally less well preserved. The second track (3/2) is slightly rotated outward from the midline and overall weakly imprinted. The second digit is only sparsely preserved but still visible. The divarication angle to the third digit is much smaller than the one between the latter and the fourth digit. The third track (3/3) is the best preserved within trackway 3 (Fig. 16.5B). All digits left imprints with clear outlines. The third digit is well separated from the others by a distinct rim. The second digit is divided in two pads. As in the other tracks, the second digit is also closer to the third digit than to the more spread-out fourth digit. The metatarsal imprint is posteriorly extended and acute. The fourth track (3/4) is less well preserved, because the fourth digit is partially overprinted by another large track. The other features are similar to the former tracks with a slender third digit, which is separated by a weak displacement rim, and an acute but rounded metatarsal imprint. The second digit and metatarsal imprint of the fifth track (3/5) are also very shallowly imprinted with the outline almost absent. The imprints of the third and fourth digits are well visible, suggesting no significant deviations from the usual features in this trackway. The sixth track (3/6) is also difficult to recognize due to its shallow impression and an open fissure grazing its almost circular metatarsal imprint (Fig. 16.5C). The relatively well-preserved imprint of the third digit is divided in two digital pads. The fourth digit reaches as far distally as the mid length of the proximal pad of the third digit. The presence of the second digit is only assumable. The last observable track (3/8) is also a right one because the seventh track is not preserved. Its metatarsus is overprinted, but the three digits are well visible (Fig. 16.4D). The length of the lateral digits is similar, but the fourth digit is wider divaricated than is the second digit as in some other tracks of this trackway. Two pads are present in the imprints of the third and fourth digits, respectively. The distal ends of all three digits are blunt.

Trackway 4 consists of three left tracks with all tracks of the right side missing. Their affiliation to the trackway is further evident by the shared orientation, size, and comparable stride lengths (Table 16.1, Fig. 16.2). However, the preservation is rather poor. The first track (4/1) is the best preserved with imprints of all three digits and the metatarsus (Fig. 16.5E). The latter is rounded and slightly acute and separated from the third and second digits by low elevations. The imprints of the digits are deepest in the distal half, respectively. The lateral track wall of the second digit is collapsed. All digits are also wider distally than proximally. The third and

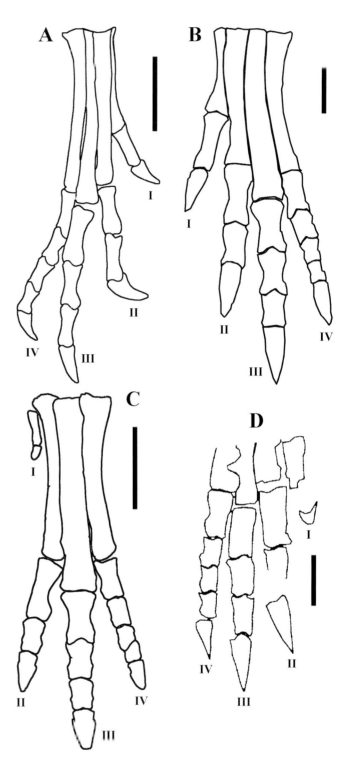

16.11. Sketches of foot skeletons representing four ornithischian clades, which are considered as possible trackmakers of the respective small tracks on the Chicken Yard Tracksite. (A) *Heterodontosaurus tucki*, scale bar = 3 cm (modified from Sereno, 2012:fig. 71D). (B) *Hypsilophodon foxii*, scale bar = 2 cm (modified from Galton, 2009:fig. 4L). (C) *Eousdryosaurus nanohallucis*, scale bar = 5 cm (modified from Escaso et al., 2014:fig. 6). (D) *Stenopelix valdensis*, scale bar = 2 cm (modified from Butler and Sullivan, 2009:fig. 6).

fourth digits taper distally, but a separate claw mark is absent. The second track (4/3) has a poorly defined outline showing all three digits with different lengths. The second digit is the shortest and the third by far the longest. The deepest parts are again the distal portions of the digits, but the interior of the track is generally more irregular in depth. The shape of the metatarsal imprint and the distal ends of the digits are similar to the first track. The third track (4/5) preserves only the imprints of the third and fourth digits, whereas the second digit is only weakly indicated and flat.

Trackway 5 consists of five tracks of different preservational conditions ranging from reasonably good to poor preservation. None of the tracks are consecutive, with one to three tracks missing in between, so that their affiliation to a single trackway is treated as tentative. However, the shared direction is conspicuous (Fig. 16.2) and the similarity in morphology and size between the individual tracks is striking. The first track (5/1) is reasonably well preserved with imprints of all three digits and of the metatarsus (Fig. 16.5F). There is even a crescent-shaped imprint visible in front of the second digit, which may represent a blunt claw mark. The fourth digit is widely separated from the third digit, whereas the second digit is only separated from the latter at its distal tip, forming a merged large imprint. The posterior outline of the metatarsal imprint is wide. The second track (5/3) is very shallowly imprinted but its outline is very similar to the first track regarding the asymmetric divarication angles of the lateral digits and the length ratio between the third and fourth digits. It is only slightly smaller in absolute dimensions. The third (5/6) is a right track located distally relative to two missing tracks. It is poorly preserved because most of it is overprinted by a large theropod track anteriorly. Thus, only the posterior outline comprising the external track walls

of the lateral digits and the metatarsus are still present (Fig. 16.2). The fourth track (5/8) is better preserved with three recognizable digits, in which the fourth digit is again more widely separated from the third digit than the second one (Fig. 16.5G). The latter is also more shallowly imprinted and more slender than the rest. The comparatively longer third digit distinguishes this track from the other ones of this trackway. The last track (5/12), a left one, skips even three missing tracks. It is remarkably similar to most of the other tracks regarding the asymmetrical divarication angles, the general outline, and the proportions (Fig. 16.5H). The three digits and the metatarsus are all preserved and well separated from each other. The lateral digits are of the same length and the distal ends of the digits as well as the metatarsal imprint are rounded.

The following tracks are isolated finds. Track 6 is a shallow imprint with three digits preserved (Fig. 16.5I). The third and fourth digits are well separated from each other. All digits are slender and the lateral digits are roughly symmetrical in length and divarication angle. The metatarsal imprint is very shallow, small, and slightly acute posteriorly. Track 7 has an incomplete second digit but the remaining parts are preserved. The track has a shallow metatarsal imprint, whereas the third and fourth digits are slightly deeper imprinted. The metatarsal imprint is acute and posteriorly slightly offset, the fourth digit is very slender, and the divarication angle is comparable to the other described tracks (Table 16.1). Track 7 is only slightly longer than wide. Track 8 has very shallow lateral digits and metatarsal imprint with very weakly preserved outlines, respectively. The third digit is slender, well separated from the lateral digits, and distally with the deepest part of the track. The track is symmetrical and has a consistent posterior metatarsal border.

REFERENCES

Alexander, R. M. 1976. Estimates of the speed of dinosaurs. Nature 261: 129–130.

Azuma, Y., R. Li, P. J. Currie, Z. Dong, M. Shibata, and J. Lu. 2006. Dinosaur footprints from the Lower Cretaceous of Inner Mongolia, China. Memoir of the Fukui Prefectural Dinosaur Museum 5: 1–14.

Barrett, P. M., R. J. Butler, R. J. Twitchett, and S. Hutt. 2011. New material of Valdosaurus canaliculatus (Ornithischia: Ornithopoda) from the Lower Cretaceous of southern England. Special Papers in Palaeontology 86: 131–163.

Belvedere, M., P. Mietto, and S. Ishigaki. 2010. A Late Jurassic diverse ichnocoenosis from the siliciclastic Iouaridène Formation (Central High Atlas, Morocco). Geological Quarterly 54: 367–380.

Brown, C. M., D. C. Evans, N. E. Campione, L. J. O'Brien, and D. A. Eberth. 2013. Evidence for taphonomic size bias in the Dinosaur Park Formation (Campanian, Alberta), a

model Mesozoic terrestrial alluvial-paralic system. Palaeogeography, Palaeoclimatology, Palaeoecology 372: 108–122.

Butler, R. J., and R. M. Sullivan. 2009. The phylogenetic position of the ornithischian dinosaur Stenopelix valdensis from the Lower Cretaceous of Germany and the early fossil record of Pachycephalosauria. Acta Palaeontologica Polonica 54(1): 21–34.

Butler, R. J., L. Jin, J. Chen, and P. Godefroit. 2011. The postcranial osteology and phylogenetic position of the small ornithischian dinosaur Changchunsaurus parvus from the Quantou Formation (Cretaceous: Aptian-Cenomanian) of Jilin Province, northeastern China. Palaeontology 54: 667–683.

Butler, R. J., L. B. Porro, P. M. Galton, and L. M. Chiappe. 2012. Anatomy and cranial functional morphology of the small-bodied dinosaur Fruitadens haagarorum from the Upper

Jurassic of the USA. PLoS One 7(4): e31556. doi:10.1371/journal.pone.0031556.

Butler, R. J., P. M. Galton, L. B. Porro, L. M. Chiappe, D. M. Henderson, and G. M. Erickson. 2010. Lower limits of ornithischian body size inferred from a diminutive new Upper Jurassic heterodontosaurid from North America. Proceedings of the Royal Society B 277: 375–381.

Carpenter, K., and Y. Wilson. 2008. A new species of Camptosaurus (Ornithopoda: Dinosauria) from the Morrison Formation (Upper Jurassic) of Dinosaur National Monument, Utah, and a biomechanical analysis of its forelimb. Annals of the Carnegie Museum 76: 227–263.

Castanera, D., B. Vila, N. L. Razzolini, P. L. Falkingham, J. I. Canudo, P. L. Manning, and A. Galobart. 2013. Manus track preservation bias as a key factor for assessing trackmaker identity and quadrupedalism in basal ornithopods.

PLoS One 8(1): e54177. doi:10.1371/journal.
pone.0054177.

Chinnery-Allgeier, R. J., and J. I. Kirkland. 2010.
An update on the paleobiogeography of cera-
topsian dinosaurs; pp. 387–404 in M. J. Ryan,
B. J. Chinnery-Allgeier, and D. A. Eberth (eds.),
New Perspectives on Horned Dinosaurs: The
Royal Tyrrell Museum Ceratopsian Symposium.
Indiana University Press, Bloomington, Indiana.

Choiniere, J. N., C. A. Forster, and W. J. de Klerk.
2012. New information on *Nqwebasaurus
thwazi*, a coelurosaurian theropod from the
Early Cretaceous Kirkwood Formation in
South Africa. Journal of African Earth Sciences
71–72: 1–17.

Cotton, W. D., J. E. Cotton, and A. P. Hunt. 1998.
Evidence for social behavior in ornithopod dino-
saurs from the Dakota Group of northeastern
New Mexico, U.S.A. Ichnos 6(3): 141–149.

Escaso, F., F. Ortega, P. Dantas, E. Malafaia, B.
Silva, J. M. Gasulla, P. Mocho, I. Narváez, and
J. L. Sanz. 2014. A new dryosaurid ornithopod
(Dinosauria, Ornithischia) from the Late Jurassic
of Portugal. Journal of Vertebrate Paleontology
34(5): 1102–1112.

Farlow, J. O., and M. G. Lockley. 1993. An
osteometric approach to the identification of
the makers of early Mesozoic tridactyl dinosaur
footprints; pp. 123–131 in S. G. Lucas and M.
Morales (eds.), The Nonmarine Triassic. Bulletin
3. New Mexico Museum of Natural History and
Science, Albuquerque, New Mexico.

Farlow, J. O., and R. E. Chapman. 1997. The scientific study of dinosaur footprints; pp. 519–553
in J. O. Farlow and M. K. Brett-Surman (eds.),
The Complete Dinosaur. Indiana University
Press, Bloomington, Indiana.

Farlow, J. O., R. E. Chapman, B. H. Breithaupt,
and N. Matthews. 2012. The scientific study of
dinosaur footprints; pp. 713–760 in M. K. Brett-
Surman, T. R. Holtz, and J. O. Farlow (eds.),
The Complete Dinosaur. 2nd edition. Indiana
University Press, Bloomington, Indiana.

Foster, J. R., and M. G. Lockley. 1995. Tridactyl
dinosaur footprints from the Morrison
Formation (Upper Jurassic) of northeast
Wyoming. Ichnos 4: 35–41.

Galton, P. M. 1974. The ornithischian dinosaur
Hypsilophodon from the Wealden of the Isle
of Wight. Bulletin of the British Museum of
Natural History. Geology 25(1): 1–152.

Galton, P. M. 1981. *Dryosaurus*, a hypsiloph-
odontid dinosaur from the Upper Jurassic of
North America and Africa. Postcranial skeleton.
Paläontologische Zeitschrift 55(3/4): 271–312.

Galton, P. M. 2009. Notes on Neocomian
(Lower Cretaceous) ornithopod dinosaurs
from England—*Hypsilophodon, Valdosaurus,
'Camptosaurus,' 'Iguanodon'*—and referred
specimens from Romania and elsewhere. Revue
de Paléobiologie 28: 211–273.

Galton, P. M. 2012. *Hypsilophodon foxi* and other
smaller bipedal ornithischian dinosaurs from
the Lower Cretaceous of southern England; pp.
225–282 in P. Godefroit and O. Lambert (eds.),
Bernissart Dinosaurs and Early Cretaceous
Terrestrial Ecosystems. Indiana University Press,
Bloomington, Indiana.

Galton, P. M., and H. P. Powell, H. 1980. The
ornithischian dinosaur *Camptosaurus prest-
wichii* from the Upper Jurassic of England.
Palaeontology 23: 411–443.

Gierlinski, G. D. 2009. A preliminary report on
new dinosaur tracks in the Triassic, Jurassic
and Cretaceous of Poland, pp. 75–90 in IV
Jornadas Internacionales sobre Paleontologia
de Dinosaurios y su Entorno, Burgos, Spain,
September 13–15, 2007.

Gierlinski, G. D., and K. Sabath. 2008.
Stegosaurian footprints from the Morrison
Formation of Utah and their implications for
interpreting other ornithischian tracks. Oryctos
8: 29–46.

Gierlinski, G. D., G. Niedzwiedzki, and P. Nowacki.
2009. Small theropod and ornithopod
footprints in the Late Jurassic of Poland. Acta
Geologica Polonica 59: 221–234.

Gilmore, C. W. 1915. Osteology of *Thescelosaurus*,
an orthopodous dinosaur from the Lance
Formation of Wyoming. Proceedings of the
United States National Museum 49: 591–616.

Hailu, Y., and P. Dodson. 2004. Basal Ceratopsia;
pp. 478–493 in D. B. Weishampel, P. Dodson,
and H. Osmólska (eds.), The Dinosauria. 2nd
edition. University of California Press, Berkeley,
California.

Hammer, Ø., and D. A. T. Harper. 2006.
Paleontological Data Analysis. Blackwell
Publishing, Oxford, U.K., 351 pp.

Hammer, Ø., D. A. T. Harper, and P. D. Ryan.
2001. PAST: Paleontological Statistics software
package for education and data analysis.
Palaeontologia Electronica 4(1): 1–19.

Heinrich, R. E., C. B. Ruff, and D. B. Weishampel.
1993. Femoral ontogeny and locomotor
biomechanics of *Dryosaurus lettowvorbecki*
(Dinosauria, Iguanodontia). Zoological Journal
of the Linnean Society 108: 179–196.

Henderson, D. M. 2006. Simulated weathering of
dinosaur tracks and the implications for their
characterization. Canadian Journal of Earth
Sciences 43: 691–704.

Hornung, J. J., and M. Reich. 2012. Excursion
Guide C2: dinosaur tracks from the Berriasian
Obernkirchen Sandstone on exhibit at the
Göttingen University Geopark; pp. 169–187
in A. Richter and M. Reich (eds.), Dinosaur
Tracks 2012. An International Symposium,
Obernkirchen, April 14–17, 2011, Abstract
Volume and Field Guide to Excursions.
Universitätsverlag, Universitätsdrucke,
Göttingen, Germany.

Hornung, J. J., A. Böhme, and M. Reich. 2012.
Excursion Guides A2: Harrl hill near Bückeburg;
pp. 101–112 in A. Richter and M. Reich
(eds.), Dinosaur Tracks 2012. An International
Symposium, Obernkirchen, April 14–17, 2011,
Abstract Volume and Field Guide to Excursions.
Universitätsverlag, Universitätsdrucke,
Göttingen, Germany.

Hornung, J. J., A. Böhme, T. van der Lubbe,
M. Reich, and A. Richter. 2012. Vertebrate
tracksites in the Obernkirchen Sandstone (late
Berriasian, Early Cretaceous) of northwest
Germany—their stratigraphical, palaeogeographical, palaeoecological, and historical
context. Paläontologische Zeitschrift 86:
231–267.

Hübner, T. R. 2011. Ontogeny in *Dysalotosaurus
lettowvorbecki*. Ph.D. dissertation, Ludwig-
Maximilians-Universität München, Munich,
Germany, 338 pp.

Jackson, S. J., M. A. Whyte, and M. Romano.
2009. Laboratory-controlled simulations of
dinosaur footprints in sand: a key to understanding vertebrate track formation and
preservation. Palaios 24: 222–238.

Kim, J. Y., M. H. Lockley, H. M. Kim, J. D. Lim,
and K. S. Kim. 2009. New dinosaur tracks from
Korea, *Ornithopodichnus masanensis* ichnogen.
et ichnosp. nov. (Jindong Formation, Lower
Cretaceous): implications for polarities in ornithopod foot morphology. Cretaceous Research
30: 1387–1397.

Li, R., M. G. Lockley, M. Matsukawa, K. Wang,
and M. Liu. 2011. An unusual theropod track
assemblage from the Cretaceous of the
Zhucheng area, Shandong Province, China.
Cretaceous Research 32: 422–432.

Li, J., M. G. Lockley, Z. Yuguang, H. Songmei,
M. Masaki, and B. Zhiqiang. 2012. An
important ornithischian tracksite in the Early
Jurassic of Shenmu region, Shaanxi, China.
Acta Geological Sinica 86(1): 1–10. doi:
10.1111/j.1755–6724.2012.00606.x.

Lockley, M. G. 1991. Tracking Dinosaurs: A
New Look at an Ancient World. Cambridge
University Press, Cambridge, UK.

Lockley, M. G. 2000. Philosophical perspectives on
theropod track morphology: blending qualities
and quantities in the science of ichnology. Gaia
15: 279–300.

Lockley, M. G. 2009. New perspectives on
morphological variation in tridactyl foot prints:
clues to widespread convergence in developmental dynamics. Geological Quarterly 53(4):
415–432.

Lockley, M. G., and M. Matsukawa. 1999.
Some observations on trackway evidence for
gregarious behavior among small bipedal
dinosaurs. Palaeogeography, Palaeoclimatology,
Palaeoecology 150: 25–31.

Lockley, M. G., R. T. McCrea, and M. Matsukawa.
2009. Ichnological evidence for small quadrupedal ornithischians from the basal Cretaceous
of SE Asia and North America: implications for
a global radiation; pp. 255–269 in E. Buffetaut,
G. Cuny, J. Le Loeuff, and V. Suteethorn (eds.),
Late Palaeozoic and Mesozoic Ecosystems in SE
Asia. Special Publications 315. The Geological
Society, London, U.K.

Lockley, M. G., V. F. Santos, C. Meyer, and A.
Hunt. 1998. A new dinosaur tracksite in
the Morrison Formation, Boundary Butte,
Southeastern Utah; pp. 317–330 in K.
Carpenter, D. Chure, and K. Kirkland (eds.),
The Upper Jurassic Morrison Formation: An
Interdisciplinary Study. Modern Geology 23(2).

Lockley, M. G., R. Li, M. Matsukawa, L. Xing,
J. Li, M. Liu, and X. Xing. 2013. Tracking
the yellow dragons: implications of China's
largest dinosaur tracksite (Cretaceous of
the Zhucheng area, Shandong Province,
China). Palaeogeography, Palaeoclimatology,
Palaeoecology 423: 62–70.

Lubbe, van der, T., A. Richter, and A. Böhme.
2009. Velociraptor's sisters: first report of
troodontid tracks from the Lower Cretaceous
of northern Germany. Journal of Vertebrate
Paleontology 29(3): 194A.

Manning, P. L. 2004. A new approach to the
analysis and interpretation of tracks: examples
from the Dinosauria; pp. 93–123 in McIlroy,
D. (ed.), The Application of Ichnology to
Palaeoenvironmental and Stratigraphic Analysis.

Special Publications 228. Geological Society, London, U.K.

Martin, A. J., T. H. Rich, M. Hall, P. Vickers-Rich, and G. Vazquez-Prokopec. 2011. A polar dinosaur-track assemblage from the Eumeralla Formation (Albian), Victoria, Australia. Alcheringa 36(2): 171–188. doi:10.1080 /03115518.2011.597564.

Marty, D., A. Strasser, and C. A. Meyer. 2009. Formation and taphonomy of human footprints in microbial mats of present-day tidal-flat environments: implications for the study of fossil footprints. Ichnos 16(1–2): 127–142.

Mateus, O., and J. Milàn. 2008. Ichnological evidence for giant ornithopod dinosaurs in the Upper Jurassic Lourinha Formation, Portugal. Oryctos 8: 47–52.

Milàn, J. 2006. Variations in the morphology of emu (Dromaius novaehollandiae) tracks, reflecting differences in walking pattern and substrate consistency: ichnotaxonomic implications. Palaeontology 49: 405–420.

Milàn, J., and D. B. Loope. 2007. Preservation and erosion of theropod tracks in eolian deposits: examples from the Middle Jurassic Entrada Sandstone, Utah, U.S.A. Journal of Geology 115: 375–386.

Milner, A. R. C., M. G. Lockley, and S. B. Johnson. 2006. The story of the St. George Dinosaur Discovery Site at Johnson Farm: an important new Lower Jurassic dinosaur tracksite from the Moenave Formation of southwestern Utah; pp. 329–345 in J. D. Harris, S. G. Lucas, J. A. Spielmann, M. G. Lockley, A. R. C. Milner, and J. I. Kirkland. (eds.), The Triassic-Jurassic Terrestrial Transition. Bulletin 37. New Mexico Museum of Natural History Science, Albuquerque, New Mexico.

Moratalla, J. J., J. Hernan, and S. Jiménez. 2003. Los Cayos dinosaur tracksite: an overview on the Lower Cretaceous ichno-diversity of the Cameros Basin (Cornago, La Rioja Province, Spain). Ichnos 10: 229–240.

Moratalla, J. J., J. L. Sanz, and S. Jiménez. 1988. Multivariate analysis on Lower Cretaceous dinosaur footprints: discrimination between ornithopods and theropods. Geobios 21(4): 395–408.

Norman, D. B. 2004. Basal iguanodontia; pp. 413–437 in D. B. Weishampel, P. Dodson, and H. Osmólska (eds.), The Dinosauria. 2nd edition. University of California Press, Berkeley, California.

Norman, D. B., and P. M. Barrett. 2002. Ornithischian dinosaurs from the Lower Cretaceous (Berriasian) of England. Special Papers in Palaeontology 68: 161–189.

Olsen, P. E., and E. Rainforth. 2003. The Early Jurassic ornithischian dinosaurian ichnogenus Anomoepus; pp. 314–368 in P. M. Letourneau and P. E. Olsen (eds.), The Great Rift Valleys of Pangea in eastern North America. Columbia University Press, New York, New York.

Pol, D., O. W. M. Rauhut, and M. Becerra. 2011. A Middle Jurassic heterodontosaurid dinosaur from Patagonia and the evolution of heterodontosaurids. Naturwissenschaften 98: 369–379.

Posmoşanu, E. 2003. Iguanodontian dinosaurs from the Lower Cretaceous bauxite site from Romania. Acta Paleontologica Romaniae 4: 431–439.

Richter, A., and A. Böhme. 2016. Too many tracks: preliminary description and interpretation of the diverse and heavily dinoturbated Early Cretaceous 'Chicken Yard' ichnoassemblage (Obernkirchen tracksite, northern Germany); chap. 17 in P. L. Falkingham, D. Marty, and A. Richter (eds.), Dinosaur Tracks: The Next Steps. Indiana University Press, Bloomington, Indiana.

Richter, A., A. Böhme, and T. van der Lubbe. 2009. 'Chicken Run': a new unusual, heavily dinoturbated tracksite from the Lower Cretaceous sandstones of Obernkirchen, northern Germany. Journal of Vertebrate Paleontology 29(3): 171A.

Richter, A., J. J. Hornung, A. Böhme, and U. Stratmann. 2012. Excursion Guide A1: Obernkirchen sandstone quarries: a natural workstone lagerstaette and a dinosaur tracksite; pp. 73–100 in A. Richter and M. Reich (eds.), Dinosaur Tracks 2012. An International Symposium, Obernkirchen, April 14–17, 2011, Abstract Volume and Field Guide to Excursions. Universitätsverlag, Universitätsdrucke, Göttingen, Germany.

Romilio, A., and S. W. Salisbury. 2011. A reassessment of large theropod dinosaur tracks from the mid-Cretaceous (late Albian-Cenomanian) Winton Formation of Lark Quarry, central-western Queensland, Australia: a case for mistaken identity. Cretaceous Research 32: 135–142.

Romilio, A., and S. W. Salisbury. 2014. Large dinosaurian tracks from the Upper Cretaceous (Cenomanian/Turonian) portion of the Winton Formation, Lark Quarry, central-western Queensland, Australia: 3D photogrammetric analysis renders the 'stampede trigger' scenario unlikely. Cretaceous Research 51: 186–207.

Romilio, A., R. T. Tucker, and S. W. Salisbury. 2013. Reevaluation of the Lark Quarry dinosaur Tracksite (late Albian-Cenomanian Winton Formation, central western Queensland, Australia): no longer a stampede? Journal of Vertebrate Paleontology 33(1):102–120.

Ruiz-Omeñaca, J. I., X. Pereda Suberbiola, and P. M. Galton. 2007. Callovosaurus leedsi, the earliest dryosaurid dinosaur (Ornithischia: Euornithopoda) from the Middle Jurassic of England; pp. 3–16 in K. Carpenter (ed.), Horns and Beaks: Ceratopsian and Ornithopod Dinosaurs. Indiana University Press, Bloomington, Indiana.

Ruiz-Omeñaca, J. I., J. I. Canudo, G. Cuenca-Bescós, P. Cruzado-Caballero, J. M. Gasca,

and M. Moreno-Azanza. 2012. A new basal ornithopod dinosaur from the Barremian of Galve, Spain. Comptes Rendue Palevol 11(6): 435–444. doi:10.1016/j.crpv.2012.06.001.

Santa Luca, A. P. 1980. The postcranial skeleton of Heterodontosaurus tucki (Reptilia, Ornithischia) from the Stormberg of South Africa. Annales of the South African Museum 79: 159–211.

Scheetz, R. D. 1999. Osteology of Orodromeus makelai and the phylogeny of basal ornithopod dinosaurs. Ph.D. dissertation, Montana State University, Bozeman, Montana, 186 pp.

Sereno, P. C. 2012. Taxonomy, morphology, masticatory function and phylogeny of heterodontosaurid dinosaurs. ZooKeys 226: 1–225.

Stanford, R., R. E. Weems, and M. G. Lockley. 2004. A new dinosaur ichnotaxon from the Lower Cretaceous Patuxent Formation of Maryland and Virginia. Ichnos 11: 251–259.

Thulborn, T. 1990. Dinosaur Tracks. Chapman and Hall, London, U.K., 411 pp.

Thulborn, R. A. 2013. Lark Quarry revisited: a critique of methods used to identify a large dinosaurian track-maker in the Winton Formation (Albian-Cenomanian), western Queensland, Australia. Alcheringa 37(3): 312–330.

Thulborn, R. A., and M. Wade. 1984. Dinosaur trackways in the Winton Formation (mid-Cretaceous) of Queensland. Memoirs of the Queensland Museum 21: 413–517.

Valais de, S. 2011. Revision of dinosaur ichnotaxa from the La Matilde Formation (Middle Jurassic), Santa Cruz Province, Argentina. Ameghiniana 48(1): 28–42.

Xing, L. D., and M. G. Lockley. 2014. First report of small Ornithopodichnus trackways from the Lower Cretaceous of Sichuan, China. Ichnos 21: 213–222.

Xing, L. D., Y. Q. Liu, H. W. Kuang, H. Klein, J. P. Zhang, M. E. Burns, J. Chen, M. W. Wang, and J. Hu. 2014. Theropod and possible ornithopod track assemblages from the Jurassic-Cretaceous boundary Houcheng Formation, Shangyi, northern Hebei, China. Paleoworld 23: 200–208.

Xing, L., M. G. Lockley, J. Zhang, H. Klein, W. S. Persons IV, and H. Dai. 2014. Diverse sauropod-, theropod-, and ornithopod-track assemblages and a new ichnotaxon Siamopodus xui ichnosp. nov. from the Feitianshan Formation, Lower Cretaceous of Sichuan Province, southwest China. Palaeogeography, Palaeoclimatology, Palaeoecology 414: 79–97.

Zheng, X-T., H.-L. You, X. Xu, and X.-M. Dong. 2009. An Early Cretaceous heterodontosaurid dinosaur with integumentary structures. Nature 458: 333–336.

Depositional Environments and Their Influence on the Track Record

4

17.1. The Chicken Yard level at night, photograph from a higher layer on the east side; artificial low-angle light from the northern margin. The surface is moderately to heavily dinoturbated. *Courtesy Tobias Landmann/Schaumburger Zeitung, 2011.*

Too Many Tracks: Preliminary Description and Interpretation of the Diverse and Heavily Dinoturbated Early Cretaceous "Chicken Yard" Ichnoassemblage (Obernkirchen Tracksite, Northern Germany)

17

Annette Richter and Annina Böhme

THE MODERATELY TO HEAVILY DINOTURBATED BERRIA-sian Chicken Yard level from the Obernkirchen tracksite (Lower Saxony, northern Germany) is preliminarily described and analyzed. Its ichnoassemblage is characterized by an extraordinary high track density composed of several different morphotypes and size classes of theropod and ornithopod true tracks with an overall similar preservation quality. The occurrence of didactyl tracks of a new, so far unnamed ichnotaxon that can be attributed to deinonychosaurian dinosaurs is particularly remarkable. Despite the high track density and associated frequent overprinting of tracks, several trackways were identified. Their orientation analysis tends toward a primarily bimodal orientation pattern despite the overall chaotic appearance. Also, the history and development of the term "dinoturbation" and its application to Mesozoic dinosaur tracksites are reviewed, and the different factors and scenarios that may have led to high dinoturbation in general and at the Chicken Yard level in particular are discussed together with some recommendations for the analysis of heavily dinoturbated paleosurfaces.

INTRODUCTION

Early Cretaceous (Berriasian) dinosaur tracksites have been known from Lower Saxony, Northwestern Germany, since the 19th century (Struckmann, 1880a, 1880b). Hornung et al. (2012) and Hornung, Böhme, and Reich (2012) have reviewed and comprehensively listed 13 localities with terrestrial deposits of Berriasian age in the County of Lower Saxony. Within a terrestrial succession of mainly limnic claystones and black shales, dinosaur tracks have so far only been discovered in sandstones, which are excellent building stones and have therefore been quarried at different locations for the last 1000 years (Graupner, 1977; Lepper, 1997; Lepper and Richter, 2010). Today, only two main areas are still exploited: the Wesling quarry at Münchehagen (Wings, Lallensack, and Mallison, 2016) and—on a larger scale—the

Obernkirchener Sandsteinbrüche quarries (Obernkirchen sandstone quarries).

The majority of the historical track discoveries are isolated natural track casts on mostly small quarried slabs (Ballerstedt, 1920). However, the modern quarrying activities are extending horizontally and have uncovered large track-bearing paleosurfaces, where dinosaur tracks and trackways are preserved as impressions or negative epirelief (Wings, Broschinski, and Knötschke, 2005).

Generally, a low to moderate dinoturbation consisting of easily identifiable trackways, with isolated tracks (that cannot be assigned to any given trackway) being the exception, is typical for the Berriasian ichnoassemblages of northwest Germany. Ornithopod and sauropod trackways are often aligned parallel and with regular intertrackway spacing (Fischer, 1998; Wings, Broschinski, and Knötschke, 2005) potentially indicating gregariousness, and preferred walking directions can be determined with only a few exceptions where tracks overprint each other. The majority of the trackways are characterized by short stride lengths and are attributed to herbivores, whereas the theropod trackways generally have longer stride lengths and are scarcer. Often the tracks have a considerable depth and most levels also exhibit ripplemarks and occasionally large fossil tree trunks.

Currently, in the Obernkirchen sandstone quarries (named hereafter Obernkirchen tracksite), two levels, an upper level with mainly ichnoassemblage and a lower level with a variety of tridactyl (chiefly theropod) tracks and even didactyl tracks, crop out and have been excavated. The latter was discovered by the authors in 2007 and briefly first mentioned by Richter, Böhme, and van der Lubbe (2009). It is an extraordinary track-bearing level that is moderately to heavily dinoturbated and was named the "Chicken Yard" level (Fig. 17.1).

The term "dinoturbation" was coined by Dodson et al. (1980) and was derived from the term "bioturbation," which is used for sediment disruption caused by diverse invertebrate burrowing organisms. Bioturbation was introduced by the German geologist Rudolf Richter (1936) for "all kinds of displacement within sediments and soils produced by the activity of organisms and plants" (English translation from Flügel, 2004:185). Bioturbation leads to alteration of bedding and lamination structures and contributes to the homogenization of the substrate (Reineck, 1963). The term is meant and used for horizontal (sediment surface) as well as vertical (cross-sections) dimensions of the substrate, the latter being widely applied to thin section analyses. The primary definition and use of the term "bioturbation" arose from marine geosciences (Flügel, 2004). Interestingly, in most of the sediment cases, the focus lies on the vertical component of the textural or fabric changes.

Droser and Bottjer (1986) additionally developed a semi-quantitative "bioturbation index," using the burrow density of substrates for classifying bioturbation fabrics, which they called "ichnofabric." They showed that a differentiation of bioturbation types is possible by defining categories between two extremes: 0% bioturbation (no evidence of biogenic mixing of sediments) and 100% (complete biogenic mixing of sediments; Flügel, 2004). Between these two end members, a continuous spectrum can be subdivided, for instance, in 10% increments.

Seen in the context of the primary term, dinoturbation can be seen as a special case of bioturbation. When introduced first by Dodson et al. (1980), it was regarded from two viewpoints: dinoturbation must have had an impact (1) on the sediment and (2) on the remains of small biota (small animals and plants) because, as the comparison to extant African biota showed, cropping and "trampling" of large vertebrates—mainly herbivores—in certain savannah areas keeps the diversity of microvertebrate remains and their tracks and traces low (Dodson et al., 1980). Accordingly, dinoturbation may also obscure or delete previously left tracks, notably those of smaller vertebrates, as well as body fossils of invertebrates (clams, snails), plants, and even vertebrates (Lockley et al., 1986; Lockley, 1993) at Mesozoic tracksites.

At the time the term was coined, the focus was on the strong horizontal impact of animal activities on the substrate surface. Thus, considering the locomotor activities of dinosaurs on a substrate, dinoturbation was applied mainly to two-dimensional bedding planes. Moderate to heavy dinoturbation is known from many sites, but the majority of ichnoassemblages are characterized by a rather low dinoturbation index.

Lockley and Conrad (1989) proposed a useful approach toward characterizing dinoturbated tracksites by estimating the percentage of tracks per surface unit, which they called the "dinoturbation index," subdivided into three major categories: lightly (0%–33%), moderately (34%–66%), or heavily (67%–100%) trampled, which is comparable to the approach of Droser and Bottjer (1986). Accordingly, the most heavily dinoturbated surfaces would correspond to complete superficial trampling or alteration in a way that most remaining structures (tracks) would be very incomplete and poorly defined, whereas only a few could easily and clearly be identified as dinosaur tracks (Fig. 17.1). Typically, such surfaces are characterized by a very high number of isolated and incomplete, overprinted tracks when compared to the total number of recognizable trackways, which is then typically low because it is so difficult to unambiguously identify single trackways.

As mentioned, a low to moderate dinoturbation is typical for German Berriasian tracksites. The Obernkirchen lower Chicken Yard level strongly differs from all of them (including the Obernkirchen upper level) because of its extraordinarily high track density, being the first site from the "German Wealden" that is heavily dinoturbated. On the excavated area (approximately 400 m²) more than 900 rather complete tracks can be counted. Comparable tracksites such as the Australian Lark Quarry site (Thulborn and Wade, 1984; Romilio, Tucker, and Salisbury, 2013) among others are significant witnesses of extraordinarily high levels of dinosaur activity within the fossil record, and possibly of a longer recording timespan.

At the time of writing, negotiations about the protection of the Chicken Yard level were still ongoing, whereas the upper track-bearing level was already protected and open to the public. A first brief description of the Chicken Yard level and the discussion about the origin of its high dinoturbation index including the importance of directional analysis is the focus of this work.

GEOLOGICAL SETTING

The Obernkirchen tracksite is located within the Obernkirchen Sandstone, a thin subunit of the Bückeberg Formation (informally described as the German Wealden) exposed in the northern and western vicinities of Hannover (Hornung et al., 2012). The absolute age of the Obernkirchen Sandstone is estimated between 142 and 138 Ma, spanning ~3.4 Ma (Hornung and Böhme, 2012).

The Lower Saxony Basin, where the Bückeberg Formation was deposited, was a southern sub-basin of the North German Basin. The prograding deltaic sediments from one of two major river systems indicate the deposition of a barrier (p. 394) and delta system (p. 397) to the West of Hannover (Pelzer, 1998; Hornung and Böhme, 2012), which is believed to have matured the sand grains of the Obernkirchen sandstones. It is likely that the formation of the barrier facies was related to landward transport of sand during storm events, and maybe even longshore currents (Pelzer, 1998). This is also confirmed by the occasional occurrence of large tree trunks in the Obernkirchen Sandstone at the Münchehagen tracksite (back-barrier storm deposits sensu Hornung et al., 2012) and in the Obernkirchen Sandstone from Obernkirchen/Bückeberg (landward deltaic succession sensu Hornung et al., 2012; Richter and Knötschke, pers. obs., August, 2007; Richter et al., 2012). In the Early Cretaceous (during the Berriasian and earliest Valanginian), the Bückeberg Formation was deposited at a paleolatitude of about 32°–33° N (Hornung and Böhme, 2012) under warm and humid subtropical to paratropical conditions with seasonal droughts (Pelzer and Wilde, 1987; Hornung and Böhme, 2012).

MATERIALS AND METHODS

The documentation of the excavated Chicken Yard level was achieved after cleaning the surface and taking off the sedimentary infillings of the majority of tracks. In 2008, high-resolution digital photographs were taken from approximately 3–4 m above the ground in cooperation with the Leibniz Universität Hannover and by using a Canon EOS 400 camera. The photogrammetry software Eos Systems PhotoModeler Pro 5.23 was used to produce (with assistance of manually referenced tie-points) a geometric transformation of 95 photographs into one reference system in order to generate an orthophoto (mosaic of photographs without distortion) as a basis for the final track map. To ensure recognition of the tracks on the orthophoto, the tracks were outlined with chalk before the photographs were taken.

Finally, a classical track outline site map was drawn on the basis of the orthophoto using Adobe Photoshop Elements 7.0, outlining the most complete tracks–approximately 900 and much the same count of partially preserved tracks (~700, only partially documented in the map). Then, 715 of the most complete tracks were selected for the total track directional analyses and are represented in fine-scaled (1°) and linear rose diagrams, constructed with Microsoft Excel and modified with Adobe Photoshop Elements 7.0. Within the 715 analyzed tracks, tridactyl tracks of theropods (424 isolated and within-trackway tracks), medium sized and large iguanodontid ornithopods (196, isolated tracks and tracks of longest trackways), indeterminate medium and large tridactyl tracks (20), deinonychosaurian tracks (55, only tracks from trackways), and small ?ornithopod tracks (20, only tracks from trackways) were considered.

Four simplified rose diagrams show the directions of the most pronounced trackways of tridactyl theropods, medium-sized and large iguanodontid ornithopods, deinonychosaurid dinosaurs, and small ?ornithopods.

Due to the high track density, many of the partially preserved, heavily overprinted tracks and those represented only

by weak toe tip impressions were not considered on the main site map and in the count, but they are given as a separate exemplified area containing all textural characters of the surface in detail.

The best-preserved "elite" tracks and trackways of tridactyl theropods and iguanodontid dinosaurs, with all three toes recognizable and in the best case the metatarsal "heel" region, were selected for measurements. Overprinting, however, leads to lacking parameters even for some of the best-preserved tracks. The following track and trackway parameters were measured in the field using standard rulers, calipers (metric units) and goniometers (e.g., following Romilio and Salisbury, 2011; Hübner, 2016): maximum track length (from central posterior border of the metatarsal imprint and the outmost distal tip of digit III), maximum track width (from outmost left and right track margins), digits I–II–III lengths (from central posterior border of the metatarsal imprint and the outmost distal tip of the digit), depth (from track floor of the anterior-most phalange of digit III to an iron bar laid out across the track), digit divarication angles (between the central posterior border of the metatarsal imprint and the outmost distal tip of each digit), pace, stride (both from corresponding points, usually from and to outmost distal tip of digit III or from and to central posterior border of the metatarsal imprint). Tracks of didactyl theropods and small ?ornithopods were not measured in detail, as these tracks are in the focus of Lockley et al. (2016) and Hübner (2016).

Outline drawings of trackways (all of tridactyl theropods and the longest ones from the medium-sized and large iguanodontid dinosaurs) and some individual elite tracks are superimposed onto the photogrammetric map. The trackways are numbered consecutively (roman numbers: tridactyl theropods; letters: iguanodontid ornithopods) including missing tracks in gaps.

Indeed, it proved problematic to number the tracks at all–the sheer mass made it very difficult to put a logical order into the number rows. The decision was made to use the natural weathering of the sandstone into large slabs (German "Blöcke," B) for numbering each of them (B1, B2, etc.) and then label the tracks within each block in a clockwise direction, starting in "North/300," therefore easily being able to redetect them, resulting in numbers such as B3/27 –block 3, track 27. These numbers will be published in a forthcoming, more detailed publication about the Chicken Yard level and are used here only for three isolated examples not belonging to trackways ("B2e/24," B1a/14," and "B8/2"), located on three different sandstone slabs.

The Eos Systems PhotoModeler Scanner computed three dimensinoal contour line graphics for selected tracks, each calculated out of five photographs taken from different positions with a Canon EOS 550D camera. The graphics were

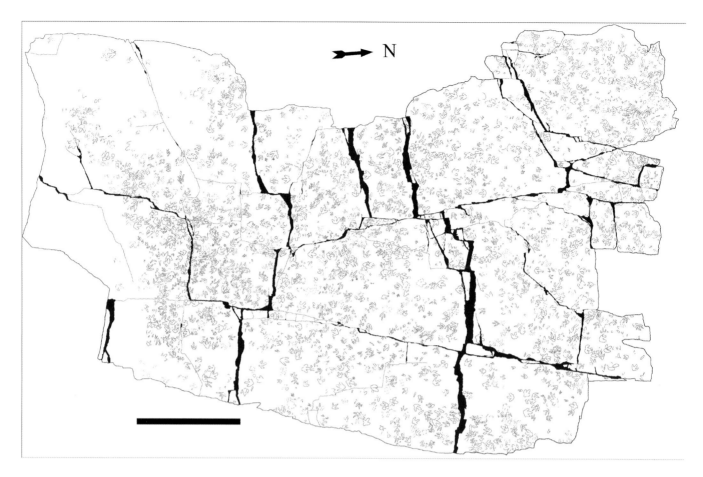

17.2. Sitemap with outline drawings of selected complete and well-preserved dinosaur tracks at the Chicken Yard level, Obernkirchen tracksite, based on the photogrammetric orthophoto. Scale bar: 5 m. *Courtesy Schaumburger Landschaft, 2011.*

revised with Corel Draw X5 and Adobe Photoshop Elements 7.0. For the work with PhotoModeler, calibrating the cameras and lenses used is obligatory.

The size classes for tracks (based on pes length [PL]) follow those cited in Marty, Strasser, and Meyer (2009): minute: PL < 10 cm; small: 10 < PL < 20 cm; medium-sized: 20 < PL < 30 cm; large: 30 < PL < 50 cm. Huge tracks (PL > 50 cm) were not found on the Chicken Yard level.

The dinoturbation index scheme developed by Lockley and Conrad (1989) was applied for the Chicken Yard level.

INTRODUCING THE CHICKEN YARD LEVEL – PRELIMINARY DESCRIPTION AND INTERPRETATION

At present, the Chicken Yard level is approximately 400 m² in size, and it is located on top of a very thick, homolithic and homogenous, fine-grained quartzose sandstone bank of a thickness of approximately 1.6 m. Most of its surface is covered with tracks (Figs. 17.1, 17.2).

Viewed from above, the edge of the overhanging sandstone cliff, which is approximately 6.5 m high, the hundreds of tridactyl tracks look like bird tracks, almost like *Gallus* imprints in a chicken yard. For this reason, one of the museum volunteers coined the nickname Chicken Yard (German: Hühnerhof) and from then on this name has been used for this level.

In the southwestern corner of the Chicken Yard level and along its southern edge, the surface is quite smooth and lacks any tracks, but some neomiodontid clams are preserved here. Neither ripple-marks nor desiccation cracks are preserved on the Chicken Yard level. Some few plant remains (stems and chaff) are scattered on the surface of the Chicken Yard.

TRACK PRESERVATION

Description

The majority of tracks from the Chicken Yard level show a similar quality of preservation. The preservation quality of the tracks is generally high, as most of the tracks preserve claw impressions and many of them even delicate and clear digital pad impressions. At low-angle light conditions, a marked fine-scale alteration of the surface can be observed.

Most of the tracks from the whole area originally had sedimentary infillings. The natural track casts had approximately the same grain size (slightly finer) than the track-bearing sandstone itself. After removal of the track infills, some track floors additionally showed tiny adhesion spikes sensu Jackson, Whyte, and Romano (2009); they will be described in detail elsewhere (Richter et al., in prep.). No mudcracks can be observed on the Chicken Yard level. Some small areas were slightly damaged by the machines of the quarry workers rolling on the surface, originating in artificial "textures."

No evidence for the former presence of microbial mats, such as calcareous "biofilm remains," microbial wrinkle marks (e.g., Dai et al., 2015), or microbial lamination (e.g., Marty, 2008), has so far been found on the Chicken Yard level and underlying and overlying sandstones.

Interpretation

As Hornung et al. (2012) pointed out, it is important to note that the vertebrate (dinosaur) track-bearing deposits of the Obernkirchen Sandstone are all located within a limnic depositional system, in contrast to former publications (e.g., Diedrich, 2004) that have suggested marine influences. Within a cyclic, limnic, depositional environment, the preservation of tracks is common (Souza Carvalho, Borghi, and Leonardi, 2013).

The complete and well-preserved tracks of the Chicken Yard preserve substantial anatomical details. Even if the Chicken Yard tracks do not preserve skin impressions, the fine anatomical details such as digital pad and claw impressions are diagnostic for true tracks, which have not suffered from significant erosion and/or diagenetic modification. Because of the numerous track interferences where complete tracks systematically overprint incomplete tracks and the abundance of distinctive sedimentary infillings even within incomplete tracks, the possibility that incomplete tracks are undertracks can be ruled out.

The sedimentary infillings are comparable to extant tracks and their infillings from Lake Manyara, especially in the vicinity of the lakeshore and the shallow water zone (Cohen et al., 1991). This corresponds well with the observed adhesion spikes and the absence of mudcracks. More or less synchronous sedimentary track infillings originate right after the foot impression when the substrate is still moist to a certain degree (up to saturated; Jackson, Whyte, and Romano, 2009). They are sometimes called 'Plomben' in historical German track literature around 1900. Literally plugs, they are sometimes incorrectly associated with the 'infillings' of layers above the track horizons, originating after the track-making time, building up the next, completely new layer.

Table 17.1. Some representative examples, with focus on track data from continuous trackways; depth from digit III (theropods: distal phalange)

Trackways (number of datasets)	Depth ranges mean: phalange III (cm)	mean digit depth (cm)
I (8)	1.8–4.3	2.7
II (17)	0.8–2.2	1.4
VII (7)	1.3–2.4	2.0
B (6)	2.3–3.5	2.9

Notes: Roman numbers indicate theropod trackways. B is the ornithopod trackmaker. For the position of trackways, see Figure 17.11.

Although there is no evidence for their presence, it cannot be excluded that microbial mats may have played a role during track formation and preservation on the Chicken Yard level, as well as those of other German Berriasian localities with less homogeneous sandstones than at the Chicken Yard level.

TRACK DEPTH

Description

All tracks from the Chicken Yard level are rather shallow (on average ranging in depth between 1.2 to 2.5 cm) (Table 17.1). There is a slight size-independent trend to deeper tracks in the outermost northwestern quadrant of the level (2.5 to 3.8 cm).

No overprinting of deeper tracks by shallower tracks occurs. The majority of the dinosaur tracks have small or no displacement rims, and only in those areas with intensive overprinting, can slight displacement rims be observed.

The track floors are rather even (anteroposteriorly as well as transversal) and show only very weakly (maximum 0.5 to 0.6 cm, mostly not at all) a "classical" deeper anterior part as interpreted as result of the third phase of walking ([1] heeldown, [2] forward rotation, and [3] toe-off, Thulborn and Wade, 1989), and as exemplified by the template experiments of Manning (2004).

Interpretation

Having such flat track floors can be interpreted in different ways. (1) a slippery surface might necessitate a deliberate flat placement of the dinosaur foot (at a high angle, approximately 90°) to prevent instability (Manning, 2004), or (2) a (semi) consolidated, firm underground will not react with a high degree of plasticity (Falkingham et al., 2011; Falkingham, 2014)

As all preserved tracks of the Chicken Yard level are comparably shallow with only a few exceptions, this can be interpreted as a proof of similar to identical substrate conditions

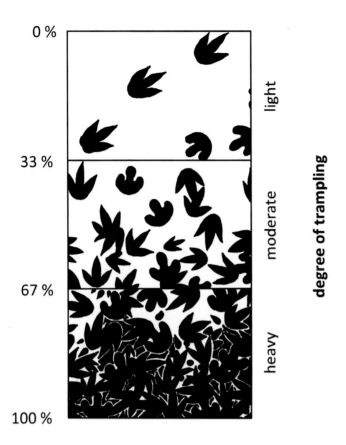

0 %

33 %

67 %

100 %

light

moderate

heavy

degree of trampling

17.3. Dinoturbation index scheme; the track density and associated degree of overprinting are estimated. Drawing after Lockley and Conrad (1989), with schematic tridactyl tracks as observed on the Chicken Yard level.

all over the level (Lockley, 1993) except in the southern part, where no tracks are preserved at all.

Even if substrate conditions were possibly more or less the same with a semiconsolidated firm underground, it should also be kept in mind that the foot impact of heavy animals leads to an additional densification of the substrate (Manning, 2004), which is of importance for areas heavily walked upon. From one walking event to the next, the underground must have become increasingly more dense and firm. Therefore, the dinosaurs must altogether have been "paving" their way through the river delta sediments. The last dinosaurs to step on that layer probably found the best and comparably most solid substrate conditions to walk on, leaving shallow tracks and involving less locomotion energy compared to wet, loose sand (Bates et al., 2013).

The tracks are neither deep enough, nor are they found in channels, to discuss avulsion ("dinovulsion") phenomena such as those known from a Jurassic locality from Utah (Jones and Gustason, 2006). Moreover, computational weathering experiments of Henderson (2006) suggest that it is most unlikely to consider very flat tracks from homolithic rocks to be undertracks. Additionally, the flatter tracks never overprint the deeper ones on the Chicken Yard level, indicating no

complete drying out (at least in the preserved "track recording window").

DINOTURBATION INDEX

Description

The dinoturbation index of the Chicken Yard level is moderate (southern third) to heavy (two northern thirds and middle part of southern area) according to the scheme (Fig. 17.3) developed by Lockley and Conrad (1989). In the southernmost parts, there are no tracks, which represents the "lightly" dinoturbated situation.

A high dinoturbation index and associated degree of overprinting characterizes most areas of the Chicken Yard level, and a small, representative area was chosen to depict the overall density of complete and incomplete (overprinted) tracks as well as impressions that cannot readily be identified as tracks (Fig. 17.4). Roughly estimated, one-third are well-preserved and complete tracks and the remainder are overprinted, incomplete tracks.

Interpretation

An exact count of the incomplete (partially overprinted) tracks was not made, but this seems to be an approach worthwhile for the future work in order to estimate a degree of overprinting. Unless this will be done, the overprinted tracks can be estimated as abundant and with high numbers. Altogether, taking the variable dinoturbation indexes into account, there seems to have been a strong lateral change of either substrate/paleoenvironment or animal behavior.

GROSS TRACK MORPHOTYPES

Due to the overall similar track preservation on the Chicken Yard level, several different morphotypes of theropod and ornithopod tracks can clearly be discriminated. The majority of the tracks are medium-sized to large and primarily tridactyl. There are no tracks from obligatory quadrupedals such as ankylosaurs (in spite of the rediscovery of the *Metatetrapodus*-type tracks by Hornung, König, and Reich (2012) in the Göttingen University collections) or of sauropods, as observed at the Münchehagen tracksite (Fischer, 1998:plate 2). The different morphotypes are briefly described hereafter.

Tridactyl Theropod Tracks

Description Many of these tridactyl tracks show the typical toe pads and some also show claw marks. Different morphotypes can be differentiated, not all of them forming trackways though. Future multivariate analysis (such as in Castanera et al., 2015) will help to show whether there is just preservational variation or taxonomic reason for the differences, and they will be described in detail elsewhere (Richter et al., in prep.). Hereafter, we give descriptions for some representative morphotypes.

Morphotype 1: Mostly represented by single tracks, characterized by a slender appearance of the digit impressions, marked but not broad digital pads, and a digit IV/III length ratio of 0.80. The length of the track figured here is 36.5 cm. (See Fig. 17.5A.)

Morphotype 2. Not only represented by single tracks, but also by one rather long trackway (no. VII) with 7 tracks. They show extremely marked, broad, and very round digital pad impressions. Thus, the trackmaker must have had rather fleshy feet. The digit IV/III length ratio of VII/6 is 0.83, of VII/5 is 0.84. Mean length is approximately 35.5 cm. (See Figs. 17.5B and 17.5E.)

Morphotype 3. Generally not found in trackways, this morphotype can be attributed to a theropod in a heavily dinoturbated zone (northwestern part of the Chicken Yard) and represents the type 3 here. There are only a few other isolated tracks of this morphotype. This is the largest and most robust morphotype. The mean digit IV/III length ratio is 0.80. Mean length is approximately 44.5 cm. (See Fig. 17.5C.)

Morphotype 4. A more slender appearing theropod morphotype with narrow digit impressions, and a digit IV/III length ratio of 0.75, indicating a comparatively longer digit III. Mean length is 30 cm. (See Fig. 17.5D.)

Interpretation For the medium-sized and some of the large theropod tracks, slightly younger theropod dinosaurs from the Isle of Wight (Wessex Formation, Valanginian and Hauterivian), such as *Neovenator* or *Baryonyx* and even the more gracile *Eotyrannus* might be considered as possible trackmakers insofar as both localities are geographically and temporally relatively close to each other.

The enigmatic *Bueckeburgichnus* from the German Berriasian (ichnotaxonomy is discussed in Hornung, Böhme, and Reich, 2012) shows hallux impressions, which seem unlikely to be preserved at all within the depositional conditions of the Chicken Yard level with its comparably shallow tracks. However, the gross overall morphology of all tracks of the basic theropod morphotype resembles *Megalosauripus* rather than *Bueckeburgichnus*, as the latter is characterized by more stout digits II and III and a less robust digit IV. A more detailed ichnotaxonomic discussion for typical Early Cretaceous medium- to large-sized theropods of Germany will be the focus of a forthcoming work about the Obernkirchen theropods (Richter et al., in prep.).

17.4. Detail of a heavily dinoturbated area from the northwestern margin of the Chicken Yard level (to identify on the trackway map, see marked zone in Fig. 17.11). The dinoturbation index can be estimated at >80%. Scale bar: 1 m.

Didactyl Theropod Tracks

Description On the Chicken Yard level, the most remarkable discovery was didactyl tracks (Fig. 17.6) left by deinonychosaurian dinosaurs (van der Lubbe, Richter, and Böhme, 2009). This striking and completely new track morphotype is characterized by the lack of a proper digit II impression, with only a small but distinctive rounded groove around the area of the toe articulation of digit II (see van der Lubbe, Richter, and Böhme, 2009; Lockley, et al., 2016). The mean pes lengths of the didactyl tracks from the seven trackways range between 13 and 24.5 cm, with a mean length of 20 cm.

Interpretation The good preservation quality and quantity (systematic occurrence in seven trackways each with

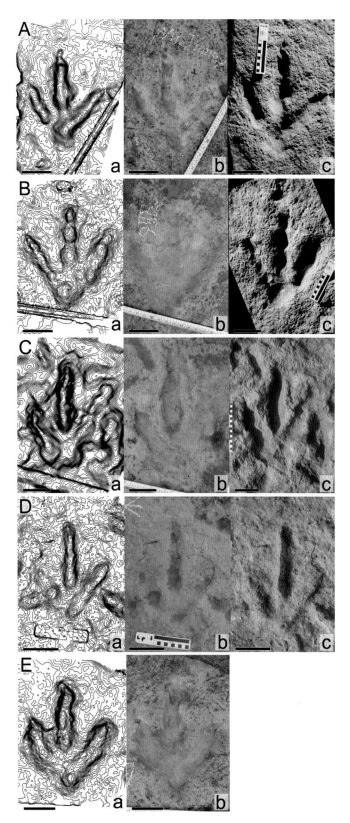

17.5. Four track morphotypes of tridactyl theropods from the Chicken Yard level. (A, B) Tracks are shown as (a) contour line models, (b) photogrammetric photographs, and (c) grazing light photographs. (C, D) Tracks are shown as (a) contour line models, (b) basic photogrammetric photographs, and (c) additional daylight photographs. (E) Track is shown as (a) a contour line model and (b) a photogrammetric photograph. (A) Morphotype 1, medium-sized "standard type" (nicknamed "Mr. Jimmy" during the Obernkirchen Symposium 2011; dextral [dex.]; B2e/24). (B) Morphotype 2: digital pad type (trackway VII/6, sinistral]). (C) Morphotype 3, large and robust theropod track (dex.; B1a/14). (D) Morphotype 4, gracile theropod (dex.; slight overprinting of digit III by another track; B8/2). (E) Morphotype 2: second specimen from the digital pad type (trackway VII/5, dex.). All tracks pictured are localized on the trackway map, Figure 17.11; all black scale bars 10 cm; all line distances 0.1 cm. *Grazing light photographs are courtesy of Oliver Gerke, 2010.*

more than three tracks) of these didactyl tracks exclude the possibility that they are all preservational artifacts of tracks left by a trackmaker that usually leaves tridactyl tracks or are due to a paleopathology on both feet of a habitually tridactyl trackmaker. The first scenario was observed in walking experiments with emus by Milàn (2006): emus tended to lower the load on digit II on firm substrates, thus producing didactyl tracks. This can also be observed on recent tidal flats with extant sea gulls when the typical stiff, hard state of sand appears during the approaching phase of the lowering tide (on a very homogenous quartz grain beach), thus allowing producing pseudodidactyl tracks with only very weak traces of the second toes by healthy, normal gull feet (Richter, pers. obs., June, 2010; Falkingham, et al., 2010:fig. 5) (Figure 17.7).

In the case of the Chicken Yard level, this scenario can be excluded, as several long trackways cross the entire area. Also, these didactyl tracks always and unambiguously exhibit a distinct, circular impression below the area of toe II (van der Lubbe, Richter, and Böhme, 2009; Lockley, et al., 2016). The didactyl tracks have been assigned to the Troodontidae by van der Lubbe, Richter, and Böhme (2009) and not to the more frequent Dromaeosauridae (see also Lockley, et al., 2016).

Although the absolute lengths of the tracks assigned to Troodontidae are much shorter than those from other theropods, their trackmakers must have been quite large by standards of their family (2–3 m body length). The tracks provide evidence for the existence of these sickle-clawed dinosaurs in the Berriasian of northwest Germany, but skeletal remains are still unknown. For didactyl dromaeosaurs, in the Early Cretaceous of Central Europe, there is only one skeletal taxon known from velociraptorine teeth from the Isle of Wight (Sweetman, 2004).

It is worth mentioning that one right track and one left track of a troodontid are located almost directly beside each other, evoking the picture of a "standing" animal. Their heel (= metatarsal pad) impressions are longer than those from

17.6. Grazing light photograph of a troodontid, didactyl, pes track (dextral) from the Chicken Yard level. Joint print at toe II marked by white arrow. Localized on the trackway map, Figure 17.11. Scale bar: 10 cm. *Courtesy Oliver Gerke, 2010.*

all other didactyl tracks. They were shortly mentioned in van der Lubbe, Richter, and Böhme (2009), and tridactyl tracks with similar, elongated "metatarsal impressions" are generally interpreted as "crouching traces" (Gierlinski, 1994; Milàn, Loope, and Bromley, 2008; Milner, et al., 2009).

Medium-Sized to Large Ornithopod Tracks

Description These are medium-sized to large tridactyl pes tracks (Fig. 17.8) with petaloid-shaped and rounded toes (broadly recognizable rounded nails) and a short to mid-sized "metatarsal heel" impression. The pes lengths tend to be slightly longer than the widths. Trackway B (Fig. B18), for example, shows a mean length of 40.5 cm and a mean width of 36.2 cm. Mean pace length is 100.7 cm and mean stride length about 203.3 cm. The pes tracks are commonly associated with oval- to kidney-shaped manus tracks, located slightly in front of the pes tracks. More tracks from the Chicken Yard level still have to be measured and additionally investigated with multivariate analysis (compare Hornung, et al., 2016).

Interpretation To date, these tracks match most closely to the ichnogenus *Iguanodontipus* that is commonly assigned to iguanodontid dinosaurs (Díaz-Martínez et al., 2015; Hornung, et al., 2016; former *Amblydactylus*, see discussion in Lockley, et al., 2014). In the Early Cretaceous, *Iguanodontipus* tracks are commonly found in Europe (England, Portugal, Spain, Switzerland). However, there is also an ongoing debate to which dinosaurs (*Iguanodon*, *Mantellisaurus*, *Delapparentia*) these tracks are best assigned to (Santos, Callapez, and Rodrigues, 2012).

Maybe, these dinosaurs needed quarters or functional This would be corroborated by the quadrupedal stance of the iguanodontians, proven by the existence of manus imprints, using their forefeet for stabilization (as described by Wright, 1999).

The presence of shallow manus tracks indicates that there was only a limited influence of erosion on the Chicken Yard level, as the usual preservational bias for ankylopollexian manus impressions would greatly hamper their existence when minimal erosion was present or as undertracks on lower levels (Castanera et al., 2013).

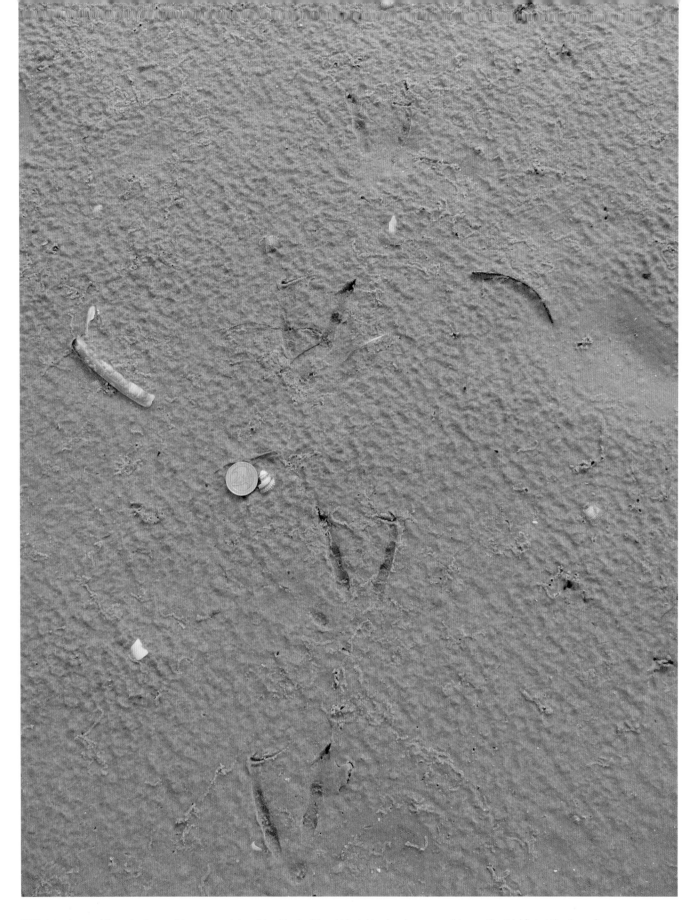

17.7. A trackway of the extant seagull *Larus argentatus*. A healthy-looking, tridactyl trackmaker was walking on the tidal flat at the time of approaching low tide, producing a pseudodidactyl trackway. Island of Langeoog, German North Sea Coast, 2010. Scale: 1-Euro coin = 2.3 cm.

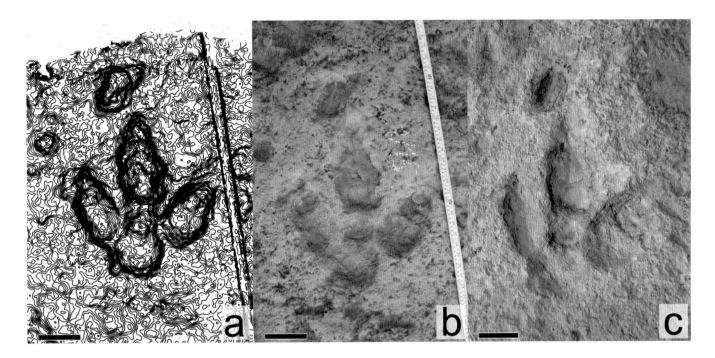

17.8. A pes and manus track of the *Iguanodontipus*-type trackway B (localized on the trackway map, Fig. 17.11; sinistral). (A) Contour line model. (B) Basic photogrammetric photograph. (C) Field photograph during dusk. Scale bar: 10 cm; distance between neighboring contour lines is 0.1 cm.

Small ?Ornithopod Tracks

Description There are a few small ?ornithopod tracks on the Chicken Yard level, but they are difficult to spot within the wealth of the larger tracks. There are short trackways with only a few consecutive tracks. The maximum length of the tracks is 16.2 cm and the maximum width is 13.5 cm. Mean length and width according to measurements of Hübner (2016) is 13.1 cm and 10.9 cm, respectively (see chap. 16 for a detailed description of the 28 tracks).

Interpretation Hübner (2016) shows that it is very difficult to assign these tracks even tentatively to either heterodontosaurids, basal ornithopods, dryosaurid ornithopods, and/or basal marginocephalians. Known skeletal genera from the slightly younger Isle of Wight are *Iguanodon*, *Valdosaurus*, and/or *Hypsilophodon* (Martill and Naish, 2001), and the enigmatic *Stenopelix* from the same horizons of the Obernkirchen Sandstone that is currently interpreted as basal Marginocephalian or even a very basal Ceratopsian (Butler et al., 2011).

TRACK SIZE CLASSES

Description

Track size classes from the Chicken Yard level include a few small (10 to 20 cm), many medium-sized (20 to 30 cm), and some larger (30 to 50 cm) tracks. Regarding the 715 most complete tracks, they consist of 67.0% theropods (including 7.7% troodontids), 27.4% medium-sized and large iguanodontid

ornithopods, 2.8% indeterminate taxa (of medium-sized and large tridactyl dinosaurs), and 2.8% small ?ornithopods.

Interpretation

On the Chicken Yard level, the track sizes are rather widely scaled. In general, track surfaces are controlled by the substrate properties (Falkingham et al., 2011), and they are rather small-scaled. However, size-class exceptions in spite of substrate-dependences are known for example from a carbonate platform tracksite, the Late Jurassic Chevenez-Combe Ronde tracksite from northwest Switzerland, with many small- and medium-sized sauropod tracks that are associated with minute and small tridactyl tracks (Marty, 2008), and this seems to be far more common in the vertebrate ichnological record so that it could be considered an exception (Marty, pers. comm., February, 2015). For widely scaled track sizes, a firm underground underlying the sand upper layer may be a possible explanation.

TRACK AND TRACKWAY ORIENTATION

At first glance, the tracks on the Chicken Yard level look as if they were randomly placed and multimodal, resulting in a chaotic pattern. However, a total directional analysis of 715 clearly distinguishable and plus/minus complete individual tracks delivers a different picture (Fig. 17.9). The majority of tracks point to the north, and a second maximum to the south with a slightly southwestward trend, resulting in an

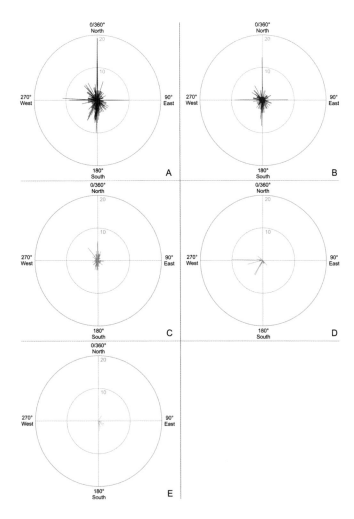

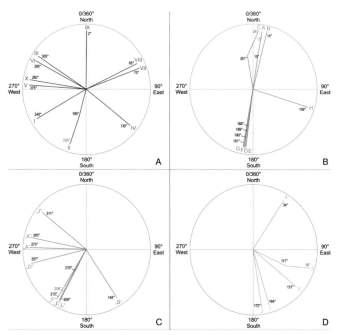

17.10. (A–D) Simplified rose diagrams with trackway directions on the Chicken Yard level. (A) Direction of tridactyl theropod trackways I–X. (B) Directions of medium to large ornithopod trackways A–H. (C) Directions of eight troodontid trackways (following the preliminary naming of van der Lubbe, Richter, and Böhme [2009], A, B, D–F, J–L). (D) Directions of five small ?ornithopod trackways 1–5 sensu Hübner (2016). Changes in direction were depicted stylized. The points of change in direction are the "zero-points" for reading the new direction (red numbers).

17.9. (A–E) Rose diagrams with track orientation (including tracks of trackways) on the Chicken Yard level; fine-scaled view in 1°-steps, linear scale. (A) Rose diagram for 715 clearly distinguishable and complete tracks (theropods, medium-sized and large iguanodontid ornithopods, few indeterminate medium-sized and large tridactyl dinosaurs, troodontids, and small ?ornithopods sensu Hübner, 2016). The marked north–south–orientation is dominant, a second east-west trend and much more scattered rest of orientations are visible. (B) Rose diagram for 424 tridactyl theropod tracks. Visible are a primary, bimodal north–south, a secondary east–west, and a tertiary, weak southeast–northwest orientation. (C) Rose diagram for 196 medium-sized and large iguanodontid ornithopod tracks. A well-pronounced, bimodal north–south pattern is visible with an additional northwest trend. (D) Rose diagram for 55 troodontid tracks from the eight trackways with marked west and southwest trend. (E) Rose diagram for 20 small ?ornithopod tracks and their mainly southeastward movements.

overall bimodal pattern. Less pronounced but still recognizable is another, bimodal east–west trend and some more separate, individual walking directions.

The picture becomes more differentiated when splitting up the tracks into tridactyl theropod (424), ornithopod (only medium-sized and large iguanodontid ornithopod, 196), deinonychosaurian dinosaur (55), and small ?ornithopod (20) sensu Hübner (2016) tracks. The iguanodontids were walking mainly north–south in both directions, delivering a strong bimodal pattern. This correlates with the "Upper Level" of the Obernkirchen tracksite walking directions.

The tridactyl theropod tracks show three trends. There seems to be a strong bimodal north–south orientation but also a marked west–east tendency and a third trend toward a west-northwest–east-southeast direction. The tracks of the deinonychosaurian dinosaurs show a marked trend to the west and southwest, those from the small ?ornithopod tracks reflect mainly southeastward movements.

Trackways

Despite the high track density, some trackways of all observed dinosaur groups can clearly be identified on the Chicken Yard level. The 10 tridactyl theropod trackways from the Chicken Yard are represented by straight trackways with only slight turns and iguanodontid ornithopods by 8 almost straight trackways (Fig. 17.10 and 17.11; Table 17.2). Only the longest trackways (minimum five tracks) of the latter category are analyzed here.

Theropod trackways V and X represent the strict "Westerners." The trackmaker of trackway II went southward but made a slight turn from 195° to 200°, and only the trackway IX points more or less exactly northward.

The longest of them all, trackway A consists of 28 tracks and is located very close and parallel to trackway B with 21 tracks. Both belong to iguanodontian trackmakers which headed toward the north.

The troodontid trackways are subject to another publication (see Lockley et al., 2016, and van der Lubbe, et al., in prep.) but will be also mentioned briefly (Fig. 17.10C). They consist of seven trackways with more than three tracks, or even eight trackways if the shortest one with just three consecutive tracks is also included. The longest one runs 273° (almost exactly west), and there is a second trend toward the southwest (208°–210°, 212°, and 209°). Opposite walking directions (e.g., east or northeast) are not represented. A comparable picture is delivered by the small ?ornithopods (Fig. 17.10D, mostly having moved in southeastern directions (117/96°, 131°, 164°, and 172°), with only one exception to the north-northeast (34°; see also Hübner, 2016). Note that directional changes of some dinosaurs (theropods: II; large ornithopods: A, C; small ?ornithopods: 3; and troodontids: E) have been worked into the same diagrams, respectively, for simplification and recognition that they belong each to one trackway.

Interpretation of Track and Trackway Directions

The marked north–south trend of especially larger tracks and trackways is unambiguous and may indicate a preferred walking direction or pathway. Besides that, each of the dinosaur groups worked out here shows astonishing details that could not be expected a priori by the superficially "chaotic" arrangement of tracks at the beginning of research. In marked contrast to the single tracks, the long theropod trackways do not represent a north–south trend but rather an east–west trend. Due to the completeness of these long trackways, this may imply that these trackways—as a result of a possible "final trackmaking phase"—represent a later change from one preferred walking direction to another over a certain time span. Also, among the larger tridactyl theropod trackmakers, the trackway directions allude that trackmakers VI and III and additionally VIII and VII could possibly have been walking parallel to each other as well as V and X, which, however, are not very close in a locality.

A much more strikingly north–south trend is represented by all single large ornithopod tracks and trackways (Figs. 17.9C and 17.10B, Table 17.2) with their north-northeast and south-southwest directions. The mostly parallel arrangement of the iguanodontid ornithopod trackways is also obvious and documents once more the social behavior of the iguanodontian dinosaurs and their roaming in groups. Three trackways (A, B, C) are oriented to the north and four (D–G) to the south, expressing a clear bimodal north–south orientation pattern. Another (H) crosses the main north–south walking

Table 17.2. Directions of trackways for the tridactyl theropods (roman numbers: trackways) and iguanodontid ornithopods (letters A–H); Chicken Yard level

Trackway Number	Orientation (0°/360° = North)
I	240°
II	195–200°
III	305°
IV	130°
V	275°
VI	300°
VII	70°
XIII	65°
IX	2°
X	280°
A	10–13°
B	15°
C	351–26°
D	189°
E	188°
F	190°
G	191°
H	109°

direction in an east–southeastern direction. Trackways A and C both make slight turns to the right.

Four of five small ?ornithopod trackways show a strong preference to southeastern directions, only one at 34° (north-northeast), and none with a north–south direction.

On the Chicken Yard level, many of the didactyl tracks are organized in trackways, too, and they also do not mirror the north–south trend of some of the other groups but depict a distinct western and southwestern trend. Three (E, F, L) of these are even parallel to each other. Parallel didactyl trackways (ichnotaxon *Dromaeopodus*) were already reported from the Early Cretaceous of Shandong Province (China) and interpreted as evidence of gregarious behavior (Li et al., 2015). The parallel didactyl trackways from the Chicken Yard level could thus be further evidence that these animals frequently occurred in groups (refer also to van der Lubbe, et al., in prep.).

The different but mostly straight walking directions and the lack of randomly arranged, curved trackways (see "Discussion") reject the presence of a shrinking water hole scenario as was proposed for instance for the Isle of Wight dinoturbation layers (Martill and Naish, 2001). The presence of straight trackways, some of which are parallel, seem to indicate a "passing-by" without much interaction between individual trackmakers. For the theropod tracks, this also excludes rambling paths such as have been suggested for the Hettangian tridactyl tracks from Dinosaur State Park (Galton and Farlow, 2003). For the iguanodontid ornithopods, the presence of manus tracks, the short pace/stride values and the gentle turns of the trackways indicate a slow walking gait.

N

17.11. Outline drawings of the 10 tridactyl theropod (I–X) and 8 ornithopod (A–H) trackways of the Chicken Yard level superimposed onto the photogrammetric map. Also, the examples for the gross track morphotypes are marked here (see Gross track types) and the exact position of "slab 8 (B8)" with the realistic textural density from Figure 17.4 is shown. Scale bar: 5 m.

Two groups of iguanodontid trackways on the upper track level of Obernkirchen (the Upper Level approximately 4 m above the Chicken Yard level) show a clear bimodal north–south orientation (Böhme et al., 2009). This corresponds very well with the directions gathered from the tracks of the Chicken Yard level and may be a first hint toward a possible longer-term trend of displacement along a stable landscape marker, such as the shoreline of a large lake.

There may have been two different dinosaur walking events altogether: a rather random primary one and then a second, with a strong primary north–south bimodal pattern and a weaker secondary east–west 'bimodal' trend, together forming almost a quadrimodal configuration. The more marked north–south orientation of the iguanodontian trackways could also indicate an additional species-related, different behavior, such as a preferred route.

The Early Cretaceous dinosaur tracks from Glen Rose/ Paluxy River (United States) serve as good comparison. Besides the large sauropod tracks, the tridactyl tracks show a mirror-image, bimodal orientation pattern, with most of them heading generally northward or southward, in approximately equal numbers. This bimodal pattern of dinosaur trackways is commonly observed; it suggests that the trackmakers were moving back and forth along the local coastline (Lockley, 1993; Farlow et al., 2011). This could fit well into a small-scale, regional wandering due to the aforementioned nutritional aspects. Dinosaurs wandering alongside a shoreline or a lakeside (or here, a large lake's delta) could also have done so in order to avoid possible geographical barriers in the hinterland (e.g., dense forests). This interpretation could also fit for the second tracksite of the Obernkirchen sandstone quarries (Upper Level), where iguanodontid dinosaurs walked in groups besides each other in a comparable bimodal pattern (Böhme et al., 2009).

Taking the bimodal trackway pattern into account, it can also be postulated that returning individuals left two trackways. Interpretations such as this have been made and seem logical (Farlow et al., 2011), but bidirectional walking zones could only be proven by unambiguous special cases of individual pathology, such as "a hurt, biped dinosaur" (Lockley, 1993:99) or the tyrannosaurid of McCrea et al. (2014).

17.12. Vertical cross-section of a theropod track from the northern continuation of the Chicken Yard level showing the homolithic texture of the fine-grained sandstone. Only small coal ribbons are recognizable on a macroscopic scale within the sandstone. Scale bar: 10 cm.

Additional Remark

In spite of the excellent and numerous horizontal data, the homolithic nature of the Chicken Yard locality level sandstones (see Fig. 17.12) does not enable the inclusion of data sets concerning the vertical aspects.

DISCUSSION

Fossil localities with a high dinoturbation index and thus high numbers of tracks received iconic nicknames such as "dinosaur disco" (Interlier trackoito, northwest Switzerland: Lockley and Meyer 2000), "the Raceway" (Fleming Fjord Formation, Jameson Land, East Greenland: Milàn, Clemmensen, and Bonde, 2004), the "Blue Hole Ballroom" (Paluxy River, United States: Farlow et al., 2011), or the controversial "Dinosaur stampede" (Lark Quarry tracksite, Australia. Thulborn and Wade, 1979, 1984; Romilio and Salisbury, 2011). The Chicken Yard level is one of the ichnoassemblages with the highest dinoturbation index worldwide (Fig. 17.13).

Important Parameters for Moderate to High Dinoturbation

The most important factors that may have led to the exceptionally high degree of dinoturbation observed on the Chicken Yard level, sensu Lockley and Conrad (1989), are paleoenvironment, duration of exposure of the sediment surface, moisture content of the substrate at time of track formation, population density, and animal activity and behavior. For the following discussion (following Marty, pers. comm., November, 2012), we propose a slightly simplified parameter list:

1. Paleoenvironment–physical barriers, such as shore-lines or water holes, renewed wetting on tidal flats, and so on.
2. Track recording–time of exposure, substrate properties sensu lato, climatic conditions, and so on.
3. Trackmakers–behavioral parameters of animals: frequency, level of activity, overprinting, carrion, feeding grounds, and so on.

17.13. Detail from the Chicken Yard level at night, photograph taken from a higher level on the east side, artificial grazing light from the northern margin (lower right in the photograph), showing the general aspect of a moderately to heavily dinoturbated ichnoassemblage. *Courtesy Tobias Landmann/Schaumburger Zeitung, 2011.*

Paleoenvironment Behrensmeyer et al. (1992:48) stated, "The margins of large, fluctuating lakes are generally poor in organic remains, although wood and small vertebrate remains may occur; but they provide an optimal context for ichnofossils, including invertebrate and vertebrate tracks, trails, fish nests, and root traces." As the Obernkirchen Chicken Yard level belonged to a large former lacustrine area, it fits well into this scenario. Driftwood, root traces, and the dinosaur tracks are the trademarks of the Berriasian deposits of Obernkirchen, although some small freshwater-to-brackish clams occur as well (Neomiodontidae; see Richter et al., 2012). For an actualistic comparison, an African lake system chosen by Cohen et al. (1991, 1993) proves ideal. Like the many known fossil track–bearing lake-margin strata, Lake Manyara in Tanzania represents an aggrading, closed depositional system with numerous and diverse trackmakers

(birds, mammals). The shores of Lake Manyara show trampled (= "heavily bioturbated") areas in different distances to the shoreline, so that the aspect of trampling and thus large-scale bioturbation delivered useful observations transferrable into dinoturbation criteria (for more details see Cohen et al., 1991, 1993).

The dinosaurs leaving their tracks on the Chicken Yard level must have crossed a large river delta system at the margin of a large lake some 140 million years ago and thus were probably only temporarily present in a paleoenvironment composed of emerged sand bars and barriers (Hornung et al., 2012; Hornung and Böhme, 2012) when they have passed alongside the southern and southeastern shore of the Lower Saxony Basin. In the hinterland of Obernkirchen, forests were present and probably sustained some of the herbivorous dinosaurs (Hornung et al., 2012). It remains questionable whether there were narrow zones, leading to migration corridors in a true sense, or if a passing-by in different directions just followed the actual paleogeographical situation. According to the paleoenvironmental interpretation of Hornung et al. (2012), a network of fluvial channels developed toward the lagoon and sandy mouthbars divided the many arms, some of which were in the process of being abandoned and reduced to oxbow lakes (see pp. 397, 398). It is likely that dinosaurs passing by had to avoid certain areas of these sandbars or deeper oxbow lakes.

Track Recording An important parameter that influences dinoturbation is time (Lockley and Conrad, 1989). In order to produce a heavily dinoturbated level, a certain time span was required for track formation before the environmental conditions changed, track formation became no longer possible, and the next depositional event covered and preserved the track-bearing surface (e.g., growth of microbial mats, tempestitic influences: see Hornung et al., 2012; and discussions herein).

However, the duration of exposure is not measurable and can greatly vary in different extant depositional systems between some days and up to several weeks, months, and even years (Cohen et al., 1991; Marty, 2008). Moreover, it is possible that due to repeated drying and wetting cycles, tracks may form at several different occasions (track recording windows) and lead to the formation of a heavily dinoturbated level that consists of several different track generations left by unrelated noncontemporaneous trackmakers.

Time periods on the order of days to weeks are often considered as track-formation windows in carbonate depositional environments (e.g., Kvale et al., 2001; Marty, 2008). Cohen et al. (1991) provided interesting "track survivorship" curves showing that in an extant lake system from Africa, for small vertebrates, a maximum of 20 days of general track persistence was valid but only in a very restricted zone. On the other hand, microbial mats that are ubiquitous on modern and ancient carbonate and siliciclastic tidal flats may lithify by the precipitation of calcium carbonate and consequently enhance the preservation potential of tracks and other traces (Marty, Strasser, and Meyer, 2009). Such consolidated mats may be exposed for longer time periods during which they may withstand trampling by large and heavy trackmakers and even resist heavy rainfall (Marty, Strasser, and Meyer, 2009). Scott et al. (2007) noted that microbial mats do temporarily protect tracks due to fluid chemistry influences allowing calcification.

Consequently, ichnoassemblages are commonly affected by a certain degree of time-averaging. Even though, the degree of time-averaging is very low when compared to most other sedimentary depositional units (e.g., a single bed), it is still high enough to make it generally impossible to

determine—based on tracks—whether on a given surface two animals were really present at the same time or not (Marty, 2008). To make that determination, other characteristics such as (very long) parallel trackways with a very regular intertrackway spacing, synchronous turns, or very similar track preservation need to be present.

The longer a drying phase (track recording window) is, the more animals can pass through a given zone, and the less that can be said about the faunal diversity, as time-averaging increases (Falkingham et al., 2011). On the other hand, it is probable that additional densifying processes such as those deriving from many animals walking upon a substrate during a certain time span simultaneously stabilizes the substrate as a whole, strengthening it against contemporary erosion, which is known from African flamingo populations and their innate substrate influence (Scott, Renaut, and Owen, 2012).

Studies about extant African land vertebrate faunas from lakes show that within transects from lake shores to landward areas, the depth of the tracks increases with increasing water content, and some "mass track occurrences" such as those from flamingos compare well to fossil dinoturbation areas concerning the density of the tracks and the interactions of tracks with the substrate (Cohen et al., 1991, 1993; Scott et al., 2007) and each other (overprinting).

The different size classes of the tracks on the Chicken Yard level indicate that quite different sizes and types of small, medium-sized, and large dinosaurs were able to walk on the surface, implying a moist thin layer that overlaid a firmer subsurface layer, resulting in a certain stability for different weight groups. The vertical cross-section shows the homolithic sandstone entity (see Fig. 17.12). The low displacement rims of the track additionally hint to this, as seemingly there was not much substance to be squeezed at the margins. Concerning the sizes, a similar situation with huge and small tracks is known from Lark Quarry, Australia (Thulborn and Wade, 1984).

Concerning the time of exposure (parameter 2) and the activity of the dinosaurs (parameter 3), it remains questionable what caused the high number of incomplete and/or overprinted tracks at the Chicken Yard. Random overprinting during a short but high-activity time period can be discussed. The shallow tracks with their flat or lacking displacement rims could point to a "short" drying out event, the surface buried before drying out completely. The marked similarity and equal preservational quality of all tracks with no shallower tracks overprinting deeper tracks might result from that short event, which necessarily implies a high biological activity at the very end of that phase.

Also, another scenario can be constructed. Several walking cycles result in several track-recording events modified afterward by alternating drying out and rewetting, along with the associated erosion. In that case, only the last "track generation" would not undergo erosion and could become ichnotaxonomically the most useful group of tracks in "elite preservation quality," including rather complete trackways. If the latter had been the case on the Chicken Yard level, it would explain a longer time span to develop bimodal traveling paths, but the preservational quality of all recognizable tracks should be quite different, assuming that there are recognizable but eroded tracks preserved from theoretical cycle(s) before the last one.

Thus, a third scenario seems more probable for the Chicken Yard. All recognizable tracks show a similar quality of preservation, which speaks against multiple drying-out cycles plus erosion. Also, there are indicators such as the lack of desiccation cracks, sedimentary infillings in the tracks (possibly originating from shallow water), and in certain tracks adhesion spikes, which give additional evidence for a high moisture or even permanent water cover of the uppermost substrate surface and no final drying out at all.

Additionally, supported by the single theropod track north–south–directional trend contrasting with the east-west one for the long theropod trackways, plus many walking directions for the four gross dinosaur track types, this most probably points at a changing environment during the track recording time (e.g., instable and migrating river delta) and thus a "longer" time span. This could have resulted in slight changes of abundantly used travel paths as for the theropods, which changed from north-south to east-west. Altogether this evidence would suggest a "longer-term" exposure (e.g., several weeks as in Farlow et al., 2006, or even more), likewise allowing for many different dinosaurs to pass by in all possible directions without the need to hypothesize an intensified biological activity during a short time. This interpretation is preferred here.

Trackmakers The behavior of animals plays a most crucial role, and even a few animals can leave many tracks if there is a driver such as a water hole or carrion. Martill and Naish (2001) reconstructed a water hole scenario from the terrestrial Valanginian Wessex formation from the Isle of Wight, representing a formerly subtropical environment with seasonal droughts: "As the [fresh] water body became smaller, more dinosaurs would have gathered around the shrinking water hole, churning up the mud, as occurs in modern-day seasonal wetlands such as the South American Pantanal. Eventually, the margin of the water body would have become an intensely churned quagmire" (319). This scenario represents an abundant environmental situation that is not restricted to South America. The shrinking, drying-up water reservoirs are suboptimal preservational zones for light-weight animals (the "Goldilocks effect;" Falkingham, et al., 2011) and also preferred ambush hunting zones for

predators and thus, mostly avoided by small animals, which consequently do not leave tracks. But all other, larger animals gather exactly the way Martill and Naish (2001) describe it and leave their tracks. Thus, for terrestrial deposits, this could be a favorable reason to explain dinoturbated areas.

Carrion scenarios can easily be observed today. Dead seals (*Phoca vitulina*) or harbor porpoises (*Phocoena phocoena*) regularly attract large groups of seagulls at the North Sea Coast. The latter produce high track densities around the carcasses (in analogy to dinoturbation, this could be named "aviturbation"). Their foot skeletons are anatomically almost identical, and the different genera and species cannot easily be distinguished (Richter, pers. obs., August, 2010) (Fig. 17.14).

Another scenario indicates that many dinosaurs could have been looking for special nourishment conditions, for instance, upon regularly emerged supratidal flats – not carrion but living fish trapped in tidal ponds (Petti et al., 2011) similarly to extant shore birds moving restlessly along the shoreline and along tide pools (Lockley and Meyer, 2000). Thus, they represent trackways of animals feeding in special environments ("foraging behavior").

For theropod-dominated heavily dinoturbated levels, carrion-feeding could be proposed, but an ichnoassemblage providing evidence for this kind of behavior has not been found so far. However, theropods can be abundant in general without dinoturbation, such as marked majorities of theropods have lately been confirmed again for the important Shandong Tianjialou Formation (?Aptian-Albian) from China (Li et al., 2015).

Judging from seagulls gathering around carrion leaving dense track zones as pictured in Figure 17.14, the majority of the tracks should be arranged radially to the (former) center of the food and just around this area. This scenario can clearly be ruled out for the Chicken Yard ichnoassemblage, because the heavily dinoturbated area has a considerable size and extension. Also, within the chaos of trampling caused by carrion-feeding, trackways should not be expected. The rather distinct trackways from the Chicken Yard clearly speak against such a scenario.

CONCLUDING REMARKS AND OUTLOOK

The Chicken Yard level at the Obernkirchen tracksite is a good example of a moderately to heavily dinoturbated paleosurface and as such delivers the typical advantages and disadvantages of mass track occurrences. The density of tracks is so high that the degree of overprinting exacerbates the general analysis, but on the other hand it exhibits a high ichnodiversity despite the rather small size of this site.

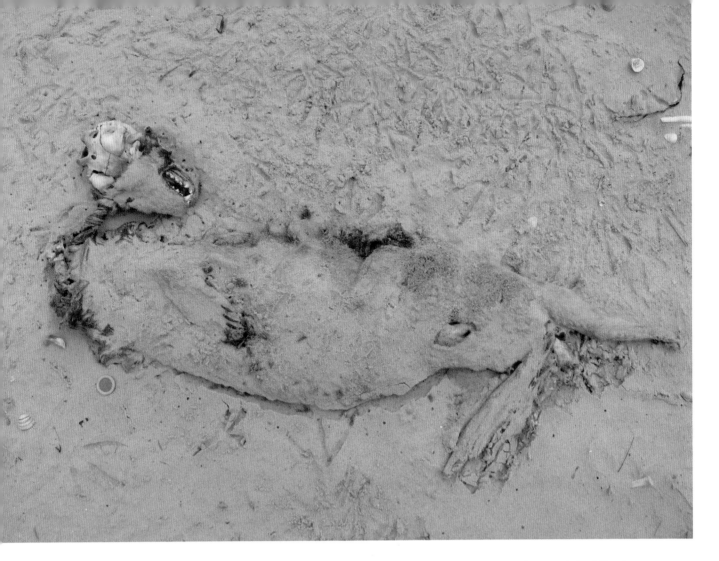

17.14. Dense track zone of extant sea gulls at a carrion feeding place, a dead subadult seal (*Phoca vitulina*). The gull *Chroicocephalus ridibundus* was observed feeding together with the *Larus*-species *L. marinus*, *L. canus*, and *L. argentatus*, and other than minor size differences, their tracks are virtually nondistinguishable. Island of Langeoog, German North Sea Coast, 2010. Below sternal region: 2-Euro coin as scale = 2.5 cm.

Although there is some mixed preservation with many incomplete and/or overprinted tracks (sensu Castanera et al., 2013), the overall track preservation quality is very good and the Chicken Yard level tracks exhibit enough anatomical details to differentiate among several, gross track morphotypes: tridactyl and deinonychosaurian theropods (the latter formerly unknown from the German Berriasian), iguanodontian ornithopods, and small basal ornithopods or basal marginocephalians (compare Fig. 17.15). Even within the tridactyl theropods, different morphotypes could be recognized, though future investigations (such as multivariate analysis) have to show whether there is a taxonomic reason for the observed differences or to what degree these morphotypes may be affected by a preservational influence. As the skeletal record of Early Cretaceous dinosaurs is poor in Germany, the tracks are an important element for reconstructing the local dinosaur fauna.

Although the precise reason for the unique preservation of this densely dinoturbated level with its strong predominance toward theropod tracks remains unknown, at least the

"track chaos" definitively has a hidden trend toward changing bimodal orientation patterns of the tracks and trackways. This and the linear trackways without any pronounced turns support a passing-by scenario.

The paleoenvironment of the Chicken Yard level is interpreted as a former delta environment at the fringe of the Lower Saxony Basin. We can imagine that oxbow lakes with calm and shallow water and eventually changing sandbars could have been the ideal place for dinosaurs leaving their tracks (following Hornung et al., 2016, see pp. 397, 398) and also for preserving them (Fig. 17.15). But also some indications such as the lacking desiccations tracks, the changing trackway directions, and others point toward this interpretation.

Different size classes and the shallow and similar preservation of the tracks fit best to the explanation of a thin moist top layer stabilized by a firm underground, perhaps further consolidated by dinoturbation.

Despite the presence of some obvious trackways, one should bear in mind that a big problem of heavily dinoturbated surfaces will always be the difficulty to unambiguously

17.15. The different dinosaurs of the Chicken Yard level. Non-naturalistic grouping to show the different gross track morphotypes. *Artwork by Frederik Spindler, Freiberg, Germany, 2014.*

recognize trackways. Especially in the case of "accelerating, decelerating or turning trackways," the trackway identification is hampered by the high track density, and only more or less regular trackways can "easily" be detected. Nonetheless, the case of the German Chicken Yard proves that it is indeed rewarding to search carefully for these few trackways and to analyze their orientations. In comparison to the "total directions analysis" with a strong north–south and weaker east–west trend, the trackways prove different behaviors of different dinosaur groups; for instance, the large iguanodontian ornithopod trackways mostly represent the marked north–south trend, whereas the orientation of the larger theropod tracks are much more variable (at least within the trackway data) and even show a change from north–south to the lately produced, last-generation east–west trackways.

Thus, for heavily dinoturbated surfaces, we recommend a detailed directional analysis for single tracks as well as for trackways. In forthcoming works, the degree of overprinting deserves a closer look and could ideally be transferred into a measurable "degree" and thus a parameter for the assessment of dinoturbated surfaces. This might allow further conclusions toward the behavior of the trackmakers. For instance, around a water hole or carrion or other feeding place, tracks may be arranged very chaotically and result in a high degree of overprinting (not many trackways visible), whereas in an area that is a regular pathway or migration corridor (even if that area is changing from time to time), the last animals that went over the surface will still leave straight trackways that can easily be recognized. The latter case would possibly be the better scenario to explain the Chicken Yard level ichnoassemblage than any "water hole" or "feeding ground" scenario would.

The Chicken Yard sandstone is homolithic and delivered no additional vertical data. Though, if the substrate of a dinoturbation layer is heterolithic and allows for vertical investigations (subsurfaces), this should be done level by level. Details of a cross-sectional analysis must be taken into account, as undertracks may contribute toward an even better understanding of the origin of the tracksite than if it were only based on the true tracks. This way, it will become clearly visible whether the dinoturbation is a true signal from only one paleosurface or whether it is an accumulated (composite signal) from tracks left on different layers with undertracks and/or overtracks, which would lead to different interpretations (see also Marty, 2008).

Hopefully, the ongoing quarrying activities will open up the connection of the current Chicken Yard level to the east in the nearby future. The Chicken Yard level should definitely be protected, which we hope to support by this research.

ACKNOWLEDGMENTS

Our sincerest and primary gratitude goes to Klaus Köster, the owner of the Obernkirchen sandstone quarries, who supported our work so generously in spite of many obstacles. His employees are thanked as well. A big thank-you to all of our good-witted volunteers who have been strong and full of enthusiasm and have never lost faith in the project. Among those, Uwe Stratmann deserves an extra thank-you for his massive support since the beginning of the project in winter 2007/2008 and the discovery of the Upper Level. We acknowledge the permanent background counseling activities of our cherished colleague Dr. Jahn Hornung and some fruitful discussions and exchange with our colleague Dr. Tom Hübner. The Lower Saxony Ministry of Science and Culture generously supported the Obernkirchen digging campaigns 2009, 2010, 2011, and donated even more money for the color illustrations and the overall quality of this "life of the past" book. Dr. Manfred "3h" Wiggenhagen taught us the primary photogrammetric methodology—his sedulous help deserves a "grand merci!" The rural county of Schaumburg, namely, Fritz Klebe and his collaborators, continuously helped through difficult political times—thank you for that! Also, the erection of the public area Upper Level Obernkirchen/"DinOS" as project from the European Union is their merit. A huge thanks goes to the foundation Schaumburger Landschaft, namely, Sigmund Graf Adelmann, who made the Dinosaur Track Symposium Obernkirchen 2011 possible and who financially supported the making of the main site map by Torsten van der Lubbe, the discoverer of the troodontid tracks, whom we also thank for his support from 2008 until 2011. The foundation Klosterkammer Hannover, namely, Dr. Stephan Lüttich, is thanked for parts of the symposium as well as for supporting the making of scientifically correct, new landscape illustrations for the Lower Saxony Lower Cretaceous. These have been brilliantly done by Frederik Spindler, Freiberg, whom we thank for his enormous skills and his patience. The foundation Stiftung Niedersachsen generously funded the Obernkirchen symposium, is thanked, and in particular, Dr. Joachim Werren. Oliver Gerke is thanked for providing three photographs of the Chicken Yard level at night. We also thank Tobias Landmann, Schaumburger Zeitung, for allowing us to use two of his Chicken Yard level night photographs. Martin G. Lockley, an anonymous reviewer, Peter Falkingham, and especially Daniel Marty deserve our deepest gratitude because their comments substantially enhanced the quality of the manuscript—Daniel, merci beaucoup! Robert Fischer helped with one of the figures. Finally, A.R. wishes to express her gratitude toward her family, who proved to be very patient from the symposium through the finishing of this book.

REFERENCES

Ballerstedt, M. 1921. Dinosaurierfährten im Wealdensandstein des Harm bei Bückeburg und eine zurzeit freiliegende Spur eines 'vierfüßigen' plumpen Dinosauriers. Zeitschrift der Deutschen Geologischen Gesellschaft 72: 8–10.

Bates, K. T., R. Savage, T. C. Pataky, S. A. Morse, E. Webster, P. L. Falkingham, L. Ren, Z. Qian, D. Collins, M. R. Bennett, J. McClymont, and R. H. Crompton. 2013. Does footprint depth correlate with foot motion and pressure? Journal of the Royal Society Interface 10: 20130009, doi: 10.1098/rsif.2013.0009.

Behrensmeyer, A. K., J. D. Damuth, W. A. DiMichele, R. Potts, H.-D. Sues, and S. L. Wing. 1992. Terrestrial Ecosystems through Time: Evolutionary Paleoecology of Terrestrial Plants and Animals. The University of Chicago Press, Chicago, Illinois, 588 pp.

Böhme, A., U. Stratmann, M. Wiggenhagen, A. Richter, and T. van der Lubbe. 2009. New tracks on the rock: parallel trackways of a new type of Iquanodontipus-Caririchnium-like morphology from the Lower Cretaceous sandstones of Obernkirchen, northern Germany. Journal of Vertebrate Paleontology 29(3, Supplement): 66A.

Butler, R. J., L. Jin, J. Chen, and P. Godefroit. 2011. The postcranial osteology and phylogenetic position of the small ornithischian dinosaur Changchunsaurus parvus from the Quantou Formation (Cretaceous: Aptian–Cenomanian) of Jilin Province, northeastern China. Palaeontology 54: 667–683.

Castanera, D., J. L. Barco, I. Díaz-Martínez, J. H. Gascón, L. Pérez-Lorente, and J. I. Canudo. 2013. Geometric morphometric analysis applied to theropod tracks from the Lower Cretaceous (Berriasian) of Spain. Palaeontology 58(1): 183–200.

Castanera, D., B. Vila, N. L. Razzolini, P. L. Falkingham, J. I. Canudo, P. L. Manning, and À. Galobart. 2013. Manus track preservation bias as key factor for assessing trackmaker identity and quadrupedalism in basal ornithopods. PLoS One 8(1): e54177.

Cohen, A. S., J. Halfpenny, M. G. Lockley, and E. Michel. 1993. Modern vertebrate tracks from Lake Manyara, Tanzania and their paleobiological implications. Paleobiology 19(4): 433–458.

Cohen, A. S., M. G. Lockley, J. Halfpenny, and A. E. Michel. 1991. Modern vertebrate track taphonomy at Lake Manyara, Tanzania. Palaios 6: 371–389.

Dai, H., L. Xing, D. Marty, J. Zhang, W. S. Persons IV, H. Hu, and F. Wang. 2015. Microbially-induced sedimentary wrinkle structures and possible impact of microbial mats for the enhanced preservation of dinosaur tracks from the Lower Cretaceous Jiaguan Formation near Qijang (Chongqing, China). Cretaceous Research 53: 98–109.

Díaz-Martínez, I., X. Pereda-Suberbiola, F. Pérez-Lorente, and J. I. Canudo. 2015. Ichnotaxonomic review of large ornithopod dinosaur tracks: temporal and geographic implications. PLoS One 10(2): e0115477.

Diedrich, C. 2004. New important iguanodontid and theropod trackways of the tracksite Obernkirchen in the Berriasian of NW Germany and the megatracksite concept of Central Europe. Ichnos 11(3–4): 215–228.

Dodson, P., A. K. Behrensmeyer, R. T. Bakker, and J. McIntosh. 1980. Taphonomy and paleoecology of the dinosaur beds of the Jurassic Morrison formation. Paleobiology 6(2): 208–232.

Droser, M. L., and D. J. Bottjer. 1986. A semiquantitative field classification of ichnofabric. Journal of Sedimentary Petrology 56: 558–559.

Falkingham, P. L. 2014. Interpreting ecology and behaviour from the vertebrate fossil track record. Journal of Zoology 292(4): 222–228.

Falkingham, P. L., K. T. Bates, L. Margetts, and P. L. Manning. 2011. The 'Goldilocks' effect: preservation bias in vertebrate track assemblages. Journal of the Royal Society Interface 8(61): 1142–1154.

Falkingham, P. L., L. D. Agenbroad, K. Thompson, and P. L. Manning. 2010. Bird tracks at the Hot Springs Mammoth Site, South Dakota, U.S.A. Ichnos 17: 34–39.

Farlow, J. O., W. Langston Jr., E. E. Deschner, R. Solis, W. Ward, B. L. Kirkland, S. Hovorka, T. L. Reece, and J. Whitcraft. 2006. Texas giants: dinosaurs of the Heritage Museum of the Texas Hill Country. The Heritage Museum of the of the Texas Hill Country, Canyon Lake, Texas, 105 pp.

Farlow, J. O., M. O'Brien, G. J. Kuban, B. F. Dattilo, K. T. Bates, P. L. Falkingham, L. Piñuela, A. Rose, C. Freels, C. Kumagai, C. Libben, J. Smith, and J. Whitcraft. 2011. Dinosaur tracksites of the Paluxy River Valley (Glen Rose Formation, Lower Cretaceous), Dinosaur Valley State Park, Somervell County, Texas; pp. 41–69 in P. Huerta, F. T. Fernández-Baldo, and J. I. Canudo Sanagustín (eds.), Proceedings of the V International Symposium about Dinosaur Palaeontology and Their Environment. Colectivo Arqueológico y Paleontológico Salense, Burgos, Spain.

Fischer, R. 1998. Die Saurierfährten im Naturdenkmal 'Saurierfährten Münchehagen'; pp. 3–60 in R. Fischer (ed.), Das Naturdenkmal 'Saurierfährten Münchehagen.' Mitteilungen des geologischen Instituts der Universität Hannover 37. Universität Hannover, Hannover, Germany.

Flügel, E. 2004. Microfacies of Carbonate Rocks: Analysis, Interpretation and Application. Springer, Berlin, Germany, 976 pp.

Galton, P. M., and J. O. Farlow. 2003. Dinosaur State Park, Connecticut, USA: history, footprints, trackways, exhibits. Zubia 21: 129–13.

Gierlinski, G. 1994. Early Jurassic theropod tracks with metatarsal impressions. Przeglad Geologiczny 42: 280–284.

Graupner, A. 1977. Natursteine an hannoverschen Bauwerken. Hannoversche Geschichtsblätter, Neue Folge 31(1/2): 105–152.

Henderson, D. M. 2006. Simulated weathering of dinosaur tracks and the implications for their characterisation. Canadian Journal of Earth Sciences 43: 691–704.

Hornung, J. J., and A. Böhme. 2012. The 'German Wealden' and the Obernkirchen sandstone: an introduction; pp. 62–72 in A. Richter and M. Reich (eds.), Dinosaur Tracks 2011. An International Symposium, Obernkirchen, April 14–17, 2011, Abstract Volume. Universitätsdrucke Göttingen, Göttingen, Germany.

Hornung, J. J., A. Böhme, and M. Reich. 2012. The type material of the theropod ichnotaxon 'Bueckeburgichnus' maximus Kuhn, 1958–reconsidered; p. 27 in A. Richter and M. Reich (eds), Dinosaur Tracks 2011. An International Symposium, Obernkirchen, April 14–17, 2011, Abstract Volume. Universitätsdrucke Göttingen, Göttingen, Germany.

Hornung, J. J., J. König, and M. Reich. 2012. The type material of the ankylosaurian ichnospecies Metatetrapous valdensis Nopcsa, 1923 (Early Cretaceous, northern Germany); p. 29 in A. Richter, and M. Reich (eds.), Dinosaur Tracks 2011. An International Symposium, Obernkirchen, April 14–17, 2011, Abstract Volume. Universitätsdrucke Göttingen, Göttingen, Germany.

Hornung, J. J., A. Böhme, N. Schlüter, and M. Reich. 2016. Diversity, ontogeny, or both? A morphometric approach to iguanodontian ornithopod (Dinosauria: Ornithischia) track assemblages from the Berriasian (Lower Cretaceous) of northwestern Germany; chap. 12 in P. L. Falkingham, D. Marty, and A. Richter (eds.), Dinosaur Tracks: The Next Steps. Indiana University Press, Bloomington, Indiana.

Hornung, J. J., A. Böhme, T. van der Lubbe, M. Reich, and A. Richter. 2012. Vertebrate tracksites in the Obernkirchen Sandstone (late Berriasian, Early Cretaceous) of northwest Germany: their stratigraphical, palaeogeographical, palaeoecological, and historical context. Paläontologische Zeitschrift 86: 231–267.

Hübner, T. 2016. Elusive ornithischian tracks in the famous Berriasian (Lower Cretaceous) 'Chicken Yard' tracksite of northern Germany: quantitative differentiation between small tridactyl trackmakers; chap. 16 in P. L. Falkingham, D. Marty, and A. Richter (eds.), Dinosaur Tracks: The Next Steps. Indiana University Press, Bloomington, Indiana.

Jackson, S. J., M. A. Whyte, and M. Romano. 2009. Laboratory-controlled simulations of dinosaur footprints in sand: a key to understanding vertebrate track formation and preservation. Palaios 24: 222–238.

Jones, L. S., and E. R. Gustason. 2006. Dinosaurs as possible avulsion enablers in the Upper Jurassic Morrison Formation, East-Central Utah. Ichnos 13: 31–41.

Kvale, E. P., G. D. Johnson, D. L. Mickelson, K. Keller, L. C. Furer, and A. W. Archer. 2001. Middle Jurassic (Bajocian and Bathonian) dinosaur megatracksites, Bighorn Basin, Wyoming, U.S.A. Palaios 16: 233–254.

Lepper, J. 1997. Naturwerksteine in Niedersachsen. Zeitschrift für angewandte Geologie 41(1): 3–10.

Lepper, J., and A. Richter (eds.). 2010. Steine an der Leine: Naturwerksteine im Stadtbild von Hannover. E. Schweizerbart'sche Verlagsbuchhandlung, Stuttgart, Germany, 90 pp.

Li, R., M. G. Lockley, M. Matsukawa, and M. Liu. 2015. Important dinosaur-dominated footprint assemblages from the Lower Cretaceous Tianjialou Formation at the Houzuoshan Dinosaur Park, Junan County, Shandong Province, China. Cretaceous Research 52: 83–100.

Lockley, M. G. 1993. Auf den Spuren der Dinosaurier. Dinosaurierfährten–Eine Expedition in die Vergangenheit. Birkhäuser Verlag, Basel, Switzerland, 313 pp.

Lockley, M. G., and C. Meyer. 2000. Dinosaur Tracks and Other Fossil Footprints from Europe. Columbia University Press, New York, New York, 323 pp.

Lockley, M. G., and K. Conrad. 1989. The palaeoenvironmental context, preservation, and palaeoecological significance of dinosaur tracksites in the western USA; pp. 121–134 in D. D. Gillette and M. G. Lockley (eds.), Dinosaur Tracks and Traces. Cambridge University Press, Cambridge, UK.

Lockley, M. G., K. J. Houck, and N. K. Prince. 1986. North America's largest trackway site: implications for Morrison Formation paleoecology. Geological Society of America Bulletin 97: 1163–1176.

Lockley, M. G., L. Xing, J. A. F. Lockwood, and S. Pond. 2014. A review of large Cretaceous ornithopod tracks, with special reference to their ichnotaxonomy. Biological Journal of the Linnean Society 113: 721–736.

Lockley, M. G., J. D. Harris, R. Li, L. Xing, and T. van der Lubbe. 2016. Two-toed tracks through time: on the trail of 'raptors' and their allies; chap. 11 in P. L. Falkingham, D. Marty, and A. Richter (eds.), Dinosaur Tracks: The Next Steps. Indiana University Press, Bloomington, Indiana.

Lubbe, T. van der, A. Richter, and A. Böhme. 2009. Velociraptor's sisters: first report of troodontid tracks from the Lower Cretaceous of northern Germany. Journal of Vertebrate Paleontology 29(3, Supplement): 194A.

Manning, P. L. 2004. A new approach to the analysis and interpretation of tracks: examples from the Dinosauria; pp. 93–123 in D. McIlroy (ed.), The Application of Ichnology to Palaeoenvironmental and Stratigraphic Analysis. Special Publication 228. Geological Society, London, UK.

Martill, D. M., and D. Naish. (eds.) 2001. Dinosaurs of the Isle of Wight. Field Guide No. 10. The Palaeontological Association, London, UK, 433.

Marty, D. 2008. Sedimentology, taphonomy, and ichnology of Late Jurassic dinosaur tracks from the Jura carbonate platform (Chevenez-Combe Runde tracksite, NW Switzerland): insights into the tidal-flat palaeoenvironment and dinosaur diversity, locomotion, and palaeoecology. PhD dissertation, University of Fribourg, Fribourg, Germany. GeoFocus 21, 278 pp.

Marty, D., A. Strasser, and C. A. Meyer. 2009. Formation and taphonomy of human footprints in microbial mats of present-day tidal-flat environments: implications for the study of fossil footprints. Ichnos 16(1–2): 127–142.

McCrea, R., L. Buckley, J. O. Farlow, M. G. Lockley, P. J. Currie, N. A. Matthews, and S. G. Pemberton. 2014. A 'Terror of Tyrannosaurs': the first trackways of tyrannosaurids and evidence of gregariousness and pathology in Tyrannosauridae. PLoS One 9(7): e103613.

Milàn, J. 2006. Variations in the morphology of emu (Dromaius novaehollandiae) tracks reflecting differences in walking pattern and substrate consistency: ichnotaxonomic implications. Palaeontology 49(2): 405–420.

Milàn, J., D. B. Loope, and R. G. Bromley. 2008 Crouching theropod and Navahopus sauropodomorph tracks from the Early Jurassic Navajo Sandstone of USA. Acta Palaeontologica Polonica 53: 197–205.

Milàn, J., L. B. Clemmensen, and N. Bonde. 2004. Vertical sections through dinosaur tracks (Late Triassic lake deposits, East Greenland): undertracks and other subsurface deformation structures revealed. Lethaia 37: 285–296.

Milner, A. R. C., J. D. Harris, M. G. Lockley, J. I. Kirkland, and N. A. Matthews. 2009 Bird-like anatomy, posture, and behaviour revealed by an Early Jurassic theropod dinosaur resting trace. PLoS One 4: e4591.

Pelzer, G. 1998. Sedimentologie und Palynologie der Wealden-Fazies im Hannoverschen Bergland. Cour. Forschungsinstitut Senckenberg 207: 211 pp., 27 app.

Pelzer, G., and V. Wilde. 1987. Klimatische Tendenzen während der Ablagerungen der Wealden-Fazies in Nordwesteuropa. Geologisches Jahrbuch, Serie A 96: 239–263.

Petti, F.M., M. Bernardi, P. Ferretti, R. Tomasoni, and M. Avanzini. 2011. Dinosaur tracks in a marginal marine environment: the Coste dell'Anglone ichnosite (Early Jurassic, Trento Platform, NE Italy). Italian Journal of Geosciences 130(1): 27–41.

Reineck, H. 1963. Parameter von Schichtung und Bioturbation. Geologische Rundschau 56: 420–438.

Richter, A., A. Böhme, and T. Van der Lubbe. 2009. 'Chicken Run': a new unusual, heavily dinoturbated tracksite from the Lower Cretaceous sandstones of Obernkirchen, northern Germany. Journal of Vertebrate Paleontology 29(3, Supplement): 171A.

Richter, A., J. J. Hornung, A. Böhme, and U. Stratmann. 2012. Excursion Guide A1: Obernkirchen sandstone quarries: a natural workstone lagerstaette and a dinosaur tracksite; pp. 73–99 in A. Richter and M. Reich (eds), Dinosaur Tracks 2011. An International Symposium, Obernkirchen, April 14–17, 2011, Abstract Volume. Universitätsdrucke Göttingen, Göttingen, Germany.

Richter, R. 1936. Marken und Spuren aus dem Hunsrückschiefer. II Schichtung und Grundleben. Senckenbergiana 18: 215–244.

Romilio, A., and S. W. Salisbury. 2011. A reassessment of large theropod dinosaur tracks from the mid-Cretaceous (late Albian-Cenomanian) Winton Formation of Lark Quarry, central-western Queensland, Australia: a case for mistaken identity. Cretaceous Research 32: 135–142.

Romilio, A., R. T. Tucker, and S. W. Salisbury. 2013. Reevaluation of the Lark Quarry dinosaur tracksite (late Albian-Cenomanian Winton Formation, central-western Queensland, Australia): No longer a stampede? Journal of Vertebrate Paleontology 33(1): 102–120.

Santos, V. F., P. M. Callapez, and N. P. C. Rodrigues. 2012. Dinosaur footprints from the Lower Cretaceous of the Algarve Basin (Portugal): new data on the ornithopod palaeoecology and palaeobiogeography of the Iberian Pensinula. Cretaceous Research 40: 158–169.

Scott, J. J., R. W. Renaut, and R. B. Owen. 2012. Impacts of flamingos on saline lake margins and shallow lacustrine sediments in the Kenya Rift Valley. Sedimentary Geology 277/278: 32–51.

Scott, J. J., R. W. Renaut, R. B. Owen, and W. A. S. Sarjeant. 2007. Biogenic activity, trace formation, and trace taphonomy in the marginal sediments of saline, alkaline lake Bogoria, Kenya Rift Valley; pp. 311–332 in SEPM (ed.), Sediment-Organism Interactions: A Multifacetes Ichnology. Special Publication 88. Society for Sedimentary Geology, Tulsa, Oklahoma.

Souza Carvalho, I. de, L. Borghi, and G. Leonardi. 2013. Preservation of dinosaur tracks induced by microbial mats in the Sousa Basin (Lower Cretaceous), Brazil. Cretaceous Research 44: 112–121.

Struckmann, C. 1880a. Die Wealden-Bildungen der Umgegend von Hannover. Eine geognostische Darstellung. Hahn'sche Buchhandlung, Hannover, Germany, 122 pp.

Struckmann, C. 1880b. Vorläufige Nachricht über das Vorkommen grosser vogelähnlicher Thierfährten (Ornithoidichnites) im Hastingssandsteine von Bad Rehburg bei Hannover. Neues Jahrbuch für Mineralogie, Geologie und Palaeontologie Jahrgang 1881 (I. Band): 125–128.

Sweetman, St. C. 2004: The first record of velociraptorine dinosaurs (Saurischia, Theropoda) from the Wealden (Early Cretaceous, Barremian) of southern England. Cretaceous Research 25: 353–364.

Thulborn, R. A., and M. Wade. 1979. Dinosaur stampede in the Cretaceous of Queensland. Lethaia 12: 275–517.

Thulborn, R. A., and M. Wade. 1984. Dinosaur trackways in the Winton Formation (mid-Cretaceous) of Queensland, Memoirs of the Queensland Museum 21: 413–517.

Thulborn, R. A., and M. Wade. 1989. A Footprint as a History of a Movement; pp. 51–56 in D. D. Gillette and M. Lockley (eds.), Dinosaur Tracks and Traces. Cambridge University Press, Cambridge, UK.

Wings, O., A. Broschinski, and N. Knötschke. 2005. New theropod and ornithopod dinosaur trackways from the Berriasian of Münchehagen (Lower Saxony, Germany). Kaupia, Darmstädter Beiträge zur Naturgeschichte 14: 105.

Wings, O., J. N. Lallensack, and H. Mallison. 2016. The Early Cretaceous dinosaur trackways in Münchehagen (Lower Saxony, Germany): 3D photogrammetry as basis for geometric morphometric analysis of shape variation and evaluation of material loss during excavation; chap. 3 in P. L. Falkingham, D. Marty, and A. Richter (eds.), Dinosaur Tracks: The Next Steps. Indiana University Press, Bloomington, Indiana.

Wright, J. 1999. Ichnological evidence for the use of the forelimb in iguanodontid locomotion; pp. 209–219 in A. R. Milner and D. J. Batten (eds.), Life and Environments in Purbeck Times. Special Papers in Palaeontology 68. The Palaeontological Association, London, UK.

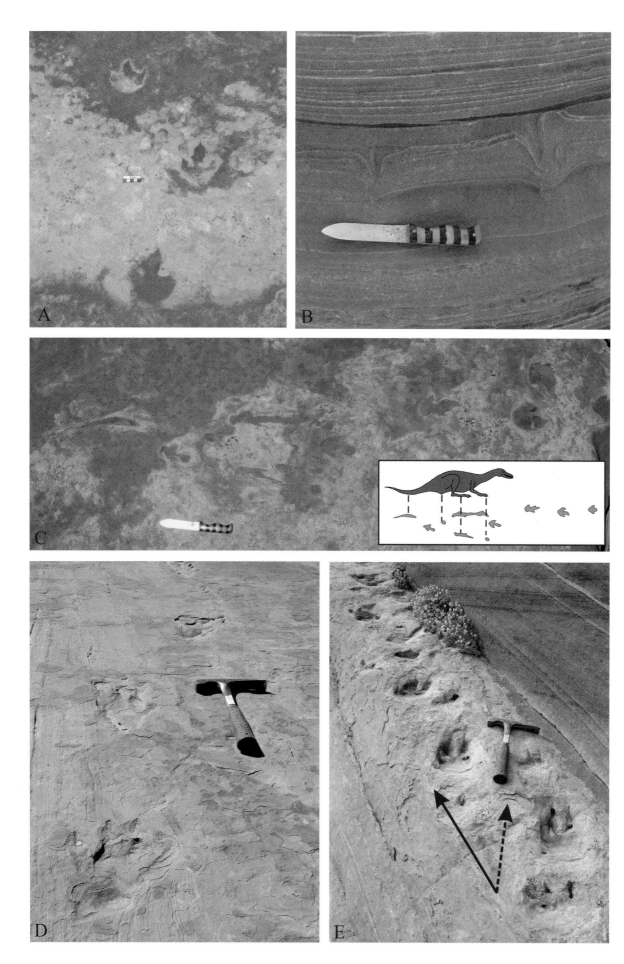

Dinosaur Tracks in Eolian Strata: New Insights into Track Formation, Walking Kinetics, and Trackmaker Behavior

18

David B. Loope and Jesper Milàn

DINOSAUR TRACKS ARE ABUNDANT IN WIND-BLOWN Mesozoic deposits, but the nature of loose eolian sand makes it difficult to determine how they are preserved. This also raises the questions: Why would dinosaurs be walking around in dune fields in the first place? And, if they did go there, why would their tracks not be erased by the next wind storm?

INTRODUCTION

Most dunes today form only in deserts and along shore-lines – the only sandy land surfaces that are nearly devoid of plants. Normally plants slow the wind at the ground surface enough that sand will not move even when the plant cover is sparse. However, some dunes, such as the Great Sand Dunes of Colorado, form in semiarid areas where deflation combines with local wind corridors to permit accumulations that cover otherwise vegetated areas. Because animals are totally dependent on plants as an energy source, it might seem that dunes would be a poor place to look for animal tracks. A short walk in a modern dune field when the sun is low and shadows are long will demonstrate that this is not the case. The scarcity of life in dune fields is actually a boon for generating distinct, recognizable tracks. In the Nebraska Sand Hills (a giant dune field in central North America), the now stabilized, grass-covered dunes are at present traversed by huge numbers of cattle, but none of their tracks will get preserved. However, thin, 800-year-old cross-beds inside the dunes were deposited while the region was a howling desert and contain large numbers of distinct bison tracks and track-ways (Loope, 1986). The scarcity of trackmakers prevented bioturbation (complete mixing of the sediment), thereby allowing full, three-dimensional preservation of the tracks and the trackways that did get made by the relatively small number of animals inhabiting or traversing the dunes. The

hooves of the bison deformed soft, laminated sediment – the perfect medium to preserve recognizable tracks. The next windstorm buried the tracks. Today, the thick cover of grasses protects the land surface so well that there are no soft, laminated sediments for cattle to step on. And, if any tracks were, somehow, to get formed, no moving sediment would be available to bury them. Mesozoic eolian sediments around the world, which have been the focus of a number of case studies in recent years, preserve the tracks of dinosaurs that walked on actively migrating sand dunes. This chapter summarizes the known occurrences of dinosaur tracks in Mesozoic eolian strata and discusses their unique modes of preservation and the anatomical and behavioral information about the trackmakers that can be deduced from them.

TRACKS IN DEPOSITS OF LOWER JURASSIC DUNES, NAVAJO SANDSTONE, UTAH, UNITED STATES

The Early Jurassic Navajo Sandstone is a thick, widespread sedimentary layer on the Colorado Plateau of Utah and Arizona. The sandstone and its correlative strata, the Nugget Sandstone, have preserved more than 60 sites with dinosaur tracks and trackways (e.g., Lockley, Hunt, and Meyer, 1994; Rainforth and Lockley, 1996a, 1996b; Milàn, Loope, and Bromley, 2008; Lockley, 2011a, 2011b; Lockley et al., 2011), and a sparse but diverse vertebrate fauna comprising tritylodonts, crocodylomorphs, and dinosaurs (Irmis, 2005).

The Navajo Sandstone was deposited by large sand dunes that migrated southward along the subsiding, western coast of Pangaea. The sloping layers (cross-beds) deposited by the migrating dunes contain thousands of tracks of small theropod dinosaurs. Many of these tracks are preserved in dry avalanches (grain flows) that were deposited at the angle of repose of dry sand (about 32°). In a few places, it is possible to

18.1. Dinosaur tracks in the Navajo Sandstone at Coyote Buttes, Utah. (A) Tracks of small theropod dinosaurs on the upper surface of an eolian grain flow (layer deposited by avalanching dry sand on the steep, downwind slope of a sand dune). (B) Close-up of dinosaur tracks in cross-section. Notice how the sharp digits have penetrated several layers of sand. Strata slope downward away from viewer. (C) Trackway of a crouching theropod on a dune slope, with interpretative drawing inset (from Milàn, Loope, and Bromley, 2008). (D) *Otozoum* trackway on a firm, wind-rippled interdune surface. (E) Trackway of sauropodomorph dinosaur adopting a sideways walking gait for the first part of the trackway. The later (upper) part shows that the animal then started moving directly up the slope. The solid arrow shows the direction of progression and the dashed arrow indicates the orientation of the animal's body.

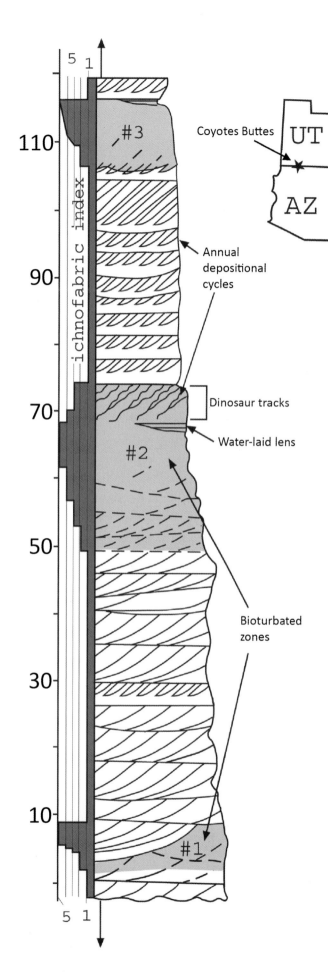

18.2. Stratigraphic column showing the distribution of tracks and burrows in the Lower Jurassic Navajo Sandstone. Tracks are restricted to one interval, but the burrows indicate there were three time periods when sufficient moisture was present in the dune field to support abundant life.

see many three-toed tracks on the upper surface of a single sandstone layer (Fig. 18.1A). Many more tracks, however, can be seen only in vertical cross-section (Loope and Rowe, 2003; Loope, 2006) (Fig. 18.1B). When viewed in cross-section, the preserved track has a U or W shape, with the top-most portion cut off by erosion. Apparently, when these animals stepped on the steep dune slopes, they created small, thin avalanches of dry sand, initiating from above the track. As an animal traversed the dune slope, each step was onto the sliding sand that it triggered by its previous step. The end result is that different tracks within the same animal's trackway are sometimes preserved in different layers of sand, giving the false impression that the tracks are emplaced at different times and not by the same animal. As each avalanche buried a track, it eroded down into the tracked surface. The tracks of very small animals did not penetrate deeply into the dune slope, so many probably were completely eroded. The theropods, although small by dinosaur standards, were sufficiently large that they deformed the layered sand deeply enough (about 10 cm) so that most of the track (but not all) escaped erosion.

Theropod tracks are not the only signs of life in the Jurassic dune deposits. There are also abundant, small burrows, and surface trails made by insects or other invertebrates (Fig. 18.2). Theropods likely fed on the burrowers, but it is a mystery what the burrowers ate. There are no traces of rooted plants in the vicinity of the tracks, and very few in the whole formation. The three intervals containing abundant traces of animal life record relatively wet climatic conditions in the dune field (Fig. 18.2), but the dunes (apparently never stabilized) continued to migrate southward during both wet and dry intervals.

In a few other places, the Navajo Sandstone contains thin, isolated limestones that are completely surrounded by sandstone. These were deposited in lakes that formed between the dunes during the wet climatic intervals that lasted thousands of years. Petrified wood and stromatolites are recorded and sometimes abundant at some of these sites, and dinosaur tracks are also found around these ancient oases (Eisenberg, 2003; Parrish and Falcon-Lang, 2007).

Among the abundant tracks and trackways in the Navajo Sandstone are rare examples of trackways that have preserved evidence of individual behavior of the trackmakers. At the Coyotes Buttes locality, one trackway has preserved tracks of a small theropod, walking directly up a sloping dune front; crouching down; making full impressions of the metatarsi, the belly, and both hands; and then continuing straight up the

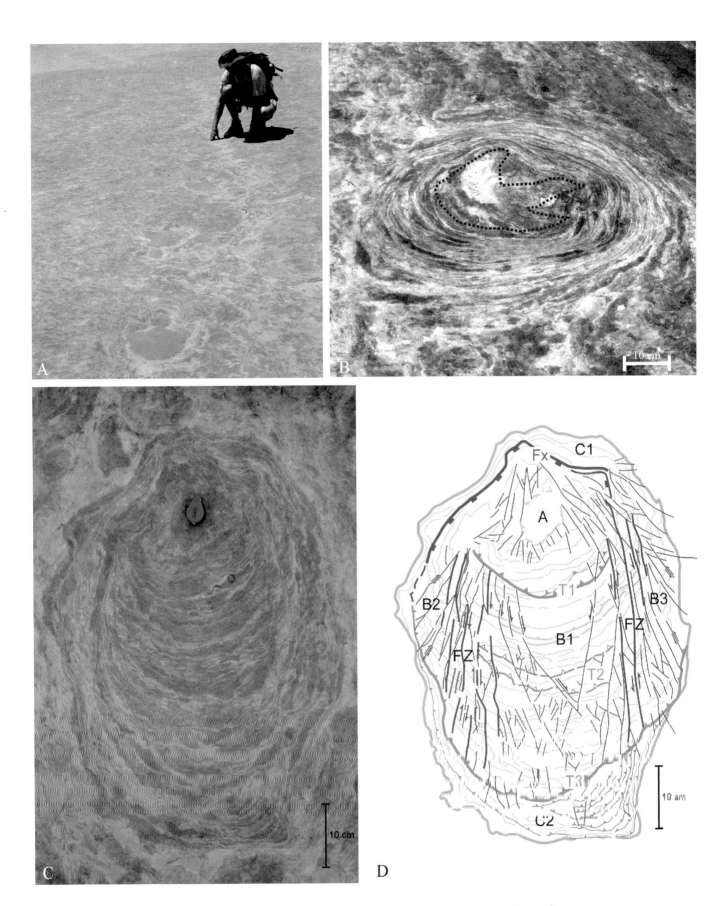

18.3. Tracks in the Entrada Sandstone at Twentymile Wash, Utah. (A) Trackway of large theropod preserved on a laminated inter-dune surface. (B) Close-up of single track with an extensive zone of disturbed sediment around it. The estimated extent of the original footprint is indicated by broken line. (C) Track where the dynamic contact between the trackmaker's foot and the substrate has caused an extensive set of faulting and rotated discs. (D) Interpretation of C (from Gravesen, Milàn, and Loope, 2007).

18.4. Cretaceous dune deposits from southern Mongolia. (A) Cross-bedded sandstone on the left contains well-preserved tracks; deposits on the right are bioturbated and have abundant dinosaur bones (man sits at bone site). (B) Cross-section of a typical dinosaur track from one of the cross-bedded parts of the formation.

dune front (Milàn, Loope, and Bromley, 2008) (Fig. 18.1C). One sauropodomorph, trackway, *Otozoum*, shows normal bipedal progression (Fig. 18.1D), whereas another sauropodomorph trackway, *Navahopus*, shows a sauropodomorph in quadrupedal stance walking up the sloping dune front. The first part of the trackway shows the animal walking at an angle upward, all the time keeping the axis of the body directed upward, before changing its mode of progression to directly up the slope (Fig. 18.1E). Trackways showing a similar mode of progression are known from Pleistocene coastal eolianites from the Mediterranean island Mallorca, where Pleistocene goats adopted a similar sideways gait when they progressed up the steep dune faces (Fornos et al., 2002), and reptile trackways in the Permian eolian Coconino Sandstone of Grand Canyon also show a similar sideways mode of progression (Brand and Tang, 1991; Loope, 1992).

DINOSAUR TRACKS IN INTERDUNE DEPOSITS, MIDDLE JURASSIC ENTRADA SANDSTONE, UTAH, UNITED STATES

Although small theropod tracks are found in cross-bedded dune facies within the Entrada Sandstone (Lockley, Mitchell, and Odier, 2007), tracks are also found preserved in flat-bedded eolian sandstones. In many dune fields, the areas between migrating dunes are covered with wind-ripples. When wind-ripples migrate, they commonly climb over one another, producing firm, flat deposits of thin-bedded sand. Animals do not sink deeply into these deposits, so tracks are shallow (Fig. 18.1D). In some interdune areas, the water table is near the surface, and the sand can get cemented by salts when moisture evaporates. In the wind-deposited, Middle Jurassic Entrada Sandstone of south-central Utah, hundreds of large dinosaur tracks are arranged in long, horizontal

trackways (Milàn and Loope, 2007) (Fig. 18.3A). The animals walked over a flat desert surface that was mantled by small dunes. Salts lightly cemented the sandy surface, so the weight of the large theropods not only depressed the material directly under their feet (the "true track"), but it also disturbed a large area around each track (Foster, Hamblin, and Lockley, 2000; Breithaupt, Matthews, and Noble, 2004; Milàn and Loope, 2007) (Fig. 18.3B). As animals walk, they first compress the sediment under their feet, then push it backward, and, as they remove the foot, stretch the sediment. The damp, thinly laminated sands of the Entrada Sandstone at the Twentymile Wash locality in Utah has captured this interaction between the trackmaker and the substrate.

Close inspection reveals that the disturbed area around each track contains several types of small faults that make it possible to interpret the walking dynamics of the animal (Gravesen, Milàn, and Loope, 2007; Milàn, Gravesen, and Loope, 2014). The sediment layers under the dinosaurs' feet were pushed down and then radially outward, moving as sheet-like slabs, not as loose sand grains. During the kickoff phase of the dinosaur's step, a plate of sand is first rotated below the foot, before being pushed backward, creating a set of parallel faults in the sediment (Fig. 18.3C). The well-preserved evidence of "dinosaur-induced tectonics" has been successfully described using terminology from structural geology, which allowed precise reconstruction of the dinosaurs' walking dynamics (Gravesen, Milàn, and Loope, 2007) (Fig. 18.3D).

SCATTERED TRACKS IN CRETACEOUS DUNES, SOUTHERN MONGOLIA

In the dune deposits of the Cretaceous Djadochta Formation of the Gobi Desert, scattered tracks are visible in cross sections of sloping dune deposits (Loope et al., 1998) (Fig. 18.4B). Ukhaa Tolgod, a site recently discovered by paleontologists from the Mongolian Academy of Sciences and the American Museum of Natural History, is famous for its dinosaur bones. The bones, however, are never found in the cross-beds with the tracks. The bones are in crudely bedded or unbedded sandstone in between the cross-bedded deposits (Fig. 18.4A). The unbedded sediment was bioturbated by abundant burrowing insects (?) and plant roots that are now replaced by calcite (rhizoliths).

The distribution of the fossils and the tracks suggests that the animals whose bones were fossilized lived in a dune field that was stabilized by plants and had abundant life (somewhat similar to the modern Nebraska Sand Hills, a grass-stabilized dune field on the North American Great Plains). When the Cretaceous dunes were active (desert conditions) only a few animals walked across the actively migrating dunes, so only a few tracks and no bones are preserved during those time intervals.

EOLIAN ICHNOFACIES

Based on characteristic assemblages of vertebrate tracks occurring together in eolian deposits, multiple eolian vertebrate ichnofacies have been erected (Lockley, Hunt, and Meyer, 1994). The term "ichnofacies" was introduced by Seilacher (1964, 1967) to cover recurring associations of trace fossils related to sedimentary facies and depositional environments. The term has later been used in many different scales from global associations to individual rock units (Bromley, 1996). Dealing with vertebrate trace fossils, Lockley, Hunt, and Meyer (1994:242) suggest defining vertebrate ichnofacies as "multiple ichnocoenosis that are similar in ichnotaxonomic composition and show recurrent association in particular definite environments."

So far, two distinct eolian vertebrate ichnofacies have been recognized. The Paleozoic *Laoporus* ichnofacies is a recurrent association of vertebrate tracks from the Permian Coconino Sandstone of Arizona and contemporary Lyons Sandstone of Colorado. The ichnofacies comprise the synapsid trackways *Laoporus*, the lizard-like ichnogenus *Dolichopodus* together with abundant invertebrate traces including *Octopodichnus* and *Paleohelcura* (Lockley, Hunt, and Meyer, 1994). As *Laoporus* is a junior synonym of *Chelichnus*, the *Laoporus* ichnofacies is now referred to as the *Chelichnus* ichnofacies (Hunt and Lucas, 2007; Lockley, 2007).

The Mesozoic *Brasilichnium* ichnofacies are confined to erg settings and found in the Lower Jurassic Navajo and Nugget sandstones of the Colorado Plateau region, the equivalent Aztec Sandstone of California, and the Botucatu Formation of South America. The trackmaker for *Brasilichnium* is considered a synapsid with mammal-like reptilian affinities (Lockley, 2011a). Apart from abundant *Brasilichnium* tracks, also the invertebrate tracks *Octopodichnus* and *Paleohelcura* are abundant, as well as theropod, prosauropod, and small mammal tracks (e.g., Lockley, Hunt, and Meyer, 1994; Rainforth and Lockley, 1996a, 1996b; Milàn, Loope, and Bromley, 2008; Lockley, 2011a, 2011b; Lockley et al., 2011). Because of the morphological similarities between Jurassic and Late Paleozoic *Octopodichnus*, *Paleohelcura* and mammaloid tracks (*Chelichnus* in the Permian and *Brasilichnium* in the Jurassic) the dune ichnofacies from eolian deposits spanning this long time interval are almost indistinguishable.

CONCLUSIONS

Recent research into tracks registered and preserved in eolian strata and the usage of the concept of eolian vertebrate ichnofacies has provided important information about animal behavior and ancient environments. Because skeletal remains are uncommon in eolian strata, tracks provide additional, and often the only, information about the fauna that inhabited the area. Tracks are easily recognized in vertical cross-sections as disturbances of the uniformly bedded eolian strata. Close examination of dinosaur tracks in finely laminated interdune strata, provides insight into the second for second interaction between an extinct animal and the sediment it walked on, allowing a very detailed reconstruction of the walking kinetics of the trackmaker.

REFERENCES

Brand, L. R., and T. Tang. 1991. Fossil vertebrate footprints in the Coconino sandstone (Permian) of northern Arizona: evidence for underwater origin. Geology 19: 1201–1204.

Breithaupt, B. H., N. A. Matthews, and T. A. Noble. 2004. An integrated approach to three-dimensional data collection at dinosaur tracksites in the Rocky Mountain west. Ichnos 11: 11–26.

Bromley, R. G. 1996. Trace Fossils: Biology, Taphonomy and Applications. 2nd edition. Chapman and Hall, London, U.K., 384 pp.

Eisenberg, L., 2003. Giant stromatolites and a supersurface in the Navajo sandstone, Capitol Reef National Park, Utah. Geology 31: 111–114.

Fornos, J. J., R. G. Bromley, L. B. Clemmensen, and A. Rodriguez-Perea. 2002. Tracks and trackways of *Myotragus balearicus* Bate (Artiodactyla, Caprinae) in Pleistocene aeolianites from Mallorca (Balearic Islands, western Mediterranean). Palaeogeography Palaeoclimatology Palaeoecology 180(4): 277–313.

Foster, J. R., A. H. Hamblin, and M. G. Lockley. 2000. The oldest evidence of a sauropod dinosaur in the western United States and other important vertebrate trackways from Grand Staircase-Escalante National Monument, Utah. Ichnos 7: 169–181.

Gravesen, O., J. Milàn, D. B. Loope. 2007. Dinosaur tectonics: a structural analysis of theropod undertracks with a reconstruction of theropod walking dynamics. Journal of Geology 115: 375–386.

Hunt, A. P., and S. G. Lucas. 2007 Tetrapod ichnofacies: a new paradigm. Ichnos 14: 59–68.

Irmis, R. B. 2005. A review of the vertebrate fauna of the Lower Jurassic Navajo sandstone in Arizona. Mesa Southwest Museum Bulletin 11: 55–71.

Lockley, M. G. 2007. A tale of two ichnologies: the different goals and missions of vertebrate and invertebrate ichnology and how they relate in ichnofacies analysis. Ichnos 14: 39–57.

Lockley, M. G. 2011a. The ichnotaxonomix status of *Brasilichnium* with special reference to occurrences in the Navajo sandstone (Lower Jurassic) in the western USA. New Mexico Museum of Natural History and Science Bulletin 53: 306–315.

Lockley, M. G. 2011b. Theropod- and prosauropod dominated ichnofaunas from the Navajo-Nugget sandstone (Lower Jurassic) at Dinosaur National Monument: implications for prosauropod behavior and ecology. New Mexico Museum of Natural History and Science Bulletin 53: 316–320.

Lockley, M. G., A. P. Hunt, C. A. Meyer. 1994. Vertebrate tracks and the ichnofacies concept; pp. 241–268 in S. K. Donovan (ed.), The Palaeobiology of Trace Fossils. Wiley, New York, New York.

Lockley, M. G., L. Mitchell, and G. Odier. 2007. Small theropod track assemblages from Middle Jurassic eolianites of eastern Utah: paleoecological insights from dune facies in a transgressive sequence. Ichnos 14: 132–143.

Lockley, M. G., A. R. Tedrow, K. C. Chamberlain, N. J. Minter, and J.-D. Lim. 2011. Footprints and invertebrate traces from a new site in the Nugget sandstone (Lower Jurassic) of Idaho: implications for life in the northwestern reaches of the great Navajo-Nugget erg system in the western USA. New Mexico Museum of Natural History and Science Bulletin 53: 344–356.

Loope, D. B. 1986, Recognizing and utilizing vertebrate tracks in cross section: Cenozoic hoofprints from Nebraska. Palaios 1: 141–151.

Loope, D. B. 1992. Comment on 'Fossil vertebrate footprints in the Coconino sandstone (Permian) of northern Arizona: evidence for underwater origin.' Geology 20: 667–668.

Loope, D. B. 2006, Dry-season tracks in dinosaur-triggered grainflows. Palaios 21: 132–142.

Loope, D. B., and C. M. Rowe. 2003. Long-lived pluvial episodes during deposition of the Navajo Sandstone. Journal of Geology 111: 223–232.

Loope, D. B., L. Dingus, C. C. Swisher III, and C. Minjin. 1998. Life and death in a Late Cretaceous dunefield, Nemegt Basin, Mongolia. Geology 26: 27–30.

Milàn, J., O. Gravesen, and D. B. Loope. 2014. Dinosaur tectonics: when biomechanics meet structural geology. Journal of Vertebrate Paleontology, Program and Abstracts 2014: 188.

Milàn, J., and D. B. Loope. 2007 Preservation and erosion of theropod tracks in eolian deposits: examples from the Middle Jurassic Entrada sandstone, Utah, USA. Journal of Geology 115: 375–386.

Milàn, J., D. B. Loope, and R. G. Bromley. 2008. Crouching theropod and *Navahopus* sauropodomorph tracks from the Early Jurassic Navajo sandstone of USA. Acta Palaeontologia Polonica 53: 197–205.

Parrish, J. T., and H. J. Falcon-Lang. 2007. Coniferous trees associated with interdune deposits in the Jurassic Navajo Sandstone Formation, Utah, USA. Palaeontology 50: 829–843.

Rainforth, E. C., and M. G. Lockley. 1996a. Tracks of diminutive dinosaurs and hopping mammals from the Jurassic of North and South America; pp. 265–269 in M. Morales (ed.), The Continental Jurassic. Bulletin 60. Museum of Northern Arizona, Flagstaff, Arizona.

Rainforth, E. C., and M. G. Lockley. 1996b. Tracking life in a Lower Jurassic desert: vertebrate tracks and other traces from the Navajo sandstone; pp. 285–289 in M. Morales (ed.), The Continental Jurassic. Bulletin 60. Museum of Northern Arizona, Flagstaff, Arizona.

Seilacher, A. 1964. Biogenic sedimentary structures; pp. 296–316 in J. Imbrie and N. Newell (eds.), Approaches to Palaeoecology. Wiley, New York, New York.

Seilacher, A. 1967. Bathymetry of trace fossils. Marine Geology 5: 413–428.

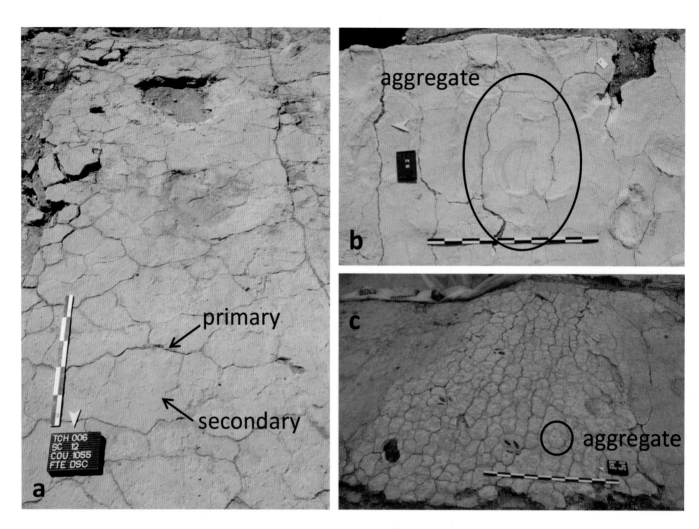

19.1. Different desiccation crack patterns as observed on Late Jurassic dinosaur tracksites excavated on Highway A16 (Canton Jura, northwest Switzerland). Differentiation of primary and secondary crack patterns (A) and different aggregate dimensions (B, C). (A) Courtedoux-Tchâfouè tracksite, level 1055 (thickness of layer 1055 around 3 cm, mudstone). Scale bar is 0.5 m. (B) Courtedoux–Bois de Sylleux tracksite, level 1040 (thickness of layer 1040 2–4 cm, mudstone), note semicircular sauropod manus and oval pes tracks partially surrounded by displacement rims and highlighted with black chalk. Scale bar is 1 m. (C) Courtedoux–Bois de Sylleux tracksite, level 1070 (thickness of layer 1070 5–6 cm, mudstone), note several tridactyl theropod tracks colored in black. Scale bar is 1 m.

Analysis of Desiccation Crack Patterns for Quantitative Interpretation of Fossil Tracks

19

Tom Schanz, Maria Datcheva, Hanna Haase, and Daniel Marty

THIS CHAPTER PRESENTS A CONCEPTUAL APPROACH TO interpret fossil track environments employing the progress made in soil mechanics regarding understanding and modeling of soil desiccation cracks. It must be emphasized that the thorough analysis of soil desiccation phenomena is crucial for the understanding of track formation and preservation processes and the interpretation of the paleoenvironmental setting associated with fossil track-bearing strata. The basis for the proposed method is the generally accepted fact that crack formation depends on the intrinsic state properties of the soil and on the boundary conditions the soil layer is exposed to during desiccation. Both the observations made during the experimental test series and results obtained from numerical simulation (mathematical modeling) show that it is possible to reconstruct crack patterns that are typically found in association with fossil tracks. It has to be admitted, however, that despite a vast literature on the examination, explanation, interpretation, and simulation of present-day soil (desiccation) crack patterns, fossil soil conditions and desiccation crack patterns are generally still difficult to assess. This study addresses experimental investigation of crack patterns on fine-grained to more coarse-grained soils with varying thickness and base layer roughness, exposed to drying, and experimental and numerical simulation of crack appearance in and around artificially designed impressions that resemble sauropod pes tracks. The results may be used to infer subsoil properties and boundary conditions in a given paleoenvironment at the time of track formation and during track preservation and to quantify the influence of the impression relief (track) on the crack pattern formation during desiccation.

INTRODUCTION

In recent years, research on the quantitative interpretation of fossil tracks has been given more and more attention involving different interdisciplinary approaches in order to reconstruct track formation and to understand the subsoil conditions at the time of track formation. This includes biomechanical, paleogeological, neoichnological, computational, and most recently soil mechanical studies, and a recent literature review can be found in Falkingham (2014). A state-of-the-art gait reconstruction via forward dynamic simulation and employing solely the animal biology and ignoring the physical properties of the substrate can be found in Sellers et al. (2009) and Sellers et al. (2013). On the other side, fossil tracks may be considered as paleopenetrometers as proposed in Falkingham, Margetts, and Manning (2010), where it is demonstrated via numerical simulations that the characteristic features of the imprints provide a wealth of information about the specific weight, duty factor, and limb kinematics of the animal, as well as about substrate (or sediment) properties. Thorough analysis on the importance of the soil condition at track site involving finite element analysis (FEA) and geotechnical interpretation is performed in Falkingham et al. (2011a, 2011b). The conclusion drawn is that the "Goldilocks principle" applies to the vertebrate track formation and preservation processes. One of the components supporting the Goldilocks effect is the substrate where the tracks are formed, and this confirms that assessing substrate mechanical properties is of major importance for reconstructing extinct animal gaits. Moreover, when interpreting fossil tracks with regard to deeper insight into the body mass and the locomotion of the trackmaker, it is essential to know the subsoil constitutive properties.

This study focuses on track interpretation using a soil mechanics approach that was recently introduced by Schanz et al. (2013). They showed that trackmakers' mass can be found with an error of about 15% by employing the track topography characteristics and if the properties of the track-bearing strata are known. Moreover, in order to successfully apply the method proposed by Schanz et al. (2013), a detailed knowledge is required about trackmaker locomotion (e.g., gait, walking speed), track geometry (e.g., imprint diameter and depth), as well as about the (sub)soil properties. It should be noted that it is a challenging task to identify the original (sub)soil mechanical and physical parameters from the present-day lithified (sub)soil constitution, and this is especially relevant for (very) fine-grained lithified soils, where even innovative methods such as micro-CT do not provide satisfactory results for the characterization of the original soil condition. The reason is in the significant influence of the diagenesis in the original soil forming the present-day clay and mudstone. This is contrary to the situation with tracks preserved in sandstone where direct micromechanical analysis such as micro-CT is applicable owing to the sandstone granulometric features. Therefore, indirect indicators gain importance for the interpretation of tracks imprinted in originally fine-grained soils. We suggest the results of the quantitative analysis of fossil wave ripple-marks, if ripples exist, be considered indirect indicators. This is done to gain insight into the hydraulic regime (water depth and flow velocity over time) as well as to aid the qualitative analysis of fossil desiccation crack patterns. By qualitatively interpreting the geometrical characteristics of the fossil crack patterns, we expect to gain information about the subsoil thickness, subsoil type, and the contact of the subsoil layer with the underlying stratum at time of fossil crack formation.

On many tracksites, drying and wetting processes resulted in the development of large-scale desiccation crack patterns in the track-bearing strata. Because the occurrence of desiccation cracks in present-day soils can be described from a soil mechanics point of view, in situ crack patterns might just as well be interpreted in terms of the original soil properties and boundary conditions at the time before and during the formation of the cracks. Here, we introduce an experimentally based approach to qualitatively relate desiccation crack patterns with track substrate boundary conditions at the time the cracks were formed. Furthermore, a numerical approach using coupled thermo-hydro-mechanical (THM) analysis of tensional stresses in a hypothetical track exposed to desiccation shrinkage is presented and the numerical predictions are validated experimentally.

The literature overview hereafter focuses on research and results most closely related to the specific purpose of our discussion: the reconstruction of the original subsoil properties around fossil tracks, based on in situ crack patterns combined with data from experiments designed to simulate the conditions at the time the foot imprints were created.

PHYSICAL ORIGIN OF CRACKS

First, we will try to reveal the physical origin of cracks in soils as it is understood in soil mechanics literature. Shrinkage of soils causes cracking when the sample is sufficiently large and the shrinkage is restrained. The physical origin of shrinkage is explained by the nature of the forces causing it, namely by mechanical, thermal, or hydrostatic forces. Therefore, shrinkage cracks differ depending on the boundary conditions under which they develop. Syneresis cracks appear in the subaqueous environment due to loss of pore water ascribed to the contraction of swelling clay lattices in response to a drastic change in the salinity of the surrounding liquid phase (Burst, 1965). Diastasis cracks result from compaction and are usually formed in stiff mud underbedded with loose sand and are triggered by seismic or other shocks. The subaerial exposure of the soil results in desiccation cracks that appear when the water evaporates from the soil pore system. An extended literature review on shrinkage cracks is given in Plummer and Gostin (1981) where the authors reveal the wide variety in crack form and structure depending on the conditions of formation. They established some basic criteria for distinguishing between cracks formed by subaerial and subaqueous processes in mudstone: "Desiccation mudcracks are generally continuous, polygonal, and often of several generations with V- or U-shaped cross-sections that are infilled from above. Syneresis cracks, on the other hand, are generally discontinuous, spindle, or sinuous in shape and of one generation only, with V- or U-shaped cross-sections that are infilled from either above or below" (Plummer and Gostin, 1981:1153). We should also acknowledge the hypothesis raised by Cowan and James (1992) that diastasis is an alternative to desiccation and syneresis to mud cracks found in carbonate and terrigenous clastic sediments. Hereafter, however, only soil shrinkage and cracking processes related to desiccation are considered, as only the paleoenvironments that may preserve tracks of terrestrial vertebrates are taken into account.

Second, it has to be answered what the basics of soil desiccation processes are. For doing so, the factors influencing desiccation cracking processes have to be ascertained. Soil is a natural material of high complexity and because of that, many factors govern soil desiccation cracking. These factors may be classified in three categories:

1. Grain size distribution, mineral composition, clay content and type, soil plasticity;
2. Boundary conditions during desiccation: relative humidity, temperature, mechanical loading; and
3. Geometry and size effects—layer thickness, size of the sample, sample base properties.

Another important question is how to model initiation, development, pattern, and depth of desiccation cracks. The essential mechanism of desiccation cracking is still not well understood and a rational model to describe this phenomenon as well as a prediction of crack initiation and the associated crack network propagation are still difficult to establish (Tang, Cui, et al., 2011). Some existing approaches to interpret the physical origin of soil cracking phenomenon are summarized herein.

Tensile Stress and Tensile Strength Approach

The most commonly used theory of desiccation cracking follows the tensile stress and tensile strength approaches. For modeling shrinkage crack initiation and propagation, it is commonly accepted that desiccation cracks initiate when the normal stress at a certain point exceeds a given threshold value (tensile strength). There are two main frameworks for modeling soil cracking based on limited tensile strength: (1) physical analogy models employing springs and dashpots to represent the interparticle forces, and (2) soil mechanics models within the continuum mechanics framework.

Hereafter, we focus our discussion on soil mechanics interpretation and modeling of desiccation crack formation. Models developed based on the soil mechanics approach may be divided into models using single effective stress (e.g., Péron, 2008; Schanz, 2009) or those employing two stress variables (e.g., Morris, Graham, and Williams, 1992; Schanz, 2009). These concepts reflect the current developments in unsaturated soil mechanics. In both cases, soil is considered to be a porous medium and desiccation is understood as a multiphase flow process generally involving coupled THM phenomena. Because fracture mechanics is the field of continuum mechanics concerned with the process of crack propagation in a deformable continuum, it was found to provide sufficiently good tools to model soil desiccation cracks. Both linear elastic and elastoplastic approaches were applied combined with the fundamental principles of unsaturated soil mechanics.

Two Stress Variables Concept The linear elastic material concept has been intensively employed to establish a relationship for the depth of the cracks, the properties of the soil, and the given suction profile expressing the unsaturated soil behavior in terms of net stress (total stress relative to air pressure), matric suction (difference between pore gas and pore fluid pressures), and void ratio. The crack propagation is commonly assessed by applying linear elastic fracture mechanics (LEFM) (Morris, Graham, and Williams, 1992; Konrad and Ayad, 1997; Prat et al., 2008). It is assumed that intergranular voids act as micro cracks and their growth under tensile load ends with producing macro cracks. There is also discussion on the different mechanisms of crack propagation depending on existing weak interfaces (e.g., between coarse and fine layers) and the development of vertical tensile stresses that cease the cracks due to horizontal cracking or layered clays and slurries where horizontal crack propagation leads to separation of the layers (Morris, Graham, and Williams, 1992).

Effective Stress Concept Konrad and Ayad (1997) also applied LEFM to analyze crack propagation in cohesive soils undergoing desiccation, employing an effective stress concept to model the unsaturated soil behavior. Their key assumptions are the following: (1) crack initiation occurs when the total horizontal stress (minor principal stress) is equal to the tensile strength of the soil; (2) crack propagation at the onset of its formation is analyzed with a trapezoidal distribution of tensile stress and the LEFM theory; (3) crack spacing is determined from a horizontal stress relief distribution, calculated by using the finite-element method with a fictitious stress distribution opposite to the actual total tensile stress distribution. A numerical model within the LEFM concept for crack initiation and propagation in drying soils is presented in Prat et al. (2008). Crack initiation and propagation criteria are based on fracture parameters determined experimentally and the process is considered to be a coupled hydromechanical one.

Energy-Based Concept Constraints in shrinkage as a reason for the appearance of desiccation macro cracks are discussed in Péron, Hueckel, et al. (2009). These constraints are found to be a frictional or any other traction or displacement boundary condition, any eigen-stress concentration within the soil, or intrinsic factors such as soil texture and soil structure. The physical variables at the macroscale investigated were water content, suction, degree of water saturation, strain, stress, and crack geometry. It was emphasized that any crack initiation model should consider the elastoplastic process that is accompanies cracking. Moreover, based on the obtained results, it was suggested that crack pattern geometry reproduces the energy redistribution. The crack orientation and spacing were studied by means of a basic fundamental energy concept, and the main result was that cracking by creating new solid surfaces is an energy-consuming process, and therefore, the geometry of the lowest crack area enclosing the maximum soil volume and consuming a minimum of surface energy per volume unit will be realized (Péron, Hueckel, et al., 2009).

Nonlinear Effects, Plastic Deformation Process The deficiency of LEFM was recently addressed in Hallett and Newson (2005). According to the authors LEFM fails in accounting for the large amount of irrecoverable energy caused by plastic processes such as particle rearrangement, destructuration, friction within the particle matrix, and bond rupture between particles not associated with fracture. The authors proposed instead a model based on elastoplastic fracture mechanics in order to better assess the irreversible processes during desiccation cracking.

Alternatives to the Conventional Tensile Strength Concept

An alternative to the tensile stress concept was introduced by Shin and Santamarina (2011), who claimed a new fundamental understanding of desiccation crack formation in soils by combining particle-level and macroscale observations. Their proposed approach explored the different stages in desiccation cracking such as (1) water evaporation and surface settlement before crack initiation—water surface above the sediment surface, (2) mobilization of the tensile membrane and development of suction, (3) membrane invasion—crack initiation, and (4) crack propagation. It was discovered that the soil is fully water saturated at the time of crack initiation. Also, Shin and Santamarina (2011) underlined the relevance of surface defects, small topographic irregularities, inclusions, and large aggregate (agglomerate of soil particles) anomalies as triggers for crack initiation and formation.

An integrated approach to soil shrinkage, cracking, and soil structure was proposed in a series of papers by Chertkov (see Chertkov, 2012). A concept of the critical sample size, lacunar factor, and crack factor was introduced for the physical description of shrinkage and cracking. However, the model requires at least 8 to 10 parameters to be independently determined for a given soil in order to make any prediction on soil cracking.

In addition, it has to be acknowledged that research on desiccation cracks is not only carried out in engineering sciences but also in geology (sedimentology). However, the approach and the terminology used to describe lithified desiccation (mud) cracks are different. A good overview is provided in Demicco and Hardie (1994)

STATE OF THE ART ON EXPERIMENTAL AND THEORETICAL MODELING OF SOIL DESICCATION CRACKING

There exists a vast literature on desiccation cracking of fine-grained soils devoted to geoenvironmental, agricultural, and soil physics applications, see, for example, Alaoui, Lipiec, and Gerke (2011).

Modeling of Soil Cracking within the Unsaturated Soil Mechanics Framework

There are several examples of modeling the soil desiccation cracking that employ the methods and the assumptions of the unsaturated soil mechanics. With respect to geoenvironmental applications, an analytical approach relating desiccation cracking potential to hydraulic and mechanical stress conditions of the soil was introduced by Heibrock (1996).

Input parameters were derived by experimental results on tensile strength with varying water content. Moreover, different macrostructure- or microstructure-based models exist allowing the modeling of changes in hydraulic properties such as suction, water content relationship, and relative permeability due to cracking. An example for a straightforward way to such modeling is the introduction of the dependence of the microvoid volume (considering saturated aggregates) on water content as shown in Romero (2013).

The shrinkage and cracking of swelling soils requires special consideration. In this case, double porosity models for expansive clays come into picture. Such a model for repeated drying-wetting cycles is discussed in Fityus and Buzzi (2009). They employed a concept that regards soil consisting of saturated, structured clay peds, separated by air-filled cracks. It is stated that after a few cycles, the soil is in a kind of equilibrium referring to purely reversible swelling and shrinkage for further cycles, and that the volume of water gained or lost from the soil is equal to the volume change of the soil peds and to the change in crack volume. The peds were considered to expand and shrink completely in a reversible way. Therefore, when the expansive soil contains cracks, the volume of water loss is equal to the volume of air gained by the shrinkage crack network, whereas the soil remains very close to a fully water-saturated condition. The main conclusion in Fityus and Buzzi (2009) is that the unsaturated soil mechanics should be applied with a special caution in case of cracked expansive soils.

Phenomenological Approach to Soil Shrinkage and Desiccation Cracking

Chertkov (2012, 2013) analyzed the influence of the water content distribution and the 'sample to layer size ratio' on the shrinkage anisotropy characteristics. Two classes were identified: small samples without cracking during shrinkage, and sufficiently large samples with internal cracking. Additionally, pore size distribution, soil texture, and structure were considered to understand possible differences in horizontal and vertical shrinkage of a sample. The main conclusion from the Chertkov concept is that the critical sample geometry for the occurrence of cracks can be shown to depend on both the soil texture and structure characteristics.

Experimental Assessment and Image-Based Modeling

Desiccation cracking of thin clay layers was experimentally studied by Tang et al. (2011a). According to these authors, evaporation during drying occurs in two phases: During the first phase the water content decreases at an almost constant rate, while it slows down in the second phase when approaching the air entry value (AEV). Most of the cracks appeared during the first phase. Prior to crack initiation, shrinkage mainly occurred in the vertical direction, and the lateral contribution to volumetric strain was minor even after the end of the cracking processes. During crack evolution, crack intersections first reached a stable value, followed by crack length and width (Tang, Shi, et al., 2011).

Concerning the effect of clay layer thickness on the crack pattern, it was found by Tang et al. (2008) that:

1. The number of intersections and crack segments decreases with an increase in thickness of soil layer;
2. The average length, width of cracks, average area of aggregates, and crack intensity factor of crack patterns increases with increase of thickness; and
3. Lower initial soil density or higher adhesion at the base tends to produce lower average aggregates area for a given layer thickness.

Nahlawi and Kodikara (2006) found that soil layers with greater thicknesses generally showed a slower water loss rate and higher water content when cracking. Chertkov and Ravina (1999) indicated the average spacing between cracks (d) as a function of soil depth (z), and there is a trend of d to increase with z.

Automatic Digital Photo Analysis For quantitative characterization of the crack inventory, the automatic digital photo analysis is extensively used, see for example Baer, Kent, and Anderson (2009) and Li and Zhang (2010). The statistical characteristics of crack networks such as soil surface crack area, fractal dimensions of crack edges, mass fractal dimensions, and lacunarity can be deduced either under field or laboratory conditions.

Regarding the digital photo analysis, Tang et al. (2008) developed special software to study the effect of temperature, thickness of soil layer, and soil type on surface shrinkage crack patterns with the help of digital photo analysis. In their experimental setup, thin layers of clay with a ratio from about 15 to 30 of length to height were placed over a smooth glass base. After performing digital photo analysis, data were collected on the number of crack segments and intersections, average crack length and width, average aggregate area, crack intensity factor, and the corresponding probability density functions. They observed that with increasing soil layer thickness, the average crack length, width, aggregate area, and crack intensity factor increased. In addition, the increasing of the number of drying and wetting cycles caused increasing in the irregularity of the observed crack patterns. The amount of cracking also was directly related to the fine fraction and the plasticity index of the soil, and the larger the fine fraction and the higher the plasticity index were, the greater became the crack width and crack intensity factor. Tang et al. (2008) observed that first appeared the main cracks starting at the soil surface and forming the main aggregates. In a second phase, the main aggregates split into several subaggregates by subcracking. Cracks may close partially or completely after subsequent wetting for soils showing sufficient swell potential and thus enabling self-healing, (e.g., see Rayhani, Yanful, and Fakher, 2008). Next, Tang et al. (2008) investigated the effects of temperature, thickness of soil layer, wetting and drying cycles, and soil type on the geometric structure of surface shrinkage cracks in clayey soils. They selected the following parameters to characterize the crack patterns: number of crack segments and intersections, average crack length and width, average aggregate area, crack intensity factor, and the corresponding probability density functions of these parameters. Tang, Shi, et al. (2011) and Tang, Cui, et al. (2011) also reported on repeated drying-wetting cycles applied to very thin clay samples that were initially at slurry condition. By image processing, different parameters such as surface crack and structure evolution were monitored. After three of five cycles, the cracking water content and the surface crack ratio reached equilibrium. Regularity of the overall crack pattern vanished with increasing the number of cycles. Void ratio changes became more and more reversible with increasing the number of cycles. However, due to the particular geometric restrictions of this study, a direct, quantitative one-to-one application to a real scale field situation is difficult, albeit Li and Zhang (2011) by studying crack patterns in a field qualitatively confirmed the results of Tang, Shi, et al. (2011) and Tang, Cui, et al. (2011).

The influence of temperature on desiccation cracking of saturated slurries is studied in Tang et al. (2010). By digital photo analysis of crack patterns, it was concluded that the initial critical water content for crack initiation during the initial drying conditions increases with temperature, and at water content lower than the shrinkage limit, crack occurrence vanished. The surface crack ratio increases with increasing temperature, and because the water content profile above the height of the shrinking layers is not constant, Tang et al. (2010) stated that direct comparison of cracking characteristics with the soil water characteristic curve (the relation between soil suction and soil water content) and shrinkage curves is not possible.

X-Ray CT Imaging The evolution of desiccation cracks in active clays under varying physicochemical initial and boundary conditions can be investigated by means of X-ray CT analysis. Using X-ray CT imaging, Gebrenegus, Ghezzehei, and Tuller (2011) demonstrated the possibility to quantify the effect of bentonite content, pore fluid chemistry, and drying rate on the desiccation crack initiation and evolution. By means of a stochastic modeling of the evolution of crack aperture, the crack distribution was successfully simulated. It was found that crack volume and crack-specific surface show a high sensitivity to temperature and bentonite content, whereas solution chemistry plays a minor role in desiccation cracking processes.

Numerical Simulation of Desiccation Cracking and Reinterpretation of Fossil Tracks

Numerical modeling and simulation of soil shrinkage and desiccation crack formation is a powerful tool for investigating the considered processes especially for gaining information beyond the experimental resources. Here we discuss applications of the finite and discrete element methods recently applied to evaluate soil cracking due to drying-wetting cycles, as well as a reinterpretation of fossil tracks.

Finite Element Analysis In almost all fields of research in engineering, FEA is a well-accepted and widely used numerical analysis technique for solving mathematical problems. FEA allows finding an approximate solution to the governing equations within the mechanics of continuous media. Soil shrinkage and desiccation cracking have been successfully analyzed via FEA as far as these processes are modeled as taking place in a continuous medium and the corresponding mathematical model was built within the framework of continuum mechanics. In most cases, soil desiccation is considered as a coupled hydromechanical problem and FEA is done accordingly. In most cases, FEA was applied to problems related to geotechnical projects where soil properties are known or are back-calculated based on measurements done shortly before, during, or after the execution of the considered project. To our best knowledge, there is no example in the literature on FEA performed to reconstruct the soil properties in the past employing contemporary information.

The importance of accounting for the fundamental processes involved in desiccation cracking was confirmed by Rodríguez et al. (2007) who performed FEA of desiccation tests. A major conclusion in their work was that the generation of desiccation cracks in soils may be regarded as a coupled hydromechanical boundary value problem employing continuum mechanics formulations.

A simple and straightforward way to interpret desiccation is to make an analogy with thermoelasticity. An example of such a modeling approach is given in Péron et al. (2007) who used two-dimensional FEA to investigate desiccation of soils, assuming the total strain rate is decomposed to a purely elastic mechanical part and a drying induced part, proportional with the opposite sign to water content increment.

When referring to the application of numerical simulations to the interpretation of fossil tracks, a series of recent papers by Falkingham and coauthors should be addressed

(Falkingham et al., 2009; Falkingham, Margetts, and Manning, 2010; Falkingham et al., 2011a, 2011b). Falkingham et al. (2009) emphasized that in order to recover information regarding the trackmakers, the interaction of several factors including substrate properties, must be taken into account. FEA has the advantage of making it possible to easily look within the sediment and view surfaces at any level within the three-dimensional volume. In Falkingham et al. (2009), a simplified von Mises elastoplasticity model was used to represent the plastic behavior of the sediment.

Discrete Element Modeling Even if the FEA and the continuum mechanics formulation of the desiccation crack formation showed their ability to tackle different problems by simulation shrinkage and desiccation processes in soils, they have some deficiencies. Recently, a discrete element modeling of drying shrinkage and cracking of soils has been successfully performed by Péron, Delenne, et al. (2009), who revealed the efficiency of the discrete element modeling, where the soil is represented by an assembly of discrete elements, as compared to results based on continuum medium and calculations via FEA.

EXPERIMENTAL APPROACH FOR INTERPRETATION OF DESICCATION CRACK PATTERNS

In this section, we present a strategy to reconstruct the soil properties and soil layer size and boundary conditions using the crack pattern caused by soil desiccation. First, we bridge the distinction in scale between laboratory experimental results and observations from fossil track-bearing surfaces in the field. Exemplarily, in Figure 19.1, crack patterns from Late Jurassic dinosaur tracksites excavated on Highway A16 (Courtedoux, Canton Jura, northwest Switzerland) by Paleontology A16 (e.g., Marty et al., 2003, 2010; Marty, 2008) are illustrating a qualitative geometrical difference even between levels that formed in the same paleoenvironment and that are located vertically close to each other (magnitude of several to several tens of centimeters). In Figure 19.1A, primary cracks after first drying and secondary cracks after consequent drying can be identified. In addition, crack patterns observed in the field differ in dimension, in other words, the size of aggregates formed by primary cracking may be different depending on soil type and thickness and other variable environmental parameters (Figs. 19.1B, 19.1C).

The experimental work presented herein focuses on the reproduction and interpretation of crack patterns caused by the initial drying process from a slurry condition, that is, primary crack patterns, as well as repeated wetting and drying cycles, that is, secondary crack patterns. In order to relate the resulting desiccation crack patterns qualitatively and

| Clay rough | Clay smooth | Silt rough | Silt smooth |

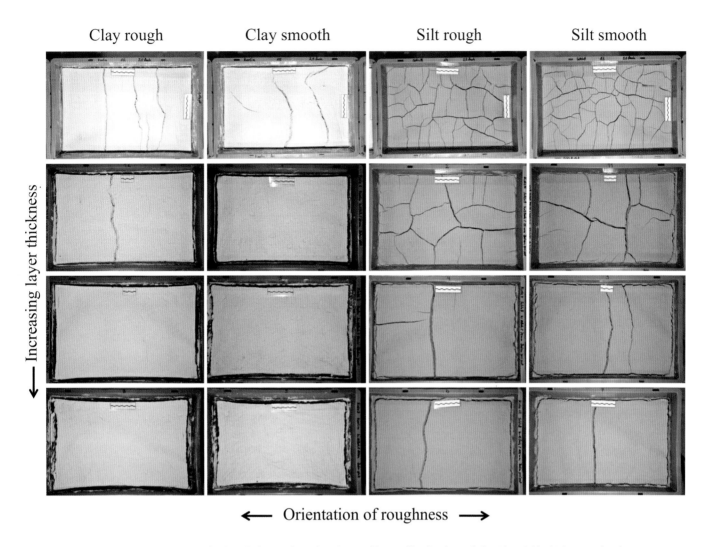

← Increasing layer thickness

←— Orientation of roughness —→

19.2. Primary desiccation crack patterns as developed after 60 days of air drying of layers of kaolin clay and silt with variable thickness and with rough and not rough (smooth) bottom. Scale bars are 5 cm and 10 cm in length. With decreasing grain size and increasing layer thickness–2.5 times (first row), 5 times (second row), 10 times (third row) and 15 times (fourth row)–the size of crack aggregates as well as crack openings increases.

quantitatively to the initial (intrinsic and state) soil properties such as fine-grained fraction amount, mineralogy, water content, tensile strength, and dry density and boundary conditions, in other words, soil layer thickness and roughness of the soil layer base, an experimental study was conducted as follows. Two types of fine-grained soils susceptible to the formation of desiccation cracks have been used, namely kaolin clay (grain size < 0.06 mm) and silt (grain size < 0.19 mm). For testing the materials both soils were placed in boxes (50 × 36.5 cm) that had been prepared with varying base roughness in advance. "Rough" bedding was simulated with oriented ripples (7 mm in height, 5 mm in width, 16 mm clear diameter) forming the box base, and "smooth" bedding with a flat box base. Embedded soil layer thickness varied between 2.5 and 15 times the roughness, that is, the ripple height. In order to induce complete drying processes, the initial water content of the materials was set to correspond to the particular liquid limit: 26.2% for silt and 48% for clay. The material dry density of 1.5 g/cm³ for silt and 1.1 g/cm³ for clay allows the

materials to have approximately saturated initial conditions. The testing materials were subjected to drying for around 60 days until no further observable changes in crack patterns occurred. Measurements of temperature and relative humidity during the whole drying period allowed for the determination of the average applied suction to the soil samples of about 130 MPa. In order to account for primary crack pattern changes due to wetting and drying cycles, that is, secondary crack formation, additional tests have been performed on the silty soil. After first drying was completed, two more wetting and drying cycles were conducted.

Figure 19.2 summarizes the final primary desiccation crack patterns developed depending on the soil type, soil layer thickness, and soil base roughness. The results indicate that the size of the formed crack aggregates as well as the crack opening width increase with increasing soil layer thickness. Qualitatively this relationship is valid for both types of soil, whereas quantitatively the size of cracking aggregates related to layer thickness differs for the two soil types. This

Table 19.1. Average diameter (in centimeters) of cracking aggregates observed after 60 days of drying (values of layer thickness: times roughness, i.e., ripple height)

Soil Type	Layer Thickness			
	2.5	**5**	**10**	**15**
Silt	6.2	11	23.4	30.5
Clay	22.3	30.8	36.9	—

may be explained by the existence of large particles in the silt samples that play the role of imperfections that trigger the crack initiation and promote the desiccation crack processes. The porosity in the larger grains environment drops faster by increasing the applied suction and this allows the menisci to disintegrate the soil and consolidating micro defects, thus causing the initiation of macro cracks. Because of this interaction between grain size and stress singularity (concentration), the wider pattern in the grain size distribution plot indicates a soil more prone to cracking. This means, the higher the amount of coarse material, the finer the structure of the crack pattern and the smaller the crack aggregate areas. The experimental results presented in Figure 19.2 also show that the dominating orientation of cracks is perpendicular to the orientation of the roughness, that is, the ripples. The quantitative characteristics that can be extracted based on these test results are summarized in Table 19.1.

Figure 19.3 summarizes the results obtained on silty soil samples subjected to repeated wetting and drying cycles. Secondary crack patterns developed within aggregates that formed during primary cracking. It is obvious that (1) primary crack patterns remain nearly unaffected by repeated wetting and drying cycles and (2) secondary crack patterns develop during the cyclic drying-wetting process and become more distinctive with increasing number of cycles.

In general, the characteristics, dependences, and correlations obtained under fixed and controlled boundary conditions in the laboratory may be used for the interpretation and evaluation of crack patterns observed in the field. Finer primary crack networks, such as those illustrated in Figure 19.1C, can most probably be attributed to a silty soil substrate with a moderate layer thickness. However, for the interpretation of coarser crack networks, such as the crack pattern shown in Figure 19.1B, there arise some difficulties and the conclusion about the type of the soil that was exposed to drying may suffer of high inaccuracy if the information on the substrate layer thickness is missing.

Based on literature and our own experience, it can be concluded that the analysis of present-day soil cracking patterns is a potential tool for the interpretation of fossil crack patterns and thus for the evaluation of the original soil conditions. If the crack patterns are associated with vertebrate tracks, they allow important conclusions to be made about

the original soil conditions at the time of track formation and preservation. Using this approach, some difficulties (e.g., insufficient resolution in micro-CT, which is a promising approach for coarse-grained materials; Schanz et al., 2012), related to the analysis of lithified fine-grained sediments can be overcome. Furthermore, mineral alterations and modification (recrystallization) as a consequence of diagenetic processes do not have to be considered for the interpretation.

HYBRID EXPERIMENTAL: NUMERICAL SIMULATION APPROACH FOR INTERPRETATION OF DESICCATION CRACK PATTERNS

Summarizing the understanding of soil cracking phenomenon and the physical origin of desiccation cracks, we can draw conclusions on the type of the numerical analysis that is adequate for realistic physical modeling. Numerical simulations based on appropriate mathematical modeling are well-accepted tools in various research fields including soil mechanics. From the current review on the current level of knowledge about cracking phenomena and modeling soil shrinkage, it can be concluded that coupled THM analysis is the most proper way to simulate the process of soil cracking under varied environmental conditions. Hereafter, we present the results of such a numerical analysis and present a comparison of these results with the experimental data in order to validate the applicability of the coupled THM approach, especially to the initial and boundary conditions close to ancient paleoenvironmental settings associated with the formation and preservation of sauropod tracks. For this reason, both experimental and numerical modeling approaches have been applied to a soil sample with a geometry representing a medium-sized sauropod pes track. The soil sample embodies an oval imprint surrounded by a circular displacement rim (raised rim of sediment) that aims to mimic a typical sauropod pes track as reported in, for example, Lockley, Farlow, and Meyer (1994); Marty (2008); and Marty et al. (2010). It is supposed that the track cracks in the field appeared due to drying the track-bearing soil after the track had been formed. In order to verify this hypothesis and to allow for further and more precise interpretation of the track formation, a series of laboratory and numerical simulation tests was carried out with special efforts being taken to reproduce most closely the shape evolution of the track during soil drying.

Both the experimental and the numerical models predicted an occurrence of typical radial and circumferential cracks as, for example, those observed in the Copper Ridge sauropod tracksite, United States (see Ishigaki and Matsumoto, 2009).

First, we will discuss the experimental setup and the observed phenomena while conducting our experimental

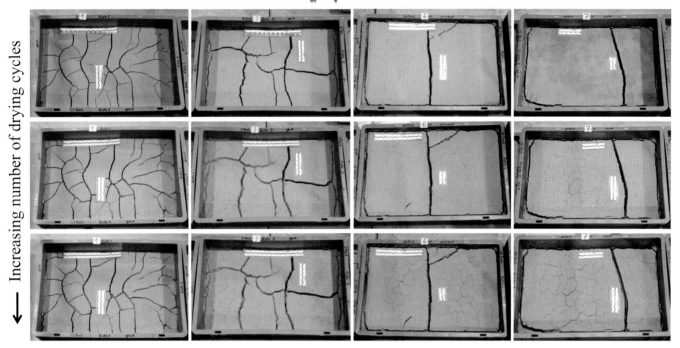

Increasing layer thickness ⟶

Increasing number of drying cycles

19.3. Primary and secondary crack patterns developed with the number of drying cycles: first (first row) to third (third row) cycles. The silt layer thickness is 2.5 times (first column), 5 times (second column), 10 times (third column), and 15 times (fourth column) the sample base roughness.

tests. The imprint diameter was 40 cm, the outer dimension (outside the displacement rim) was 80 cm, and the height of the wall was 6 cm. We are aware that this displacement rim is exceptionally wide and it is not to be considered as a typical displacement rim around sauropod tracks, even if it may occur in particular substrate. We have chosen this geometry to be able to prepare it in a homogenous way not introducing artificial disturbance causing unphysical crack initiation. In this sense, our experiment is of qualitative nature not meant to derive quantitative results.

The material used was clayey silt, fully water-saturated at the beginning of the test. The environmental humidity and temperature was changed from 62% and 37°C to initiate soil drying. The experimental results are presented in Figure 19.4. Significant radial and circumferential cracks on and around the displacement rim can clearly be identified. Radial shrinkage strains in our case may reach up to 2% 4%. During the drying process, the soil water was released by evaporation, and the reduction of the soil pore water induced an overall shrinkage of the specimen. Due to the boundary conditions chosen (rough base of the sample), the magnitude of shrinkage is different along the specimen height, in other words, the shrinkage intensity at the top of the track and at the base of the track is not the same. This difference in shrinkage intensity induces tension stresses on the surface of the track. Thus, it may be stated that if the tension stress is higher than the tensile strength of the soil, cracks will occur.

Hereafter, a short description of the numerical simulation procedure is provided, and the interpretation of the obtained results is discussed. The THM analysis of the soil behavior and the simulation of the conducted tests were performed using the finite element code CODE BRIGHT (DIT-UPC, 2009). The equations that govern the THM response can be categorized into four main groups, namely, balance equations, constitutive equations, equilibrium restrictions, and definition constraints. The balance equations are given in Olivella et al. (1994, 1996). The unknown variables in the applied numerical model are displacements, pore air pressure, water pressure, and temperature. For the mechanical stress-strain analysis, we employ the two stress variables concept in unsaturated soil mechanics with constitutive laws based on nonlinear elasticity and plasticity with hardening dependent on both stress level and suction (the difference between pore air and water pressures). The details on the mathematical model can be found in Nguyen-Tuan et al. (2013). For numerical simulations, the silt material parameters were collected from experimental data reported in the literature, dedicated to soil materials with similar properties (e.g., Alonso, Gens, and Josa, 1990; Geiser, 1999).

The simulation procedure is divided in two stages corresponding to the change in the sample environment during the experiment. For the first stage, the applied suction at the upper face was increased gradually from initially 15 MPa to 30 MPa corresponding to 75% ambient room relative humidity.

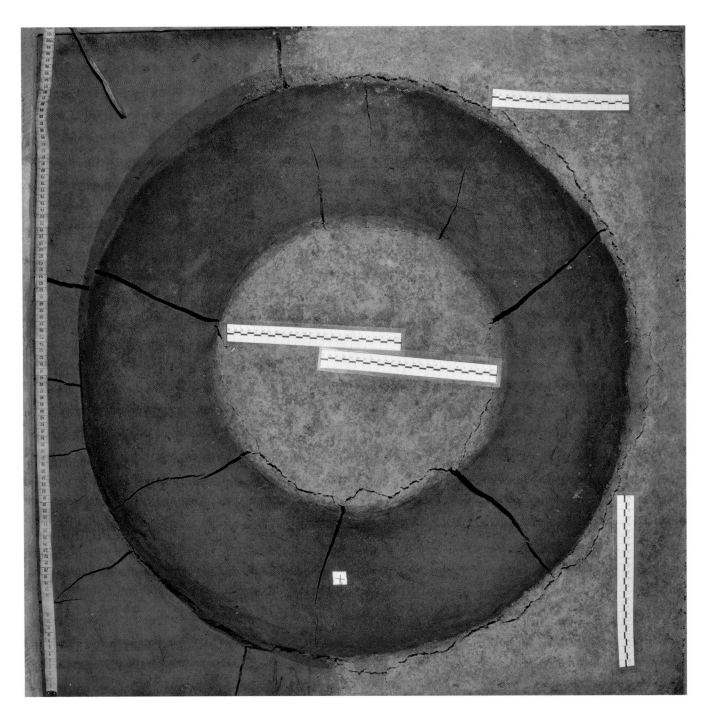

19.4. Final crack pattern around a circular depression surrounded by a broad displacement rim as obtained in the experiment simulating drying of a sauropod pes track. Scale bars are 25 cm in length.

The temperature on the top surface of the sample was kept constant (T = 20°C) and equal to the initial temperature of the sample. Within the second stage, the applied suction on the upper face was increased gradually from 30 MPa to 45 MPa, corresponding to 62% ambient relative humidity that applies after initiating drying in the test. The temperature on the upper face of the soil sample was increased from 20°C to 37°C in this stage. This way, the first drying of the sample with the initially formed imprint was simulated.

The obtained tensile stress distribution is shown in Figure 19.5. It can be seen that the tensile stress value is more pronounced at the inside and outside edges of the impression ring (track wall). The stress contour plot reveals that there are lines at the inner and the outer edge of the track wall as well as between them, along which the tensile stress is higher than on the outside of the rim. This indicates that radial and circumferential cracks will most probably appear along these lines. Thus, it can be affirmed that the results from the numerical analysis agree with the observations in the experiment.

CONCLUSIONS AND PERSPECTIVES

Commonly, fossil tracks may either be found in sediments of granular origin as sandstone, or they are preserved in sediments formed from fine-grained soils as clay- or mudstones. Our objective was to gain information about the initial state of these sediments at the moment the tracks were generated. It is nowadays straightforward to assess the sediment properties of sandstone by granulometric analysis (grain size distribution), for example, via micro-CT approach. However, for sediments formed by diagenesis from fine-grained soils, such a direct procedure is not applicable. Therefore, indirect methods become important and such methods are employed by the approach suggested here. The main properties of the sediment bed include among others geometric and constitutive parameters. The thickness of the compressed layer and the substrate characteristics ensure more or less frictional contact to the underlying sediments and have a major impact on the track formation. It is obvious that the initial stiffness of the sediment is the most significant material parameter to be known when interpreting fossil tracks. The initial soil stiffness may be retrieved with little uncertainty from the knowledge on the original soil plasticity. Therefore, another important factor is the soil type, particularly whether it is clay, silt, or mud. The approach suggested here aims to determine the soil parameters by interpretation of desiccation crack patterns, which formed more or less coevally with tracks formation. It is evident from the laboratory and field observations that the geometry of primary and secondary desiccation crack patterns includes implicit information about the original soil properties. Namely, the examination of single fossil track cross-sections gives information on the thickness of the compressed layer and on the roughness of the contact of this layer to the underlying soil layer. It has to be noted that the absence of desiccation crack patterns at a fossil track site does not imply that the sediment holding the track had not been subject to subaerial exposure. The reason for the absence of soil cracking during desiccation might just be the combination of soil layer thickness and soil type, which is not promoting cracking. Another alternative is the underwater formation of ripple-marks in the first period. Later these surfaces may fall dry and the ripple-marks may be overlaid by desiccation cracks in the second period. The next observation is that the track as a trigger for local crack development has minor significance compared to other heterogeneities in the subsoil such as large inclusions that may facilitate the crack initiation during hydromechanical

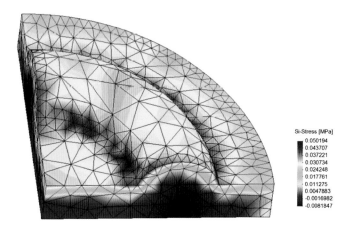

19.5. Final tensile stress distribution (in megapascals [MPa]) obtained via numerical simulation of the drying experiment on a sample with a circular depression (40 cm in diameter) surrounded by a broad displacement rim (sauropod track analogue).

loading. It has to be pointed out that reasonable correlations exist between soil type, typical soil consistency and plasticity, and soil stiffness based on a natural analogue, whose behavior can be studied nowadays.

In summary, we can conclude that by qualitative analysis of the geometrical features of desiccation crack patterns in clay- and mudstones it is possible to gain insight into paleogeoenvironmental situation at the time when the material was still fine-grained soil (e.g., clay or silt). In particular, primary and secondary crack patterns can be classified according to the soil type, the relevant thickness of the imprinted soil layer, and the substrate properties. To visualize the different categories of crack patterns and to facilitate their classification, we introduce a pattern plot matrix where the fossil crack pattern plots are compared with those obtained via laboratory experiments. Depending on the constitutive and geometric properties of the soil, it is expected that there be a significant difference in the geometric characteristics of the resulting crack patterns. Thus, the approach suggested here will bring additional information for quantitative analysis of fossil tracks left and reserved in formerly fine-grained soils. Considering that desiccation cracks appear after soil dry-wetting cycles, we have foreseen to concentrate our research on the statistical properties of secondary crack patterns whose evolution converges after a low number of cycles into a steady state. Moreover, we intend in the near future to establish a kind of catalog with different categories of crack patterns that will allow us to identify the original substrate properties of lithified track-bearing sediments via the pattern plot matrix. In conclusion, we are able to estimate subsoil stiffness for the time the fossil tracks were formed and therefore fill an important gap in order to further assess prehistoric animal weight and locomotion.

Alaoui, A., J. Lipiec, and H. H. Gerke. 2011. A review of the changes in the soil pore system due to soil deformation: a hydrodynamic perspective. Soil and Tillage Research 115–116: 1–15.

Alonso, E. E., A. Gens, and A. Josa. 1990. A constitutive model for partially saturated soils. Géotechnique 40(3): 405–430.

Baer, J. U., T. F. Kent, and S. H. Anderson. 2009. Image analysis and fractal geometry to characterize soil desiccation cracks. Geoderma 154: 153–163.

Burst, J. F. 1965. Subaqueously formed shrinkage cracks in clay. Journal of Sedimentary Petrology 35(2): 348–353.

Chertkov, V. Y. 2012. An integrated approach to soil structure, shrinkage, and cracking in samples and layers. Geoderma 173–174: 258–273.

Chertkov, V. Y. 2013. Shrinkage anisotropy characteristics from soil structure and initial sample/layer size. Geoderma 200–201: 1–8.

Chertkov, V. Y., and I. Ravina. 1999. Morphology of horizontal cracks in swelling soils. Theoretical and Applied Fracture Mechanics 31: 19–29.

Cowan, C. A., and N. James. 1992. Diastasis cracks: mechanically generated synaeresis-like cracks in Upper Cambrian shallow water oolite and ribbon carbonates. Sedimentology 39: 1101–1118.

Demicco, R. V., and L. A. Hardie. 1994. Sedimentary Structures and Early Diagenetic Features of Shallow Marine Carbonate Deposits. SEPM Atlas Series 1. Society for Sedimentary Geology, Tulsa, Oklahoma, 265 pp.

DIT-UPC. 2009. CODE_BRIGHT User's Guide. Universitat Politècnica de Catalunya, Barcelona, Spain, 265 pp.

Falkingham, P. L. 2014. Interpreting ecology and behaviour from the vertebrate fossil track record. Journal of Zoology 292(4): 222–228.

Falkingham, P. L., L. Margetts, and P. L. Manning. 2010. Fossil vertebrate tracks as paleopenetrometers: confounding effects of foot morphology. Palaios 25: 356–360.

Falkingham, P. L., K. T. Bates, L. Margetts, and P. L. Manning. 2011a. The Goldilocks effect: preservational bias in vertebrate track assemblages. Journal of the Royal Society Interface 8(61): 1142–1154.

Falkingham, P. L., K. T. Bates, L. Margetts, and P. L. Manning. 2011b. Simulating sauropod manus-only trackway formation using finite-element analysis. Biology Letters 7(1): 142–145.

Falkingham, P. L., L. Margetts, I. M. Smith, and P. L. Manning. 2009. Reinterpretation of palmate and semi-palmate (webbed) fossil tracks; insights from finite element modelling. Palaeogeography, Palaeoclimatology, Palaeoecology 271: 69–76.

Fityus, S., and O. Buzzi. 2009. The place of expansive clays in the framework of unsaturated soil mechanics. Applied Clay Science 43: 150–155.

Gebrenegus, T., T. A. Ghezzehei, and M. Tuller. 2011. Physicochemical controls on initiation and evolution of desiccation cracks in sand-bentonite mixtures: X-ray CT imaging and stochastic modeling. Journal of Contaminant Hydrology 126: 100–112.

Geiser, F. 1999. Comportement mécanique d'un limon non saturé–Étude expérimentale et modélisation constitutive. PhD dissertation, École Polytechnique Fédérale de Lausanne, Lausanne, Switzerland, 244 pp.

Hallett, P. D., and T. A. Newson. 2005. Describing soil crack formation using elastic-plastic fracture mechanics. European Journal of Soil Science 56: 31–38.

Heibrock, G. 1996. Zur Rissbildung durch Austrocknung in mineralischen Abdichtungsschichten an der Basis von Deponien. Heft 26, Schriftenreihe des Instituts für Grundbau, Ruhr-Universität Bochum, Bochum, Germany.

Ishigaki, S., and Y. Matsumoto. 2009. 'Off-tracking'-like phenomenon observed in the turning sauropod trackway from the Upper Jurassic of Morocco. Memoir of the Fukui Prefectural Dinosaur Museum 8: 1–10.

Konrad, J.-M., and R. Ayad. 1997. An idealized framework for the analysis of cohesive soils undergoing desiccation. Canadian Geotechnical Journal 34: 477–488.

Li, J. H., and L. M. Zhang. 2010. Geometric parameters and REV of a crack network in soil. Computers and Geotechnics 37: 466–475.

Li, J. H., and L. M. Zhang. 2011. Study of desiccation crack initiation and development at ground surface. Engineering Geology 123: 347–358.

Lockley, M. G., J. O. Farlow, and C. A. Meyer. 1994. Brontopodus and Parabrontopodus ichnogen. nov. and the significance of wide- and narrow-gauge sauropod trackways. Gaia 10: 135–146.

Marty, D. 2008. Sedimentology, taphonomy, and ichnology of Late Jurassic dinosaur tracks from the Jura carbonate platform (Chevenez–Combe Ronde tracksite, NW Switzerland): insights into the tidal-flat palaeoenvironment and dinosaur diversity, locomotion, and palaeoecology. PhD dissertation, University of Fribourg, Fribourg, Switzerland. GeoFocus 21: 278 pp.

Marty, D., W. A. Hug, A. Iberg, L. Cavin, C. A. Meyer, and M. G. Lockley. 2003. Preliminary report on the Courtedoux rinosaur tracksite from the Kimmeridgian of Switzerland. Ichnos 10: 209–219.

Marty, D., M. Belvedere, C. A. Meyer, P. Mietto, G. Paratte, C. Lovis, and B. Thüring. 2010. Comparative analysis of Late Jurassic sauropod trackways from the Jura Mountains (NW Switzerland) and the central High Atlas Mountains (Morocco): implications for sauropod ichnotaxonomy. Historical Biology 22: 109–133.

Morris, P. H., J. Graham, and D. J. Williams. 1992. Cracking in drying soils. Canadian Geotechnical Journal 29: 263–277.

Nahlawi, N., and J. K. Kodikara. 2006: Laboratory experiments on desiccation cracking of thin soil layers. Geotechnical and Geological Engineering 24: 1641–1664.

Nguyen-Tuan, L., V. Viefhaus, M. Datcheva, and T. Schanz. 2013. Coupled thermo-hydro-mechanical modelling of crack development along fossil dinosaur's footprints in soft cohesive sediments; p. 555, 13 pp. in S. Idelsohn, M. Papadrakakis, and B. Schrefler (eds.), V Int. Conference on Computational Methods for Coupled Problems in Science and Engineering, Coupled Problems 2013. CD-ROM Proceedings ISBN: 978-84-941407-5-4, E-book ISBN: 978-84-941407-6-1.

Olivella, S., J. Carrera, A. Gens, and E. E. Alonso. 1994. Non-isothermal multiphase flow of brine and gas through saline media. Transport in Porous Media 15: 271–293.

Olivella, S., J. Carrera, A. Gens, and E. E. Alonso. 1996. Porosity variations in saline media caused by temperature gradients coupled to multiphase flow and dissolution/precipitation. Transport in Porous Media 25(1): 1–25.

Péron, H. 2008. Desiccation cracking of soils. PhD dissertation, École Polytechnique Fédérale de Lausanne, Lausanne, Switzerland, 392 pp.

Péron, H., J. Y. Delenne, L. Laloui, and M. S. El Youssoufi. 2009. Discrete element modelling of drying shrinkage and cracking of soils. Computers and Geotechnics 36: 61–69.

Péron, H., L. B. Hu, L. Laloui, and T. Hueckel. 2007. Numerical and experimental investigations of desiccation of soil; pp. 391–396 in Z. Z. Yin, Y. P. Yuan, and A. C. F. Chiu (eds.), Proceedings of the 3rd Asian Conference on Unsaturated Soils. Science Press, Princeton Junction, New Jersey.

Péron, H., T. Hueckel, L. Laloui, and L. B. Hu. 2009. Fundamentals of desiccation cracking of fine-grained soils: experimental characterisation and mechanisms identification. Canadian Geotechnical Journal 46: 1177–1201.

Plummer, P. S., and V. A. Gostin. 1981. Shrinkage cracks: desiccation or synaeresis? Journal of Sedimentary Petrology 51(4): 1147–1156.

Prat, P. C., A. Ladesma, M. R. Lakshmikantha, H. Levatti, and J. Tapia. 2008. Fracture mechanics for crack propagation in drying soils.; 8 pp. in Proceedings of the 12th International Conference of the International Association for Computer Methods and Advances in Geomechanics (IACMAG), Goa, India, October 1–6.

Rayhani, M. H. T., E. K. Yanful, and A. Fakher. 2008. Physical modeling of desiccation cracking in plastic soils. Engineering Geology 97: 25–31.

Rodríguez, R., M. Sánchez, A. Ledesma, and A. Lloret. 2007. Experimental and numerical analysis of desiccation of a mining waste. Canadian Geotechnical Journal 44: 644–658.

Romero, E. 2013. A microstructural insight into compacted clayey soils and their hydraulic properties. Engineering Geology 165: 3–19.

Schanz, T. 2009. Zu Verformungsverhalten und Rissgefährdung einer mineralischen Abdichtung im Deponiebau infolge gekoppelter zyklischer hydraulischer und mechanischer Beanspruchung. Bautechnik 86(2): 111–119.

Schanz, T., Y. Lins, H. Viefhaus, and P. M. Sander. 2012. Quantitative interpretation of dinosaur tracks revisited. Journal of Vertebrate Paleontology 32(Supplement): 166A.

Schanz, T., Y. Lins, H. Viefhaus, T. Barciaga, S. Läbe, H. Preuschoft, U. Witzel, and P. M. Sander. 2013. Quantitative interpretation of tracks for determination of body mass. PLoS One 8(10): e77606.

Sellers, W. I., L. Margetts, R. A. Coria, and P. L. Manning. 2013. March of the titans: the locomotor capabilities of sauropod dinosaurs. PLoS One 8(10): e78733.

Sellers, W. I., P. L. Manning, T. Lyson, K. A. Stevens, and L. Margetts. 2009. Virtual palaeontology: gait reconstruction of extinct vertebrates using high performance computing. Palaeontologia Electronica 12(3): 11A.

Shin, H., and J. C. Santamarina. 2011. Desiccation cracks in saturated fine-grained soils: particle level phenomena and effective stress analysis. Géotechnique 61: 961–972.

Tang, C.-S., Y.-J. Cui, A.-M. Tang, and B. Shi. 2010. Experiment evidence on the temperature dependence of desiccation cracking behaviour of clayey soils. Engineering Geology 114: 261–266.

Tang, C.-S., B. Shi, C. Liu, W.-B. Suo, and L. Gao. 2011. Experimental characterization of shrinkage and desiccation in thin clay layer. Applied Clay Science 52: 69–77.

Tang, C., B. S. Shi, C. Liu, L. Zhao, and B. Wang. 2008. Influencing factors of geometrical structure of surface shrinkage cracks in clayey soils. Engineering Geology 101: 204–217.

Tang, C.-S., Y.-J. Cui, B. Shi, A.-M. Tang, and C. Liu. 2011. Desiccation and cracking behaviour of clay layer from slurry state under wetting-drying cycles. Geoderma 166(1): 111–18.

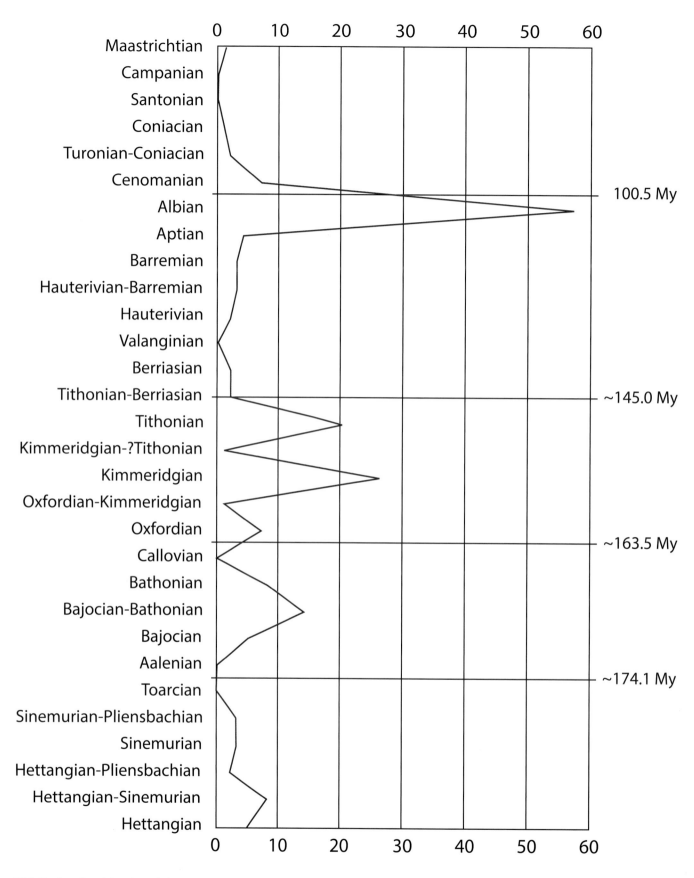

20.1. Number of tracksites through time. Note the Late Jurassic and the Early Cretaceous peaks related to the high number of tracksites from the Reuchenette Formation (Late Jurassic, Switzerland) and the Glen Rose Formation (Early Cretaceous, United States). No record for the Pliensbachian, Toarcian (Early Jurassic), the Aalenian, Callovian (Middle Jurassic), and for the Santonian and Campanian stages (Late Cretaceous) is known. Number of tracksites is on the y-axis, the Jurassic and Cretaceous stages are on the x-axis.

A Review of the Dinosaur Track Record from Jurassic and Cretaceous Shallow Marine Carbonate Depositional Environments

20

Simone D'Orazi Porchetti, Massimo Bernardi, Andrea Cinquegranelli, Vanda Faria dos Santos, Daniel Marty, Fabio Massimo Petti, Paulo Sá Caetano, and Alexander Wagensommer

AN EXTENSIVE LITERATURE ON DINOSAUR ICHNOLOGY is available today, with hundreds of papers describing dinosaur tracks in different depositional settings. In recent years, it has become common practice in paleontology to gather data in databases to ease organization, managing, and analysis of large amounts of information. A review of the occurrences of dinosaur tracks in shallow marine carbonate depositional environments is presented here, based on 131 papers published between 1917 and 2013 and describing a total of 212 tracksites. The raw data set, which adopts the tracksite as the basic unit, reveals an abundance of saurischian footprints and a relative paucity of ornithischian tracks in these depositional environments. Theropods are dominant during most of the Jurassic and the Cretaceous period, whereas sauropods are always well represented and dominant at least in the Late Jurassic. A complementary database for siliciclastic coastal and inland depositional settings is required before testing for trends and patterns to investigate the relationships between dinosaur tracks and facies/environment in general. This contribution intends to promote new large-scale evolutionary studies based on track data.

INTRODUCTION

Intrinsically, the nature of vertebrate tracks deals with in situ trace fossils that can be unambiguously linked to the environment in which they were left (Lockley 1991, 1998). Particular depositional environments are interpreted as having been preferred by specific groups of dinosaurs and other vertebrates, thus giving rise to the tetrapod ichnofacies concept (Lockley et al., 1994; Lockley, Hunt, and Meyer, 1994; Hunt and Lucas, 2007; Lockley, 2007).

Dinosaur tracks are often recorded in coastal, shallow marine environments of Europe and North America (e.g., Farlow, Pittman, and Hawthorne, 1989; Farlow, 1992; Dalla Vecchia, 1994, 1998; Lockley et al., 1994; Meyer and Pittman, 1994; Avanzini et al., 1997, 2006; Kvale et al., 2001; Marty

et al. 2003, 2010), and more recently also from the southern hemisphere in South America (Moreno and Pino, 2002; Moreno and Benton, 2005; Pazos et al., 2012) and Madagascar (Wagensommer, Latiano, and Nicosia, 2010; Wagensommer et al., 2012).

Lockley, Hunt, and Meyer (1994) first noticed the recurrent, statistically significant association of sauropod tracks, especially wide-gauge tracks, together with less-abundant theropod tracks, in low latitude carbonate-dominated systems, and introduced the term *Brontopodus* ichnofacies for this association. The occurrence of dinosaur tracks in these environments also led to some important paleoenvironmental and paleogeographical reinterpretations, as in the case of the Tethyan, peri-adriatic carbonate platforms (Bosellini, 2002; Dalla Vecchia, 2003, 2005; Nicosia et al., 2007; Zarcone et al., 2010), and the Jura carbonate platform (e.g., Meyer, 1990, 1993; Meyer and Hauser, 1994; Marty and Meyer, 2013), which prior to the discovery of dinosaur tracks were thought to be devoid of terrestrial biota. However, it is still rather poorly understood whether dinosaurs used these environments only occasionally as migration corridors, if they were present seasonally for feeding or even breeding, or if at least some of them continuously inhabited these environments (see also, Lopez-Martínez, Moratalla, and Sanz, 2000; Marty, 2008; Myers and Fiorillo, 2009; Diedrich, 2011; Fricke, Hencecroth, and Hoerner, 2011; Marty and Meyer, 2013; Rowe et al., 2013).

This contribution reviews the Jurassic and Cretaceous dinosaur track record of shallow marine carbonate depositional systems, briefly "carbonate platforms," and presents a database (http://dx.doi.org/10.6084/m9.figshare.1348706) that was built for this purpose. This database should serve as a reference for future (statistical) analyses based on this coastal dinosaur track record, and it may also be completed with complementary data from other (siliciclastic) environments and time periods. For the time being, Triassic dinosaur track occurrences, including those usually labeled as "dinosauroid" or "dinosauromorph" tracks (e.g., Marsicano, Domnanovich, and Mancuso, 2007; Brusatte, Niedźwiedzki, and Butler, 2011; Kubo and Kubo, 2013; Niedźwiedzki, Brusatte, and Butler, 2013), are not included as they are often not easy to distinguish from other nondinosaurian archosaur footprints or are hard to refer to one dinosaur group or another (i.e., Saurischia, Ornithischia). However, recently more and more Triassic dinosauroid and dinosaur ichnites are discovered in shallow marine carbonate deposits of Central Europe (e.g., Gand and Demathieu, 2005) and notably the Alps (e.g., Avanzini, 2002; Furrer and Lozza, 2008; Meyer et al., 2009, 2010, 2013; Bernardi et al., 2013), and once these tracks are better understood, they will certainly become an important addition to the database.

MATERIALS AND METHODS

The database was built on data gathered from the literature, and to the best of our knowledge, all available reports of dinosaur tracksites from carbonate-dominated shallow marine deposits of Jurassic and Cretaceous age have been included. In total, 131 publications published between 1917 and 2013 are considered in the database. Apart from some data of tracksites that the authors have studied themselves, the database only includes published data. Not all of this published data can still be verified today, as dinosaur tracks are generally not collected and deposited in a museum collection. This is related to the fact that dinosaur tracksites are generally large, form part of the landscape, and thus cannot easily be relocated to museum collections. Moreover, most tracksites are not protected against erosion and disintegration, and for this reason, some previously available data may today be distorted, destroyed, or not be accessible.

In order to avoid subjective filtering at the data-entry level, the basic unit adopted in the database is the tracksite, which refers to a natural, in situ area composed of one or several bedding planes (levels, horizons) with tracks. Accordingly, at a given tracksite, tracks may not occur in just one single ichnoassemblage but at multiple, superimposed levels, and true tracks may also be associated on a single surface with undertracks and/or overtracks (Marty, 2008). Even though the entire track-bearing interval of a given tracksite may usually be considered as penecontemporaneous in geological terms, such intervals may represent up to several tens of thousands of years (Marty, 2008), or one to several elementary sequences in the concept of cyclostratigraphy (Strasser et al., 1999).

All tracksites considered in the database share some basic environmental parameters such as carbonate deposition, close vicinity to the marine depositional paleoenvironments,

warm (sub)tropical climate, and low input of sediment from the continent. Several tracksites are from peripheral continental (e.g., Texas, United States) or intracontinental (lowland coastal, e.g., Wyoming, United States) settings, whereas others are from intraoceanic (e.g., southern Italy, Croatia) and epeiric (Jura Mountains, Switzerland) carbonate platforms located far away from emergent mainland. Accordingly, the database includes tracksites from a range of different marginal marine environmental settings, and at a single tracksite, several subenvironments of proximal to distal peritidal facies (e.g., sabkha, algal marsh, mudflat, tidal channel) may also be observed in close lateral and/or vertical vicinity. However, a detailed description of the spatial distribution of shallow water carbonate depositional environments during the Jurassic and the Cretaceous is outside the scope of this chapter, and the reader is referred to the sedimentological literature on this topic (e.g., Hardie and Shinn, 1986; Tucker et al., 1990; Demicco and Hardie, 1994; Wright and Burchette, 1996; Schlager, 2002; Kiessling, Flügel, and Golonka, 2003).

In track-rich regions, numerous tracksites may occur at short distances from one another and even belong to the same stratigraphic interval (e.g., Marty et al., 2007). As we will show in more detail later, peaks in the number of tracksites from a given region (Fig. 20.1) may be a function of intensive studying (sampling bias). The quality of the fossil (ichnological) record or the discrepancy in the number of entries and amount of data from different areas and hence the quality of the biological signal may depend on the presence of research institutions close to the study areas, number of specialists for selected branches (ichnology, in this case), outcrop area and territorial accessibility of potential data-yielding rock units, and so on (e.g., Alroy et al., 2001; Smith, 2001; Benson and Upchurch, 2012; Dunhill, 2012; Dunhill et al., 2012; Smith and Benson, 2013). A good example are the systematic track excavations carried out prior to the construction of Highway A16 in northwest Switzerland (e.g., Marty et al., 2003, 2007; Marty, 2008), which have led to the documentation of a total of nearly 14,000 dinosaur tracks including 656 trackways from six tracksites, which are located geographically very close one to another (within a few kilometers) and which constitute one of the most abundant data entries in the database.

DATABASE STRUCTURE

Chrono- and Lithostratigraphy

The three categories "epoch," "stage," and "formation" are related to the age of the tracksite. For some localities, no precise dating is available, and two or more successive stages bracket the best available range estimate. Where a formal lithostratigraphic designation is lacking, "no data" figures as an entry.

Geographical and General Setting

The five categories "tracksite name," "number of levels," "locality," "country," and "outcrop area" provide general information about the tracksite. "Number of levels" indicates how many different levels with dinosaur tracks are present at any given tracksite. "Locality" is specified by the name of a nearby locality or a toponym, as reported in the original publication. The average dimension of the outcrop area is expressed in square meters, but in many cases, there is no precise information available about the tracksite dimensions.

References

This category comprises the references that were used as data sources. A specific bibliography for the database is included.

Tracks, Trackways, and Trackmakers

Related information is provided in nine categories. The category "total number of tracks and trackways" reports the total number of tracks or trackways at a single tracksite. The category "inferred trackmaker(s)" is the trackmaker attribution based on the identification given by original authors. When conflicting attributions are found in different papers on the same material, we report the attribution with the largest consensus (see *Deltapodus* attribution). If no consensus is found on the attribution, we report conflicting hypotheses in the supplementary notes, supplying the original attribution in the "trackmakers" column.

The three categories "herbivores (H)," "carnivores (C)," and "ratio H/C" are related to the diet of the inferred trackmakers and the ratio between these two categories. All nonavian dinosaurs are divided in two broad ecologic categories, "herbivores," including Sauropodomorpha and all ornithischians, and "carnivores," referring to all Theropoda. We consider the Theropoda as largely carnivorous, following the consensus on most groups of nonavian theropods (Fastovsky and Smith, 2004), even though some of them, especially among coelurosaurs, are considered as herbivores (Weishampel and Norman, 1989; Barrett, 2005; Butler and Barrett, 2008; Zanno and Makovicky, 2011). The ratio (H/C) is expressed exclusively in relative terms. H = C is reported when these two categories are equally abundant. We adopted the mathematical symbols >, < (greater than, lesser than), or even >>, << (much greater than, much lesser than) in case one category is much more abundant than another. In the case of monotypic tracksites (in terms of dietary type), H (or C) = 100% is reported.

Track sizes

Four categories named "average dimensions (pes track length) theropods/sauropods/ornithopods/other ornithischians" comprise information about track size. As all the data is derived from literature, we assume that the pes length is usually measured from the rear of the foot to the tip of the most protruding toe (which is generally digit III), but there are many exceptions to this general rule, depending on the type of track (e.g., sauropod or tridactyl tracks) and on track preservation (e.g., completeness). For instance, most of the tridactyl tracks described from the Middle Jurassic of Wyoming by Kvale et al. (2001) are preserved as isolated toe prints. In this case, we only reported the length of digit III from some selected tracks (accompanied by an asterisk in the database). Furthermore, many tracks are not true tracks but undertracks, overtracks, or penetrative tracks, and this may distort the original dimensions and shape of the foot (Milàn and Bromley, 2006, 2008). The uniformity of the available measurements is another general issue for our database. Many papers report discrete measurements for each track encountered at a single tracksite. In other cases, the original work includes only mean values, with no information on the distribution of these parameters. Moreover, some papers give no absolute measurements but only qualitative assessment (e.g., small, medium, large). In some cases, no data were available at all. In the "track size" column, we have tried to mitigate these different approaches as follows: if no measurements were available, the corresponding box has no entry (e.g., no data). When no numerical data were available, the original qualitative term was reported (small or large, for instance). If an average length is the only available information, it is reported in the appropriate box. In several cases, a size range is reported. When the original article gave a complete list of measurements for several footprints, we calculated a mean value and added the minimum and maximum values, as appropriate.

Depositional Setting

This category comprises all relevant information (e.g., supratidal/intertidal flat, sabkha) about the depositional paleoenvironment based on the indicated references. However, original works are not always accompanied by a detailed analysis of the geologic framework and often only provide general information such as "shallow marine depositional environment."

Supplementary Notes

A further column is dedicated to supplementary notes. Information is reported here that cannot be addressed in previous categories, such as the current conditions of the outcrop (destroyed, eroded) or specific information on the history of the sites.

DISCUSSION

Dinosaur tracksites used for this study belong to distinct regions, but these may also have changed their paleogeographic position and environmental setting between the Jurassic and the Cretaceous. North America yields tracks from lowlands bordering an epicontinental sea (Sundance Formation, Wyoming) during the Middle Jurassic and from coastal environments facing the Western Interior Seaway (Glen Rose Formation, Texas) during the Early Cretaceous. Many tracksites are from the so-called European archipelago (e.g., Iberian Massif, Armorican Massif, and Massif Central), where sea-level rise and fall (either linked to seafloor spreading or to climatic changes) continuously and strongly modified the paleogeographical and paleoenvironmental setting throughout the Mesozoic. A third region encompasses the Periadriatic area where an alternation of carbonate platforms and pelagic basins formed a complex patchwork during most of the Jurassic and the Cretaceous. In this scenario, track-bearing localities are from carbonate platform environments (Trento Platform, Apulian Platform, Apennine Platform, Adriatic-Dinaric Platform). Central Asia has yielded large dinosaur tracksites from Turkmenistan and Uzbekistan, representing lowland shorelines of a Late Jurassic large epicontinental sea. From Africa, a good, new sample comes from Madagascar, where tracks are found on the western coast of the Eastern Gondwana continent. Coastal deposits from northern Africa (e.g., Morocco, Tunisia) in a mixed carbonate and siliciclastic facies are excluded because they do not match the paleoenvironmental requirements of this review. To the best of our knowledge, the Far East and Australia are devoid of dinosaur tracks from shallow marine carbonate environments.

As opposed to a considerable amount of data on dinosaur tracks from shallow marine carbonates, reports of extant large tetrapods within marginal marine environments are rather rare. This may depend on two factors: first, there may be no exact homology between Mesozoic and modern carbonate depositional settings, especially for large epeiric seas that are absent today (Hallam, 1992; Harries, 2009), and second, there might be no extant ecological homologues for

nonavian dinosaurs. This creates a severe limiting factor: the virtual absence of present-day possibilities of comparison on this topic.

Taphonomy is a further relevant issue that must be considered. Chemical-physical and microbiological factors may play a strong role in early consolidation of shallow marine carbonate deposits, thus helping to preserve vertebrate bioturbation. Microbial mats are especially relevant in this process (Marty, Strasser, and Meyer, 2009) and may have acted like flypaper for dinosaur tracks. Early dolomitization of the trampled substrate has also been suggested as an important preservational variable (Avanzini et al., 1997). However, dinosaur tracks are also abundant in siliciclastic coastal plain facies, thus suggesting the need for further clastic-carbonate facies comparisons before the preservational role of microbial mats is fully evaluated and understood.

Trackmaker Attribution

Ascribing tridactyl tracks to either theropods or basal ornithopods always involves a certain degree of uncertainty (e.g., Farlow et al., 2006; Castanera et al., 2013; Hübner, 2016). To some extent, uncertainties in trackmaker attribution may also affect the tracks of large and obligate quadrupedal herbivores (saurischians vs. ornithischians). A good example is the case of *Deltapodus*, an ichnotaxon first attributed to a sauropod (Whyte and Romano, 1994) but today considered as a stegosaurian ichnotaxon by most ichnologists (Whyte and Romano, 1994, 2001; Lockley and Meyer, 2000; Cobos et al., 2010; Li et al., 2012; Xing et al., 2013, and references therein). This ambiguity in trackmaker attribution is potentially an intrinsic issue in ichnology and is also linked to preservational quality, as a result of dynamic interactions between anatomical, kinematic, and sedimentological factors (e.g., Padian and Olsen, 1984; Gatesy, 2003).

Relative Timing of Trackway Registration and Assessing Trackmaker Frequency

Determining the relative timing of trackway registration in a given ichnoassemblage is a difficult task. A (parallel) series of trackways each with similar track dimensions may either result from a single individual passing several times or from several individuals of similar size passing together in a group or a herd (Ostrom, 1972; Lockley, 1986, 1989; Lockley and Hunt, 1995; Myers and Fiorillo, 2009). However, if compared to many body fossil deposits, the temporal window for track-making and preservation of tracks is comparatively short and may vary between a few hours and up to several days, weeks, or months (Cohen et al., 1993; Marty, 2008; Marty, Strasser, and Meyer, 2009). Trackways are often used in census studies as the unit for counting individuals on each individual track-site (Lockley, 1998), and the same approach was adopted in this work. Unfortunately, identifying trackways on a heavily dinoturbated surface is sometimes impossible because of too many interfering tracks. An alternative possibility may be a classification based on a tracks/area ratio, the dinoturbation index (Lockley and Conrad, 1989), but this ratio is generally impossible to assess if based on literature data only. Sometimes tracksites are not located very conveniently or, when cleaned up in order to expose tracks, the travel direction may be chosen as the rationale under which a site is expanded by excavation of overlying rocks. By doing so, an outstanding long trackway may be the result, but another important pattern (e.g., several parallel trackways, herding) may not be detected (Marty et al., 2012).

Jurassic and Cretaceous Tracksites at a Glance: Distribution over Time, Space, and Taxonomic Groups

To the best of our knowledge, a total of 212 dinosaur tracksites have been reported from Jurassic and Cretaceous shallow marine carbonate systems worldwide. Almost every stage of the Jurassic and the Cretaceous includes occurrences of dinosaur tracks in shallow marine carbonate systems. Exceptions are the Pliensbachian and Toarcian (Early Jurassic), the Aalenian and Callovian (Middle Jurassic), and the Santonian and Campanian stages (Late Cretaceous).

Some 129 tracksites are known from the Jurassic and 83 from the Cretaceous. However, this number is somewhat misleading, as some formations contain many individual tracksites all located approximately at the same stratigraphic level and yielding an almost identical ichnofauna. There are also some formations with several tracksites at different stratigraphic levels, but the track-bearing layers still are very close in space and time and are similar in composition. For instance, almost all known dinosaur tracksites from the Early Jurassic are from the Causses (France) and from the Province of Trento (northern Italy), whereas half of all Middle Jurassic entries belong to the Bemaraha Formation in western Madagascar. Not less than 49 tracksites (23% of our total database) have been reported from the Glen Rose Formation of Texas (Albian, Early Cretaceous), which is the most prolific formation in our database, followed by the Reuchenette Formation in northwest Switzerland (Kimmeridgian, Late Jurassic), which yielded 25 individual tracksites. If we consider tracksites from the same geological formation as a single entry, 49 occurrences remain worldwide, with the bulk of them (31 occurrences, or 63%) being quite evenly distributed over the Late Jurassic and Early Cretaceous. Referring for statistical purposes to either these 49 occurrences by formation or all

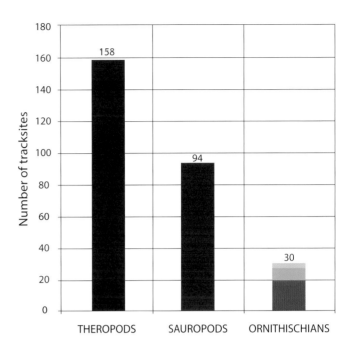

20.2. Total occurrences for each dinosaur group. Theropods are the most abundant, with 158 entries (left column), and sauropods follow with 94 (central column). Ornithischians are poorly represented in the record with respect to Saurischia. A total of 30 entries (right column) is the sum of 19 entries for ornithopods (dark lower block), 8 thyreophorans (middle block), and 3 for other ornithischians (pale upper block).

212 known individual tracksites is mainly a matter of what kind of question is investigated. On the other hand, abundances of dinosaur groups in the database are easily visualized when data from different outcrops (tracksites) are separated, so this approach is followed herein (Fig. 20.2). In this regard, it can be argued that we must utilize the raw data on tracksites and not infer (interject our own bias) by concluding that some formations are "oversampled" when they may in fact genuinely represent depositional systems where track registration is naturally greater due to the intrinsic dynamics that create the sedimentary record: in other words, there is no objective way to pool some data sets and not others.

Nevertheless, the geographic distribution of the tracksites depends on the outcrop area of potential data-yielding rock units (i.e., shallow marine deposits), but it also strongly reflects the intensity with which different geographic areas have been studied (sampling bias). Of the 212 tracksites, 170 are located in Europe, 59 in North America (United States only), 13 localities in Africa (Madagascar only), 7 localities in Asia (including the Arabian Peninsula), and 3 tracksites in South America. No occurrences are known from Australia or Antarctica to date. The United States (59 tracksites) is the richest country in terms of single track-bearing localities, followed by France (46) and Switzerland (30). In all these countries, potential data-yielding rock units crop out over large areas and have been studied intensively since at least the 1980s

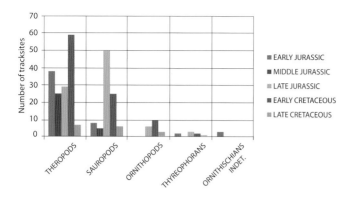

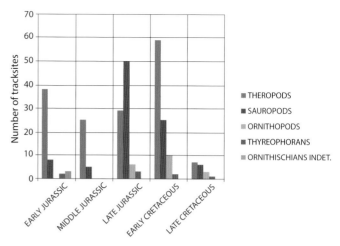

20.3. Distribution through time for each trackmaker group. The y-axis is for the number of tracksites, whereas trackmaker groups are reported on the x-axis. Each column has a different color depending on the time sampled (see legend on the right). Theropods are the most common group, except for the Late Jurassic, where sauropod tracks became dominant. Theropod track occurrences have a peak in the Early Cretaceous, but they fall down in the next stage (Late Cretaceous), where the record is generally low for all groups. Ornithopods are best represented in the Early Cretaceous, whereas thyreophorans and other ornithischians have a very scarce record throughout the Jurassic and the Cretaceous.

20.4. Distribution of trackmaker groups over time (epochs on the x-axis). Theropods are well represented throughout the Jurassic and the Cretaceous. The absolute peak in theropod tracks is in the Early Cretaceous, possibly in correspondence with highly prolific track-bearing formations. Sauropods are dominant trackmakers in the Late Jurassic, being less represented before and after this epoch. Ornithopod tracks appear in the Late Jurassic with few entries until the end of the Cretaceous. Thyreophoran and generic ornithischian tracks appear as soon as the Early Jurassic, but they are very poorly represented in the record.

Of the 212 tracksites, 158 yield tridactyl tracks assigned to theropods. This is the dominant trackmaker group in almost every time slice of the Jurassic and Cretaceous periods except for the Late Jurassic where they are outnumbered by sauropods. A total of 103 tracksites yielded theropod tracks only.

Overall, sauropods are the second most common group, and they outnumber all other ornithischian herbivores in each of the five time slices we have considered (Early, Middle, and Late Jurassic, and Early and Late Cretaceous). Among ornithischians, ornithopods are the most common group, with 19 entries, although almost half of them (9) are reported as dubious, presumably due to the danger of confusing them with theropod tracks. Thyreophorans are a rare but recurrent element, with eight entries quite evenly distributed over the considered time interval. Three sites, all from the Early Jurassic of France, have yielded tracks attributed to undetermined ornithischians (Fig. 20.3).

There are 55 localities where herbivores and carnivores co-occur, of which 36 yield a theropod/sauropod association and 8 a theropod/ornithischian association. Co-occurrences of theropods with both sauropods and ornithischians have only been reported from 11 tracksites. Of 212 tracksites, 44 contain sauropod tracks only. Herbivores-only associations with both saurischian and ornithischian dinosaurs are known from 3 localities only. At 14 tracksites out of 36, where sauropods and theropods co-occur, sauropods are more abundant than theropods. This latter group is dominant at 12 tracksites, whereas sauropods and theropods are equally abundant at 2 localities only. The remaining 10 tracksites yield no data on the H/C ratio. Theropods co-occur with sauropods and other herbivores (e.g., ornithopods and other ornithischians) at 47 localities. We observed a majority of herbivores at 20 tracksites, whereas

at 17 localities, theropods dominate in number. At 2 localities, herbivores and carnivores were equally abundant. No data were available for 8 tracksites (Fig. 20.4).

When the data is sorted by lithostratigraphic units (formations), the results are slightly different, because of fewer "raw data issues" (i.e., analysis based on the number of tracksites). As stated earlier, this is related to the presence of multiple tracksites from the same formation that may be geographically located close to each other and thus better be considered as one entry only (data sorted by lithostratigraphic unit). The general dominance of saurischians over ornithischians is confirmed for all time units, but an important reduction in the relative abundance of each taxonomic group (i.e., theropods, sauropods, other ornithischians) is detected. For instance, during the Early Jurassic, the count of "other ornithischian" is equal to theropods and greater than sauropods (Fig. 20.5). This is slightly in contrast with the results obtained from the investigation based on tracksites.

Observation on the Brontopodus Ichnofacies Concept

Lockley, Hunt, and Meyer (1994) first stated the presence of recurrent ichnocoenoses with sauropod (originally widegauge sauropods, i.e., *Brontopodus* and cf. *Brontopodus*) and large theropod tracks in shallow marine and lacustrine carbonate depositional environments. Most of the original data set (Lockley, Hunt, and Meyer, 1994) was recovered from North America (Texas) and Europe (Switzerland and Portugal), but occurrences from South America, eastern Asia, and

Africa were also considered. The original definition of the *Brontopodus* ichnofacies was later widened by including ichnoassemblages yielding *Parabrontopodus* and similar forms, with dominance or general abundance of sauropod tracks (e.g., Meyer and Hauser, 1994; Meyer and Pittman, 1994; Lockley and Meyer, 2000). In later revisions of the ichnofacies concept, Hunt and Lucas (2007) proposed conceptual redefinition of the categories and of their ranking. In their interpretation, all tetrapod ichnofacies defined prior to 2005 should be considered as ichnocoenoses, whereas the ichnofacies label was used to characterize only five very broadly defined universal or archetypal ichnofacies. For example, their *Brontopodus* ichnofacies actually included clastic as well as carbonate facies-related ichnoassemblages (their ichnocoenoses, or ichnofacies of other workers) specifically stating that they were usually herbivore dominated with a low (about 10%) number of carnivores and high (4–8) ichnogeneric diversity. This definition is not in the least consistent with the carbonate platform ichnofaunas described here, which are heavily theropod and sauropod dominated (usually only 2 ichnogenera) as is characteristic of the *Brontopodus* ichnofacies as originally defined (Lockley, Hunt, and Meyer, 1994). In x-axis subsequent works, Lockley (2007) and Lockley and Gierlinski (2014) reconsidered both these approaches, distinguishing an old and a new paradigm for the ichnofacies concept, referring the former to the definition of Lockley, Hunt, and Meyer (1994) and the latter to that of Hunt and Lucas (2007). A recent evaluation on the meaning and usefulness of vertebrate ichnofacies was presented by Santi and Nicosia (2008).

As we limited our analysis to a specific depositional setting within marine carbonate systems, we can test whether correlations between a depositional environment and the vertebrate track record occur and whether any pattern is retrievable. We can primarily state that saurischian dinosaur tracks are dominant in shallow marine/coastal deposits, and theropod tracks are the most abundant feature.

Consequently, the question arises whether the *Brontopodus* ichnofacies is a specific subset in the larger scenario of vertebrate tracks associated to shallow marine carbonate facies. The difficulty to assign some dinosaur track assemblage to one or another of the meanings of the *Brontopodus* ichnofacies was stated by Marty (2008) who noticed that the Transjurane ichnocoenosis was different from previously defined ichnocoenoses, and specifically from the *Brontopodus* ichnocoenosis as defined by Hunt and Lucas (2007). Marty (2008) interpreted the Transjurane as a new ichnocoenosis of the *Brontopodus* ichnofacies (in the definition of Lockley, 2007).

Limiting our analysis to the ichnocoenoses where theropods are associated to sauropods (exclusively, or with a small number of ornithischians), we can state that carnivores are

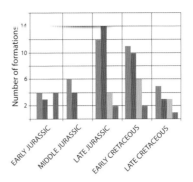

20.5. Occurrences of trackmaker groups over the Jurassic and the Cretaceous, sampled by formations (FM). Differences in relative abundance (i.e., theropods vs. sauropods) are reduced if compared to the graphs based on the tracksite unit (compare to Figs. 20.3 and 20.4). A total of 14 formations are reported on the y-axis, time is on the x-axis. Saurischians are dominant over ornithischians in all time units, but a reduction in the relative abundance of each taxonomic group is evident. For instance, during the Early Jurassic, the number of "other ornithischians" equals theropods and is greater than sauropods.

almost as common as herbivores and, in general, localities where H>>C are extremely rare. In any case, our data gives no numerical quantification on the relative abundance of one group with respect to another.

In the light of our investigation, it is of interest to observe that Hunt and Lucas (2007) assumed for the *Brontopodus* ichnocoenosis a pattern of abundance for theropod tracks below the threshold of 10%. Our results do not confirm this conclusion and in fact theropod footprints are usually abundant also when associated to herbivores tracks. Our data are more, if not fully, in agreement with the conclusions of Lockley, Hunt, and Meyer (1994) as far as theropod and sauropod tracks, when in co-occurrence, show near-equal abundance.

However, any further comment on the *Brontopodus* ichnofacies must take into account those coastal environments where mixed and siliciclastic deposition took place. The preference of one group for a specific environment must therefore be tested on a more complete data set.

CONCLUSIONS

Our data show that theropod tracks are dominant in shallow marine carbonate systems. Actually, theropods appear to be over-represented if compared with herbivores, the bulk being made up by small to medium-sized animals, and this may be related to a higher activity level of (small) carnivores than herbivores.

Sauropod footprints are the second most abundant ichnites. There is a strong affinity of some sauropod groups to coastal environments.

Ornithischians are relatively rare in comparison with saurischians. Most ornithischians appear to have entered these environments only occasionally.

The reasons for these habitat preferences are manifold and might be related to differences in physiology, different behaviors, and ecological/trophic needs that were more or less adapted for these coastal environments. In order to test our conclusions, in the future, these data should be plotted against the general pattern of distribution based on the body fossil record. This process might take into account the estimated abundance of these groups in terms of biomass, and their relative distribution during the Jurassic and the Cretaceous. The relative scarcity of data we recorded from the Late Cretaceous may be critical in this process, and the body fossil record could help drawing a more complete scenario.

This work is a first step toward the compilation of a database on the global distribution of dinosaur tracks in coastal environments, including those in widespread siliciclastic facies (Lockley et al., 2006, 2010, 2014; McCrea et al., 2014). In the future, a possible extension to inland depositional settings is envisaged. This database should be a core for further implementation, and should also include the Triassic period.

A further step would be the statistical elaboration of the data set, as far as a greater amount of data becomes available.

ACKNOWLEDGMENTS

M. Avanzini (Museo delle Scienze, Trento, Italy), A. Cobos (Fundacion Conjunto Paleontologico de Teruel-Dinopolis, Spain), C. A. Meyer (Naturhistorisches Museum Basel, Switzerland), G. Gierlinski (Polish Geological Institute, Warsaw, Poland), I. de S. Carvalho (University of Rio de Janeiro, Brazil) are kindly acknowledged for bibliographic input and comments. M. G. Lockley and an anonymous reviewer are acknowledged for their insightful comments and suggestions on an earlier version of the manuscript. We would also like to thank the editors for their constant and valuable assistance. S. D'Orazi Porchetti's research is supported by a Fundação de Amparo à Pesquisa do Estado de São Paulo (FAPESP) postdoctoral grant (process number 2013/01930-1).

REFERENCES

Alroy, J., C. R. Marshall, R. K. Bambach, K. Bezusko, M. Foote, F. T. Fürsich, T. A. Hansen, S. M. Holland, L. C. Ivany, D. Jablonski, D. K. Jacobs, D. C. Jones, M. A. Kosnik, S. Lidgard, S. Low, A. I. Miller, P. M. Novack-Gottshall, T. D. Olszewski, M. E. Patzkowsky, D. M. Raup, K. Roy, J. J. Sepkoski Jr., M. G. Sommers, P. J. Wagner, and A. Webber. 2001. Effects of sampling standardization on estimates of Phanerozoic marine diversification. PNAS 98(11): 6261–6266.

Avanzini, M. 2002. Dinosauromorph tracks from the Middle Anisian (Middle Triassic) of the Southern Alp (Valle di Non–Italy). Bolletino della Societa Paleontologica Italiana 41(1): 37–40.

Avanzini, M., S. Frisia, K. Van Den Driesche, and E. Keppens. 1997. A dinosaur tracksite in an Early Liassic tidal flat in northern Italy: paleoenvironmental reconstruction from sedimentology and geochemistry. Palaios 12: 538–551.

Avanzini, M., D. Piubelli, P. Mietto, G. Roghi, R. Romano, and D. Masetti. 2006. Lower Jurassic (Hettangian-Sinemurian) dinosaur track megasite, Southern Alps, northern Italy; pp. 207–216 in J. D. Harris, S. G. Lucas, J. A. Spielman, M. G. Lockley, A. R. C. Milner, and J. I. Kirkland (eds.), The Triassic-Jurassic Terrestrial Transition. Bulletin 37. New Mexico Museum of Natural History and Science, Albuquerque, New Mexico.

Barrett, P. M. 2005. The diets of ostrich dinosaurs (Theropoda: Ornithomimosauria). Palaeontology 48: 347–358.

Benson, B. J., and P. Upchurch. 2012. Diversity trends in the establishment of terrestrial vertebrate ecosystems: interactions between spatial and temporal sampling biases. Geology 41: 43–46.

Bernardi, M., F. M. Petti, S. D'Orazi Porchetti, and M. Avanzini. 2013. Large tridactyl footprints associated with a diverse ichnofauna from the Carnian of Southern Alps; pp. 48–54 in L. H. Tanner, J. A. Spielmann, and S. G. Lucas (eds.), Triassic Geology and Paleontology. Bulletin 61. New Mexico Museum of Natural History and Science, Albuquerque, New Mexico.

Bosellini, A. 2002. Dinosaurs 're-write' the geodynamics of the eastern Mediterranean and the paleogeography of the Apulian Platform. Earth-Science Reviews 59: 211–234.

Brusatte, S. L., G. Niedźwiedzki, and R. J. Butler. 2011. Footprints pull origin and diversification of dinosaur stem-lineage deep into Early Triassic. Proceedings of the Royal Society of London B 278: 1107–1113.

Butler, R. J., and P. M. Barrett. 2008. Palaeoenvironmental controls on the distribution of Cretaceous herbivorous dinosaurs. Naturwissenschaften 95: 1027–1032.

Castanera, D., B. Vila, N. L. Razzolini, P. L. Falkingham, J. I. Canudo, P. L. Manning, and À Galobart. 2013. Manus track preservation bias as a key factor for assessing trackmaker identity and quadrupedalism in basal ornithopods. PLoS One 8(1): e54177. doi:10.1371/journal. pone.0054177.

Cobos, A., R. Royo-Torres, L. Luque, L. Alcalá, and L. Mampel. 2010. An Iberian stegosaurs paradise: the Villar del Arzobispo Formation (Tithonian-Berriasian) in Teruel (Spain). Palaeogeography Palaeoclimatology Palaeoecology 293: 223–236.

Cohen, A., J. Halfpenny, M. G. Lockley, and E. Michel. 1993. Modern vertebrate tracks from Lake Manyara, Tanzania and their paleobiological implications. Paleobiology 19: 443–458.

Dalla Vecchia, F. M. 1994. Jurassic and Cretaceous sauropod evidence in the Mesozoic carbonate platforms of the Southern Alps and Dinarids; pp. 65–74 in M. G. Lockley, V. F. dos Santos, C. A. Meyer, and Hunt, A. P. (eds.), Aspects of Sauropod Paleobiology. Special issue, Gaia 10.

Dalla Vecchia, F. M. 1998. Theropod footprints in the Cretaceous Adriatic-Dinaric carbonate platform (Italy and Croatia). Gaia 15: 366–367.

Dalla Vecchia, F. M. 2003. Observations on the presence of plant-eating dinosaurs in an oceanic carbonate platform. Natura Nascosta 27: 14–27.

Dalla Vecchia, F. M. 2005. Between Gondwana and Laurasia: Cretaceous sauropods in an intraoceanic carbonate platform; pp. 395–429 in V. Tidwell and K. Carpenter (eds.), Thunder-Lizards. Indiana University Press, Bloomington, Indiana.

Demicco, R. V., and L. A. Hardie. 1994. Sedimentary structures and early diagenetic features of shallow marine carbonate deposits. SEPM Atlas of Sedimentary Structures 1. Society for Sedimentary Geology, Tulsa, Oklahoma, 270 pp.

Diedrich, C. G. 2011. Upper Jurassic tidal flat megatracksites of Germany: coastal dinosaur migration highways between European islands, and a review of the dinosaur footprints. Palaeobiodiversity and Palaeoenvironments 91: 129–155.

Dunhill, A. M. 2012. Problems with using rock outcrop area as a paleontological sampling proxy: rock outcrop and exposure area compared with coastal proximity, topography, land use, and lithology. Paleobiology 38(1): 840–857.

Dunhill, A. M., M. J. Benton, R. J. Twitchett, and A. J. Newell. 2012. Completeness of the fossil record and the validity of sampling proxies at outcrop level. Palaeontology 55(6): 1139–1175.

Farlow, J. O. 1992. Sauropod tracks and trackmakers: integrating the ichnological and skeletal records. Zubia 10: 89–138.

Farlow, J. O., J. G. Pittman, and J. M. Hawthorne. 1989. *Brontopodus birdi*, Lower Cretaceous sauropod footprints from the U.S. Gulf Coastal Plain; pp. 371–394 in D. D. Gillette and M. G. Lockley (eds.), Dinosaur Tracks and Traces. Cambridge University Press, Cambridge, UK.

Farlow, J. O., W. Langston Jr., E. E. Deschner, R. Solis, W. Ward, B. L. Kirkland, S. Hovorka, T. L. Reece, and J. Whitcraft. 2006. Texas Giants: Dinosaurs of the Heritage Museum of the Texas Hill Country. Heritage Museum of the Texas Hill Country, Canyon Lake, Texas, 105 pp.

Fastovsky, D. E., and J. B. Smith. 2004. Dinosaur paleoecology; pp. 614–626 in D. B. Weishampel, P. Dodson, and H. Osmólska (eds.), The Dinosauria. 2nd edition. University of California Press, Berkeley, California.

Fricke, H. C., J. Hencecroth, and M. E. Hoerner. 2011. Lowland-upland migration of sauropod dinosaurs during the Late Jurassic epoch. Nature 480: 513–515.

Furrer, H., and H. Lozza. 2008. Neue Funde von Dinosaurierfährten im Schweizerischen Nationalpark. Cratschla, Ediziuns specialas 1: 4–24.

Gand, G., and G. Demathieu. 2005. The French Middle Triassic dinosauroid trackways: palaeontological result and nomenclatural re-evaluation. Geobios 38: 725–749.

Gatesy, S. 2003. Direct and indirect track features: What sediment did a dinosaur touch? Ichnos 10(2–4): 91–98.

Hallam, A. 1992. Phanerozoic Sea-Level Changes. Columbia University Press, New York, New York, 224 pp.

Hardie, L. A., and E. A. Shinn. 1986. Carbonate depositional environments, modern and ancient. Part 3: tidal flats. Colorado School of Mines, Quarterly 81: 1–74.

Harries, P. J. 2009. Epeiric seas: a continental extension of shelf biotas. Earth System: History and Natural Variability 4: 138–156.

Hübner, T. 2016. Elusive ornithischian tracks in the famous Berriasian (Lower Cretaceous) 'Chicken Yard' tracksite of northern Germany: quantitative differentiation between small tridactyl trackmakers; chap. 16 in P. L. Falkingham, D. Marty, and A. Richter (eds.), Dinosaur Tracks: The Next Steps. Indiana University Press, Bloomington, Indiana.

Hunt, A. P., and S. G. Lucas. 2007. Tetrapod ichnofacies: a new paradigm. Ichnos 14: 59–68.

Kiessling, W., E. Flügel, and J. Golonka. 2003. Patterns of Phanerozoic carbonate platform sedimentation. Lethaia 36: 195–226.

Kubo, T., and M. O. Kubo. 2013. Analysis of Triassic archosauriform trackways: difference in stride/foot ratio between dinosauriforms and other archosauriforms. Palaios 28: 259–265.

Kvale, E. P., G. D. Johnson, D. L. Mickelson, K. Keller, L. C. Furer, and A. W. Archer. 2001. Middle Jurassic (Bajocian and Bathonian) dinosaur megatracksites, Bighorn Basin, Wyoming. Palaios 16: 233–254.

Li, J., M. G. Lockley, Y. G. Zhang, M. Matsukawa, S. Hu, and Z. Q. Bai. 2012. An important ornithischian tracksite in the Early Jurassic of the Shenmu Region, Shaanxi, China. Acta Geological Sinica 86. 1–10.

Lockley, M. G. 1986. The paleobiological and paleoenvironmental importance of dinosaur footprints. Palaios 1(1): 37–47.

Lockley, M. G. 1989. Tracks and traces: new perspectives on dinosaurian behavior, ecology and biogeography; pp. 134–145 in K. Padian and D. J. Chure (eds.), The Age of Dinosaurs. Short Courses in Paleontology 2. Paleontological Society, Knoxville, Tennessee.

Lockley, M. G. 1991. Tracking Dinosaurs: A New Look at an Ancient World. Cambridge University Press, Cambridge, UK, 252 pp.

Lockley, M. G. 1998. The vertebrate track record. Nature 396: 429–432.

Lockley, M. G., 2007. A tale of two ichnologies: the different goals and potentials of invertebrates and vertebrate (tetrapod) ichnotaxonomy and how they relate to ichnofacies analysis. Ichnos 14(1–2): 39–57.

Lockley, M. G., and A. P. Hunt. 1995. Dinosaur Tracks and Other Fossil Footprints of the Western United States. Columbia University Press, New York, New York, 338 pp.

Lockley, M. G., and C. A. Meyer. 2000. Dinosaur Tracks and Other Fossil Footprints of Europe. Columbia University Press, New York, New York, 360 pp.

Lockley, M. G., and G. D. Gierlinski. 2014. Jurassic ichnofaunas and ichnofacies of the western interior USA. Volumina Jurassica 12(2): 133–150.

Lockley, M. G., and K. Conrad. 1989. The paleoenvironmental context, preservation and paleoecological significance of dinosaur tracksites in the western USA; pp. 121–134 in D. D. Gillette and M. G. Lockley (eds.), Dinosaur Tracks and Traces. Cambridge University Press, Cambridge, UK.

Lockley, M. G., A. P. Hunt, and C. A. Meyer. 1994. Vertebrate tracks and the ichnofacies: implications for paleoecology and palichnostratigraphy; pp. 241–268 in S. K. Donovan (ed.), The Palaeobiology of Trace Fossils. John Wiley and Sons, Chichester, UK.

Lockley, M. G., C. A. Meyer, A. P. Hunt, and S. G. Lucas. 1994. The distribution of sauropod tracks and trackmakers; pp. 233–248 in M. G. Lockley, V. F. dos Santos, C. A. Meyer, and A. P. Hunt (eds.), Aspects of Sauropod Paleobiology. Special issue, Gaia 10.

Lockley, M. G., J. Holbrook, R. Kukihara, and M. Matsukawa. 2006. An ankylosaur-dominated dinosaur tracksite in the Cretaceous Dakota Group of Colorado and its paleoenvironmental and sequence stratigraphic context. New Mexico Museum of Natural History and Science Bulletin 35: 95–104.

Lockley, M. G., D. Fanelli, K. Honda, K. Houck, and N. A. Matthews. 2010. Crocodile waterways and dinosaur freeways: implications of multiple swim track assemblages from the Cretaceous Dakota Group (Dakota Group) area, Colorado. New Mexico Museum of Natural History and Science Bulletin 51: 137–156.

Lockley, M. G., K. Cart, J. Martin, R. Prunty, K. Houck, K. Hups, J.-D. Lim, K.-S. Kim, K. Houck, and G. Gierlinski. 2014. A bonanza of new tetrapod tracksites from the Cretaceous Dakota Group, western Colorado: implications for paleoecology. New Mexico Museum of Natural History and Science Bulletin 62: 393–409.

Lopez-Martínez, N., J. J. Moratalla, and J. L. Sanz. 2000. Dinosaur nesting on tidal flats. Palaeogeography, Palaeoclimatology, Palaeoecology 160: 153–163.

Marsicano, C. A., N. S. Domnanovich, and A. C. Mancuso. 2007. Dinosaur origins: evidence from the footprints record. Historical Biology 19: 83–91.

Marty, D. 2008. Sedimentology, taphonomy and ichnology of Late Jurassic dinosaur tracks from the Jura carbonate platform (Chevenez-Combe Ronde tracksite, NW Switzerland): insights into the tidal-flat palaeoenvironment and dinosaur diversity, locomotion and palaeoecology. PhD dissertation, University of Fribourg, Fribourg, Switzerland. GeoFocus 21, 278 pp

Marty, D., and C. A. Meyer. 2013. A bridge over troubled water: the continuous record of terrestrial vertebrates from the Oxfordian to the Berriasian in the Jura Mountains. 57th Annual Meeting of the Palaeontological Association, Zürich, Switzerland, December 13–16. Palaeontology Newsletter 84.

Marty, D., A. Strasser, and C. A. Meyer. 2009. Formation and taphonomy of human footprints in microbial mats of present-day tidal-flat environments: implications for the study of fossil footprints. Ichnos 16: 127–142.

Marty, D., C. A. Meyer, G. Paratte, C. Lovis, and M. Cattin. 2012. Spatial relationships in dinosaur ichnoassemblages and its influence on paleobiological interpretations: a case study from the Late Jurassic of NW Switzerland. Qijiang International Dinosaur Track Symposium, Chongqing Municipality, China, November 28–30. Abstract volume.

Marty, D., L. Cavin, W. A. Hug, C. A. Meyer, M. G. Lockley, and A. Iberg. 2003. Preliminary report on the Courtedoux dinosaur tracksite from the Kimmeridgian of Switzerland. Ichnos 10: 209–219.

Marty, D., M. Belvedere, C. A. Meyer, P. Mietto, G. Paratte, C. Lovis, and B. Thüring. 2010. Comparative analysis of Late Jurassic sauropod trackways from the Jura Mountains (NW Switzerland) and the central High Atlas Mountains (Morocco): implications for sauropod ichnotaxonomy. Historical Biology 22(1–3): 109–133.

Marty, D., J. Ayer, D. Becker, J.-P. Berger, J.-P. Billon-Bruyat, L. Braillard, W. A. Hug, and C. A. Meyer. 2007. Late Jurassic dinosaur tracksites of the Transjurane highway (Canton Jura, NW Switzerland): overview and measures for their protection and valorization. Bulletin für Angewandte Geologie/Bulletin of Applied Geology 12: 75–89.

McCrea, R. T., L. G. Buckley, P. J. Currie, A. G. Plint, J. W. Haggart, C. W. Helm, and S. G. Pemberton. 2014. A review of vertebrate track-bearing formations from the Mesozoic and earliest Cenozoic of western Canada with a description of a new theropod ichnospecies and reassignment of an avian ichnogenus. New Mexico Museum of Natural History and Science Bulletin 62. 5–93.

Meyer, C. A. 1990. Sauropod tracks from the Upper Jurassic Reuchenette Formation (Kimmeridgian; Lommiswil, Kt. Solothurn) of northern Switzerland. Eclogae Geologicae Helvetiae 83(2): 389–397.

Meyer, C. A. 1993. A sauropod dinosaur megatracksite from the Late Jurassic of northern Switzerland. Ichnos 3: 29–38.

Meyer, C. A., and M. Hauser. 1994. Additional theropod and sauropod prints from the Upper

Jurassic megatracksite of northern Switzerland; pp. 49–56 in M. G. Lockley, M. G., V. F. dos Santos, C. A. Meyer, and A. P. Hunt (eds.), Aspects of Sauropod Paleobiology. Special issue, Gaia 10.

Meyer, C. A., and J. G. Pittman. 1994. A comparison between the *Brontopodus* ichnofacies of Portugal, Switzerland, and Texas; pp. 125–133 in M. G. Lockley, V. F. dos Santos, C. A. Meyer, and A. P. Hunt (eds.), Aspects of Sauropod Paleobiology. Special issue, Gaia 10.

Meyer, C. A., B. Thüring, L. Costeur, and S. Thüring. 2009. The early rise of sauropods: evidence from the Late Triassic of the eastern Swiss Alps; p. 48 in D. Schwarz-Wings, O. Wings, and F. Sattler (eds.), 7th Annual Meeting of the European Association of Vertebrate Palaeontologists, July 20–24. Abstract Volume. Museum für Naturkunde, Berlin, Germany.

Meyer, C. A., B. Thüring, D. Marty, L. Costeur, and S. Thüring. 2010. Tracking early dinosaurs: new discoveries from the Upper Austroalpine Nappes of eastern Switzerland (Hautpdolomit, Norian); p. 58 in 8th Meeting of the European Association of Vertebrate Palaeontologists: Abstract Volume, June 7–12, Aix-en-Provence, France.

Meyer, C. A., D. Marty, B. Thüring, R. Stecher, and S. Thüring. 2013. Dinosaurierspuren aus der Trias der Bergüner Stöcke (Parc Ela, Kanton Graubünden, SE-Schweiz). Mitteilungen der Naturforschenden Gesellschaft beider Basel 14: 135–144.

Milàn, J., and R. G. Bromley. 2006. True tracks, undertracks and eroded tracks, experimental work with tetrapod tracks in laboratory and field. Palaeogeography, Palaeoclimatology, Palaeoecology 231: 253–264.

Milàn, J., and R. G. Bromley. 2008. The impact of sediment consistency on track–and undertrack morphology: experiments with emu tracks in layered cement. Ichnos 15: 18–24.

Moreno, K., and M. J. Benton. 2005. Occurrence of sauropod dinosaur tracks in the Upper Jurassic of Chile (redescription of *Iguanodonichnus frenki*). Journal of South American Earth Sciences 20: 253–257.

Moreno, K., and M. Pino. 2002. Huellas de dinosaurios en la Formación Baños del Flaco (Titoniano–Jurasico Superior), VI Región, Chile: paleoetología y paleoambiente. Revista Geológica de Chile 29(2): 151–165.

Myers, T. S., and A. R. Fiorillo. 2009. Evidence for gregarious behavior and age segregation in sauropod dinosaurs. Palaeogeography, Palaeoclimatology, Palaeoecology 274: 96–104.

Nicosia, U., F. M. Petti, G. Perugini, S. D'Orazi Porchetti, E. Sacchi, M. A. Conti, N. Mariotti, and A. Zarattini. 2007. Dinosaur tracks as palaeogeographic constraints: new scenarios for the Cretaceous geography of the periadriatic region. Ichnos 14: 69–90.

Niedźwiedzki, G., S. L. Brusatte, and R. J. Butler. 2013. *Prorotodactylus* and *Rotodactylus* tracks: an ichnological record of dinosauromorphs from the Early-Middle Triassic of Poland; pp. 319–351 in S. J. Nesbitt, J. B. Desojo, and R. B. Irmis (eds.), Anatomy, Phylogeny and Palaeobiology of Early Archosaurs and Their Kin. Special Publication 379. Geological Society of London, London, UK.

Ostrom, J. H. 1972. Were some dinosaurs gregarious? Palaeogeography, Palaeoclimatology, Palaeoecology 11: 287–301.

Padian, K., and P. E. Olsen. 1984. The fossil trackways *Pteraichnus*: not pterosaurian, but crocodilian. Journal of Paleontology 58: 178–184.

Pazos, P. J., D. G. Lazo, M. A. Tunik, C. A. Marsicano, D. E. Fernández, and M. B. Aguirre-Urreta. 2012. Paleoenvironmental framework of dinosaur tracksites and other ichnofossils in Early Cretaceous mixed siliciclastic-carbonate deposits in the Neuquén Basin, northern Patagonia (Argentina). Gondwana Research 22: 1125–1140.

Rowe, M., G. Bakken, J. Ratliff, and V. Langman. 2013. Heat storage in Asian elephants during submaximal exercise: behavioral regulation of thermoregulatory constraints on activity in endothermic gigantotherms. Journal of Experimental Biology 216: 174–178.

Santi, G., and U. Nicosia. 2008. The ichnofacies concept in vertebrate ichnology. Studi Trentini di Scienze Naturali, Acta Geologica 83: 223–229.

Schlager, W. 2002. Sedimentology and Sequence Stratigraphy of Carbonate Rocks. Vrije Universiteit/Earth and Life Sciences, Amsterdam, the Netherlands, 146 pp.

Smith, A. B. 2001. Large-scale heterogeneity of the fossil record: implications for Phanerozoic diversity studies. Philosophical Transactions Royal Society of London B 356: 351–367.

Smith, A. B., and B. J. Benson. 2013. Marine diversity in the geological record and its relationship to surviving bedrock area, lithofacies diversity, and original marine shelf area. Geology 41: 171–174.

Strasser, A., B. Pittet, H. Hillgärtner, and J.-B. Pasquier. 1999. Depositional sequences in shallow carbonate-dominated sedimentary systems: concepts for a high-resolution analysis. Sedimentary Geology 128: 201–221.

Tucker, M. E., J. L. Wilson, P. D. Crevello, J. R. Sarg, and F. Read. 1990. Carbonate Platforms: Facies, Sequences and Evolution. Blackwell Science, Oxford, U.K., 328 pp.

Wagensommer, A., M. Latiano, and U. Nicosia. 2010. First report of dinosaur footprints from Madagascar: two tracksites from the Middle Jurassic Bemaraha Formation. Ichnos 17(2): 127–136.

Wagensommer, A., M. Latiano, G. Leroux, G. Cassano, and S. D'Orazi Porchetti. 2012. New dinosaur tracksites from the Middle Jurassic of Madagascar: ichnotaxonomical, behavioural and palaeoenvironmental implications. Palaeontology 55(1): 109–126.

Weishampel, D. B., and D. B. Norman. 1989. Vertebrate herbivory in the Mesozoic; jaws, plants and evolutionary metrics. Special Papers Geological Society America 238: 87–100.

Whyte, M. A., and M. Romano. 1994. Probable sauropod footprints from the Middle Jurassic of Yorkshire, England; pp. 15–26 in M. G. Lockley, V. F. dos Santos, C. A. Meyer, and A. P. Hunt (eds.), Aspects of Sauropod Paleobiology. Special issue, Gaia 10.

Whyte, M. A., and M. Romano. 2001. Probable stegosaurian dinosaur tracks from the Saltwick Formation (Middle Jurassic) of Yorkshire, England. Proceedings of the Geological Association 112: 45–54.

Wright, V. P., and T. Burchette. 1996. Shallow-water carbonate environments; pp. 325–394 in H. G. Reading (ed.), Sedimentary Environments: Processes, Facies and Stratigraphy. Wiley, New York, New York.

Xing, L., M. G. Lockley, R. McCrea, G. Gierlinski, L. G. Buckley, J. Zhang, L. Qi, and C. Jia. 2013. First record of *Deltapodus* tracks from the Early Cretaceous of China. Cretaceous Research 42: 55–65.

Zanno, L. E., and P. J. Makovicky. 2011. Herbivorous ecomorphology and specialization patterns in theropod dinosaur evolution. PNAS 108: 232–237.

Zarcone, G., F. M. Petti, A. Cillari, P. Di Stefano, D. Guzzetta, and U. Nicosia. 2010. A possible bridge between Adria and Africa: new palaeobiogeographic and stratigraphic constraints on the Mesozoic palaeogeography of the Central Mediterranean area. Earth-Science Reviews 103: 154–162.

APPENDIX

Supplemental files can be found at http://dx.doi.org/10.6084/m9.figshare.1348706.

Paleoenvironment Reconstructions of Vertebrate Tracksites in the Obernkirchen Sandstone, Lower Cretaceous of Northwest Germany

Jahn J. Hornung, Annette Richter, and Frederik Spindler
Artwork by Frederik Spindler, Freiberg, 2013

INTRODUCTION

These four color plates show reconstructions of the Obernkirchen/Bückeberge and Münchehagen areas in Lower Saxony during the deposition of the late Berriasian Obernkirchen Sandstone (Bückeberg Formation), with a focus on some of the localities visited during the Dinosaur Tracks 2011 symposium in Obernkirchen.

The sandstone was deposited marginally to a large inland lake, filling the Lower Saxony Basin. Its interpretation as sandy barrier deposits (Münchehagen area), fringing distally a lagoon and delta system (Obernkirchen and Bückeberg area) was based upon the seminal work by Pelzer (1998), with amendments by Hornung et al. (2012), and still unpublished data. The trackmakers and other biota were based mainly upon references in Hornung et al. (2012, and many citations therein), Pelzer (1998, vegetation), Richter et al. (2012), Wings et al. (2012), and Hornung and Reich (2013).

REFERENCES

Hornung, J. J., and M. Reich. 2013. The first record of the pterosaur ichnogenus *Purbeckopus* in the late Berriasian (Early Cretaceous) of northwest Germany. Ichnos 20: 164–172.

Hornung, J. J., A. Böhme, N. Schlüter, and M. Reich. 2016. Diversity, ontogeny, or both? A morphometric approach to Iguanodontian ornithopod (Dinosauria: Ornithischia) track assemblages from the Berriasian (Lower Cretaceous) of northwestern Germany; chap. 12 in P. L. Falkingham, D. Marty, and A. Richter (eds.), Dinosaur Tracks: The Next Steps. Indiana University Press, Bloomington, Indiana.

Hornung, J. J., A. Böhme, T. van der Lubbe, M. Reich, and A. Richter. 2012. Vertebrate tracksites in the Obernkirchen Sandstone (late Berriasian, Early Cretaceous) of northwest Germany–their stratigraphical, palaeogeographical, palaeoecological, and historical context. Paläontologische Zeitschrift 86(3): 231–267.

Pelzer, G. 1998. Sedimentologie und Palynologie der Wealden-Fazies im Hannoverschen Bergland. Courier Forschungsinstitut Senckenberg, 207: 1–211.

Richter, A., J. J. Hornung, A. Böhme, and U. Stratmann. 2012. Obernkirchen Sandstone Quarries–a natural workstone lagerstaette and a dinosaur tracksite; 73–100 in A. Richter and M. Reich (eds.), Dinosaur Tracks 2011: An International Symposium, Obernkirchen, April 14–17, 2011. Abstract Volume and Field Guide to Excursions. Universitätsverlag Göttingen, Göttingen, Germany.

Wings, O., D. Falk, N. Knötschke, and A. Richter. 2012. The Early Cretaceous Dinosaur Trackways in Münchehagen (Lower Saxony, Germany)–The Natural Monument "Saurierfährten Münchehagen and the adjacent Wesling Quarry. In: Richter, A. & Reich, M. (eds.): Dinosaur Tracks 2011: An International Symposium, Obernkirchen, April 14–17, 2011. Abstract Volume and Field Guide to Excursions: 113–142, Göttingen (Universitätsverlag Göttingen).

(*left*) An aerial look southward from the present-day Münchehagen area, where a small group of titanosauriform sauropods wander along the sandy barrier, bypassing narrow inlets via shallows created by washover and breach fans. The thin band of sand is only sparsely vegetated by horse-tails and other pioneering vegetation. Barren parts of the barrier are quickly shaped into dune fields by the wind. At the horizon, across the approximately 5–10-km-wide back-barrier lagoon, the delta fringe in the Bückeberg area is visible, passing further south into the gently sloping heights of the largely eroded upland of the Rhenish Massif.

(*following page, left*) Following a small flock of pterosaurs on their fast flight, we come close to the small herd of titanosauriform sauropods, who leave their trackways on the wave ripple-marked, shallowly submergent sand flats on the landward side of the small barrier islands. In the background, two iguanodontian ornithopods also use the migratory route along the barrier. It is still a bright day, but the sky begins to darken, as the first rainstorm of the upcoming rainy season looms over the northwestern horizon.

(*following page, right*) A network of fluvial channels untolds toward the lagoon, as the large river draining the southern uplands terminates in the lake near present-day Oberkirchen. Sandy mouthbars divide the many arms, some of which are in the process to be abandoned and being reduced to oxbow lakes. The older parts of the deltaic plain are forested by conifers and other dense vegetation which becomes lower, shrub-like and sparser toward the young mouthbars at the delta front. The rise of the lake level was outpaced by progradation of the delta for some time; however, during phases of increased rate of lake-level rise, the low-lying islands and their vegetation will be drowned to become swamps, forming the coal seams that have been mined in the area for centuries. Far to the north we see the band of the sandy barrier near present-day Münchehagen. It is the end of the dry season and the river is calm and low standing, its waters being relatively clear. Two different species of iguanodontian ornithopods (see Hornung et al., 2016, for track-based iguanodontian diversity) have gathered in groups at the waterways.

Zooming in to the Bückeberg delta, we observe an area that will eons later become a place colloquially called "Chicken Yard" in the main Obernkirchen quarry. An abandoned channel, which became a shallow oxbow lake, is fringed by a typical vegetation, consisting of conifer trees with an underwood of Ginkgoales, Cycadales, Bennettitales, and ferns. A fringe of broken shoots of horse-tails have been left by the last storm-tide at the flood line, together with the empty shell of a large pleurosternid turtle and a dead lepisosteiform fish (*Scheenstia* sp.). The shallow bottom of the oxbow lake is covered by dinosaur tracks, mostly left by a diversity of theropods. Currently, two troodontids wade through the calm water while a small allosauroid has passed the water to feed on the carcass of a basal ankylosaur on the shore. A single, subadult iguanodontian adds its track to the ichnocoenosis. The preservation of the tracks is favored by an algal-microbial matt that gives the subaqueous sediment surface a greenish hue, especially in very shallow areas. On a sandbank in the background, a group of pholidosaurid crocodyliforms is basking in the sun, not tempering with two large pterosaurs, who are going to leave the typical track *Purbeckopus* cf. *pentadactylus*.

Dinosaur Track Terminology: A Glossary of Terms

Daniel Marty, Peter L. Falkingham, and Annette Richter

G

Despite the fact that many efforts have been undertaken to define and harmonize terminology for vertebrate tracks (Leonardi, 1987; Gamez Vintaned and Liñán, 1996; Allen, 1997; Fornos et al., 2002; Romano and Whyte, 2003; Manning, 2004; Marty, Strasser, and Meyer, 2009) as well as for arthropod tracks (Davis, Minter, and Braddy, 2007; Minter, Braddy, and Davis, 2007), inconsistent or different usage of track terminology has led to much confusion in dinosaur (and more generally vertebrate) ichnology. Obviously, a clear and concise terminology is essential for the future study of dinosaur tracks. Here we present a guide to common terms used in vertebrate ichnology. In some cases, we have tried to synonymize multiple terms that have the same meaning. In order to avoid confusion, we strongly suggest not to use the synonyms that are provided in parentheses after each of the terms.

Deep track A deep track is formed when a foot enters a soft substrate and penetrates to depth, leaving the true track (the surface having been touched by the foot) within the substrate. After withdrawal, the track is (at least partially) closed (collapsed, sealed), forming a deep track, and there is no clear true track visible on the tracked surface (Gatesy et al., 1999). Often, this may lead, on the tracked surface, to the appearance of a track with very thin digits and sharply dipping track walls (Gatesy, 2003). In cross-sections, such tracks may reveal information about how the foot was placed (Gatesy et al., 1999).

Dinoturbation (trampling) Derived from the term "bioturbation," which is widely used in invertebrate ichnology, dinoturbation is a synonym of "trampling by dinosaurs." Originally defined by Lockley and Conrad (1989) as light, moderate, or heavy, depending on the surface percentage of substrate impacted. For a detailed review on dinoturbation, see Richter and Böhme (2016).

Displacement rim (raised rim, expulsion rim, bourrelet) Depending on the substrate consistency, a marginal rim of displaced material may form on the tracked surface around a track. This may be associated with marginal upward thrusts in the rim and with radial fractures in the substrate around the track (Allen, 1997; Manning, 2004).

Elite track Elite track refers to a well-preserved true track with clear impressions of the toes (phalangeal pads, phalanges), claws, or even skin, and it reveals information about the anatomy of the trackmaker's foot (Lockley and Hunt, 1995; Lockley and Meyer, 2000). Only well-preserved true tracks or elite tracks should be used for ichnotaxonomic purposes.

Heteropody Lockley, Farlow, and Meyer (1994) defined heteropody as the difference in area (total track area) between the pes and manus tracks in a given trackway of a quadrupedal animal. Heteropody is often used in the classification of sauropod trackways, but it is generally not (clearly) stated how it is precisely quantified.

Ichnoassemblage A single, spatially restricted track-bearing level (stratal horizon) (Fig. 0.3). According to Hunt and Lucas (2007), an ichnoassemblage is one of three "fundamental terms" in ichnology, the other two being ichnocoenosis and ichnofacies. The three terms form a distinct hierarchy.

Ichnocoenosis This term refers to recurrent ichnoassemblages (Hunt and Lucas, 2006, 2007) and is not confined to an assemblage of tracks of a single horizon. Ichnoceonosis may be used in a general sense to imply ichnoassemblages that represent part of a once-living community. However, most vertebrate ichnocoenoses discussed in the literature have an adjectival prefix, often an ichnogenus name: for example, *Brontopodus* ichnocoenosis (sensu Hunt and Lucas, 2007) or a more general adjectival prefix such as "shorebird." According to Hunt and Lucas (2007), vertebrate ichnocoenoses may compose larger ichnofacies, which they label as archetypal.

Ichnofacies This term, originally introduced in invertebrate ichnology by Seilacher (1964, 1967), was first defined for vertebrate ichnology by Lockley, Hunt, and Meyer (1994) as recurrent, regionally extensive ichnoassemblages or ichnocoenoses associated with similar sedimentary facies. However, the term ichnofacies has proved difficult to define unambiguously and Hunt and Lucas (2007) regarded some, but not all, previously defined ichnofacies as ichnocoenoses. Hunt and Lucas (2007) again argued against applying the same approach to both vertebrate and invertebrate ichnology and advocated less broadly defined ichnofacies that more precisely represent the facies preferences of characteristic trackmakers. This latter approach advocates that vertebrate ichnofacies have adjectival prefixes (such as *Brontopodus* or "shorebird" sensu Lockley, Hunt, and Meyer (1994), which refer more directly to the dominant or characteristic track types associated with particular sedimentary facies.

Ichnofauna A widely used but rather vague term to denote any ichnoassemblage, ichnocoenosis, or ichnofacies, whether of local or regional extent. Often, used generally to refer to ichnoassemblages, ichnocoenoses, or other categories of trace fossils from particular geographical areas or rock units.

Modified track Tracks having been modified by physiochemical (e.g. weathering) and/or by biological influences after they were made. Biological influences include modification through the growth of microbial mats, which do not cover the whole tracked surface or the whole track bottom (Marty, Strasser, and Meyer, 2009). The terms "slightly," "moderately," and "strongly modified" may be used to characterize the degree of modification. Modified tracks differ from unmodified true tracks in that they will not reveal details of the anatomy of the foot anymore and from overtracks in so far that the tracked surface is not covered but modified. However, identifying a track as modified may not always be possible.

Natural track cast (mold, track fill) Lithified sediment filling a true track and forming a negative replica (in convex hyporelief) of the track (Lockley, 1991).

Overall track If the track walls are not vertical, the track intersection with the tracked surface is larger than the dimensions of the trackmaker's foot and is termed the overall track (Brown, 1999). See Falkingham (2016).

Overprinting/overprint (overstepping, overstep) When a quadrupedal animal overprints the track of its front foot with the track left by its hind foot. Overprinting may be partial or complete, either obscuring the manus impression or obliterating it completely.

Overstepping (overstep, overprinting, overprint) When in the trackway of a quadrupedal animal the (right or left) pes track of the hind foot is placed in front of the preceding (right or left) manus track of the forefoot.

Overtrack (overprint) and internal overtrack An overtrack appears in a horizon above the tracked surface. It is the covering of the true track (the track bottom) and the tracking surface with sediment and/or by repeated growth of microbial mats and associated binding of sediment particles (Marty, Strasser, and Meyer, 2009). A rapid and important sedimentation event may bury the tracked surface without leaving behind any overtrack, merely a homogeneous natural track cast.

Paleosurface A paleosurface is the original surface of an ancient landform that is preserved in the rock record with definable topography and surface sculpture intact (Smith, 1993; Widdowson, 1997). Most paleosurfaces with tetrapod tracks are related to (at least partial) subaerial exposure, even though some paleosurfaces with tetrapod tracks (notably such with deep tracks, swim tracks, or ripplemarks) may never or only partially (or after track formation) have been subaerially exposed.

Striations (striation marks) Produced during foot movement, striations refer to impressions left by skin tubercles or claws. They can be used for the reconstruction of foot movement (Avanzini, Piñuela, and García-Ramos, 2012; Romano and Whyte, 2012; Xing et al., 2015).

Substrate (medium, media, soil, sediment) The substrate upon which the trackmaker moved and in which tracks were formed. May be any combination of sediment (sand, clay, carbonate mud) and grain size and can be observed in fossil tracks by different lithologies. Must be compliant enough to deform underfoot but firm enough to support the trackmaker (Falkingham et al., 2011a; Falkingham, Hage, and Bäker, 2014).

Substrate consistency The consistency of the substrate depends on its physical properties (nature and grain size of sediment, presence of microbial mat) and its water content (dry, moist/damp, water unsaturated, and water saturated/wet). Accordingly, substrate consistency may vary from firm or stiff to semifluid at the time of track formation (Manning, 2004; Falkingham et al., 2011a).

Track (footprint, print, footmark, impression) Track is the preferred synonym of "footprint" as a track may not be strictly a "print" of a "foot," but rather a more complex feature. The term "track" is used in this book in the context of tracks of dinosaurs and their descendants, the birds. Track remains the singular (one track, or the track in question) when discussing a single unit from within a trackway. A track is a feature formed by (a part of) the autopodium of the animal (Leonardi, 1987). If the animal leaves an impression of another part of the body such as the tail, then this impression should be named in combination with "trace," for example, tail trace.

Track consolidation and lithification The term "consolidation" is used in neoichnology to describe tracks that are cohesive or hard and difficult to disintegrate. Often, this is related to desiccation or the presence of a dry, coherent microbial mat. A consolidated layer is rigid and hard but not necessarily lithified. Lithification means that the substrate has been (at least partially) stabilized by cementation (e.g., precipitation of carbonates).

Track preservation To facilitate the descriptions of tracks, qualitative terms such as "poorly," "reasonably well-preserved," and "well-preserved" can be used. The first refers to true tracks that only show the typical gross outline of the foot and lack any further anatomical details. The latter refers to "perfect" true (elite) tracks with a well-defined outline and impressions of the toes (phalangeal pads, phalanges), claws, or eventually of the skin. Reasonably well-preserved may be applied for true tracks exhibiting some anatomical details and which

can therefore be identified as true tracks. Poorly preserved true tracks are typically difficult to identify as true tracks and may easily be mistaken as undertracks or vice versa. Belvedere and Farlow (2016) introduce a numerical scale, based on morphological details, to quantify the quality of preservation of vertebrate tracks. Note that "preservation" as used to describe tracks does not necessarily refer to how well the fossil matches the original–a "messy" track may be well-preserved geologically but not be considered a "well-preserved track."

Tracked surface (tracking surface) The tracked surface is the surface (paleosurface), on which the animal (trackmaker) interfaces (Fornos et al., 2002), in other words, the surface, exposed at the time of track formation (the immediate sediment/foot interface). By their very nature, a track will deform the tracked surface such that preformational and postformational surfaces do not correspond.

Trackmaker (track maker) The animal that produced the track.

Tracksite (track site) A tracksite is a natural in situ area composed of one or several bedding planes (levels, horizons) with tracks. If a tracksite is composed of different bedding planes, this must be clearly stated and the different bedding planes must be numbered.

Track wall Vertical or inclined parts of a track linking the base of the track with the tracked surface (Brown, 1999).

Trackway (track) A series of tracks (commonly a minimum of three consecutive hind foot tracks in the case of a bipedal animal), left by a single animal (Leonardi, 1987; Thulborn, 1990).

Trackway configuration In quadrupedal animals, trackway configuration refers to the general trackway characteristics, such as the arrangement of pes and manus tracks (e.g., whether the pes or manus tracks are closer to the trackway midline), their orientation (degree of rotation), or the trackway width (gauge). See also overprinting and overstepping above.

Trackway pattern Refers to the general pattern of a trackway of a quadrupedal animal. Depending on the presence or absence of pes and manus tracks, any given quadrupedal trackway can be described with five different trackway patterns: quadrupedal (pes and manus always present), pes-only (primary overstep of Peabody, 1959), manus-only, pes-dominated (some manus tracks absent, either completely overprinted and/or [partially] deformed), manus-dominated (manus tracks occasionally associated with some shallow and barely visible pes tracks) (Marty, Meyer, and Billon-Bruyat, 2006; Marty, 2008; Falkingham et al., 2011b; Falkingham, Bates, and Mannion, 2012).

Trackway and track parameters These terms are used for the description of trackway characteristics (Romano, Whyte, and Jackson, 2007). Even though the most important track and trackway parameters are commonly measured in a similar way, many different terminologies

exist for the measuring of track and trackway parameters of tetrapods and for the reference points used (Peabody, 1948, 1959; Baird, 1952; Lessertisseur, 1955; Haubold, 1971; Leonardi, 1987; Thulborn, 1990; Lockley, 1991). The most important parameters include track length, width, and depth; track rotation; interdigital divarication angles; stride and pace length; pace angulation; width of the angulation pattern; and trackway width (gauge).

true track (surface print, direct track) A true track is the track bottom respective to the substrate, which was in direct contact with the trackmaker's foot (Gatesy, 2003). True tracks may also be observed as underprints. If the foot is penetrating deep into the substrate, the true track may correspond to a complex wrapped surface reflecting the morphology of the foot and natural (or artificial) casts may reveal very fine anatomical details such as toes and claws in three dimensions (3-D) (Avanzini, Piñuela, and García-Ramos, 2012; Huerta et al., 2012).

Underprint (subtrace) Underprints are true tracks, which are produced when the foot penetrates or breaks through one (the uppermost) or several layer(s) of substrate and leaves the true track on a substrate layer below the uppermost substrate surface penetrated by the foot. Note that underprints are not a subclass of undertracks, as the substrate is penetrated and not deformed as is the case in undertracks. Underprints may also be described as unsealed deep tracks and accordingly they are often complex 3-D structures and together with their natural track casts may record important information about foot anatomy and kinematics (Gatesy et al., 1999; Avanzini, Piñuela, and García-Ramos, 2012; Falkingham and Gatesy, 2014).

Undertrack (transmitted print/track/ relief, ghost print) A track that is formed in laminated and plastic substrate by the transmission of force and deformation beyond the foot-sediment interface. Often creates a miniature stratigraphic sequence or stack of undertracks (Thulborn, 1990; Lockley, 1991). After consolidation, the (laminated/bedded) substrate package may split at successively deeper bedding planes (which may not necessarily correspond to paleosurfaces) and reveal changing track dimensions (i.e., length, width, depth) at different horizons. Thulborn (2012) proposed the term "transmitted relief" instead of "undertrack" and "transmitted track." This term "transmitted relief" is not adopted here.

Weathered track (altered track, eroded track) This term refers to tracks that have been modified by erosional processes (e.g., wind, rain, tides, root growth). Weathering processes may take place on different geological times scales, and it is important to distinguish recent weathering (due to the natural exposition of a tracksite) from ancient weathering and taphonomic processes having occurred between track formation and track burial and preservation.

REFERENCES

Allen, J. R. L. 1997. Subfossil mammalian tracks (Flandrian) in the Severn Estuary, S. W. Britain: mechanics of formation, preservation and distribution. Philosophical Transactions of the Royal Society B: Biological Sciences 352: 481–518.

Avanzini, M., L. Piñuela, and J. C. García-Ramos. 2012. Late Jurassic footprints reveal walking kinematics of theropod dinosaurs. Lethaia 45: 238–252.

Baird, D. 1952. Revision of the Pennsylvanian and Permian footprints *Limnopus*, *Allopus* and *Baropus*. Journal of Paleontology 26: 832–840.

Belvedere, M., and J. O. Farlow. 2016. A numerical scale for quantifying the quality of preservation of vertebrate tracks; chap. 6 in P. L. Falkingham, D. Marty, and A. Richter (eds.), Dinosaur Tracks: The Next Steps. Indiana University Press, Bloomington, Indiana.

Brown, T. J. 1999. The Science and Art of Tracking. Berkley Books, New York, New York, 240 pp.

Davis, R. B., N. J. Minter, and S. J. Braddy. 2007. The neoichnology of terrestrial arthropods. Palaeogeography, Palaeoclimatology, Palaeoecology 255: 284–307.

Falkingham, P. L. 2016. Applying objective methods to subjective track outlines; chap. 4 in P. L. Falkingham, D. Marty, and A. Richter (eds.), Dinosaur Tracks: The Next Steps. Indiana University Press, Bloomington, Indiana.

Falkingham, P. L., and S. M. Gatesy. 2014. The birth of a dinosaur footprint: subsurface 3D motion reconstruction and discrete element simulation reveal track ontogeny. Proc Natl Acad Sci USA 111: 18279–18284.

Falkingham, P. L., J. Hage, and M. Bäker. 2014. Mitigating the Goldilocks effect: the effects of different substrate models on track formation potential. Royal Society Open Science 1: 140225–140225.

Falkingham, P. L., K. T. Bates, and P. D. Mannion. 2012. Temporal and palaeoenvironmental distribution of manus- and pes-dominated sauropod trackways. Journal of the Geological Society, London 169: 365–370.

Falkingham, P. L., K. T. Bates, L. Margetts, and P. L. Manning. 2011a. The 'Goldilocks' effect: preservation bias in vertebrate track assemblages. Journal of the Royal Society: Interface 8: 1142–1154.

Falkingham, P. L., K. T. Bates, L. Margetts, and P. L. Manning. 2011b. Simulating sauropod manus-only trackway formation using finite-element analysis. Biology Letters 7: 142–145.

Fornós, J. J., R. G. Bromley, L. B. Clemmensen, and A. Rodríguez-Perea. 2002. Tracks and trackways of *Myotragus balearicus* Bate (Artiodactyla, Caprinae) in Pleistocene aeolianites from Mallorca (Balearic Islands, western Mediterranean). Palaeogeography, Palaeoclimatology, Palaeoecology 180: 277–313.

Gamez Vintaned, J. A., and E. Liñán. 1996. Revision de la terminología icnológica en español. Revista Española de Paleontología 11: 115–176.

Gatesy, S. M. 2003. Direct and indirect track features: What sediment did a dinosaur touch? Ichnos 10: 91–98.

Gatesy, S. M., K. M. Middleton, F. A. Jenkins, and N. H. Shubin. 1999. Three-dimensional preservation of foot movements in Triassic theropod dinosaurs. Nature 399: 141–144.

Haubold, H. 1971. Ichnia amphibiorum et reptiliorum fossilium; part 18 in O. Kuhn (ed.), Handbuch der Paläoherpetologie. Gustav Fisher Verlag, Stuttgart, Germany.

Huerta, P., F. T. Fernández-Baldor, J. O. Farlow, and D. Montero. 2012. Exceptional preservation processes of 3D dinosaur footprint casts in Costalomo (Lower Cretaceous, Cameros Basin, Spain). Terra Nova 24: 136–141.

Hunt, A. P., and S. G. Lucas. 2006. Tetrapod ichnofacies of the Cretaceous. New Mexico Museum of Natural History and Science Bulletin 35: 61–67.

Hunt, A. P., and S. G. Lucas. 2007. Tetrapod ichnofacies: a new paradigm. Ichnos 14: 59–68.

Leonardi, G. 1987. Glossary and Manual of Tetrapod Footprint Palaeoichnology. República Federativa do Brasil, Ministério das Minas e Energia, Departamento Nacional da Produção Mineral, Brasilia, Brazil, 137 pp.

Lessertisseur, J. 1955. Traces fossiles d'activité animale et leur signification paléobiologique. Mémoires de la Société Géologique de France 74: 150.

Lockley, M. G. 1991. Tracking Dinosaurs. Cambridge University Press, Cambridge, UK, 252 pp.

Lockley, M. G. 2007. A tale of two ichnologies: the different goals and potentials of invertebrate and vertebrate (tetrapod) ichnotaxonomy and how they relate to ichnofacies analysis. Ichnos 14: 39–57.

Lockley, M. G., and A. P. Hunt. 1995. Dinosaur Tracks and Other Fossil Footprints of the Western United States. Columbia University Press, New York, New York, 338 pp.

Lockley, M. G., and C. A. Meyer. 2000. Dinosaur Tracks and Other Fossil Footprints of Europe. Columbia University Press, New York, New York, 360 pp.

Lockley, M. G., and K. Conrad. 1989. The palaeoenvironmental context, preservation, and palaeoecological significance of dinosaur tracksites in the western USA; pp. 121–134 in D. D. Gillette and M. G. Lockley (eds.), Dinosaur Tracks and Traces. Cambridge University Press, Cambridge, UK.

Lockley, M. G., A. P. Hunt, and C. A. Meyer. 1994. Vertebrate tracks and the ichnofacies concept: implications for paleoecology and palichnostratigraphy; pp. 241–268 in S. Donovan (ed.), The Paleobiology of Trace Fossils. Wiley and Sons, New York, New York.

Lockley, M. G., J. O. Farlow, and C. A. Meyer. 1994. Brontopodus and Parabrontopodus ichnogen. nov. and the significance of wide- and narrow-gauge sauropod trackways. Gaia 10: 135–145.

Manning, P. L. 2004. A new approach to the analysis and interpretation of tracks: examples from the Dinosauria. Geological Society, London, Special Publications 228: 93–123.

Marty, D. 2008. Sedimentology, taphonomy, and ichnology of Late Jurassic dinosaur tracks from the Jura carbonate platform (Chevenez-Combe Ronde tracksite, NW Switzerland): insights into the tidal flat palaeoenvironment and dinosaur diversity, locomotion, and palaeoecolog. GeoFocus 21: 278.

Marty, D., A. Strasser, and C. A. Meyer. 2009. Formation and taphonomy of human footprints in microbial mats of present-day tidal-flat environments: implications for the study of fossil footprints. Ichnos 16: 127–142.

Marty, D., C. A. Meyer, and J.-P. Billon-Bruyat. 2006. Sauropod trackway patterns expression of special behaviour related to substrate consistency: an example from the Late Jurassic of northwestern Switzerland. Hantkeniana 5: 38–41.

Minter, N. J., S. J. Braddy, and R. B. Davis. 2007. Between a rock and a hard place: arthropod trackways and ichnotaxonomy. Lethaia 40: 365–375.

Peabody, F. E. 1948. Reptile and Amphibian Trackways from the Lower Triassic Moenkopi Formation of Arizona and Utah. University of California Publications Bulletin of the Department of Geological Sciences 27(8): 25–467.

Peabody, F. E. 1959. Trackways of living and fossil salamanders. University of California Publications in Zoology 63: 1–72.

Richter, A., and A. Böhme. 2016. Too many tracks: preliminary description and interpretation of the diverse and heavily dinoturbated Early Cretaceous 'Chicken Yard' ichnoassemblage (Obernkirchen Tracksite, northern Germany); chap. 17 in P. L. Falkingham, D. Marty, and A. Richter (eds.), Dinosaur Tracks: The Next Steps. Indiana University Press, Bloomington, Indiana.

Romano, M., and M. A. Whyte. 2003. Jurassic dinosaur tracks and trackways of the Cleveland Basin, Yorkshire: preservation, diversity and distribution. Proceedings of the Yorkshire Geological Society 52: 361–369.

Romano, M., and M. A. Whyte. 2012. Information on the foot morphology, pedal skin texture and limb dynamics of sauropods: evidence from the ichnological record of the Middle Jurassic of the Cleveland Basin, Yorkshire. Zubia 30: 45–92.

Romano, M., M. A. Whyte, and S. J. Jackson. 2007. Trackway ratio: a new look at trackway gauge in the analysis of quadrupedal dinosaur trackways and its implications for ichnotaxonomy. Ichnos 14: 257–270.

Seilacher, A. 1964. Biogenic sedimentary structures; pp. 296–316 in J. Imbrie and N. Newell (eds.), Approaches to Palaeoecology. Wiley, New York, New York.

Seilacher, A. 1967. Bathymetry of trace fossils. Marine Geology 5: 413–428.

Smith, R. M. H. 1993. Sedimentology and ichnology of floodplain paleosurfaces in the Beaufort Group (Late Permian), Karoo sequence, South Africa. Palaios 8: 339–357.

Thulborn, R. A. 1990. Dinosaur Tracks. Chapman and Hall, London, UK, 410 pp.

Thulborn, T. 2012. Impact of sauropod dinosaurs on lagoonal substrates in the Broome Sandstone (Lower Cretaceous), Western Australia. PLoS One 7: e36208.

Widdowson, M. 1997. Palaeosurfaces: recognition, reconstruction and palaeoenvironmental interpretation. Geological Society, London, Special Publications 120: 330.

Xing, L., D. Li, M. G. Lockley, D. Marty, J. Zhang, W. Scott Persons IV, H.-L. You, C. Peng, and S. B. Kümmell. 2015. Dinosaur natural track casts from the Lower Cretaceous Hekou Group in the Lanzhou-Minhe Basin, Gansu, Northwest China: ichnology, track formation, and distribution. Cretaceous Research 52: 194–205.

Index

Contributors

Luis Alcalá, Fundación Conjunto Paleontológico de Teruel-Dinópolis/Museo Aragonés de Paleontología, Spain.

Matteo Belvedere, Museum für Naturkunde—Leibniz institute for Evolution and Biodiversity Research, Germany.

Massimo Bernardi, Museo delle Scienze–Trento, Italy and University of Bristol, UK.

Annina Böhme, Niedersächsisches Landesmuseum Hannover, Germany.

Brent Breithaupt, Bureau of Land Management, United States.

Lisa G. Buckley, Peace Region Palaeontology Research Centre/University of Alberta, Canada.

José I. Canudo, Universidad de Zaragoza, Spain.

Diego Castanera, Universidad de Zaragoza, Spain.

Andrea Cinquegranelli, San Felice Circeo, Italy.

Alberto Cobos, Fundación Conjunto Paleontológico de Teruel-Dinópolis/Museo Aragonés de Paleontología, Spain.

Maria Datcheva, Bulgarian Academy of Sciences, Bulgaria.

Simone D'Orazi Porchetti, Universidade de São Paulo, Brazil.

Richard G. Ellis, Boston University School of Medicine, U.S.

Scott Ernst, Office de la culture Paléontologie A16, Switzerland.

Peter L. Falkingham, Liverpool John Moores University, U.K.

James O. Farlow, Indiana University–Purdue University, United States.

Denver W. Fowler, Museum of the Rockies and Montana State University, United States.

Ashley E. Fragomeni, Cleveland Museum of Natural History, United States.

Francisco Gascó, Fundación Conjunto Paleontológico de Teruel-Dinópolis/Museo Aragonés de Paleontología, Spain.

Stephen M. Gatesy, Brown University, United States.

Hanna Haase, Ruhr-Universität Bochum, Germany.

Lee E. Hall, Cleveland Museum of Natural History, United States.

Jerry D. Harris, Dixie State College, United States.

Jahn J. Hornung, Fuhlsbüttler Straße 611, 22337 Hamburg, Germany.

Tom Hübner, LWL-Museum für Naturkunde Münster, Germany.

Jens N. Lallensack, Steinmann-Institut für Geologie, Mineralogie und Paläontologie, Universität Bonn, Germany.

Rihui Li, Qingdao Institute of Marine Geology, China.

Martin G. Lockley, University of Colorado, Denver, United States.

David B. Loope, University of Nebraska, United States.

Heinrich Mallison, Museum für Naturkunde–Leibniz institute for Evolution and Biodiversity Research, Germany.

Luis Mampel, Fundación Conjunto Paleontológico de Teruel-Dinópolis/Museo Aragonés de Paleontología, Spain.

Daniel Marty, Office de la culture–Paléontologie A16, Porrentruy Switzerland and Naturhistorisches Museum Basel, Switzerland.

Neffra Matthews, Bureau of Land Management, United States.

Richard T. McCrea, Peace Region Palaeontology Research Centre, Canada.

Jesper Milàn, Geomuseum Faxe/Østsjællands Museum, Denmark.

Andrew R. C. Milner, St George Dinosaur Discovery Site at Johnson Farm, United States.

José Joaquin Moratalla, Instituto Geológico y Minero de España, Spain.

Tommy Noble, Bureau of Land Management, United States.

Carlos Pascual, Soria, Spain.

Fabio Massimo Petti, Museo delle Scienze–Trento, Italy.

Laura Piñuela, Museo del Jurásico de Asturias, Spain.

Mike Reich, Bayerische Staatssammlung für Paläontologie und Geologie, München, Germany.

Annette Richter, Niedersächsisches Landesmuseum Hannover, Germany.

Rafael Royo-Torres, Fundación Conjunto Paleontológico de Teruel-Dinópolis/Museo Aragonés de Paleontología, Spain.

Paulo Sá Caetano, Universidade Nova de Lisboa, Portugal.

Vanda Faria dos Santos, Museu Nacional de História Natural e da Ciência, Portugal.

Tom Schanz, Ruhr-Universität Bochum, Germany.

Nils Schlüter, Georg-August Universität, Göttingen, Germany.

Frederik Spindler, Freiberg, Germany

Kent A. Stevens, University of Oregon, United States.

Torsten van der Lubbe, Hannover, Germany.

Bernat Vila, Universidad de Zaragoza, Spain.

Alexander Wagensommer, Gruppo di Ricerca sulle Impronte di Dinosauro, Italy.

Oliver Wings, Niedersächsisches Landesmuseum Hannover/Museum für Naturkunde Berlin, Germany.

Lida Xing, China University of Geosciences, China.

PETER FALKINGHAM is Senior Lecturer in Vertebrate Biology in the School of Natural Sciences and Psychology at Liverpool John Moore's University, United Kingdom. In 2010, he completed a PhD dissertation at Manchester University titled "Computer Simulation of Dinosaur Tracks" and has continued in that field to date, combining advanced imaging and simulation techniques to model footprint formation in dinosaurs and living birds.

DANIEL MARTY is a Research Paleontologist at the Paleontology A16 (Office de la culture, Canton Jura, Switzerland), a paleontological service founded in the year 2000, which is in charge of the excavation, documentation, and safeguarding of the paleontological heritage along the future course of Swiss federal Highway A16. He is responsible for the excavation, documentation, and scientific research of dinosaur tracksites that were uncovered prior to the construction of Highway A16. He is involved in several research projects related to these dinosaur track discoveries in collaboration with researchers from Europe and abroad. Furthermore, he is working part-time at the Natural History Museum Basel, where he is editor of the Swiss *Journal of Palaeontology* and Swiss *Journal of Geosciences (Paleontology)*, and he teaches paleontology at the University of Basel.

ANNETTE RICHTER is Senior Custodian of Earth Sciences and head of the Natural History Department at the Lower Saxony State Museum at Hannover. She teaches at Gottfried Wilhelm Leibniz University and the University of Applied Sciences Hannover. She specializes in the field of fossil and extant reptiles and combines this with aspects of regional geology within northern Germany.

This book was designed by Jamison Cockerham and set in type by Tony Brewer at Indiana University Press, and printed by Sheridan Books.

The fonts are Electra, designed by William A. Dwiggins in 1935, Frutiger, designed by Adrian Frutiger in 1975, and Futura, designed by Paul Renner in 1927. All were published by Adobe Systems.